Lecture Notes in Computer Science 4537

Commenced Publication in 1973
Founding and Former Series Editors:
Gerhard Goos, Juris Hartmanis, and Jan van Leeuwen

Kevin Chen-Chuan Chang Wei Wang
Lei Chen Clarence A. Ellis
Ching-Hsien Hsu Ah Chung Tsoi
Haixun Wang (Eds.)

Advances in Web and Network Technologies, and Information Management

APWeb/WAIM 2007 International Workshops:
DBMAN 2007, WebETrends 2007, PAIS 2007 and ASWAN 2007
Huang Shan, China, June 16-18, 2007
Proceedings

Volume Editors

Kevin Chen-Chuan Chang
University of Illinois at Urbana-Champaign, USA
E-mail: kcchang@cs.uiuc.edu

Wei Wang, University of New South Wales, Australia
E-mail: weiw@cse.unsw.edu.au

Lei Chen, Hong Kong University, Hong Kong
E-mail: leichen@cs.ust.hk

Clarence A. Ellis, University of Colorado at Boulder, USA
E-mail: skip@cs.colorado.edu

Ching-Hsien Hsu, Chung Hua University, Taiwan
E-mail: chh@chu.edu.tw

Ah Chung Tsoi, Monash University, Australia
E-mail: ahchung.tsoi@arc.gov.au

Haixun Wang, T. J. Watson Research Center, USA
E-mail: haixun@us.ibm.com

Library of Congress Control Number: 2007928328

CR Subject Classification (1998): H.2-5, C.2, I.2, K.4, J.1

LNCS Sublibrary: SL 3 – Information Systems and Application, incl. Internet/Web and HCI

ISSN 0302-9743
ISBN-10 3-540-72908-9 Springer Berlin Heidelberg New York
ISBN-13 978-3-540-72908-2 Springer Berlin Heidelberg New York

Springer is a part of Springer Science+Business Media

springer.com

Printed in Germany

Typesetting: Camera-ready by author, data conversion by Scientific Publishing Services, Chennai, India
Printed on acid-free paper SPIN: 12073054 06/3180 5 4 3 2 1 0

APWeb/WAIM 2007 Workshop Chair's Message

As an important part of the joint APWeb/WAIM 2007 conference, continuing both conferences' traditional excellence on theoretical and practical aspects of Web-based information access and management, we were pleased to have four workshops included in this joint conference. This proceedings volume compiles, the technical papers selected for presentation at the following workshops, held at Huang Shan (Yellow Mountains), China, June 16 – 18, 2007.

1. International Workshop on DataBase Management and Application over Networks (DBMAN 2007)
2. International Workshop on Emerging Trends of Web Technologies and Applications (WebETrends 2007)
3. International Workshop on Process Aware Information Systems (PAIS 2007)
4. International Workshop on Application and Security service in Web and pervAsive eNvirionments (ASWAN 2007)

These four workshops were selected from a public call-for-proposals process. The workshop organizers put a tremendous amount of effort into soliciting and selecting research papers with a balance of quality, novelty, and application relevance. We asked all workshops to follow a rigid paper selection process, including the procedure to ensure that any Program Committee members (including workshop Program Committee Chairs) were excluded from the paper review process of any papers they were involved in. A requirement about the overall paper acceptance ratio was also imposed on all the workshops.

I am very grateful to Jeffrey Xu Yu, Xuemin Lin, Zheng Liu, Wei Wang and many other people for their great effort in supporting the conference organization. I would like to take this opportunity to thank all workshop organizers and Program Committee members for their great effort in putting together the workshop program. I hope you enjoy these proceedings in.

June 2007 Kevin C. Chang

International Workshop on DataBase Management and Application over Networks (DBMAN 2007) Chairs' Message

With the increasing ubiquity of personal computing devices, such as mobile phones and PDAs, and the increasing deployment of sensor networks, new distributed applications are developed over networked databases posing interesting challenges. The 1st International Workshop on Database Management and Applications over Networks (DBMAN 2007) was held in Huanshan, China on June 18, 2007, in conjunction with the WAIM/APWeb 2007 conference. It aimed to bring together researchers in different fields related to database management and application over networks and to provide a forum where researchers and practitioners could share and exchange their knowledge and experience. In response to the call for papers, DBMAN attracted 105 submissions. The submissions are highly diversified, coming from Australia, China, France, Japan, Korea, New Zealand, Norway, Spain, Thailand, Taiwan and the USA, resulting in an international final program. All submissions were peer reviewed by at least two Program Committee members. The Program Committee selected 14 full papers and 14 short papers for inclusion in the proceedings. The competition was keen, with an overall acceptance rate of around 27%. The accepted papers covered a wide range of research topics and novel applications on database management over networks.

The workshops would not have been successful without the help of many organizations and individuals. First, we would like to thank the Program Committee members and external reviewers for evaluating the assigned papers in a timely and professional manner. Next, the tremendous efforts put forward by members in the Organization Committee of WAIM/APWeb in accommodating and supporting the workshops are appreciated. Of course, without support from the authors with their submissions, the workshop would not have been possible.

June 2007

Lei Chen
M. Tamer Özsu

International Workshop on Emerging Trends of Web Technologies and Applications (WebETrends 2007) Chairs' Message

The 1st Workshop on Emerging Trends of Web Technologies and Applications was held in Huangshan, China, on June 16, 2007, in conjunction with AP-Web/WAIM 2007: the Ninth Asia-Pacific Web Conference (APWeb) and the Eighth International Conference on Web-Age Information Management (WAIM).

The explosive growth of the World Wide Web is revolutionizing how individuals, groups, and communities interact and influence each other. It is well recognized that the Web has transformed everyday life; however, its deepest impact in science, business, and society is still in the making. As technologies mature and communities grow, a new world that is fundamentally revolutionary is just beginning. The goal of this workshop is to bring together researchers to share visions for the future World Wide Web.

The workshop solicted papers in a wide range of fields with a focus on emerging Web technologies and applications. We welcome innovative ideas from traditional topics such as Web wrapping, Internet crawling, search engine optimization, current hot areas such as Wiki, blogosphere, social network, as well as promising future directions such as Web services and the Semantic Web. Our aim is to promote the awareness of the potential impact of the Web, discuss research directions and agendas, share experience and insights, and build a joint community across disciplines for creating enabling technologies for the future Web.

The workshop received 40 submissions from many countries and regions. All submissions were peer reviewed by at least two Program Committee members. The Program Committee selected 12 full papers for inclusion in the proceedings.

We are grateful to members of the Program Committee who contributed their expertise and ensured the high quality of the reviewing process. We are thankful to the Workshop Chair Kevin Chang and the PC Chair Jeffrey Xu Yu for their support.

June 2007

Haixun Wang
Chang-shing Perng

International Workshop on Process Aware Information Systems (PAIS 2007) Chairs' Message

A process-aware information system (PAIS) is a software system that manages and executes operational processes involving people, applications, and/or information sources on the basis of an explicit imbedded process model. The model is typically instantiated many times, and every instance is typically handled in a predefined way. Thus this definition shows that a typical text editor is not process aware, and likewise a typical e-mail client is not process aware. In both of these examples, the software is performing a task, but is not aware that the task is part of a process. Note that the explicit representation of process allows automated enactment, automated verification, and automated redesign, all of which can lead to increased organizational efficiency. The strength and the challenge of PAISs are within the CSCW domain where PAISs are used by groups of people to support communication, coordination, and collaboration. Even today, after years of research and development, PAISs are plagued with problems and pitfalls intertwined with their benefits. These problems are frequently elusive and complex due to the fact that "PAISs are first and foremost people systems." Conclusively, we were able to address some of the people issues and frameworks for research in this domain through the workshop, as we expected.

Although the workshop was the first in the series, we received a reasonable number of submissions, 60 papers from about 20 regions and countries. All submissions were peer reviewed by at least two or three Program Committee members and five external reviewers. After a rigorous review process, the Program Committee selected 8 full papers and 11 short papers among the 60 submissions. The acceptance rates of full paper and short paper submissions are 13.3% and 18.3%, respectively. Also, the accepted papers covered a wide range of research topics and novel applications related to PAISs.

We are grateful to the Kyonggi University and the University of Colorado at Boulder for sponsoring the workshop and would like to express thanks to the research members of the CTRL/CTRG research groups for developing the workshop's Web site and review-processing system. We would also like to thank all authors who submitted papers and all the participants in the workshop program. We are especially grateful to members of the Program Committee who contributed their expertise and ensured the high quality of the reviewing process. We are thankful to the APWeb/WAIM organizers for their support and local arrangement.

June 2007

Clarence A. Ellis
Kwanghoon Kim

International Workshop on Application and Security Service in Web and Pervasive Environments (ASWAN 2007) Program Chairs' Message

We are proud to present the proceedings of the 2007 International Workshop on Application and Security Service in Web and Pervasive Environments, held at HuangShan, China during June 16–18, 2007.

Web and pervasive environments (WPE) are emerging rapidly as an exciting new paradigm including ubiquitous, Web, grid, and peer-to-peer computing to provide computing and communication services any time and anywhere. In order to realize their advantages, security services and applications need to be suitable for WPE. ASWAN 2007 was intended to foster the dissemination of state-of-the-art research in the area of secure WPE including security models, security systems and application services, and novel security applications associated with its utilization. The aim of ASWAN 2007 was to be the premier event on security theories and practical applications, focusing on all aspects of Web and pervasive environments and providing a high-profile, leading-edge forum for researchers and engineers alike to present their latest research.

In order to guarantee high-quality proceedings, we put extensive effort into reviewing the scientific papers. We received 61 papers from Korea, China, Hong Kong, Taiwan and the USA, representing more than 50 universities or institutions. All submissions were peer reviewed by two to three Program or Technical Committee members or external reviewers. It was extremely difficult to select the presentations for the workshop because there were so many excellent and interesting ones. In order to allocate as many papers as possible and keep the high quality of the workshop, we decided to accept 16 papers for oral presentations. We believe all of these papers and topics will not only provide novel ideas, new results, work in progress and state-of-the-art techniques in this field, but will also stimulate future research activities.

This workshop would not have been possible without the support of many people to make it a success. First of all, we would like to thank the Steering Committee Chair, Laurence T. Yang, and Jong Hyuk Park for nourishing the workshop and guiding its course. We thank the Program Committee members for their excellent job in reviewing the submissions and thus guaranteeing the quality of the workshop under a very tight schedule. We are also indebted to the members of the Organizing Committee. Particularly, we thank Byoung-Soo Koh, Jian Yang, Xiaofeng Meng, Yang Xiao and Yutaka Kidawara for their devotions and efforts to make this workshop a real success. Finally, we would like to take this opportunity to thank all the authors and participants for their contributions, which made ASWAN 2007 a grand success.

June 2007

Laurence T. Yang
Sajal K. Das
Eung Nam Ko
Ching-Hsien Hsu
Djamal Benslimane
Young Yong Kim

Organization

International Workshop on DataBase Management and Application over Networks (DBMAN 2007)

Program Co-chairs

Lei Chen, Hong Kong University of Science and Technology, Hong Kong, China
M. Tamer Özsu, University of Waterloo, Waterloo, Canada

Program Committee

Gustavo Alonso, ETH, Switzerland
Luc Bouganim, INRIA, France
Klemens Böhm, Universität Karlsruhe (TH), Germany
Ilaria Bartolini, University of Bologna, Italy
Ahmet Bulut, Citrix Systems, USA
Selcuk Candan, Arizona State University, USA
Yi Chen, Arizona State University, USA
Reynold Cheng, Hong Kong Polytechnic University, China
Mitch Cherniack, Brandeis University, USA
Khuzaima Daudjee, University of Waterloo, Canada
Ada Waichee Fu, Chinese University of Hong Kong, China
Björn THór Jónsson, Reykjavik University, Iceland
Wang-Chien Lee, Penn State University, USA
Chen Li, University of California, Irvine, USA
Mingjin Li, Microsoft Research Asia, China
Alexander Markowetz, Hong Kong University of Science and Technology, China
Vincent Oria, New Jersey Institute of Technology, USA
Sule Gunduz Oguducu, Istanbul Technical University, Turkey
Jian Pei, Simon Fraser University, Canada
Peter Triantafillou, University of Patras, Greece
Anthony Tung, National University of Singapore, Singapore
Özgür Ulusoy, Bilkent University, Turkey
Patrick Valduriez, INRIA, France
Jari Veijaleinen, University of Jyvaskyla, Finland
JianLiang Xu, Hong Kong Baptist University, China

External Reviewers

Yu Li, Jinchuan Chen, Xiang Lian, Yingyi Bu, Qiang Wang, Yingying Tao, Huaxin Zhang, Oghuzan Ozmen, Lukasz Golab, Weixiong Rao, Shaoxu Song, Rui Li, Lei Zou, Yongzhen Zhuang, Xiaochun Yang, Bin Wang.

International Workshop on Emerging Trends of Web Technologies and Applications (WebETrends 2007)

Program Co-chairs

Haixun Wang, IBM T. J. Watson Research Center, USA
Chang-shing Perng, IBM T. J. Watson Research Center, USA

Program Committee

Jeff Wei-shinn Ku, University of Southern California, USA
Yijian Bai, University of California, Los Angeles, USA
Jian Yin, IBM T. J. Watson Research Center, USA
Tao Li, Florida International University, USA
Hui Xiong, Rutgers University, USA
Charles Perng, IBM T. J. Watson Research Center, USA
Haixun Wang, IBM T. J. Watson Research Center, USA

International Workshop on Process Aware Information Systems (PAIS 2007)

Program Co-chairs

Clarence A. Ellis, University of Colorado at Boulder, USA
Kwang-Hoon Kim, Kyonggi University, Korea

Program Committee

Hayami Haruo, Kanagawa Institute of Technology, Japan
Jintae Lee, University of Colorado at Boulder, USA
George Wyner, Boston University, USA
Jorge Cardoso, University of Madeira, Portugal
Yang Chi-Tsai, Flowring Technology, Inc., Taiwan
Michael zur Muehlen, Stevenson Institute of Technology, USA
Dongsoo Han, Information and Commnications University, Korea
Ilkyeun Ra, University of Colorado at Denver, USA
Taekyou Park, Hanseo University, Korea
Joonsoo Bae, Chonbuk National University, Korea

Yoshihisa Sadakane, NEC Soft, Japan
Junchul Chun, Kyonggi University, Korea
Luis Joyanes Aguilar, Universidad Pontificia de Salamanca, Spain
Tobias Rieke, University of Muenster, Germany
Modrak Vladimir, Technical University of Kosice, Slovakia
Haksung Kim, Dongnam Health College, Korea
Yongjoon Lee, Electronics and Telecommunications Research Institute, Korea
Taekyu Kang, Electronics and Telecommunications Research Institute, Korea
Jinjun Chen, Swinburne University of Technology, Australia
Zongwei Luo, The University of Hong Kong, China
Peter sakal, Technical University Bratislava, Slovakia
Sarka Stojarova, Univeristy of Brno, Czech Republic
Boo-Hyun Lee, KongJu National University, Korea
Jeong-Hyun Park, Electronics and Telecommunications Research Institute, Korea
Yanbo Han, Chinese Academy of Sciences, China
Jacques Wainer, State University of Campinas, Brazil
Aubrey J. Rembert, University of Colorado at Boulder, USA

External Reviewers

Aubrey Rembert, Hyongjin Ahn, Minjae Park, Jaekang Won, Hyunah Kim

Workshop Liaison

Taewook Kim

International Workshop on Application and Security Service in Web and Pervasive Environments (ASWAN 2007)

Steering Chair

Laurence T. Yang, St. Francis Xavier University, Canada

General Co-chairs

Sajal K. Das, University Texas at Arlington, USA
Eung Nam Ko, Baekseok University, Korea

Program Co-chairs

Ching-Hsien Hsu, Chung Hua University, Taiwan
Djamal Benslimane, University Claude Bernard, France
Young Yong Kim, Yonsei University, Korea

Publicity Co-chairs

Jian Yang, Macquarie University, Australia
Xiaofeng Meng, Renmin University of China, China
Yang Xiao, University of Alabama, USA
Yutaka Kidawara, NICT, Japan

Web Management Chair

Byoung-Soo Koh, DigiCAPS Co., Ltd, Korea

Program Committee

Andrew Kusiak, The University of Iowa, USA
Anna Cinzia Squicciarini, Purdue University, USA
Antonio Coronato, SASIT-CNR, Italy
Apostolos N. Papadopoulos, Aristotle University, Greece
Aris M. Ouksel, The University of Illinois at Chicago, USA
Barbara Catania, Università di Genova, Italy
Cho-Li Wang, The University of Hong Kong, China
Christoph Bussler, Cisco Systems, USA
Christophides Vassilis, Foundation for Research and Technology-Hellas, Greece
Claudio Sartori, Università di Bologna, Italy
Deok Gyu Lee, ETRI, Korea
Dimitris Papadias, Hong Kong University of Science and Technology, China
Do van Thanh, NTNU, Norway
Dongseop Kwon, Myungji University, Korea
Emmanuelle Anceaume, IRISA, France
Evi Syukur, Monash University, Australia
Fangguo Zhang, Sun Yat-sen University, China
Gianluca Moro, University of Bologna, Italy
Huafei Zhu, Institute for Infocomm Research, Singapore
Ilsun You, Korean Bible University, Korea
Javier Garcia Villalba, Complutense University of Madrid, Spain
Jong Hyuk Park, Hanwha S&C Co., Ltd., Korea
Jean-Henry Morin, Korea University, Korea
Karl M. Goeschka, Vienna University of Technology, Austria
Katsaros Dimitrios, Aristotle University, Greece
Marco Aiello, University of Trento, Italy
Massimo Esposito, ICAR-CNR, Italy
Massimo Poncino, Politecnico di Torino, Italy
Mirko Loghi, Politecnico di Torino, Italy
Naixue Xiong, JAIST, Japan
Nicolas Sklavos, Technological Educational Institute of Messolonghi, Greece
Ning Zhang, University of Manchester, UK

Paolo Bellavista, University of Bologna, Italy
Paris Kitsos, Hellenic Open University, Greece
Qi Shi, Liverpool John Moores University, UK
Rodrigo Fernandes de Mello, University of Sao Paulo, Brazil
Sheng-De Wang, National Taiwan University, Taiwan
Stavros Christodoulakis, MUSIC/TUC, Crete, Greece
Tetsu Iwata, Nagoya University, Japan
Tom Gross, University of Weimar, Germany
Tore Jonvik, Oslo Unversity College, Norway
Traore Jacques RD-MAPS-CAE, France Telecom R&D, France
Vesna Hassler, European Patent Office, Austria
Weisong Shi, Wayne State University, USA
Wenfei Fan, University of Edinburgh, UK
Yeong Deok Kim, Woosong University, Korea
Klaus R. Dittrich, Universität Zurich, Switzerland
Trevor Jim, AT&T Labs Research, USA
Wen Ouyang, Chung Hua University, Taiwan

Table of Contents

International Workshop on DataBase Management and Application over Networks (DBMAN 2007)

Information Access and Dissemination I

Data Mining

Sensor, P2P, and Grid Networks I

Information access and Dissemination 2

Stream Data Management

Sensor, P2P, and Grid Networks 2

Potpourri

International Workshop on Emerging Trends of Web Technologies and Applications (WebETrends 2007)

Keynote Talk

Session 1

Session 2

International Workshop on Process Aware Information Systems (PAIS 2007)

Full Paper

Short Paper

International Workshop on Application and Security Service in Web and Pervasive Environments (ASWAN 2007)

WPE Models and Applications

Security and Services of WPE

WSN/RFID/Web Services

Data Management and Access Control for WPE

Cost Framework for a Heterogeneous Distributed Semi-structured Environment

Tianxiao Liu[1], Tuyêt Trâm Dang Ngoc[2], and Dominique Laurent[3]

[1] ETIS Laboratory - University of Cergy-Pontoise & XCalia S.A, France
Tianxiao.Liu@u-cergy.fr, Tianxiao.Liu@xcalia.com
[2] ETIS Laboratory-University of Cergy-Pontoise, France
Tuyet-Tram.Dang-Ngoc@u-cergy.fr
[3] ETIS Laboratory-University of Cergy-Pontoise, France
Dominique.Laurent@u-cergy.fr

Abstract. This paper proposes a generic cost framework for query optimization in an XML-based mediation system called XLive, which integrates distributed, heterogeneous and autonomous data sources. Our approach relies on cost annotation on an XQuery logical representation called Tree Graph View (TGV). A generic cost communication language is used to give an XML-based uniform format for cost communication within the XLive system. This cost framework is suitable for various search strategies to choose the best execution plan for the sake of minimizing the execution cost.

Keywords: Mediation system, query optimization, cost model, Tree Graph View, cost annotation.

1 Introduction

The architecture of mediation system has been proposed in [Wie92] for solving the problem of integration of heterogeneous data sources. In such an architecture, users send queries to the mediator, and the mediator processes these queries with the help of wrappers associated to data sources. Currently, the semi-structured data model represented by XML format is considered as a standard data exchange model. XLive [NJT05], mediation system based on XML standard, has a mediator which can accept queries in the form of XQuery [W3C05] and return answers. The wrappers give the mediator an XML-based uniform access to heterogeneous data sources.

For a given user query, the mediator can generate various execution plans (referred to as "plan" in the remainder of this paper) to execute it, and these plans can differ widely in execution cost (execution time, price of costly connections, communication cost, etc. An optimization procedure is thus necessary to determine the most efficient plan with the least execution cost. However, how to choose the best plan based on the cost is still an open issue. In relational or object-oriented databases, the cost of a plan can be estimated by using a cost model. This estimation is processed with database statistics and cost formulas

K.C. Chang et al. (Eds.): APWeb/WAIM 2007 Ws, LNCS 4537, pp. 1–11, 2007.

for each operator appearing in the plan. But in a heterogeneous and distributed environment, the cost estimation is much more difficult, due to the lack of underlying databases statistics and cost formulas.

Various solutions for processing the overall cost estimation at the mediator level have been proposed. In [DKS92], a calibration procedure is described to estimate the coefficients of a generic cost model, which can be specialized for a class of systems. This solution is extended for object database systems in [GGT96][GST96]. The approach proposed in [ACP96] records cost information (results) for every query executed and reuses that information for the subsequent queries. [NGT98] uses a cost-based optimization approach which combines a generic cost model with specific cost information exported by wrappers. However, none of these solutions has addressed the problem of overall cost estimation in a semi-structured environment integrating heterogeneous data sources.

In this paper, we propose a generic cost framework for an XML-based mediation system, which integrates distributed, heterogeneous and autonomous data sources. This framework allows to take into account various cost models for different types of data sources with diverse autonomy degrees. These cost models are stored as annotations in an XQuery logical representation called Tree Graph View (TGV) [DNGT04][TDNL06]. Moreover, cost models are exchanged between different components of the XLive system. We apply our cost framework to compare the execution cost of candidate plans in order to choose the best one.

First, we summarize different cost models for different types of data sources (relational, object oriented and semi-structured) and different autonomy degrees of these sources (proprietary, non-proprietary and autonomous). The overall cost estimation relies on the cost annotation stored in corresponding components TGV. This cost annotation derives from a generic annotation model which can annotate any component (i.e. one or a group of operators) of a TGV.

Second, in order to perform the cost communication within the XLive system during query optimization, we define an XML-based language to express the cost information in a uniform, complete and generic manner. This language, which is generic enough to take into account any type of cost information, is the standard format for the exchange of cost information in XLive.

The paper is organized as follows: In Section 2, we introduce XLive system with its TGV modeling of XQuery and we motivate our approach to cost-based optimization. In Section 3, we describe the summarized cost models and show how to represent and exchange these cost models using our XML-based generic language. Section 4 provides the description of TGV cost annotation and the procedure for the overall cost estimation at the mediator level. We conclude and give directions for future work in Section 5.

2 Background

XQuery Processing in XLive. A user's XQuery submitted to the XLive mediator is first transformed into a canonical form. Then the canonized XQuery is modeled in an internal structure called TGV. We annotate the TGV with information on evaluation, such as the data source locations, cost models, sources functional capabilities of sources, etc. The *optimal* annotated TGV is then selected based on a cost-based optimization strategy. In this optimization procedure, TGV is processed as *the logical execution plan* and the cost estimation of TGV is performed with cooperation between different components of XLive. This *optimal* TGV is then transformed into an execution plan using a physical algebra. To this end, we have chosen the XAlgebra [DNG03] that is an extension to XML of the relational algebra. Finally, the physical execution plan is evaluated and an XML result is produced, Fig.1 depicts the different steps of this processing.

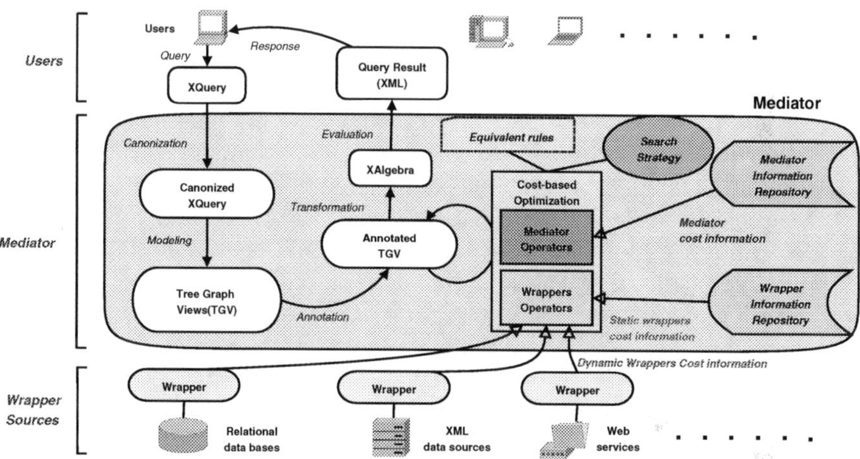

Fig. 1. Cost-based optimization in processing of XQuery in the XLive system

Tree Graph View. TGV is a logical structure model implemented in the XLive mediator for XQuery processing, which can be manipulated, optimized and evaluated [TDNL06]. TGV takes into account the whole functionality of XQuery (collection, XPath, predicate, aggregate, conditional part, etc.) and uses an intuitive representation that provides a global view of the request in a mediation context. Each element in the TGV model has been defined formally using Abstract Data Type in [Tra06] and has a graphical representation. In Fig. 2 (a), we give an example of XQuery which declares two FOR clauses ($a and $b), a join constraint between authors and a contains function, then a return clause projects the title value of the first variable. This query is represented by a TGV in Fig. 2 (b). We can distinguish the two domain variables $a and $b of the XQuery, defining each nodes corresponding to the given XPaths. A join hyperlink links the two *author* nodes with an equality annotation. The *contains* function is linked to the $b "author" node, and a projection hyperlink links the node *title* to the ReturnTreePattern in projection purposes.

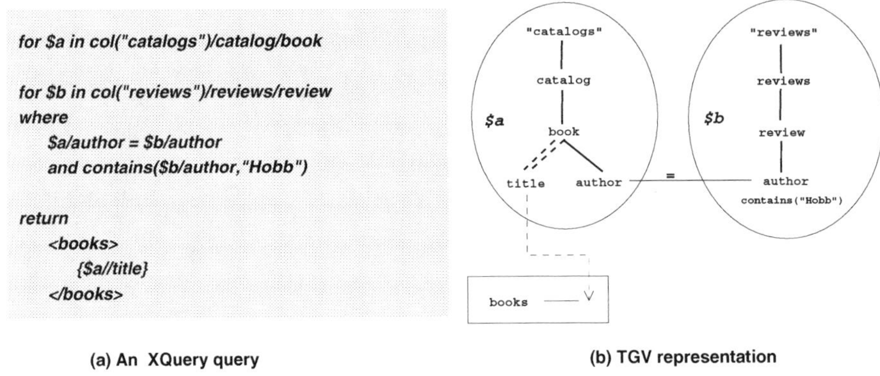

Fig. 2. An example of XQuery and its TGV representation

TGV Generic Annotation. The motivation to annotate a TGV is to allow annotating subsets of elements of a TGV model with various information. Precisely, for each arbitrary component (i.e. one or a group of operators of TGV), we add some additional information such as cost information, system performance information, source localization, etc. Our annotation model is generic and allows annotation of any type of information. The set of annotation based on the same annotation type is called an *annotated view*. There can be several annotated views for the same TGV, for example, time-cost annotated view, algorithm annotated view, sources-localization annotated view, etc.

3 Heterogeneous Cost Models and Cost Communication Within XLive

3.1 Cost Models for Heterogeneous Autonomous Data Sources

Cost Models Summary. We summarize different existing cost models for various types of data sources in Fig. 3. This summary is not only based on types of data sources but also on autonomy degrees of these sources. In addition, this summary gives some relations between different works on cost-based query optimization. The cost models with the name "operation" contain accurate cost formulas for calculating the execution cost of operators appearing in the plan. Generally, cost information such as source statistics is necessary for these cost models, because these statistics are used to derive the value of coefficients in cost formulas. It is often data sources implementers who are able to give accurate cost formulas with indispensable sources statistics.

When the data s ources are autonomous, cost formulas and source statistics are unavailable. For obtaining cost models we need some special methods that vary with the autonomy degree of data sources. For example, the method by *Calibration* [DKS92] estimates the coefficients of a generic cost model for each type of relational data sources. This calibration needs to know access methods

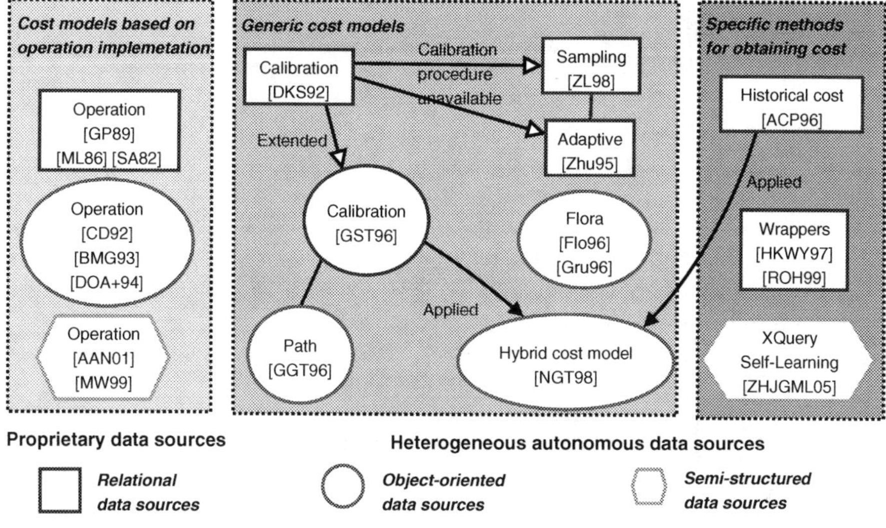

Fig. 3. Cost models for heterogeneous sources

used by the source. This method is extended to object-oriented databases by [GST96]. If this calibration procedure can not be processed due to data source constraints, a sampling method proposed in [ZL98] can derive a cost model for each type of query. The query classification in [ZL98] is based on a set of common rules adopted by many DBMSs. When no implementation algorithm and cost information are available, we can use the method described in [ACP96], in which cost estimation of new queries is based on the history of queries evaluated so far.

Generic Cost Model. Here, we show how to reuse the summary in Fig. 3 to define our generic cost model used for XQuery optimization in the XLive system. First, a cost model is generally designed for some type of data source (but there are also some methods that can be used for different types of sources, for example, the method by history [ACP96]). Second, this cost model can contain some accurate cost formulas with coefficients' value derived from data sources statistics, or a specific method for deriving the cost formulas. This cost model may also have only a constant value for giving directly the execution cost of operators. The possible attributes of our generic cost model are described in Table 1. This descriptive definition of cost model is used for TGV cost annotation for the purpose of overall cost estimation in the mediator level (ref. Section 4).

For a cost model, all attributes are optional by reason of generality. We apply the principle *as accurate as possible*. For example, the method by *calibration* can normally provide more accurate cost models than the method based on *historical costs*, but it has a lower accuracy level than cost models based on operation implementation. That means if the cost models based on operations implementation are available, we use neither the method by *calibration* nor *history*.

Table 1. Definition of generic cost model

Attribute	Description
Data source type	This type can be relational, object-oriented, semi-structured, files, Web services, etc.
Method	The specific method stored in this field can be used to derive the practicable cost formulas. These cost formulas may be inaccurate, but can at least roughly estimate the execution cost. This respect our *as accurate as possible* principle. Generally, some APIs corresponding to the specific method are available in this field, these APIs are implemented by XLive system and can give some useful services such as "provide the value of coefficients in the formulas".
Formulas	This is the core of a cost model, but they are often unavailable in a heterogeneous environment. These formulas are given in form of equations. The values of coefficients appearing in the formulas can also be represented in form of equations, for example, *Cardinality = 10000*. All these formulas forms an equations system. For some cost models, only a constant cost value is available. This value can be provided by data source (stored in *wrapper information repository*), or derived from results of executed queries (*historical cost*)

3.2 Generic Language for Cost Communication (GLCC)

XML-Based Generic Language. To perform cost communication within our XLive system, we define a language to express the cost information in a uniform, complete and generic manner. This language fits to our XML environment, to avoid costly format converting. It considers every cost model type and allows wrappers to export their specific cost information. In our XLive context, this language is generic enough to express cost information of different parts of a TGV and is capable to express cost for various optimization goals, for example, response time, price, energy consummation, etc.

Our language extends the MathML language [W3C03], which allows us to define all mathematical functions in XML form. MathML fits to cost communication within XLive due to its semi-structured nature. We use the *Content Markup* in MathML to provide explicit encoding for cost formulas. We just add some rules to MathML to define the grammar of our language. Furthermore, this grammar is extensible so that users can always define its own tags for any type of cost.

Cost formulas are represented in the form of equations set. Each equation corresponds to a cost function that may be defined by the source or by the mediator. Each component of TGV is annotated with an equation set in which the number of equations is undefined. One function in a set may use variables defined in other sets. We define some rules to ensure the consistency of the equations system. First, every variable should have somewhere a definition. Second, by

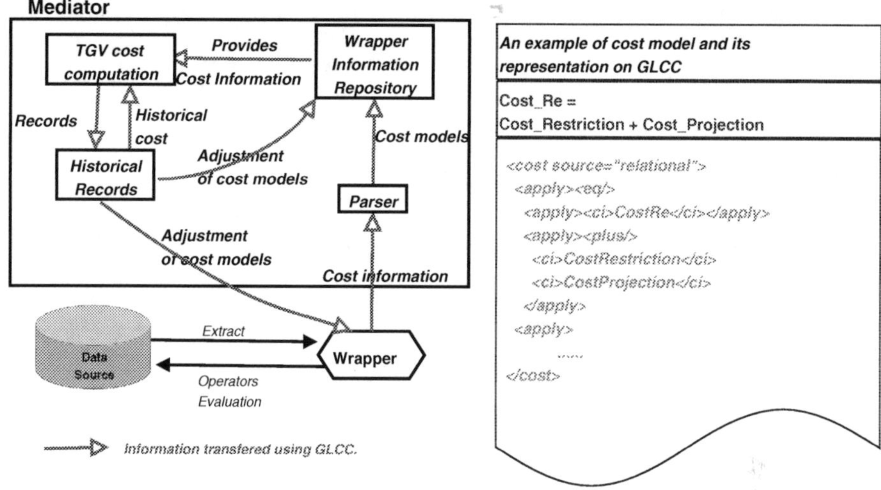

Fig. 4. Dynamic cost evaluation with GLCC in XLive system

reason of generality, there are no predefined variable names. For example, in the grammar, we do not define a name "time" for a cost variable because the cost metric can be a price unit. It is the user of the language who gives the specific significant names to variables. This gives a much more generic cost definition model compared to the language defined in [NGT98].

Dynamic Cost Evaluation. Fig. 4 gives a simple example for the expression of a cost model and shows the role of our language in cost communication. After extracting cost information from data source, the wrapper exports that information using our language to the parser, which derives cost models that will be stored in the wrapper information repository. When the mediator needs to compute the execution cost of a plan (TGV), the wrapper information repository provides necessary cost information for operators executed on wrappers. We have a cache for storing historical execution cost of queries evaluated, which can be used to adjust the exported cost information from the wrapper. All these communications are processed in the form of our language. Our language completes the interface between different components of XLive.

4 Overall Cost Estimation

4.1 TGV Cost Annotation

As mentioned in Section 2, the TGV is the logical execution plan of XQuery within the query processing in XLive. The purpose of our query optimization is to find the *optimal* TGV with the least execution cost. For estimating the overall cost of a TGV, we annotate different components (one or a group of operators) of

TGV. For an operator or a group of operators appearing in a TGV, the following cost information can be annotated:

- Localization: The operator(s) can be executed on the mediator or on the wrappers (data sources).
- Cost Model: Used to calculate the execution cost of the component.
- Other information: Contains supplementary information that is useful for cost estimation. For example, several operators' (such as join operator) implementation allows parallel execution between its related operators.

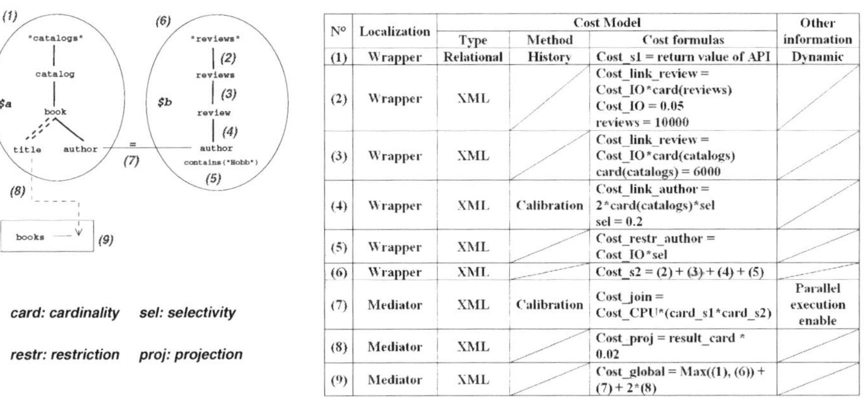

N°	Localization	Cost Model			Other information
		Type	Method	Cost formulas	
(1)	Wrapper	Relational	History	Cost_s1 = return value of API	Dynamic
(2)	Wrapper	XML		Cost_link_review = Cost_IO*card(reviews) Cost_IO = 0.05 reviews = 10000	
(3)	Wrapper	XML		Cost_link_review = Cost_IO*card(catalogs) card(catalogs) = 6000	
(4)	Wrapper	XML	Calibration	Cost_link_author = 2*card(catalogs)*sel sel = 0.2	
(5)	Wrapper	XML		Cost_restr_author = Cost_IO*sel	
(6)	Wrapper	XML		Cost_s2 = (2) + (3) + (4) + (5)	
(7)	Mediator	XML	Calibration	Cost_join = Cost_CPU*(card_s1*card_s2)	Parallel execution enable
(8)	Mediator	XML		Cost_proj = result_card * 0.02	
(9)	Mediator	XML		Cost_global = Max((1), (6)) + (7) + 2*(8)	

Fig. 5. An example for TGV cost annotation

Fig. 5 gives an example for TGV cost annotation. In this example, different components of the TGV introduced in Fig. 2 (Ref. Section 2) are annotated. We can see for the operators executed on Source1(S1), we have only the historical cost to use for estimate the total execution cost of all the these operators; in contrast, for each operator executed on Source2(S2), we have a cost model for estimating its execution cost. For the join operator(numbered (7)) executed on the mediator, the operators linked to it can be executed in parallel.

4.2 Overall Cost Estimation

Cost Annotation Tree (CAT). We have seen how to annotate a TGV with cost information. Now we are concentrated on how to use this cost annotation for the overall cost estimation of a TGV. As illustrated in Fig. 5, the cost of an annotated component of TGV generally depends on the cost of other components. For example, for the cost formula annotated in (6), we see that it depends on the cost of (2), (3), (4) and (5). From the cost formulas annotated for each component of TGV, we obtain a *Cost Annotation Tree (CAT)*. In a CAT, each node represents a component of TGV annotated by cost information and this CAT describes the hierarchical relations between these different components. Fig. 6 (a) illustrates the CAT of the TGV annotated in Fig. 5.

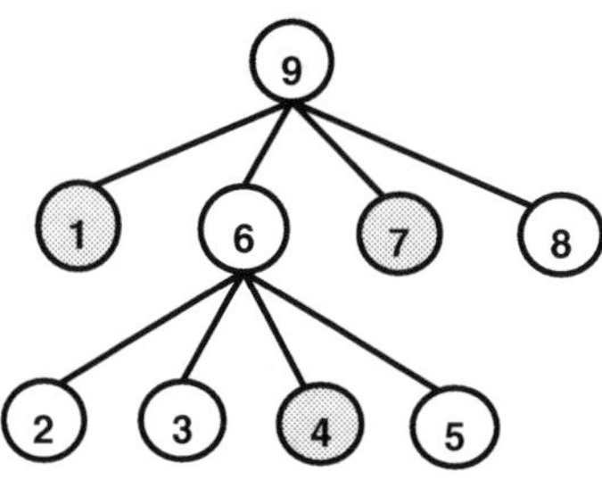

(1) *Node that needs to call APIs for obtaining the necessary coefficients' value*

(a) Cost Annotation Tree (CAT)

```
1  associateCost (node) {
2      node.analyzeCostModel ( );
3      if (node.hasSpecialMethod( )) {
4          node.callAPI( );
5      }
6      for (each child of node) {
7          associateCost(child);
8      }
9      node.configCostFormula( );
10     node.calculateCost( );
11 }
```

(b) Overall cost estimation algorithm

Fig. 6. Cost Annotation Tree and the algorithm for overall cost estimation

Overall Cost Estimation Algorithm. We now show how to use the CAT of a TGV to perform the overall cost estimation. We use the recursive breadth-first search algorithm of a tree for performing cost estimation of each node. For each node of CAT, we define a procedure called *associateCost* (Fig. 6 (b)) for operating the cost annotation of a node. This procedure first analyzes the cost annotation of the node and derives its cost model (line 2); If a specific cost method is found, it calls an API implemented by XLive for obtaining the necessary values of coefficients or cost formulas for computing the cost (line 3-5); if the cost of this node depends on the cost of its child nodes, it executes recursively the *associateCost* procedure on its child nodes (line 6-8). When these 3 steps are terminated, a procedure *configCostFormula* completes the cost formulas with obtained values of coefficients (line 9) and execution cost of this node will be calculated (line 10). By using this algorithm, we can obtain the overall cost of a TGV, which is the cost of the root of CAT.

4.3 Application: Plan Comparison and Generation

It has been shown in [TDNL06] that for processing a given XQuery, a number of candidate plans (i.e. TGV) can be generated using transformation rules that operate on TGVs. These rules have been defined for modifying the TGV without changing the result. The execution cost of a TGV can be computed by using our generic cost framework and thus we can compare the costs of these plans to choose the best one to execute the query.

However, as the number of rules is huge, this implies an exponential blow-up of the candidate plans. It is impossible to calculate the cost of all these candidate plans, because the cost computation and the subsequent comparisons will be even more costly than the execution of the plan. Thus, we need a search strategy to reduce the size of the search space containing candidate execution plans. We note in this respect that our cost framework is generic enough to be applied to various search strategies such as exhaustive, iterative, simulated annealing, genetic, etc.

5 Conclusion

In this paper, we described our cost framework for the overall cost estimation of candidate execution plans in an XML-based mediation system. The closest related work is DISCO system [NGT98], which defines a generic cost model for an object-based mediation system. Compared to DISCO work and other mediation systems, we have the following contributions: First, to our knowledge, our cost framework is the first approach proposed for addressing the costing problem in XML-based mediation systems. Second, our cost communication language is completely generic to express any type of cost, which is an improvement compared to the language proposed in DISCO. Third, our cost framework is generic enough to fit to overall cost computation within various mediation systems.

As futur work, we plan to define a generic cost model for XML sources with cost formulas that can compute the cost with given parameters that are components in TGV. This cost model would be generic for all types of XML sources. We will also concentrate on the design of an efficient search strategy that will be used in our cost-based optimization procedure.

Acknowledgment

This work is supported by *Xcalia S.A.* (France) and by ANR *PADAWAN* project.

References

[AAN01] Aboulnaga, A., Alameldeen, A., Naughton, J.: Estimating the Selectivity of XML Path Expressions for Internet Scale Applications. In: VLDB (2001)

[ACP96] Adali, S., Candan, K., Papakonstantinou, Y.: Query Caching and Optimization in Distributed Mediator Systems. In: ACM SIGMOD (1996)

[BMG93] Blakeley, J.A., McKenna, W.J., Graefe, G.: Experiences Building the Open OODB Query Optimizer. In: ACM SIGMOD (1993)

[CD92] Cluet, S., Delobel, C.: A General Framework for the Optimization of Object-Oriented Queries. In: ACM SIGMOD (1992)

[DKS92] Du, W., Krishnamurthy, R., Shan, M.C.: Query Optimization in a Heterogeneous DBMS. In: VLDB (1992)

[DNG03] Dang-Ngoc, T.T., Gardarin, G.: Federating Heterogeneous Data Sources with XML. In: Proc. of IASTED IKS Conf. (2003)

[DNGT04] Dang-Ngoc, T.T., Gardarin, G., Travers, N.: Tree Graph View: On Efficient Evaluation of XQuery in an XML Mediator. In: BDA (2004)

[DOA+94] Dogac, A., Ozkan, C., Arpinar, B., Okay, T., Evrendilek, C.: Advances in Object-Oriented Database Systems. Springer-Verlag (1994)

[Flo96] Florescu, D.: Espace de Recherche pour l'Optimisation de Requêtes Objet PhD thesis, University of Paris IV (1996)

[GGT96] Gardarin, G., Gruser, J.R., Tang, Z.H.: Cost-based Selection of Path Expression Algorithms in Object-Oriented Databases. In: VLDB (1996)

[GM93] Graefe, G., McKenna, W.J.: The Volcano Optimizer Generator: Extensibility and Efficient Search. In: ICDE (1993)

[GP89] Gardy, D., Puech, C.: On the Effects of Join Operations on Relation Sizes. ACM Transactions on Database Systems (TODS) (1989)

[Gru96] Gruser, J.R.: Modéle de Coût pour l'Optimisation de Requêtes Objet. PhD thesis, University of Paris IV (1996)

[GST96] Gardarin, G., Sha, F., Tang, Z.H.: Calibrating the Query Optimizer Cost Model of IRO-DB. In: VLDB (1996)

[HKWY97] Haas, L.M., Kossmann, D., Wimmers, E.L., Yang, J.: Optimization Queries Across Diverse Data Sources. In: VLDB (1997)

[ML86] Mackert, L.F., Lohman, G.M.: R* Optimizer Validation and Performance Evaluation for Local Queries. In: ACM SIGMOD (1986)

[MW99] McHugh, J., Widom, J.: Query Optimization for Semistructured Data. Technical report, Stanford University Database Group (1999)

[NGT98] Naacke, H., Gardarin, G., Tomasic, A.: Leveraging Mediator Cost Models with Heterogeneous Data Sources. In: ICDE (1998)

[NJT05] Dang Ngoc, T.T., Jamard, C., Travers, N.: XLive: An XML Light Integration Virtual Engine. In: Bases de Données Avancées (BDA) (2005)

[ROH99] Roth, M.T., Ozcan, F., Haas, L.: Cost Models DO Matter: Providing Cost Information for Diverse Data Sources in a Federated System. In: VLDB (1999)

[RS97] Tork Roth, M., Schwarz, P.M.: Don't Scrap It, Wrap it! A Wrapper Architecture for Legacy Data Sources. In: VLDB (1997)

[SA82] Selinger, P.G., Adiba, M.E.: Access path selection in distributed database management systems. In: ICOD (1982)

[TDNL06] Travers, N., Dang-Ngoc, T.T., Liu, T.: TGV: An Efficient Model for XQuery Evaluation within an Interoperable System. Int. Journal of Interoperability in Business Information Systems (IBIS), vol. 3 (2006)

[Tra06] Travers, N.: Optimization Extensible dans un Mèdiateur de Donnèes XML, PhD thesis, University of Versailles (2006)

[W3C03] W3C. Mathematical Markup Language (Mathml TM) Version 2.0 (2003)

[W3C05] W3C. An XML Query Language (XQuery 1.0) (2005)

[Wie92] Wiederhold, G.: Mediators in the Architecture of Future Information Systems. Computer 25(3), 38–49 (1992)

[ZHJGML05] Zhang, N., Haas, P.J., Josifovski, V., Zhang, C., Lohman, G.M.: Statistical Learning Techniques for Costing XML Queries. In: VLDB (2005)

[Zhu95] Zhu, Q.: Estimating Local Cost Parameters for Global Query Optimization in a Multidatabase System. PhD thesis, University of Waterloo (1995)

[ZL98] Zhu, Q., Larson, P.A.: Solving Local Cost Estimation Problem for Global Query Optimization in Multidatabase Systems. Distributed and Parallel Databases (1998)

Cost-Based Vertical Fragmentation for XML

Sven Hartmann, Hui Ma, and Klaus-Dieter Schewe

Massey University, Department of Information Systems & Information Science Research Centre, Private Bag 11 222, Palmerston North, New Zealand
{s.hartmann,h.ma,k.d.schewe}@massey.ac.nz

Abstract. The Extensible Markup Language (XML) has attracted much attention as a data model for data exchange, data integration and rich data representation. A challenging question is how to manage native XML data in distributed databases. This leads to the problem of how to obtain a suitable distribution design for XML documents. In this paper we present a design approach for vertical fragmentation to minimise total query costs. Our approach is based on a cost model that takes the complex structure of queries on XML data into account. We show that system performance can be improved after vertical fragmentation using our approach, which is based on user access patterns.

1 Introduction

Vertical fragmentation is an important database distribution design technique for improving system performance. It has been widely studied in the context of the relational model, cf. [6, 10, 18, 19, 20, 21, 23] and object oriented data model, cf. [3, 7, 11]. With the emergence of XML as a standard format for data exchange, data integration and rich data representation, distribution design for XML becomes a highly relevant topic, in particular as data shared over the web is naturally distributed to meet the needs of the users who are physically distributed. Due to the particularities of the XML data model, the adaption of distribution design principles for relational or object-oriented data to XML poses a real challenge for database research. In [12], horizontal, vertical, and split fragmentation techniques for XML are studied. In [13], a horizontal fragmentation algorithm for XML is proposed. In [5], a fragmentation method for XML is proposed together with an allocation model for distributed XML fragments. For that, local index structures are presented that allow efficient storage of global context for local fragments, facilitate the local execution of queries, and support the reconstruction of fragments distributed over multiple sites. In [2], horizontal, vertical, and hybrid fragmentation techniques for XML are surveyed. In addition, correctness rules evaluating a fragmentation schema are presented and experimental results are given to emphasise the beneficial effects of fragmentation.

To the best of our knowledge no work so far has discussed how to perform fragmentation in the context of XML procedurally using an top-down design approach based on user queries as input information. In the paper at hand, we

K.C. Chang et al. (Eds.): APWeb/WAIM 2007 Ws, LNCS 4537, pp. 12–24, 2007.

will study this problem for the vertical fragmentation of XML data. Most vertical fragmentation algorithms for relational data presented in the literature are affinity-based, that is, they use attribute affinities as input. Affinity-based fragmentation approaches have also been adopted to the object oriented data model, using different kinds of affinities such method affinities, attribute affinities, or instance variable affinities. One disadvantage of affinity-based approaches is that they do not reflect local data requirements and therefore cannot improve the local availability of data and reduce data transportation between sites to improve system performance. To overcome this deficiency of affinity-based algorithms, a cost-based vertical fragmentation approach for relational data has recently been proposed in [14].

In this paper, we will discuss cost-based vertical fragmentation in the context of XML, thus extending earlier work presented in [16, 14]. In particular, we will outline a cost model that incorporates available information about user queries issued at various sites of the system. Using such a cost model it is possible to handle vertical fragmentation and fragment allocation simultaneously with low computational complexity and resulting high system performance.

The remainder of this paper is organised as follows. In the remainder of this section we start with a brief review of the XML data model, discuss query algebra for XML, and give a definition of vertical fragmentation of XML data. In Section 2, we present a cost model for XML. In Section 3, we propose a cost-based design approach for vertical fragmentation and illustrate it by an example. We conclude with a short summary in Section 4.

XML as Native Data Type. With the emergence of XML, most providers of commercial database management systems made an effort to incorporate some support for XML data management into their products. In the beginning, this was typically achieved by shredding XML data to object-relational structures or by mapping it to existing data types such as LOBs (CLOBs, BLOBs). The limitations of this approach are well-known. Since then, much effort has been spent to optimise database management systems for the handling of XML data. Only recently, native XML support has been implemented, e.g., in the latest version of IBM DB2 [22] where XML can now be included into tables using the new XML data type. That is, tables may hold relational data and native XML data simultaneously. In SQL table declarations, it is now possible to specify that a column should be of type XML, cf. Example 1.

Example 1. With the XML data type, we may for example define a table with the table schema DEPARTMENT = (Name : *STRING*, Homepage : *STRING*, Address : *STRING*, Lecturers: XML, Papers: XML).

Internally, XML and relational data are stored in different formats, which match their corresponding data models. XML columns are stored on disk pages in tree structures matching the XML data model. Using the common XML data model, XML documents are regarded as *data trees*. XML documents are frequently described by XML schemas. Most database management systems support XSDs

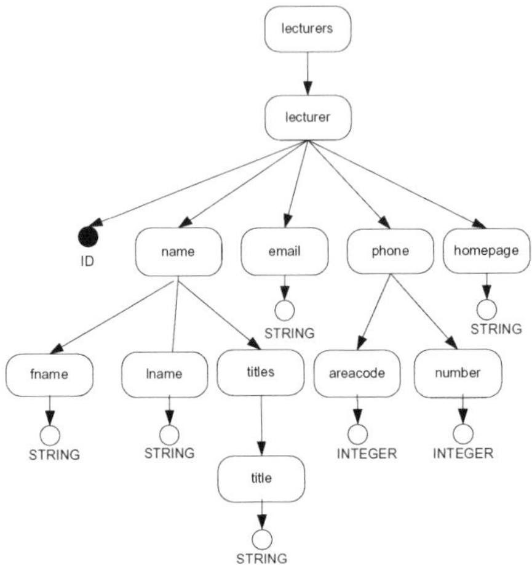

Fig. 1. XML Tree for the Lecturer Data Type

(XML Schema Definitions) as XML schemas. XML schemas can be visualised as *schema trees*, too. This often helps imagining the nested structure of the corresponding XML documents. In our example, we might be given an XML schema for the Lecturers attribute with the schema tree in Figure 1.

When declaring a column as XML, each cell in this column may hold an XML document. That is, the domain of the XML data type is actually complex-valued. For the sake of simplicity, we assume that the XML documents in such a column are described by an XML schema. In DB2, for example, XSDs (XML Schema Definitions) can be used for this purpose. To facilitate the adaption of the vertical fragmentation technique from the relational model to XML, it is useful to assume that each XML element e has a label ℓ and a content type t. In particular, each column ℓ of type XML is thus associated with the root element e of an XML schema with label ℓ and content type t. For the expressiveness of different XML schema languages and content types of elements, we refer to [4]. In the paper at hand, we will also use the label-extended type system defined in [16] where content types are called *representation types*.

Example 2. Consider LECTURERS and PAPERS that have been declared as XML. From the XML schema we obtain their representation types which may look for examples as follows:

$t_{\text{LECTURERS}}$ = lecturers: {lecturer: (id: *INTEGER*, name: (fname: *STRING*, lname: *STRING*, titles:{title: *STRING*}), homepage: *STRING*, email: *STRING*, phone: (areacode: *INTEGER*, number: *INGEGER*))}

t_{PAPERS} = papers: {paper: (no : *STRING*, title: *STRING*, taught : *IDREF*, description: *STRING*, points: *STRING*, campus: *STRING*, literature: [(article : *STRING*) $\oplus$ (book: *STRING*)])}

Query Algebra. Database management systems exploit query algebra for evaluating user queries. For that, user queries are transformed into query trees and further optimised during query optimisation. The query tree specifies which query plan is actually executed inside the database management system to compute the query result. Query algebras for XML have been studied extensively in the literature. Prominent examples include the XQuery algebra, the IBM and Lore algebra [9], and XAL [8]. In [1] query algebras for XML are surveyed, while in [24] query algebras for nested data structures are surveyed. In [16], a generic query algebra for complex value databases is presented. However, currently no complete theory of query optimisation based on structural recursion is known. For our purposes here, we restrict the discussion to a simpler query algebra that provides at least standard operations such as projection, selection, join, union, difference, and intersection.

For projection, for example, we can use an operator $\pi_{p'_1,\ldots,p'_i}(e)$ with path expressions $p'_1,\ldots,p'_i$. When applied to an XML element e the tree rooted at e will be reduced to the subtrees rooted at the subnodes selected by the specified path expression $p'_1,\ldots,p'_k$ plus the paths from e to these subnodes. The label of e will be the same as before, while the content type can be updated to reflect the reductions.

Each node in an XML tree can be reached from the root node by a unique path. The labels along this path give rise to a simple path expression $\ell_0/\ell_1/\ldots/\ell_k$. In DB2, for example, all labels that are used in XML documents are stored in a global string table [22]. As common, we use XPath expressions to access nodes in XML trees. To support query processing we need a path language that is expressive enough to be practical, yet sufficiently simple to be reasoned about efficiently. The computational tractability of query containment and query equivalence has been widely studied in the literature, cf. [17]. Here we use path expressions p which may include node labels, the context node (.), the child axis (/), the descendant-or-self axis (//), and the label wildcard (*). That is, we use the XPath fragment $XP^{[//,*]}$ for which query equivalence can be decided in polynomial time [17].

Example 3. The algebraic query expression $\pi_{\text{id, name/fname}}(\text{lecturers})$ projects the lecturer elements to the lecturers' id and first name.

Vertical Fragmentation. Consider a table schema with a column ℓ declared as XML, and let $t = t_e$ be the representation type of the root element e of the corresponding XML schema in the label-extended type system [16]. Vertical fragmentation will replace the column ℓ by new columns $\ell_{v1},\ldots,\ell_{vi}$, where each ℓ_{vj} is again declared as XML. Let t_j denote the representation types of the root elements of the corresponding XML schemas for the new columns. Vertical fragmentation of e can be achieved by projection operations using path expressions that are evaluated from the root.

The general rule of performing fragmentation is that it should be lossless, that is, there should not be any loss or addition of data after the fragmentation. In the RDM the correctness rules for fragmentation include three criteria, *completeness, disjointness* and *reconstruction*. We now adapt them to vertical fragmentation of XML documents. *Completeness* requires that each leaf node (that is, element node with pure text content or attribute node) together with its attached data item in an XML document must appear in at least one of the fragments. *Disjointness* requires that each leaf node together with its attached data item in an XML document occur no more than one of the fragments. *Reconstruction* requires that it is possible to reconstruct the original XML document by using join operations. In other words, one has to ensure that the fragmentation is reversible.

2 A Query Processing Cost Model

In practice there are usually several query trees that can be used to compute the result of a query: queries may be rewritten in various equivalent forms, and intermediate results may be executed at different sites. In this section we look at a cost model for the queries that can be used to evaluate and compare query trees. For that, we suggest to estimate the sizes of intermediate results for all intermediate nodes in the query tree. These sizes determine the costs for retrieving data for the next step, i.e., the operation associated with the predecessor in the query tree, and the costs for the transportation of data between nodes. Afterwards, we present a cost model for measuring system performance.

Size Estimation. We first approach an estimation of the sizes intermediate query results. In order to do so, we first look at elements and attributes defined in the XML schema. Then, if r is the root element, the estimated size $s(r)$ is the size estimated for the entire document rooted at r.

Let s_i be the average size of values of base type b_i. In particular, let s_0 be the average size of values of base types ID and $IDREF$, that is, of identifiers. From these assumptions we can compute estimates for the average size of elements. Let $att(\ell)$ denote the set of attributes defined on element ℓ. We can proceed inductively to define the size $s(e)$:

$$s(e) = \begin{cases} s_i & \text{if } e = b_i \\ \sum_{a \in att(\ell)=1} s(a) & \text{if } e = \epsilon \\ s(n) + \sum_{a \in att(\ell)} s(a) & \text{if } e = n \\ q \cdot s(e') + \sum_{a \in att(\ell)} s(a) & \text{if } e = (e')^* \\ \sum_{i=1}^{n} s(e_i) + \sum_{a \in att(\ell)} s(a) & \text{if } e = e_1, \dots, e_n \\ \sum_{i=1}^{n} p_i \cdot s(e_i) + \sum_{a \in att(\ell)} s(a) & \text{if } e = e_1, \dots, e_n \\ \frac{1}{n} \sum_{i=1}^{n} s(e_i) & \text{if } e = (a_1 : e_1) \oplus \cdots \oplus (a_n : e_n) \\ r \cdot s(e') & \text{if } e = \{e'\} \text{ or } e = [e'] \text{ or } e = \langle e' \rangle \end{cases}$$

where q is the average number of successor elements with e', p_i is the probability that the successor of ℓ is defined by e_i. In particular, we have $\sum_{i=1}^{n} p_i = 1$.

- The size of a projection node $\pi_{p'}$ is $(1-p) \cdot s \cdot \frac{s_e}{s_p}$ where s is the size of the successor node in the query tree, s_p is the average size of an element reached by p', s_e is the average size of an element defined by e, and p is the probability that two element x_1, x_2 matching p' coincide on their projection to e, i.e.,$\pi_e^{e'}(x_1) = \pi_e^{e'}(x_2)$.
- For a join node $\bowtie_{p'}$ the size is $\frac{s_1}{s_1'} \cdot p \cdot \frac{s_2}{s_2'}(s_1' + s_2' - s)$, where the $s_i (i = 1, 2)$ are sizes of the successors in query tree, $s_i'(i = 1, 2)$ are the sizes of the elements, e_1, e_2, p is the probability that such two elements match, and s is the size of common leaf nodes e.

Query Processing Costs. Taking the cost model in [15] we now analyse the query costs in the case of vertical fragmentation. For the convenience of discussion we briefly present the cost model in the following. The major objective is to base the fragmentation decision on the efficiency of the most frequent queries.

Fragmentation results in a set of fragments $\{F_1, \ldots, F_n\}$ of average sizes $s_1, \ldots, s_n$. If the network has a set of nodes $N = N_1, \ldots, N_k$ we have to allocate these fragments to one of the nodes, which gives rise to a mapping $\lambda : \{1, \ldots, n\} \to \{1, \ldots, k\}$, which we call a *location assignment*. This decides the allocation of leaves of query trees, which are fragments. For each intermediate node v in each relevant query tree, we must also associate a node $\lambda(v)$, i.e., $\lambda(v)$ indicating the node in the network that the intermediate query result corresponding to v will be stored at.

Given a location assignment λ we can compute the total costs of query processing. Let the set of queries be $Q^m = \{Q_1, \ldots, Q_m\}$, each of which is executed with a frequency f_j. The total costs of all the queries in Q^m are the sum of the costs of each query multiplied by its frequency. The cost of each query are composed of two parts, the storage costs and transportation costs. Storage costs measure the costs of retrieving the data back from secondary storage, depend on the size of the intermediate results and on the assigned locations, which decide the storage cost factors. The transportation costs provide a measure for transporting between two nodes of the network, depend on the sizes of the involved sets and on the assigned locations, which decide the transport cost factor between every pair of sites.

$$
\begin{aligned}
Costs_\lambda(Q^m) &= \sum_{j=1}^{m} (stor_\lambda(Q_j) + trans_\lambda(Q_j)) \cdot f_j \\
&= \sum_{j=1}^{m} (\sum_h s(h) \cdot d_{\lambda(h)} + \sum_h \sum_{h'} c_{\lambda(h')\lambda(h)} \cdot s(h')) \cdot f_j
\end{aligned}
$$

where h ranges over the nodes of the query tree for Q_j, $s(h)$ are the sizes of the involved sets, and d_i indicates the storage cost factor for node N_i $(i = 1, \ldots, k)$, h' runs over the predecessors of h in the query tree, and c_{ij} is the transportation cost factor for data transport from node N_i to node N_j $(i, j \in \{1, \ldots, k\})$.

3 Cost-Based Approach for Vertical Fragmentation

The aim of vertical fragmentation is to improve the system performance. In this section we will adapt the cost-efficient vertical fragmentation approach in [14] to XML. We start with some terminology, then present a vertical fragmentation design methodology, and illustrate it with an example.

Assume an XML document with a representation type t_e being accessed by a set of queries $Q^m = \{Q_1, \dots, Q_j, \dots Q_m\}$ with frequencies $f_1, \dots, f_m$, respectively. To improve the system performance, element e is vertically fragmented into a set of fragments $\{e_{V1}, \dots, e_{Vu}, \dots, e_{Vk_i}\}$, each of which is allocated to one of the network nodes $N_1, \dots, N_\theta, \dots, N_k$. Note that the maximum number of fragments is k, i.e.,$k_i \leq k$. We use $\lambda(Q_j)$ to indicate the site that issues query Q_j, and use $elem_j = \{e_i | f_{ji} = f_j\}$ to indicate the set of leaf elements that are accessed by Q_j, with f_{ji} as the frequency of the query Q_j accessing e_i. Here, $f_{ji} = f_j$ if the element e_i is accessed by Q_j. Otherwise, $f_{ji} = 0$.

The term *request* of an element at a site θ is used to indicate the sum of frequencies of all queries at the site accessing the element or attribute:

$$request_\theta(e_i) = \sum_{j=1,\lambda(Q_j)=\theta}^{m} f_{ji}.$$

With the size ℓ_i of an element e_i, we can calculate the *need* of an element or attribute as the total data volume involved to retrieve a_i by all the queries:

$$need_\theta(e_i) = \sum_{j=1,\lambda(Q_j)} f_{ji} \cdot \ell_i.$$

Finally, we introduce a term *pay* to measure the costs of accessing an element once it is allocated to a network node. The *pay* of allocating an element e_i to a site θ measures the costs of accessing element e_i by all queries from the other sites θ', which is different from *theta*. It can be calculated using the following formula:

$$pay_\theta(e_i) = \sum_{\theta'=1,\theta\neq\theta'}^{k} \sum_{j=1,\lambda(Q_j)=\theta}^{m} f_{ji} \cdot c_{hh'}.$$

A Design Methodology for Vertical Fragmentation. Following [20] we assume a simple transaction model for database management systems where the system collects the information at the site of the query and executes the query there. Under this assumption we can evaluate the costs of allocating a single element and then make decisions by choosing a site that leads to the least query costs. Following the discussion of how fragmentation affects query costs [16], the allocation of fragments to sites according to the cost minimisation heuristics already determines the location assignment, provided that an optimal location assignment for the queries was given prior to the fragmentation.

We take a two-step approach for vertical fragmentation of a table. For columns of some conventional (relational) data type we use the heuristic [14] to assign attributes to fragments, such that they share the primary key attributes. For columns of XML type we adapt the heuristic approach in [14] to perform vertical fragmentation with the following steps. Note that, for the sake of simplicity, we do not consider replication at this stage.

To record query information we use a *Element Usage Frequency Matrix* (EUFM) which records frequencies of queries, the subset of leaf nodes accessed by the queries and the sites that issue the queries. Each row in the EUFM represents one query Q_j; the head of column is the set of nodes that are covered by the representation type t_e, listed in some hierarchical order from top to bottom. When considering nodes we do not distinguish between attribute and element nodes, but record them in the same matrix. In addition, there are two further columns with one column indicating the site that issues the queries and the other indicating the frequency of the queries. The values on a column indicate the frequencies f_{ji} of the queries Q_j that use the corresponding leaf node e_i grouped by the site that issues the queries. Note that we treat any two queries issued at different sites as different queries, even if the queries themselves are the same.

The EUFM is constructed according to the optimised queries to record all the element requirements returned by queries as well as all the elements used in some join predicates. If a query returns all the information of an element then every descendant element are accessed by the query. As a general pragmatic guideline we follow the recommended rule of thumb to consider the 20% most frequent queries, as these usually account for most of the data access [23]. This procedure can be described as follows and implemented as in Table 1.

1. Take the most frequently processed 20% queries Q^N, which retrieve data from XML documents.
2. Optimise all the queries and construct an *EUFM* for each XML document E based on the queries.
3. Calculate the *request* at each site for each leaf element.
4. Calculate the *pay* at each site for each element.
5. Cluster all leaf elements to the site which has the lowest value of the *pay* to get a set of rooted label paths $RP(e_{V\theta})$ for each site.
6. Perform vertical fragmentation using the sets of rooted label path and allocate the fragments to the corresponding site.

The algorithm first finds the site that has the smallest value of *pay* and then allocates the element to the site. A vertical fragmentation and allocation schema are obtained simultaneously.

An Example. We now illustrate the algorithm using an example. Assume there are five queries that constitute the 20% most frequently queries accessing an XML document from three different sites.

1. $\pi_{\text{lecturers}}(\text{LECTURERS} \bowtie_{\text{lecturers//@id= papers//taught}} \text{PAPERS})$ is issued at site 1 with frequency as 20.
2. $\pi_{\text{titles, homepage}}(\text{LECTURERS})$ is issued at site 2 with frequency as 30.
3. $\pi_{\text{name/lname, phone}}(\text{LECTURERS})$ is issued at site 3 with frequency as 100.
4. $\pi_{\text{fname, email}}(\text{LECTURERS})$ is issued at site 1 with frequency as 50.
5. $\pi_{\text{titles, areacode}}(\text{LECTURERS})$ is issued at site 2 with frequency as 70.

Table 1. Algorithm for Vertical Fragmentation

Input:	$Q^M = \{Q_1, \dots, Q_m\}$ /* a set of queries $elem(e) = \{e_1, \dots, e_n\}$ /*a set of all the leaf nodes of e $RP(e) = \{rp_1, \dots, rp_n\}$ /*a set of rooted label path of all leaf nodes a set of network nodes $N = \{1, \dots, k\}$ the $EUFM$ of e
Output:	fragmentation and fragment allocation schema $\{e_{V1}, \dots, e_{Vk}\}$

Method:	`for each` $\theta \in \{1, \dots, k\}$ `let` $elem(e_{V\theta}) = \emptyset$ `endfor`
	`for each` element $e_i \in elem(e), 1 \leq i \leq n$ `do`
	  `for each` node $\theta \in \{1, \dots, k\}$
	    `do` calculate $request_\theta(e_i)$
	  `endfor`
	  `for each` node $\theta \in \{1, \dots, k\}$
	    `do` calculate $pay_\theta(e_i)$
	  `endfor`
	  `choose` w such that $pay_w(e_i) = \min_{\theta=1}^{k} pay_\theta(e_i)$
	  $elem(e_{Vw}) = elem(e_{Vw}) \cup e_i$ /* add e_i to e_{Vw}
	  $RP(e_{Vw}) = RP(e_{Vw}) \cup rp_i$ /* add rp_i to $RP(e_{Vw})$
	`endfor`
	`for each` $\theta \in \{1, \dots, k\}$, $e_{Vw} = \pi_{RP(e_{Vw})}(e)$ `endfor`

To perform vertical fragmentation using the design procedure introduced in 3 we first construct an Element *Usage* Frequency Matrix as in Table 2. Secondly, we compute the *request* for each element at each site, the results of which are shown in the Element *Request and Pay* Matrix in Table 3. Thirdly, assuming the values of transportation cost factors are: $c_{12} = c_{21} = 10$, $c_{13} = c_{31} = 25$, $c_{23} = c_{32} = 20$, we can now calculate the *pay* of each attribute at each site using the values of the *request* in Table 3. The results are shown in a Attribute *Request and Pay* Matrix in Table 3.

Table 2. Element *Usage* Frequency Matrix

Site	Query	Frequency	lecturers							
			lecturer							
			ID	name			email	phone		homepage
				fname	lname	titles		areacode	number	
						title				
Length			20	$20 \cdot 8$	$20 \cdot 8$	$2 \cdot 15 \cdot 8$	$30 \cdot 8$	10	20	$50 \cdot 8$
1	Q_1	20	20	20	20	20	20	20	20	20
	Q_4	50	0	50	0	0	50	0	0	0
2	Q_2	30	0	0	30	30	0	0	0	30
	Q_5	70	0	0	0	70	0	70	0	0
3	Q_3	100	0	0	100	0	0	100	100	0

Table 3. Element *Request and Pay* Matrix

request/pay	lecturers							
	lecturer							
	ID	name			email	phone		homepage
		fname	lname	titles title		areacode	number	
$request_1(e_i)$	20	70	20	20	70	20	20	20
$pay_1(e_i)$	0	0	2800	1000	0	3200	2500	300
$request_2(e_i)$	0	0	30	100	0	70	0	30
$pay_2(e_i)$	200	700	2200	200	700	2200	2200	200
$request_3(e_i)$	0	0	100	0	0	100	100	0
$pay_3(e_i)$	500	1750	1100	2500	1750	1900	500	1100
site	1	1	3	2	1	3	3	2

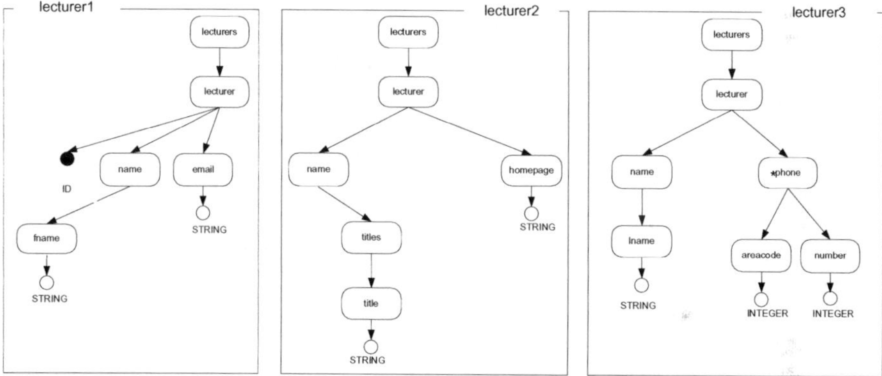

Fig. 2. XML Tree for the Fragments of Lecturers

Grouping the leaf elements to the site of the lowest *pay* we get allocations of elements as shown as the last row in Table 3. Correspondingly, we get three sets of rooted label paths that can be used to perform vertical fragmentation:

- $RP(e_{V1}) = \{$lecturers/lecturer/(ID, name/fname, email)$\}$
- $RP(e_{V2}) = \{$lecturers/lecturer/(name/titles/title, homepage)$\}$
- $RP(e_{V3}) = \{$lecturers/lecturer/(name/lname, phone)$\}$

When applying projection to the Lecturer column in our table schema, we get three new columns of type XML. The XML schema trees corresponding to these vertical fragments are shown in Figure 2.

We now look at how the system performance is changed due to the outlined fragmentation by using the cost model presented above. Assume that the average number of lecturers is 20 and the average number of titles for each lecturer is 2. With the average length of each element given in Table 3, we can compute the total query costs. Two different ways of distributed query evaluation are considered. If distributed XML query processing and optimisation are supported then selection and projection should be processed first locally to reduce the size of data transported among different sites. In this case, the optimised allocation of document LECTURERS is site 2 (see Figure 3) which leads to total query costs of 16,600,000 while the total query costs after the vertical fragmentation and allocation are 4,474,000, which is about one fourth of the costs before the fragmentation. If queries are evaluated in a way that whole documents are shifted to the site issuing the queries to execute the queries over there, then the improvement of the system performance after fragmentation is even more obvious. Before fragmentation, the optimised allocation of XML document LECTURERS is site 2, which leads to total query costs of 67,500,000. After vertical fragmentation and fragment allocation, the total query costs are 9,780,000, which is only about one seventh of the costs before fragmentation. This shows that vertical fragmentation can indeed improve system performance.

Discussion. In distributed databases, costs of queries are dominated by the costs of data transportation from a remote site to the site that issued the queries. To compare different vertical fragmentation schemata we would like to compare how it affect the transportation costs. Due to the complexity of vertical fragmentation it is practically impossible to achieve an optimised vertical fragmentation schema by exhaustedly comparing different fragmentation schema using the cost model. However, from the cost model above, we observe that the less the value of the pay of allocating an element to a site the less the total costs will be to access it [14]. This explains that the design procedure above can at least achieve semi-optimal vertical fragmentation schema.

Using the above algorithm we can always guarantee that the resulting vertical fragmentation schema meet the criteria of correctness rules. Disjointness and completeness are satisfied because all leaf nodes occur and only occur in one of the fragments. Reconstruction is guarantied because all fragments are composed of a set of rooted label path of leaves. Because we use rooted label paths for fragmentation, so that the upper common label path between a pair of vertical fragments serves as a hook for fragments [5]. Using the identification schemata in [5, 25], which use reasonable storage consumption, fragments reconstruction can be in performed in constant time.

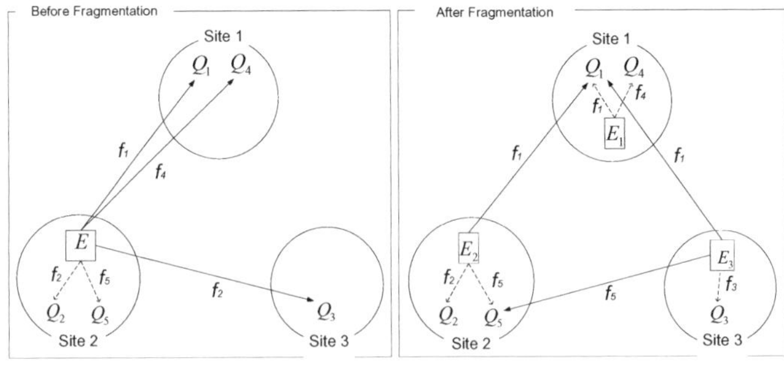

Fig. 3. Allocation before and after Fragmentation

4 Conclusion

In this paper we presented a vertical fragmentation design approach for XML based on a cost model for measuring total query costs of accessing XML documents. This approach takes into consideration of hierarchical structure of XML documents which are supported by some DBMS as a native data type. Furthermore, a design procedure for vertical fragmentation is presented for XML distribution design. An related problem left for future work is an integrated approach of horizontal and vertical fragmentation.

References

[1] Abraham, J., Chaudhari, N., Prakash, E.: XML query algebra operators, and strategies for their implementation. In: IEEE Region 10 Conference, pp. 286–289. IEEE Computer Society Press, Los Alamitos (2004)

[2] Andrade, A., Ruberg, G., Baião, F.A., Braganholo, V.P., Mattoso, M.: Efficiently processing xml queries over fragmented repositories with partix. In: Ioannidis, Y., Scholl, M.H., Schmidt, J.W., Matthes, F., Hatzopoulos, M., Boehm, K., Kemper, A., Grust, T., Boehm, C. (eds.) EDBT 2006. LNCS, vol. 3896, pp. 150–163. Springer, Heidelberg (2006)

[3] Bellatreche, L., Simonet, A., Simonet, M.: Vertical fragmentation in distributed object database systems with complex attributes and methods. In: Thoma, H., Wagner, R.R. (eds.) DEXA 1996. LNCS, vol. 1134, pp. 15–21. Springer, Heidelberg (1996)

[4] Bex, G.J., Neven, F., Van den Bussche, J.V.D.: Dtds versus xml schema: a practical study. In: WebDB, pp. 79–84. ACM Press, New York (2004)

[5] Bremer, J.-M., Gertz, M.: On distributing xml repositories. In: WebDB, pp. 73–78 (2003)

[6] Cornell, D., Yu, P.: A vertical partitioning algorithm for relational databases. In: ICDE, pp. 30–35 (1987)

[7] Ezeife, C.I., Barker, K.: Vertical fragmentation for advanced object models in a distributed object based system. In: ICCI, pp. 613–632 (1995)

[8] Frasincar, F., Houben, G.-J., Pau, C.: Xal: an algebra for xml query optimization. In: ADC, pp. 49–56 (2002)

[9] Goldman, R., McHugh, J., Widom, J.: From semistructured data to xml: Migrating the lore data model and query language. In: WebDB

[10] Hoffer, J.A., Severance, D.G.: The use of cluster analysis in physical database design. In: VLDB, pp. 69–86 (1975)

[11] Karlapalem, K., Navathe, S.B., Morsi, M.M.A.: Issues in distribution design of object-oriented databases. In: IWDOM, pp. 148–164 (1992)

[12] Ma, H.: Distribution design in object oriented databases. Master's thesis, Massey University (2003)

[13] Ma, H., Schewe, K.-D.: A heuristic approach to cost-efficient horizontal fragmentation of XML documents. In: Pastor, Ó., Falcão e Cunha, J. (eds.) CAiSE 2005. LNCS, vol. 3520, pp. 131–136. Springer, Heidelberg (2005)

[14] Ma, H., Schewe, K.-D., Kirchberg, M.: A heuristic approach to vertical fragmentation incorporating query information. In: Baltic Conference on Databases and Information Systems, pp. 69–76 (2006)

[15] Ma, H., Schewe, K.-D., Wang, Q.: A heuristic approach to cost-efficient fragmentation and allocation of complex value databases. In: ADC, pp. 119–128 (2006)
[16] Ma, H., Schewe, K.-D., Wang, Q.: Distribution Design for Higher-Order Data Models. Data. and Knowledge Engineering 60, 400–434 (2007)
[17] Miklau, G., Suciu, D.: Containment and equivalence for a fragment of XPath. J. ACM 51 1, 2–45 (2004)
[18] Muthuraj, J., Chakravarthy, S., Varadarajan, R., Navathe, S.B.: A formal approach to the vertical partitioning problem in distributed database design. In: International Conference on Parallel and Distributed Information Systems, pp. 26–34 (1993)
[19] Navathe, S., Karlapalem, K., Ra, M.: A mixed fragmentation methodology for initial distributed database design. Journal of Computer and Software Engineering 3 (1995)
[20] Navathe, S.B., Ceri, S., Wiederhold, G., Dour, J.: Vertical Partitioning Algorithms for Database Design. ACM TODS 9, 680–710 (1984)
[21] Navathe, S.B., Ra, M.: Vertical Partitioning for Database Design: A Graphical Algorithm. SIGMOD Record 14, 440–450 (1989)
[22] Nicola, M., van der Linden, B.: Native xml support in db2 universal database. In: VLDB, pp. 1164–1174 (2005)
[23] Özsu, M., Valduriez, P.: Principles of Distributed Database Systems (1999)
[24] Schewe, K.-D.: On the unification of query algebras and their extension to rational tree structures. In: ADC, pp. 52–59 (2001)
[25] Weigel, F., Schulz, K.U., Meuss, H.: Node identification schemes for efficient xml retrieval. In: Foundations of Semistructured Data (2005)

Efficiently Crawling Strategy for Focused Searching Engine

Liu Huilin, Kou Chunhua, and Wang Guangxing

College of Information Science and Engineering, Northeastern University
Shenyang 110004, China

Abstract. After the analysis and study of focused search engine, we design an efficiently crawling strategy, which include two parts: topic filter and links forecast. The topic filter can filter the web pages having already been fetched and not related to topic; and the link forecaster can predict topic links for the next crawl, which can guide our crawler to fetch topic pages as many as possible in a topic group and when all the topic pages are fetched, it will try to look for another topic group by traversing the unrelated links. This strategy considers not only the precision of fetching web pages but also the recall. With the strategy, the focused search engine can get expected pages without browsing large amount of unnecessary pages, and the user can get the useful information without browsing the huge list which is high recall and low precision. Our extensive simulation studies show that the new strategy has a good efficiency and feasibility.

1 Introduction

Information on the Internet is in disorder and be redundant, which limited search efficiency seriously. The appearance of general search engines alleviates the state. However, it is not enough to solve the problems. One side, as the amount of web sites is growing rapidly, the number and size of stored documents is growing even faster and site contents are getting updated more and more often. Although great efforts in hardware and software are done, it is hard to keep up with the growing amount of information available on the web. Study shows, the largest crawls cover only 30–40% of the web, and refreshes take weeks to a month. Another side, the general search engines which are designed for all fields, can fulfill the all-sided search demand. When users input the query, they will return a huge ranked list of the resultant web pages that match the query. The high recall and low precision [1] coupled with this huge list make it difficult for the users to find the information that they are looking for. Such search engine can not meet people's demand more and more. Another kind of search engine—focused search engine [2] appears. Focused search engine is a new kind of search engine service mode, compared with low queue precision, large amounts information, and not enough depth of general search engine. It provides valuable information and related services for a specific area, a specific group or a specific demand. Its features is "special, refined, deep", and with industry characteristic which makes focused search engine more focused, specific and in-depth compared

K.C. Chang et al. (Eds.): APWeb/WAIM 2007 Ws, LNCS 4537, pp. 25–36, 2007.

with general search engines massive information disorder. It supplies users with not hundreds of or thousands of related web pages, but extremely narrow and highly specific and targeted information. Therefore it attracts many users.

The key part for a good focused search engine is an appropriate crawling strategy [3], by which the crawler will only collect focused information without being bothered by the unnecessary web pages. There are two problems in the crawling strategy:

1. One problem is the precision of calculating the relevance between web pages and topic. Some methods have a high precision, but the strategies are so complex that the systems need many resources to run. Although it is no need of high real-time, it is not worth exchanging the benefit with so much time and space. Otherwise, some strategies have low calculation, but the systems have to sacrifice the precision. How to design an appropriate algorithm is a problem.
2. The other problem is how to predict the potential Url of the links for the next crawl. In the process of fetching the web pages, the existence of web tunnel characteristic [4] impacts on the quality of collection. To improve the accuracy of fetching web pages, we need to increase the threshold, which will lose topic pages group, and the recall decreases. Otherwise, to improve the recall will result in low precision. Therefore, how to balance the recall and precision is really a question.

In this paper, we propose an efficiently crawling strategy, which is consist of topic filter and link forecaster. The topic filter is used for filtering the web pages which is not related to the topic, and the link forecaster is used for predicting the potential Urls of the links for the next crawl. The benefit brought by the strategy is much more than the cost, because of no needing to craw the entire Internet. So energy consumption is saved. The contributions of this paper are summarized as follows:

1. We propose an efficient approach which combining the topic filter and the links forecaster. The topic filter can be used to filter the unrelated web pages and save the related web page. The forecaster can be used to forecaster the topic links and find topic links first and consider other links then.
2. Last but not the least, our extensive simulation studies show that the strategy maintains pages on a specific set of topics that represent a relatively narrow segment of the web. It only needs a very small investment in hardware and network resources and yet achieves respectable coverage at a rapid rate. The focused crawling strategy will be far more nimble in detecting changes to pages within its focus than the general search engine's strategy that is crawling the entire Web.

The rest of the paper is arranged as follows. Section 2 briefly reviews the previous related work. And the Section 3 introduces the architecture of focused crawler in which our strategy is applied. Our algorithm is introduced in detail in Section 4. The extensive simulation results to show the effectiveness of the proposed algorithm are reported in Section5. Finally we conclude in Section 6.

2 Related Work

The concept of focused search engine has appeared since the general search engine turned up. Therefore, there was focused crawl strategy called Fish-Search [5] designed by De Bra et al. at that time. In Fish-Search, the Web is crawled by a team of crawlers, which are viewed as a school of fish. If the ''fish'' finds a relevant page based on keywords specified in the query, it continues looking by following more links from that page. If the page is not relevant, its child links receive a low preferential value. But it is difficult to assign a more precise potential score to the pages which haven't been yet fetched. Shark-Search [6] by Hersovici et al. did some improvements to the original Fish-Search algorithm. Shark-Search is a modification of Fish-search which differs in two ways: a child inherits a discounted value of the score of its parent, and this score is combined with a value based on the anchor text that occurs around the link in the Web page.

Besides the kind of methods above, there is another kind of methods which use a baseline best-first [7] focused crawling strategy combined with different heuristics. For example, C. Aggarwal, F. Al-Garawi and P. Yu designed an intelligent crawling with arbitrary predicates [8]. They propose the novel concept of intelligent crawling which actually learns characteristics of the linkage structure of the World Wide Web while performing the crawling. Specifically, the intelligent crawler uses the inlinking web page content, candidate URL structure, or other behaviors of the inlinking web pages or siblings in order to estimate the probability that a candidate is useful for a given crawl. Also, the system has the ability of self-learning, i.e. to collect statistical information during the crawl and adjust the weight of these features to capture the dominant individual factor at that moment.

There are other methods used to capture path information which is help lead to targets such as genetic programming. For example, a genetic programming algorithm is used to explore the space of potential strategies and evolve good strategies based on the text and link structure of the referring pages. The strategies produce a rank function which is a weighted sum of several scores such as hub, authority and SVM scores of parent pages going back k generations [9].

Currently, the one thing people focus on is how to judge whether a web page is a topic page. The method designed can judge the web pages at certain accuracy, and massive time and space is not needed. And the other thing people focus on is how to recognize the potential Urls for the next crawl. Many methods designed before, think the links of a related page tend to topic. It is true, but not all the links is topic-related. If fetched all the pages corresponding to the links, a lots of unrelated pages will be fetched back, which will affect the precision. The links of an unrelated page shouldn't be discarded entirely, because there are some links which may point to a topic page. Even if the links don't point to a topic page, traversing some unrelated links can reach a topic group, which plays an important role in ensuring the recall. Therefore, the current research is the appropriate topic filter and a good link forecaster which forecast the potential Urls, and ensure at certain precision and recall.

3 Architecture of Crawling for Focused Search Engine

The web crawling for focused search engine is done by a focused crawler, which adds a topic relevance determinator, based on a general crawler [10]. Its structure can be described by figure 1.

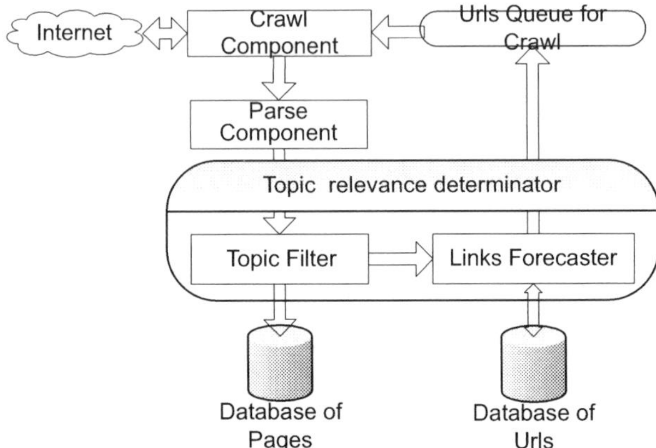

Fig. 1. Architecture of Crawling

From figure1, we can know the focused crawler is constituted of several important parts such as crawl component, parse component and topic relevance determinator. The crawl component fetches the web pages according to the Urls queue coming from the database of Urls. The parse component parses these pages. The topic relevance determinator analyze the pages which have been parsed, and determine which pages should be preserved and which links should be added into the Urls queue for the next crawl. The database of pages is used for saving the topic pages, and the database of Urls is used for saving all the Urls of links. With the architecture it works like this: First, inject a set of Urls selected from the authority sites of certain special fields by manual into the database of Urls, and links forecaster will add these Urls to the Urls queue. Then the crawl component fetches the corresponding pages. After that, these pages will be parsed by the parse component. The html page information like title, anchor, and text and so on, is available from the parsing results. Making full use of the information, system will analyze the pages' relevance, and save in database of pages or throw some of them by the different relevance through the topic filter, one part of topic relevance determinator. Then update the database of Urls and forecast the potential Urls for the queue for crawl through the other part of topic relevance determinator. A new cycle begins.

4 Crawling Strategy of Focused Search Engine

From the architecture of the focused crawler, the topic relevance strategy [11] contains the topic filter and the links forecaster. The topic filter calculates topic relevance by classifying based on vector space model, and then decides whether web pages will be preserved or not, according to the value of the topic relevance. The links forecaster predicts topic relevance of linked web pages, when no previous knowledge of link structure is available, except that found in web pages already fetched during a crawling phase. So the links forecaster needs to get the useful information from the web pages already fetched. For example, the forecaster can get some apocalypse from the topic relevance of the page in which links are and from the existing link structure in the database of Urls. Then the focused crawler can be guided by such a strategy that predicts first, and judges later. And the strategy can fetch topic pages at certain precision and recall, and not need to crawl the entire web.

4.1 The Topic Filter

The task of the topic filter is to judge the similarity between the topic and the page having already been fetched. In fact, the calculation of the content topic relevance also can be considered as a process of classification to pages. The first thing that must be done is to convert the web page into an expression form that makes the web pages can be recognized by computers. There are some models based on keywords such as Boolean model, vector space model and probability model. So far, the vector space model is applied widely and that's the reason why we adopt it. The vector space model (VSM) [12]uses the characteristic item set which appears in the web pages to replace them, so the original pages can be expressed by the vector. The strategy designed by us just expresses the web page texts using this kind of model. Then, a text D can be represented as:

$$D=\{(T1,W1), (T2,W2), (T3,W3)......(Tn,Wn)\}$$

In the formula, Ti is the characteristic item, Wi is the weight of ith characteristic item (n=1, 2 ....).

Words appearing on one page are considered as the characteristic item. Therefore, the first important work is extracting the characteristic item from web pages. The focused search engine aims at crawling web pages in certain specific domain, and characteristic items in this kind of web pages generally are the technical terms. So we do not need to do segmentation and frequency statistics [13] to the ordinary words. In order to improve the efficiency of whole system and the accuracy of segmentation, we use the topic dictionary to help finish segmentation of web pages. The topic dictionary is determined by experts in the field.

The different logic division of the text will make the characteristic items play a different role. When one characteristic item vector is used to describe the content of the text logic passage for every text logic passage, matching search considers not only the number of the characteristic items but also the location. The method can improve the precision of matching results greatly. Therefore, since the web pages have been expressed in different logic part, we can't only adopt the frequency of characteristic item as the weight. In one web page, text in Head label, the anchor text and the ordinary text in Body label nearly show the whole information, so they can be used while calculating the weight of characteristic items. We design the following algorithm to get the weight.

Hweight is the weight of Head label in web page; Aweight is the weight of Anchor label in web page; Bweight is the weight of Body label in web page; and satisfy Hweght+Aweight+Bweight=1; If the characteristic item appears in a certain label, it will be recorded as Focused (label) =true, otherwise recorded as Focused (label) =false. Therefore, the weight of a characteristic item is calculated as follows:

```
Initialize weight=0;
If(Focused(Head)==true)      Then weight+=Hweight;
If(Focused(Anchor)==true)    Then weight+=Aweight;
If(Focused(Body)==true)      Then weight+=Bweight;
```

After transforming the web page text into the vector space model using the topic dictionary and the segmentation technology, we'll select a typical topic web page as the benchmark-vector. Then it is feasible to calculate the similarity [14] of web pages. For example, similarity between the web page text i and the base-vector j can be expressed as:

$$DSim(D_i, D_j) = \frac{\sum_{k=1}^{n}(W_{ik}, W_{jk})}{\sqrt{\sum_{k=1}^{n} W_{ik}^2 * \sum_{k=1}^{n} W_{jk}^2}} \tag{1}$$

According to formula (1) we can calculate the similarity between one web page and benchmark-vector. We can get a threshold value through experiments. The page whose value is higher than this threshold belongs to this topic, and will be preserved. Otherwise, the page will be discarded. The value of similarity will also provide important information for the links forecaster.

4.2 The Links Forecaster

The topic filter can filter the web pages which are not related to the topic. However, if we capture all pages down from the web and then judge whether we will save the page, we'll waste massive time and space. Therefore, it is necessary to forecast topic relevance according to the web pages which have been fetched. Our goal is to crawl related web pages as much as possible and to crawl unrelated web pages as little as possible. Therefore, we need analyze the features of web pages which will be helpful for predicting the topic links, and then design a better strategy.

4.2.1 The Analysis of Web Features

First, the labels of a web page include the topic features. An important label is the text around links such as anchor text. They are the descriptions to pointed web pages. If the topic word turns up in these texts, the page that the link points to is likely the topic page. For example, if a web page includes such a label: <a href="content.HTM" > agriculture</a>, the pointed page is related to agriculture possibly.

Second, web is a hypertext system based on Internet. The biggest difference between hypertext system and ordinary text system is that the former has a lot of hyperlinks. Study shows, hyperlinked structure on the web has contained not only the characteristic of page importance, but also the topic relevance information. Because of the distribution characteristic of pages on the Web, we can have such conclusions: (1) Page A pointed by Page B has a good relevance with the topic of A. (2) Page linked by a topic page will be a worthy one. Therefore, we calculate the score of a page by PageRank [15, 16]. The Algorithm has an effective use of huge link relations in the Web [17].It can be described as: one link from A pointing to B is considered as an A' providing vote to B, then we can judge the page importance according to the number of vote. The formula [18] is:

$$PR(A) = (1-d) + d\,(PR(T1)/C(T1) + \ldots + PR(Tn)/C(Tn)) \tag{2}$$

In formula (2),

PR(A): The PageRank value of page A;

PR(Ti): ThePageRank value of Ti Pointing to page A;

C(Ti): The number of out links of page Ti;

d: Damping ($d \in (0,1)$)

4.2.2 The Prediction Algorithm

Definition: There are three attributes in every Url in the database of Urls. One is score which is used for storing the value calculated by the topic filter. The value of score represents the extent similar to the topic. The bigger the value is, the web page tends to the topic more. Otherwise, the smaller the value is, the web page is further from the topic. Another is pscore which is used for storing the value calculated by PageRank according to the score. The value of pscore represents the linked page's extent similar to the topic. The bigger the value is, the linked page tends to the topic more. Otherwise, the smaller the value is, the linked page is further from the topic. The last one is fetch flag which is used for the information whether the page has been fetched. If the value of the fetch flag is true, it represents that the page corresponding to the Url has already been fetched, otherwise, it hasn't been fetched.

Initiation: Inject the topic Urls into the database of Urls and initiate all the score=0, all the pscore=0, all the fetch flag=false. Initiate fetching turn=n, topic threshold=t, number of urls in Urls queue for crawl=N. Then add all the Urls in the database of Urls to the Urls queue for crawl.

Description: First, judge whether the fetching turn is reached, if not reached, fetch the web pages corresponding to the Urls of the Urls queue. The topic filter will calculate the similarity value of every page. Then the forecaster will do two things: one is to update the database of Urls, the other thing is to forecast the potential links for the next crawl. During updating the database, there are four things to do:

1. The fetch flag of the Url whose pages having already been fetched should be set true. The purpose to do so is that the system won't fetch these pages later.
2. The Urls extracted from the fetched page by the parser should be added into the database of Urls and the link relation between each other should also be stored. The purpose to do so is that it can form a hyperlinked structure that will be used while forecasting the potential links.
3. If the anchor text of the link includes topic words, the pscore of the Url of the link should be set 1.0. Because once the topic words turn up in anchor text, the link may point to a topic page, the pscore of the Url should be set as high as 1.0. The Url will likely exist in the Url queue in the next cycle.
4. If the similarity value larger than the topic threshold, the score of the Url should be set the value of the similarity. Otherwise, the pscore of the Url of every link in the page should be set 0.5*p. Here p is a random probability. The links in unrelated page shouldn't be discarded, because we may traverse some such links and find many topic web pages which get together. The benefit is so large that we must consider.

During the forecasting, there are four things to do:

1. According to the links relation that has existed in the database of Urls, compute the PageRank value. Every page whose topic relevance value is higher than the threshold does a division to the topic relevance value. Give every share to the links of the page. Iterating in such way, every Url will have a PageRank value.
2. Normalize the PageRank by $e^{PageRank\_value} - e^{-PageRank\_value} / e^{PageRank\_value} + e^{-PageRank\_value}$. If the normalization larger than the original pscore of the Url, replace it.
3. Sort the Urls according to the value of pscore, top N Urls are considered to be the Urls that tends to the topic and added to the queue for crawl.

The steps are shown in the Algorithm 1.

Algorithm 1. Forecast links

```
1:  While (fetching turn is not reached)
2:  Fetch the web pages corresponding to the Urls.
3:  The similarity value of every page is calculated by the topic filter.
4:  Update the database of Url, do 5, 6, 7, 8, 9;
5:  Set the fetch flag of the Url whose pages having been fetched=true.
6:  Add the Urls of the links in the page to the database and store the link relation.
7:  If the anchor text of the link include topic words, set the pscore of the Url of the
    link=1.0.
8:  If the similarity value larger than topic threshold, set the score of the Url =the
    similarity.
9:  Else set pscore of the Url of the link=0.5*p (p is random probability)
10: Forecast the potential links, do 11, 12, 13,14;
11: According to the link relation, compute the PageRank (PR (T) is the score) value
    of every Url.
12: Normalize the PageRank value, compute the normalization by
    e^PageRank_value - e^-PageRank_value / e^PageRank_value + e^-PageRank_value
13: If the pscore larger than the original pscore of the Url, replace it.
14: Sort the Urls according to the value of pscore, top N Urls are added to the queue
    for crawl.
15: EndWhile
```

5 Experiments

The algorithm is implemented with Eclipse and running Linux Professional Edition. The experiments are conducted on a PC with InterPentium4 2.8GHz CPU, 1G main memory and 80G hard disk. We conducted experiments with the real pages on Internet.

To examine the effects of topic-first strategy, we use two evaluation benchmarks [19]: precision, and recall. Precision is the ratio of information from given algorithm and result by manual categorizing. Recall is the ratio of correct information that classifies by an algorithm and the precise information with reality.

For this algorithm:

$$\text{Precision} = \text{expPage}/\text{toreturnPage}; \quad (3)$$

$$\text{Recall} = \text{expPage}/\text{existPage}; \quad (4)$$

In the formula (3) and (4),
expPage: the number of correct expected page,
toreturnPage: the total number of returned pages actually,
existPage: the number of expected pages.

In this experiment, 1000 agricultural web pages are disposed for text training, in order to extract and choose characteristic as the base-vector of this topic. The number of topic web pages, which exist on the Internet, is estimated by the method of statistical sampling. Then we crawl the Internet respectively with common search engine and focused search engine for agriculture. The common search engine adopts the breadth-first as the crawling strategy, while the focused search engine for agriculture adopts the topic-first as the crawling strategy.

Experiment 1: We start with 100 seed Urls which are all topic Urls, set the threshold 0.3, and crawl with the increase of link-depth .Then we can draw the figures below:

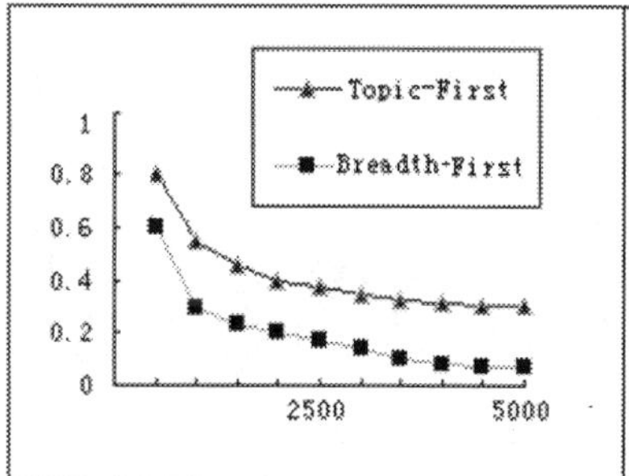

Fig. 2. Precision of topic-first and Breadth-first with topic seed Urls

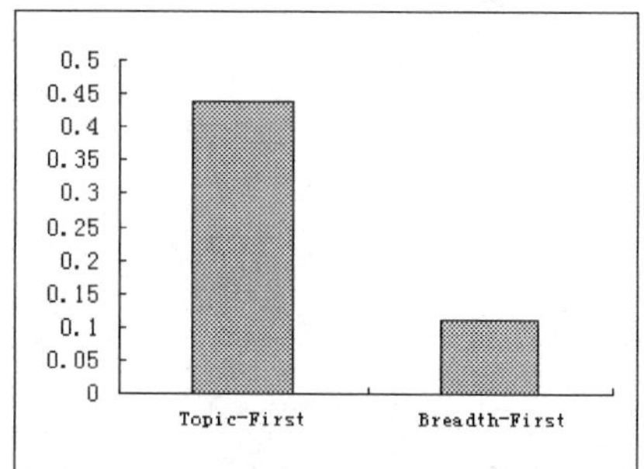

Fig. 3. Recall of topic-first and breadth-first with topic seed Urls

The Carves in Fig. 2 show the precision of two different strategies. From this figure we can know, the topic-first strategy has a high precision in the first round, due to the Urls we have injected are all about agriculture. In the process of crawling, the topic strategy will filter unrelated web pages, while the breadth-first strategy will crawl all pages, so it has a low precision compared with topic-first. With the increase of link-depth, precision of two methods decline and tend to be stable. Comparing the two carves, we can easily find out that the agriculture focused crawling strategy is superior to breadth-first obviously in precision.

And Fig.3 compares the recall value of two strategies. From Fig.3, we can know after crawling 5000 pages, the recall value of agricultural focused crawling strategy ups to 44.1%, and the breadth-first only is about 11%.

Experiment 2: We start with 100 seed Urls which inlcluding 50 topic Urls and 50 unrelated Urls, set the threshold 0.3, and crawl with the increase of link-depth .Then we can draw the figures below:

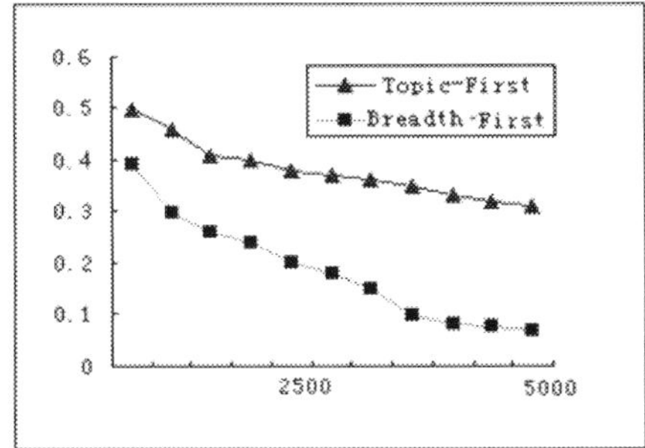

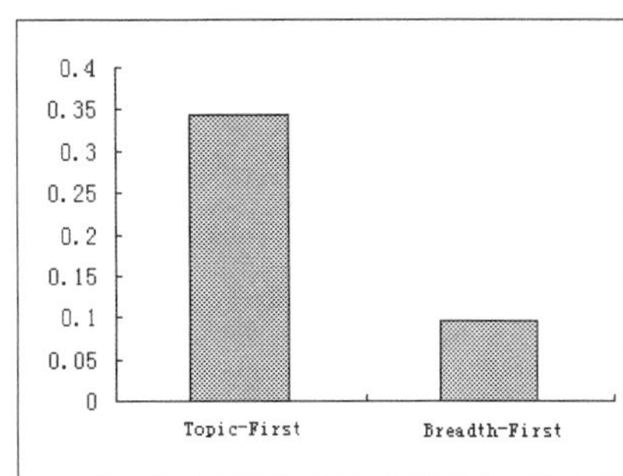

Fig. 4. Precision of topic-first and breadth-first with not all topic seed Urls

Fig. 5. Recall of topic-first and breadth-first with not all topic seed Urls

The Carves in Fig. 4 show the precision of two different strategies with seed Urls not all being topic Urls. They all decline until stable state. Compared with Fig.3, the precision is apparently lower, because the initiation seed Urls is not all topic Urls, the forecaster will add unrelated Urls to the Urls Queue when there are no topic Urls. In this way, the effect of our topic-first strategy will be a little worse. Therefore, to make the strategy work well, better seed Url should be selected.

Fig.5 shows the recall of topic-first and breadth-first with seed Urls not all being topic Urls, we can know after crawling 5000 pages, the recall value of agricultural focused crawling strategy ups to 34.6%, and the breadth-first only is about 9.7%. Although lower than the topic-first in Fig.5, the precision of the topic-first is still better than the breadth-first.

Experiment 3: We start with 100 seed Urls which are all topic Urls, set the threshold 0.2, and crawl with the increase of link-depth .Then we can draw the figures below:

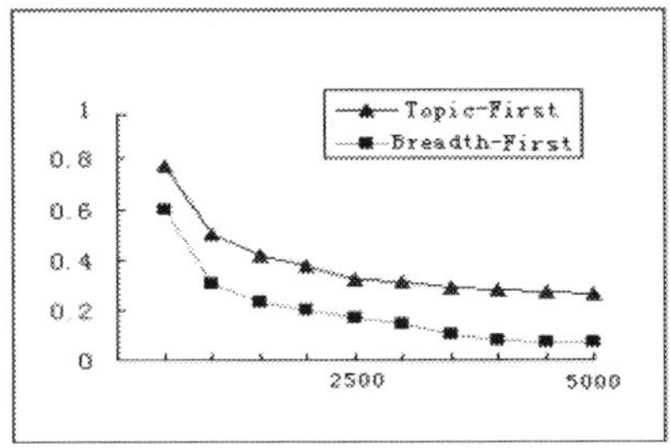

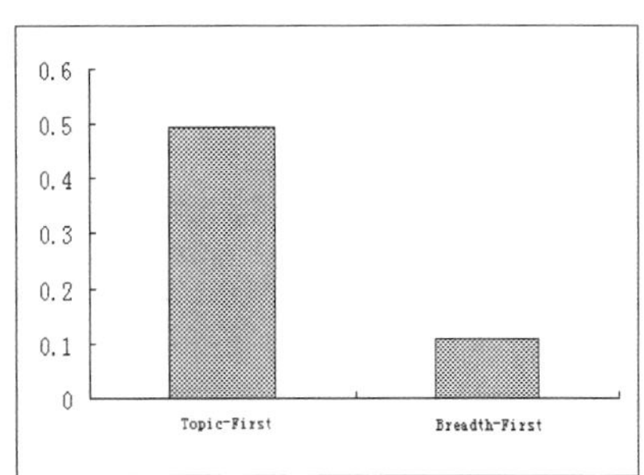

Fig. 6. Precision of topic-first and breadth-first with a different threshold

Fig. 7. Recall of topic-first and breadth-first with a different threshold

The Carves in Fig. 6 show the precision of two different strategies with a different threshold. They all decline until stable state. Compared with Fig.3, the precision is a little lower, because the threshold is set lower, a lot of unrelated web pages may be judged as the topic pages, which will misdirect the links forecaster to add the unrelated Urls to the Urls queue.

Fig.7 shows the recall of topic-first and breadth-first with a different threshold, we can know after crawling 5000 pages, the recall value of our topic-first strategy ups to 48.9%, and the breadth-first is only about 11%. The reason why the recall value of our topic-first strategy is higher than that in Fig.3 is the lower threshold. Although lower threshold results in lower precision, it can improve the recall.

In these experiments, after the same time, the number of the topic pages having already been fetched with our topic-first strategy is more than that of the breadth-first with the fewer web pages fetched. It means a lot of time and space is saved. Therefore, it is obvious that the topic-first strategy works better than breadth-first in both precision and recall. All these results reflect the great advantage of topic-first crawling strategy. And to keep the effect of our topic-first strategy, good seed Urls and appropriate threshold should be selected.

6 Conclusion

Crawling strategy of focused search engine combines the topic filter and the links forecaster. In the part of topic filter, we uses the method of classification based on the vector space model, which has many good point such as clear expression, convenient, smaller computation complexity. While extracting the characteristic items, a certain domain topic dictionary can cover some topic well use few words, and make the words quantity and complexity reduce greatly. While computing the weight of the characteristic items, we consider the feature of the web page and improve the precision. Therefore, with the process of topic filter, the system can filter out the web pages which have nothing to do with the topic. In the part of the links forecaster, we have used the anchor text and the link structure analysis method, which allows us to get the topic web pages without crawling entire Internet. This strategy has been used in the focused search engine for agriculture designed by us; at present it works well, and has an appropriate precision and recall ratio for search results. Our following work is: determine the best parameter and the threshold by the massive experiments, do further research and continually optimize the strategy.

Acknowledgement

This work is supported by the national natural science foundation under the grant No. 60573089.

References

1. Crestani, F., Lalmas, M., van Rijsbergen, C.J.: Information Retrieval: Uncertainty and Logics[M]. Kluwer Academic Publishers, Boston, Massachusetts (1998)
2. Chakrabarti, S., van den Berg, M., Dom, B.: Focused crawling a new approach to topic-specific Web resource discovery[J]. Computer Networks 31(11-16), 1623–1640 (1999)

3. Pant, G., Tsioutsiouliklis, K., Johnson, J., Giles, C.: Panorama: extending digital libraries with topical crawlers[C]. In: Proceedings of ACM/IEEE Joint Conference on Digital Libraries, Tucson, Arizona, June 2004, pp. 142–150 (2004)
4. Bergmark, D., Lagoze, C., Sbityakov, A.: Focused crawls, tunneling, and digital libraries[C]. In: Proceedings of the 6th European Conference on Digital Libraries, Rome, Italy (2002)
5. Bra, P.D., Post, R.: Information retrieval in the World Wide Web: making client-base searching feasible[C]. In: Proceedings of the 1st International WWW Conference, Geneva, Switzerland (1994)
6. Hersovici, M., Jacovi, M., Maarek, Y., Pelleg, D., Shtalhaim, M., Ur, S.: The Shark-search algorithm-an application: tailored Web site mapping[C]. In: Proceedings of the 7th International WWW Conference, Brisbane, Australia (1998)
7. Menczer, F., Pant, G., Srinivasan, P.: Evaluating Topic-Driven Web Crawlers[C]. In: Proceedings of the 24th annual international ACM SIGIR conference on Research and development in information retrieval, (September 2001)
8. Aggarwal, C., Al-Garawi, F., Yu, P.: Intelligent crawling on the World Wide Web with arbitrary predicates[C]. In: Proceedings of the 10th International WWW Conference, Hong Kong (2001)
9. Johnson, J., Tsioutsiouliklis, K., Giles, C.L.: Evolving strategies for focused Web crawling [C]. In: Proceedings of the 20th International Conference on Machine Learning (ICML-2003), Washington, DC, USA (2003)
10. Liang, Xiaoming, L., Xing, L.: Discussion on theme search engine[A]. In: Advances of Search Engine and Web Mining[C], pp. 34–40. Higher Education Press, Beijing (2003)
11. Can, F., Nuray, R., Sevedil, A.B.S.: Automatic performance evaluation of Web search engines. Information Processing & Management[J] 40(3), 495–514 (2004)
12. Ming, L., et al.: Improved Relevance Ranking in Web-Gather[J]. Comput. Sci. & Technol. 16(5), 410–417 (2001)
13. Hong, C., Jinsheng, Y.: Search on Forestry Focused Search Engine [J]. Computer Applications, 24 (2004)
14. Gianluigi, G., Sergio, G., Ester, Z.: A Probabilistic Approach for Distillation and Ranking of Web Pages[C]. World Wide Web, pp. 1386–145X (2001)
15. HuanSun, Wei, Y.: A note on the PageRank algorithm. Applied Mathematics and Computation[J] 179(2), 799–806 (August 2006)
16. Higham, D.J.: Google PageRank as mean playing time for pinball on the reverse web. Applied Mathematics Letters[J] 18(12), 1359–1362 (December 2005)
17. Almpanidis, G., Kotropoulos, C., Pitas, I.: Combining text and link analysis for focused crawling-An application for vertical search engines[M]. Information Systems Article in Press, Uncorrected Proof, Available online November 7, 2006 (2006)
18. Brin, S., Page, L.: The anatomy of a large-scale hypertextual web search engine[C]. In: Proceedings of the Seventh International World Wide Web Conference,Brisbane, Australia (1998)
19. Chekuri, C., Goldwasser, M., Raghavan, P., Upfal, E.: Web search using automatic classification [C]. In: Proceedings of the Sixth International WWW Conference (1997)

QoS-Guaranteed Ring Replication Management with Strong Consistency*

Wei Fu, Nong Xiao, and Xicheng Lu

School of Computer, National University of Defense Technology, Changsha, P.R. China
lukeyoyo@tom.com

Abstract. As an important strategy for reducing access time, balancing overload, and obtaining high data availability, replication is widely used in distributed system such as Data Grids. Currently grid researchers often consider replica placement problem from the view of whole systems, i.e, ignoring individual requirements. With the rapid growth of Data Grid applications, fulfilling requirements of every request becomes a big challenge, especially in Data Grids with massive data, multiple users and geographically distribution. We build up a ring-based Data Grid model and define a QoS-guaranteed Replica Management Problem. A QoS-guaranteed replica placement algorithm qGREP is proposed, which takes individual QoS requirement into account and achieves tradeoff between consistency cost and access efficiency. Also a ring-base electing algorithm VOSE is presented to solve update conflicting while minimizing propagating costs. Analysis and experiments show that the model and algorithm are feasible in complex Data Grid environment and achieve good performance under different simulating conditions.

1 Introduction

Data Grid[2, 3] employs Grid technologies to aggregate massive storage capacity, network bandwidth and computation power across widely distributed network environment to provide pervasive access to very large quantities of valuable data resources. As an important strategy for reducing access time, balancing overload, and obtaining high data availability, replication is widely used in distributed system such as distributed file system[4-6], Web caching[7-9], database management and Data Grids[2, 10-13].

Replica placement problem[14] is an important function of replica management, which defined as: given M sites and a data objects, find a subset of sites R to place a replica of the data onto each of them, obtaining some benefits. For example, traditional researches aimed at minimizing the total/average costs as much as possible. The cost can be accessing delay, communication message, or response time. However, this criterion cannot be valid in some circumstance. In bank or telecom systems, the total/average response time is required to be as low as possible. At the same time, the response time of EACH request is also required below some service quality requirements. These two

* This paper is supported by National Basic Research Program (973) NO. 2003CB317008, NO. 2005CB321801, and Chinese NSF NO. 60573135, NO. 60503042.

K.C. Chang et al. (Eds.): APWeb/WAIM 2007 Ws, LNCS 4537, pp. 37–49, 2007.

ways have different objectives to place replica, although in many cases they have almost the same solutions. In this paper we will consider the replica placement from the viewpoint of individuals, each holding a pre-defined QoS requirement.

Data objects are often supposed read-only to cater to the requirement of today's grid applications. Unfortunately writing in grid has not been well solved yet and still remains as an unavoidable challenge. In a writable system, consistency is very important for the correctness and efficiency of data access. We would like concentrate our discussion on maintaining the consistency of data content while guaranteeing the access efficiency.

Generally speaking, access efficiency and consistency cost are two conflicting goals. More replicas obtain better access efficiency by bringing data to the vicinity of more users, but the cost of update will increase correspondingly. And vice versa. Therefore a suitable trade-off should be made for both satisfactions. In a writable Data Grid environment, we consider both query & update request and present a ring based model to manage replica consistently. With a pre-description of QoS, a ring structure with sufficient replicas will be produced to meet individual access requirements, meanwhile minimizing the cost of update. An electing algorithm is proposed to harmonize simultaneous update to achieve strong consistency.

The paper is organized as follows. In Section 2 we propose a ring-based model to depict a QoS-guaranteed replica management. In Section 3, an algorithm is presented to solve the replica placement problem under the proposed model. In Section 4, a varietal electing algorithm is provided for strong consistency. Section 5 includes experiments which show some characteristic of our model. After a introduction of related work in Section 6, we give a brief conclusion and some future work at Section 7.

2 Ring-Based Data Grid Model

Typically, Data Grid is described as a combination of Computational Elements (CE) and Storage Elements (SE) [10]. Every data access will be directed to some SE eventually, either from CE or SE. We regard the access from CE to SE as coming directly from the SE being accessed. Thus all CEs can be omitted and all SEs make up an overlay storage graph. We propose a ring-based Data Grid model as follows:

Definition 1. *A Ring-based Data Grid* RDG={G, DM, D, ξ,s, R}, where

- G: is a network of storage nodes, represented by G=(V, E), where V is the set of all storage element nodes, and $E \in V \times V$ is the set of direct links between two nodes. Each edge has communication delay as weight;
- DM: is a matrix holding shortest communication costs of all node pairs in G;
- D: represents the diameter of G, measured by network delay;
- ξ: stands for a proportion of update $\xi(0 \le \xi < 1)$, this value is a statistics from accessing history in a given time window. Here we assume that accesses are partitioned into query from ordinary node to replica node and update from one replica node to all the other replica nodes;
- s: is the node id where the object residing on, here we assume each node has a global unique distinguished id;
- R: represents for a set of nodes, each of them replicate data from s to its local disk. Nodes in R are organized as a unidirectional ring call *replica ring*;

Definition 2. $\forall$ u, v$\in$V, the *cost* between u and v is defined as:

- If (u,v) $\in$E, then d(u, v) equals to the delay of the link between u and v.
- Otherwise, d(u, v) equals to the total delay of the shortest path between u and v.

Note: A given cost can be considered as relatively stable. If not, it can be replaced by an average value of historical observations.

A storage graph is illustrated as follows by Fig. 1.

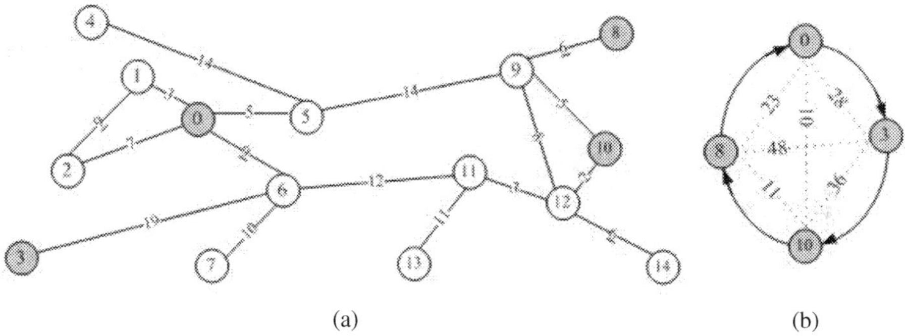

Fig. 1. (a) is a storage graph of 15 nodes with random distribution topology, D = 48 and ξ= 0.2. Source data resides on site labeled by 0. (b) is a replica ring generated from (a). Dashed lines in (b) represent the propagating paths when updates take place. It guarantees that query cost from any node is no more than 16, while update cost is also minimal.

In Fig. 1.(a) each grey node represents for a replica node with a replica residing on it. Let R be the set of replicas. White node represents for ordinary node, which can turn into a replica node by fetching data to its local disks. Fig. 1.(b) shows a unidirectional replica ring composed by three replica nodes. There are logical communication path between two neighbor node pair. The number of nodes in the ring is constricted by the QoS requirement, and this will be discussed below. In reality, the total cost of a typical cost includes the delay of network, overhead of data transferring and other factors. In order to simplify the problem, we only consider the communication delay.

In our model, any two nodes can communicate with each other directly or indirectly, i.e., there exists no isolated node. Otherwise, it can be removed from the graph easily. The storage capacity of each node is assumed to be infinite, due to the lower and lower price of disks. To say the least, a replica replacement policy can partly solve the problem of volume exceeding. Besides the infinite capability of storage, each node has little but enough computational capability to run some simple protocols.

Through some sort of grid monitoring and/or information service[1], global information can be sent to every node for the preparation of our algorithms VOSE and qGREP, as to be discussed below.

3 QoS-Guaranteed Replica Placement and qGREP Algorithm

Replica promotes access efficiency and decrease total query cost by placing data to users' vicinity. With the storage graph and the ring-based replica model, in this section we discuss about replica placement with QoS guarantee. We want to find a tradeoff between query and update cost, while guaranteeing that every node could obtain data within a QoS restriction and overall accessing cost can be decreased as much as possible This deducts the following QoS-guaranteed Replica Management Problem:

Definition 3. Given a ring-based Data Grid RDG with G, DM, ξ, and s, a *QoS-guaranteed Replica Management Problem (QRMP)* is to find a set of nodes *R* to replicate data from s to R, promising that each query can be fulfilled within a pre-defined QoS restriction *Q*, while update cost is minimized.

We present a central algorithm to find a minimal replica set for a given storage graph and QoS restriction, as depicted by the pseudo-code in Fig. 2.

```
Input:  G =(V, E); DM; ξ; Q; s
Output: R: init {s};
1   Begin
2   D = MAX_{1≤i,j≤n}{DM[i,j]} ;// the diameter of G
3   If Q<ξ*D/(1-ξ) // qualification inequation, see Sec. 3.2
4         Warning!     //run a risk, set Sec. 3.2
5   D_query=Q + 1;       //current max query cost, initially greater than Q
6   CV=DM[s];            //vector of the minimum cost to replica ring
7   While D_query > Q, loop:  // see Sec. 3.1
8         For each row vector DM[i]=DM[i,.] in DM, loop
9               Profit Vector:PV[i] = GetProfit(CV, DM[i]);
10        Find a row making the biggest PV[i], let it be i';
11        Add2Ring(R, i');
12        CV = Merging(CV, DM[i']);
13        D_query= MAX_{1≤i≤n} {CV[i]};
14     End of loop
15  End    //normal terminating
```

Fig. 2. Centralized algorithm: QoS-Guaranteed Replica Placement (qGREP)

Line 9: GetProfit(V1,V2) returns the sum of positive differences of V1 and V2. If V1[i]> V2[i], the minus of (V1 [i]- V2[i]) is added into PV[i]; otherwise omits it.
Line 11: Add2Ring(R, i') adds the selected node i' into the replica ring R. First it computes the sum of costs to any two neighbored node pair in the ring. Then it chooses the pair with the smallest sum and inserts itself between them.
Line 12: Merging(V1,V2) picks out the smaller element of V1[i] and V2[i] as the new V1[i].

It can be easily deduced that qGREP has a computational complexity of $\Theta(n^3)$ at the worst case, and a space complexity of $\Theta(n^2)$.

3.1 Origination and Principle

qGREP considers both query and update cost at the same time. Distinctly, more replicas mean less total/individual query cost. But more replicas mean more update cost, including higher probability of update conflict and more propagation overhead. Therefore the basic strategy or principle is to give a optimal tradeoff between them.

As illustrated in Figure 1, each node in V will find a nearest replica in R for better access latency. After a new replica u is inserted between a pair (i, i+1) in the ring, query cost decrement is calculated as follows:

$$\Delta\text{Query(u)}= \sum_{v\in(V-R)} \underset{r\in R}{MIN}(d(v,r)) - \sum_{v\in(V-R)} \underset{r\in R\cup\{u\}}{MIN}(d(v,r)) \quad (1)$$

Meanwhile, the update cost will increase at two aspects. Firstly, the chance of update conflict will increase. Secondly, the propagation of update will suffer longer latency among more replica nodes. Let replicas number be **n**, the probability of update conflict should be:

$$P_n=1-C_n^0(1-\xi)^n - C_n^1\xi(1-\xi)^{n-1}=1-(1-\xi)^n - n\xi(1-\xi)^{n-1} \quad (2)$$

After u added, the probability should be:

$$P_{n+1}=1- C_{n+1}^0(1-\xi)^{n+1} - C_{n+1}^1\xi(1-\xi)^n=1-(1-\xi)^{n+1} - (n+1)\xi(1-\xi)^n \quad (3)$$

Hence we calculate the difference of update cost as:

$$\Delta\text{Update(u)}=P_{n+1}\times \underset{r\in R\cup\{u\}}{MAX}(d(r,v)) - P_n\times \underset{v\in R}{MAX}(d(r,v)) \quad (4)$$

Where v is the update initiator. Thus the total changing of access cost should be:

$$\Delta\text{Cost} = (1-\xi)\times \Delta\text{Query} - \xi\times\Delta\text{Update} \quad (5)$$

Positive ΔCost means that adding replica induces more query cost decrement than update cost increment. Therefore the total access cost decreases. And vice verse. When adding replica does harm to the overall performance, no more replica should be added. In Sec. 3.2 we will prove that qGREP always make positiveΔCost under some conditions. Therefore, qGREP should stop as soon as every query can be served within a pre-defined QoS restrict.

3.2 Correctness and Qualification

In this section we first give proof that with correct QoS constrict for a given storage graph, qGREP will produce a solution with low time complexity.

Recall the pseudo-code of qGREP. Essentially, AV records the smallest costs from every node to access replica ring, and D_{query} is the biggest one. When qGREP terminated normally, any element of AV must be smaller than Q. Thus the QoS restrict is fulfilled. On the other hand, once qGREP notices that QoS restrict is fulfilled, it will terminate immediately, and the dead lock of no-stop-looping would never occur. Furthermore, PV[i] stands for the possible decrement of query cost if node i becomes a replica node. Thus qGREP always promotes the most "valuable" node to hold new replica.

Next we will prove that under suitable conditions, qGREP will always decrease total access cost when adding new replica.

Lemma 1. Add a new replica node u, the difference of total query cost should be: Δ Query(u) $\geq \underset{r\in R}{MIN}(d(u,r)) > 0$.

Nodes in node u's vicinity may turn to access this new replica if their query costs are less than before, thus the total cost will be decreased. The worst case is that no such node exists, however this will still eliminate the query latency from u itself, i.e. $\underset{r\in R}{MIN}(d(u,r))$.

Lemma 2. $\forall$ $0<\xi<1$, the conflict probability will increase: $P_{n+1} > P_n > 0$. Furthermore, ΔUpdate(u)>0.

Proof. From the meaning of P_n we know $P_n > 0$. For any integer n>1, $0<\xi<1$,

$$P_{n+1} - P_n = (1-\xi)^n + n\,\xi\,(1-\xi)^{n-1} - (1-\xi)^{n+1} - (n+1)\,\xi\,(1-\xi)^n$$
$$= n\,\xi^2\,(1-\xi)^{n-1} > 0, \text{ so it can be deduced easily that } P_{n+1} > P_n > 0. \quad (6)$$

From (4) and (6), ΔUpdate(u)> Pn ($\underset{r\in R\cup\{u\}}{MAX}(d(r,v)) - \underset{v\in R}{MAX}(d(r,v))$). Since for any update initiator v, $\underset{r\in R\cup\{u\}}{MAX}(d(r,v)) > \underset{v\in R}{MAX}(d(r,v))$, thus ΔUpdate(u)>0. ■

Theorem of Qualification. If QoS restrict Q meets the inequation: $Q > \xi D/(1-\xi)$, where D is the network diameter of storage graph G, adding a node to replica ring will always decrease the total access cost, i.e., ΔCost > 0.

Proof. Substitute ΔQurey and ΔUpdate in (5) with (1) and (4) :

$$\Delta\, Cost = (1-\xi)\left(\sum_{v\in(V-R)} \underset{r\in R}{MIN}(d(v,r)) - \sum_{v\in(V-R)} \underset{r\in R\cup\{u\}}{MIN}(d(v,r))\right) - \xi\,\left(P_{n+1} \underset{r\in R\cup\{u\}}{MAX}(d(r,v)) - P_n \underset{v\in R}{MAX}(d(r,v))\right)$$
$$> (1-\xi)\,\underset{r\in R}{MIN}(d(u,r)) - \xi\, P_n \left(\underset{r\in R\cup\{u\}}{MAX}(d(r,v)) - \underset{v\in R}{MAX}(d(r,v))\right)$$

Since $\underset{r\in R}{MIN}(d(u,r)) \geq Q$, and $\underset{r\in R\cup\{u\}}{MAX}(d(r,v)) - \underset{v\in R}{MAX}(d(r,v)) \leq D$, $Q > \xi D/(1-\xi)$, Δ Cost $>(1-\xi)\,Q - \xi D > \xi D - \xi D = 0$ ■

Theorem of Qualification reveals the internal relationship between network topology and access model. It stands for the trade-off point of QRMP problem. Therefore, we can use the inequation to judge the existence of feasible solution. When it is violated, qGREP runs a risk of producing bad solution. To avoid this, some mechanism should be introduced, for example, adjusting the value of Q.

3.3 A Simple Example of qGREP

Here we present an example of qGREP to show its workflow as well as verify its correctness. A distance matrix DM can be generated from a storage graph illustrated by Fig. 1. Since the graph has bidirectional links, the matrix must be symmetric.

$$DM = \begin{bmatrix}
0 & 3 & 7 & 28 & 14 & 5 & 9 & 19 & 23 & 19 & 16 & 11 & 14 & 22 & 20 \\
3 & 0 & 9 & 31 & 17 & 8 & 12 & 22 & 28 & 22 & 19 & 14 & 17 & 25 & 23 \\
7 & 9 & 0 & 35 & 21 & 12 & 16 & 26 & 32 & 26 & 23 & 18 & 21 & 29 & 27 \\
28 & 31 & 35 & 0 & 42 & 33 & 19 & 29 & 48 & 42 & 36 & 31 & 34 & 42 & 40 \\
14 & 17 & 21 & 42 & 0 & 19 & 23 & 33 & 39 & 33 & 38 & 25 & 28 & 36 & 34 \\
5 & 8 & 12 & 33 & 19 & 0 & 14 & 24 & 20 & 14 & 11 & 6 & 9 & 17 & 15 \\
9 & 12 & 16 & 19 & 23 & 14 & 0 & 10 & 29 & 23 & 17 & 12 & 15 & 23 & 21 \\
19 & 22 & 26 & 29 & 33 & 24 & 10 & 0 & 39 & 33 & 27 & 22 & 25 & 33 & 31 \\
23 & 28 & 32 & 48 & 39 & 20 & 29 & 39 & 0 & 6 & 11 & 17 & 14 & 28 & 20 \\
19 & 22 & 26 & 42 & 33 & 14 & 23 & 33 & 6 & 0 & 5 & 11 & 8 & 22 & 14 \\
16 & 19 & 23 & 36 & 38 & 11 & 17 & 27 & 11 & 5 & 0 & 5 & 2 & 16 & 8 \\
11 & 14 & 18 & 31 & 25 & 6 & 12 & 22 & 17 & 11 & 5 & 0 & 3 & 11 & 9 \\
14 & 17 & 21 & 34 & 28 & 9 & 15 & 25 & 14 & 8 & 2 & 3 & 0 & 14 & 6 \\
22 & 25 & 29 & 42 & 36 & 17 & 23 & 33 & 28 & 22 & 16 & 11 & 14 & 0 & 20 \\
20 & 23 & 27 & 40 & 34 & 15 & 21 & 31 & 20 & 14 & 8 & 9 & 6 & 20 & 0
\end{bmatrix}$$

Q=16, ξ=0.2, R={0}, AV=DM[0]=$[0\ 3\ 7\ 28\ 14\ 5\ 9\ 19\ 23\ 19\ 16\ 11\ 14\ 22\ 20]$, D= MAX(DM[• , •])=48, Q>48*0.2/0.8=12, so this graph has a feasible solution. The workflow looks like this:

1) Since AV[4]=28>Q, qGREP enter the loop, calculating PV of each row and then choose node 10 obtaining the biggest Profit (79) to join into the replica ring, thus R={0, 10}. After the first loop, AV=$[0\ 3\ 7\ 28\ 14\ 5\ 9\ 19\ 11\ 5\ 0\ 5\ 2\ 16\ 8]$

2) Since AV[4]=28>Q again, qGREP calculates PV of each row and then choose node 3 obtaining the biggest Profit (28) to join into the replica ring, thus R={0, 3, 10}. After the second loop, AV=$[0\ 3\ 7\ 0\ 14\ 5\ 9\ 19\ 11\ 5\ 0\ 5\ 2\ 16\ 8]$.

3) Since AV[8]=19>Q, qGREP calculates PV of each row and then choose node 8 obtaining the biggest Profit (19) to join into the replica ring ,thus R={0, 3, 10, 8}. Because d(0,8)+d(8,3) > d(3,8)+d(8,10) > d(10,8)+d(8,0), 8 is inserted between 0 and 10. After the third loop, AV=$[0\ 3\ 7\ 0\ 14\ 5\ 9\ 0\ 11\ 5\ 0\ 5\ 2\ 16\ 8]$.

4) Now all elements in AV are less than Q, so the algorithm ends with replicas at node of 0, 3, 8 and 10.

4 Ring-Based Election and Consistency Maintaining

In our model both query and update requests can be forwarded to any replica node, thus achieving a good balance of overload. When several updates arrive at different sites simultaneously, good consistency mechanism is in urgent need to handle the conflict problem. We use a unidirectional ring to run election algorithm. Winner updates its replica and propagates it to all the other replica sites, but losers have to wait for the next round of election. Obviously, our model behaves a strong consistency to data requests.

A unidirectional ring structure is a graph that in-degree and out-degree of all nodes are both 1, as depicted in Fig. 1. Many graceful ring-based electing algorithms can be employed here. We present a special vote-alike algorithm called VOSE. Every node maintains a list of update initiators named *elector*. Each update initiator sends out a message *vote* to announce himself as an elector, and other nodes are electorates. Once receiving a vote, a node adds it to the list. Then it chooses a node from the list and sends its id to its next neighbor. Once an elector receives a vote including his own id, it wins the elections and the election terminates. A pseudo-code is illustrated as Fig. 3

```
Var: List_p;
Begin
1   If p is an elector, then
2     Add p into List_p, status = wait, send<vote, p>;
3     Receive<vote, q>;
4     If p≠q, then
5          List_p=List_p∪[1], let r=GetMin(List_p), send <vote, r>;
6       Else //p=q
7           p wins, the election terminates.
8   //otherwise, p must be an electorate
9   Else looping until the election terminates
10    Receive <vote, q>, add p into List_p;
11    let r=GetMin(List_p), send <vote, r>; //r has a minimal cost
End
```

Fig. 3. Distributed algorithm: VOte Swallow Electing (VOSE)

Line 5/11: $\forall$ u$\in$List, let Pro(u)= $\underset{v\in R}{MAX}(d(u,v))$, GetMin(List) chooses the id of some node who makes the minimal Propagating cost Pro(u). If several nodes have the same value of Pro(u), the one with smallest label will be selected. Since every node has information about the topology and DM, this can be done within $\Theta(n^2)$.

After a round of election, the winner terminates the election and performs updating using broadcasting. From PR2 every node knows the position of other replica nodes and communication cost between them, so broadcasting can be employed to propagate the update. Furthermore, the cost should be the largest delay between winner and any node in the replica ring. As illustrated in Fig. 1.B, propagating cost of node 0 is Pro(0)=max{10, 23, 28}=28. Since GetMin(List) selects an elector with minimal Pro(u), the update cost is minimized. Note that the broadcasting way is not constricted along the ring.

Votes from electors with smaller Pro(u) value always "swallow" those with bigger value, hence it is named VOSE. It has a message complexity of $\Theta(n^2)$ in the worst cases, and time complexity is $\Theta(n^2)$. Additionally, G, DM and $List_p$ requires a cost complexity of $\Theta(n^2)$. Detail information please refer to [15].

5 Experiments and Evaluation

Methodology of our Experiments

In order to simulator the real internet environment, we adopt the famous network topology generator BRITE to produce networks with different size. The information about interconnecting and delays are extracted from *output.brite* files to build up adjacency matrix *AM*. Then the Floyd-Warshall algorithm is employed to generate distance matrix *DM* from AM, with time complexity$\Theta(n^3)$. This cost can be awful when n is big enough. Fortunately, in Grid system, nodes are often made up of trustful servers with powerful capacity and strong resistibility to failure, as declared in PR1. So the effectiveness of such a DM can last a long time. Once RDG notices that the topology's changing, it collects new information and generates DM again.

In BRITE, firstly N nodes are placed into a square plane ordered by HS and LS, either randomly or referring to the heavy tail phenomenon. Then edges are added to interconnect nodes with a probability p. Before outputting, attributes such as delay and bandwidth are assigned to topology components. In our experiments, HS=10*N, LS=N and p= 0.45. In our experiments, the effects of QoS restrict *Q* is replaced by a scale of QoS restrict and network diameter *D*, i.e., Q=QoS/D for the sake of practicability. Every experiment repeats 10 times with random selected original site. Query cost T is substituted by D*rn/n, where rn is the replica number, and n is the total number. Thus total access cost C equals to $(1-\xi)* Q + \xi*T$.

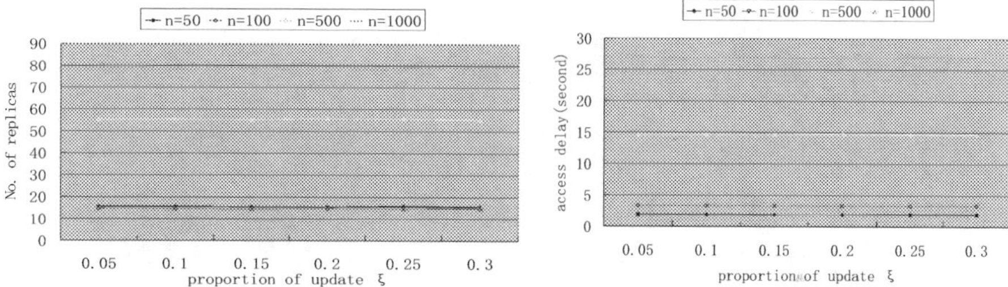

Fig. 4, 5. Given Q=D*0.1, with the growth of nodes, both replica number and access cost keep stable as the proportion of update increases

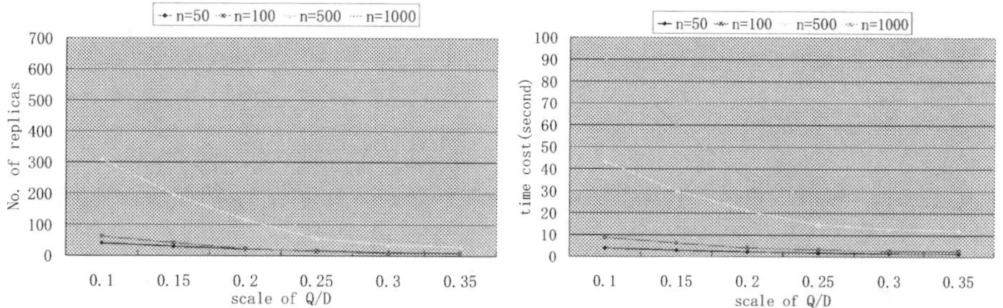

Fig. 6, 7. Given ξ=0.1, n and C decrease fast with the expanding of Q/D

Above all, we study how the access pattern, i.e. proportion of update ξ, impacts the system performance and the replica placement. The QoS Q is set to be D/10. Results are illustrated in Fig. 4 and 5. It is obvious that rn and C both keep steady even if n increases a lot. There are two explanations for this phenomenon. Firstly, we suppose that query happens much more frequently than update, so the weight of update ξ is relatively small. Secondly, qGREP gains enough query benefits to counteract increment of costly updating operations.

Next we exchange the role of Q andξ. When$\xi$is fixed as 0.1, results can be depicted as Fig. 6 and 7. Totally different from Fig. 4 and 5, Fig. 6 and 7 declare that Q is one of the most important factors in RDG model. If the QoS restrict is relaxed below about 1/3 of the network diameter D, qGREP only needs a small number of replica nodes for better performance as well as less storage consumption.

Furthermore, Fig. 8 illustrates the relationship between replica percentages and Q. It shows that network with more nodes generate less replica percentages. It is make sense because in BRITE generated topology, as number of nodes increases, the number of edges increases correspondingly. Thus a replica node has more opportunities to serve other non-replica nodes in his vicinity. As a result, a comparatively smaller replica percentage can fulfill the QoS restrict.

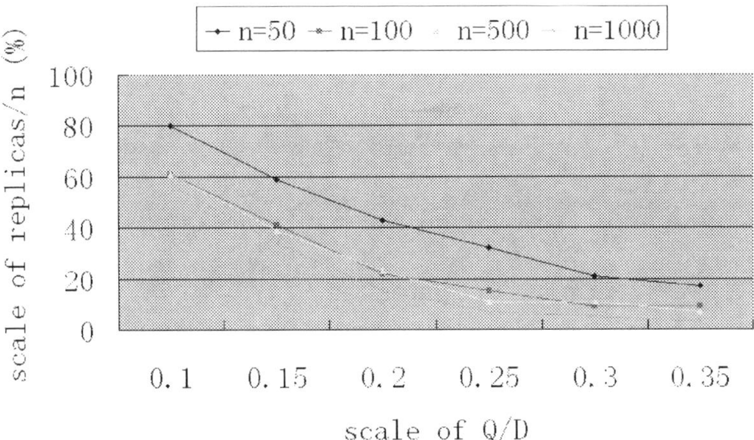

Fig. 8. Another profile of the same experiments, except Y axis's substituting by a percentage

6 Related Work

Our work has close relationship with replica management, QoS, and replica consistency problem. We will talk about them respectively.

Replica Placement and QoS Problem

Replica Placement and QoS problem have both been widely studies in grid researches[12, 13, 16]. Almost all of the replica placement pay more attention to optimize the overall performance, such as the mean access latency or TTL hops. They seldom think about each part individually, such as QoS requirement. Literature [12] associates replica placement with workload. It selects optimal locations for replica so that the workload of each node becomes balance.

Recently some inspirations emerged to meet the individual requirement. Literature [14] proposed QoS-Aware Replica Placement Problem for replica-aware service and replica-blind service. It promises that any quest can be served within a given QoS requirement, and the replica-aware problem is proved to be NP-complete. Two heuristic algorithms named *l-greedy-insert* and *l-greedy-delete* is provided for optimal solution. On the basis of the work, [17] gave another proof of NP-hard property of this problem as well as two distributed algorithms. The newest research [13] proposes another heuristic algorithm that determines the positions of replicas with minimized costs.

All of them only care about query for read-only objects. Instead, we study the placement policy considering both query cost and update cost. While QoS of query is similar to [13], we turn the QoS of overall cost to be a question under the concept of probability, associating with a proportion of update. We believe this has never been carefully studied before.

Ring-based Replica Consistency Problem

Most of the literatures[2, 3, 11] about Data Grid always regard data as read-only objects. Due to the consideration of data update, consistency becomes an unavoidable problem in our model. Literature[18] gave some models for replica synchronization and consistency in Data Grid and classified 5 level of Consistency Levels. However, it presented no practical solutions.

As for the structure of replica, Literature[5] presented a chain-based replication for high throughput and availability while maintaining strong consistency in cluster computing. In his model, every query request is directed to the tail, while every update request is directed to the head of the chain. This brings problem of overload and hot accessing. Furthermore, the update cannot be initiated at multiple nodes simultaneously. Additionally, Literature [19] gave a tree structure of replica management, which is based on the tree-like structure of the system it concerns about.

In our ring-based Data Grid model, every node can make response for query or update request, which achieves good balance of overload. Ring-based electing algorithm is presented to negotiate update conflicting.

7 Conclusion and Future Work

Achieving strong consistency produces considerable additional overhead for distributed system, especially Data Grid with multiple writable data objects. In this paper, we exploit the management capability of Grid system and propose a novel ring-based replica consistency maintaining mechanism to negotiate update conflicting, supporting simultaneous query & update access to any replica site. Comprehensive consideration of individual QoS requirement has not been emphasized before. We present a QoS-guaranteed replica placement algorithm to make an ideal trade-off between query cost decrement and update cost increment while promising each user can obtain what he want in a pre-defined time limit. A qualification inequation reveals the internal relationship between the diameter, QoS and access pattern.

It must be pointed out that our algorithms depend on the functionality of Grid heavily. The topology and distance information should be collected and delivered to every participant. If the underlying network changes all the time, the additional cost of DM reproducing will become unacceptable. Furthermore, data update frequency, represented by the proportion of update ξ, should not be very frequent. From the Qualification inequation Q>ξ* D/(1-ξ), if ξ = 0.2, then Q>D/4; if ξ = 0.8, then Q >4D. This can be intolerant in Data Grid where D is very big.

Till now, we discuss the problem in a static network. In the future we will study the influence of node adding/quitting, as well as the failure-tolerance of replica nodes. We believe that a lazy algorithm should be developed to obtain the DM efficiently. Also relatively relax consistency model for RDG is under consideration.

References

1. Wei Fu, Q.H.: GridEye: A Service-oriented Grid Monitoring System with Improved Forecasting Algorithm. In: The 5th International Conference on Grid and Cooperative Computing, Changsha, China (2006)
2. Ann Chervenak, I.F., Kesselman, C., Salisbury, C., Tuecke, S.: The Data Grid: Towards an Architecture for the Distributed Management and Analysis of Large Scientic Datasets. Journal of Network and Computer Applications 23, 187–200 (2001)
3. Hoschek, W., F.J.J.-M., Samar, A., Stockinger, H., Stockinger, K.: Data Management in an International Data Grid Project. In: Proceedings of the 1st IEEE/ACM International Workshop on Grid Computing. 2000, Bangalore, India, Springer-Verlag, Heidelberg (2000)
4. Dahlin, M.D., R.Y.W., Anderson, T.E., Patterson, D.A.: Cooperative Caching: Using Remote Client Memory to Improve File System Performance. In: Proceedings of Operating Systems Design and Implementation (OSDI'94) (1994)
5. Schneider, R.v.R.a.F.B.: Chain Replication for Supporting High Throughput and Availability. In: 6th Symposium on Operating Systems Design and Implementation(OSDI'04) (2004)
6. Foster, L.W.D.a.D.V.: Comparative models of the file assignment problem. ACM Computing Surveys 14(2), 287–313 (1982)
7. Satyanarayanan, M.: Coda: A Highly Available File System for a Distributed Workstation Environment. IEEE Trans. on Computers 39(4), 447–459 (1990)
8. Jia, X., D.L., Hu, X.-D., Du, D.-Z.: Placement of read-write web proxies in the internet. In: ICDCS'01 (2001)
9. Jianliang, X., B.L., Lee, D.L.: Placement Problems for Transparent Data Replication Proxy Services. IEEE JOURNAL ON SELECTED AREAS IN COMMUNICATIONS 20(7), 1383–1398 (2002)
10. David, W.B., D.G.C., Capozza, L., Millar, A.P., Stocklinger, K., Zini, F.: Simulation of dynamic grid rdeplication strategies in optorsim. In: Proceedings of 3rd Intl IEEE Workshop on Grid Computing (Grid'02) (2002)
11. Stockinger, H., A.S., Allcock, B., Foster, I., Holtman, K., Tierney, B.: File and object replication in data grids. In: 10th IEEE Symposium on High Performance and Distributed Computing (HPDC'01) (2001)
12. Pangfeng, L., J.-J.W.: Optimal Replica Placement Strategy for Hierarchical Data Grid Systems. In: Proceedings of the Sixth IEEE International Symposium on Cluster Computing and the Grid (CCGRID'06) (2006)
13. Hsiangkai, W., P.L., Jan-Jan W.: A QoS-Aware Heuristic Algorithm for Replica Placement. in Grid Computing Conference 2006 (2006)
14. Xu, X.T.a.J.: Qos-aware replica placement for content distribution. IEEE TRANSACTIONS ON PARALLEL AND DISTRIBUTED SYSTEMS 16(10), 921–932 (2005)

15. Tel, G.: Intorduction to Distributed Algorithms, 2nd edn. Publishing House of Electronics Industry. pp. 227–267 (2003)
16. Kavitha Ranganathan, I.F.: Identifying Dynamic Replication Strategies for a High-Performance Data Grid. In: Grid Computing Conference 2001 (2001)
17. Jeon, W.J., I.G., Nahrstedt, K.: QoS-aware Object Replication in Overlay Networks. In: IPTPS 2005 (2005)
18. D"ullmann, D., W.H., Jaen-Martinez, J., Segal, B., et al.: Models for Replica Synchronisation and Consistency in a Data Grid. In: IEEE Symposium on High Performance Distributed Computing HPDC', 2001. Redondo Beach, California, USA (2001)
19. Kalpakis, K., K.D., Wolfson, O.: Optimal Placement of Replicas in Trees with Write, and Storage Costs. IEEE TRANSACTIONS ON PARALLEL AND DISTRIBUTED SYSTEMS 12(6), 628–637 (2001)

A Mobile Database Sharing Protocol to Increase Data Availability in Mobile Environments

Hien Nam Le[1] and Mads Nygård[2]

[1] Department of Telematics
hiennam@item.ntnu.no
[2] Department of Computer and Information Science,
Norwegian University of Science and Technology (NTNU)
mads@idi.ntnu.no

Abstract. This paper presents a mobile database sharing protocol to improve data availability in mobile environments. Long disconnection periods of mobile hosts could cause data inconsistency and blocking of mobile transactions. The proposed mobile database sharing protocol allows two conflict data caching modes to reduce the blocking of mobile transactions that access to shared data. Furthermore, data can be directly shared among mobile hosts that are being disconnected from the database servers. Data consistency is ensured via the execution constraints and dependency among transactions.

Keywords: Mobile database, mobile data management, conflicting operations.

1 Introduction

The rapid growth of wireless network technologies and capacity of portable computing devices has promoted a new mobile environment. People can access to information and carry out their transactions while being on the move. The environment for accessing and processing information is changing from stationary to mobile and location independent.

A mobile environment is characterized by unique characteristics: the mobility of users or computers, the frequent and unpredictable disconnection of wireless networks, and the limitation of computing capacity of mobile devices [1]. These characteristics pose many challenges for mobile transaction processing. Long disconnection periods in communication could cause data inconsistency or blocking of other transactions. For example, if a data item is locked at a disconnected mobile computer for a long period, this could delay the progress of transactions on other mobile hosts. The duration of disconnected or connected periods is not always as planned, i.e., varying in time, due to many factors, for example, the limited battery-life of mobile computers, the unstable bandwidth of wireless networks. This might result in longer delay time or interruption periods for database operations.

In this paper, we present a mobile database sharing protocol to increase data availability in mobile environments. The main contributions of our research include: (1) two conflict caching modes to allow accessing to shared data objects that have been locked by mobile hosts that are being disconnected from the database servers for

K.C. Chang et al. (Eds.): APWeb/WAIM 2007 Ws, LNCS 4537, pp. 50–61, 2007.

long periods; and (2) mobile data sharing protocols to support direct data sharing among mobile hosts that are being disconnected from the database servers.

We consider a large database *DB* that is distributed among several fixed servers S_i. The correctness criterion of the system is the serializable schedule of transactions [4]. Data items are cached at mobile hosts for disconnected transaction processing. To avoid data inconsistency among transactions at different mobile hosts, the database servers grant read or write locks on data items that are being downloaded into the mobile hosts (before that mobile hosts are disconnected from the database servers). When mobile host MH_i is being disconnected from the database servers, it can share data with other mobile hosts MH_j via the support of mobile affiliation workgroups (reviewed in the next section). We also distinguish two classes of transactions in mobile environments: *online transaction* and *offline transaction*. An *online transaction* is a transaction that interacts directly with the transaction manager at the fixed database servers to perform read or write operations on shared data. An *offline transaction is* a transaction that is executed and managed by the mobile transaction manager at the disconnected mobile host. Offline transactions are first locally committed at the disconnected mobile host, and then finally be verified and committed at the database servers. A locally committed offline transaction may be finally aborted if there is a non-serializable execution occurred. We use the following notations: T_i^k denotes a transaction T^k at mobile host MH_i. A transaction T_i^k that shares data to or obtains data from other transactions is called a *delegator* or *delegatee* transaction, respectively. In some cases, a transaction can play roles as both delegator and delegatee transaction. The read and write locks on shared data item X are denoted by X_R and X_W, respectively; R_X and W_X represent the read and write operations on the shared data item X.

The rest of the paper is organized as follows. Section 2 discusses the related research. Section 3 presents the conflict caching modes. The mobile database sharing protocols is addressed in Section 4, and Section 5 concludes the paper.

2 Related Research

In this section, first we briefly recap the concepts of mobile affiliation workgroups and mobile data sharing workspace that support data sharing among mobile hosts. Later, we review several related work.

2.1 Mobile Affiliation Workgroup Model and Mobile Sharing Workspace

Our mobile database sharing protocol is developed based on the concepts of mobile affiliation workgroups and mobile data sharing workspace (as illustrated in Figure 1). The objective of the mobile affiliation workgroup model is to support information sharing among transactions at different disconnected mobile hosts in mobile environments. This can be done by allowing such disconnected mobile hosts to form dynamic mobile affiliation workgroups (MAW) [11] by taking advantages of wireless communication technologies, i.e., the ability of direct communication among mobile hosts within a limited area. Sharing of information within a mobile affiliation workgroup is carried out through a dynamic mobile sharing workspace created by participating mobile hosts. Further detail about the mobile affiliation workgroup can be found in [12].

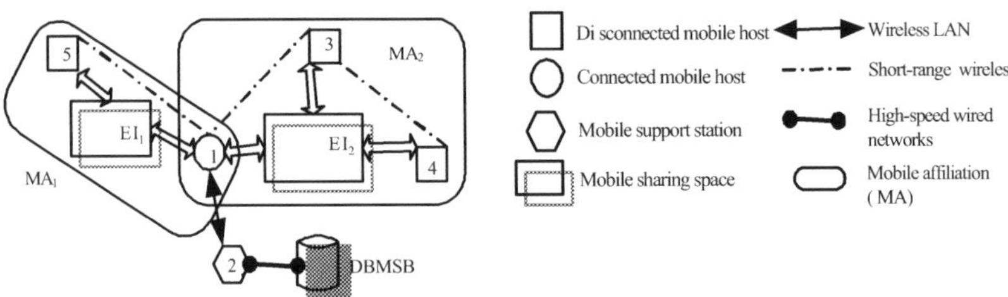

Fig. 1. Mobile affiliation workgroup and mobile sharing workspaces

2.2 Related Research

Mobile transactions and mobile databases have been wisely applied to cope with the challenges of sharing data among mobile transactions. However, there are still some major limitations. The architecture of mobile transaction environments [5] relies too much on mobile support stations that are stationary and connected.

Current approaches that tackle long locking periods among transactions may not be suitable in mobile environments. For example, altruistic locking protocols [9] could cause cascading aborts; speculative locking protocols [8] might require extra computing resources at mobile hosts. The delegation approach [7,10] that supports sharing information among transactions may not be adequate because it requires tight cooperation, i.e., connectivity, among delegator and delegatee transactions. [6] presents a multi-version transaction model for improving data availability in mobile environments. A mobile transaction in this model has three different states of execution: start, commit and termination. The shared data updated by a transaction T_i is made available to another transaction T_j immediately after T_i commits. In this model, however, a data item could be locked for long periods of time at a mobile host until the mobile host releases that locks at mobile support stations.

In [3], four different disconnection modes for mobile databases are presented: *basic sign-off*, *check-out*, *relaxed check-out* and *optimistic check-out*. This model only supports planned disconnected database operations for one mobile host, while our mobile data sharing protocol can deal with both planned and unplanned disconnections of a single host or a group of mobile hosts. To support dynamic database sharing in a mobile database community, the Accessing Mobile Database architecture [2] based on the concept of mobile agents is proposed. The limitation of this architecture is that the role of a mobile host is restrained to either the database server or a database client. Our mobile data sharing mechanism does not discriminate between a mobile database server and a mobile database client.

3 Conflict Caching Modes

In this section, we present the two conflict caching modes that play important roles in the mobile database sharing protocols. According to [4], two database operations Op_i and Op_j of two transactions T^i and T^j are in conflict if they are: (1) accessing the same data item, (2) one of them is a write operation. The conflict of database operations is denoted by $Conflict(Op_i,Op_j)$.

Definition (*directly conflicting transactions*). *Two transactions T^i and T^j are in direct conflict, denoted by $T^i C_d T^j$, if there is an operation Op_i of transaction T^i that conflicts with an operation Op_j of transaction T^j.*

Definition (*indirectly conflicting transactions*). *Two transactions T^i and T^j are in indirect conflict, denoted by $T^i C_{id} T^j$, if there is a transaction T^k that T^i either develops a direct conflict or an indirect conflict with, and T^k develops either a direct conflict or an indirect conflict with T^j, i.e.,*

$$T^i C_{id} T^j \text{ if } \exists T^k, (T^i C_d T^k \vee T^i C_{id} T^k) \wedge (T^k C_d T^j \vee T^k C_{id} T^j)$$

If the database servers will make use of the standard compatibility between read and write locks to ensure data consistency, there is no conflict lock request is allowed. However, in mobile environments, this seems to be too restricted to be useful, due to, long disconnection periods of mobile hosts. In our mobile transaction processing system, there are two different data caching modes: read-write conflict and write-read conflict. These conflict mobile data caching modes are discussed below.

3.1 Read-Write Conflict Data Caching Mode

Fig. 2 illustrates the read-write caching conflict scenario. Suppose that an online transaction T_1^i at connected mobile host MH_1 is holding a read lock X_R on a shared data item X, and an offline transaction T_2^j at mobile host MH_2 requests a write lock X_W on this shared data. The write lock request can be granted to the offline transaction T_2^j because the write operation W_X by transaction T_2^j is not immediately carried out at the database servers, even after the online transaction T_1^i has committed. And the transaction T_1^i is scheduled to execute before transaction T_2^j, i.e., $T_1^i \rightarrow T_2^j$ (the execution constraints are discussed in next section). To handle this limitation, the mobile transaction processing system will allow these conflict lock requests to be compatible.

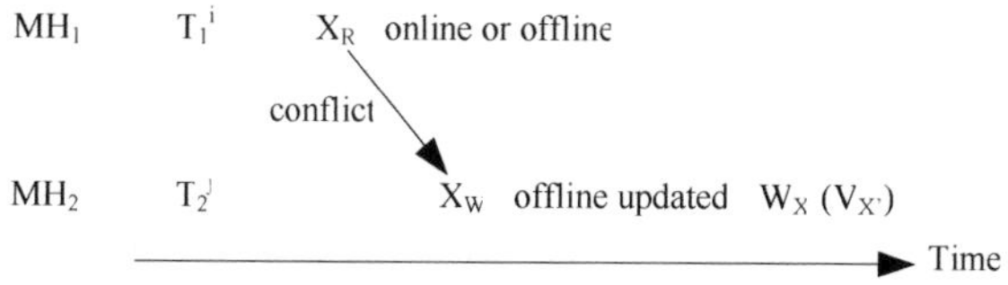

Fig. 2. Read-write conflict data caching mode

Definition (*read-write conflict*). *If an online transaction T_i^k holds a read lock on data item X, and an offline transaction T_j^l requests a write lock on data item X, the database server grants the write lock to T_j^l. We call this conflict mode a read-write (RW) conflict and denote it $X_{RW}(T_i^k, T_j^l)$.*

Our read-write conflict mode focuses on supporting offline transactions at the disconnected mobile hosts. The read-write conflict provides the mobile transaction processing system the ability to avoid blocking of the execution of an offline updating transaction, i.e., if the shared data item X is read locked by an online transaction T_i^k, the write lock request from an offline transaction T_j^l will be granted. In Figure 2, when the mobile host MH_2 reconnects to the database servers, the write (offline) transaction T_2^j will be converted to an online transaction (i.e., with online write lock on the shared data item X) so that the updated data value $V_{X'}$ will be integrated into the database servers. At this time, any on-going online transaction T^p that currently holds read lock on the shared data item X is either allowed to commit (given that the final commitment of the transaction T_2^j will be delayed) or aborted. We define the execution constraint among transactions that access shared data in conflict modes as follows:

Definition (*execution constraint*). *A transaction T^i is said to be scheduled before a transaction T^j, denoted by $T^i \rightarrow T^j$, if all the conflicting operations Op_i of transaction T^i is executed before the conflicting operations Op_j of transaction T^j, i.e.,*

$$T^i \rightarrow T^j \Leftrightarrow (\forall Op_i \in T^i, Op_j \in T^j, Conflict(Op_i, Op_j) \Rightarrow Op_i \rightarrow Op_j)$$

Rule 1 (*execution constraint of rw-conflict*). *If transaction T_i^k develops a read-write conflict with transaction T_j^l on shared data X, i.e., transaction T_j^l will modify the shared data X offline after it is being read by transaction T_i^k, transaction T_i^k will be scheduled before transaction T_j^l, i.e., $T_i^k \rightarrow T_j^l$.*

3.2 Write-Read Conflict Data Caching Mode

In read-write conflict data caching mode, a write lock request on the shared data item of an offline transaction is granted even if the shared data is currently being read lock by other transactions. On the other hand, an online transaction or an offline transaction can be allowed to read a shared data item while another offline transaction holds a write lock on the same shared data item, as long as these transactions can be serialized with the offline updating transaction.

Figure 3 illustrates the write-read conflict scenario. The offline transaction T_2^j at disconnected mobile host MH_2 holds a write lock X_W on the shared data item X. However, this data item is not being immediately modified at the database servers because the mobile host MH_2 that executes transaction T_2^j is currently being disconnected. When an (online or offline) transaction T_1^i at mobile host MH_1 requests a read lock X_R on the data item X, this read lock will conflict with the write lock on X held by transaction T_2^j. In this case, the database server can grant a read lock on X (and consequently allow the read operation to be executed) for transaction T_1^i, given the original value V_X of the data item X is returned (this value might be inconsistent with the value of X that is stored and being modified at the disconnected mobile host MH_2). In fact, at the database servers, the original data value V_X is the most up-to-date and consistent data. Consequently, to ensure that the involved transactions are serializable, transaction T_1^i must be scheduled before transaction T_2^j, i.e., $T_1^i \rightarrow T_2^j$. Note that the offline transaction T_2^j may not know about this conflict that is happening at the database servers. To handle this limitation, the mobile transaction processing system will allow these conflict lock requests to be compatible.

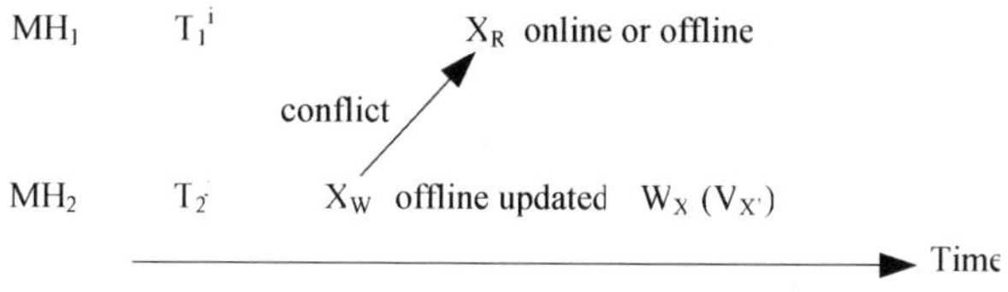

Fig. 3. Write-read conflict data caching mode

Definition (*write-read conflict*). *If an offline transaction T_i^k holds a write lock on data item X, and a transaction T_j^l requests a read lock on data item X, the database server grants the read lock request and the un-modified value of X is returned. We call this conflict mode a write-read (WR) conflict and denote it $X_{WR}(T_i^k, T_j^l)$.*

The write-read conflict mode allows read operations to be executed when there is a write operation that is being executed at the disconnected mobile host, i.e., avoids blocking of the execution of the read operations on the shared data item. In Figure 3, when the mobile host MH_2 reconnects to the database servers, the write (offline) transaction T_2^j will be converted to an online updating transaction with an online write lock on the shared data item X so that the updated data value $V_{X'}$ will be integrated into the database servers. At this time, any on-going online transaction T^p that currently holds a read lock on the shared data item X is either allowed to commit (given that the final commitment of the transaction T_2^j will be delayed) or aborted.

Rule 2 (*execution constraint of wr-conflict*). *If transaction T_i^k develops a write-read conflict with transaction T_j^l on shared data X, i.e., transaction T_i^k will read the shared data X after it is being modified offline by transaction T_j^l, transaction T_i^k will be scheduled before transaction T_j^l, i.e., $T_i^k \rightarrow T_j^l$.*

4 Mobile Database Sharing Protocol

In this section, we present the mobile database sharing protocols that support sharing of information among transactions at different mobile hosts. Mobile hosts will cache shared data item before disconnecting from the database servers. While being disconnected from the database servers, the mobile hosts can directly shared its cached data to other hosts.

For a shared data item, it can be cached either with a read lock or a write lock at a mobile host. If a shared data item X is cached with read lock X_R at the mobile host MH_i, an offline transaction T_i^k can only perform read operation R_X on X. On the other hand, if X is cached with a write lock X_W, the offline transaction T_i^k can modify X, i.e., write operation W_X. Hence, depending on a status of a cached data item, a delegator transaction T_i^k can shared either an *original* shared data or an *updated* shared data to a delegatee transaction T_j^l (Table 1 summaries these sharing data options). If a delegator transaction T_i^k at mobile host MH_i holds a read lock on a data item X, the shared data value V_X will be identical to the value cached at the mobile host, i.e., the original data state is shared. However, if the delegator T_i^k at mobile host MH_i holds a write lock on data item X, the shared value V_X can be either an old value V_X (i.e., before the delegator transaction updates X) or an updated value $V_{X'}$ (i.e., after the delegator transaction has updated X).

Table 1. Locks and equivalent shared data states

		Lock on X	
		Read	Write
Shared data	Original value V_X	Relevant	Relevant
	Modified value $V_{X'}$	N/A	Relevant

Delegatee transactions T_j^l at other mobile hosts MH_j are only allowed to read these shared values, i.e., to the target mobile hosts, the shared data item are read-only. By this way, a local transaction at the disconnected mobile host can obtain the needed data without being connected to the database servers, i.e., reducing the blocking period. This is one of the novel advantages of our mobile data sharing mechanism to increase the data availability in mobile environments.

4.1 Sharing Original Data

In this sub-section, we discuss the execution constraints when a delegator transaction T_i^k shares an original cached data item to a delegatee transaction T_j^l. The following rule defines the execution constraint between these mobile transactions:

Rule 3 (*execution constraint of sharing original data state*). *If delegator transaction T_i^k shares an original data state to delegatee transaction T_j^l, transaction T_j^l must be scheduled before transaction T_i^k, i.e., $T_j^l \rightarrow T_i^k$.*

This rule describes the mobile sharing data states scenario in which a delegator transaction T_i^k shares an original data state V_X of the data item X to a delegatee transaction T_j^l. The delegator transaction T_i^k can hold a read lock, or a write lock on the shared data item but the shared data state has not been modified. In this scenario, both the delegator T_i^k and delegatee T_j^l transactions read the same value V_X of data item X. If the delegator transaction T_i^k reads a consistent data value of X, then the delegatee transaction T_j^l will be assured to read the same consistent data value as the delegator transaction T_i^k.

If the delegator transaction T_i^k holds a read lock X_R on X and there is another transaction T_x^y (at a different mobile host) with which the delegator transaction T_i^k holds a read-write conflict or a write-read conflict, i.e., $T_i^k \rightarrow T_x^y$, this rule ensures that $T_j^l \rightarrow T_i^k \rightarrow T_x^y$, i.e., both transactions T_i^k and T_j^l read consistent data values in relation to the transaction T_x^y.

If the delegator transaction T_i^k holds a write lock X_W on X, and there is another transaction T_x^y (at a different mobile host) with which the delegator transaction T_i^k holds a read-write conflict or a write-read conflict, i.e., $T_x^y \rightarrow T_i^k$, this rule ensures that either $T_j^l \rightarrow T_x^y \rightarrow T_i^k$ or $T_x^y \rightarrow T_j^l \rightarrow T_i^k$, i.e., both transactions T_j^l and T_x^y read consistent data values in relation to the transaction T_i^k.

In Figure 4(a), an example of sharing original values with a read lock is shown. Time proceeds from left to right. The delegator transaction $T_1^{\,1}$ at the mobile host MH_1 holds a read lock on the shared data item X and shares the value V_X to the delegatee transaction $T_2^{\,1}$ at the mobile host MH_2. If these two transactions $T_1^{\,1}$ and $T_2^{\,1}$ finally commit when the mobile hosts reconnect to the database servers, the transaction $T_2^{\,1}$ must be scheduled before the transaction $T_1^{\,1}$, i.e., $T_2^{\,1} \rightarrow T_1^{\,1}$. In Figure 4(b), an example of sharing original values with a write lock is shown. Time proceeds from left to right. The delegator transaction $T_1^{\,1}$ at the mobile host MH_1 holds a write lock on data item Y and shares the original (i.e., non-modified) value V_Y to the delegatee transaction $T_2^{\,1}$ at the mobile host MH_2. In this case, the final transaction schedule will again be $T_2^{\,1} \rightarrow T_1^{\,1}$.

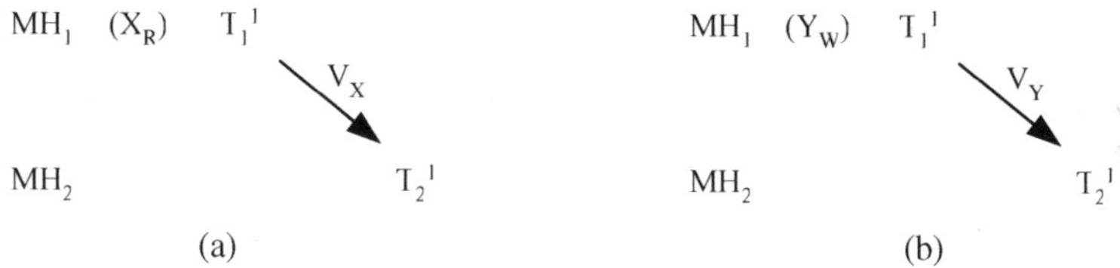

Fig. 4. Execution constraints with sharing original data

4.2 Sharing Modified Data

When a delegator transaction shares an updated data state to a delegatee transaction, the following rule is applied:

Rule 4 (*execution constraint of sharing updated data state*). *If delegator transaction T_i^k shares an updated data state to delegatee transaction T_j^l, transaction T_j^l must be scheduled after transaction T_i^k and before any transaction T_i^n that is scheduled - due to another update - after transaction T_i^k in the locally committed transaction set LC_i at the same mobile host, i.e.,*

$$\forall T_i^n \in LC_i, T_i^k \in LC_i, T_i^k < T_i^n \Rightarrow T_i^k \rightarrow T_j^l \rightarrow T_i^n$$

This rule is denoted by $T_i^k \rightarrow\!\bullet\, T_j^l$.

This mobile sharing data scenario happens when a delegator transaction T_i^k holds a write lock X_W on the shared data item X, and the shared data item has been modified. If the delegatee transaction T_j^l were only to be scheduled after the delegator transaction T_i^k, and the shared data item is later modified again by another transaction T_i^n (the transaction T_i^n is executed at the same mobile host as the delegator transaction and also scheduled after T_i^k), the execution schedule $T_i^k \rightarrow T_i^n \rightarrow T_j^l$ will not be correct. Instead, the correct execution schedule must be $T_i^k \rightarrow T_j^l \rightarrow T_i^n$, i.e., with the above rule $T_i^k \rightarrow\!\bullet T_j^l$ being met.

In Figure 5, an example of sharing updated values with a write lock is shown. The transaction $T_1^{\,1}$ shares the new value $V_{Y'}$ to the delegatee transaction $T_2^{\,1}$. At mobile host MH_1, there is another transaction $T_1^{\,2}$ that later updates it to a new value $V_{Y''}$. The transaction $T_1^{\,2}$ is scheduled after the transaction $T_1^{\,1}$. Rule 4 ensures that the transaction T_2^1 will be scheduled between transactions T_1^1 and T_1^2. This means that the final global transaction schedule is $T_1^1 \rightarrow T_2^1 \rightarrow T_1^2$. When the mobile hosts MH_1 and MH_2 reconnect to the database servers, this transaction execution constraint will be used to support the transaction integration process.

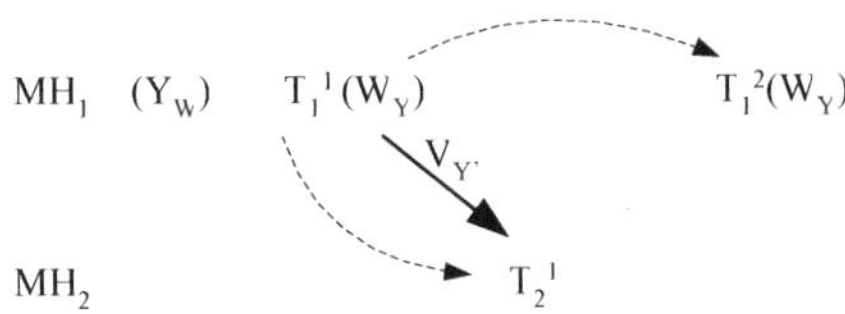

Fig. 5. Execution constraints with sharing modified data

Rule 5 (*transaction dependency*). *If a delegator transaction T_i^k shares a modified data item X to a delegatee transaction T_j^l, the delegator transaction T_i^k will hold an abort-dependency with transaction T_j^l on shared data X, i.e., if the transaction T_i^k aborts, then the transaction T_j^l aborts.*

The above rule will ensure that if an offline transaction is operating on an invalid shared data item, this offline transaction will eventually abort when it reaches transaction integration stage (explained in the next section).

4.3 Transaction Integration

The transaction integration stage is carried out when the mobile host reconnects to the database servers. In this stage, locally committed transactions, which have been disconnectedly processing at the mobile host, will be validated against other transactions to ensure that the states of the database servers are consistent.

In mobile environments, there is no guarantee that all the mobile hosts will synchronously connect to the database servers to integrate the locally committed transactions at the same time. Therefore, there is no guarantee that a delegator transaction will be integrated into the database servers before a delegatee transaction or via versa. Furthermore, a local transaction can play roles as both the delegator and delegatee transactions. Consequently, the database servers must keep track of the commit or abort state of both delegator and delegatee transactions in order to determine the effect of one transaction on the others (i.e., both the transaction execution constraints and dependencies must be checked).

Figure 6 presents examples of these effects. In Figure 6(a), the delegator transaction T_1^i and the delegatee transaction T_2^j, which belong to different mobile hosts MH_1 and MH_2 respectively, develop an abort-dependency (T_1^i *AD* T_2^j) and a commit-dependency (T_1^i *CD* T_2^j). If the delegator transaction T_1^i commits or aborts before the delegatee transaction T_2^j, the final state of the delegatee transaction T_2^j can be determined normally. However, if the delegatee transaction T_2^j requests to finally commit before the delegator transaction T_1^i (as shown in Figure 6(b)), the final state of the delegatee transaction T_2^j will not be determined until the state of the delegator transaction T_1^i is known. In this case, the commit of the delegatee transaction T_2^j will be delayed, i.e., resulting in a pending commit.

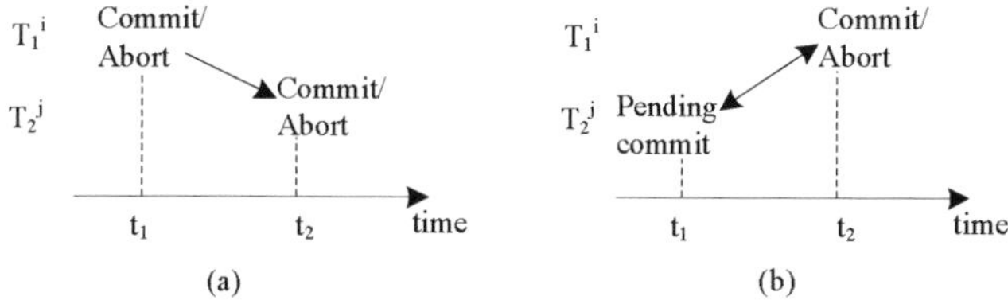

Fig. 6. The effect of the order of transaction termination requests

The database servers will manage two transaction sets: (1) *GlobalAbortedDelegator GAD* and (2) *PendingCommit PC*. The *GAD* set will keep track of the abort of delegator transactions. If a delegator transaction is aborted, it will be added to the *GAD* set so that the associated delegatee transactions will be notified. For a delegatee transaction that holds an abort-dependency with a delegator transaction that has not been integrated yet, the commitment of the delegatee transaction will be delayed and added to the *PC* set.

Handling the abort of delegator transactions

The following algorithm takes care of the final aborts of delegator transactions T_i^{Dor} and verifies the abort dependencies of associated delegatee transactions T_j^l which are queued in the *PendingCommit* (*PC*) set.

```
(1) For each delegator transaction T_i^Dor that has aborted
      Add T_i^Dor to the GlobalAbortedDelegator set

(2)   For each T_j^l in the PendingCommit set that holds
      an abort-dependency with transaction T_i^Dor
           Abort T_j^l
           If T_j^l is a delegator transaction
               Add T_j^l to the GlobalAbortedDelegator set
```

Fig. 7. Handling the abort and abort dependency of delegator transactions

The algorithm is presented in Figure 7 and is explained as follows:

(1) Each of the aborted delegator transaction T_i^{Dor} will be added to the *GAD* set. This will trigger a separate verification of the abort dependencies of the associated delegatee transactions.
(2) Any associated delegatee transaction T_j^l in the *PC* set holding an abort-dependency with the delegator transaction T_i^{Dor} will be aborted. If the aborted transaction T_j^l is a delegator transaction (i.e., this transaction plays roles a both delegator and delegatee transactions), the transaction T_j^l will be added to the *GAD* set.

Verify transaction dependency and execution constraint of offline transactions
For an offline transaction T_i^k which has been locally committed at the mobile host, the offline transaction will be verified in two steps. First, any abort-dependency with a delegator transaction T_j^{Dor} will be evaluated. Second, the execution constraints between transaction T_i^k with other transactions T_j^l will be checked to ensure the correctness of transactions, i.e., serializable execution.

```
(1) For each abort-dependency with T_j^Dor
       If T_j^Dor is in the GlobalAbortedDelegator set
            Abort T_i^k
            If T_i^k is a delegator transaction
              Add T_i^k to the GlobalAbortedDelegator set
       Else If T_j^Dor has not committed
              Add T_i^k to the PendingCommit set

(2) For all transactions T_j^l that hold execution
    constraint with T_i^k
       Check the execution constraints
       If T_j^l and T_i^k are non-serializable and T_j^l has
       committed
            Notify the transaction manager for manual
            handling
       Else If T_j^l has not committed
            Choose T^m = T_j^l | T_i^k
            Abort T^m
            If T^m is a delegator transaction
              Add T^m to the GlobalAbortedDelegator set
            If T^m = T_i^k
              Exit the algorithm
    Commit T_i^k
```

Fig. 8. Verifying transaction dependencies of a locally committed transaction

The algorithm of this step is presented in Figure 7 and is explained as follows:

(1) For each abort-dependency between the locally committed transaction T_i^k and a delegator transaction T_j^{Dor}, if the delegator transaction T_j^{Dor} has aborted, the transaction T_i^k must abort too. Otherwise, if the delegator transaction T_j^{Dor} has not reached the transaction integration stage, the locally committed transaction T_i^k will be added to the *PC* set. In this case, the abort-dependency will be re-evaluated when the termination state of the delegator transaction T_j^{Dor} is known (see point (2) in Figure 7 above).

(2) The execution constraints between transactions T_i^k and other transactions T_j^l will be determined based on Rules 1-4. If transactions T_i^k and T_j^l end up being non-serializable and T_j^l has committed, a notification will be sent to transaction manager so that it can be handled separately, e.g., by compensating T_j^l. If T_j^l is alive, one of the transactions T_j^l and T_i^k must be aborted. Then, we have to make a choice between it and T_i^k - which one depends on the policy to be used in a specific system. If the aborted transaction T^m is a delegator transaction, it will be added to the *GAD* set so that related pending transactions T^p in the *PC* set may

be re-evaluated. After all the execution constraints are checked, transaction T_i^k will be finally committed at the database servers.

5 Conclusions

In this paper, we present a mobile database sharing protocol to increase data availability in mobile environments. To deal with the problem of long disconnection periods, the mobile database sharing protocol support two conflict caching modes to allow accessing to shared data objects that have been conflicting locked by disconnected mobile hosts. Furthermore, the data sharing mechanism also support direct data sharing among mobile hosts that are being disconnected from the database servers. The data consistency is ensured via the execution constraints and dependency among delegator and delegatee transactions. Finally, a prototype of the mobile database sharing protocol has been successfully implemented [13]. Future work is to investigate the performance of the data sharing protocol.

References

1. Pitoura, E., Samaras, G.: Data Management for Mobile Computing. Kluwer Academic Publishers, Boston (1998)
2. Brayner, A., Filho, J.M.: Sharing Mobile Databases in Dynamically Configurable Environments, CAiSE 2003. In: Eder, J., Missikoff, M. (eds.) CAiSE 2003. LNCS, vol. 2681, pp. 724–737. Springer, Heidelberg (2003)
3. Holliday, J., Agrawal, D., Abbadi, A.E.: Disconnection Modes for Mobile Databases. Wireless Networks 8(4), 391–402 (2002)
4. Garcia-Molina, H., Ullman, J., Widom, J.: Database Systems: The Complete Book. Prentice Hall, Englewood Cliffs (2001)
5. Serrano-Alvarado, P., Roncancio, C., Adiba, M.E.: A Survey of Mobile Transactions. Distributed and Parallel Databases 16(2), 193–230 (2004)
6. Madria, S.K., Baseer, M., Bhowmick, S.S.: A Multi-version Transaction Model to Improve Data Availability in Mobile Computing, CoopIS/DOA/ODBASE, pp. 322–338 (2002)
7. Ramampiaro, H.: CAGISTrans: Adaptable Transactional Support for Cooperative Work, Dr.ing Thesis 2001:94, Norwegian University of Science and Technology (2001)
8. Reddy, P.K., Kitsuregawa, M.: Speculative Locking Protocols to Improve Performance for Distributed Database System. IEEE Transactions on Knowledge and Data. Engineering 16(2), 154–169 (2004)
9. Salem, K., Garcia-Molina, H., Shands, J.: Altruistic Locking. ACM Transactions on Database Systems 19(1), 117–165 (1994)
10. Chrysanthis, P.K., Ramamritham, K.: Synthesis of Extended Transaction Models Using ACTA. ACM Transactions on Database Systems 19(3), 450–491 (1994)
11. Le, H.N., Nygård, M.: A Mobile Affiliation Model for Supporting Collaborative Work in Mobile Environments, the 3rd Workshop on Ubiquitous Mobile Information and Collaboration Systems, pp. 649–660 (2005)
12. Le, H.N., Nygård, M.: Mobile Transaction System for Supporting Mobile Work, the 8th Workshop on Mobility in Databases and Distributed Systems, pp. 1090–1094 (2005)
13. Skjerven, S.: Adaptive Locking Protocols for Mobile Databases, Master Thesis, Dept. of Computer and Information Science, Norwegian University of Science and Technology (2006)

Mining Recent Frequent Itemsets over Data Streams with a Time-Sensitive Sliding Window

Long Jin[1], Duck Jin Chai[1], Jun Wook Lee[2,*], and Keun Ho Ryu[1]

[1] Database/Bioinformatics Laboratory, Chungbuk National University, Korea
{kimlyong,djchai,khryu}@dblab.chungbuk.ac.kr
[2] Telematics Research Division, ETRI, Daejeon, Korea
junux@etri.re.kr

Abstract. A data stream is a massive unbounded sequence of data elements continuously generated at a rapid rate. Consequently, the knowledge embedded in a data stream is more likely to be changed as time goes by. Mining frequent patterns is the one of them and has been widely studied over the last decade. There are several models and approaches, but there is only one study on the time-sensitive sliding window model. This study spends much memory and has a low accuracy. In this paper, we propose an efficient discounting method and a Sketch data structure for solving these problems. This approach has several advantages. (i) The accuracy is increased compared with that of previous techniques. The efficient discounting method not only loses the information about accumulated count but also decrease many missing true answers. (ii) The memory is saved. The Sketch data structure saves much space. (iii) It is not necessary to have the discount table and reduce significantly the computing time of discounting table. Experiment results show that our proposed method exactly increases the accuracy and saves the memory and the computing time.

1 Introduction

A data stream is a massive unbounded sequence of data elements continuously generated at a rapid rate. Due to this reason, it is impossible to maintain all elements of a data stream. As a result, data stream processing should satisfy the following requirements [1]. First, each data element should be examined at most to analyze a data stream. Second, memory usage for analyzing data stream should be restricted finitely although new data elements are continuously generated in a data stream. Third, newly generated data elements should be processed as fast as possible. Finally, the up-to-date analysis result of a data stream should be instantly available when requested.

Frequent itemset mining is a core data mining operation and has been extensively studied over the last decade. There are two groups: finding frequent items and finding frequent itemsets on data stream. For finding these, there are four types of models. (i) Most of them [2, 3, 4] utilize all the data between a particular point of time (called landmark) and the current time for mining. The landmark usually refers to the time

[*] Corresponding author.

K.C. Chang et al. (Eds.): APWeb/WAIM 2007 Ws, LNCS 4537, pp. 62–73, 2007.

when the system starts. Moreover, the support count of an itemset in this model is the number of transactions containing it between the landmark and the current time. (ii) The landmark model is not aware of time and therefore cannot distinguish between new data and old ones. To overcome this difficulty, the time-fading model has been presented recent works [5, 6, 7]. It assigns different weights to transactions such that new ones have higher weights than old ones. All these approaches provide approximate answers for long-term data and adjust their storage requirement based on the available space. Therefore, they satisfy the two requirements – approximation and adjustability. (iii) In the sliding-window model, users can only be interested in the data recently arriving within a fixed time period. As a transaction arrives, the oldest transaction in the sliding window is expired. Therefore, under this model, the methods for finding the expired transaction and for discounting the support counts of the itemsets involved are required [8, 9, 10, 11]. (iv) In the time-sensitive sliding-window model [12], it is natural to specify a time period as the basic unit and regards a fixed time period as the basic unit for mining. A data stream is decomposed into a sequence of basic units, which are assigned with serial numbers starting at 1. This model slides the sliding window over this sequence to see a set of overlapping sequences.

Lin et al. [12] proposed a time-sensitive sliding window model. They use the accumulated count (Acount) and the potential count (Pcount) concepts. Acount accumulates its exact support counts in the subsequent basic blocks. Pcount estimates the maximum possible sum of its support counts in the past basic blocks. There maintains a potential frequent itemsets pool (PFP) and deletes the itemset if the sum of its Acount and Pcount is less than the support count threshold ε. And they output the frequent itemsets only if these Acount is bigger than the threshold ε. There are many problems. First, the information about Acount is lost. There deletes the itemset c that doesn't satisfy the threshold. The itemset c appears again at the after blocks. Although Acount is already computed, there is estimated Pcount and substituted Pcount for Acount. So the accuracy of Acount is decreased. Second, many true answers are missed. When filtering the frequent itemsets, they only use Acount. So there occur many missing true frequent itemsets by decreasing the accuracy of Acount. Also it needs to minimize the missing of true answers and guarantee that the no false answers are outputted, e.g. no-false-alarm.

In this paper, we propose two kinds of solutions. First, we propose an efficient discounting method for increasing the accuracy of Acount. This efficient discounting method not only loses the information about Acount but also decrease many missing true answers. Second, we propose a Sketch data structure for decreasing the memory space. This structure saves much space than that of [12]. And it doesn't need to have the discount table (DT) as in [12] and significantly reduces the computing time of DT. Third, we propose the detail of efficient algorithm.

The remaining of this paper is organized as follows. Section 2 defines some symbols and definitions. The detail of efficient discounting method and Sketch data structure is shown in Section 3. Section 4 develops the detail of algorithm and shows the examples. Section 5 presents our performance of our proposed method. Section 6 summarizes this paper.

2 Preliminaries

Given a time point t and a time period p, the set of all the transactions arriving in [t-p+1, t] will form a basic block. The number of basic block's ciphers is $|p|$. A data stream is decomposed into a sequence of basic blocks, which are assigned with serial numbers starting at 1. Given a window with length $|\omega|$, we slide it over this sequence to see a set of overlapping sequences, where each sequence is called the time-sensitive sliding window abbreviated as *TS*. Let the basic block numbered i be denoted as B_i. The number of transactions in B_i is denoted as $|B_i|$, which is not fixed due to the variable data arrival rate. For each B_i, the *TS* that consists of the $|\omega|$ consecutive basic blocks from $B_{i-|\omega|+1}$ to B_i is denoted as TS_i. Let the number of transactions in TS_i be denoted as Σ_i.

For efficiently mining the frequent itemsets in TS_i, we design the data structure that is consisted of one *SA* and one *PFP* as follows.

1. *SA* is abbreviated the Support Array. The basic blocks may have different numbers of transactions and we store it into an entry in the *SA*. In there, only $|\omega|$+1 entries are maintained in *SA*. As B_i arrives, $SA[j]$ keeps the support count of B_{i-j} for $1 \leq j \leq |\omega|+1$ and i-j>0. After B_i is processed, the last entry $SA[|\omega|+1]$ is ignored and the others are moved to the next positions, i.e., $SA[j] \rightarrow SA[j+1]$ for $1 \leq j \leq |\omega|$. Finally, the support count of B_i is put into $SA[1]$.
2. *PFP* is abbreviated the potential frequent itemsets pool. *PFP* includes the itemsets that are frequent in B_i but not frequent in TS_{i-1} in *PFP* because they are possibly frequent in TS_i. *PFP* is consisted of an ID, an itemset, an accumulated count Acount, and a potential count Pcount. Where, Acount is the type of Sketch data structure that will describes in Section 3.2. This kind of Acount stores all accumulated counts from $B_{i-|\omega|+1}$ to B_i. So there doesn't need the discounting table (DT) in [12].

This structure efficiently stores the data and minimizes the memory space. We will describe how to efficiently store the data and minimize the memory space in Section 3.

3 Our Proposed Method

This section shows how to increase the accuracy and save the memory. For these ones, we propose an efficient discounting method in Section 3.1 and a Sketch data structure in Section 3.2.

3.1 Efficient Discounting Method

For increasing the accuracy of Acount, we use a error rate r and control the accuracy of Acount. For deleting the infrequent itemsets and maintaining the potential frequent itemsets, there are two kinds of deletions. First is the logical deletion that the sum of Acount and Pcount is less than the support count threshold ε. The logical deletion doesn't delete the itemset from PFP. And we update Acount and Pcount of the itemset if the itemset appears after blocks. Second is the physical deletion that Acount is less than the threshold (ε-r). The physical deletion deletes the itemset from PFP.

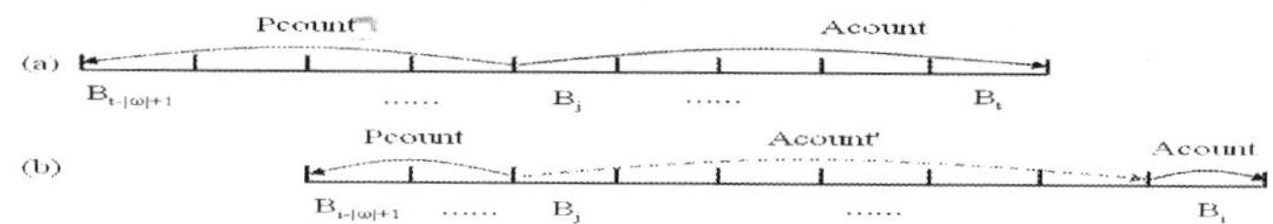

Fig. 1. Before and after sliding window

Definition 1. Let the minimum support be ε and the error rate be *r*. the itemset c is discarded from the potential frequent itemsets only if it satisfies the following formula.

$$L = \{c \mid c.Pcount + c.Acount < \varepsilon \times \Sigma_i,\ c \in PFP\}$$
$$P = \{g \mid g.Acount < (\varepsilon - r) \times \Sigma_i,\ g \in PFP\}$$
$$where,\ i > 0$$

If there are a set of itemsets *L* that doesn't satisfy the logical deletion, some itemsets are not frequent at this window, but frequent at the future window. So we must filter these itemsets from *L*. We filter the frequent itemsets from the potential frequent itemsets, we only use Acount and that is bigger than the threshold ε. Then we only delete the itemsets that whose Acount is less than the threshold (ε-*r*). Then we can guarantee to minimize the missing true answers and dynamically control the level of the infrequent itemsets by user.

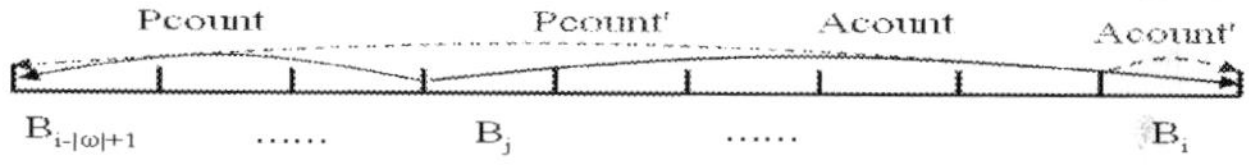

Fig. 2. The count state in two cases

Theorem 1. Let AC be the accurate count between our proposed method and time-sensitive sliding window model [12]. And *j* is the block point that the itemset first appears at TS_i. Then the boundary of AC is as follows.

$$0 < AC < r \times \Sigma_i - \sum_{k=i-|\omega|}^{j-1} \varepsilon |B_k|$$

Proof

1) Our proposed method

$$If\ C_i \geq \varepsilon |B_i|,\ then\ Acount = \sum_{k=j}^{i} C_k\ and\ Pcount = \sum_{k=i-|\omega|+1}^{j-1} \varepsilon |B_k|$$

2) Mining and Discounting in [12]

$$If\ C_i \geq \varepsilon |B_i|,\ then\ Acount' = C_i\ and\ Pcount' = \sum_{k=i-|\omega|+1}^{i-1} \varepsilon |B_k|$$

3) The accurate count AC is as following.

$$AC = (Acount' + Pcount') - (Acount + Pcount)$$
$$= (C_i + \sum_{k=i-|\omega|+1}^{i-1} \varepsilon |B_k|) - (\sum_{k=j}^{i} C_k + \sum_{k=i-|\omega|+1}^{j-1} \varepsilon |B_k|) = \sum_{k=j}^{i-1} \varepsilon |B_k| - \sum_{k=j}^{i-1} C_k$$

4) From Definition 1, the boundary of Acount is as follows.

$$\begin{cases} Acount + Pcount < \varepsilon \times \Sigma_{i-1} & (1) \\ Acount \geq (\varepsilon - r) \times \Sigma_{i-1} & (2) \end{cases}$$

$$From\ (1),\ Acount < \varepsilon \times \Sigma_{i-1} - \sum_{k=i-|\omega|}^{j-1} \varepsilon |B_k| = \sum_{k=j}^{i-1} \varepsilon |B_k|$$

$$\therefore (\varepsilon - r) \times \Sigma_{i-1} \leq Acount < \sum_{k=j}^{i-1} \varepsilon |B_k|$$

5) From 4), the boundary of AC is as follows.

$$0 < \sum_{k=j}^{i-1} \varepsilon |B_k| - \sum_{k=j}^{i-1} C_k \leq \sum_{k=j}^{i-1} \varepsilon |B_k| - (\varepsilon - r) \times \Sigma_{i-1} = \sum_{k=j}^{i-1} \varepsilon |B_k| - \varepsilon \times \Sigma_{i-1} + r \times \Sigma_{i-1}$$

$$= r \times \Sigma_{i-1} + \sum_{k=j}^{i-1} \varepsilon |B_k| - \sum_{k=i-|\omega|}^{i-1} \varepsilon |B_k| = r \times \Sigma_{i-1} - \sum_{k=i-|\omega|}^{j-1} \varepsilon |B_k|$$

$$\therefore 0 < AC \leq r \times \Sigma_{i-1} - \sum_{k=i-|\omega|}^{j-1} \varepsilon |B_k|$$

The efficient discounting method has two advantages. First, the information about Acount is maintained. Acount is already computed and substituted Acount for Pcount. So the accurary of Acount is increased, e.g. Acount' at the (b) of Fig.1. Second, many missing true answers are decreased. For maintaining the information about Acount before blocks, we compute Acount more accurately and decrease the information about Acount at the blocks that doesn't satisfy the minimum threshold. Filtering the frequent itemsets, we only use Acount and minimize many missing true frequent itemsets.

The efficient discounting method has the disadvantages, too. There are many potential frequent itemsets by deleting the itemsets that satisfies the physical deletion. So there spend much memory space than [12]. For solving this problem, we propose a Sketch data structure and describe it in next section.

3.2 Sketch Data Structure

The size of sliding window must have at least 2 and have more than several basic blocks at general case. The count of any frequent itemset is at most p at each basic block. The time period p is not too big, because the input stream is divided into several basic blocks and this is the basic unit. So the size of the basic block couldn't be very big, e.g. the number of ciphers is not bigger than 4~6 in general case.

Definition 2. Let the number of basic block's ciphers be $|p|$, $p \geq 1$, and $|\omega|>1$. Then the following formula is satisfied.

$$\omega \leq p \Rightarrow 1 \leq |p| \leq |\omega| \qquad (1)$$

Proof. The stream data is continuously inputted and the size of basic block couldn't be very big. So the time period of basic block is as follows in a general case. (i) The time period of basic block is 1 minute, then p=60 and $|p|$=2. (ii) The time period of basic block is 1 hour, then p=3600 and $|p|$=4. (iii) The time period of basic block is 1 day, then p=86400 and $|p|$=5. So the window size ω is more than 4~6 in general case and the formula (1) is materialized. The memory space is saved more and more according to increasing the size of sliding window ω.

Definition 3. Let the frequent itemset count be C and the number of ith cipher on C is a_i. Then the array of any frequent itemset count $A[]$ is as follows:

$$A_i = \begin{cases} A_i + (C_t / 10^{i-1}) \% 10 * 10^{|\omega|-t}, & 0 < t \le |\omega| \\ A_i \% 10^{|\omega|-1} * 10 + (C_t / 10^{i-1}) \% 10, & t > |\omega| \end{cases}$$

$$where,\ 1 \le i \le |p|\ and\ 1 \le k \le |\omega|$$

Proof

1) The number of ith cipher is as follows.

$$a_i = (C / 10^{i-1}) \% 10,\ 1 \le i \le |p|$$

$$C = \overbrace{a_{|p|}\, a_{|p|-1} \cdots a_i \cdots a_2\, a_1}^{|p|}$$

2) The number of end cipher is as follows at time t.

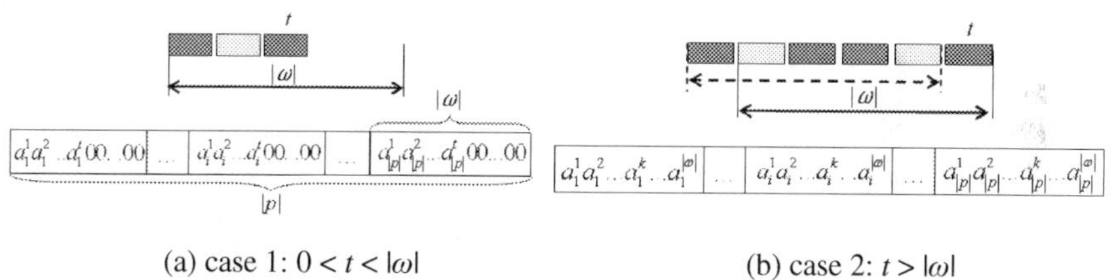

(a) case 1: $0 < t < |\omega|$ (b) case 2: $t > |\omega|$

Fig. 3. The state of accumulated count

$$Case\ 1: 0 < t \le |\omega|$$

$$a_i^t = (C_t / 10^{i-1}) \% 10 * 10^{|\omega|-t},\ 1 \le i \le |p|$$

$$Case\ 2: t > |\omega|$$

$$a_i^{|\omega|} = (C_t / 10^{i-1}) \% 10,\ 1 \le i \le |p|\ and\ 1 \le k \le |\omega|$$

3) The array of any frequent itemset count $A[]$ is as follows by from 2).

$$A_i = A_i + (C_t / 10^{i-1}) \% 10 * 10^{|\omega|-t},\ 0 < t \le |\omega|$$

$$A_i = A_i \% 10^{|\omega|-1} * 10 + (C_t / 10^{i-1}) \% 10,\ t > |\omega|$$

Let $|p|$ be 3 and $|\omega|$ be 5. Fig. 4 is the example of definition 3. The (a) of Fig. 4 is the example that the *buffer* isn't full. And the (b) of Fig. 4 is the example that the sliding window starts to slide on the *buffer*.

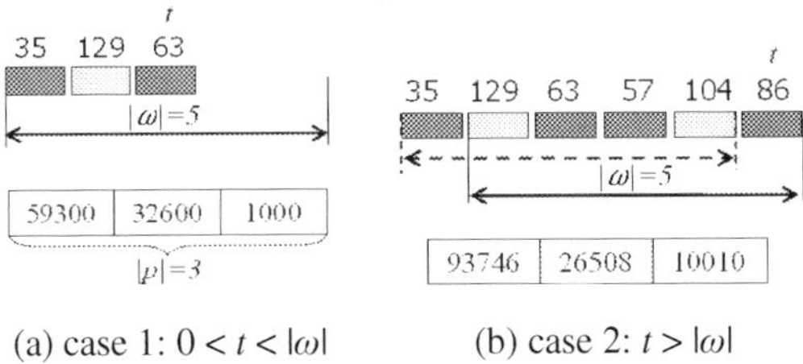

(a) case 1: $0 < t < |\omega|$ (b) case 2: $t > |\omega|$

Fig. 4. The example of definition 3

The detailed algorithm of definition 3 is shown as follows in Fig. 5. This structure has several advantages. (i) The size of count space only depends on $|p|$ and isn't on $|\omega|$. So much space is saved than that of [12]. (ii) It doesn't need to have the discount table (DT) as in [12]. (iii) It saves the computing time of DT.

```
Algorithm 1 (AC_List Algorithm)
Input: count C, potential frequent item cout PFI_Count[], ω, p, t
Output: PFI_Count[]
Method:
1. If t < |ω| then //didn't full the basic block
     For each i from 0 to |p|-1
       PFI_Count[i] ← PFI_Count[i]+(C/10^i)%10*10^(|ω|-t);
2. Else //general case
     For each i from 0 to |p|-1
       PFI_Count[i] ← PFI_Count[i]%10^(|ω|-1)*10+(C/10^i)%10;
3. Return PFI_Count[];
```

Fig. 5. AC_List Algorithm

4 Algorithms and Examples

Mining frequent itemsets in TS_i consists of three steps. First, the frequent itemsets in B_i are mined by using FP-growth[18] and added into *PFP* with their potential counts computed. If the itemset exists in *PFP*, it is added into *PFP* and discounted from its potential count in $B_{i-|\omega|}$. Second, for each itemset that is in *PFP* but not frequent in B_i, we scan B_i to accumulate its support count and discount the potential count of each itemset in $B_{i-|\omega|}$. And we delete it from *PFP* if it is not frequent in TS_i. Last, the frequent itemsets are outputted. The detail of algorithm is shown in Fig. 6. There are two cases: first time and other time. If first block is inserted, we only insert the frequent itemset into *PFP*. Other time, we not only insert the frequent itemset but also update the itemset that is not frequent in this block but already exist in *PFP*.

```
Algorithm 2 (Main Algorithm)
Input: Stream S, Parameters ε, P, |ω|, M
Output: All the frequent itemsets satisfying M
Method:
1. Let TA and PFP be empty
2. While B_i comes
3.   If (i=1)
       New_itemset_insertion
4.   Else //i>1)
       New_itemset_insertion
       Old_itemset_update
6.   SA_update
7.   Frequent_itemset_output
```

Fig. 6. Main Algorithm

4.1 Main Operations

Fig. 7 is the detail of the main operations of main algorithm. First procedure is New_item_insertion. First step is that we adopt the FP-growth algorithm to mine all the frequent itemsets from B_i. Let F_i denote this set. Next, we check each itemset in F_i to see whether it has been kept in *PFP* and then either update or create an entry. There are Acount and Pcount. Acount. We compute Acount by AC_List algorithm and estimate Pcount as follows.

$$Pcount\ at\ TS_i = \lceil \varepsilon \times \Sigma_{i-1} \rceil - 1 = \left\lceil \sum_{j=1}^{|\omega|} \varepsilon \times SA[j] \right\rceil - 1 \tag{2}$$

Second procedure is Pcount_discounting. We use this procedure only if the time point i is greater than $|\omega|$ and slide on *buffer*. Each itemset uses Pcount to keep its maximum possible count in the past basic blocks before it is inserted into *PFP*. By formula (2), since B_i comes, Pcount is computed by including the support count threshold of an extra basic block, i.e., $B_{i-|\omega|}$. As B_{i+1} comes, if Pcount is nonzero, we subtract the support count threshold of $B_{i-|\omega|}$ from Pcount. If Pcount is smaller than the support count threshold of $B_{i-|\omega|+1}$, Acount should have the exact support counts from $B_{i-|\omega|+2}$ to B_{i-1}. In this case, we set Pcount to 0.

Third procedure is Old_itemset_update. For each itemset g that has been kept by *PFP* but not in F_i, we compute its support count in B_i to increase its Acount. If g.Acount is zero, we insert it, too. Suppose that g was inserted into *PFP* when B_k comes ($k < i$). At this point, we have g.Acount, the exact support count of g in [B_k, ..., B_i], and g.Pcount, the maximum possible support count of g in [$B_{i-|\omega|+1}$, ..., B_{k-1}]. If the sum is less than the support count threshold ε, g may be frequent at the future *buffer* and we don't delete g from *PFP*. And we continue to check that if g.Acount is less than the support count threshold (ε-r), g must not be frequent in TS_i and can be safely deleted from *PFP*.

Forth procedure is Frequent_itemset_output. It supports two kinds of outputs: NFD and NFA. NFD is the no-false-dismissal model and it guarantees that no true answer is missed. In this mode, an itemset that is frequent in B_i but not in TS_i is still outputted. Sometimes user hopes that all the itemsets outputted are real answers. Therefore, we also provide the no-false-alarm mode (denoted as NFA), which outputs only the itemsets with Acount satisfying the support count threshold. Since Acount accumulates the support counts of an itemset in the individual basic blocks after that itemset is inserted into *PFP*, this mode guarantees that no false answer is outputted.

```
Procedure 1 (New_itemset_insertion)
Input: B_i
Output: F_i, updated PFP
1. Discover F_i from B_i
2. For each itemset f in F_i
   If (f ∈ PFP) //Increase f.Acount;
     PFP[f].Acount ← AC_List(f.Acount, PFP[f].Acount, ω, p, i);
     If (i > |ω|) Pcount_discounting;
   Else //Insert f into PFP
     Create a PFP parameter temp and initializes it;
     temp.itemset ← f;
     temp.Acount ← AC_List(f.Acount, temp.Acount, ω, p, i);
     Estimate temp.Pcount;
     PFP ← PFP ∪ temp;
```

```
Procedure 2 (Pcount_discounting)
Input: g, i
Output: updated PFP
If (g.Pcount > 0)
  g.Pcount = g.Pcount – ε*SA[|ω|+1]
  If (g.Pcount < ε*SA[|ω|])
    g.Pcount = 0
```

```
Procedure 3 (Old_itemset_update)
Input: F_i, B_i, PFP
Output: updated PFP
1. For each itemset g in PFP but not in F_i
   //Increase g.Acount by scanning B_i once
   If (g.Acount > 0)
     PFP[g].Acount ← AC_List(g.Acount, PFP[g].Acount, ω, p, i);
   Else PFP[g].Acount ← AC_List(0, PFP[g].Acount, ω, p, i);
   If(i > |ω|) Pcount_discounting;
2. If (PFP[g].Acount < ε x Σ_i) //Logical discard
     If (PFP[g].Acount < (ε-r) x Σ_i) //Physical discard
       Delete g from PFP
```

```
Procedure 4 (Frequent_itemset_output)
Input: PFP
Output: The set of frequent itemsets O
1. If (M = NDF)
     For each itemset f in PFP
       If (f.Acount + f.Pcount ≥ ε x Σ_i)
         O = O + {f}
2. Else
     For each itemset f in PFP
       If (f.Acount ≥ ε x Σ_i)
         O = O + {f}
```

Fig. 7. Main Operations

4.2 Examples

Let ε, $|\omega|$, r, and p be 0.4, 4, 0.1, and 60 minutes, respectively. Consider the stream of transactions shown in Table 1.

Table 1. A stream of transactions

	Time period	Number of transactions	Time Stamp
B_1	**09:00 ~ 09:59**	**27**	**a(11), b(20), c(9), d(9), ab(6)**
B_2	**10:00 ~ 10:59**	**20**	**a(20), c(15), d(6), ac(15)**
B_3	**11:00 ~ 11:59**	**27**	**a(19), b(9), c(7), d(9), ac(7), bd(9)**
B_4	**12:00 ~ 12:59**	**23**	**a(10), c(7), d(15)**
B_5	**13:00 ~ 13:59**	**30**	**a(20), b(19), c(20), d(19), ac(20), bd(19)**
B_6	**14:00 ~ 14:59**	**22**	**a(9), b(12), c(6), d(12), ab(3), bd(12)**

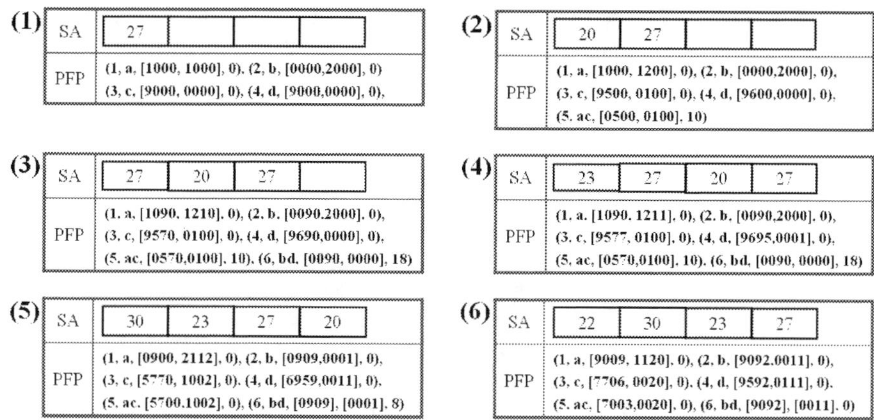

Fig. 8. The snapshots after processing each basic block

Fig. 8 is the illustration of mining the frequent itemsets using our proposed method in Table 1 and shows the snapshot of the potential frequent itemsets and the support array after processing each basic block.

5 Experiment and Evaluation

In this section, we perform a series of experiments for evaluating the performance of TFP-tree with the classical frequent pattern mining algorithms on synthetic data. All the experiments are performed on a Windows XP with Pentium PC 2.8GHz and 512 Mbytes of main memory. Also we use JDK 1.4, MS-SQL 2000 database and JDBC driver for connecting MS-SQL 2000.

We generate three dataset, DT1, DT2, and DT3, by using our data generation program. The parameters of the number of distinct itemset, the average transaction length, the average length of the maximum pattern, and total number of transaction are respectively DT1(40, 8, 4, 10000), DT2(55, 12, 6, 10000), and DT3(45, 10, 5, 10000). For examining the performance, the basic block is 60, the minimum support is 0.6, and the error rates are 0.1, 0.2, and 0.3. The mining algorithm we applied to find frequent itemsets in a block is the FP-growth. There are two kinds of the experiments: the evaluation of our proposed method and the comparing our proposed method with others. There are two compared algorithm: TSM(Selective adjustment) and TSM-S(Self-adjustment) in [12].

Fig. 9 is the first experiment. The (a) of Fig. 9 is the experiment of accuracy at the minimum support 0.6 and the window size 4. The numbers of frequent itemsets is increasing according to increasing of the error rate, but some dataset doesn't increase sharply like DT2. This reason is that the frequent itemsets is close to the round of the minimum support. The (b) of Fig. 9 is first experiment of saving the memory at the minimum support 0.6 and the error rate 0.1. The usage of memory is increasing slowly according to increasing of the window size. And total usage of memory is smooth. The (c) of Fig. 9 is second experiment of saving the memory at the minimum support 0.6 and the window size 4. The usage of memory is increasing smoothly according to increasing of the error rate. This proves that our proposed method doesn't use much memory and the total memory is close to first block at some times.

Fig. 10 is the experiment of comparing our proposed method with others. The (a) of Fig. 10 is the evaluation of accuracy. From the experiment, our propose method is finding the frequent itemsets more than TSM. The (b) of Fig. 10 is the evaluation of using the memory. Our propose method is saving the memory more than others. The (c) of Fig. 10 is the evaluation of the execution time. From the figure, the execution of our proposed method is smaller than others, too.

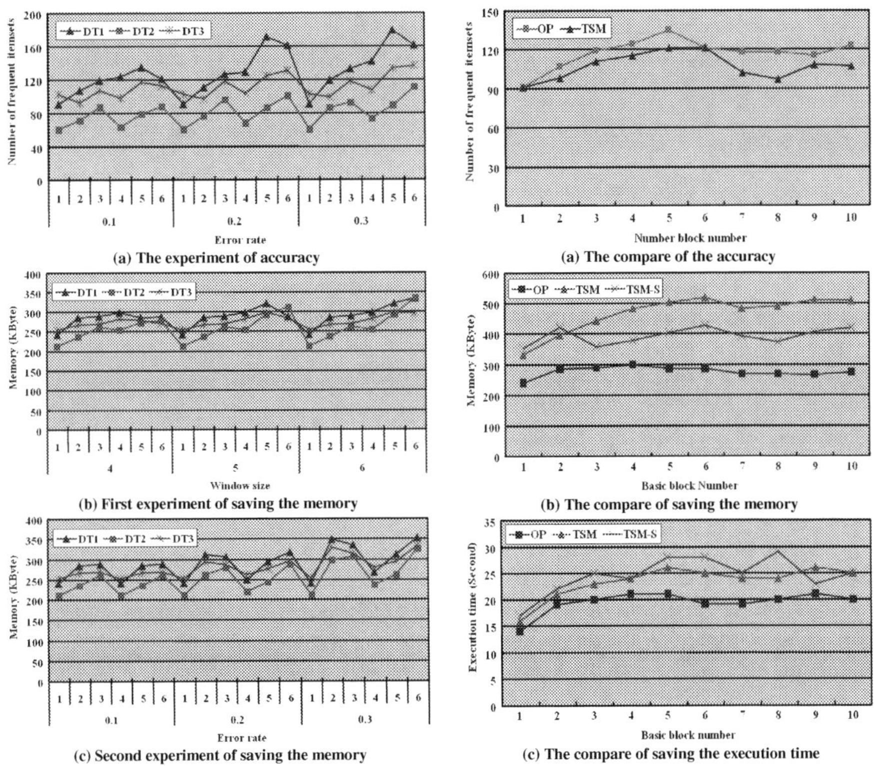

(a) The experiment of accuracy

(b) First experiment of saving the memory

(c) Second experiment of saving the memory

Fig. 9. The evaluation of our proposed method

(a) The compare of the accuracy

(b) The compare of saving the memory

(c) The compare of saving the execution time

Fig. 10. The comparing our proposed method with others

Through the above evaluation, our proposed method is more accurate and saves the memory and the execution time.

6 Conclusion

We have proposed an efficient discounting method and a novel Sketch data structure for efficient mining of frequent itemsets in time-sensitive sliding-window model. And we proposed the detail of efficient algorithm. Through the experiments, our proposed method has several advantages. (i) The accuracy is increased compared with that of previous techniques. From the experiment, the efficient discounting method not only loses the information about Acount but also decreases many missing of true answers. (ii) The memory is saved. The Sketch data structure saves much space than that of [12]. (iii) It doesn't need to have the discount table (DT) and saves the computing time of DT.

In our future work, we are going to apply the proposed method to the real application systems and evaluate our proposed method with other algorithms.

Acknowledgment

This work was partially supported by the RRC program of MOCIE and ITEP, the Korea Research Foundation Grant by the Korean Government (the Regional Research Universities Program/Chungbuk BIT Research-Oriented University Consortium), and by Telematics · USN Division of ETRI in Korea.

References

1. Garofalakis, M., Gehrke, J., Rastogi, R.: Querying and mining data streams: you only get one look. In: the tutorial notes of the 28th Int'l Conf. on Very Large Data Bases (VLDB) (2002)
2. Emaine, E., Lopez-Ortiz, A., Munro, J.I.: Frequency estimation of internet packet streams with limited space. In: Proc. of European Symp. On Algorithms (ESA) (2002)
3. Karp, R.M., Papadimitriou, C.H., Shenker, S.: A simple algorithm for finding frequent elements in streams and bags, ACM Trans. On Database Systems (TODS) (2003)
4. Manku, G., Motwani, R.: Approximate frequency counts over data streams. In: Proc. of the Int'l Conf. on Very Large Data Bases (VLDB) (2002)
5. Chang, J.H., Lee, W.S.: Finding recent frequent itemsets adaptively over online data streams. In: Proc. of ACM SIGKDD Conf. (2003)
6. Cohen, E., Strauss, M.: Maintaining time decaying stream aggregates. In: Proc. of PODS Symp. (2003)
7. Giannella, C., Han, J., Pei, J., Yu, P.S.: Mining frequent patterns in data streams at multiple time granularities. In:Kargupta, H., Joshi, A., Sivakumar, K., Yesha, Y. (eds.) Next Generation Data Mining (2003)
8. Babcock, B., Datar, M., Motwani, R., O'Callaghan, L.: Maintaining variance and k-Medians over data stream windows. In: Proc. of ACM PODS Symp (2003)
9. Charikar, M., Chen, K., Farach-Colton, M.: Finding frequent items in data streams. In: Proc. of ICALP (2002)
10. Cormode, G., Muthukrishnan, S.: What's hot and what's not: tracking most frequent items dynamically. In: Proc. of ACM PODS Symp. (2003)
11. Jin, C., Qian, W., Sha, C., Yu, J.X., Zhou, A.: Dynamically maintaining frequent items over a data stream. In: Proc. of ACM CIKM Conf. (2003)
12. Lin, C.H., Chiu, D.Y., Wu, Y.H., Chen, A.L.P.: Mining frequent itemsets from data streams with a time-sensitive sliding window, SIAM Inter'l Conf. on Data Mining (2005)
13. Han, J., Pei, J., Yin, Y., Mao, R.: Mining frequent patterns without candidate generation. In: Proc. of ACM SIGMOD Int'l Conf. Management of Data (2000)
14. Jin, L., Lee, Y., Seo, S., Ryu, K.H.: Discovery of Temporal Frequent Patterns using TFP-tree. In: Yu, J.X., Kitsuregawa, M., Leong, H.V. (eds.) WAIM 2006. LNCS, vol. 4016, Springer, Heidelberg (2006)
15. Seo, S., Jin, L., Lee, J.W., Ryu, K.H.: Similar Pattern Discovery using Calendar Concept Hierarchy in Time Series Data. In: Proc. of APWeb Conf. (2004)

Mining Web Transaction Patterns in an Electronic Commerce Environment

Yue-Shi Lee and Show-Jane Yen

Department of Computer Science and Information Engineering, Ming Chuan University
5 The-Ming Rd., Gwei Shan District, Taoyuan County 333, Taiwan
{leeys,sjyen}@mcu.edu.tw

Abstract. *Association rule mining* discovers most of the users' purchasing behaviors from transaction database. Association rules are valuable for cross-marking and attached mailing applications. Other applications include catalog design, add-on sales, store layout, and customer segmentation based on buying patterns. *Web traversal pattern mining* discovers most of the users' access patterns from web logs. This information can provide navigation suggestions for web users such that appropriate actions can be adopted. *Web transaction pattern mining* discovers not only the pure navigation behaviors but also the purchasing behaviors of customers. In this paper, we propose an algorithm *IWA* (Integrating Web traversal patterns and Association rules) for mining web transaction patterns in the electronic commerce environment. Our *IWA* algorithm takes both the traveling and purchasing behaviors of customers into consideration at the same time. The experimental results show that *IWA* algorithm can simultaneously and efficiently discover traveling and purchasing behaviors for most of customers.

1 Introduction

As the *electronic commerce* (*EC*) activities become more and more diverse, it is very critical to provide the right information to the right customers. *Web mining* [4, 6, 7, 14, 15, 16, 17] refers to extracting useful information and knowledge from web logs, which applies data mining techniques [11, 12, 13] in large amount of web data to improve the web services. Mining *web traversal patterns* [16] is to discover most of users' access patterns from web logs. These patterns can be used to improve website design, such as provide efficient access between highly correlated objects, and better authoring design for web pages, and provide navigation suggestions for web users. However, this information just considered the navigation behaviors of users.

In EC environment, it is very important to find user purchasing behaviors. Association rule mining [1, 2, 3, 10] can discover the information about user purchasing behaviors. There are many researches worked on this field, e.g., *Apriori* algorithm [1], *DHP* (Direct Hashing and Pruning) algorithm [2], *FP-Growth* (Frequent Pattern-Growth) algorithm [3] and *H-mine* (Hyper-Structure Mining) algorithm [10]. However, association rule mining just purely considers the purchasing behaviors of customers. It is important to consider navigation behaviors and purchasing behaviors of customers simultaneously in EC environment. *Web transaction patterns* [9]

K.C. Chang et al. (Eds.): APWeb/WAIM 2007 Ws, LNCS 4537, pp. 74–85, 2007.

combine web traversal patterns and association rules to provide more information for web site managers, such as putting advertisements in proper places, better customer classification, and behavior analysis, etc.

In the following, we describe the definitions about web transaction patterns: Let $W = \{x_1, x_2, \ldots, x_n\}$ be a set of all web pages in a website and $I = \{i_1, i_2, \ldots, i_m\}$ be a set of all purchased items. A *traversal sequence* $S = <w_1, w_2, \ldots, w_p>$ $(w_i \in W, 1 \leq i \leq p)$ is a list of web pages which is ordered by traversal time, and each web page can repeatedly appear in a traversal sequence, that is, backward references are also included in a traversal sequence. For example, if there is a path which visit web page *B*, and then go to web page *G* and *A* sequentially, and come back to web page *B*, and then visit web page *C*. The sequence <*BGABC*> is a traversal sequence. The *length* of a traversal sequence *S* is the total number of visited web pages in *S*. A traversal sequence with length *l* is called an *l-traversal sequence*. For example, if there is a traversal sequence $\alpha = <AGDFAB>$, the length of α is 6 and we call α a 6-traversal sequence. Suppose that there are two traversal sequences $\alpha = <a_1, a_2, \ldots, a_m>$ and $\beta = <b_1, b_2, \ldots, b_n>$ $(m \leq n)$. If there exists $i_1 < i_2 < \ldots < i_m$, such that $b_{i1} = a_1$, $b_{i2} = a_2$, $\ldots b_{im} = a_m$, then β *contains* α. For instance, if there are two traversal sequences $\alpha = <BEA>$ and $\beta = <ABCEA>$, then α is a sub-sequence of β and β is a super-sequence of α.

A *transaction sequence* is a traversal sequence with purchasing behaviors. A transaction sequence is denoted as $<w_1\{i_1\}, w_2\{i_2\}, \ldots w_q\{i_q\}>$, in which $w_k \in W$ $(1 \leq k \leq q)$, $i_j \subset I$ $(1 \leq j \leq q)$ and $<w_1, w_2, \ldots, w_q>$ is a traversal sequence. If there is no purchased item in a web page, then the purchased items need not be attached with the web pages. For example, for a traversal sequence <*ABCDE*>, if item *i* is purchased on web page <*B*>, and there is no purchased items on the other web pages, then the transaction sequence can be denoted as <*AB*{*i*}*CDE*>. Suppose that there are two transaction sequences $\alpha = <a_1\{p_1\}, a_2\{p_2\}, \ldots, a_m\{p_m\}>$ and $\beta = <b_1\{q_1\}, b_2\{q_2\}, \ldots, b_n\{q_n\}>$ $(m \leq n)$. If there exists $i_1 < i_2 < \ldots < i_m$, such that $b_{i1} = a_1$, $b_{i2} = a_2$, $\ldots b_{im} = a_m$, and $p_1 \subseteq q_{i1}, p_2 \subseteq q_{i2}, \ldots, p_m \subseteq q_{im}$, then β *contains* α.

Table 1. User transaction sequence database *D*

TID	User Transaction Sequence
1	*BECAF{1}C*
2	*DBAC{2}AE{3}*
3	*BDAE*
4	*BDECAF{1}C*
5	*BAC{2}AE{3}*
6	*DAC{2}*

A *user transaction sequence database D*, as shown in Table 1, contains a set of records. Each record includes *traversal identifier (TID)* and *user transaction sequence*. A user transaction sequence is a transaction sequence, which stands for a complete browsing and purchasing behavior by a user. For example, the first user transaction sequence means that the user purchased item 1 on page *F* when he/she traversed the sequence *BECAFC*.

The *support* for a traversal sequence α is the ratio of user traversal sequences which contains α to the total number of user traversal sequences in *D*, which is denoted as *Support* (α). The *support count* of α is the number of user traversalsequences which contain α. The *support* for a transaction sequence β is the ratio of user transaction sequences which contains β to the total number of user transaction sequences in *D*. It is denoted as *Support* (β). The *support count* of β is the number of user transaction sequences which contain β. For a traversal sequence $<x_1, x_2, \ldots, x_l>$, if there is a link from x_i to x_{i+1} (for all i, $1 \le i \le l-1$) in a web site structure, then the traversal sequence is a *qualified traversal sequence*. A traversal sequence α is a *web traversal pattern* if α is a qualified traversal sequence and *Support* $(\alpha) \ge min\_sup$, in which the *min_sup* is the user specified *minimum support* threshold. For instance, in Table 1, if we set *min_sup* to 80%, then $Support\ (<BE>) = 5/6 = 83.33\% \ge min\_sup = 80\%$, and there is a link from web page *B* to web page *E* in the web site structure shown in Figure 1. Hence, $<BE>$ is a web traversal pattern. If the length of a web traversal pattern is *l*, then it can be called an *l-web traversal pattern*.

A *qualified transaction sequence* is a transaction sequence in which the corresponding traversal sequence is qualified. For instance, $<BAC\{2\}>$ is a qualified transaction sequence according to Figure 1. A transaction sequence α is a web transaction sequence if α is a qualified transaction sequence and *Support* $(\alpha) \ge min\_sup$. For example, in Table 1, suppose *min_sup* is set to 50%, then $Support\ (<AC\{2\}>) = 3/6 = 50\% \ge min\_sup = 50\%$, and there is a link from page A to page C in the web site structure in Figure 1. Hence, $<AC\{2\}>$ is a web transaction sequence. The length of a transaction sequence is the number of web pages in the transaction sequence. A transaction sequence of length *k* is called a *k-transaction sequence*, and a web transaction sequence of length *k* is called a *k- web transaction sequence*.

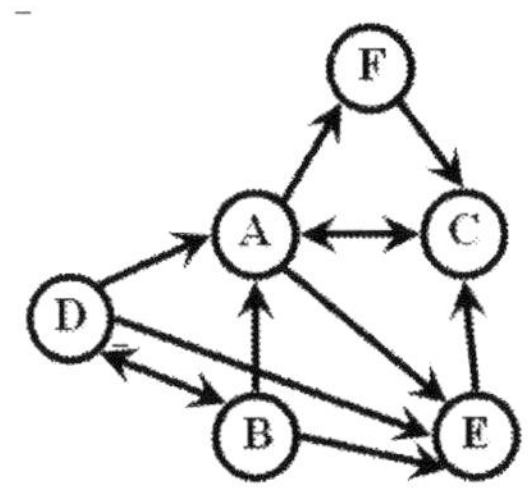

Fig. 1. Web site structure

A *transaction pattern* can be denoted as $<x_1, x_2, \ldots, x_i', \ldots, x_j', \ldots, x_n>: x_i\{I_i\} \Rightarrow x_j\{I_j\}$. The confidence of a pattern $<x_1, x_2, \ldots, x_i', \ldots, x_j', \ldots, x_n>: x_i\{I_i\} \Rightarrow x_j\{I_j\}$ is the ratio of support for transaction sequence$<x_1, x_2, \ldots, x_i\{I_i\}, \ldots, x_j\{I_j\}, \ldots, x_n>$ to support for transaction sequence $<x_1, x_2, \ldots, x_i\{I_i\}, \ldots, x_j, \ldots, x_n>$. For a web transaction sequence $<x_1, x_2, \ldots, x_i\{I_i\}, \ldots, x_j\{I_j\}, \ldots, x_n>$, if the confidence of transaction pattern $<x_1, x_2, \ldots, x_i', \ldots, x_j', \ldots, x_n>: x_i\{I_i\} \Rightarrow x_j\{I_j\}$ is no less than a user-specified minimum confidence threshold, then this transaction pattern is a web transaction pattern, which means that when the sequence <x_1, x_2, ..., x_i,..., x_j,..., x_n> is traversed, users purchased item I_i on page x_i, item I_j on page x_j was purchased too.

2 Related Work

Association rule mining [1, 2, 3, 10] is to find associations among purchased items in EC environment. *Path traversal pattern* mining [4, 6, 7, 8, 9, 17] discover *simple path traversal patterns* which there is no repeated page in the pattern, that is, there is no backward reference in the pattern. These algorithms just consider the forward references in the traversal sequence database. Hence, simple path traversal patterns are not fit in real web applications. *Non-simple path traversal pattern*, i.e., web traversal pattern, contains not only forward references but also backward references. This information can present user navigation behaviors completely and correctly. The related researches are *MFTP (Mining Frequent Traversal Patterns)* algorithm [16], and *FS-Miner* algorithm [15].

However, web site managers would like to know not only pure navigation behaviors but also purchasing behaviors of customers. Yun and Chen [9] proposed *MTS* algorithm, which takes both the traveling and purchasing behaviors of customers into consideration at the same time. Figure 2 shows a user transaction sequence, which the user traversed from page *A* to page *B* and purchased item 1 on page B. Next, he went to page *C*, and then went to page *D*. Thereafter, the customer went back to page *C* and then went to page *E* on which items 3 and 4 were purchased. And then, he went back to page *C* and bought item 2, and so on.

MTS algorithm cuts each user transaction sequence into several web transaction records, that is, when a customer has backward behavior, a web transaction record is generated. *MTS* algorithm discovers web transaction patterns from all the web transaction records. Besides, if there is no item purchased on the last page of a web transaction record, the web transaction record does not be generated. Table 2 shows all the web transaction records generated from the user transaction sequence in Figure 2. Because there is no item bought on pages *D* and *I*, the web transaction records <*ABCD*> and <*ASI*> cannot be generated.

However, cutting user transaction sequences into web transaction records may lead to record and discover incorrect transaction sequences and lose the information about backward references. For example, from Figure 2, we can see that the customer visited page *E* and purchased items 3 and 4, and then went back to page *C* and purchased item 2 on page C. However, the first transaction record in Table 2 shows that the customer purchased item 2 on page *C* and then bought items 3 and 4 on page *E*, which is incorrect since backward information has been lost. Thus, *MTS* algorithm cannot generate correct web transaction patterns based on these web transaction records.

Hence, we propose an algorithm *IWA* for mining web transaction patterns without dividing any user transaction sequences. *IWA* algorithm also takes both traveling and purchasing behaviors of customers into consideration at the same time, and considers not only forward references but also backward references. Besides, *IWA* algorithm allows noises which exist in user transaction sequences.

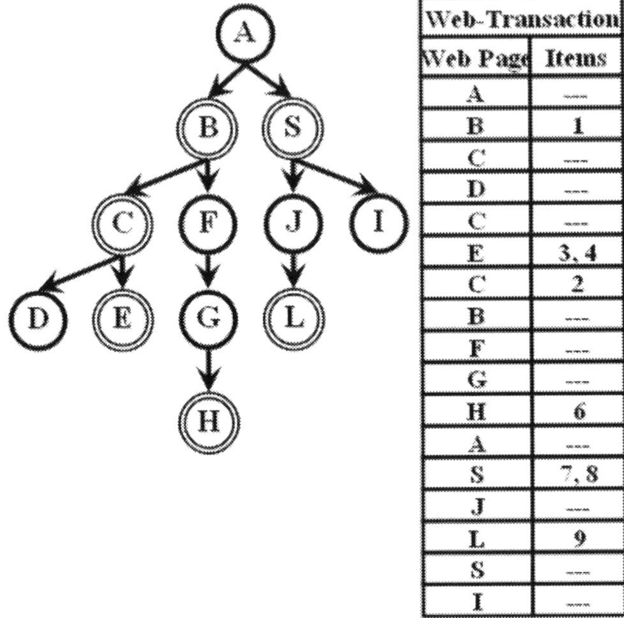

Web-Transaction	
Web Page	Items
A	---
B	1
C	---
D	---
C	---
E	3, 4
C	2
B	---
F	---
G	---
H	6
A	---
S	7, 8
J	---
L	9
S	---
I	---

Fig. 2. A user transaction sequence

Table 2. Web transaction records

Path	Purchases
ABCE	$B\{1\}, C\{2\}, E\{4\}, E\{3\}$
ABFGH	$B\{1\}, H\{6\}$
ASJL	$S\{7\}, S\{8\}, L\{9\}$

3 Mining Web Transaction Patterns

In this section, we describe our web transaction pattern mining algorithm *IWA* according to a web site structure (e.g., Figure 1). The reason for using web site structure is to avoid the unqualified web traversal sequences to be generated in our mining process. Our algorithm *IWA* discovers web traversal patterns and web transaction patterns from a user transaction sequence database *D* (e.g., Table 1).

IWA algorithm first scans user transaction sequence database *D* once to obtain all the 1-web traversal patterns and 1-web transaction sequences. A two-dimensional matrix *M* for generating 2-web traversal patterns and 2-web transaction sequences is constructed. For matrix *M*, the rows and columns are all the 1-web traversal patterns and 1-web transaction sequences, and each entry records support count for 2-web traversal patterns and 2-web transaction sequences from row to column. From this matrix, we can obtain all the 2-web traversal patterns and 2-web transaction sequences. In order to generating *k*-web traversal patterns and *k*-web transaction sequences ($k \geq 3$), all the (*k*-1)-web traversal patterns and (*k*-1)-web transaction sequences are joined to generate candidate *k*-traversal (transaction) sequences. After counting the supports for candidate *k*-traversal (transaction) sequences, *k*-web traversal patterns and *k*-web transaction sequences can be discovered and user transaction sequences whose length is less than (*k*+1) can be deleted from user transaction sequence database *D*.

The candidate generation method for web traversal pattern is like the join method proposed in [4]. For any two distinct web traversal patterns , say $<s_1, \ldots, s_{k-1}>$ and $<u_1, \ldots, u_{k-1}>$, we join them together to form a *k*-traversal sequence only if either $<s_2, \ldots, s_{k-1}>$ exactly is the same with $<u_1, \ldots, u_{k-2}>$ or $<u_2, \ldots, u_{k-1}>$ exactly the same with $<s_1, \ldots, s_{k-2}>$. For example, candidate sequence <*BDAC*> can be generated by joining the two web traversal patterns <*BDA*> and <*DAC*>. For a candidate *l*-traversal sequence α, if a qualified length (*l*-1) sub-sequence of α is not a web traversal pattern, then α must not be web traversal pattern and α can be pruned. Hence, we also check all of the qualified web traversal sub-sequences with length *l*-1 for a candidate *l*-traversal sequence to reduce some unnecessary candidates.

The candidate generation method for web transaction sequence is similar to the join method for generating web traversal pattern, which is divided into two parts. For the first part, a (*k-1*)-web traversal pattern $<s_1, \ldots, s_{k-1}>$ and a (*k-1*)-web transaction sequence $<u_1, \ldots, u_{k-1}\{j_{k-1}\}>$ can be joined to form a *k*-web transaction sequence $< s_1, u_1, \ldots, u_{k-1}\{j_{k-1}\}>$ only if $<s_2, \ldots, s_{k-1}>$ is exactly the same with $<u_1, \ldots, u_{k-2}>$. A (*k-1*)-web traversal pattern $<s_1, \ldots, s_{k-1}>$ and a (*k-1*)-web transaction sequence $<u_1\{j_1\}, \ldots, u_{k-1}>$ can be joined to form a *k*-web transaction sequence $< u_1\{j_1\}, \ldots, u_{k-1}, s_{k-1}>$ only if $<s_1, \ldots, s_{k-2}>$ is exactly the same with $<u_2, \ldots, u_{k-1}>$. In the second part, for any two distinct (*k-1*)-web transaction sequences $<s_1\{i_1\}, \ldots, s_{k-1}\{i_{k-1}\}>$ and $<u_1\{j_1\}, \ldots, u_{k-1}\{j_{k-1}\}>$ which the set of purchased items can be empty, we join them together to form a *k*-transaction sequence only if either $<s_2\{i_2\}, \ldots, s_{k-1}\{i_{k-1}\}>$ exactly is the same with $< u_1\{j_1\}, \ldots, u_{k-2}\{j_{k-2}\}>$ or $<u_2\{j_2\}, \ldots, u_{k-1}\{j_{k-1}\}>$ exactly the same with $<s_1\{i_1\}, \ldots, s_{k-2}\{i_{k-2}\}>$.

For a candidate *l*-traversal (transaction) sequence α, if a qualified length (*l*-1) sub-sequence of α is not a web traversal pattern (or web transaction sequence), then α must not be web traversal pattern (or web transaction sequence) and α can be pruned. Hence, we also check all of the qualified web traversal (transaction) sub-sequences with length *l*-1 for a candidate *l*-traversal (transaction) sequence to reduce some unnecessary candidates. For the above example, we need to check if <*BAC*> is a web traversal patterns, which is a qualified traversal sequence. If <*BAC*> is not a web traversal pattern, <*BDAC*> is also not a web traversal pattern. We do not need to check <*BDC*>, because <*BDC*> is an unqualified traversal sequence (i.e., there is no link from web page *D* to web page *C* in Figure 1).

In the following, we use an example to explain our algorithm *IWA*. The web site structure *W* and user transaction sequence database is shown in Figure 1 and Table 1, respectively. Suppose that minimum support *min_sup* is set to 30%, that is, minimum support count is $\lceil 6\times30\% \rceil = \lceil 1.8 \rceil = 2$.

After scanning user transaction sequence database *D*, 1-web traversal patterns are <*A*> , <*B*>, <*C*>, <*D*>, <*E*> and <*F*>, and 1-web transaction sequences are <*C{2}*>,

<E{3}> and *<F{1}>*. The two-dimensional matrix *M* is shown in Table 3, in which the notation "X" represents that there is no connection in web site structure W. For example, for entry <AB>, there is no connection from web page *A* to web page *B*. From matrix *M*, we can easily obtain all the 2-web traversal patterns *<AC>*, *<AE>*, *<AF>*, *<BA>*, *<BD>*, *<BE>*, *<CA>*, *<DA>*, *<DE>*, *<EC>* and *<FC>*, and 2-web transaction sequences *<AC{2}>*, *<AE{3}>*, *<AF{1}>*, *<BE{3}>*, *<C{2}A>* and *<F{1}C>*.

Table 3. Matrix M

	A	B	C	D	E	F	C{2}	E{3}	F{1}
A	--	X	5	X	3	2	3	2	2
B	5	--	X	2	5	X	X	2	X
C	4	X	--	X	X	X	--	X	X
D	4	1	X	--	3	X	X	1	X
E	X	X	2	X	--	X	0	---	X
F	X	X	2	X	X	--	0	X	---
C{2}	2	X	--	X	X	X	---	X	X
E{3}	X	X	0	X	--	X	0	---	X
F{1}	X	X	2	X	X	X	0	X	---

After generating all the 2-web traversal patterns and 2-web transaction sequences, the candidate 3-traversal (transaction) sequences can be generated by applying the join method on 2-web traversal patterns and 2-web transaction sequences. The generated candidate 3-traversal sequences are *<ACA>*, *<AEC>*, *<AFC>*, *<BAC>*, *<BAE>*, *<BAF>*, *<BDA>*, *<BDE>*, *<BEC>*, *<CAC>*, *<CAE>*, *<CAF>*, *<DAC>*, *<DAE>*, *<DAF>*, *<DBE>*, *<DEC>*, *<ECA>* and *<FCA>*, and the generated candidate 3-transaction sequences are *<BAC{2}>*, *<CAC{2}>*, *<DAC{2}>*, *<C{2}AC{2}>*, *<BAE{3}>*, *<CAE{3}>*, *<DAE{3}>*, *<C{2}AE{3}>*, *<BAF{1}>*, *<CAF{1}>*, *<DAF{1}>*, *<C{2}AF{1}>*, *<BDE{3}>*, *<AC{2}A>*, *<C{2}AC>*, *<C{2}AE>*, *<C{2}AF>*, *<AF{1}C>* and *<F{1}CA>*. Because there is a direct link from web page *D* to web page *E*, and transaction sequence *<DE{3}>* is not a 2-web transaction sequence, transaction sequence *<DAE{3}>* can be pruned. After the pruning, most of the candidates can be pruned.

IWA scans *D* to count supports for all the remaining candidates and find 3-web traversal patterns and 3-web transaction sequences. The user transaction sequences whose length is less than 4 are also deleted from *D*. The 3-web traversal patterns are *<ACA>*, *<AFC>*, *<BAC>*, *<BAE>*, *<BAF>*, *<BDA>*, *<BDE>*, *<BEC>*, *<CAC>*, *<CAE>*, *<CAF>*, *<DAC>*, *<DAE>* and *<ECA>*, and the 3-web transaction sequences are *<BAC{2}>*, *<DAC{2}>*, *<BAE{3}>*, *<CAE{3}>*, *<C{2}AE{3}>*, *<BAF{1}>*,

<CAF{1}>, *<AC{2}A>*, *<C{2}AE>* and *<AF{1}C>*. The user transaction sequence TID 6 in *D* is deleted.

After applying candidate generation method, the generated candidate 4-traversal sequences are *<ACAC>*, *<ACAE>*, *<ACAF>*, *<BACA>*, *<BAFC>*, *<BDAC>*, *<BDAE>*, *<BECA>*, *<CACA>*, *<CAFC>*, *<DACA>*, *<ECAC>*, *<ECAE>* and *<ECAF>*, and the generated candidate 4-transaction sequences are *<ACAE{3}>*, *<ACAF{1}>*, *<BDAC{2}>*, *<ECAE{3}>*, *<ECAF{1}>*, *<BAC{2}A>*, *<DAC{2}A>*, *<BAF{1}C>*, *<CAF{1}C>*, *<AC{2}AE{3}>*, *<AC{2}AE>*}. After scanning database *D* and count supports for all the candidates, the discovered 4-web traversal patterns are *<ACAE>*, *<BACA>*, *<BAFC>*, *<BECA>*, *<CAFC>*, *<ECAC>* and *<ECAF>*, and 4-web transaction sequences are *<ACAE{3}>*, *<ECAF{1}>*, *<BAC{2}A>*, *<BAF{1}C>*, *<CAF{1}C>*, *<AC{2}AE{3}>* and *<AC{2}AE>*. The user transaction sequence TID 3 in *D* is deleted.

Finally, the 5-web transaction sequences are *<BACAE{3}>*, *<BECAF{1}>*, *<ECAF{1}C>* and *<BAC{2}AE{3}>*, and 6-web transaction sequences is *<BECAF{1}C>*. Because there is no candidate 7-web transaction sequences generated, the mining process is terminated. Suppose that the minimum confidence is 50%. The generated web transaction patterns are *<BAC'AE'* >: $C\{2\} \Rightarrow E\{3\}$, *<AC'AE'* >: $C\{2\} \Rightarrow E\{3\}$ and *<C'AE'* >: $C\{2\} \Rightarrow E\{3\}$.

4 Experimental Results

We use a real dataset which the information about renting DVD movies is stored. There are 82 web pages in the web site. We collect user traversing and purchasing behaviors from 02/18/2001 to 03/24/2001 (seven days). There are 428,596 log entries in the original dataset. After preprocessing web logs, we reorganize the original log entries into 12,157 user transaction sequences. The maximum length of these user transaction sequences is 27, the minimum length is 1, and the average length is 3.9. Table 4 and Table 5 show the execution times and the number of web transacttion patterns generated by *IWA* on the real dataset for different minimum supports, respectively.

We also generate synthetic dataset to evaluate the performance of our algorithm. Because there is no algorithm for mining web transaction patterns without cutting user transaction sequences, that is, both forward and backward traversing and purchasing behaviors are considered, we evaluate performance of our *IWA* algorithm by comparing with *MTS* algorithm [9] which cannot deal with backward references. For synthetic dataset, the number of web pages is set to 100 and the average number of out-links for each page is set to 100×30%=30. The purchasing probability for a customer in a web page is set to 30%. We generate five synthetic datasets in which the numbers of user transaction sequences are set to 200K, 400K, 600K, 800K and 1,000K, respectively.

Figure 3 and Figure 4 show the execution times and number of generated web transaction patterns for *MTS* and *IWA*, respectively, when minimum support is set to 5%. From Figure 3, we can see that *IWA* outperforms *MTS* although the number of web transaction patterns generated by *IWA* is more than that of *MTS*, which is shown

Table 4. Execution times (seconds) for *IWA* algorithm on real dataset

k-web transaction sequences (Lk)	Minimum Support			
	5%	10%	15%	20%
L1	0.03	0.02	0.02	0.02
L2	1.091	1.131	1.061	1.041
L3	0.02	0.01	0.01	0.01
L4	0.03	0.02	0.01	0.00
L5	0.04	0.02	0.00	0.00
L6	0.04	0.00	0.00	0.00
Total	1.251	1.201	1.101	1.071

Table 5. The number of web transaction patterns generated by *IWA* algorithm on real dataset

k-web transaction sequences (Lk)	Minimum Support			
	5%	10%	15%	20%
L1	14	4	4	4
L2	13	6	4	3
L3	8	6	2	1
L4	5	4	0	0
L5	3	1	0	0
L6	1	0	0	0
Total	44	21	10	8

in Figure 4. This is because *MTS* need to take more time to cut each user transaction sequences into transaction records, such that the web transaction patterns about backward references cannot be generated. For *IWA* algorithm, it is not necessary to take time to cut user transaction sequences and the complete user behaviors can be retained. Hence, *IWA* algorithm can discover correct and complete web transaction patterns. In Figure 3, the performance gap between *IWA* and *MTS* increases as the number of user transaction sequences increases.

Figure 5 shows the execution times for *MTS* when minimum support is 0.05% and execution times for *IWA* when minimum support is 5%. Owing to *MTS* algorithm cuts user transaction sequences into a lot of transaction records, it need to take time to do

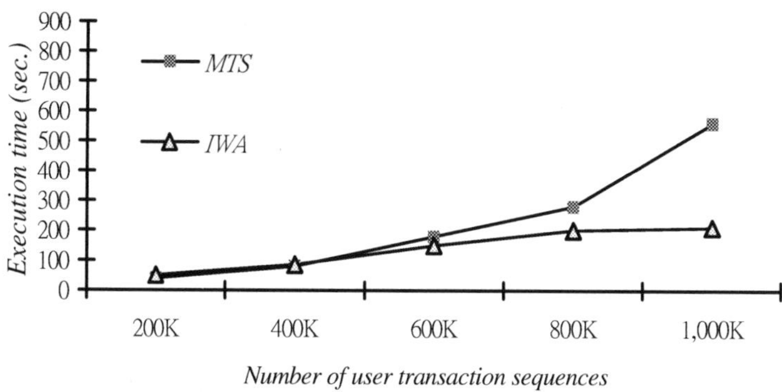

Fig. 3. Execution times for *MTS* and *IWA* algorithms

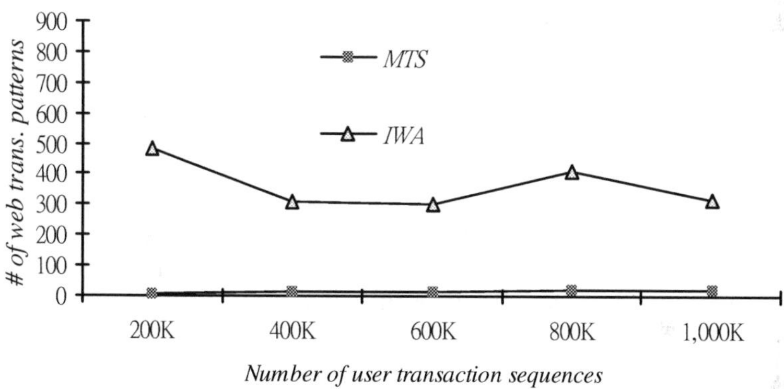

Fig. 4. Number of web transaction patterns generated by *MTS* and *IWA*

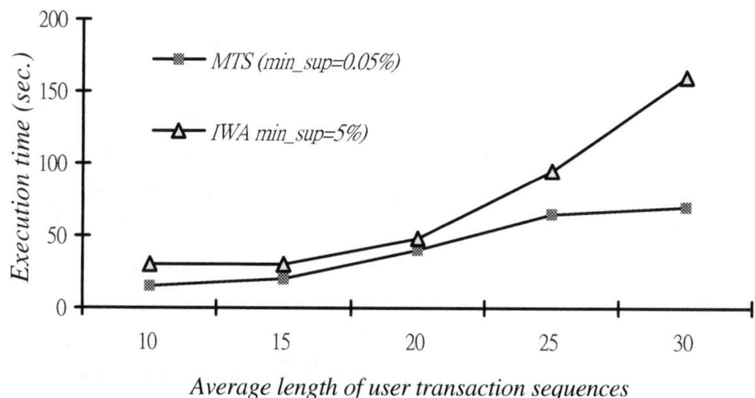

Fig. 5. Execution times for *MTS* and *IWA* algorithms

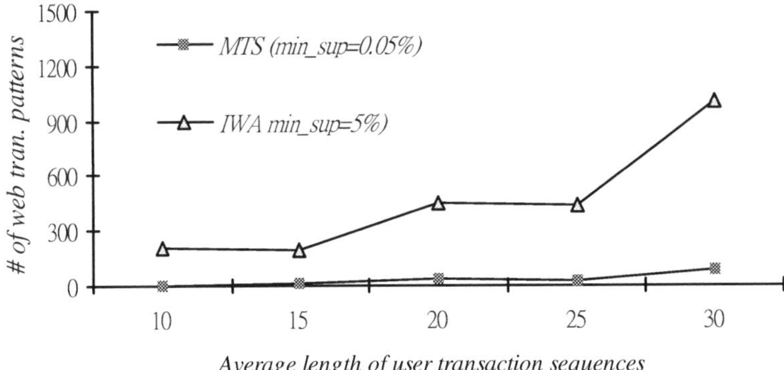

Fig. 6. Number of web transaction patterns generated by *MTS* and *IWA*

the preprocessing and mining we transaction patterns from a large number of transaction records. Hence, the execution time for *IWA* is slightly more than that of *MTS* although the number of web transaction patterns generated by *MTS* is much less than that of *IWA*, which is shown in Figure 6.

Figure 6 shows the number of web transaction patterns generated by *MTS* and *IWA* when minimum supports are set to 0.05% and 5%, respectively. Although minimum support for *MTS* is much lower than minimum support for *IWA*, the number of web transaction patterns generated by *MTS* is still less than that of *IWA*, since *MTS* cuts user transaction sequences into many short transaction records such that many web transaction patterns about backward references cannot be generated.

5 Conclusions

Mining association rules [1, 2, 3, 10] only discovers associations among items. Mining web traversal patterns [15, 16] just finds navigation behaviors for most of the customers. However, we cannot acquire the information about traveling and purchasing behaviors simultaneously. Therefore, this paper proposes an algorithm *IWA* for mining web transaction patterns which provide not only navigation behaviors but also the purchasing behaviors for most of the customers. Our algorithm retains complete user navigation and purchasing behaviors and discovers web transaction patterns including forward and backward references. The experimental results also show that our algorithm outperforms *MTS* algorithm which just considers forward references and incorrect web transaction patterns may be generated.

References

1. Agrawal, R., et al.: Fast Algorithm for Mining Association Rules. In: Proceedings of the International Conference on Very Large Data Bases, pp. 487-499 (1994)
2. Park, J.S., Chen, M.S., Yu, P.S.: Using a Hash-Based Method with Transaction Trimming for Mining Association Rules. IEEE Transaction on Knowledge and Data. Engineering 9(5), 813–825 (1997)

3. Han, J., Pei, J., Yin, Y., Mao, R.: Mining Frequent Patterns without Candidate Generation: A Frequent- Pattern Tree Approach. Data. Mining and Knowledge Discovery 8(1), 53–87 (2004)
4. Chen, M.S., Park, J.S., Yu, P.S.: Efficient Data Mining for Path Traversal Patterns in a Web Environment. IEEE Transaction on Knowledge and Data. Engineering 10(2), 209–221 (1998)
5. Yen, S.J.: An Efficient Approach for Analyzing User Behaviors in a Web-Based Training Environment. International Journal of Distance Education Technologies 1(4), 55–71 (2003)
6. Chen, M.S., Huang, X.M., Lin, I.Y.: Capturing User Access Patterns in the Web for Data Mining. In: Proceedings of the IEEE International Conference on Tools with Artificial Intelligence, pp. 345–348 (1999)
7. Pei, J., Han, J., Mortazavi-Asl, B., Zhu, H.: Mining Access Patterns Efficiently from Web Logs. In: Proceedings of the Pacific-Asia Conference on Knowledge Discovery and Data Mining, pp. 396–407 (2000)
8. Xiao, Y., Dunham, M.H.: Efficient Mining of Traversal Patterns. IEEE Transaction on Data. and Knowledge Engineering 39(2), 191–214 (2001)
9. Yun, C.H., Chen, M.S.: Using Pattern-Join and Purchase-Combination for Mining Web Transaction Patterns in an Electronic Commerce Environment. In: Proceedings of the COMPSAC, pp. 99–104 (2000)
10. Han, J., Pei, J., Lu, H., Nishio, S., Tang, S., Yang, D.: H-Mine: Hyper-Structure Mining of Frequent Patterns in Large Databases. In: Proceedings of 2001 International Conference on Data Mining (ICDM'01), San Jose, CA (November 2001)
11. Chen, S.Y., Liu, X.: Data Mining from 1994 to 2004: an Application-Orientated Review. International Journal of Business Intelligence and Data. Mining 1(1), 4–21 (2005)
12. Ngan, S.C., Lam, T., Wong, R.C.W., Fu, A.W.C.: Mining N-most interesting itemsets without support threshold by the COFI-tree. International Journal of Business Intelligence and Data. Mining 1(1), 88–106 (2005)
13. Xiao, Y., Yao, J.F., Yang, G.: Discovering Frequent Embedded Subtree Patterns from Large Databases of Unordered Labeled Trees. International Journal of Data. Warehousing and Mining 1(2), 70–92 (2005)
14. Cooley, R., Mobasher, B., Srivastava, J.: Web Mining: Information and Pattern Discovery on the World Wide Web. In: Proceedings of IEEE International Conference on Tools with Artificial Intelligence (1997)
15. EL-Sayed, M., Ruiz, C., Rundensteiner, E.A.: FS-Miner: Efficient and Incremental Mining of Frequent Sequence Patterns in Web logs, In: Proceedings of 6th ACM International Workshop on Web Information and Data Management, pp.128–135 (2004)
16. Yen, S.J.: An Efficient Approach for Analyzing User Behaviors in a Web-Based Training Environment. International Journal of Distance Education Technologies 1(4), 55–71 (2003)
17. Yen, S.J., Lee, Y.S.: An Incremental Data Mining Algorithm for Discovering Web Access Patterns. International Journal of Business Intelligence and Data Mining (2006)

Mining Purpose-Based Authorization Strategies in Database System*

Jie Song, Daling Wang, Yubin Bao, Ge Yu, and Wen Qi

School of Information Science and Engineering, Northeastern University
Shenyang 110004, P.R. China
sy_songjie@163.com, {dlwang,baoyubin,yuge}@mail.neu.edu.cn,
ddqiwen@126.com

Abstract. With the development of computer and communication technology, access control of the resources in databases has become an issue focused by both consumers and enterprises. Moreover, the new concept of purpose-based authorization strategies is widely used instead of the traditional one of role-based strategies. The way of acquiring the optimal authorization strategies is an important problem. In this paper, an approach of mining authorization strategies based on purpose in database system is proposed. For obtaining the optimal authorization strategies of the resources in databases for supporting various purposes, an algorithm of clustering purposes is designed, which is based on the inclusion relationship among resources required by the purposes. The resultant purpose hierarchy is used for guiding the initial authorization strategies. The approach provides valuable insights into the authorization strategies of database system and delivers a validation and reinforcement of initial strategies, which is helpful to the database administration. The approach can be used not only in database system, but also in any access control system such as enterprise MIS or web service composing system. Theories and experiments show that this mining approach is more effective and efficient.

1 Introduction

With the rapid development of computer network and communication techniques, the access control of system resources, especially the resources in databases, has become an issue focused by both consumers and enterprises [15, 10]. Therefore, a new concept of purpose-based authorization strategies is widely used instead of the traditional one of role-based strategies. The purpose is adapted because access control strategies are concerned with which resources is used for which purposes, rather than which users are performing which actions on which resources [5, 16].

In database systems, authorizations were commonly predefined by the system administrators [9]. However, the predefined authorizations may be not accurate because of the complication of gigantic database systems with thousands of tables. Take querying request for an example, there are numerous of tables and columns, some of which are

* This work is supported by National Natural Science Foundation of China (No. 60573090, 60673139).

K.C. Chang et al. (Eds.): APWeb/WAIM 2007 Ws, LNCS 4537, pp. 86–98, 2007.

joined for individual query, and different queries may be considered as different performing ways of one purpose. In this case, for a purpose, which table should be assigned to it is illegible. According to this condition, the authorization strategies are often changed when the systems run because some purposes are authorized too deficient to be performed while others are excessively authorized. How to constitute the purpose-based authorization strategies in database systems is an important problem.

In this paper, a new approach is proposed to settle the problem of the purpose-based authorization strategies. Consider a database system of super market as a motivation example, where system purposes are organized according to a hierarchical structure (see Fig. 1) based on the principles of generalization and specialization. The different purposes represent the different system functions and every purpose has been authorized to some tables in the database.

In this structure, the relationship among purposes and that between purposes and database resources are predefined by the database administrators when the database is ready to be on line. The purposes are invariable except that the functions of the system are extended, but the relationships among them should be adjusted according to actual status when the database system runs, so as the relationship between purposes and database resources. These modifications are ineluctable, costly and taken place more than once, the times of such modifications should be reduced. Since any modification is based on the actual running statuses, which are logged in the access logs, a mining algorithm can be used for discovering the strategies from these logs instead of artificial analysis, and thus provides the guidance for initial authorization strategies and database privacy administration. In implementation of the algorithm, a hierarchical clustering method of data mining [7, 14], which is based on the inclusion relationship among resources required by the purposes, is used in database system for mining authorization strategies to support the initialization of them. The experiment results show that with a good mining performance, this approach can provide a nice support to purpose-based authorization strategies for database systems.

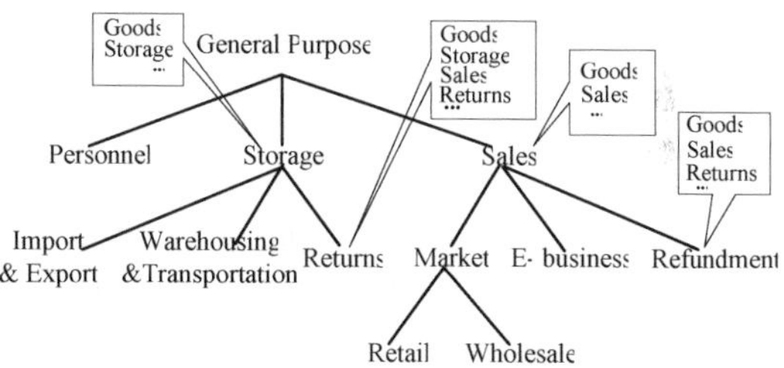

Fig. 1. A Purpose Hierarchy Structure

The rest of this paper is organized as follows. Section 2 provides a precise definition and description of the purpose-based access control model. Section 3 sets out the algorithm of authorization strategies of mining and Section 4 describes its performance. After discussing related works in section 5, Section 6 concludes the paper with an overview of current and future research.

2 Model Definition

In this section, the model of purpose-based authorization proposed in this paper is well defined using formalization definitions. In this model, the most important notions are resources and purposes. The all purposes are initialized by a set of the accessible resources, and all of them are organized as a hierarchical structure, which is revised by strategies mining results.

The notion of purpose plays a central role because the purpose is the basic concept on which access decisions are made. For simplifying the management and appropriate for common business environments, the purposes are organized according to a hierarchical structure based on the principles of generalization and specialization. There is one-to-multi relationship between the purpose and resources.

Definition 1 (Resources). A resource indicates the data of a system and the way to access the data. In a broad sense, resources indicate tables, views in a database, or even file, operation, function, service, sub-system or something which can be requested, ignored there is a containment relationship. So a request is for one or multi resources. In this case, whatever requests, each of them is regarded as requesting a resource set. Let set R present the set of all resources in the system. $R = \{r_i \mid 0 < i < n\}$ Where n means the number of resources in the system.

Definition 2 (Purpose and Purpose Tree). A purpose describes the intention(s) for accessing some resources. A purpose set is organized as a hierarchical structure, referred to as Purpose Tree, where each node represents a purpose in the purpose set and each edge represents a hierarchical relationship (specialization and generalization) between two purposes. Let *PTree* be a Purpose Tree and P be the purpose set in *PTree*. (Where n means the number of purposes).

$$PTree = P \cap \{\text{relationship between } p_i \text{ and } p_j \mid p_i, p_j \in P\} \ (0 < i, j < n)$$

In the purpose-based authorization model, an accessible resource set is assigned to one purpose for supporting the positive privacy strategies of the purpose, which means the resources that the purpose can access are included in this set. This mechanism is similar to the famous RBAC model [8, 12]. The relative definitions are given as follows.

Definition 3 (Accessible Resources). Accessible Resources (AR) defines the resources one purpose can access. By using Accessible Resources, one can guarantee that accesses to particular resources are allowed and on the contrary the others are rejected according to a certain purpose. Two AR sets may consist of the same resources, and thus provide greater flexibility. Evidently, $AR \subseteq R$:

$$AR = \begin{cases} \{r_i \mid r_i \in R\} & (0 < i < n) \\ \varnothing & (\text{ no resource is permited}) \end{cases}$$

Where n is the number of accessible resource. In purpose-based authorization model, one purpose corresponds with one AR. Intuitively, based on the hierarchical structure of purposes, AR_p of one purpose includes not only the AR set assigned to itself, but also the AR sets assigned to its ancestors. If AR_p is an Accessible Resources that purpose p is authorized, then:

$$AR_p = (\bigcup p_{i \in P} \, AR \text{ of } Ancestor(p_i)) \ (p \in Ancestor(p))$$

The strategies of verification are based on what kind of AR the purpose has, so the verification strategies are straightforward. $\forall\ p \in P,\ r \in R,\ AR_p$ is Accessible Resources of p, the condition that purpose p can access resource r is $r \in AR$.

3 Authorization Strategies Mining

As discussed in the previous section, a purpose-based authorization model needs to be built up. In this section, the main idea of this paper, authorization mining is formally described. There are two steps of this approach: analyzing permission and mining purpose hierarchy. And the result of the strategies should be well used as a guidance of initial strategies.

- **Permission Analyzing**

Any person acting with a purpose has to access some resources in order to perform it, and there are many ways to fulfill the same purpose. However, the best way is the one that needs the least resources according to information hiding principle, and these resources are actual *AR* of the purpose contrast to initial *AR*. In this section, the approach of extracting the best *AR* set by analyzing access logs is described.

Definition 4 (Logs and Record). The purposes performing statuses are recorded in *logs*, which contain many *records*. Each *record* has three parts: purpose, requested resources and their access statuses.

$$Logs = \{ record_i \} \qquad record = < p_j, \{ < r_k, status > \} >$$
$$(p_j \in P, r_k \in R, 0 < i, j, k < n, staus \in \{true, false\})$$

Here r_k is one of the resources used to perform the purpose, and a status has the alternative values: true or false, which means p_j can or cannot access r_k. For example, a user chooses several resources to accomplish a purpose, and then these resources are verified. If the purpose does not have permission to access the resource, a tuple <*resource*, *false*> is recorded in *logs*. Since the way of a user to perform one purpose is different, the resource he chooses is also different.

Definition 5 (Record Frequency). Record Frequency (*RF*) is the frequency of a specified *record* occurs within the *records*, which belong to the same purpose. If *p* is a purpose and $p \in P$, *re* is one of the *records* in *Logs*, $re \in Logs$:

$$RF_p = \frac{|\{record_i \mid record_i = re\}|}{|\{record_i \mid \text{purpose of } record_i \text{ is } p\}|}$$

Definition 6 (Permission Set). Permission Set is extracted from *Logs* as the result of permission analyzing, which is the best mapping relationship between purposes and resources, and is used in next purpose hierarchy mining. Let *PS* be the Permission Set, then ps_i is a two-tuple $<p_i, AR_{pi}>$, *PS* satisfies:

$$PS = \{ps_i\} = \{< p_i, AR_{p_i} >\}$$
$$AR_{p_i} \text{ is resurce set of reasonable } record \text{ of } p_i \qquad |PS| = |P|$$

There are several steps to generate permission set. First, rational *records* that can fulfill a certain purpose are extracted from logs. Of these *records*, those happened seldom or *RF*s less than a certain value are omitted. Table 1 shows the relationship between *records* and resources for a certain purpose as example. In the table, "√" means the resource has been assigned to the purpose, and "×" means the resource

Table 1. All Records of A Certain Purpose

	r_1	r_2	r_3	r_4	r_5	r_6	r_7	***RF***
record$_1$	√	√	×					rf_1
record$_2$	√	√		×				rf_2
record$_3$	√	√	×	×				rf_3
record$_4$	√	√			√	√	√	rf_4

needs to be performed for the purpose but has not been authorized. The values in the last row of the table show the *RF*s of each *record*.

Then, the analyzing algorithm is following several rules to extract the permission set of each purpose:

Rule 1: According to the information hiding principle, find the *record* which needs the least resources. If several *record*s are selected, then follow the rule 2.

Rule 2: Because of minimizing resource allocation, resource retraction is advocated while re-assignment should try to be avoided, so the record which can fulfill the purpose is selected first. That is, all the resources it needs are marked "√". If there is no purpose satisfied this condition, then follow the rule 3, or more purposes satisfied, follow the rule 4.

Rule 3: If resource re-assignment is ineluctable, select the *record* with least "×" in order to furthest restrict the authorization. If an exclusive *record* cannot be selected by this rule, then follow rule 4.

Rule 4: The higher the *RF* of *record* is, the more often the *record* is chosen to perform the purpose. For adapting the habit of most users, the *record* with maximal *RF* is selected. If there are several *record*s with equal *RF*s, then select one randomly.

In table 1, *record*$_1$ and *record*$_2$ are selected by rule 1. Since rule 2 and rule 3 cannot find the result, an exclusive *record*$_1$ is select at last by rule 4 (RF_1>RF_2). The detail of permission analyzing algorithm is shown as algorithm 1.

Algorithm 1. Permission Analyzing Algorithm

```
Input: Logs and P
Output: PS
Analyzing():
PS = ∅
For each p of P
    record_p = all records of p in Logs;
    Delete the records from record_p
which RF of them < min_value;
    Delete the records from record_p
which resources of them > max_value;
    If one or more records in record_p can fulfill the purpose
        Find the record with the largest RF;
        Add record to PS;
        Break;
    End If
    Find records that have the least "×";
    Find the record with the largest RF;
    Add record to PS;
End For
Return PS ;
```

- **Purpose Hierarchy Mining**

When permission set of each purpose is extracted from a number of logs by the algorithm proposed in previous section, this *PS* is prepared for the next purpose hierarchy mining. In this section, a purposes hierarchical clustering method based on inclusion relationship among the resources required by the purposes is proposed for mining the purpose hierarchy.

In detail, the idea of the algorithm is that the less resource a purpose requires, the higher level the purpose is in the hierarchy. The resources required by the same level purposes have not inclusion relationship. According to the idea, some definitions related with the clustering and the algorithms are given here.

Definition 7 (Overlap). $\forall p_i, p_j \in P \ \ (i \neq j)$, then p_i overlap p_j iff $ARp_j \subset ARp_i$. The overlap is signed as "≥", so p_i overlap p_j means $p_i \geq p_j$.

Definition 8 (Purpose Cluster Set). The *Purpose Cluster Set* is the set of some purpose clusters. The purposes in one cluster have the relationship "≥". The set is signed as *PC* and its every purpose cluster is signed as p_ic_i ($0<i<n$). Every p_ic_i can be ∅ or compounded by purpose and p_ic_i. In this case:

$$p_ic_i = \begin{cases} \varnothing \\ (p_j, p_ic_i) \end{cases} \quad \text{and} \quad PC = \{p_ic_i \mid 0 < i < n\}$$

Definition 9 (*PY* and *PN*). Let *PY* be the set of purposes which have been clustered, and *PN* be the set of purposes which haven't been clustered. Then:

$$PY \subseteq P \ \text{ and } \ PN \subseteq P \ \text{ and}$$
$$PY \cap PN = \varnothing \ \text{ and } \ PY \cup PN = P$$

Obviously, because of $p_i \geq p_j$, there is such purpose cluster as $(p_i,(p_jc_j))$. The "⊂" between resources required by some purposes results in the "≥" between the purposes, and the "≥" generates the hierarchy of the purposes. According to the relationship, the purpose hierarchy can be built by mining purpose clusters based on the "⊂" between resources required by the purposes in every cluster.

Followings are the steps of the purpose hierarchy mining algorithm.

Step 1. Go through all *PS* to see if there are several purposes that be assigned the same resources. If there are, combine the purposes as one purpose;

Step 2. Initialize the variables: *PY*=∅, *PN*=*P*, *PC*=∅;

Step 3. From all purposes in *PN*, find one that doesn't overlap any purpose in *PY* while requires the least resources, and put it into *PY*. Delete that purpose from *PN* and add a new corresponding cluster to the set of all clusters. After this step:

$$PY = PY \cup$$
$$\{p_i \mid p_i \in PN \text{ and } p_i \text{ doesn't overlap the elements in } PN$$
$$\text{while } |AR_{p_i}| \text{ is minimal }\}$$
$$PN = PN \cap \overline{\{p_i\}}$$
$$PC = PC \cup \{p_ic_i \mid 0 < i < n\}$$

Step 4. Repeat step 3 until there is no purpose in *PN* that doesn't overlap any purpose in *PY*;

Step 5. Now, each purpose in *PN* should overlap at least one of the purposes in *PY*. Find one purpose (p_i for example) in *PY* that requires the least resources, then traverse all purposes in each cluster from lower level to higher level until finding the first one (p_j for example) which is overlapped by p_i. If p_j is at the lowest level, add a new lower level to the corresponding cluster (pc_j) in *PC*. And if p_j isn't at the lowest level, then add p_i to p_j's sub-level. Add p_i to *PY* and remove it from *PN*. After this step:

$$PY = PY \bigcup \{p_i \mid p_i \in PN \text{ and } \mid AR_{p_i} \mid \text{is minimal}\}$$

$$PN = PN \bigcap \overline{\{p_i\}}$$

$$p_j c_j = (p_j (p_i c_i))$$

Step 6. Repeat step 5 until there is no element in *PN*.

Step 7. Now the *PTree* can be constructed according to *Purpose Cluster Set*. First, add a general purpose to the *PTree* as a root node, and then every *pc* in *PC* is a subtree of it. For each *pc*, while *pc* is not empty and there is a level, add the first purpose of this level to its super-node as a sub-node and the purposes of the same level are its brother-node. When all *pc* in *Purpose Cluster Set* have been added to the root node, the *PTree* is constructed.

The detail of purpose hierarchy mining algorithm is shown as algorithm 2 and algorithm 3.

Algorithm 2. Mining Cluster Algorithm

```
Input: P
Output: PTree
Mining():
Sort P set by AR of each p;
Combine purposes whose AR is the same as one
purpose;
PTree = ∅ , PY= ∅ , PN= P, PC = ∅
For each p of PN from begin to end // PN or P
              //have be sorted, so AR of p_i in PN is minimal
   If AR of p_i in temp is minimal,
      Add p_i to PY and delete p_i from PN and add p_ic_i to PC;
   End if
   Else
      For each c_j in PC
         While p_ic_j is not empty
            templevel = the lowest level of p_ic_j;
            If p_j is at templevel and p_j is overlapped by p_i
               Add p_i to p_j's lower level;
               Add p_i to PY, delete p_i from PN;
            End If
            templevel = templevel's super-level;
         End While
      End For
   End Else
Return buildingTree(PC);
```

Algorithm 3. Building Purpose Hierarchy Algorithm

```
Input: PC
Output: PTree
buildingTree():
Add a general purpose to PTree as root node;
For each pc in PC
  While pc ≠ ∅
    If p is each level's first purpose
        add a sub-node to its super-node of PTree
      End If
    Else
        add a brother-node to the same level's purposes;
      End Else
  End While
End For
Return PTree;
```

Example (Supermarket Database). In this example, the purpose tree of the motivation example is shown as the left part of Fig. 2, in which the true name of each purpose is replaced by p_i in order to present tersely. And table 2 shows the best *record* of each purpose as the result analyzed by algorithm 1. Now apply the algorithm 2 to construct the cluster hierarchy as following items:

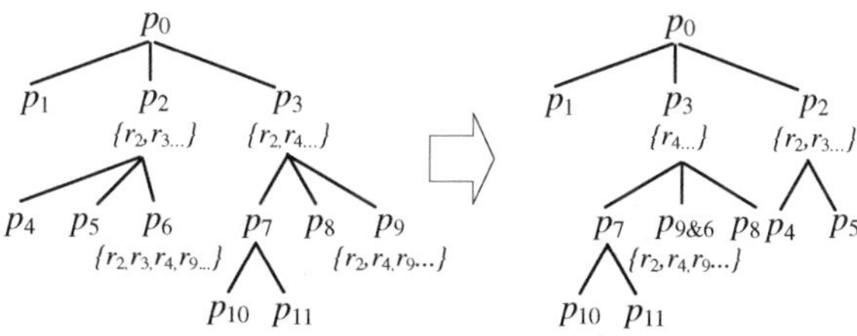

Fig. 2. Purpose Hierarchy of the Example

Table 2. Analyzed Result of Motivation Example

	r_1	r_2	r_3	r_4	r_5	r_6	r_7	r_8	r_9	r_{10}	r_{11}
p_1	√										
p_2		√	√								
p_3				√							
p_4		√	√		×						
p_5		√	√			√					
p_6		√		×					√		
p_7				√			√				
p_8		×		√				√			
p_9		√		√					√		
p_{10}				√			√			√	
p_{11}		×		√			√				√

1. In the first two steps, variables are initialized:
$PN=\{p_1,p_2,p_3,p_4,p_5,p_{6\&9},p_7,p_8,p_9,p_{10},p_{11}\}$
Here, since p_6 and p_9 require the same resources, they are combined as $p_{6\&9}$;

2. Step 3 adds the first cluster(pc) to the PC, and because p_1 and p_3 only require one recourse, they are the first elements that be added to PY. And the rest are remained in PN:

$PY = \{p_1, p_3\}$
$PN = \{p_2, p_4, p_5, p_{6\&9}, p_7, p_8, p_9, p_{10}, p_{11}\}$
$PC = \{(p_1), (p_3)\}$;
3. After the iteration of step 4:
$PY = \{p_1, p_3, p_2\}$
$PN = \{p_4, p_5, p_{6\&9}, p_7, p_8, p_9, p_{10}, p_{11}\}$
$PC = \{(p_1), (p_3), (p_2)\}$;
4. In step 5, p_7 requires two resources and is the one that requires the least. And p_3 in *PY* is overlapped by it, so p_7 is added to *PC* as p_3's sub-level's purpose. Then add p_7 to *PY* and delete it from *PN*:
$PY = \{p_1, p_3, p_7, p_2\}$
$PN = \{p_4, p_5, p_{6\&9}, p_8, p_9, p_{10}, p_{11}\}$
$PC = \{(p_1), (p_3\,(p_7)), (p_2)\}$;
5. The iteration of step 6, the result is:
$PY = \{p_1, p_3, p_7, p_{10}, p_{11}, p_{6\&9}, p_8, p_2, p_4, p_5\}$
$PN = \varnothing$
$PC = \{(p_1), (p_3\,(p_7(p_{10}, p_{11}), (p_{6\&9}), (p_8)), (p_2(p_4, p_5))\}$;

Since *PN* has no element, no cluster is created. So, *PC* is stable and algorithm stops. Now the cluster hierarchy is got and a purpose hierarchy is derived from it. The right part of Fig. 2 shows the *PTree* of the example.

- **Result Explaining**

The purpose hierarchy shows that purpose can access to all the resources of its ancestor. It means that a specialized purpose requires more resources to be performed than a generalized one. After a careful comparing between the result tree and initial tree shown as Fig. 2, some apparent changes are made evidently. For example, the order of the sub-tree of p_0 (General Purpose) is changed, while these changes do nothing with the authorization strategies. However, the other changes imply the real strategies. Some purposes are excessively authorized, such as p_3 (Sales), while some purposes which can be fulfilled by the less resources have to make a detour, because they are authorized irrationally, and p_6 (Returns) is an instance. Purposes those require the same resources are regarded as different because they are authorized different resources which are not appropriate, p_6 (Returns) and p_9 (Refundment) are of such cases. All of these against the information hiding principle cause a waste of resources and the increase of management cost.

With the purpose hierarchy, the resources which a purpose actually requires are obvious. Then it can be used in database system's authorization strategies, and make the strategies more reasonable and effective.

4 Experiments and Performance

Now, the purpose-based authorization strategies mining approach is well used in National Ocean Environment Database System, which has more than 1000 tables (about ten million records each table) and 1000 purposes. All the visited SQL are use a key word "for" to indicate its purpose, and be executed by a special SQL syntax analyzer. Take a querying for example (updating and deleting are similar with querying), the SQL syntax is like following:

SELECT <column_name> [,<column_name>]
FROM <table_or_view_name>[,<table_or_view_name>]
[WHERE <condition>]
FOR <purpose_name>

Authorization strategies mining can be done based on the log table which is chosen as a container of logs, some practical results about this database are given to prove that strategy mining is effective and efficient.

The storage overhead of proposed approach is also inconspicuous. Some storage is spent on archiving logs. But the experiment in Fig.3 shows that it only costs about 1MB storage per ten thousand records. Comparing with the tremendous database with the storage level of GBs or TBs, it may as little as the storage cost of an index of small table, it can be ignored. And as for strategies mining, better result is more important than mining performance. But the performance is also gratifying as shown in Fig. 4 and Fig. 5 under variety number of records (see definition 4) and purpose involved. The results of tests prove that the performance of analyzing and mining algorithm is very good (the mining performance shown in Fig. 4 doesn't include I/O cost for result output).

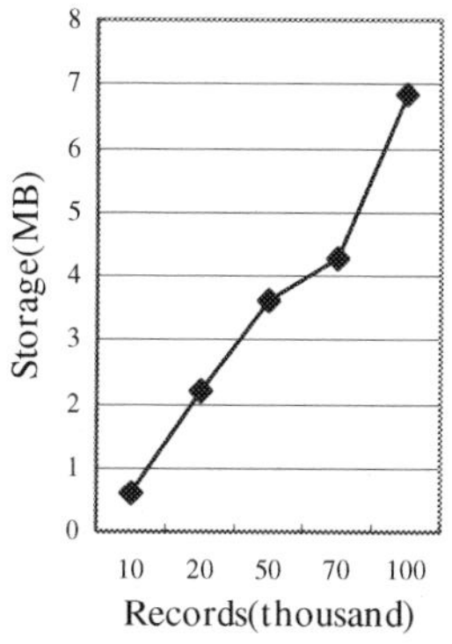

Fig. 3. Results of Storage Evaluation

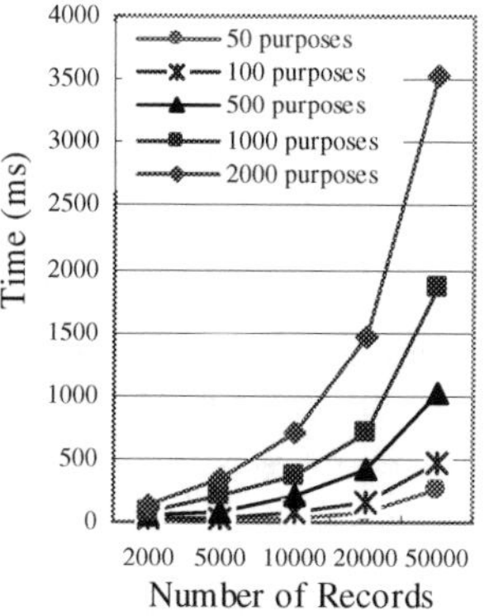

Fig. 4. Analyzing Performance

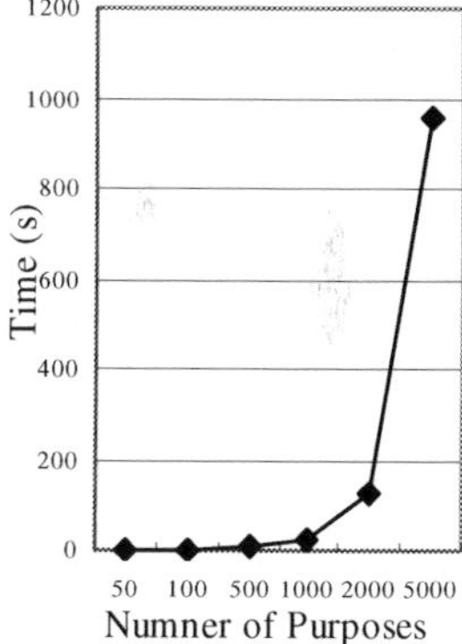

Fig. 5. Mining Performance

Maybe the most important thing concerned by the users is the effect of strategy mining. Table 3 shows some qualitative comparisons between the way with and without authorization strategies mining.

The experiment results show that this approach can reduce the times, number of purposes involved, cost and side-effect of strategies modification, so can meet the requirements of users and implement a good degree of information hiding. Moreover, with the increasing of purposes, resources, and query requests of users in database, the advantages of purpose-based authorization will more obvious.

Table 3. Comparisons Between the Way with or without Mining

Aspect	With Strategies mining	Without Strategies mining
Times of strategies modification to achieve reasonable one	2 or 3 times	More than 20 times
The average number of purposes involved in once strategies modification	About 10 purposes	1 or 2 purposes
The user satisfaction after once strategies modification	Very good	Part of them are satisfied
Time cost of one strategies modification	More times is needed	Quickly
Side-effect of one strategies modification	No side-effect	Uncertain side-effects
Degree of information hiding	Very good	Overage resources may be authorized

5 Related Work

There are some researches about the authorization strategies in database system though they are not based on data mining technique. Some of them are helpful to this research. Bertino and Samarati proposed two extensions to the authorization model for relational databases, and gave the conversion from an authorizations status to another one [3]. Bertino and Jajodia presented an authorization model that can be used to express a number of discretionary access control policies for relational data management systems [4]. Afinidad and Levin proposed a model for representing temporal access control policies. In this model, temporal authorizations are represented by time attributes associated with subjects and objects [1]. Keahey and Welch, et al. proposed a fine-grain authorization for resource management[11,17]. Take Elisa Bertino [5] and Ji-Won Byun [6] built a purpose-based access control model for not only database system, but also other privacy system with complex data. In this paper, the purpose-based access control model is based on [5] and [6]. However, the issue of this paper is how to support and consummate the purpose model with data mining technology different from [5] and [6].

Data mining techniques has been widely used in database systems [2]. Since traditional top-down approach to role engineering does not analyze the permission enter in the role definition process, J¨urgen Schlegelmilch proposed RCA to solve this problem [15]. Whereas, role mining focuses mostly on the rights that a performance needs, rather than the resources a performance needs. What's more, different role may have the same purpose, while role mining cannot figure out this deep-seated relation. Purpose mining delves into the relationship between resources and purposes, and optimizes the authorization strategies in database systems. A hierarchy mining algorithm has been proposed to commendably find out that relationship, and a purpose tree is constructed to motivate the authorization strategies.

Other mining applications also support the strategies mining algorithm. Qiang Yang presented an approach of case mining to automatically discover case bases from large datasets in order to improve both the speed and the quality of case based reasoning [18]. Case mining constructed a case base from a large raw dataset with an objective to improve the case-base reasoning systems' efficiency and quality. Jie Zhang presented an approach that exploits the data dimension by mining user query log to glorify the ontology learning process [19]. Since these works both propose some kinds of mining algorithm, but the strategies mining algorithm is brand-new and dose not strictly match any of them.

6 Conclusions and Future Work

In this paper, an approach of purpose-based authorization strategies mining is presented and implemented, which is primarily used in database access control. Generally, compared with previous studies, purpose-based authorization strategies are more flexible and efficient. And mining algorithm operates on a more microcosmic level, reduces the waste of resources and saves the management cost, when database systems are getting larger and larger.

Future work of this research includes a weighted purposes hierarchy mining algorithm, which provides a more flexible model of authorization strategies. And the ignored containment relation between resources should be well considered. And the implementation of the mining algorithm to the other privacy system such as web service, composing system, and work flow system, which need to be well considered to enhance the usage of purpose-based authorization strategies mining algorithm.

References

[1] Afinidad, F., Levin, T., Irvine, C., Nguyen, T.: A Model for Temporal Interval Authorizations. HICSS (2006)

[2] Apté, C., Liu, B., Pednault, E.P.D., Smyth, P.: Business applications of data mining. Commun. ACM 45(8), 49–53 (2002)

[3] Bertino, E., Samarati, P., Jajodia, S.: An Extended Authorization Model for Relational Databases. IEEE Trans. Knowl. Data Eng. 9(1), 85–101 (1997)

[4] Bertino, E., Jajodia, S., Samarati, P.: A Flexible Authorization Mechanism for Relational Data Management Systems. ACM Trans. Inf. Syst. 17(2), 101–140 (1999)

[5] Bertino, E.: Purpose Based Access Control for Privacy Protection in Database Systems. In: Zhou, L.-z., Ooi, B.-C., Meng, X. (eds.) DASFAA 2005. LNCS, vol. 3453, 2, Springer, Heidelberg (2005)

[6] Byun, J., Bertino, E., Li, N.: Purpose based access control of complex data for privacy protection. SACMAT 2005, pp. 102–110 (2005)

[7] Du, Z., Lin, F.: A novel parallelization approach for hierarchical clustering. Parallel Computing 31(5), 523–527 (2005)

[8] Ferraiolo, D., Sandhu, R., Gavrila, S., Kuhn, D., Chandramouli, R.: Proposed NIST Standard for Role-Based Access Control. ACM Transactions on Information and ystemsSecurity, vol. 4(3) (August 2001)

[9] Griffiths, P., Wade, B.: An Authorization Mechanism for a Relational Database System. ACM TODS 1(3), 242–255 (September 1976)

[10] Hitchens, M., Varadarajan, V.: Tower: A Language for Role-Based Access Control. In: Proceedings of the Strategies Workshop, Bristol, UK (2001)

[11] Keahey, K., Welch, V.: Fine-Grain Authorization for Resource Management in the Grid Environment. In: Parashar, M. (ed.) GRID 2002. LNCS, vol. 2536, pp. 199–206. Springer, Heidelberg (2002)
[12] Kim, D., Ray, I., France, R., Li, N.: Modeling Role-Based Access Control Using Parameterized UML Models. In: Wermelinger, M., Margaria-Steffen, T. (eds.) FASE 2004. LNCS, vol. 2984, pp. 180–193. Springer, Heidelberg (2004)
[13] Kumar, A., Karnik, N., Chafle, G.: Context sensitivity in role-based access control. In: ACM SIGOPS Operating Systems Review (July 2002)
[14] Lee, J., Yeung, D., Tsang, E.: Hierarchical clustering based on ordinal consistency. Pattern Recognition 38(11), 1913–1925 (2005)
[15] Schlegelmilch, J.: Role mining with ORCA. SACMAT 2005, pp. 168–176 (2005)
[16] Tachikawa, T., Higaki, H., Takizawa, M.: Purpose-Oriented Access Control Model in Object-Based Systems. In: Mu, Y., Pieprzyk, J.P., Varadharajan, V. (eds.) ACISP 1997. LNCS, vol. 1270, pp. 38–49. Springer, Heidelberg (1997)
[17] Thompson, M., Essiari, A., Keahey, K., Welch, V., Lang, S., Liu, B.: Fine-Grained Authorization for Job and Resource Management Using Akenti and the Globus Toolkit. CoRR cs.DC/0306070 (2003)
[18] Yang, Q., Cheng, H.: Case Mining from Large Databases. In: Ashley, K.D., Bridge, D.G. (eds.) ICCBR 2003. LNCS, vol. 2689, pp. 691–702. Springer, Heidelberg (2003)
[19] Zhang, J., Xiong, M., Yu, Y.: Mining Query Log to Assist Ontology Learning from Relational Database. In: Zhou, X., Li, J., Shen, H.T., Kitsuregawa, M., Zhang, Y. (eds.) APWeb 2006. LNCS, vol. 3841, pp. 437–448. Springer, Heidelberg (2006)

RSP-DS: Real Time Sequential Pattern Analysis over Data Streams*

Ho-Seok Kim[1], Jae-Jyn Shin[1], Yong-Il Jang[1], Gyoung-Bae Kim[2], and Hae-Young Bae[1]

[1] Department of Computer Science and Information Engineering, Inha University
Yonghyun-dong, Nam-gu, Incheon, 402-751, Korea
{hskim, jaejyn, yijang}@dblab.inha.ac.kr, hybae@inha.ac.kr
[2] Department of Computer Education, Seowon University, 231 Mochung-dong
Heung-duk-gu Cheongju-si Chungbuk, 361-742, Korea
gbkim@seowon.ac.kr

Abstract. The existing pattern analysis algorithms in data streams environment have only focused on studying performance improvement and effective memory usage. But when new data streams come, existing pattern analysis algorithms have to analyze patterns again and have to regenerate pattern tree. This approach needs many calculations in real time environments having real time pattern analysis needs. This paper proposes a method that continuously analyzes patterns of incoming data streams in real time. The proposed method analyzes patterns first, and then after obtains real time patterns by updating previously analyzed patterns. The patterns form a pattern tree, and freshly created new patterns update the pattern tree. In this way, real time patterns are always maintained in the pattern tree and old patterns in the tree are deleted easily using FIFO method. The advantage of our algorithm is proved by performance comparison with existing methods, MILE, with a condition that pattern is changed continuously.

1 Introduction

Many applications worthy of being noticed, including network packet monitoring, measurement of scientific experimental performance, financial management and other applications in ubiquitous computing, requiring real-time processing deal with new data type called data stream. In traditional databases the data does not change so much in time. On the other hand, in data streams data is inputted continuously with differing input volume and speed different from the previous data at every instant of time. It also may be generated from several sources at the same time[3]. The existing data mining techniques should be redefined depending on data stream's specific characteristics such as input volume and frequency. It is impossible to store all data from the stream as enormous data is generated continuously. Therefore data mining operations

* This research was supported by the MIC (Ministry of Information and Communication), Korea, under the ITRC (Information Technology Research Center) support program supervised by the IITA (Institute of Information Technology Assessment).

K.C. Chang et al. (Eds.): APWeb/WAIM 2007 Ws, LNCS 4537, pp. 99–110, 2007.

should not use all of the data but the recent data. Also since the characteristic of data stream changes continuously, the result of mining operation should be re-analyzed. But, these absolute conditions require much memory and calculation time to analyze the pattern of large volume of data stream that is inputted with fast speed once more[10].

Sequential pattern analysis method finds the set of frequent sub-sequences is a field of data mining[11]. GSP(Generalized Sequential Patterns) and PrefixSpan are two typical methods for sequential pattern analysis[1,4]. GSP discovers sequential pattern based on the part of the pattern that happens frequently if some patterns happen frequently. But, GSP requires so much time to calculate how many times sub-sequential pattern occur, it can't be adapted to real time data streams. PrefixSpan doesn't generate the candidate patterns and discovers sequential pattern faster than GSP, so it can be adapted to data stream analysis.

StreamPath and DSM-TKP(Data Stream Mining for Top-k Path Traversal Patterns) were proposed for sequential pattern analysis of data stream[6,9]. But, they have a weak point that these methods should search common short-term pattern in long-term patterns repeatedly. MILE(MIning from muLtiple strEams) is the algorithm to analyze sequential pattern in multiple streams, based on PrefixSpan[2]. MILE recursively utilizes the knowledge of existing patterns to avoid redundant data scanning, and can therefore effectively speed up the new patterns' discovery process.

As mentioned above, because of characteristic of data streams, method of sequential pattern analysis over data streams should discover pattern which change continuously. In this paper, we propose the algorithm that analyzes and discovers pattern over data streams which changes continuously. The proposed algorithm divides data streams into several sequences and makes several sequences. These sequences are managed by hash table that have data as key. In case that the regular number of sequences is entered, pattern analysis is processed in hash table in each time. Patterns that are analyzed continuously are set to pattern-tree and old pattern in pattern-tree is deleted to keep present pattern.

The rest of this paper is organized as following. Section 2 reviews related work: PrefixSpan and MILE. The proposed algorithm and performance evaluation is described in Section 3. A conclusion is presented in the last section.

2 Related Work

2.1 PrefixSpan

PrefixSpan is a method that discovers sequential pattern. The merits of PrefixSpan are that it recursively projects the original dataset into smaller and smaller subsets, from which patterns can be progressively mined out. PrefixSpan does not need to generate candidate patterns and identify their occurrences but grows patterns as long as the current item is frequent in the projected dataset as depicted figure 1. Therefore this algorithm is not proper for data stream environment that require fast processing time. Also, it wastes memory to store redundant patterns in pattern tree.

For example, given initial value of processing pattern and final pattern set is empty and min_sup is equal to 2 and 3 sequences are <b,c,g,a,b,a,d>, <b,c,h,f,a,b,d>, <f,c,d,b,e,a,g>,

First, it finds the set of frequent items over min_sup, frequent items are listed as following: a=3, b=3, c=3, d=3, e=1, f=2, g=2, h=1.

To make new processing pattern, each data that is selected is merged with processing pattern. But, because now frequent data is empty, each data can be processing pattern <a>, <b>, <c>, <d>, <f>, <g>. Processing pattern is added to final pattern set. In case of <a>, Suffix of processing pattern in each sequence, it is <a,b,a,d>, <d,b>, <g>. Patterns as a result of recursive call functions are depicted as pattern tree as below.

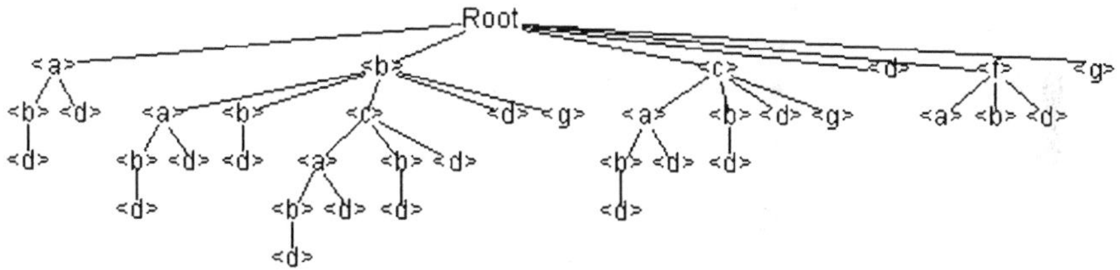

Fig. 1. Pattern Tree by PrefixSpan

2.2 MILE (MIning from muLtiple strEams)

In PrefixSpan, it is a demerit that a pattern already found will be generated once more. For example, in figure 1, left most pattern <a> and its child patterns <a,b>, <a,b,d>, <a,d> are listed Suffix of <b>, <b, c> and <c>. As the size of patterns increases more, some more redundant pattern search increase, so the cost of pattern search is an expense.

MILE divides a large number of data stream source into tuple based windows and forms several sequences. Unlike PrefixSpan, in MILE given pattern is linked with previous pattern if a current Suffix of pattern is already listed in the pattern tree. It prevents redundant pattern search, improves the speed of patter generation.

Given sequence seq1=<b,c,g,a,b,a,d>, seq2=<b,c,h,f,a,b,d>, seq3=<f,c,d,b,e,a,g> in figure 1, it forms pattern by PrefixSpan from data that is inputted later relatively to other patterns to prevent redundant pattern search.

When <c,a> pattern is formed, <a,b>, <a,d> are already listed including <a> as Prefix. So, there is no pattern generation any more and <b>, <d> including <a> as Prefix are pointed in <c,a>. After forming entire patterns, pattern tree is generated like in figure 2. There are fewer patterns in pattern tree than figure 1. Pattern <c,a,b,d> and <c,a,d> have a pointer of patterns to begin by <a> that are already found.

MILE based on PrefixSpan adapts simple sequential pattern analysis to pattern analysis of data stream. But, since pattern of each data stream changes continuously, pattern frequency of each data stream may be different. MILE divides data stream into tuple based window and analyzes it.

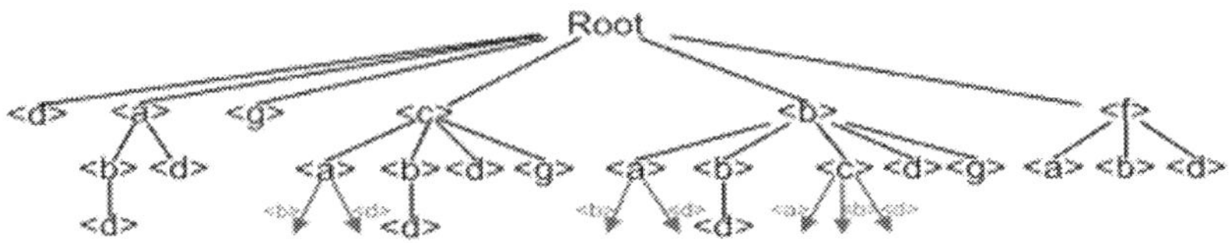

Fig. 2. Pattern Tree by MILE

3 RSP-DS (Real Time Sequential Patterns Analysis in Data Streams)

In RSP-DS, patterns of several sequences Si, Si+1, ..., Si+n divided by time based sliding window against previous patterns Si-1, Si-2, ... ,Si-(n-1) is analyzed. Terms used in pattern analysis is defined as followings.

Definition 1: old_seqs
It means the count of sequences inputted previously to discover pattern. Sequences can be inputted frequently or rarely. If patterns are discovered in sequences that are inputted continuously, we can discover patterns using only new sequences. But, in case that some patterns are discovered rarely, we have to compare with old sequences to discover those patters.

Definition 2: new_seqs
It means the count of new sequences to be used in pattern analysis. As comparing previous sequence to new sequence when analyzing pattern, it is new_seqs that decides how many new sequences are required. The number of (old_seqs + new_seqs) sequences is necessary for pattern analysis. And, after analyzing the pattern, old_seqs is set to old_seqs exclusive of the oldest number of new_seqs sequences.

3.1 Pattern Analysis and Addition

RSP-DS uses MILE to discover patterns over a data stream. MILE is an algorithm that discovers patterns over several data stream, on the other hand, RSP-DS divides a data stream into a fixed sequence and discovers pattern over those sequences.

Data Structure
It is assumed that a pattern tree is made from 3 windows (<b,c,g,a,b,c,d>, <b,c,f,a,b,d>, <f,c,d,b,a,g>). When <b,d,f,c,a,e> sequence is inputted, we should analyze pattern again including a newly inputted sequence <b,d,f,c,a,e> exclusive of <b,c,g,a,b,c,d> that is inputted for the first time. That is, it is assumed that old_seqs be 2 and new_seqs be 1. Figure 3 depicts pattern analysis when old_seqs=2, new_seqs=1. At time t1, <b,c,g,a,b,c,d>, <b,c,f,a,b,d>, <f,c,d,b,a,g> are used for pattern analysis. But, at time t2, the rest of the sequences including the newly inputted <b,d,f,c,a,e> are done exclusive of <b,c,g,a,b,c,d> that is inputted for the first time.

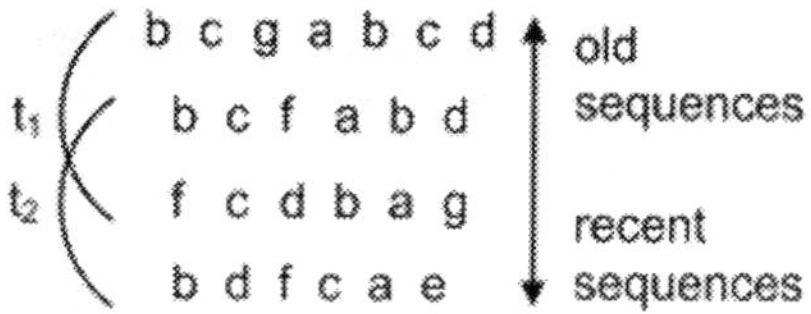

Fig. 3. The change of sequences using pattern analysis according to time

When discovering new patterns, much time is demanded to compare previous pattern tree and pattern tree that is made by current sequences. Therefore, we should make new pattern tree by updating previous pattern tree partially. If we know that the Suffix of current pattern is redundant compared to previous pattern, it can prevent us from extracting redundant pattern to link pointer to previous pattern. For example, lets assume there are pattern <b,c,a> and pattern <c,a> correctly existing. If Pattern <c,a> is found earlier than pattern <b,c,a>, when pattern <c> is found later than <b>, simply the pointer can be linked from <b> to <c,a>. At this time, we should know which pattern is entered as Suffix of another pattern to maximize it by removing redundant trace pattern.

Sequences used for analyzing pattern are managed by hash table to analyze pattern quickly. In figure 4, old_seqs is 2 and new_seqs is 1 in hash table. Sequence 3 is the last entered sequence. After sequences including seq1=<b,c,f,a,b,d>, seq2=<f,c,d,b,a,g>, seq3=<b,d,f,c,a,e> is analyzed, hash index is constructed. Data Information Table(DIT) pointed to by hash index stores statistics information about entire sequences, and Order Information Table(OIT) stores information data of each sequence.

Hash Index	nFreq	nSuffix	Seq_1	Seq_2	Seq_3
a	3	4	4 | -	5 | -	5 | -
b	3	12	1 | f,c,a; 5 | d	4 | a	1 | d,f,c,a
c	3	10	2 | a	2 | a	4 | a
d	3	7	6 | -	3 | a	2 | a
e	1	0			6 | -
f	3	11	3 | a	1 | c,a	3 | c,a
g	1	0		6 | -	

Fig. 4. The Example of Hash Table

Once new sequence is entered in OIT, the row in OIT corresponding to oldest sequence is deleted and a new row is inserted. OIT is composed of nOrder and pNext. The nOrder points to the order of data in a sequence and the pNext is the pointer that points to Suffix when this data in a sequence is the Prefix. It is to look up nOrder corresponding to each sequence <a> and <b> which is the way to decide whether Sequence <b,a> is pattern or not. Because Seq1, Seq2, Seq3 sequences appear later than a and b, these sequences are turned out pattern. The pNext is used for determining whether a data is linked to any other pattern when analyzing pattern. Therefore it can prevent redundant pattern analysis easily. DIT is composed of bFreq and nSuffix. The bFreq denotes the frequency of data appearing in sequences. The bFreq is used for determining whether a data appears frequently or not. The redundancy of pattern analysis can be minimized by first searching short data that is quite probable to be entered as Suffix of other pattern, by using nSuffix.

Algorithm

The algorithm to discover new pattern using hash table is as following.

Algorithm 2 (FindNewPattern)

```
Input: Pattern Tree(PT), Hash Table(HT), New Entered Sequence(s_new)
Output: Updated Hash Table(HT)
Delete old sequence as many as old_seqs times in OIT and then update DIT.
Input s_new into Hash Table in sequence and update OIT and DIT. pNext is null value.
foreach data D := s_new, in ascending order of nSuffix  do
  if nFreq of D >= min_sup then
    Add D to PT
  end if
  foreach Data pre_D := s_new in descending order of nSuffix in Suffix of D do
    if nFreq of pre_D >= min_sup then
      Link D and pre_D, then add to PT
      pNext of D := pre_D
      if pNext of pre_D != null then
        point the pattern <D, pre_D> in PT to the pattern begun by pNext of pre_D, then
       don't trace pNext
      end if
    end if
  end for
end for
```

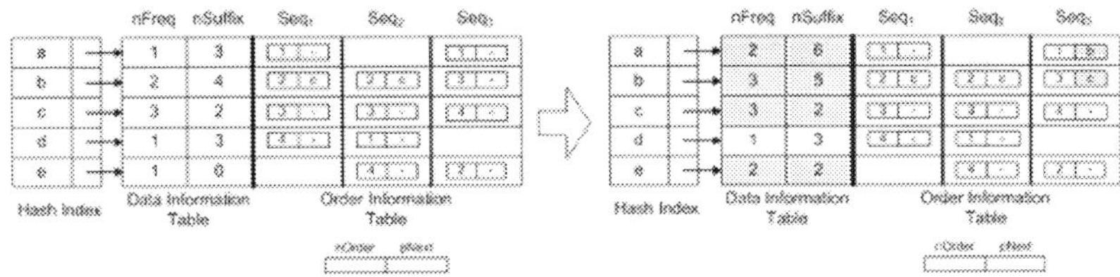

Fig. 5. Hash Table modified by FindNewPattern algorithm

Figure 5 describes hash table as an example of our algorithm. min_sup is equal to 2. Seq1=<a,b,c,d> and Seq2=<d,b,c,e> are entered previously, Seq3=<a,e,b,c> is the newest entry. Left side in Figure 5 describes when Seq2 is entered, and right side describes hash table modified after pattern is analyzed. In the right side, the gray color part is the part that is modified. The process that modifies hash table is as follows.

Seq2 is entered and nFreq, nSuffix are updated. We analysis processes in descending order of nSuffix among data in Seq3. Although <c>, <e> have same nSuffix value, <c> entered later is analyzed first. Because nFreq of <c> is equal to 2, it becomes pattern. And search is terminated because there is no Suffix of <c>.

Next is the turn of <e>. Because the nFreq of <e> is equal to 2, it also becomes a pattern. Then Suffix of <e> in each sequence is analyzed. That is, the nSuffix of data entered in Seq3 become standard between <> of Seq2 and <b,c> of Seq3. But, <b>, <c> is entered just one time, so it can't be a pattern. Therefore, only <e> is discovered as pattern.

Then, <b> whose nSuffix is equal to 5, is analyzed. The nFreq is equal to 3, it become pattern. Suffix of <b> is analyzed in each sequence. <c,d> of Seq1, <c,e> of Seq2, <c> of Seq3 is analyzed according to nSuffix of data in Seq3. Because <c> appears three times, it becomes a pattern. Because the nFreq of <a> is 2 in the last data <a> in Seq3, it become pattern. Suffix of <a> in each sequence is then analyzed. <b,c,d> of Seq1, <> of Seq2, <e,b,c> of Seq3 are analyzed according to nSuffix of data <e,b,c> in Seq3. Because <b> appears two times and is linked <a>, <a,b> become patterns. Then the pNext of <b> is not null but <c>. So, <a,b,c> becomes pattern. However, <e> appears just one time, so it can't be a pattern.

3.2 Pattern Deletion

If patterns are added continuously, there are too many patterns in a pattern tree. Moreover, all of the sequences entered may become pattern in the worst case. So the patterns that don't appear well should be deleted in spite of addition of new patterns.

Data Structure

The way of deleting patterns can be adapted from the Buffer Replacement Policy[12]. The difference between Buffer Replacement Policy and deletion of pattern is that the number of Buffer items used in Buffer Replacement Policy is regular, but the number of patterns in Deletion of pattern isn't. That is, in case that very few patterns appear frequently, there are fewer patterns in the pattern tree. But, in case when there are many patterns that appear rarely, there are a lot of pattern in pattern tree. Therefore, if there are no more entered patterns over a regular period, it is possible to delete specific patterns.

Comparing LRU(Least-Recently Used) Buffer Replacement Policy can be adapted for deletion method of RSP-DS. LRU policy processes when buffer is full, on the other hand, it is necessary to delete buffer when the scheduled time expires regularly in RSP-DS. LRU is not suitable for RSP-DS.

There is another Buffer Replacement Policy, FIFO(First In, First Out) Queue that deletes the oldest block. But, It considers only time, not access frequency.

We modify FIFO Queue algorithm to delete patterns easily. First, FIFO is generated by the time that pattern is entered. Then if that pattern is entered again, we input it in FIFO. We use pattern tree for access purpose, index of FIFO queue. Figure 6 describes modified pattern tree and FIFO structure.

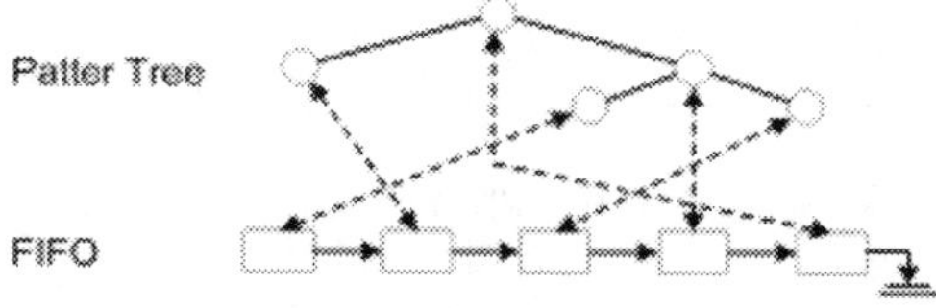

Fig. 6. Pattern Tree with FIFO for deleting pattern

Algorithm

The following is the algorithm to add pattern and delete pattern considering deletion of a pattern.

Algorithm 3 (AddPattern)

Input: Pattern Tree(PT), New Entered Sequence(s_new), FIFO
Output: Null

```
if s_new ∈ PT then
  Find s_new in FIFO, then move s_new to end of FIFO
else
  Add s_new to PT, then add s_new and the entered time in FIFO
end if
```

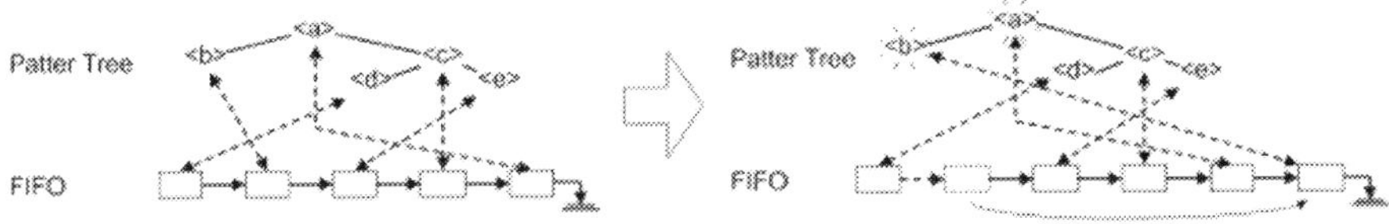

Fig. 7. Pattern Tree and FIFO modified by AddPattern()

Figure 7 describes the feature when pattern that already exists in pattern tree is entered. Simply, search the entered pattern in tree and input it to FIFO node again.

Algorithm 4 (DeletePattern)

Input: FIFO, pattern tree PT, Time Threshold t
Output: Null

```
pat_set(Pattern Set) := Pattern over time t in FIFO
foreach pat := a pattern in pat_set do
  if there is sub-pattern of pat in PT then
    Delete pat and sub-pattern of pat in PT. If there is the pointer that points to sub-pattern,
    do not trace pointer.
  else
    Delete pat in PT
  end if
end for
```

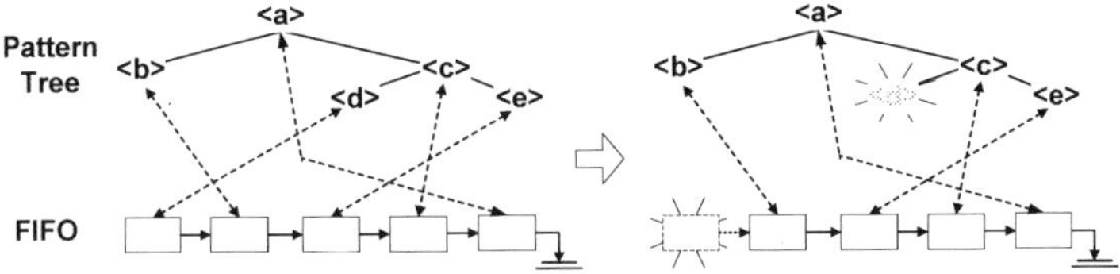

Fig. 8. Pattern Tree and FIFO modified by DeletePattern()

Figure 8 shows the process of deletion of a pattern. Even if <a,c,b> is not frequent, it does not mean that <a,c> is not frequent. Also, pattern <a,c,e> that shares same

Prefix can be frequent. So, <a,c,b> can be deleted, but not <a,c>. If <a,c> is not frequent, also <a,c,d> and <a,c,e> are not frequent. Therefore, when <a,c> is deleted, its Suffix should be deleted. Two cases can be considered when a pattern should be deleted is a Suffix or a Prefix of other pattern. In the case of deletion of <c,e> that is a Suffix of <a,e,c>, <c,e> in the tree and a pointer from <a> is deleted. And, when deleting <a,e> that is a Prefix, if some patterns are not frequent then all patterns that include that pattern are not frequent. So, <a,e> and its child patterns can be deleted in pattern tree. Otherwise, if some of the child patterns have pointers that references another patterns, these patterns can be referenced from other patterns by its Suffix. So, these patterns are not deleted. Figure 9 shows an updated tree of Figure 2 when new sequence is inserted. These patterns that are not discovered from new pattern extraction process are excluded from pattern tree and newly discovered patterns are included in the pattern tree. Patterns in <<>> are newly included patterns that are not extracted from the previous tree.

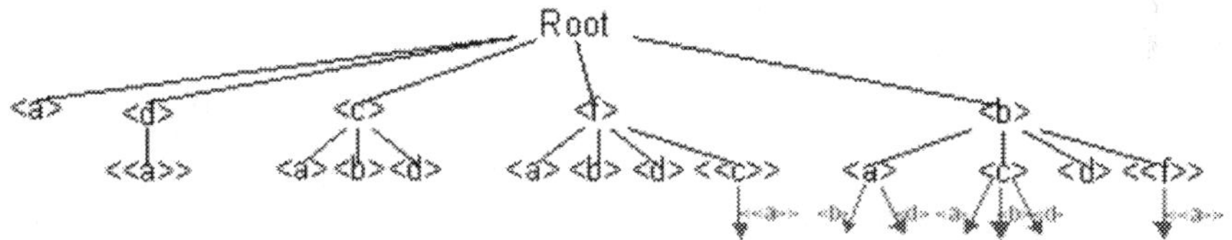

Fig. 9. Pattern Tree resulted by RSP-DS

4 Performance Evaluation

In this paper the presented algorithms have various performance measurements taken against four kind of variables. Next is the explanation of these variables.

- *min_sup*: A basis to estimate data frequency. In case of large min_sup, the calculation cost is reduced because the number of data is decreased. However, too small number of patterns makes it difficult to get appropriative results.
- *dif_items*: A kind of data stream. Many kinds of inserted data mean that there exist various data types. The value of this variable is changed by stream environment and user cannot modify it.
- *new_seqs*: The number of new sequence that will be inserted in hash table. To analyze fixed number of sequence, the last of new_seqs sequences are deleted from hash table whenever new_seqs sequences are inserted. A large size of new_seqs increases delay time and analyzing frequency. A small size of new_seqs decreases it, but calculation cost can grow up.
- *old_seqs*: The number of old sequences in hash table. This is the number of old sequences to be compared with new_seqs to analyze patterns. If old_seqs are larger than new_seqs then patterns are more easily classified whatever there is new_seqs. For example, let new_seqs is 3, old_seqs is 7 and min_sup is 6. Then data that appears after five times is decided as frequent data from new_seqs + old_seqs. At this time, if there is data D that is n times smaller than 3 exists in hash table, it is not selected to frequent data even if all of new three

sequences have D, since n+3 is lower than 6. Therefore, if data D' that satisfies n>=5 exists in the hash table, D is selected as frequent data no matter what the new sequence is.

- *data_rate*: The insertion ratio of sequence per time. If the basis of measurement is second and data_rate is 10, 10 sequences are inserted per second. This means that a unit of time_stamp is 1/10. In case when data_rate goes to excess, pattern analyzing speed cannot keep up with insertion speed. Therefore, if data_rate is lengthy, pattern analyzing frequency should be long to decrease the frequency.

Figure 10 is a building time comparison of MILE with data stream analysis that is divided by time window and RSP-DS tree. The horizontal axis represents data_rate per second and vertical axis represents average time of pattern analysis since first insertion of sequence. In the experiment, the value of dif_item, min_sup, new_seqs and old_seqs is 10. Table 1 lists update frequency of pattern tree through data_rate. In the case of MILE, tree should be built again whenever 10 sequences are inserted. Also, if the value of data_rate exceeds, update frequency is shortened and pattern calculation speed cannot keep up with data insertion speed. Therefore, processing time per sequence is extremely increased because of a lot of pending sequences for insertion in hash table. However, RSP-DS inserts new sequences in hash table, and compares with previous sequences, and then updates pattern tree. It demonstrates faster processing time than MILE.

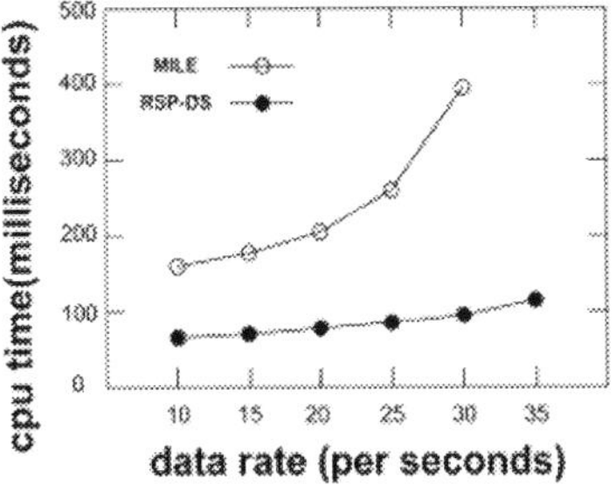

Fig. 10. A comparison of RSP-DS and MILD

Table 1. Update frequency of pattern tree with various value of data_rate

Data_rate	10	15	20	25	30	35
Update Frequency (sec)	1	2/3	1/2	2/5	1/3	2/7

Figure 11 is a performance result about RSP-DS with various values of min_sup. The value of dif_item, new_seqs, old_seqs is 10. min_sup is a ratio of new_seqs to have general value. As min_sup increases, processing cost is reduced and performance is increased by selecting frequent data from a small number of data. But, selecting too small number min_sup causes incorrect results, since it decreases the number of pattern being analyzed.

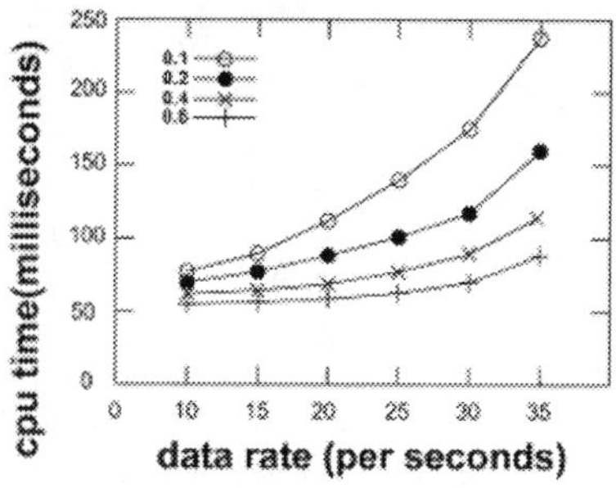

Fig. 11. A performance evaluation of min_sup

Figure 12 is a performance result as a result of variation of new_seqs. The value of dif_item, min_sup and old_seqs is 10. Table 2 shows the update frequency variation of new_seqs. If new_seqs increases, also update frequency is increased. So, increment of data_rate doesn't affect performance that significantly. However, it means that analysis performance decreases. Decreasing new_seqs reduces pattern analyzing time because it has a small number of sequences. Otherwise, high value of data_rate makes a short pattern analyzing frequency. So, pattern analyzing time is extremely increased because of low data insertion speed.

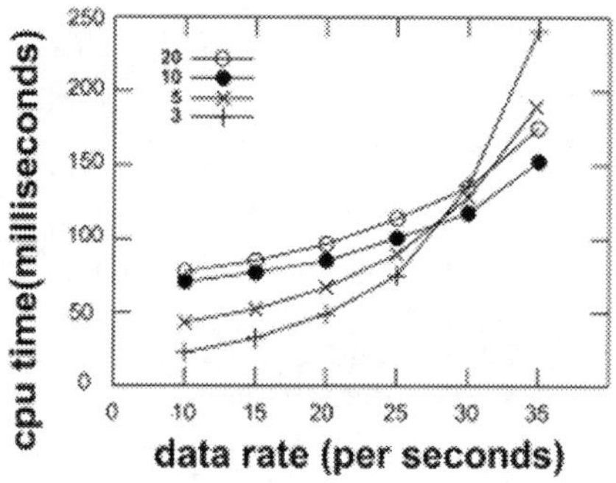

Fig. 12. A performance evaluation of new_seqs

Table 2. Update frequency of pattern tree of data_rate and new_seqs (second)

data_rate	10	15	20	25	30	35
new_seqs=3	3/10	1/5	3/20	3/25	1/10	3/35
new_seqs=5	1/2	1/3	1/4	1/5	1/6	1/7
new_seqs=10	1	2/3	1/2	2/5	1/3	2/7
new_seqs=20	2	4/3	1	4/5	2/3	4/7

5 Conclusion

Recently, data stream that has a new attribute and different from other data is being carefully studied. The pattern of data stream always changes. But, existing pattern analyzing algorithms have considered only single number of pattern analysis. This process has severe disadvantage that they repeat overall process again to analyze the patterns.

The proposed RSP-DS method makes pattern tree by discovered pattern and patterns that gradually changes is analyzed at the branch of pattern tree in real time. RSP-DS is based on MILE that is a data stream pattern analyzing algorithm.

But, the proposed method has better performance than MILE in environments with continuously pattern changes, and this is demonstrated by performance evaluation. This is achieved as MILE always creates a new tree whenever new sequences are inserted, but RSP-DS keeps current patterns of tree by editing them.

And, experiments show that varying min_sup value and the number of sequences regulates performance of the proposed algorithm. Therefore, performance optimization can be done by manipulating these two variables. Hence, a research of pattern analysis of data stream whose frequency changes continuously should be done for RSP-DS, as currently it only considers fixed frequency. And, in environment where patterns rarely change, RSP-DS algorithm processes unnecessary operations. So, study of discovering pattern variation is required which is left as a future work.

References

[1] Pei, J., Han, J., Mortazavi-Asl, B., Wang, J., Pinto, H., Chen, Q., Dayal, U., Hsu, M.C.: Mining sequential patterns by pattern-growth: The prefixspan approach. IEEE Trans. Knowl. Data Eng. 16(11), 1424–1440 (2004)

[2] Chen, G., Wu, X., Zhu, X.: Sequential Pattern Mining in Multiple Streams In: ICDM, pp. 585–588 (2005)

[3] Babcock, B., Babu, S., Datar, M., Motwani, R., Widom, J.: Models and Issues in Data Stream Systems. In: PODS, pp. 1–16 (2002)

[4] Srikant, R., Agrawal, R.: Mining sequential patterns: Generalizations and performance improvements. In: Apers, P.M.G., Bouzeghoub, M., Gardarin, G. (eds.) EDBT 1996. LNCS, vol. 1057, pp. 3–17, Springer, Heidelberg (1996)

[5] Li, H.F., Lee, S.Y., Shan, M.K.: On Mining Webclick Streams for Path Traversal Patterns. WWW, pp. 404–405 (2004)

[6] Li, H.F., Lee, S.Y., Shan, M.K.: DSM-TKP: Mining Top-K Path Traversal Patterns over Web Click-Streams Web Intelligence, pp. 326–329 (2005)

[7] Das, G., Lin, K.I., Mannila, H., Renganathan, G., Smyth, P.: Rule Discovery from Time Series, KDD, pp. 16–22 (1998)

[8] Oates, T., Cohen, P.R.: Searching for Structure in Multiple Streams of Data. In: ICML, pp. 346–354 (1996)

[9] Li, H.F., Lee, S.Y., Shan, M.K.: On Mining WebClick Streams for Path Traversal Patterns, WWW, pp. 404–405 (2004)

[10] Gaber, M.M., Krishnaswamy, S., Zaslavsky, A.: Ubiquitous Data Stream Mining. In: Dai, H., Srikant, R., Zhang, C. (eds.) PAKDD 2004. LNCS (LNAI), vol. 3056, Springer, Heidelberg (2004)

[11] Zhao, Q., Bhowmick, S.S.: Sequential Pattern Mining: A Survey, Technical Report Centre for Advanced Information Systems, School of Computer Engineering, Nanyang Technological University, Singapore (2003)

Investigative Queries in Sensor Networks

Madhan K. Vairamuthu, Sudarsanan Nesamony, Maria E. Orlowska, and Shazia W. Sadiq

School of ITEE, The University of Queensland, Australia
{madhan,sudar,maria,shazia}@itee.uq.edu.au

Abstract. Sensor networks have widespread range of applications these days. New computing locations, new needs and fresh applications have instigated new problems and exciting challenges. The view of sensor network as a distributed database and the sensor node as a table is being widely accepted by the database community. There now exists wide variety of data models and query processing methods to acquire data from sensor networks. This paper identifies a new class of query, named the investigative query, owing to needs of few special applications. Classifying sensor network query processing from communication perspective, defining the new class of query and architecture requirements for processing the same, this paper gives an application of those queries in urban water distribution management and concludes with future directions.

Keywords: Investigative Query Processing, Sensor Networks.

1 Introduction

Sensor network applications have brought a phenomenal change in how this world views Information Technology as a field of science. In the event of taking 'computing' from labs' desks to farms & roads, computer science has made huge sacrifices to cope up with energy constraints in the new little computing machines. The leap is huge, the paradigm shift to real world is brilliant but in the process we have lost much of what we used to call the power of computing. Low power, low communication range, micro storage, tiny processor, tiny operating system and tiny database are the typical characteristics of today's sensor node in a typical sensor network. Sensor networks have thus taken us on a beautiful journey on a time machine, few decades backwards in terms of computing. The same old problems of memory management, communication, computations on a low capability processor and many more such classical problems have been revisited in the recent sensor network literature.

On the other side, emergence of sensor networks in new domains has excited some fresh applications and new problems to the research community. They actually have injected fresh blood into the veins of distributed computing. Distributed data management and query processing were enlivened by the advent of these distributed sensors. The sensor network research community has two major views of sensor

K.C. Chang et al. (Eds.): APWeb/WAIM 2007 Ws, LNCS 4537, pp. 111–121, 2007.

networks. One half of the community views sensor nodes as tiny computers networked together in a huge space. Another half views the tiny sensors as small relational tables and the entire network as an active database. A huge portion of research is devoted to processing of sensor network queries.

This paper observes a rise of a new class of queries in sensor networks owing to the demand of some new applications. We call them investigative queries. Investigative queries are active recursive queries that probe into a network seeking some application data. They are executed in hierarchical fashion. The data collected by one level of this query will decide on the execution of the next level of the same query. This paper aims at defining investigative queries after classifying the list of queries available in sensor network literature. Dissemination of investigative queries into the network requires some special communication architecture to provide the hierarchy levels needed for the application. The requirements are discussed in detail. We provide a strong motivating application to demonstrate the need of investigative queries.

The paper is organised into five major sections. Section 2 aims at classifying the sensor network query routing protocols and provides different types of queries that exists in sensor network literature. In section 3 a sensor network application is described in detail where a new type of query called the investigative query is introduced. Formal definition of investigative queries is provided in section 4, which is the main contribution of this paper. The advantages of investigative queries over traditional methods are discussed along with other potential applications of sensor networks in the same section.

2 Classification of Queries in Sensor Networks

In this section we classify query processing in sensor networks in terms of their communication architecture and then by the data processing methodology. Query processing in sensor networks can be broadly classified based on their query dissemination and data gathering approaches into four major classes,

- Data Flooding
- Tree Path routing
- Multi Path routing
- Clustered routing

Each of these approaches can be sub classified based on their data processing approaches.

- Centralised
- In-network

In order to achieve this classification, we make the following reasonable assumptions.

- Existence of a base station that can issue queries and collect and store results
- Capability of sensors to sense, store, transmit and route data.

2.1 Data Flooding

Data flooding is the simplest of all the approaches, yet powerful. A query issued by base station is flooded into the network by, every sensor node transmitting the query to every node in its neighbourhood until the query is propagated to the entire network. The result is propagated in the similar fashion outwards [1-2]. The advantage of such a technique is its reliability, but the disadvantage being overall energy consumption of the network even for processing simple queries.

2.2 Tree Path Routing

Tree path routing is another simple technique where a tree is constructed out of the available dense network and the tree is maintained throughout for disseminating queries and collecting results [3-4]. Contrasting to the data flooding approach, this approach suffers from single node failure. Considering the cost of constructing a new tree, the maintenance of the tree becomes inefficient. The major advantage of this method is that it conserves over all network energy.

2.3 Multi Path Routing

Multi path routing can be considered as an approach in between data flooding and tree path routing [5-6]. The challenge in this approach has been to join the data from multiple paths and eliminating redundancy from multiple paths.

2.4 Clustered Routing

Numerous clustering algorithms [7-9] group nodes that are spatially close into clusters. The advantage of clustering is that it provides a hierarchical structure for executing queries and the disadvantage being the cost associated with clustering and electing cluster heads [8].

2.5 Centralised vs. In-Network Processing

Centralised data processing is a very common approach where using any of the four classified approaches, query is propagated through the network and result from every node is sent back to the base station in a central location and the data is processed centrally to obtain result for given query.

The advantage of this method is that once data is collected, multiple queries can be run over the available data and the main disadvantage being energy spent by individual nodes to transmit their data and route their neighbours data to the central station [10].

In-network processing is an alternative technique that minimises a node's energy consumption by performing computation at node level, combine the results and transmit just the aggregate result to the base station. This approach requires the nodes to have some basic data processing capabilities [10].

2.6 Query Types in Sensor Networks

Numerous types of queries have been introduced in sensor network literature. Simple aquisitional queries are used for acquiring data from specific nodes in a sensor network. They are very similar to simple distributed queries, which perform similar action. These queries are mainly executed with a tree path routing communication architecture or the flooding architecture [11]. Acquiring the average value of nodes or set of nodes in a network is a very common query in sensor networks. Such queries are termed as aggregate queries and have a wide variety of applications [12]. They are executed on most of the architectures specified. Tree path routing with in-network processing has been the most popular architecture to run these aggregate queries.

Queries that are targeted at nodes with a certain range of values fall under the class of range queries[13]. Range queries are very handy in temperature and pressure monitoring applications. Clustering nodes based on the their location and previous data range has proven to be suitable for these kind of queries. Window queries are very specific to spatial data. A user specified window selects the window-contained nodes alone to respond to the given query. These queries are very popular in numerous applications [14]. Tree-path and multi-path routing is proven to be efficient than flooding in many applications. Considering the poor reliability of weak sensor nodes, the notion of probability is introduced in to queries [15]. Probabilistic queries are not specific to any architecture. The probability takes into account the error in results owing to frequent node failure. Stream queries [16] and active queries [17] are being used in few real time monitoring applications, which require real time action.

2.7 Need for Advanced Query Types

Advanced applications of sensor networks demand architecture and means to support advanced queries. Queries issued by a central base station to identify location of a leak in water distribution pipes' sensors and queries issued to smart buildings to identify unoccupied rooms where the electricity usage is high are new to the database community. The demand of such applications brings in the need for new query types. The following section discusses one such application in detail.

3 A Sensor Network Application with Advanced Queries

In [18] we identified the potential benefits of deploying sensor networks for urban water management especially to identify leaks in the distribution pipes. We had explained the need to segment the network in order to hierarchically query the sensors for existence of a leak. In the following sections, we present a water network model, a sensor network model and communication model similar to [18]. The need for investigative queries is discussed at end of this section.

3.1 Urban Water Distribution Network Model

In this section first we present an abstraction of the real pipe network by the following mathematical model. We begin with introduction of essential notation.

Let the water distribution network W = (S, C, P) consists of a set of supply sources $S = \{s_1, s_2, ..., s_s\}$, a set of consumption units $C = \{c_1, c_2, ..., c_q\}$, and a set of connected pipes $P = \{p_1, p_2, ..., p_m\}$.

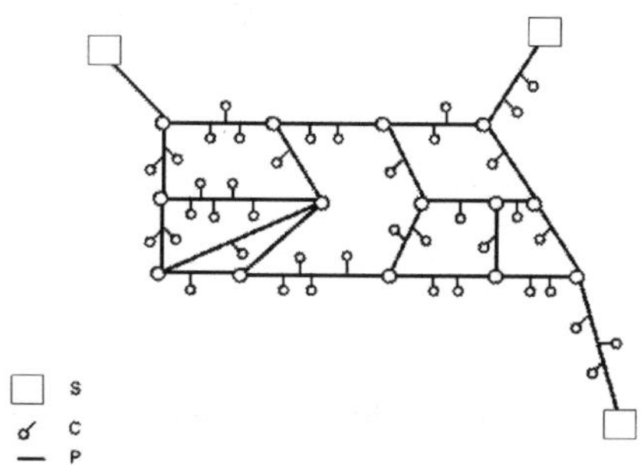

Fig. 1. A Water distribution network W

3.2 Sensor Network Model

We define three types of sensor nodes to be deployed for urban water management.

- Consumption Sensors
- Distribution Sensors
- Fragmentation Sensors

A consumption sensor is placed in front of every consumption unit for the purpose of measuring the total water consumption of a consumption unit over a monitored time period. Without any loss of specification precision, we use the same notation for these sensors as for corresponding water network consumption units, we call them c_i, where i = (1, 2, …, q).

A distribution sensor is placed at entry/exit of every pipe segment in the water network. Essentially every pipe segment would have two distribution sensor nodes, one on each end. These distribution nodes monitor the total inflow and outflow of that particular pipe over a monitored time period.

We thus extend our collection of sensors deployed by defining distribution nodes $D = \{d_1^1, d_1^2, d_2^1, d_2^2, ..., d_m^1, d_m^2\}$ where d_j^1 and d_j^2 denotes the pair of sensor nodes placed one each at both the ends of the pipe segment $p_j \in P$ in W.

Fragmentation sensors are positioned at certain position in the network such that they fragment the network into required number of connected segments and required levels of hierarchy. This hierarchical segmentation is abstracted in section 6.

We now have a third set of sensors that fragment the network as $F = \{f_1, f_2, ..., f_h\}$ where a fragmentation sensor f_i monitors the total inflow and outflow of one or two segments over a monitored time period.

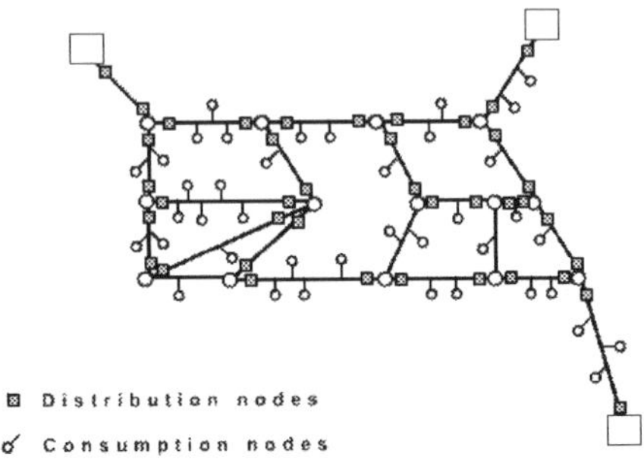

Fig. 2. SN(W) for w in figure 1

With this arrangement we have a mundane way of identifying leaks in every pipe by querying the distribution sensor at the end of every pipe segment. We have a set of points from which water flows into the whole network of pipes. We also will have the water consumption details at every consumption point in the whole network for a specific period of time. If the amount of water flowing into the network is more than the total consumption of water within the network, then it can be concluded that leak(s) are present within the network, based on various practical and reasonable assumptions discussed in [18]. Thus, if the presence of leaks is detected, then every single pipe within the network will have to be examined for a leak. In practice and in theory, this is a very ineffective and uninteresting method to solve the problem. In real world, the number of pipe segments is very high in number and the percentage of pipes containing leaks is very small. Hence, we seek a better method that does not involve unnecessary querying of all the nodes.

In [18], we proposed a technique by which, we divide the network into various partitions and perform operations at the partition level rather than the entire network. Conceptually, a fragmentation of the network is constructed with many hierarchical levels. Initially, the entire network is broken down into a set of connected fragments, with each segment containing a portion of the network. The collection of all the segments will comprise the initial network. The segmentation nodes are placed at the end points of the constructed segments, which measure the net inflow of the water into every segment. This is compared against the actual consumption of water by the end units that are located within that particular segment. Thus, instead of querying every pipe in the current segment, it is sufficient to pass on the query to the sub-segments that were created.

The levels of hierarchy that is needed for the segmentation of the entire network will depend on the application requirement and the characteristics. Eventually, the lowest level will be pipe level segments. We now have segment level distribution nodes apart from the pipe level distribution nodes. A segment level distribution node monitors the inflow and outflow volume of water pertaining to a segment over a monitored time period. It is to be noted that such a distribution node could belong to two different segments and still has capability to maintain data pertaining to both the segments.

3.3 Communication Model

The central processing station should be aware of the readings of consumption sensors over the entire network, as the fundamental requirement of the application is to measure the water intake by every consumption unit. The readings are taken for a period of time and considering the scope of the network size, the ideal communication method will be to include gateway nodes that are powerful enough to reach the base station through wireless signals and the consumption sensor nodes should be able to communicate with the gateway nodes. The communication architecture of the distribution nodes can be the same as that of the consumption nodes. For the current problem, we assume that the communication method is reliable and functional. Taking into account the scope of the network span, it is very reasonable to assume the presence of gateway nodes to facilitate communication amongst the various types of sensor nodes that prevail in the network.

3.4 Fragmentation Model

Let $F_0 \subset F$ be the level zero fragmentation sensors that defines the entire network. Similarly $\forall_{i=1..l}\ F_i \subset F$ denotes a set of subsets of ith level fragmentation sensors, where l denotes the number of levels of segmentation. There by, $F_i = \{F_{i1}, F_{i2}, ..., F_{ip}\}$ where p denotes the number of fragments in level i. So, a fragment of network W can be obtained by cutting the network W at the position of $\forall_i f_i \subset F_k \subseteq F$ for any subset of fragmentation sensors that define a fragment.

In this paper, we presume the existence of a good fragmentation of the network that satisfies the constraints associated with it. We proceed with the result of [] that has provided a fragmentation of the given network. We use this fragmentation as an input to the Investigative queries that have been described earlier. For facilitating this purpose, the fragmentation of the network has to be abstracted. We provide the abstraction of the hierarchical fragmentation as a tree.

Let A be the entire fragment representing W defined by fragmentation sensors in F_0. Let $A_1, A_2, ..., A_p$ be the first level fragments of A into sub-fragments of A. Each A_i is a fragment of W defined by the fragmentation sensors F_{1i}. We obtain the segments in the next levels in similar fashion. Therefore, $\bigcup_{i=1..p} A_i = A$ denotes that the union of fragments, in each level with the corresponding fragmentation sensors, will be the entire fragment representing W.

Let T be a tree representing the fragmentation of W as shown in figure 3. This tree is the result of the network fragmentation process that is detailed in [18]. This representation will capture the level of positioning and hierarchy of the individual segments in the network. The tree structure suits the best for manipulating hierarchical data. The investigative queries will operate on this structure as they execute themselves over a recursive hierarchy.

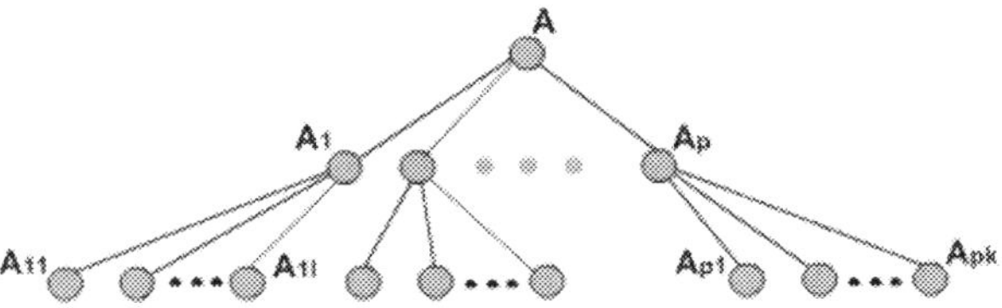

Fig. 3. Fragmentation tree T

The tree T is rooted at a node that denotes the topmost level of hierarchy of the segmentation, which is the entire network itself. All the leaf nodes of T represent the lowest level of hierarchy of segmentation, which will be all the individual pipe segments present in the whole network. The conceptual meaning of the leaf nodes carries significance as the investigative queries, when detected mismatch between the net inflow and the total consumption in a segment, execute recursively within the subsequent level of sub segments and finally terminate at one or more leaf nodes meaning that the pipe segments denoted by those leaf nodes contain the leaks.

The internal nodes of the tree will symbolise the intermediate levels of segments that are neither the entire network nor the individual pipe segments. Since T is a rooted tree, we can also define the notion of a child node of a node in the tree. The child node of a given node will be any node that is directly connected to the given node, which has got the path distance from the root node that is one greater than the path distance between the root node and the given node. Conceptually, a child node denotes a sub-segment or one component of a partition of the given segment represented by the given node.

3.5 Need for Investigative Queries in UWD Management

In the described scenario, querying the sensor nodes for identifying location of leak will be a periodical operation. One way of locating leaks would be to query every distribution sensor and consumption sensor in the network. With Investigative queries we minimise the number of distribution sensors being queried. The fragmentation of the water network gives us a hierarchy as described in the previous section. This hierarchy provides efficiency in processing the query by reducing the number of distribution sensors that need to be queried. The following section formally defines investigative queries in general.

4 Investigative Queries

4.1 Definition

Let Q(s) be an investigative query issued over a segment of interest, in the given network, denoted by node ***s*** of segmentation tree ***T***, such as the one mentioned in section 4.4. Now, Q(s) can be defined as follows.

$$Q(s) = \begin{cases} s & leaf(s) = true \wedge \delta(s) > \theta \\ Q(s_c) \forall s_c \in child(s) & leaf(s_c) = false \wedge \delta(s) > \theta \\ null & otherwise \end{cases}$$

The segment of interest denoted by node **s** is usually the root of **T**. But in general, **s** can be any node in **T** denoting any segment that needs to be investigated.

Here, *leaf(s),* a Boolean function, will return true if **s** is a leaf node or false otherwise. Function *child(s)* will return a set of nodes from **T** containing the children nodes of s.

The investigative function $\delta(s)$ and threshold θ are provided by the application on which these queries are run.

4.2 Investigative Queries in Urban Water Management

In the urban water management scenario $\delta(s)$ balances the inflow volume of water into segment denoted by s against the consumption volume in the same segment. This function is explained in detail in [ref]. The threshold limit for $\delta(s)$ is given by θ. The value of θ is determined by the organisation that controls water distribution. The query returns the nodes, corresponding to the pipe segments that contain the leak or returns null if there exists no leak in the network. The efficiency benefits of using investigative queries over simple aquisitional queries are explained in the following section.

4.3 Advantage over Traditional Queries

Let m denote the number of pipes in the water network thus $2m$ is the number of distribution nodes installed. Let l denote the level of fragmentation and maximum number of fragmentation sensors in any level be denoted by $f_s << 2m$.

To Identify leak using a simple acquisitional query, we need to query $2m$ distribution nodes.

With an Investigative Query, the number of fragmentation nodes and distribution nodes that need to be queried in the worst case, where every level has f_s nodes, is given by lf_s+2 which includes the two distribution nodes in the pipe that has the leak.

So, for a 75,000-pipe network with 150,000 distribution nodes, we would need 150,000 readings from distribution nodes to locate a leak with an acquisitional query. Whereas an investigative query, on a 10-level fragmentation hierarchy with value of f_s being 100, requires 1002 readings to locate a leak.

4.4 Other Possible Applications of Investigative Queries

Investigative queries can be applied in many applications where the data is needs to be sought in a hierarchical fashion. The same leak detection can be used in pipes applied for petroleum industries. Supply chain management with RFIDs can be a potential application for Investigative queries. Item dislocation or path of an item can be queried in such applications. A group of buildings, say a University, where there exists a natural hierarchy as streets, buildings, floors and rooms where sensors are deployed to sense occupancy, sound, temperature and light exposure. We can have a series of Investigative queries here such as locating rooms in buildings with high electricity usage with no occupants. Locating toilets in buildings with water dripping taps will be another application of investigative querying. So, a wide range of applications exists to deploy investigative queries. Different application needs different architecture to support investigative querying.

5 Conclusion

In this paper, we classified the query processing strategies and the types of queries in sensor network literature. We identified newly evolving query types due to deployment of sensors in diverse applications. Urban water distribution with sensor networks was discussed as a motivating application where investigative queries are introduced. The advantages of investigative queries were discussed after formally defining the new class of queries. Few applications that can deploy investigative queries were also listed for further research and investigation under this topic.

References

1. Maróti, M.: Directed flood-routing framework for wireless sensor networks. In: Proceedings of the 5th ACM/IFIP/USENIX international Conference on Middleware, Toronto, Canada, October 18 - 22, 2004. Middleware Conference, vol. 78, pp. 99–114. Springer, New York (2004)
2. Mehta, S., Lee, J., Kim, J.: IS-MAC based flooding protocol for sensor networks. In: Proceedings of the 2nd ACM international Workshop on Performance Evaluation of Wireless Ad Hoc, Sensor, and Ubiquitous Networks, PE-WASUN '05, Montreal, Quebec, Canada, October 10 - 13, 2005, pp. 79–83. ACM Press, New York (2005)
3. Lian, J., Chen, L., Naik, K., Özsu, M.T., Agnew, G.: Localized routing trees for query processing in sensor networks. In: Proceedings of the 14th ACM international Conference on information and Knowledge Management. CIKM '05, Bremen, Germany, October 31 - November 05, 2005, pp. 259–260. ACM Press, New York (2005)
4. Moh, M., Dumont, M., Moh, T., Hamada, T., Su, C.: Brief announcement: evaluation of tree-based data gathering algorithms for wireless sensor networks. In: Proceedings of the Twenty-Fourth Annual ACM Symposium on Principles of Distributed Computing. PODC '05, Las Vegas, NV, USA, July 17 - 20, 2005, p. 239. ACM Press, New York, (2005)
5. Zhu, J., Hung, K., Bensaou, B.: Tradeoff between network lifetime and fair rate allocation in wireless sensor networks with multi-path routing. In: Proceedings of the 9th ACM international Symposium on Modeling Analysis and Simulation of Wireless and Mobile Systems. MSWiM '06, Terromolinos, Spain, October 02-06, 2006, pp. 301–308. ACM Press, New York (2006)
6. Dong, Q., Banerjee, S., Adler, M., Misra, A.: Minimum energy reliable paths using unreliable wireless links. In: Proceedings of the 6th ACM international Symposium on Mobile Ad Hoc Networking and Computing, Urbana-Champaign, IL, USA, May 25-27, 2005, pp. 449–459. ACM Press, New York (2005)

7. Arboleda, C.L.M., Nasser, N.: Cluster-based routing protocol for mobile sensor networks. In: Proceedings of the 3rd international Conference on Quality of Service in Heterogeneous Wired/Wireless Networks. QShine '06, Waterloo, Ontario, Canada, August 07-09, 2006, vol. 191, p. 24. ACM Press, New York (2006)
8. Moussaoui, O., Naïmi, M.: A distributed energy aware routing protocol for wireless sensor networks. In: Proceedings of the 2nd ACM international Workshop on Performance Evaluation of Wireless Ad Hoc, Sensor, and Ubiquitous Networks. PE-WASUN '05, Montreal, Quebec, Canada, October 10-13, 2005, pp. 34–40. ACM Press, New York (2005)
9. Luo, H., Luo, J., Liu, Y.: Energy efficient routing with adaptive data fusion in sensor networks. In: Proceedings of the 2005 Joint Workshop on Foundations of Mobile Computing. DIALM-POMC '05, Cologne, Germany, September 02-02, 2005, pp. 80–88. ACM Press, New York (2005)
10. Madden, S.R., Franklin, M.J., Hellerstein, J.M., Hong, W.: TinyDB: an acquisitional query processing system for sensor networks. ACM Trans. Database Syst. 30(1), 122–173 (March 2005)
11. Madden, S., Franklin, M.J., Hellerstein, J.M., Hong, W.: The design of an acquisitional query processor for sensor networks. In: Proceedings of the 2003 ACM SIGMOD international Conference on Management of Data. SIGMOD '03, San Diego, California, June 09-12, 2003, pp. 491–502. ACM Press, New York (2003)
12. Sharifzadeh, M., Shahabi, C.: Supporting spatial aggregation in sensor network databases. In: Proceedings of the 12th Annual ACM international Workshop on Geographic information Systems. GIS '04, Washington DC, USA, November 12-13, 2004, pp. 166–175. ACM Press, New York (2004)
13. Kalpakis, K., Puttagunta, V., Namjoshi, P.: Accuracy vs. lifetime: linear sketches for approximate aggregate range queries in sensor networks. In: Proceedings of the 2004 Joint Workshop on Foundations of Mobile Computing. DIALM-POMC '04, Philadelphia, PA, USA, October 01-01, 2004, pp. 67–74. ACM Press, New York (2004)
14. Coman, A., Nascimento, M.A., Sander, J.: A framework for spatio-temporal query processing over wireless sensor networks. In: Proceeedings of the 1st international Workshop on Data Management For Sensor Networks: in Conjunction with VLDB 2004. DMSN '04, Toronto, Canada, August 30-30, 2004. DMSN '04, vol. 72, pp. 104–110. ACM Press, New York (2004)
15. Lam, K., Cheng, R., Liang, B., Chau, J.: Sensor node selection for execution of continuous probabilistic queries in wireless sensor networks. In: Proceedings of the ACM 2nd international Workshop on Video Surveillance &Amp; Sensor Networks. VSSN '04, New York, NY, USA, October 15-15, 2004, pp. 63–71. ACM Press, New York (2004)
16. Srivastava, U., Munagala, K., Widom, J.: Operator placement for in-network stream query processing. In: Proceedings of the Twenty-Fourth ACM SIGMOD-SIGACT-SIGART Symposium on Principles of Database Systems. PODS '05, Baltimore, Maryland, June 13-15, 2005, pp. 250–258. ACM Press, New York (2005)
17. Zoumboulakis, M., Roussos, G., Poulovassilis, A.: Active rules for wireless networks of sensors & actuators. In: Proceedings of the 2nd international Conference on Embedded Networked Sensor Systems, SenSys '04, Baltimore, MD, USA, November 03-05, 2004, pp. 263–264. ACM Press, New York (2004)
18. Nesamony, S., et al.: On Sensor Network Segmentation for Urban Water Distribution Monitoring. In: Zhou, X., Li, J., Shen, H.T., Kitsuregawa, M., Zhang, Y. (eds.) APWeb 2006. LNCS, vol. 3841, Springer, Heidelberg (2006)

A Taxonomy-Based Approach for Constructing Semantics-Based Super-Peer Networks

Baiyou Qiao, Guoren Wang, and Kexin Xie

School of Information Science and Engineering, Northeastern University, Shenyang, 110004, China
{qiaobaiyou, wangguoren}@ise.neu.edu.cn

Abstract. Clustering a peer-to-peer (P2P) network into distinct semantic clusters can efficiently improve the search efficiency and enhance scalability of the network. This paper considers P2P systems in which peers employ taxonomy hierarchy to describe the contents of their objects, and presents a taxonomy-based approach for constructing semantics-based super-peer networks. By dynamically clustering peers in taxonomy-based semantic space based on the semantics of their data and organizing the clusters into semantic routing overlays, an efficient query-routing algorithm can be used among these clusters. Preliminary evaluation indicates that our approach achieves a competitive trade-off between search latencies and overheads, and load-balancing among super-peers is well maintained.

1 Introduction

A super-peer network is a hierarchy structure formed by adding super-peer nodes to an unstructured P2P system. It combine elements of both pure and hybrid P2P systems, it has the potential to combine the efficiency of a centralized search with autonomy, load balancing, robustness to attacks and at least semantic interoperability provided by distributed search. Therefore super-peer networks, such as KaZaA [7] and Gnutella's new version [1], have been one of the most popular network applications.

Taxonomy hierarchy is widely used in the real world, for its well conceptual hierarchy structure and supporting for logical reasoning. At present there are many distributed data sources which employ taxonomy hierarchy to describe the contents of their objects, such as PC files, web directories, and digital books and so on. And how can we efficiently exchange and share information among these large-scale distributed data sources which employ a domain taxonomy hierarchy to describe the contents of their objects? An effective solution is to construct a semantics-based super-peer network by taking advantage of semantic information of taxonomy hierarchy and benefits of super-peer networks. Paper [12] proposes a simple method based on unstructured P2P networks, i.e. peers are classified into concepts in the taxonomy hierarchy according to the content of their objects, and each semantic cluster is associated with a concept. This static clustering approach, however, for a super-peer network, can result in severe load imbalance among super-peers, thus

K.C. Chang et al. (Eds.): APWeb/WAIM 2007 Ws, LNCS 4537, pp. 122–134, 2007.

inevitably reducing performance of the system. Paper [15] proposes an approach based on DHT, hashing the whole taxonomy hierarchy to super-peers, and super-peers are connected according to CHORD protocol. However, this approach results in loose semantics of data within a super-peer and makes the communications between super-peers increasing, and thus affects network performance. Other approaches, such as Edutella [6], Hypercup [16], etc. do not consider load-balancing among super-peers. Here this paper proposes a Taxonomy-based Approach (TBA), which considers the characteristics of data semantic space consisting of taxonomy hierarchy, dynamically clusters peers into many semantic clusters based on the semantics of their data and load, and organizes the clusters into semantic routing overlays. Each semantic cluster is composed of data and peers in its semantic sub-space and managed by a super-peer, and each cluster is only responsible for answering queries in its semantic sub-space.

Compared to traditional super-peer networks, the approach employs a source locating strategy based on taxonomy hierarchy, and send queries only to the semantic clusters that possibly have results. Thus, peers involved and messages to send are reduced and network performance is greatly enhanced as a result. Preliminary evaluation shows that the approach achieves a competitive trade-off between search latencies and overheads, load-balancing among clusters and data semantics within a cluster are all well maintained.

The remainder of the paper is organized as follows. System model is provided in section 2, section 3 describes the approach of constructing semantic based super-peer networks in detail, and section 4 gives a search and routing algorithm. Preliminary performance evaluation is presented in section 5. Related works are presented in section 6. Finally we conclude this paper.

2 System Model

2.1 Definition of Taxonomy Hierarchy and Encoding

Taxonomy hierarchy is a model effectively describing data semantics and used widely in various fields of computer science, which has good conceptual hierarchy structure and support for logical reasoning. In the paper, we suppose that there is a common taxonomy hierarchy, and all data are classified and organized according to it. The taxonomy hierarchy represents the semantic space of data on the network, and hence a sub-taxonomy hierarchy stands for a semantic sub-space. A taxonomy hierarchy can be expressed with a tree, called a taxonomy tree, defined as follows.

Definition 1. A taxonomy tree T=<N, E>, where N is a set of all nodes, each node represents a concept, E is the set of directed edges, and an edge represents ISA relationship. Nodes on the tree can have arbitrary fan-outs and leaf nodes can be on different levels.

Each node represents a concept. A concept in a higher layer is more general and contains more children concepts, and a concept in a lower layer is part of those in a higher layer. Each node has a unique ID, and the ID encoding is as follows.

The ID of the root node of the taxonomy tree is prescribed to be 0, and the ID of the parent node is the prefix of its children nodes. Fig.1 is an example for encoding a taxonomy tree. The ID of the root node is 0, and the IDs of its children are 0.1, 0.2, and 0.3. IDs of other nodes are also shown in Fig.1.

The advantage of the encoding is that: when the IDs of two nodes are given, it is very easy to judge their relationship. When one ID is a prefix of another, the two nodes have the ascendant-descendant relationship. The encoding is used later.

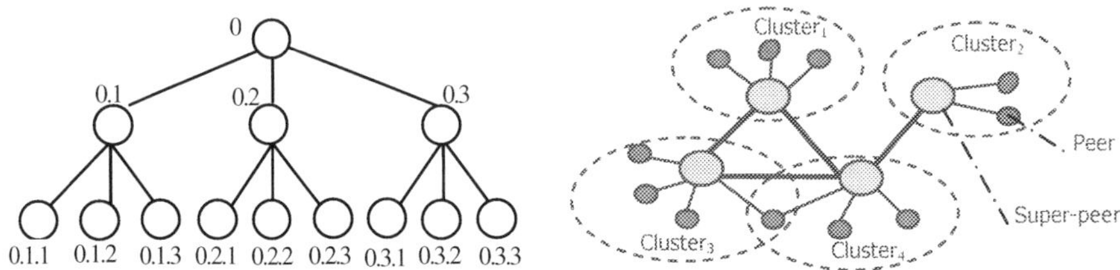

Fig. 1. An example of encoding a taxonomy tree

Fig. 2. The semantics-based super-peer network model

2.2 Semantics-Based Super-Peer Network Model

A semantics-based super-peer network is a hierarchy structure consisting of two layers, the bottom layer is composed of peers and the top layer is composed of super-peers, super-peers are connected into an overlay structure according to their semantics. A set of peers with their super-peer together forms a semantic cluster. Data are stored on peers with data index and routing information of neighbor super-peers stored on super-peers. Intra semantic cluster data communication takes place via direct peer-to-peer links; inter semantic clusters communication takes places via links between super-peers. Each semantic cluster uses a common taxonomy tree to describe its data semantics, and takes charge of a semantic sub-space, comprised of one or multi sub-trees of the taxonomy tree. If the data on each peer are classified into n nodes of the taxonomy sub-trees, there are n index items, and all index items are organized and stored in the form of one or multi sub-trees. Each cluster is only responsible for answering queries in its semantic sub-space. Each query is associated with a node of the taxonomy tree and the node express semantics of the query. The query is routed according to its semantics, so the peers involved and messages to send are reduced and network performance is greatly enhanced. Fig. 2 is the model of semantics-based super-peer network.

3 Approach for Constructing Semantics-Based Super-Peer Networks

In this section the approach for constructing semantics-based super-peer network (TBA) is described in detail from four aspects.

3.1 Constructing Strategy

We apply a partitioning strategy of the semantic space to construct the network, which a preset maximal cluster load size M is used to determine the clusters ranges. The chief idea is: suppose that there is one semantic cluster in the network initially, whose semantic space is the whole taxonomy tree; the load size of the cluster increases with peers joining, if the load size of the cluster exceeds M, the cluster will automatically partition its semantic sub-space according to the data semantics and the load size, and a new semantic cluster is generated from the original one. The new semantic cluster selects appropriate semantic clusters as its neighbors, and answers queries in its semantic sub-space. In the way each cluster continuously partitions its semantic sub-space and a semantics-based adaptive super-peer network forms. The advantage of the approach is that the load size within a cluster is less than the maximal load size M, and hence search efficiency of the system is guaranteed. Therefore, the approach is adaptive to difference of density of peers in the semantic space, and ensures stability and adaptability of the system.

3.2 Clustering Algorithm Based on Taxonomy Hierarchy

In a super-peer network, any super-peer, as an index server, takes charge of vast work such as maintaining index data, answering and routing queries, etc., and thus the capability of super-peers decides the network performance. Therefore, we regard the load of a super-peer as the load of the cluster, which is the important basis of clustering. Since maintenance and search take a super-peer a lot of time, the load of a super-peer can be measured by the number of index items. Thus we can express a semantic cluster with a weighted taxonomy sub-tree, the weight of a node representing the number of index items, and topological relationship between nodes represents semantic relationship between the corresponding index items. Now clustering problem can be regarded as partitioning a set of weighted taxonomy sub-trees. When clustering, we should consider not only the semantics between index items but also the load-balancing among super-peers, and in fact, it is a tradeoff between load-balancing among clusters and semantics within a cluster. Aimed at the problem, we propose a Self-organized Semantic Clustering Algorithm, i.e. LBFCA (Load Balance First Semantic Cluster Algorithm) in paper [19], which achieves a good trade-off between data semantics within a cluster and load-balancing among clusters. LBFCA algorithm obeys the following three rules:

1. If the original cluster is composed of a sub-tree, the new cluster generated consists of one or multi sub-trees of a node on the original tree, which makes the difference of load sizes between the two clusters minimal;
2. If the original cluster is composed of multi sub-trees, the sub-trees are split into two groups, either of which forms a cluster, which makes the difference of load sizes between the two clusters minimal;
3. If the original cluster is composed of one node on the taxonomy tree, the new cluster consists of the same node, and the two clusters share the load of the original one.

As shown in Fig. 3, SC_0 is a semantic cluster, which can be expressed with a weighted taxonomy tree, where a weight of a node represents the number of index items of the corresponding concept and the sum of weights of all nodes is the load size of the cluster SC_0, i.e. 120. If we assume that maximal cluster load size M is 120, the load size of SC_0 reached M, and SC_0 should automatically execute clustering algorithm, after applying LBFCA, a new cluster SC_8 is produced. As shown in Fig. 4,

SC_0 is composed of a sub-tree whose root ID is 0, and SC_8 is composed of two sub-trees whose root IDs are 0.2 and 0.3 respectively. Detailed description of LBFCA is in paper [19].

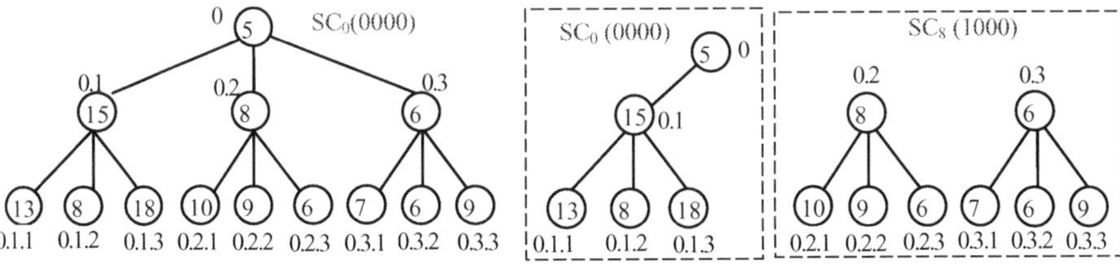

Fig. 3. An example of a semantic cluster

Fig. 4. The clustering result by LBFCA

3.3 Cluster Encoding Strategy

On the network, each cluster has a unique ID, representing the location of the cluster in the whole semantic space and being the basis of establishing neighborhood. This paper offers a cluster encoding algorithm, which automatically generates an ID for a new cluster. The system uses a binary number of *m* bits as a cluster ID, and each cluster maintains a variable Par_times to record the partitioning time (used for generating a new ID). The first cluster ID is predefined as 0 and Par_times as 0. When the cluster is partitioned with peers joining, the newly generated cluster obtains a new ID that equals the ID of the original cluster plus $2^{m\text{-Par\_times-1}}$, and Par_times is increased by one. When Par_times is more than *m*, partitioning can no longer be done, and therefore *m* should be large enough. The cluster encoding algorithm is Algorithm 1.

Algorithm 1. Algorithm for cluster encoding

Input: ID_{old}: ID of the original cluster; Par_times : partitioning times;
m: length of a cluster ID

Output: ID_{new}: ID of the new cluster

if Par_times<*m* then
  $ID_{new} = ID_{old} + 2^{m\text{-Par\_times-1}}$
  Par_times= Par_times+1
End if

For example, we assume that the length of cluster codes m = 4, i.e. binary numbers have 4 bits. At the beginning there is one cluster in the system shown in Fig. 3, whose semantic space is the whole taxonomy tree. According to the cluster encoding strategy, its cluster ID is 0, denoted by SC_0, Par_times = 0, the ID of the new cluster generated by the first time partitioning is $0 + 2^{4-1} = 8$, denoted by SC_8 shown in Fig. 4, and Par_times changes into 1, kept in the two clusters prepared for the next partitioning.

3.4 Neighborhood Establishment Among Clusters

Neighborhood establishment among clusters exerts significant effluence on P2P network performance and is directly related to query efficiency and network scalability. On a super-peer network, super-peers are responsible for maintaining and dealing with routing among clusters. In the system, each super-peer maintains a cluster information table that stores relevant status and routing information. The table is shown in Table 1, there are five fields: Cluster_ID is the cluster ID, Par_times is the partitioning time, Cluster_size is the load size of the cluster, Cluster_range is current semantic sub-space of the cluster, represented by a set of IDs of root nodes of taxonomy sub-trees, and Network_range is initially semantic sub-space of the cluster when the cluster is generated, represented by a set of IDs of root nodes of taxonomy sub-trees, representing semantic sub-spaces of the cluster and those partitioned apart from it. Obviously Cluster_range is a subset of Network_range. Neighbor_list is a list of neighbor clusters, comprised of Network_range's and Cluster_ID's of neighbor clusters. In the system, routing information among super-peers is dynamically built when partitioning clusters according to cluster IDs. If there is one and only one bit difference between two cluster IDs, the two clusters are neighbors. A cluster and a new cluster partitioned from it must be neighbors according to cluster encoding, it can effectively reduce routing hops when searching, which decreases communicating cost of the system and enhances search performance as a result.

Table 1. Information table of a cluster

Cluster_id	Network_range
Cluster_range	Cluster_size
Par_times	Neighbor_list: { Network_range, Cluster_id}

The process of adding neighbors is: when a cluster is partitioned, the new cluster and the original one are neighbors, corresponding information is added to their cluster information tables, and the new cluster broadcasts its ID to network, and finds clusters whose IDs has one and only one bit difference from its ID. If such clusters are found within a given hops, corresponding information is added to the respective cluster information tables and thus neighborhood is established.

For example, as shown in Fig. 4, clusters SC_0 and SC_8 continue being partitioned with load increase, SC_{12} is partitioned from SC_8, SC_4 from SC_0, SC_6 from SC_4, and finally obtains the result (Fig. 5). After several times of partitioning, the original cluster is partitioned into five clusters. Accordingly a taxonomy tree is partitioned into several taxonomy sub-trees. According to the mechanism of neighborhood establishment, the neighborhood among the five clusters and their cluster information are shown in Fig. 6. For instance, the cluster SC_8, whose Cluster_ID is 1000_2, Par_times is 2, and Cluster_size is 52. At the time the cluster is generated, the cluster consists of two taxonomy sub-trees whose root IDs are 0.2 and 0.3, and thus its Network_range is the set {0.2, 0.3}, and now the cluster is comprised of the sub-tree whose root ID is 0.2, so its Cluster_range is {0.2}. Since SC_0 generates SC_8 and SC_8 generates SC_{12}, SC_8 has two neighbor clusters SC_{12} and SC_0. Cluster information of other clusters is shown in Fig. 6.

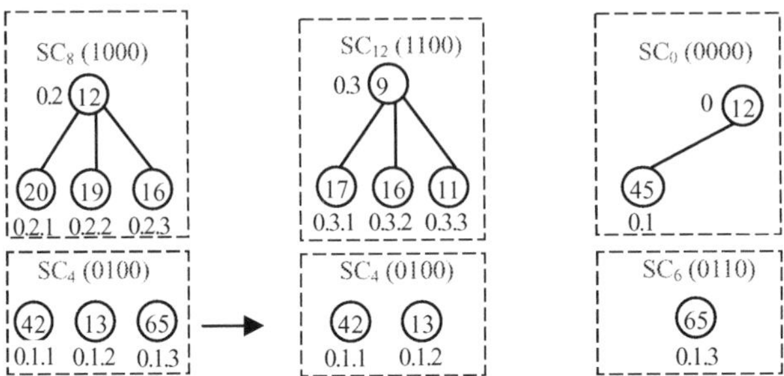

Fig. 5. Results after re-clustering based on the results in Fig. 4

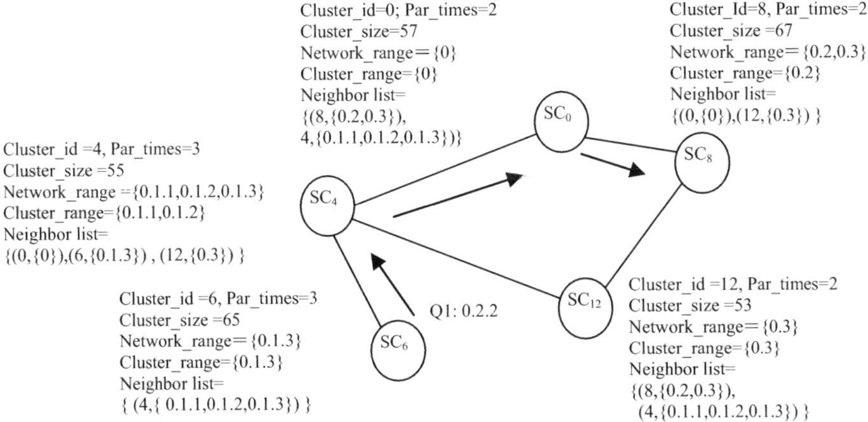

Fig. 6. Neighborhood of the five clusters

4 Query and Network Maintenance

4.1 Query-Routing Algorithm

Since semantics of data and query is described with taxonomy hierarchy, each query is associated with a concept of the taxonomy hierarchy, which is corresponding to a node on the taxonomy tree, the node express semantics of the query; we name ID of the node as Search ID (SID). According to structural characteristics of the taxonomy tree, when a node is the searching target of a query, its ascendants and descendants are searching targets as well, and hence the query is sent to the clusters that contain the target nodes. In this paper, the ascendant-descendant relationship is judged by node IDs of the taxonomy tree. The query-routing process is as follows.

When a peer in the cluster A sends a query Q to its connected super-peer, the super-peer obtains the SID from Q's semantic information. If SID has ascendant-descendant relationship with some ID(s) in A's Cluster_range, Q is executed by finding all peers in A's index that can answer Q, sending Q to these peers to execute and returning results to the peer that initially sends Q. At the same time, it is checked whether in each Network_range in A's Neighbor_list, there is some ID of ascendant-descendant relationship with SID. If so, Q is sent to these neighbor clusters. If in A's Cluster_range and any Network_range in the Neighbor_list, there is no node whose ID has ascendant-descendant relationship with SID, Q is forward to the neighbor cluster B that generates A, and then B continues to process Q. Algorithm 2 shows how to query and route.

As shown in Fig. 6, a peer in the cluster SC_6 sends a query Q_1 whose SID is 0.2.2. According to Algorithm 2, the routing path is shown in the form of arrows in Fig. 6. Since no ID in SC_6's Cluster_range and any of Network_range in its Neighbor_list have ascendant-descendant relationship with the SID, Q_1 is routed to SC_4, which generates SC_6. The ID 0 in Network_range in SC_4's Neighbor_list has ascendant-descendant relationship with the SID, and then Q_1 is forward to the neighbor SC_0. According to its Cluster_range, SC_0 is recognized as one of destinations of Q1, and meantime Q_1 is forwarded to SC_8 according to the Network_range's in SC_0's Neighbor_list, and thus SC_8 is another destination of Q_1.

Algorithm 2. Query-routing algorithm

Input: Cid: cluster ID, SID

```
If received Search(SID) before then
  Drop Search(SID);
  Return;
end if
Search_token=0;
If exist a ascendant-descendant
relationship between SID and a ID in
cluster_range then
  local_index_search(SID);
  Search_token=1;
End if
For any Network_range NR in
neighbor_list do
  If exist a ascendant-descendant
  relationship between SID and a ID in
  NR then
    Forward Search(SID) to neighbor
    Cluster_id
    Search_token=1;
  End if
End for
If Search_token!=1 then
  X=Cid XOR (2^m-1);
  J=[x OR (x+1)] XOR (2^m-1);
  Forward Search(SID) to neighbor
  cluster J;
End if
```

4.2 Network Maintenance

1. **Peer join.** When a peer P joins, he should find an existing cluster C in the network firstly, and send semantic index of its Data to C, C routes the semantic index in the overlays, and finally the semantic index is stored on the cluster B according to its semantics. B is responsible for saving connection and other relevant information of P, and meantime B joins P. This is how peer joins
2. **Peer leave.** When a peer P leaves, P sends a leaving request to its connected super-peer and then it leaves. The super-peer deletes relevant information and modifies Cluster_size value in the cluster information table.
3. **Peer failure.** The super-peer of a cluster periodically checks availability of peers it manages. If a peer fails, it is labeled with temporary failure status in the corresponding index, and if temporary failure status remains over a certain period of time, the peer is regarded to leave and the same thing is done as peer leave.

5 Performance Evaluation

5.1 Experimental Environment Setup

In order to verify validity of TBA, we evaluate the performance using simulating experiment. The experimental environment is P4 3.0GHz, 512M memory, and 80G hard disk, the operating system is Windows server 2003 and JDK 5.0 as the programming environment to develop the stimulating test software. In the experiment synthesized data are employed to test performance of the algorithm. There are many taxonomy trees of 2-10 layers composed of 200-100K concepts (nodes) of 2-30 fan-outs. For calculation convenience, it is assumed that each peer relates to only one index item, and thus the load size of a cluster can be viewed as the number of peers. For universality, distribution of peers on the taxonomy tree follows the Zipf-like (α=0.8) distribution. At the beginning there is one cluster and one super-peer in the system. With peers continuously joining, the cluster is automatically partitioned when the maximal cluster load size M is exceeded.

5.2 Experimental Results

In the paper, experimental research focuses on three aspects: network scalability, average similarity of data within a cluster and load skew degree. Comparison with a representative constructing approach, the DHT-based approach [15] is made. Experimental results are detailed as follows.

(1) Network scalability

Network scalability mainly evaluates how much query efficiency and cost change when the network size varies. In this paper the average query hops (AQH) is used to evaluate query efficiency and average message number per query (AMNPQ) to evaluate query cost. We construct a 4-layered taxonomy tree composed of 2808 concepts. When the maximal cluster load size M is 100 and 10 queries are sent randomly from each peer, the AQH with network size varying are shown in Fig. 7. It can be seen AQH increases slowly with network size increasing, and the AQH of the TBA is lower than that of DHT-based method. For TBA clusters peers based on their data semantics and routes semantically, and thus reduces the communication times inter clusters, the AQH consequently decreases. The DHT-based method can not

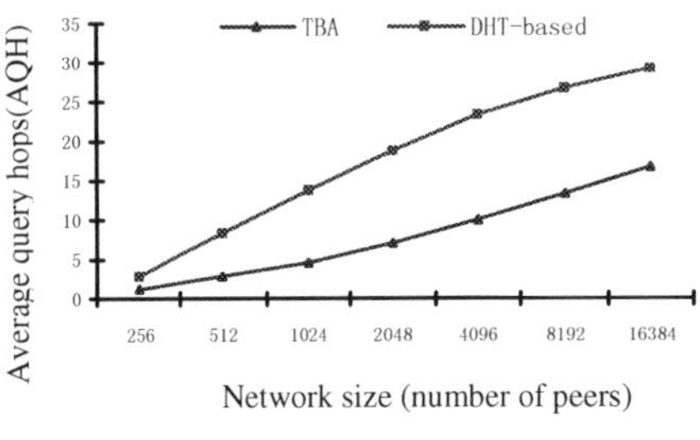

Fig. 7. Comparison of AQH with the network size varying based on two different approaches

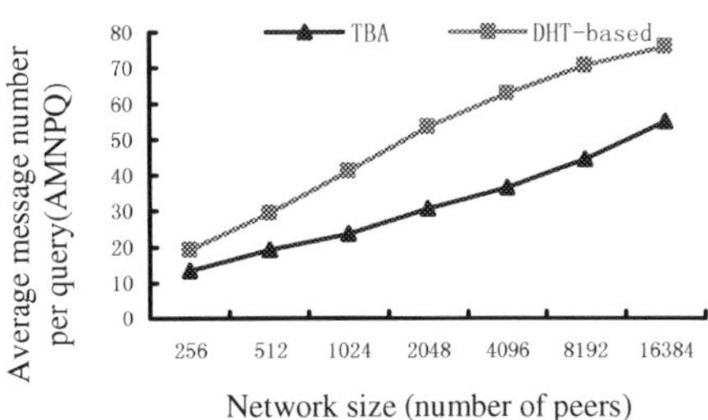

Fig. 8. Comparison of AMNPQ with the network size varying based on two different approaches

support semantic clustering and thus data semantics in a cluster is loosen, so the AQH is higher. Fig. 8 shows how the AMNPQ change in the same condition as Fig. 7. It can be seen that the AMNPQ slowly increase with peers increasing almost in consistent with changing of AQH. From the two aspects, it can be seen that TBA is better.

(2) Average similarity of data within a cluster
In this paper average similarity of data within a cluster is used to evaluate semantic similarity of data in a cluster after partitioning. Average similarity of data is denoted by Sim, shown in Formula (1)

$$\mathrm{Sim} = \frac{1}{n}\sum_{i=1}^{n} 1/D_i \quad (1)$$

Where n is the number of clusters on the network, D_i is the average distance of roots of all taxonomy sub-trees in the i-th cluster. The distance of two nodes on a taxonomy tree equals the number of the edges from the two nodes to their first common ascendant node. It is obvious that the larger D_i is, the larger the distance of the nodes is, the less the similarity is, and the smaller Sim is. On the other hand, the smaller D_i is, the less the distance of the nodes is, the more the similarity is, and the larger Sim is. For a cluster composed of one taxonomy sub-tress, since there is only one root, we predefine D_i is 1. Fig.9 shows comparison of average similarities of data within a cluster generated in two approaches in the same condition as Fig.7. It can be seen that the average similarity of data within clusters by TBA is higher than that by the DHT-based approach. With the network size varying, the two average similarities grow larger and closer to each other, both approximating to 1. This indicates that TBA can well maintain semantic relationship of data within a cluster and the average similarity grows more with the network size increasing.

(3) Load skew degree
Load skew degree represents how load skews among clusters, calculated by the standard deviation of loads among clusters, denoted by LSD in Formula (2)

$$\mathrm{LSD} = \sqrt{\sum_{1}^{n} (x_i - \overline{x})^2 / n} \quad (2)$$

Where x_i is the load of the i-th cluster and $\overline{x}$ is the average load of all clusters. Obviously the larger LSD is, the larger the difference of cluster loads is, indicating the less balance clusters have, whereas the less LSD is, the less the difference of cluster loads is, indicating the more balance clusters have. Fig.10 shows comparison of Load skew degrees of clusters by the two approaches in the same condition as Fig. 7. Load skew degrees by the two approaches fluctuate regularly and both relatively slightly, and with network size increasing, the two LSDs fluctuate gradually less and approach each other. In view of load skew degree, both approaches maintain load-balancing among clusters, TBA is better than DHT-based approach, but the advantage of TBA over DHT-based approach becomes less and less with network size increasing.

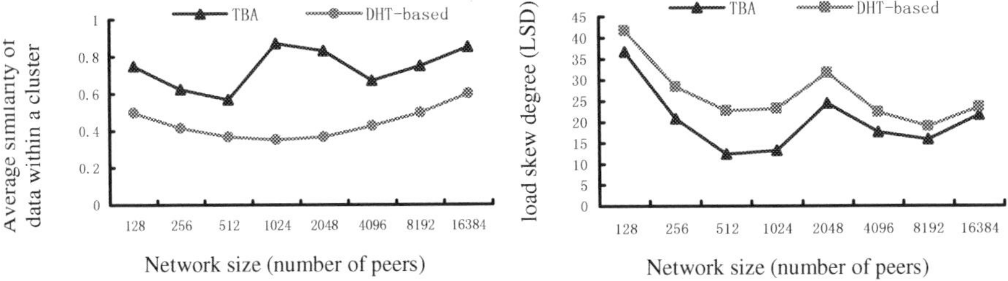

Fig. 9. Comparison of average similarity of data within a cluster based on two approaches

Fig. 10. Comparison of load skew degrees (LSD) based on two approaches

6 Related Works

Much research has been focused on improving search efficiency and scalability in P2P systems. Several techniques have been proposed to improve search in unstructured P2P systems, such as constructing routing indices of information of related neighbors on each peer [8,9], using of cache [10,11], using histograms to direct queries [2,18], etc. In structured P2P systems, most of research constructs a system based on distributed hash tables, such as CHORD[3], PASTY[4], CAN[5] , etc. Another effective method to improve query performance is partitioning the network into distinct semantic clusters and constructing SONs (Semantic Overlay Networks), each of which is a semantic cluster. Queries are merely forwarded to the SON that probably contains answers. The advantage of this approach is that it reduces the search cost and enhances search efficiency and scalability of the system. The concept of SONs is firstly proposed in paper [12], where SONs are constructed by taxonomy hierarchy, peers are distributed on a concept in taxonomy hierarchy and the concept is regarded as a SON, however, without providing how to manage these SONs; paper [13] proposes a dynamic clustering method in unstructured systems based on multidimensional semantic space, not suitable for taxonomy hierarchy semantic space; paper [14] proposes a method of creating SONs in structured P2P networks, which uses LSI to produce semantic vectors of documents, which are stored in an CAN, and uses nearest neighbor query technology; paper [15] proposes an education resource sharing model based on super-peer, in which peers are classified by taxonomy hierarchy, and taxonomy hierarchy is distributed on super-peers by hash functions, which results in loose semantics of peers in one cluster and hence increases search cost; paper [16] proposes a routing and clustering strategy based on routing indices in super-peer networks, in which concepts and schemas are used to cluster peers without considering load-balancing among super-peers; paper [17] proposes a clustering architecture based on super-peer networks, in which the semantic schema of every cluster is predefined and peers are distributed into several fixed clusters according to the match between peers and clusters.

7 Conclusions

In this paper, we focus on how to construct a semantics-based P2P network using semantic information contained by a domain taxonomy hierarchy, and propose a self-organized approach based on the taxonomy tree. We also give relevant algorithms and make comparison with the DHT-based approach. Experiments show that our approach has low search latencies and overheads, and load-balancing among super-peers is well maintained. Future work focuses on the dynamical load-balancing algorithms among clusters and maintenance approaches in the case of super-peer leave and failure based on improving clustering and the query-routing algorithm.

Acknowledgments. This research was supported by National Basic Research Program of China (2006CB303103), the National Natural Science Foundation of China (Grant No. 60473074 and 60573089), Cultivation Fund of the Key Scientific and Technical Innovation Project (Grant No. 706016) and Natural Science Foundation of Liaoning Province (Grant No. 20052031).

References

1. Gnutelliums, LLC.: Gnutella Protocol Specification Version 0.4, 2004-02-18 (2004), http://www.limewire.com/developer/gnutella-protocol-0.4.pdf
2. Löser, A., Tempich, C.: On Ranking Peers in Semantic Overlay Networks. In: Althoff, K.-D., Dengel, A., Bergmann, R., Nick, M., Roth-Berghofer, T.R. (eds.) WM 2005. LNCS (LNAI), vol. 3782, Springer, Heidelberg (April 2005)
3. Stoica, I., Morris, R., Karger, D., Kaashoek, M.F.: Chord: A scalable peer-to-peer lookup service for internet applications. In: Proc. ACM SIGCOMM (August 2001)
4. Rowstron, A., Druschel, P.: Pastry: Scalable, distributed object location and routing for large-scale peer-to-peer systems. In: Proc. of the 18th IFIP/ACM International Conference on Distributed Systems Platforms (Middleware 2001) (November 2001)
5. Ratnasamy, S., Francis, P., Handley, M., Karp, R., Shenker, S.: A scalable content-addressable network. In: Proc. ACM SIGCOMM (August 2001)
6. Nejdl, W., Wolf, B., Qu, C., et al.: EDUTELLA: a P2P Networking Infrastructure based on RDF. In: WWW 11 Conference Proceedings, Hawaii, USA, (May 2002)
7. KaZaA., http://www.kazaa.com
8. Pireddu, L., Nascimento, M.: Taxonomy-based routing indices for peer-to-peer networks. In: Proceedings of the SIGIR Workshop on P2P IR (2004)
9. Tempich, C., Staab, S., Wranik, A.: REMINDIN': Semantic Query Routing in Peer-to-Peer Networks based on Social Metaphors. In: Proceedings of the 13th International Conference on World Wide Web (WWW'2004) (2004)
10. Bhattacharjee, et al.: Efficient Peer-to-Peer Searches Using Result-caching. In: Proceeding of the 2nd IPTPS. (2003)
11. Wang, C., Xiao, L., Liu, Y.: Distributed Caching and Adaptive Search in Multilayer P2P Networks. In: Proceedings of ICDCS'04 (2004)
12. Crespo, A., Garcia-Molina, H.: Semantic Overlay Networks for P2P Systems. Technical report, Stanford University (2002)

13. Li, M., Lee, W.C., Sivasubramaniam, A.: Semantic Small World: An Overlay Network for Peer-to-Peer Search. In: Proceedings of the International Conference on Network Protocols (October 2004)
14. Tang, C., Xu, Z., Dwarkadas, S.: Peer-to-Peer Information Retrieval Using Self-Organizing Semantic Overlay Networks. In: Proceedings of SIGCOMM'03 (2003)
15. Loeser, A.: Taxonomy-based Overlay Networks for P2P Systems. In: Proceedings of IDEAS'04 (2004)
16. Nejdl, et al.: Super-Peer-based Routing and Clustering Strategies for RDF-based P2P Networks. In: Proceedings of WWW'03 (2003)
17. Eisenhardt, M., Mueller, W., Henrich, A.: Classifying Documents by Distributed P2P Clustering. GI Jahrestagung, pp. 286–291 (2003)
18. Petrakis, Y., Koloniari, G., Pitoura, E.: On Using Histograms as Routing Indexes in Peer-to-Peer Systems. In: DBISP2P (2004)
19. Qiao, B., Wang, G., Xie, K.: A self-organized Semantic Clustering Approach for Super-peer Networks. In: WISE2006 (October 2006)

A Comparative Study of Replica Placement Strategies in Data Grids

Qaisar Rasool[1], Jianzhong Li[1,2], George S. Oreku[1], Ehsan Ullah Munir[1], and Donghua Yang[1]

[1] School of Computer Science and Technology, Harbin Institute of Technology, China
[2] School of Computer Science and Technology, Heilongjiang University, China
qrasool@yahoo.com, lijzh@hit.edu.cn, gsoreku@yahoo.com, ehsan_munir@yahoo.com, yang.dh@hit.edu.cn

Abstract. Data Grids are today's emerging infrastructure providing specialized services on handling large datasets that needs to be transferred and replicated among different grid sites. Data replication is an important technique for data availability and fast access. In this paper we present a comparison of various replication models and techniques employed by some major topologies used in data grid environment. We focus on dynamic strategies for replica placement in tree, Peer-to-Peer (P2P) and hybrid architectures. Beside tree model which is being implemented for many Data Grid applications, hybrid and P2P grid models of replication are also emerging for providing scientific communities with better availability and efficient access to massive data.

1 Introduction

Grid computing is a wide-area distributed computing environment that involves large-scale resource sharing among collaborations, referred to as Virtual Organizations, of individuals or institutes located in geographically dispersed areas. A Data Grid [1] provides services that help users discover, transfer and manipulate large datasets stored in distributed repositories and also, create and manage copies of these datasets. Examples of the scientific applications dealing with huge amounts of data and the potential beneficiaries of Data Grid technology are high energy physics, astronomy, bioinformatics, and earth sciences.

Replication of data means the creation and maintenance of copies of data at multiple computers. Replication is a key to the effectiveness of distributed systems in that it can provide enhanced performance, high availability and fault tolerance. Grid being a specialized form of distributed system is going to get benefit from replication and possibilities are continually explored by the research community. In this paper we present a brief review of current and past research on replication techniques. Specifically, we focus on data replica placement policies proposed for use in the data grid environment. For each replica placement technique, we consider its methodology, objective (or performance metric), and results. In the presence of diverse and varying characteristics of tree and other architectures it is difficult to create a common ground for comparison of different replication strategies. We,

K.C. Chang et al. (Eds.): APWeb/WAIM 2007 Ws, LNCS 4537, pp. 135–143, 2007.

therefore, group topologies into tree and hybrid/P2P architectures and analyze the impact of replica placement policies in each one. A hybrid topology can carry features of both tree and P2P architectures and thus can be used for better performance of a replication strategy.

The paper is organized as follows. Section 2 describes some widely used grid topologies and their implications. In section 3, we take a closer look on data replication in grid environment and how different replica placement strategies work under analogous topologies. Section 4 concludes the paper.

2 Grid Topologies

In this section we present an overview of major grid topologies. We use terms tree, hierarchical, and multi-tier to refer the same topology. Other topologies considered are P2P and hybrid.

A hierarchical or tree model is used where there is a single source for data and the data has to be distributed among collaborations worldwide. For instance, the data grid envisioned by the GriPhyN project is hierarchical in nature and is organized in tiers [2]. The source where the data is produced is denoted as Tier 0 (e.g. CERN). Next are the Tier 1 national centers, the Tier 2 regional centers (RC), the Tier 3 workgroups and finally Tier 4, which consists of thousands of desktops. Such a multi-tier Data Grid has many advantages [3]. Firstly, it allows the scientific community to access the resources in a common and efficient way. Secondly, the datasets can be distributed to appropriate resources and accessed by multiple sites. The network bandwidth can be used efficiently because most of the data transfers involve local or national network resources, hence alleviating the workload of international network links. Thirdly, with the support of the Grid middlewares, the resources located in different centers can be utilized to support data-intensive computing. The multi-tier Data Grid model is being implemented for many scientific applications, the High Energy Physics (HEP) being the most prominent one.

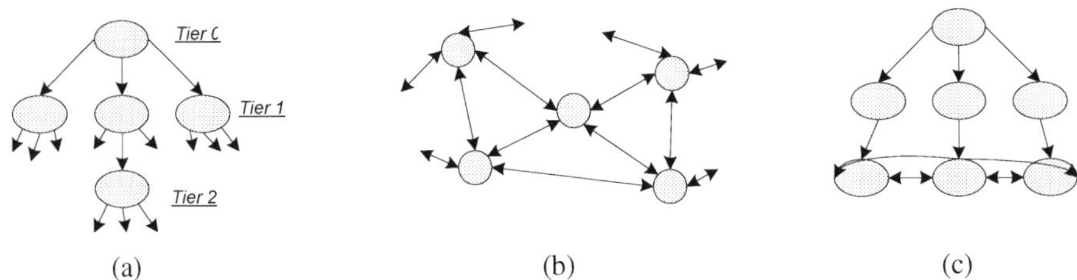

Fig. 1. Data grid topologies (a) tree, (b) P2P, and (c) hybrid

A tree topology also has shortcomings. The tree structure of the grid means that there are specific paths the messages and files can travel to get to the destination. Further, data transference is not possible among sibling nodes or nodes situated on the same tier. Peer-to-Peer (P2P) systems overcome these limitations and offer flexibility in communication among components. A P2P system is characterized by the applications that employ distributed resources to perform functions in a decentralized manner. From the viewpoint of resource sharing, a P2P system overlaps a grid system. The key characteristics that distinguishes a P2P system from other resource sharing systems is its symmetric communication model between peers, each of which acts as both a server and a client. The unique characteristics of P2P make the Data Grid more scalable and flexible to deal with the distributed data and replication.

With the needs for researchers to collaborate and share products of their analysis, new hybrid models are emerging for Data Grids which utilize the combined benefits of tree and other architectures such as ring and P2P. A hybrid model of a hierarchical Data Grid with peer linkages at the edges is shown in Fig.1(c).

3 Data Replication in Grid

Replication is used in distributed systems to improve data access efficiency and fault tolerance. Fig.2 shows pros and cons of data replication. The general idea is to store copies of data in different locations so that data can be easily recovered if one copy of data is lost. Also, placing data replicas closer to user can improve data access performance significantly. A replica is an exact copy of a file that is linked to the original file through some well-defined mechanisms [4]. In Grid environment, replica management services do not enforce any semantics to (multiple) replicas of a file and thus a replica takes the form of a user-asserted correspondence between two physical files [1]. Data replication is characterized as an important optimization technique in Grid for promoting high data availability, low bandwidth consumption, increased fault tolerance, and improved scalability.

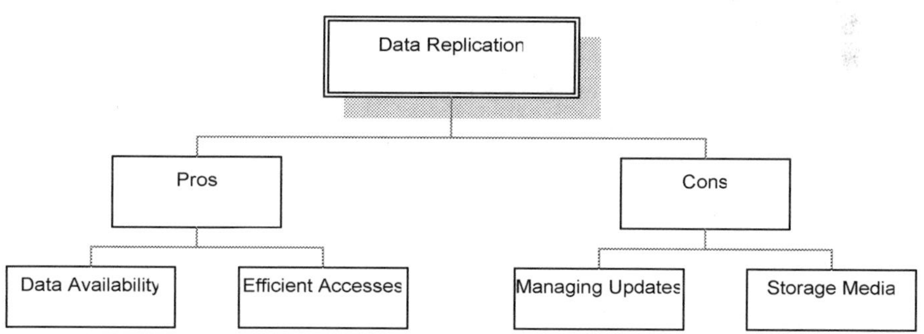

Fig. 2. Pros and cons of data replication in a generic distributed system

The goals of replica optimization is to minimize file access times by pointing access requests to appropriate replicas and pro-actively replicating frequently used files based on access statistics gathered. In general, replication mechanism determines which files should be replicated, when to create new replicas, and where the new replicas should be placed.

Replication schemes can be classified as static and dynamic. In static replication, a replica persists until it is deleted by users or its duration is expired. The drawback of static replication is evident, when client access patterns change greatly in the Data

Grid, the benefits brought by replica will decrease sharply. In contrast to static replication, dynamic replication automatically creates and deletes replicas according to changing access patterns, and thus ensures that the benefits of replication continue even if user behavior changes. Dynamic replication strategies tend to reduce hotspots created by popular data and facilitate load sharing. The benefits achieved, however, incur extra overhead costs in the form of real-time assessment of network traffic and maintaining efficient system logs to handle the dynamicity of grid environment.

3.1 Replica Placement in Tree Topology Grids

The manner in which replicas are placed in the grid nodes follows two approaches. The first approach is to distribute data from top to bottom starting from root node to lower tier nodes. The other technique is to place replicas in bottom up fashion. There is a variation in the consideration of read/write costs before making replication decision. We take both approaches into consideration in our discussion. As most of the data in Data Grids are read-only [5, 6], we ignore replica update and consistency issue.

We start with the initial work [7] on replication strategies that are proposed and simulated for a hierarchical Data Grid similar to Fig.1(a). These techniques are briefly described as follows: 1) No Replication or caching: all data held at root (the base case against which other strategies are evaluated); 2) Best Client: The access history of files is maintained and a replica of a file is created at the client node who has generated most number of requests for that file; 3) Cascading Replication: when file access exceeds a certain threshold, a replica of file is created at immediate lower tier node on the path to the best client; 4) Plain Caching: the client that requests a file stores a copy locally; 5) Fast Spread: When a client requests a file, a copy (replica) is stored at each node on the path to the client. Another policy named Cascading plus Caching was also proposed in which a client node can act as server for its siblings. We discuss this strategy in the next section. It is important to note the difference between replication and caching. In case of replication, the decision to make a replica and sending it to other node is solely taken by the server. In case of caching, a client requests a file and server sends a copy of the file to be stored in the local cache of client. We can say that caching is a client-side phenomenon whereas replication is a server-side phenomenon.

The above algorithms consider three different patterns of data access: 1) Random (no locality in data); 2) Access pattern with assumption that popular files at one site are popular at others (temporal locality); 3) Access pattern with little temporal locality and assumption that files recently accessed by a client are likely to be accessed by nearby clients (geographical locality). In order to capture the dynamics of the access pattern, the history statistics (popularity logs) are cleared at specified time intervals. The effectiveness of a replication strategy is related to how well the time interval is tuned to the access behavior (i.e. interval checking); another parameter is the threshold (only if number of requests exceed this threshold is the file replicated).

Among these strategies, Fast Spread shows a relatively consistent performance and best in case of random access pattern. When locality is introduced, Cascading offers good results. In contrast, Best Client presents the worst case among all. For locating the nearest replica each model/algorithm selects the replica server site that is at the least number of hops from the requesting client.

An improvement in Cascading technique is proposed namely Proportional Share Replica policy [8]. The method is a heuristic one that places replicas on the optimal locations by assuming that number of sites and the total replicas to be distributed are already known. Firstly an ideal load distribution is calculated and then replicas are placed on candidate sites that can service replica requests slightly greater than or equal to that ideal load. The formula for calculating ideal load is:

$$Load = Totalrequests/(Originalcopy + Replicas)$$

Table 1. Merits and demerits of replica placement techniques in tree topology Grids

Authors	Technique	Methodology/ Realization	Results/Comments	TD /BU*	Objective/ Performance metric
Kavitha and Foster [7] 2001	Best client	A replica created at node who generates max requests	- worst performance - not suitable for grid	TD	Faster avg response time and bandwidth conservation
	Cascading	Replica trickles down to lower tier if no. of requests exceeds threshold	+ better for small degree of locality - Bad for random access pattern		
	Caching	A requesting client receives and stores a copy locally	+ almost similar performance as cascading - relatively high response time		
	Fast Spread	A replica is stored at each node along the path toward client	+ consistent performance + best for random access pattern - high storage req, I/O, CPU load		
Abawajy [8] 2004	Proportional Share Replication	Calculating an ideal workload and distributing replicas	+ better results over Cascading + load sharing among replica servers	TD	Mean response time
Ming Tang et al. [3] 2005	Dynamic Replication Algorithms	Simple Bottom Up (SBU), Aggregate Bottom Up (ABU)	+ better results over Fast Spread strategy	BU	Response time, bandwidth cost, replication freq
Rahman et al. [9,10] 2005, 2006	Multi-objective approach	Solved as p-facility problem to get sites for replica placement	+ user requests and currentnetwork status + dynamic maintainability when performance metric degrades	TD	Request-weighted average response time

*TD = Top-down, BU = Bottom-up.

An intensive work in the form of dynamic replication algorithms is presented by Tang et al [3]. Nodes at the intermediate tiers (i.e. tiers between root and client nodes)

act as the replica servers, each running a Local Replica Manager (LRM). A Dynamic Replication Scheduler (DRS) makes the replication decision after gathering file access statistics. The DRS uses two algorithms, Simple Bottom-Up (SBU) and Aggregate Bottom-Up (ABU). With SBU, replicas are created as close as possible to the clients with requests rates exceeding the predefined threshold. The algorithm ABU takes into account the access history of files used by the sibling nodes and aggregates the access record of similar files so that these frequently accessed files are replicated first. The performance of algorithms was evaluated and improvements shown against Fast Spread dynamic replication strategy. The values for interval checking and threshold were based on data access arrival rate, data access distribution and capacity of the replica servers.

A method exploiting Operations Research techniques is proposed in [9,10] for replica placement. In this method, named multi-objective approach, replica placement decision is made considering both the current network status and data request pattern. The problem is formulated in *p*-median and *p*-center models to find the *p* replica placement sites. The *p*-center problem targets to minimize the max response time between user site and replica server whereas the *p*-median model focuses on minimizing the total response time between the requesting sites and the replication sites. The dynamic maintainability is achieved by considering the replica relocation cost. The decision of relocation is made when performance metric degrades significantly in last *K* time periods. The threshold value is varied proportionally to response time in each time interval.

Table 1 summarizes the major research work done on replica placement in tree topology Data Grids.

3.2 Peer-to-Peer and Hybrid Solutions for Replica Placement

In large decentralized systems where the reliability of storage elements is low and bandwidth is limited, data availability is a serious challenge. Replication can be an effective technique for improving data availability in such unreliable systems. A strategy for creating replicas automatically in a generic decentralized peer-to-peer network is proposed in [11]. The goal of this model-driven approach is to maintain a specific level of availability at all times. To achieve this goal, the optimal number of replicas for each file is calculated and then best nodes are identified to host these replicas. This is done for files that have no replica or have replicas less than an availability threshold. The approach relies on a resource discovery service to find available storage and the Network Weather Service [14] to get network status information. The availability of a file depends on the node failure rate and the accuracy of the Replica Location Service (RLS). The amount of replica availability needed for a given file is modeled as:

$$RL_{acc} * (1-(1-p)^r) \geq Avail$$

Where r is the total number of replicas for a given file, p is the probability of a node to be up, RL_{acc} is the accuracy of the replica location service, and *Avail* is the required amount of availability for the file. For a certain availability threshold, the value of r can be calculated from the above function. For example, if the probability of a node to be up is 30% and the RLS accuracy is 80% then for the availability threshold of 75%, the model recommends a minimum of 8 replicas. After calculating ideal number r of replicas, the system consults RLS to know how many replicas M are actually present. If M is less than r the system creates $(r - M)$ copies of file and distribute them to best candidate nodes. The best nodes are those which maximize the difference *replicationBenefits – replicationCosts*. The benefit is the reduction in transfer time to the potential users. The replication costs are the storage costs at the remote site and the transfer time from the current location to the new location. Let F be a file stored at a location N_1 then cost of creating its replica at a new location N_2 is:

$$replicationCosts = trans(F, N_1, N_2) + s(F, N_2)$$

where function *trans()* shows the transfer cost of F from N_1 to N_2 and function *s()* represents the storage cost of F at N_2. The benefit of creating a replica of F at N_2 is given by:

$$replicationBenefits = trans(F, N_1, User) - trans(F, N_2, User)$$

where *User* is the location from which we expect the most number of future requests. In this way, the net benefits of replicating F at N_2 are calculated and the best candidate identified to place replica.

Table 2. Merits and demerits of replica placement techniques in P2P & hybrid topology Grids

Authors	**Technique**	**Methodology/ Realization**	**Results/Comments**	**Objective/ Performance metric**
Kavitha et al [11] 2002	Model-driven P2P Replication	Computing ideal no. of replicas for a file; placing replicas at best nodes by doing cost/benefit analysis	+ No single point of failure - possibility of excessive replicas as each peer takes own decision	Replica availability
Hai Jin et al [12] 2006	Decentralized P2P Strategy	Modeling costs b/w peers using adjacency graph; 0/1 vector for replica placement	+ better performance than static technique	Data transfer cost
Kavitha and Foster [7] 2001	Cascading plus Caching	Joining the cascading and caching methodologies	+ Client can act as Server for siblings + better performance than cascading or caching alone	Response time and bandwidth conservation
Lameha -medi etal [13] 2002	Hybrid Replica Connection Service	A cost model evaluates access costs and performance gains of creating & placing replicas	+ Favorable results when replicas placed closer to users	Response time

In [12], a P2P network cost model is constructed along with a decentralized algorithm to dynamically optimize the replica placement. The optimization strategy is based on the status of the grid system. Though all the status info is obtained dynamically in a decentralized way and all decision-making of replica movement is independent among the peer nodes, the data locating and storage resource management are centralized. System scale is controlled by deploying super peers or brokers. To make the effective use of grid storage resources, the lifetimes of replicas are set. A replica will be deleted from the system if it is not accessed for a certain time period.

Among replication techniques for tree topology discussed in previous section, authors [7] also presented Caching plus Cascading replication strategy by joining Plain Caching and Cascading. A copy of the requested file is saved at the client local storage and periodically replicas of popular files are sent down to lower tiers. Any node in the hierarchy can be server. Specifically, a client can act as a Server to its siblings. This constitutes a hybrid topology in which client nodes having the same parent are linked in P2P-like structure. The strategy gives better performance than Cascading or Caching alone.

A hybrid topology is used in [13] where ring and fat-tree replica organizations are combined into multi-level hierarchies. Each site or node maintains an index of the replicas it hosts and the other locations of these replicas that it knows. Replication of a dataset is triggered when requests for it at a site exceed some threshold. A cost model evaluates the data access costs and performance gains for creating replicas. The replication strategy places a replica at a site that minimizes the total access costs including both read and write costs for the datasets.

Table 2 highlights some major work on replica placement in P2P and hybrid Data Grid topology.

4 Conclusions

Comparing different replication policies working under different or even similar topologies is complicated due to diverse nature of assumptions made regarding topology, data access patterns and policy objectives. The hierarchical model of Data Grid, used in many high energy physics projects, has motivated the development of tree-structured replication mechanisms. The decision of replica placement is very important for the effective performance of any replication scheme. For our study, we consider tree and P2P/hybrid structures separately and present the merits and demerits of different replica placement strategies in each one. A Data Grid tree provides a relatively simple top-down structure for data propagation, albeit data movement remains confined to specific paths. In contrast, Peer-to-Peer systems present a more flexible layout but at the cost of additive complexity in the replication schemes. There are proposals for P2P and hybrid replication solutions in research but practically they have not yet found a place in Data Grids. We foresee that like tree structure, P2P and other replication frameworks shall become prevalent to fulfill the needs of scientific communities for accessing huge amounts of data efficiently and securely.

Acknowledgements

We would like to thank the National Natural Science Foundation of China under Grant No. 60533110 and 60473075, the Program for New Century Excellent Talents in University under Grant No.NCEF-05-0333, the Key National Science Foundation of Heilongjiang Province under Grant zjg03-05 and the National Science Foundation of Heilongjiang Province under Grant F0208 for their support.

References

1. Chervenak, A., Foster, I., Kesselman, C., Salisbury, C., Tuecke, S.: The Data Grid: Towards an Architecture for the Distributed Management and Analysis of Large Scientific Datasets. In. Journal of Network and Computer Applications (23), 187–200 (2001)
2. The GriPhyN Project, http://www.griphyn.org
3. Tang, M., Lee, B.-S., Yeo, C.-K., Tang, X.: Dynamic Replication Algorithms for the Multi-tier Data Grid. In Future Generation Computer Systems 21, 775–790 (2005)
4. Guy, L., Kunszt, P., Laure, E., Stockinger, H., Stockinger, K.: Replica Management in Data Grids. Technical Report, Global Grid Forum Information Document, GGF5, Edinburgh, Scotland (July 2002)
5. Stockinger, H., Samar, A., Allcock, B., Foster, I., Holtman, K., Tierney, B.: File and Object Replication in Data Grids. In: 10th IEEE Symposium on High Performance and Distributed Computing. pp. 305–314 (2001)
6. Hoschek, W., Janez, F.J., Samar, A., Stockinger, H., Stockinger, K.: Data Management in an International Data Grid Project. In: proceedings of GRID Workshop. pp. 75–86 (2001)
7. Ranganathan, K., Foster, I.: Identifying Dynamic Replication Strategies for a High Performance Data Grid. In: proceedings of the Int'l Grid Computing Workshop, Denver, Colorado, USA (November 2001)
8. Abawajy, J.H.: Placement of File Replicas in Data Grid Environments. In: Bubak, M., van Albada, G.D., Sloot, P.M.A., Dongarra, J.J. (eds.) ICCS 2004. LNCS, vol. 3038, pp. 66–73. Springer, Heidelberg (2004)
9. Rahman, R.M., Barker, K., Alhajj, R.: Replica Placement in Data Grid: A Multi-objective Approach. In: Zhuge, H., Fox, G.C. (eds.) GCC 2005. LNCS, vol. 3795, pp. 645–656. Springer, Heidelberg (2005)
10. Rahman, R.M., Barker, K., Alhajj, R.: Replica Placement Design with Static Optimality and Dynamic Maintainability. In: Proceedings of the Sixth IEEE Int'l Symposium on Cluster Computing and the Grid (CCGrid'06) (2006)
11. Raganathan, K., Lamnitchi, A., Foster, I.: Improving Data Availability through Model-Driven Replication for Large Peer-to-Peer Communities. In: proceedings of Global and Peer-to-Peer Computing on Large-Scale Distributed Systems Workship, Berlin, Germany, (May 2002)
12. Jin, H., Mao, F., Chen, H., Wu, S., Zou, D.: P2P Based Decentralized Dynamic Replica Placement Strategy in Data Grid Environment. Downloaded on October 2006 (2006) from http://grid.hust.edu.cn/fmao/papers/new/replica%20placement-0224IE%20fianl%20.pdf
13. Lamehamedi, H., Szymanski, B., Shentu, Z., Deelman, E.: Data Replication Strategies in Grid Environments. In: proceedings of 5th Int'l Conference on Algorithms and Architecture for Parallel Processing, pp. 378–383 (2002)
14. Wolski, R.: Forecasting Network Performance to Support Dynamic Scheduling Using the Network Weather Service. In: proc. 6th IEEE Symposium on High Performance Distributed Computing. Portland, Oregon (1997)

A Workload Balancing Based Approach to Discourage Free Riding in Peer-to-Peer Network*

Yijiao Yu and Hai Jin

Services Computing Technology and System Lab
Cluster and Grid Computing Lab
Huazhong University of Science and Technology, Wuhan, 430074, China
yjyu@mail.ccnu.edu.cn, hjin@mail.hust.edu.cn

Abstract. We propose a workload balancing based approach to discourage free riding in *Peer-to-Peer* (P2P) network. The heart of our mechanism is to migrate some shared files from the overloaded peers to the neighbored free riders automatically and transparently, which enforces free riders to offer services when altruistic peers are heavily overloaded. The implementation techniques are discussed in detail. We design a simulation to verify this approach, and the results show that it can not only decrease the degree of free riding, but also improve the *Quality of Service* (QoS) and robustness of P2P network efficiently.

1 Introduction

In the P2P philosophy, all peers in P2P network are volunteers, which should try to share files and offer services for other peers. In 2000, Adar *et al.* measured the Gnutella network, however, they found that only 10 percent peers provide almost 90 percent services in the P2P network [1]. Most of the peers share no more than 10 files or some unpopular files, which are not queried or downloaded by others at all. That is to say, most of the peers in the P2P network consume far more files than they offer. This selfish behavior of peers is called free riding and the selfish peers are regarded as free riders [1]. Papers [2][3] report the traffic measurement results of Gnutella in 2001 and 2005 respectively, which show that free riding has increased significantly since 2000. Moreover, free riding also prevails in Fast-Track, DirectConnect [4].

Free riding is a serious problem in P2P network for both ISPs and users. From an ISP's point of view, free riding leads to the mismatch between the overlay network and the physical network, which makes the in-through traffic of physical network asymmetric to the out-through traffic of physical network [2]. To users, the response time of query increases significantly because a few altruistic peers have to handle too many requests. Especially, when a hot peer crashes due to heavy overload, the shared files of the whole P2P network will decrease sharply. Also, the topology of P2P

* This paper is supported by National Basic 973 Research Program of China under grant No.2003CB317003, and the Cultivation Fund of the Key Scientific and Technical Innovation Project, Ministry of Education of China under grant 705034.

K.C. Chang et al. (Eds.): APWeb/WAIM 2007 Ws, LNCS 4537, pp. 144–155, 2007.

network will change dramatically, and even the connected P2P network will be divided into multi-isolated networks [3][5]. In a word, the "tragedy of the commons" has now emerged, therefore the rational P2P application software developers should develop some anti-free riding measures to deal with it.

Some mechanisms have been proposed to address free riding, such as incentive mechanisms, game theoretic approaches, economics or social network based methods. There is a general principle of the previous work, which is to decrease the QoS of free riders consuming. Unfortunately, most of users enjoy high QoS of download, but they do not contribute to the P2P network. ISPs are interested in the number of P2P users. If the number of users decreases sharply due to poor QoS, it is difficult to persuade ISPs to accept the anti-free riding measures. In addition, there are some guarded hosts behind of the *Network Access Translation* (NAT) or firewall [6], which cannot provide services to P2P network. A lot of peers cannot offer as much as they consume due to the asymmetric physical network connection, such as the ADSL. Therefore, the anti-free riding method should consider the real network environments and the benefits of both the ISPs and the users.

The objective of this paper is to find an anti-free riding measure, which can not only guarantee the P2P networks run stably with high QoS but also attract a great many users. In addition, the novel approach should be easily implemented in practice. Therefore, a workload monitoring and balancing based approach is proposed and simulations are designed to verify it. On one hand, we expect all peers can enjoy high QoS of consuming, and the free riders are not punished too strictly. On the other hand, the P2P network should be robust and stable.

The paper is organized as follows. Related work is discussed in section 2. In section 3, we present the overview of the workload balancing based approach, and analyze its specific features. The detailed implementation techniques of the mechanism are discussed in section 4. In section 5, we design a simulation to demonstrate the performance of our approach. Section 6 concludes this paper with a discussion of future work.

2 Related Work

Ramaswamy *et al.* [5] made the first effort to avoid free riding in P2P network with the incentive mechanisms. Utility functions are proposed to evaluate the contribution of each peer, which are based on the number of shared files, the account of files size, and the contribution of upload. The QoS of each peer consuming is computed according to the utility functions. Some different utility functions and implementation protocols are defined in [7][8], which are more reasonable and complicated. For example, the contribution is not computed by the absolute size of upload, but the ratio between the traffic and the upload bandwidth of peer [7]. Since the utility functions and reputations are evaluated by the peer itself, the whitewashing behavior is employed by malicious peers. Ref. [9] addresses the dilemma about how to deal with whitewashing and gives some clues.

Karakaya *et al.* [10] proposed a distributed and measurement-based framework against free riding in P2P networks. There is an important difference between [10] and [7][8] that every peer is measured and evaluated by neighbors. The requests from free rider will be discarded or with lower priorities, or even the link to free rider will be broken. The punishment is executed by neighbors, which can avoid the cheating behavior efficiently.

Gupta *et al.* [11] used the game theory to analyze the action strategies of each peer. All peers on line are regarded as a set of players, and all players want to find the best action preferences for themselves. At the same time, the P2P network should get maximum benefits from the action strategies. A game theoretic approach to provide incentive and service differentiation in P2P Networks is presented in [12]. The analysis of the actions strategies are smart, however, they are difficult to be applied into engineering due to the high complexity and the requirement of the real time status information of the P2P network. Courcoubetis *et al.* [13] prove that more complex models are likely to lead to intractable game design problems and simple policies are asymptotically optimal.

Krishnan *et al.* [14] analyze the impact of free riding to P2P network with the social network theory. They present that free riding is an instinctive behavior of human, and the P2P network designers should accept this behavior. In their opinion, a little free riding can improve the efficiency of public resources sharing. Economic models are employed in [15]-[17].

3 The Workload Balancing Based Approach

In this section, we describe our mechanism, compare it to the previous mechanisms, and list its specific features.

3.1 Overview of the Novel Approach

An important assumption in this paper is that the QoS of P2P network will decrease if the overloaded peers crash or exit the P2P network, because usually most of P2P connections and shared files are kept by overloaded peers. If they exit, a lot of peers will lose the connections to the P2P network, and even the whole P2P network becomes multi-isolated sub-networks. In our opinion, the protection of hub peers (peers which are extremely highly connected by others) is one of the most important concerns to sustain the P2P network and guarantee the QoS of all peers.

To attract more users, free riding is allowed if there are no overloaded peers in the P2P network. If an overloaded peer appears, it decreases its workload as soon as possible to protect itself. The overloaded peers try to migrate some files to other neighbored free riders, and these free riders are enforced to offer services to the P2P network. Therefore, the motivation of anti-free riding is realized. It is notable that the negotiation of migration is executed through software automatically in the backend, and users are not aware of file migration. The report of status should be honestly, and the file receivers should provide services to other peers as soon as possible.

The pseudo-code of workload balancing based approach (referred to as WBBA in the rest of this paper) to discourage free riding is illustrated as follows. It is clear that WBBA is a daemon thread, which starts as the peer joins in the P2P network, executes all the time, and ends until the peer exits. With the daemon thread, each peer can reduce overloaded workload to protect itself actively and automatically.

```
WBBA
{
  Step1: Get the thresholds of overloaded status;
  Setp2: Monitor the peer's status periodically;
  Step3: Judge whether the peer is overloaded.
            If (overloaded==true)
               goto step 4;
            else
               goto step9;
  Step4: Compute the set of destination peers to accept
            the migrated files
  Step5: Compute the set of migrated files;
  Step6: Assign migrated files to the destinations;
  Step7: Migrate shared files by FTP or HTTP;
  Step8: The overloaded peer updates the routing of the
            P2P network;
  Step9: Sleep a sample interval;
            goto setp2.
}
```

It should be noted that WBBA is executed in every peer and the overloaded thresholds are only about a peer. WBBA need not the global status information of the P2P networks. Connection control is a popular and efficient way to avoid free riding in P2P networks, e.g. [3][10], and it is also adopted in the Step 8 in WBBA.

3.2 The Specific Features of WBBA

Compared WBBA to previous solutions, there are distinct differences.

First, there are no utility functions and no limitations about the QoS a peer consumes. All peers are eager to get files and services from the P2P network, and all of them want to enjoy high QoS greedily. Only when the QoS meets most peers' requirement, the number of peers will increase and the shared files will become abundant. The research of public files management has shown that a lot of users are of the selfish instinct. If users are required to know and select the policy of consuming and offering, it not only decreases the usability of software, but also leads some users to reject the P2P application. The benefits of ISPs will be damaged. In a word, with the novel mechanism, the P2P network tries its best efforts to provide services to all peers, both altruistic peers and selfish peers.

Second, WBBA is a distributed algorithm and easily implemented in engineering. Each peer only focuses on its status of workload. Central monitor and global status information of P2P network are not required in WBBA. Moreover, each peer samples the status information at its sampling point, thus synchronization is not needed. The time complexity of monitoring is very low, which does not bring too much burden.

Last but not least, the peers behind firewall and NAT can enjoy high QoS of download services. With the popular incentive mechanisms, these peers are regarded as free riders and offered poor download bandwidth. In fact, the special physical network environment forbids them to provide services to users outside. The P2P network tries its best to serve all peers. Therefore, the average QoS of all peers enjoy can be improved.

4 The Implementation Techniques

In this section, we will introduce the detail implementation techniques of WBBA. There are nine steps in WBBA. Some of them are easy to be realized, such as the *Step7* and *Step9*. For simplicity, the implementation of *Step7* and *Step9* are not discussed in this paper. We categorize the other steps in WBBA into three issues, namely the status monitoring and judgment of overload, the selection of the migration destination peers, and selection and assignment of the migration files.

4.1 Status Monitoring and Judgment of Overload

From *Step1* to *Step3*, a peer monitors its real time workload status and judges whether it is overloaded and which status information should be monitored. From our perspective, three types of status information should be monitored and considered with high priorities, namely the *P2P application resources type* (P2PART), the *hardware resources type* (HRT), and the *physical network resources type* (PNRT).

Table 1 lists the most necessary monitoring objects of the status. The left column lists the variable name, and the second column illustrates the type of the monitored information. The values of these variables are categorized into *integer type* and *ratio type*, which are shown in the right column. The variables listed in Table 1 are very important and necessary to WBBA, and P2P application designers can add other variables to satisfy with its special requirements.

Table 1. The most necessary monitoring variables of the peer status

Monitoring variables	Type	Value type
The number of files	P2PART	Integer
The access frequency	P2PART	Integer
The number of P2P connections	P2PART	Integer
The frequency of each file in a peer	P2PART	Integer
The number of forwarding P2P packages	P2PART	Integer
The CPU usage	HRT	Ratio
The memory usage	HRT	Ratio
The bandwidth usage	PNRT	Ratio
The number of TCP active connections	PNRT	Integer
The number of TCP passive connections	PNRT	Integer

The status of P2PART variables can be obtained with the P2P application directly, by adding some log operations, such as recording the access frequency of each file. The status of HRT variables can be obtained with *Application Programming Interface* (API) provided by the operating system. Due to the heterogeneity of APIs provided by different operating systems, the *Simple Network Management Protocol* (SNMP) can be utilized to sample both the HRT information and the PNRT information. Here, we do not discuss the detail implementation about the sampling.

The definition of overloaded thresholds of monitoring variables is the premise of the judgment of overload. From Table 1, we can see that the threshold of ratio type variable is easier to be defined. For example, most of users do not expect the CPU usage larger than 60%, otherwise the response of human-machine is very slow. Therefore, the default threshold of CPU usage is 60%. In this way, we can define all default thresholds of ratio type variables. However, due to the heterogeneity of software and hardware configuration of peers, the service abilities are distinct. The threshold of the integer type variables should be defined according to the configuration environments of each peer. Fortunately, either with the API or SNMP, we can get the hardware and software configuration information automatically.

The daemon thread samples the status information periodically. There is no unique rule for all P2P applications because the focuses of different applications are different. For a P2P high performance computing system, CPU usage is the most important variable in the judgment rule. While in a file sharing P2P system, the bandwidth usage is the more important. A reasonable one is usually a combination of multiple variables listed in Table 1. In section 4.2 and 5.1, we illustrate two judgment rules.

4.2 The Selection of Migration Destination Peers

When a peer detects itself being overloaded, it will migrate some shared files to the neighbored free riders immediately. The *step4* selects the destination peers for migration. According to the measurement results in [1][2][4], there are hundreds or thousands of peers connecting with an altruistic peer. It is a combinatorial problem of selecting the appropriate destination peers from thousands of peers.

There are a lot of strategies in selecting destination peers. For example, if the selection is based on fairness, we should select peers with the shortest queues. If the data access patterns are considered to improve the migration efficiency, we should copy the frequent accessed files in overloaded peer to free riders near the service requesters. To describe the main ideas of WBBA, we only illustrate a simple way, based on the number of connections of each peer, to select the destination peers. Most peers keep no more than 4 connections [2]. The number of connections indicates the number of services a peer provides and enjoys currently. If the number is too small, the peer is impossible to offer lots of services to other peers and the contribution to P2P network is also small. The number of connections is explicit and easily measured by each peer. Therefore, it is an appropriate measurement of the contribution of each peer and is used in the destination selection in this paper.

Fig. 1 illustrates a topology of a local network in a large P2P system. Every peer is modeled as a node and the logic links between peers are modeled as undirected edges. There are 9 peers in Fig.1, marked as 1, 2, ..., 9. The number of simultaneous connections of each peer is defined as the degree of the node in this paper. There are more than one link attached to peer 1, 2 and 3, and these three peers are easy to become the bottleneck of the P2P system. In this paper, altruistic peer means a special peer, which try to offer services for other peers in the P2P network. Overloaded peer is the peer whose workload is too heavy. Altruistic peer is easier to be overloaded, because it has too many workloads. However, not all altruistic peers are overloaded, if the altruistic peers are powerful enough.

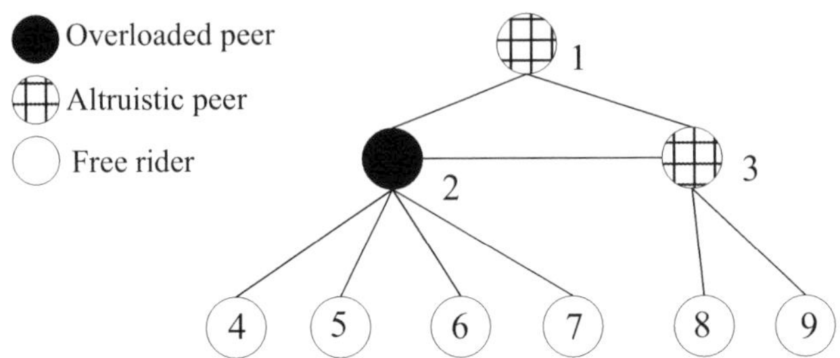

Fig. 1. A topology of a P2P network

Suppose that peer 2 reports overload because too many peers connect to it in *Step3*. Then the *Step4* is executed as follows by peer 2.

First, peer 2 broadcasts the status query messages to all neighbored peers, namely peer 1, 3, 4, 5, 6, and 7. Second, all neighbors reply the query message and report their latest status information. Third, peer 2 receives these feedback messages and judges who will be the destination peer. Given that the migration of shared files also costs a lot of computer resources and network bandwidth, we do not encourage the frequent migration of shared files between peers. Especially, the migration between busy peers and overloaded peers cannot reduce the service overload, and even makes the network become unstable. We define a simple rule for this case, the degree of destination node is no more than half of the degree of the overloaded node. In Fig.1, the degree of peer 2 is 6, and the degree of peer 3 is 4. Therefore, peer 3 is regarded as busy and will not be selected as the destination peer.

We define the *maximum set of destination peers* (MSDP), in which all elements comply with the above rule, e.g. MSDP of peer 2 is {1, 4, 5, 6, 7}. The number of final destination peers, marked as m ($1 \le m \le |MSDP|$), is generated randomly.

Finally, select m different peers from MSDP randomly. For instance, m is 3 and {4, 5, 6} is selected as the final destination peers in Fig.1. We should explain why we do not select the neighbors with the least connections as the destination peer. In a large P2P network, the degree of hub nodes can reach at more than 20,000 or 30,000. If the number of elements in MSDP is too large, sorting will cost a lot of time, which will delay the procedure of decreasing workload. Generating m different random numbers are very quick and does not cost a lot of memory resources. Using MSDP as the final set of destination peers is unacceptable in engineering, because physical network will be congested if sending too many files to many neighbored peers in a short time. We suggest that m should be no more than 50 in real environments. It should be noted that the overloaded peer should stop some most frequently accessed files services immediately when either all neighbored peers are busy, or it cannot find an appropriate destination peer to migrate files.

4.3 The Selection and Assignment of Migration Files

After the set of destination peers is chosen, *Step5* will be executed. According to the measurement results in [1][3], there are thousands of files shared by altruistic peers and the size of them is also very large. It is unnecessary and impossible to copy all shared files to other peers. The basic principle of the migration is to decrease the workload sharply by transferring files as few as possible.

There is an efficient and feasible way applied in our simulation, which only transfers the most frequently accessed shared files to free riders. To realize it, the P2P application system should record the access frequency of each file. Fortunately, this requirement is not difficult to be realized in engineering, and some P2P applications have provided this function. We only need to sort the most frequent files from large to small, and get the top n files to be migrated (suppose n ($m \le n$) files will be migrated).

There are m^n kinds of different assignments solutions of migration files. For instance, if peer 2 wants to migrate the top 4 files to the three destination peers in Fig. 1, there are 3^4 (81) kinds of assignment solutions. The number of assignment solutions increases exponentially with n. Researchers are interested in the optimal solution among the m^n kinds of solutions. However, we do not encourage spending too much time on looking for the optimal solution. Assigning the most frequently accessed files to the peers with least connections in the set of destination peers is an approximate optimal assignment in the simulation.

After the assignment of migration, the overloaded peer migrates shared files to free riders with FTP or HTTP protocols. There may be problems in the middle of the file migration process, since the network may be unreliable or the file may be too large. In this case, both the migration and the services of the unsuccessfully migrated files in the source peer should be stopped. When the migration is finished, the overloaded peer updates *Distributed Hash Table* (DHT) or routing table immediately. When the new requirements to the migration files come, it only forwards the requirement to the destination peer and stops download service.

5 Experiments

In this section, we design a simulation to illustrate the performance of WBBA.

5.1 Simulation Setup

All simulation programs are designed with Java and executed in a personal computer, with 1.7GHz Intel processor and 256MB memory. In the simulation, the first task is to make a virtual P2P network environment similar to the real P2P network environment. Especially, the network topology, the distribution of the number of files shared by each peer, and the distribution of access frequency of every file in each peer must comply with that of the real P2P network. The Pajek software [18] is employed in our simulation to obtain network topology with the scale free feature, which can generate different kinds of network topology, such as random network, scale free and Erdos-Renyi. Some peers generated by Pajek have no connections with other peers. From our perspective, all peers should connect to the P2P network. Otherwise, they cannot get services from the P2P network. Therefore, the original topology generated by Pajek is preprocessed to make it to be a connected graph. In real P2P environments, some peers join and depart frequently. However, it is not easy to describe the probability of peer's dynamic behaviors, and in many simulations the dynamic behaviors are not discussed [5]. Thus, join and departure of peers are not considered in this paper.

The number of files shared by peer i is generated with Equation (1), where d_i is the degree (the number of simultaneous connections) of peer i. The function $random(n)$ generates a pseudorandom, uniformly distributed integer value between 0 and n-1. In Equation (1), a is a constant, which represents the degree of cooperation of peers. In our simulation, a is set as 3. Since the network topology is generated with the scale free network model, the distribution of degrees is similar to the real network. The previous measurement results show that peers with large connections usually share more files. In general, Equation (1) can generate large number of shared files for these peers with large degrees. However, if the degree of peer i is large and the number of files shared is small, peer i will be regarded as routing peers in P2P network. These routing peers devote to forwarding P2P packages as the router in Internet. Equation (1) is only executed once for a peer while initializing the network environment.

$$sharedFiles_i = 1 + random(a \times d_i) \tag{1}$$

The accessed frequency in the tth iteration of the jth file in peer i is generated with Equation (2), where b is a nonnegative constant representing the popularity of files in P2P system. $d_{i,t}$ represents the degree i in the tth iteration. It is clear that the access frequency in the tth iteration of the jth file in peer i is related to that of the t-1th iteration, and c is a constant between 0 and 1. This assumption is reasonable in real P2P network. In our simulation, c is set as 0.6 and b is set as 20. In every iteration period, the access frequency of every files will be computed with Equation (2).

$$f_{i,j,t} = c \times f_{i,j,t-1} + (1-c) \times random(b \times d_{i,t}) \tag{2}$$

In the simulation, we compute the workload of each peer only with some P2PART variables. Equation (3) is defined to compute the workload of peer i in the tth iteration, which is based on the access frequencies, file size and the degree of peer. Due to the overloaded peer updates routing table, the degree of peer i is changed with t. $size_{i,j}$ is the file size of the jth file in peer i. $f_{i,j,t}$ is the access frequency of the jth file in peer i in the tth interval. $workload_{i,t}$ denotes the average traffic of each link in the tth iteration. $sharedFiles_i$ is also changed in different iteration periods, because the shared files are migrated from the overloaded peer to free riders. After the migration of shared files, $sharedFiles$ of the overloaded peer will be decreased by 1 and that of the destination peer will be increased by 1.

$$workload_{i,t} = \frac{\sum_{i=1}^{sharedFiles_{i,t}} f_{i,j,t} \times size_{i,j}}{d_{i,t}} \tag{3}$$

Moreover, considering the heterogeneity of computers, two overloaded thresholds of workloads are used in simulation to make the judgment of overload fair and similar to the real environment to represent computers of different processing capabilities, which are 20000 and 4000. If $workload_{i,t}$ is larger than the threshold of each peer, peer i will be regarded as overloaded.

5.2 Simulation Results

We evaluate the performance of the novel anti-free riding measure with three aspects, including the distribution of the degree of each peer, the distribution of the number of files shared by each peer, and the distribution of the access frequency of each peer. In addition, we are interested in how many peers have to do the migration operation, and how long the P2P network will be stable and robust. There are 2000 peers and 7827 logic links in the network topology graph. The detailed results are illustrated as follows. It is notable that lg(x) in all figures in this paper denotes the base 10 logarithm.

Fig. 2 illustrates the distribution of the degree of each peer. In Fig. 2, the largest degree of peers is no more than 105, and the smallest one is 1. Due to the heterogeneity of degrees, the *numOfPeers* is accounted with the step of 5. For instance, all peers with the degree between 1 and 5 are accounted as *numOfPeers* at *degree* = 5, and *numOfPeers* at *degree* = 10 is the number of peers whose degree is between 6 and 10. It is clear that the number of peers, whose degree is no more than 10, decreases after the anti-free riding control, and some peers have more neighbors. Moreover, the number of hub peers, with very high degrees, declines.

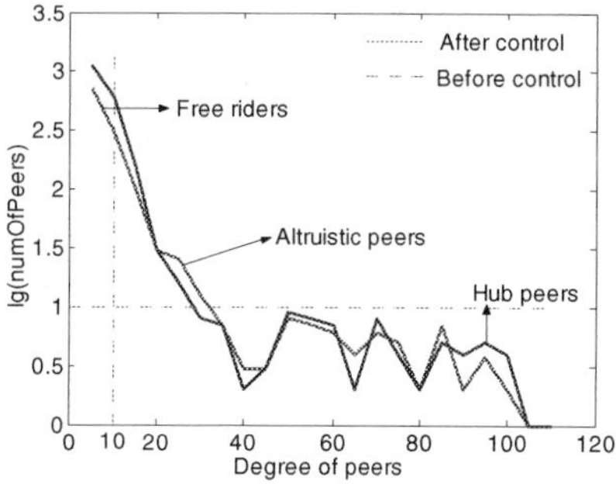

Fig. 2. Distribution of the degrees

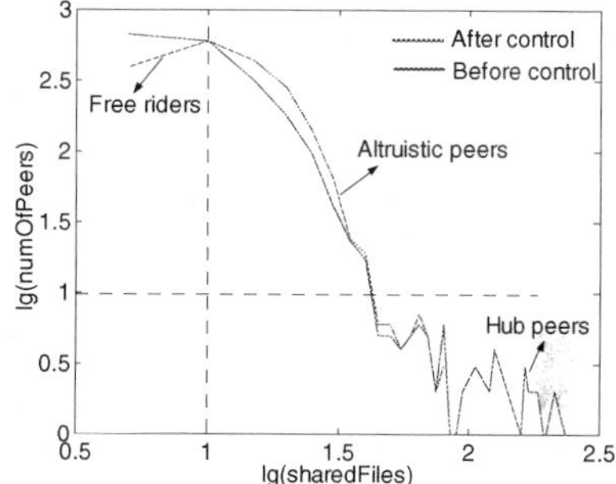

Fig. 3. Distribution of the number of shared files

The distribution of the number of files shared by each peer is shown in Fig.3. The dot line illustrates the distribution status before anti-free riding control, and the solid line describes the distribution after control. It is clear that the number of free riders, sharing no more than 10 files, decreases, and more peers share more than 10 files. However, the shared files are no longer in several hub peers, because these files in overloaded peers are migrated to neighbored free riders. From Fig.3, we can see that the free riding phenomenon is discouraged, and the overloaded peers or hub peers are removed successfully.

Fig. 4 illustrates the distribution of the access frequency of each peer during an iteration period. We use 100 as a basic step to compute *numOfPeers*. For example, all peers, the access frequency between 101 and 200 in an iteration period, are accounted as the *numOfPeers* of frequency=200. The dot line is the distribution of the frequencies of peers before control, and the solid line is the distribution of frequencies after control. The number of peers, offering service no more than 300 times in an iteration period, decreases, and the number of altruistic peers with frequency between 301 and 7,000 increases. Moreover, the largest access frequency decreases sharply after the control, which shows the altruistic peers provide more services.

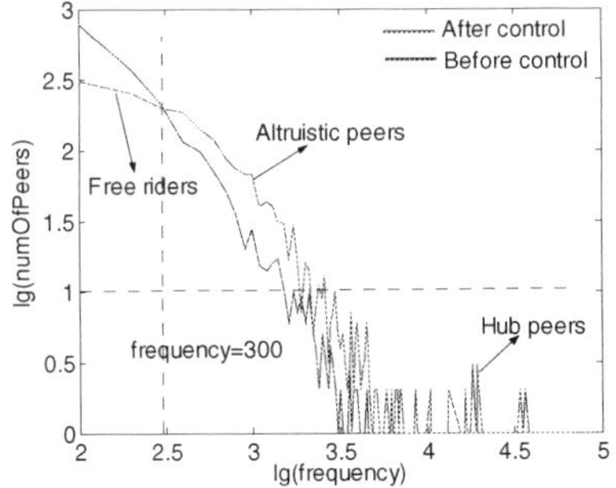

Fig. 4. Distribution of the accessed frequency of peers

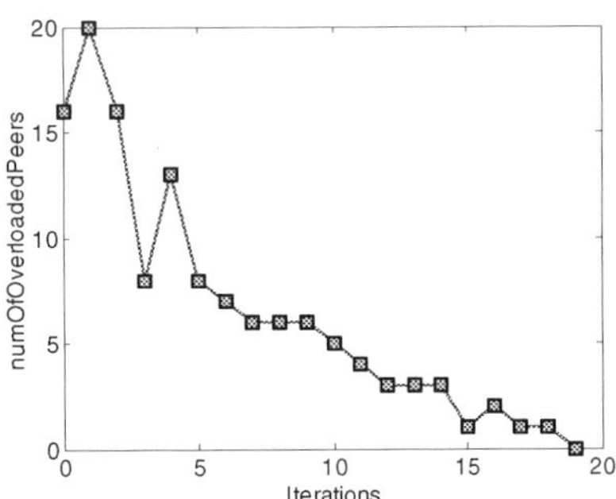

Fig. 5. The number of overloaded peers

We are interested in the time to get rid of all overloaded peers in a P2P network. Fig. 5 shows the changes in the number of overloaded peers over the iteration from 0th to 19th in the simulation. There is no overloaded peer after 19 iterations and the P2P network becomes robust. The maximum number of overloaded peers is 20, which is no more than 1% of the number of all peers in the P2P network. It illustrates that only few peers in P2P network execute the migration operation, which is feasible to practical applications. The phenomenon is also similar to the real P2P network environment that about 1% peers are very busy to provide 90% services. The number of overloaded peers does not always decrease with the increase of iterations, because new requests come to each peer in every iteration period.

6 Conclusion

In this paper, we present a workload balancing based anti-free riding approach, which is easily implemented in engineering. The simulation results show that the simple way suffices to control of the overall system to a nearly optimal objective, in which the P2P network is sustained as a connected graph and all peers can consume QoS as high as possible. The most specific feature of this way is that it not only discourages free riding behaviors of peers strictly, but also guarantees the QoS of all peers, including the free riders and guarded peers behind firewall and NAT. This feature attracts a great many peer users and is helpful to maintain the business model.

There are many ways to extend the approach in the future. For instance, we might discuss the selection strategies of the destination peers to migrate files, especially when there are too many free riders connecting to the overloaded peers. Different strategies should be considered, such as the fairness, and the data access pattern. Moreover, it is desired to discuss the impacts of join and departure behaviors of peers.

References

1. Adar, E., Huberman, B.: Free Riding on Gnutella. First Monday, 10 (2000)
2. Ripeanu, M., Foster, I., Iamnitchi, A.: Mapping the Gnutella Network. IEEE Internet Computing 1, 50–57 (2002)
3. Hughes, D., Coulson, G., Walkerdine, J.: Free Riding on Gnutella Revisited: The Bell Tolls? IEEE Distributed Systems Online, 6 (2005)
4. Sen, S., Wang, J.: Analyzing Peer-to-Peer Traffic Across Large Networks. IEEE/ACM Transactions on Networking 2, 219–232 (2004)
5. Ramaswamy, L., Liu, L.: Free Riding: A New Challenge to Peer-to-Peer File Sharing Systems. In: Proceeding of the 36th Hawaii International Conference on System Sciences, 220-229 (2003)
6. Wang, W.J., Chang, H., Zeitoun, A., Jamin, S.: Characterizing Guarded Hosts in Peer-to-Peer File Sharing Systems. In: Proceedings of Globecom 1539–1543 (2004)
7. Ma, R.T.B., Lee, S.C.M., Lui, J.C.S., Yau, D.K.Y.: An Incentive Mechanism for P2P Networks. In: Proceedings of the 24th International Conference on Distributed Computing Systems, 516–523 (2004)
8. Anceaume, E., Gradinariu, M., Ravoajia, A.: Incentive for P2P Fair Resource Sharing. In: Proceedings of the Fifth IEEE International Conference on Peer-to-Peer Computing, 253–260 (2005)
9. Feldman, M., Papadimitriou, C., Chuang, J., Stoica, I.: Free-Riding and Whitewashing in Peer-to-Peer Systems. IEEE Journal of Selected Areas in Communications 5, 1010–1019 (2006)
10. Karakaya, M., Korpeouglu, I., Ulusoy, O.: A Distributed and Measurement-based Framework Against Free Riding in Peer-to-Peer Networks. In: Proceedings of the Fourth International Conference on Peer-to-Peer Computing, 276–277 (2004)
11. Gupta, R., Somani, A.K.: Game Theory as a Tool to Strategize As Well As Predict Nodes Behavior in Peer-to-Peer Networks. In: Proceedings of the 11th International Conference on Parallel and Distributed Systems, 244–249 (2005)
12. Ma, R.T.B., Lee, S.C.M., Lui, J.C.S., Yau, D.K.Y.: A Game Theoretic Approach to Provide Incentive and Service Differentiation in P2P Networks. In: Proceedings of SIGMETRICS/ Performance'04, 189–198 (2004)
13. Courcoubetis, C., Weber, R.: Incentive for Large Peer-to-Peer Systems. IEEE Journal of Selected Areas in Communications 5, 1034–1050 (2006)
14. Krishnan, R., Smith, M.D., Tang, Z.L., Telang, R.: The Impact of Free-Riding on Peer-to-Peer Networks. In: Proceeding of the 37th Hawaii International Conference on System Sciences, 199–208 (2004)
15. Strulo, B., Smith, A., Farr, J.: An Architecture for Peer-to-Peer Economies. In: Proceedings of the Third International Conference on Peer-to-Peer Computing, 208–209 (2003)
16. Gupta, R., Somani, A.K.: Pricing Strategy for Incentive Selfish Nodes to Share Resources in Peer-to-Peer (P2P) Networks. In: Proceedings of the 12th IEEE International Conference on Networks, 624–629 (2004)
17. Ham, M.J., Agha, G.: ARA: A Robust Audit to Prevent Free-Riding in P2P Networks. In: Proceedings of the Fifth IEEE International Conference on Peer-to-Peer Computing, 125–132 (2005)
18. Pajek, Available at http://vlado.fmf.uni-lj.si/pub/networks/pajek/

QoS-Based Services Selecting and Optimizing Algorithms on Grid

Qing Zhu, Shan Wang, Guorong Li, Guangqiang Liu, and Xiaoyong Du

School of Information, Renmin University of China, Beijing 100872, P.R. China
Key Laboratory of Data Engineering and Knowledge Engineering
(Renmin University of China), Beijing 100872, P.R. China
zq@ruc.edu.cn

Abstract. QoS-Based Services Selecting and Optimizing Composition between the peers play an increasingly important role to ensure interoperability on Grid environment. However, the prohibitive cost of selecting, matching, mapping and composing algorithm has now become a key bottleneck hindering the deployment of a wide variety of Grid services. In this paper, we present QoS-Based Services Selecting and Optimizing Composition on Grid. First, it checks requesters' semantic in order to form candidate service graph. Second, it designs service selecting and mapping algorithms for optimizing the model. Third, it creates an executed plan of optimum composition on Grid. We conducted experiments to simulate and evaluate our approach.

Keywords: Information Grid, SOA, Service Composition, Selecting Algorithm.

1 Introduction

Grid has been developed to support for solving large-scale problems not only in science, but also in engineering and commerce. Grid supports for the sharing and coordinated use of diverse resources in dynamic, we call distributed virtual organizations.

The complexity of service composition in semantic Grid [3] includes three main factors: (1) different disciplines have different problems, each dependent on different aspects of domain-specific knowledge; (2) new services can be flexibly composed from available service components based on the user's function requirement and quality-of-service (QoS), and (3) both the underlying computing resources and the information input for the process are dynamic. So it's important how to understand precisely meaning of requirement and how to solve a QoS engineering problem since the service selection must select the best services to compose an efficient complex service with QoS assurance.

However QoS-based service composition presents significant challenges and requires addressing a number of critical issues such as discovering and identifying relevant services, formulating semantics, selecting algorithm and creating composition plans using current context, goals, constraints and costs, binding to and invoking composition instances and checking their validity.

K.C. Chang et al. (Eds.): APWeb/WAIM 2007 Ws, LNCS 4537, pp. 156–167, 2007.

In this paper, a key contribution is a dynamic composition model based on semantics[4], graph theory and service selecting algorithm. Candidate Service Generator is a key component of the dynamic composition framework and selects an optimum composition by considering requesters' Quality of Services (QoS). The design of QoS-Based Services Selecting and Optimizing Algorithms is a key improvement to existing Grid computing and runtime services to support the execution of applications.

The rest of the paper is organized as follows. Section 2 describes an overview of system architecture and presents the semantic interpreting according to the user's requirement and QoS-based criterion. Section 3 presents key technology of selecting and mapping algorithms. Section 4 presents simulation and evaluation. Section 5 presents related work. Finally, the paper summarizes and concludes in Section 6.

2 System Architecture

To achieve both flexibility and simplicity, we propose architecture for QoS-Based Services Selecting and Optimizing Composition in Grid Environment. The system is called GridSC (Grid Service Composition System), which can be seen as a middle control layer in an active grid environment that offers a generic service abstraction and automates mapping of processing resources to grid services.

The architecture of GridSC system consists of four components running in two different phases that are semantic Interpreter phase and composition plan generate phase. Grid Service Composition system includes other four key components: Semantic Interpreter, Candidate Service Generator, Prediction Combiner and Plan Execution as shown in Fig 1. Candidate Service Generator, Prediction Combiner and Plan Execution belong to composition plan generate phase.

Semantic Interpreter receives client's request of services, and formally describe service composition specification according to the service interface. In the meantime, semantic interpreter understands and translates the semantic meaning of client's request, then extracts semantics information.

Candidate Service Generator identifies the location and capabilities of processing resources to build a candidate service resource graph that describes the physical network topology. It translates the service specification onto the physical resource graph while taking into account all service-specific constraints and the lists of peers' resource. Finally, it provides common, reusable service candidates to prediction combiner.

Prediction Combiner includes prediction evaluating and service selecting algorithms which needs to select and map a specific service and service level along an optimal path in the execution plan. The selection is based on a user's QoS requirements and Multiple QoS constraints. It provides a new service plan that involves several basic service components through QoS-based services selecting and optimizing algorithms.

Plan execution reserves and allocates appropriate physical resources as determined by the service execution plan and resource lists on globe peers. Once the

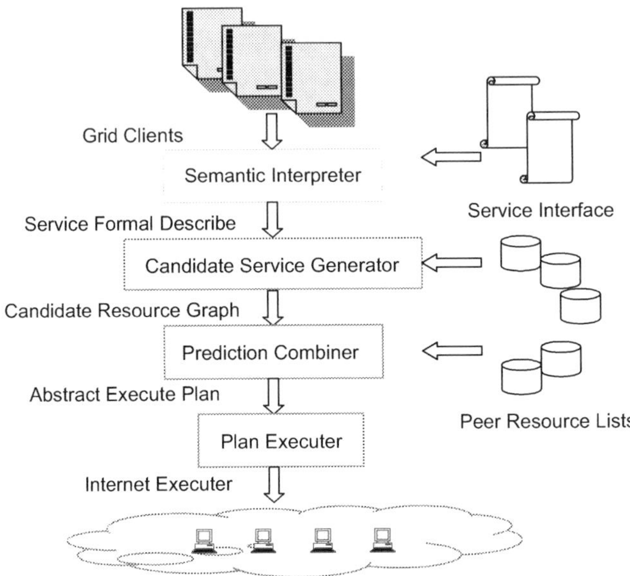

Fig. 1. Service Composition System Architecture

service has been deployed, Plan execution is the final task that maps the execution plan and combines all required components to provide an operational service composition for users. If the available resources are changed, plan execution adopts a proactive approach (e.g. failure recovery, service plan reconstruction) to maintain the quality of composed service during runtime.

From the view of the point, service specification is the foundation of service composition which can describe semantic explanation of user requisition and create requiting goal. Each service owner retains control over the services that they make available to others. They determine how the service is realized and set the policy for accessing the service. It is the key problem that consists of a new service composition QoS-based by using available web services on Grid environment. This paper focuses on the service semantic interpreter and services selecting and optimizing composition algorithms part.

2.1 Semantic Interpreter

The semantic interpreter is critical aspect with respect to service composition which is supported semantic translation, service probe, service selection, composition and monitoring. Automating these steps in Grid service usage life cycle is the aim of semantic Grid services composition. However, the semantic interpreter is the first step of understanding the meaning of service requirement on Grid.

In Grid services domain, semantics represented by the semantic metadata can be classified into the following types, namely, Functional Semantics, Data

Semantics, QoS Semantics and Execution Semantics. These different types of semantics can be used to represent the capabilities, requirements, effects and execution pattern of a Grid service. The semantic interpreter focuses to formalization expression as focused on the functional semantics. Research on Grid services composition on the other hand is based on the findings and results from the semantic Grid research to apply for services that perform some action producing an effect.

The Semantic Interpreter defines the components required by a service and describes how components are interrelated and constrained after user requesting. Lexical and syntactic information is the easiest to represent and process in a parser. But it is not sufficient to resolve all the ambiguities in understanding language of machine; and by its nature, it cannot determine what a sentence means. Therefore, we use Semantic Interpreter that can extract semantics information from Client's Request or interface.

Semantic information allows rich descriptions of Grid services and processes that can be used by computers for automatic processing in various applications. At the meanwhile, the deployment of ontology help understand a well-defined set of common data elements or vocabulary that can support communication across multiple channels, accelerate the flow of information, and meet customer needs.

When user provide m key Requirement Services, semantic interpreter abstracts m keywords, R_1, R_2, ..., R_m and the Requirement function Ψ (R_1, R_2, ..., R_m). Semantic Requirement Function Ψ (R_1, R_2, ..., R_m) is a Boolean function with parameters of key services. The two legal Boolean operations are defined as below: R_i AND R_j: Both keywords R_i and R_j are contained in the answer; R_i OR R_j: Either keyword R_i or R_j is contained in the answer. To handle any of these cases, the semantic interpreter must generate a Composition Candidate Graph that states a logical proposition. Now, let us introduce the formational notions associated with Semantic Requirement Services.

Definition 1. *Semantic Requirement is represented:*

$\Psi(R_1, R_2, ..., R_m) = \{V_t, V_n, Rule, Select\text{-}Sentence\}$,

There are four important components in a formalization description. Where:

1. Terminal symbols $V_t = \{AND, OR, R_i\}$; R_i *: the description of i-th Service Requirement;*

2. Nonterminals $V_n = \{Keyservice, Select\text{-}Sentence, Segment\}$;

3. Start Symbol: Select-Sentence;

4. Productions or Rule consisted of :

Select-Sentence $\rightarrow$ *Segment,*

Segment $\rightarrow$ *Keyservice AND Segment,*

Segment $\rightarrow$ *Keyservice OR Segment,*

Segment $\rightarrow$ *Keyservice,*

Keyservice $\rightarrow R_1|R_2|R_3|\cdots|R_i|\cdots$;

First, the Semantic Interpreter parses user's service request to a set of select keywords and computes the final states once a keyword selects. Just as above described, it receives m (m>0) keywords R_1, R_2, ..., R_m and a Boolean function Ψ (R_1, R_2, ..., R_m) as one whole select command at the start of the user Requirement Services session. Second, Semantic Interpreter translates them into a sequence of effective keywords and does group some or all of them according to the AND/OR semantics included in Boolean function. The groups of keywords are further constructed to Parser Trees. The semantic interpreter uses syntactic rules to generate parse trees and translates those trees into composition candidate service graphs.

2.2 Candidate Service Graph

Grid computing can provide ubiquitous resource and service availability. Grid is defined to be a dynamic and open environment where the availability and state of these services and resources are constantly changing. Therefore, Grid applications are similarly complex, dynamic and heterogeneous. The primary focus of the service composition model presented in this paper is to evaluate and select a specific service to find an optimal path in the execution plan. Prediction Combiner produces Composition execution plans from the pool of available services to satisfy QoS-based defined composition objectives, policies and constraints.

In GridSC system, the user's functional requirements are given in the form of a composition function graph by the semantic interpreter. The function graph consists of required service functions $F_1, \cdots, F_k$ that are connected by dependency links and commutation links. The dependency link indicates that the output of one function is used as the input by its successor.

Definition 2. *Composition Function Graph is defined as a directed weighted graph G= (F, E, C), where, the user's functional requirements* $F=\{F_1, F_2, \cdots, Fm\}$ *represents the set of* $|F|$ *nodes,* $E=\{e_1, e_2, \cdots, e_n\}$, *represents the set of* $|E|$ *edges, a weight function* $C=\{C_1, C_2, \cdots, C_t\}$ *is defined represents the set of QoS-based constraints condition,* $e_i=(F_i, F_j, C) \in E$, *represents the function relation that function* F_j *is a direct successor of state* F_i *by constraints* C_m.

In the model, composition objectives, and composition policies and constraints ($\{C_i\}$) can be dynamically defined as simple semantic information statements. The available service pool is represented as a candidate service graph CSGraph (S, E), where the nodes represents services, $S=\{s_i\}$, in the pool and the links, $E=\{s_i, s_j\}$, can be modeled as possible interaction. GridSC system can discover and locate the service resources to produced a candidate service graph is present in Figure 2.

Definition 3. *Candidate Service Graph is a state graph CSGraph = (S, E), where S represents the set of* $|S|$ *peers, denoted by composite services* S_i, $1 \leq j \leq |S|$, *and E represents the set of* $|E|$ *overlay links, denoted by* e_j , $1 \leq j \leq |E|$.

Definition 4. *Composite Service* $S_i = \{s_{i1}, s_{i2}, \cdots, s_{im}\}$, *represents the set of* $|S_i|$ *compound state of service* $S=\{S_1, S_2, \cdots, S_k\}$; *denoted by* s_{ij} ; s_{ij}: *represents basic service;* S_i *is matching with the service function set* F_i *in Candidate Service Graph CSGraph.*

Definition 5. *Overlay Links E=* $\{e_1, e_2, \cdots, e_t\}$ *represents the set of* $|E|$ *edges;* e_i*=*< s_{ij}*,* s_{km}*, C* >*: if valid (service* s_{ij} *and (condition C)) then activation (*s_{km}*).*

Let's give an example. The four viable service compositions are selected from Candidate Service Graph on Fig 3.

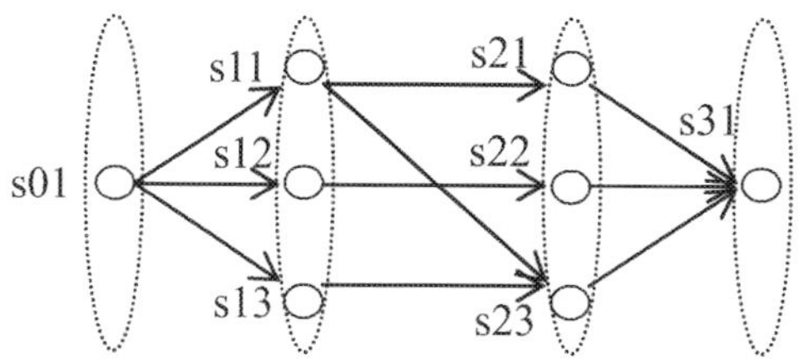

Fig. 2. Candidate Service Graph

Composition 1:	s01, s11, s21, s31;
Composition 2:	s01, s11, s22, s31;
Composition 3:	s01, s12, s22, s31;
Composition 4:	s01, s13, s23, s31;

Fig. 3. Service request compositions

Service descriptions are augmented with semantic information in the form of keywords and context information. This semantic information along with QoS-based polices and constraints are used to select applicable services (s_0) and interactions (e_{ij} where Valid (e_{ij}, C_k) = True). Candidate composition plans can be represented as paths in this graph G (S, E). Execution plans may be evaluated and ranked based on different QoS-based cost factors. In the candidate service graph, G(S, E), the available services are vertices and interaction are edges. The edges are created at runtime using a relational join operation. Service Composition can be defined as finding a path from initial to final in G_0 (S, E).

2.3 Execution Plan of Composition

Execution Plan of Service Composition is presented while the QoS-based Services Selecting and Optimizing Algorithms are discussed. We make precise some of the informal arguments and descriptions that we'll meet the terms and conceptions followed in this paper.

Definition 6. *Service Consistency Relation (*$s_i > s_j$*). Give two service components* S_i *,* $S_j \in S$ *in Candidate Service Graph CSGraph (S, E), if Composite Service* S_i*,* S_j *are each matching with Service function* F_i*,* $F_j \in F$*, in Composition Function graph G(F, E), and* $\exists e \in E$ *,* $e=(F_i, F_j, C)$*,and* $In(F_j)=1$ *Valid (Service* S_i *and* S_j*) = True, and Output(*s_i*)*$\supseteq$ *Input(*s_j*), then the relation of service* s_i *and* s_j *is called Service Consistency Relation,denoted by* $s_i > s_j$*.*

Definition 7. *Service Mapping is defined as Composition Function Model G= (F, E) is mapping into Candidate Service Graph CSGraph = (S, E), and the set of Candidate Service* S_i *is matching with Function* F_i *in order to satisfy user's requirements.*

Definition 8. *Execution Planning is defined as the path, denoted as* $P=\{p_1,p_2,\cdots,p_m\}$*, in the Candidate Service Graph CSGraph (S, E),which are selected the basic services to produce Execution Planning of service composition.*

We give the example of service composition, from Function Model graph of services mapping into Candidate Service Graph, and service Execution Planning $P_0,P_1,\cdots,P_n$. Fig 4.

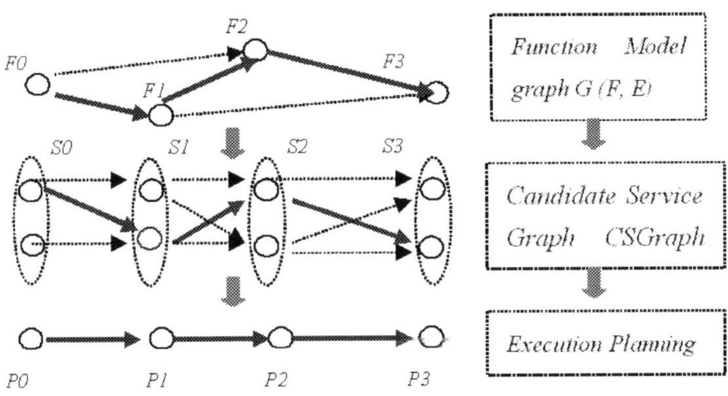

Fig. 4. Composition Mapping

The Semantic Interpreter specifies a composition request as a set of constraints, keywords, input service and output services. Candidate Service Generator discovers the participating services, S_i, generate the set of associated interactions E_i, and the composition graph G_i on Candidate Service according to the keyword set and constraint set. In the service selection step, the services in the current service pool are parsed to generate service set S by Prediction Combiner. A relational join operation is then used to construct the set of ad-hoc interactions, E, by matching interfaces, and to create service graph G(S, E). Cost associated with each C_{ij} is calculated and evaluated by Prediction Combiner. Candidate composition Execution plans can now be generated as paths in G between initial and final using graph path algorithms. The composition Execution plans can be ranked based on costs. These costs could reflect QoS-based factors, operational environments and/or user defined factors. Constraints can belong to different categories and can control aspects of both services and compositions. New services can be flexibly composed from available service components based on the user's function and quality-of-service (QoS) requirements.

2.4 QoS Criteria for Composite Services

In GridSC system, a QoS-based approach to service composition is the challenge, which maximizes the QoS of composite service executions by taking into account the constraints and preferences of the users. Traditionally, QoS[2] problem has been studied within different domains such as network layers and multimedia systems. However, we are faced with new challenges in QoS-based service composition because it requires an integrated solution considering multi-dimensional requirements (i.e., function, resource and QoS requirements) at the same time. On the other hand, the QoS of the resulting composite service executions is a determinant factor to ensure customer satisfaction, and different users may have different requirements and preferences regarding QoS. Moreover, the candidate service graph could be a state graph instead of a linear path in order to accommodate parallel execution of services.

In the following, we describe key tasks involved in generic quality criteria for basic services: (1) Execution price (2) Execution duration (3) Availability (4) Successful execution rate (5) Reputation Each basic service may provide different service levels; each level is associated with a QoS vector parameters QoS = $(Q_1,\cdots,Q_n)$, for example: Q_1=Q.price, Q_2= Q.duration, Q_3=Q.availability, Q_4=Q.succeed-rate, Q_5=Q.reputation. The quality criteria defined above in the context of basic Grid services, are also used to evaluate the QoS of composite services. If service composition satisfies all the user constraints for that task, it has the maximal score. If there are several services with maximal score, one of them is selected randomly. If no service satisfies the user constraints for a given task, an execution exception will be raised and the system will propose the user to relax these constraints.

3 Selection Algorithm of Service Composition

In this section, we present the service selection algorithms used by the Prediction Combiner for service composition with two or more QoS constraints. We use two algorithms: the SA algorithm and the Heuristic algorithm to solve the problem.

First, we present the QoS[9] quality criteria in the context of basic services, indicate its granularity and provide rules to compute its value for a given service. Second, we assume that the same service definition is used by all basic service candidates for a specific service component on Candidate Service Graph. So we are concerned about the compatibility issue among services and focus on the QoS service selection problem.

3.1 Advanced Simulated Annealing (Advanced-SA) Algorithms

Currently, we have implemented the following more general optimization algorithms in our prototype.

Simulated Annealing (SA): The simulated annealing heuristic is based on the physical process of "annealing". We use the temperature reduction ratio R as a parameter to control the cost/optimality trade-off.

SA is a search technique based on physical process of annealing, which is the thermal process of obtaining low-energy crystalline states of a solid. The temperature is increased to melt solid. If the temperature is slowly decreased, particles of the melted solid arrange themselves locally, in a stable "ground" state of a solid. SA theory states that if temperature is slowed sufficiently slowly, the solid will reach thermal equilibrium, which is an optimal state. By analog, the thermal equilibrium is an optimal task-machine mapping (optimization goal), the temperature is the total completion time of a mapping (cost function), and the change of temperature is the process of mapping change. If the next temperature is higher, which means a worse selecting and mapping, the next state is accepted with certain exponential probability. The acceptance of "worse" state provides a way to escape local optimality which occurs often in service selecting.

3.2 Max-Min Exhaustive Algorithm

Max-min Exhaustive Algorithm (Max-min): This algorithm always yields the actual optimal configuration, but the optimization cost grows rapidly with the problem size. The Max-min heuristic selects a "best" (with minimum completion time) machine for each task. Then, from all tasks, send the one with minimum completion time for execution. The idea is to send a task to the machine which is available earliest and executes the task fastest, but send the task with maximum completion time for execution. This strategy is useful in a situation where completion time for tasks varies significantly.

3.3 Heuristic Greedy Algorithm (HG)

A greedy algorithm means if it builds up a solution in small steps, choosing a decision at each step myopically to optimize some underlying criterion. There are many different greedy algorithms for the different problems. In this paper, we designed Heuristic Greedy Algorithm (HG) to optimize the cost of execution time.

Currently, we have implemented the more general optimization algorithms in our prototype. The algorithms discussed above are suitable under different circumstances. Therefore, for each physical mapping problem, the most appropriate algorithm needs to be selected. Therefore, we propose that the Prediction Combiner should be able to choose the best optimization technique to solve the problems.

4 Simulation and Evaluation

In this section, we will evaluate performance of different parameters on Max-min Exhaustive Algorithm (Max-min), Heuristic Greedy algorithm and Advanced Simulated Annealing (Advanced-SA) Algorithms. Service composition processing time includes three parts (1) Create candidate service graph time: according to user's requirement. (2) Selecting time: executing selecting algorithm to better service. (3) Execution plan time: realizing physical mapping. Our experiments mainly evaluate (2) selecting time, which actually is the major part of service composition processing time.

```
void HeuristicAlgorithm::Select()
Let S be the set of selected service nodes
    For each Si∈S, we store a Qos-based number Q(Si)
Initially S={s} and Q(s) =0
find best Qos-based service from first set of top graph
while S<>V   //For each of no-selecting service set do
    Select a service node v ∉ S with at least one edge from S for which
    CountCost(); min e=(u,v),u∈S Q(u) is as small as possible
    Add v to S and define
    If select[i] success then break;
    If select set finished then exit(0)
endwhile
```

Fig. 5. Heuristic Greedy algorithm

In the following experiments, we assume an equal-degree random graph topology for the 1-8 services of service composition candidate graph. For simplicity, we only consider one process plan with the two service composition algorithms. We produced random numbers of QoS vector parameters, QoS $= (Q_1,\cdots,Q_n)$, for example price, duration, availability, succeed-rate, and reputation. The number of service class and candidates in each service class involved in the process plan range from 5 to 40.

We run our system experiments using the simulation Grid environment on several Pentium(R)4 CPU 2.4GHZ PC with 1GB of RAM. We implement simulation experiment in C++ and both use other development tool.

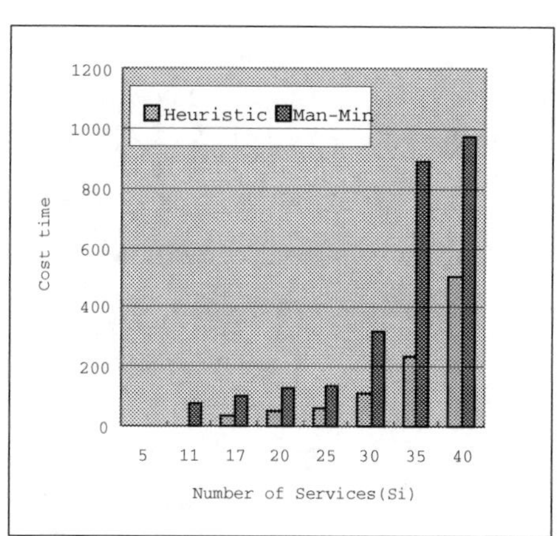

Fig. 6. Cost time of Max-min & Heuristic

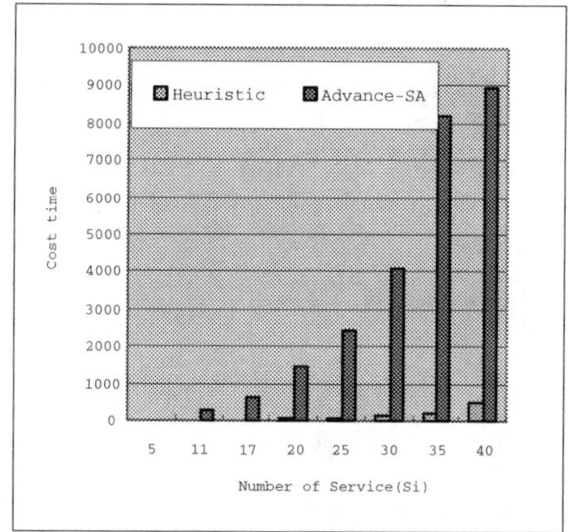

Fig. 7. Cost time of SA & Heuristic

Fig. 6 shows the selecting time with the number of service nodes increasing from 5 to 40. This experiment runs two algorithms: Heuristic Greedy algorithm and Max-min Exhaustive Algorithm. Fig.7 shows the selecting time with the number of service nodes increasing from 5 to 40. This experiment runs two algorithms: Heuristic Greedy algorithm and Advanced Simulated Annealing (Advanced-SA) Algorithms. This experiment shows Heuristic Greedy algorithm is efficient QoS-Based Services Selecting and Optimizing Algorithms about this service composition on Grid.

5 Related Work

Many projects have studied the service composition problem. SpiderNet[1] is the service composition middleware by both user's need for advanced application services and newly emerging computing environments such as smart rooms and peer-to-peer networks. SpiderNet only researched QoS-Assured Service Composition in Managed Service Overlay Networks and no semantic issue has been addressed. The SWORD project [12] and eFlow project [11] proposed a developer toolkit for the web service composition. It uses a rule-based expert system to check whether a composite service can be realized by existing services and generate the execution plan given the functional requirements for the composed application. SWORD only addressed the on-line and no QoS issue has been addressed. Our main contribution is to use semantic knowledge to interpreter user requirement of service composition, to take QoS-driven composition goal into account to find best quality composition by using selecting algorithms on Grid.

6 Conclusion

In this paper, we study the problem of QoS-Based Services Selecting and Optimizing Algorithms on Grid. Two problem Algorithms are proposed: Advanced Simulated Annealing (Advanced-SA) Algorithms, and Heuristic Greedy Algorithm (HG) Algorithms. Semantic Interpreter, Candidate Service Generator, Prediction Combiner and Plan Execution, four components consist GridSC (Grid Service Composition System) system on the Grid environment that offers service composition middle control layer. In the paper we discussed client QoS requirement and QoS constraint. We have presented two algorithms, both optimal and heuristic, to compose and select services QoS-based constraints as well as to achieve the maximum utility.

Acknowledgements

This work is supported by National Natural Science Foundation of China under Grant No. 60473069, Science Foundation of Renmin University of China (No. 30206102.202.307), and the project of China Grid (No. CNGI-04-15-7A).

References

1. Gu, X., Nahrstedt, K., Chang, R.N., Ward, C.: QoS-Assured Service Composition in Managed Service Overlay Networks, In: Proc. of The IEEE 23rd International Conference on Distributed Computing Systems (ICDCS 2003), Providence, Rhode Island, May, pp. 19-22 (2003)
2. Zeng, L., Benatallah, B., Dumas, M.: Quality Driven Web Services Composition[A]. In: Proceedings of the 12th International Conference on World Wide Web (WWW) [C], Budapest,Hungary, pp. 411–421. ACM Press, New York (2003)
3. Zhuge, H.: Semantic Grid. Scientific Issues, Infrastructure, and Methodology, Communications of the ACM 48(4), 117–119 (2005)
4. Berardi, D., Calvanese, D., De Giacomo, G., Hull, R., Mecella, M.: Automatic Composition of Transition-based Semantic Web Services with Messaging. In: Proceedings of the 31st International Conference on Very Large Data Bases (VLDB, Trondheim, Norway, 2005, pp. 613-624 (2005)
5. Arpinar, I.B., Zhang, R., Aleman, B., Maduko, A.: Ontology-Driven Web Services Composition. IEEE E-Commerce Technology, July 6-9, San Diego, CA (2004)
6. Majithia, S., Walker, D.W., Gray, W.A.: A framework for automated service composition in service-oriented architecture, in 1st European Semantic Web Symposium (2004)
7. Berardi, D., Calvanese, D., Giacomo, G.D., Lenzerini, M., Mecella, M.: Automatic composition of e-services that export their behavior. In: Proc. 1st Int. Conf. on Service Oriented Computing (ICSOC), LNCS, vol. 2910, pp. 43-58 (2003)
8. Carman, M., Serafini, L., Traverso, P.: Web service composition as planning, in proceedings of ICAPS03 International Conference on Automated Planning and Scheduling, Trento, Italy, June 9-13 (2003)
9. Chen, H., Jin, H., Ning, X.: Q-SAC: Toward QoS Optimized Service Automatic Composition. In: Proceedings of 5th IEEE/ACM International Symposium on Cluster Computing and the Grid (CCGrid05), May, pp. 623-630 (2005)
10. Benatullah, B., Dumas, M., Shang, Q.Z., et al.: Declarative composition and peer-to-peer provisioning of dynamic web services[A]. Proceedings of the 18th International Conference on Data Engineering(C).Washington: IEEE, pp. 297-308 (2002)
11. Casati, F., Ilnicki, S., Jin, L., Krishnamoorthy, V., Shan, M.: Adaptive and dynamic service composition in e-flow. Technical Report, HPL-200039, Software Technology Laboratory, Palo Alto, CA, (March 2000)
12. Ponnekanti, S.R., Fox, A.: Sword: A developer toolkit for Web service composition. In: 11th World Wide Web Conference (Engineering Track), Honolulu, Hawaii, (May 2002)

Bayesian Method Based Trusted Overlay for Information Retrieval over Networks

Shunying Lü[1], Wei Wang[2], and Yan Zhang[1]

[1] Faculty Mathematics and Computer Science, Hubei University, 430062, Wuhan, China
[2] Department of computer Science and Technology, Tongji University, 201804, Shanghai, China
willtongji@gmail.com

Abstract. Peer-to-peer (P2P) overlay networks provide a new way to retrieve information over networks. How to assurance the reliability of the resource is the crucial issue of security. This paper proposes a trustworthy overlay based on the small world phenomenon that facilitates efficient search for information retrieval with security assurance in unstructured P2P systems. Each node maintains a number of short-range links to the trusted other nodes, together with a small collection of long-range links that help increasing recall rate of information retrieval, and the trust degree of each node is evaluated by Bayesian method. Simulation test shows that the proposed model can not only increase the ratio of resource discovery, improve the interaction performance of the entire network, but also assurance the reliability of resource selection.

Keywords: Peer-to-peer; trustworthy; small world, information retrieval; Bayesian method.

1 Introduction

P2P overlay networks provide a new way to use Internet and are useful for many purposes that need constructions of large-scale, robust and distributed resource sharing systems. In overlay networks, self-organization is handled through protocols for nodes arrival and departure, based either on a fault-tolerant overlay network, such as in CAN [1], Chord [2], and Pastry [3]. However, P2P based information retrieval (IR) systems still remain a challenging problem: how to select appropriate peers (or nodes) to cooperate with. Since the candidate peers are autonomous and may be unreliable or dishonest, this raises the question of how much credence to give each resource, and we cannot expect each user to know the trustworthiness of the resource.

On the other hand, social networks [4] exhibit the small-world phenomenon [5], in which people are willing to have friends with similar interests as well as friends with many social-connections. In light of this, we propose a trustworthy small world overlay for information retrieval in P2P systems. Each node maintains a number of short-range links to the trusted other nodes, together with a small collection of long-range links that help increasing recall of information retrieval. In addition, we develop a Bayesian method based trust model which can evaluate the trust degree of nodes efficiently.

K.C. Chang et al. (Eds.): APWeb/WAIM 2007 Ws, LNCS 4537, pp. 168–173, 2007.

The rest of this paper is organized as follows. We review some related work in Section 2. In section 3, we propose the trustworthy small world overlay that facilitates efficient search for information retrieval with security assurance in P2P systems. And then we introduce the trust model based on Bayesian method in detail. The evaluation of our approach by simulations is given in section 4 and finally we conclude this paper in section 5.

2 Related Work

A P2P overlay connects peers in a scalable way and defines a logical network built on top of a networking infrastructure. Improvements to Gnutella's flooding mechanism have been studied along two dimensions. First, query caching exploits the Zipf-like distribution of popularity of content to reduce flooding. Second, approaches based on expanding ring searches, which are designed to limit the scope of queries, and random walks [6], where each peer forwards a query message to a randomly chosen neighbor, in place of flooding are also promising at improving Gnutella's scalability. Such approaches are effective at finding popular content, whereas interest-based shortcuts can find both popular and unpopular content.

There is a trend to complement the basic overlay, irrespective of its structure, with additional connections to friends in the network [7]. Proximity routing where overlays links reflect the underlying network topology also falls into this category. The whole idea behind the friendly links is to cluster peers according to some criteria, such as interest. One of the advantages of such an approach is that the notion of friends is orthogonal to the structure of the underlying overlay. And these overlays show the characteristic of complex networks [8], which can increase the degree of the availability, the potential for recovery or the efficiency for some applications where friends share some interest. But, in order to use it well, it should build a trust environment in P2P networks. In light of these, we proposed trust-based overlay approach to improve search performance and scalability of P2P networks with security assurance.

3 Trustworthy Small World Model

3.1 Relative Definitions

Small world networks can be characterized by average path length between two nodes in the network and cluster coefficient defined as the probability that two neighbors of a node are neighbors. A network is said to be small world if it has small average path length and large cluster coefficient [8]. Studies on a spectrum of networks with small world characteristics show that searches can be efficiently conducted when the network exhibits the following properties: each node in the network knows its local neighbors, called short range contacts; each node knows a small number of randomly chosen distant nodes, called long range contacts. The constant number of contacts and small average path length serve as the motivation for trying to build a small world overlay network [9].

3.2 Construction of Trustworthy Small World Overlay

We now discuss how to construct a trustworthy small world network depicted above. The construction of the small world overlay involves two major tasks: setting up short-range links and establishing long-range links [9].

In short-range links, when a peer joins the network, it first establishes its trust summarization and then pulls trust summarizations from neighbors and chooses those peers which it trusted as short-range links. In long-range links, only trusted peers can be taken as long-range links.

Faloutsos et al. [10] considers the neighborhood as an N-dimensional sphere with radius equal to the number of hops where N is the hop-plot exponent. We generalize a P2P network with the average degree k to an abstract multidimensional network and determine the dimension of the network as $N = k/2$. Thus we define the distance $dis(P, P_i)$ between peer P and trusted peer P_i as follows:

$$dis\left(P, P_i\right) = H \times e^{-Trust\left(P, P_i\right)} \tag{1}$$

In the above formula, H is the hops from peer P_i to peer P and $Trust(P, P_i)$ is the trust degree between P and P_i which defined lately. To establish long-range links, these trusted peers will actively broadcast their trust evaluation at a large interval time in the network.

The main idea of search is through short-range links and long-range links to intelligently guide the search operation to those appropriate peers which are mostly trusted.

3.3 Bayesian Trust Model

In order to make this method work well, evaluating the trust degree of the peers is the key. In this paper, we present a Bayesian trust model. Our idea is to find an important feature of trust within P2P networks, that is the successful cooperation probability between two peers, and try to estimate it, as the Bayesian method supports a statistical evidence for trust analysis. The proposed Bayesian trust model is based on our previous work [11], [12].

For the sake of simplicity, we only considered a system within the same context during a period of time. For two peers x and y, the successful cooperation probability between them is denoted by θ. They may have direct interactions between them, and they may also have other intermediate peers and each of them has direct experiences with x and y. On one hand, if there are direct interactions between x and y, we can obtain direct probability of successful cooperation, which is called *direct trust degree*, and denoted by θ_{dt}. On the other hand, if there is an intermediate node z between x and y, and there are interactions between x and z, z and y, then, we can also obtain an indirect probability of successful cooperation between x and y, which is called *recommendation trust degree*, and denoted by θ_{rt}. So, there are two kinds of probabilities of successful cooperation, which can be aggregated into global successful cooperation probability as follows:

$$\hat{\theta} = f(\lambda_0 \cdot \theta_{dt} + (1-\lambda_0) \cdot \theta_{rt}),\ \lambda_0 \in (0, 1) \tag{2}$$

where $f(\cdot)$ is trust degree combination function, satisfying the property of convex function, that is let $S \subset R^n$ is a nonempty convex set, and f is a function defined on S. f is a convex function on S if for every θ_{dt}, $\theta_{rt} \in S$, $\lambda \in (0, 1)$, we have.

$$f(\lambda \cdot \theta_{dt} + (1-\lambda) \cdot \theta_{rt}) \le \lambda f(\theta_{dt}) + (1-\lambda) f(\theta_{rt}) \quad (3)$$

$f(\cdot)$ is decided by the subject factors of x, such as personality and emotion. For example, a common trust degree combination function is $\hat{\theta} = \lambda\ \theta_{dt} + (1\text{-}\lambda)\ \theta_{rt}$, $\lambda \in (0, 1)$, and a peer will choose $\lambda > 0.5$ if it trusts more his direct experiences rather than others' recommendations.

In light of this, we analyze how to obtain these two kinds of trust degree by Bayesian method. Let x and y be two nodes in P2P networks, and their interaction results are described by binomial events (successful / failure). When there are n times interactions between them, u times successful cooperation, v times failure cooperation, and define $\hat{\theta}_{dt}$ as the probability successful cooperation at n+1 times. Then, the posterior distribution of successful cooperation between x and y is a *Beta* distribution with the density function:

$$Beta(\theta \mid u, v) = \frac{\Gamma(u+v+2)}{\Gamma(u+1)\Gamma(v+1)} \theta^u (1-\theta)^v \quad (4)$$

$$\text{and } \hat{\theta}_{dt} = E(Beta(\theta \mid u+1, v+1)) = \frac{u+1}{u+v+2} \quad (5)$$

where, $0<\theta<1$, and u, $v>0$.

Direct trust degree reflects the ability of reliable service a target node provides in the network.

With respecting to recommendation trust, we also use the approach above to evaluate it, as the recommendation is formed by several direct interactions. The selection of recommend nodes can also be decided by the trust degree of them.

Let the interactions between x and y, z and y be independent, and the number of interactions between them be n_1 and n_2 separately, in which the successful cooperation be u_1 and u_2, failure cooperation v_1 and v_2. Then the trust degree of x to y by z can be modeled as:

$$\hat{\theta}_{rt} = E(Beta(\theta \mid u_1 + u_2 + 1, v_1 + v_2 + 1)) = \frac{u_1 + u_2 + 1}{n_1 + n_2 + 2} \quad (6)$$

When there are several recommendation nodes, it is easy to extend formula (6). In addition, considering the global trust degree is affected by positive and negative feedbacks separately, the value of $\hat{\theta}_{rt}$ can be mapped onto [-1, 1]. So the formula (6) can be modified as:

$$\hat{\theta}_{rt} = \frac{\sum w \cdot (u-v)}{\sum w \cdot (u+v) + 2} \quad (7)$$

4 Simulation Results and Performance Evaluation

All experiments in the paper are performed in a simulation platform we developed under PlanetLab [13], which is suit for large scale and dynamic P2P networks simulation. Without losing generation, we first implement a Gnutella style P2P network and then apply our algorithm on it. And then we compared with some interest-based clustering model.

We setup our simulations based on the observations on P2P networks stated in [17]. Each node is assigned some amounts of contents at the start time. All contents are grouped into categories. Each node first selects a couple of content categories and takes contents only from these categories. The number of categories follows by Zipf law. Each node issues queries probabilistically during its living time to search contents that the node is interested in. For interest-based model, we configure the overlay network similar to [18], where we place the first copy of content at the peer who makes the first request for it, and subsequent copies of content are placed based on accesses.

Based on the experiments above, we compare the proposed trustworthy overlay networks with current popular P2P networks such as Gnutella and other interest-based self-clustering model. Here, λ and η are both set to 0.8.

Figure 1 shows the distribution of the search length of the network, most of which is between 2 and 4 hops. The average of the path length is short because the peers clustered around the powerful peers which are more trustworthy. The proposed trust schema can increase the degree of the availability, the potential for recovery or the efficiency for some resources.

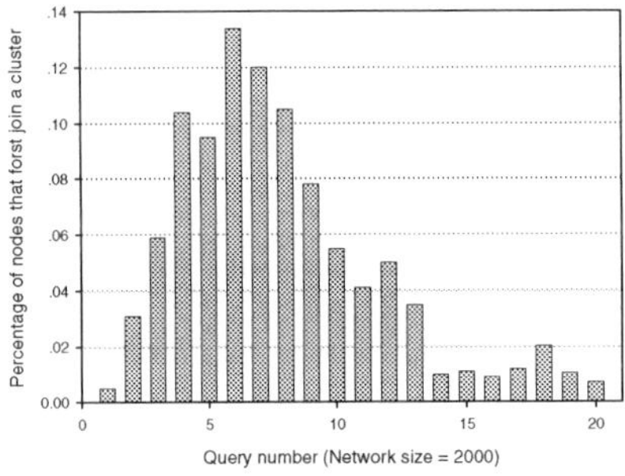

Fig. 1. The number of queries

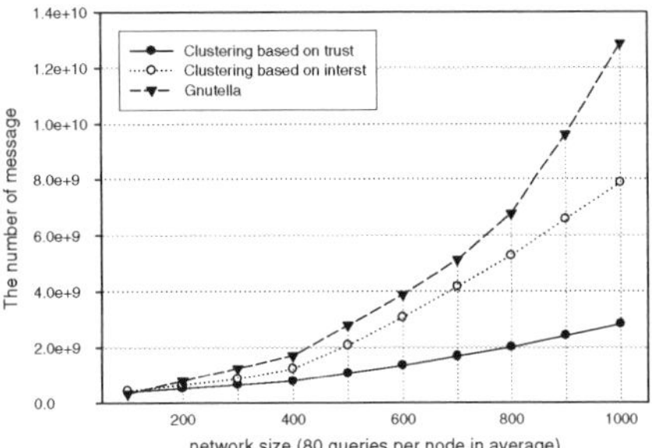

Fig. 2. Performance with varying network size

Then we evaluate the searching performance of our clustering approach. Figure 2 shows the number of messages transmitted in the P2P network with increasing network size. By applying our algorithm, the number of messages only increases linearly with the network size compared to the exponential-like increase of Gnutella style P2P network, which shows great scalability. Although the number of message increases linearly-like in interest-based clustering, it is still reaches at a high level comparing to the trust-based mechanism.

These experiments show that our proposed TSW overlay is suitable for large scale and dynamic P2P networks with trust assurance and its performance outperforms current popular P2P networks. It is promising to use it in the real world.

5 Conclusions

In this paper we present a trustworthy small world overlay model that facilitates efficient search for P2P information retrieval. In the trustworthy overlay, each node maintains a number of short-range links to the trusted other nodes, together with a small collection of long-range links that help increasing recall rate of information retrieval and reduce network traffic as well, and the trust degree of each node is evaluated by Bayesian method. Experiments have shown the following results: establishing a semantic small world overlay is feasible and it performs well, and search for IR in the overlay is efficient.

References

1. Ratnasamy, S., Francis, P., Handley, M., et al.: A scalable content-addressable network. In: Proceedings of the Conference on Applications, Technologies, Architectures and Protocols for Computer Communications, August 27–31, 161–172 (2001)
2. Stoica, I., Morris, R., Liben, N.D., et al.: Chord: A Scalable Peer-to-peer Lookup Service for Internet Applications. IEEE/ACM Transactions on Networking 11(1), 17–32 (2003)
3. Zhao, B., Kubiatowicz, J., Joseph, A.: Tapestry: an infrastructure for fault-tolerant wide-area location and routing, Report No. UCB/CSD-01-1141 (2001)
4. Newman, M.E.J.: The Structure and Function of Complex Networks. SIAM Review 45(2), 167–256 (2003)
5. Kleinberg, J.: The Small-World Phenomenon: an Algorithm Perspective. In: Proceedings of ACM Symposium on Theory of Computing (2000)
6. Cohen, E., Fiat, A., Kaplan, H.: A case for associative peer-to-peer overlays. In: Proceedings of Workshop on Hot Topics in Networks (2002)
7. Sripanidkulchai, K., Maggs, B., Zhang, H.: Efficient content location using interest-based locality in peer-to-peer systems. In: Proceedings of ACM INFOCOM Conference (2003)
8. Wang, F., Moreno, Y., Sun, Y.: Structure of Peer-to-peer Social Networks. Physical Review, 73(3) (2006)
9. Jin, H., Ning, X.M., Chen, H.H.: Efficient search for peer-to-peer information retrieval using semantic small world. In: Proceedings of ACM WWW Conference (2006)
10. Faloutsos, M., Faloutsos, P., Faloutsos, C.: On Power-Law Relationships of the Internet Topology. In: Proceedings of ACM SIGCOMM Computer Communication Review 29(4), 251–262 (1999)
11. Wang, W., Zeng, G.S., Yuan, L.L.: A Semantic Reputation Mechanism in P2P Semantic Web. In: Mizoguchi, R., Shi, Z., Giunchiglia, F. (eds.) ASWC 2006. LNCS, vol. 4185, pp. 682–688. Springer, Heidelberg (2006)
12. Wang, W., Zeng, G. S., Liu T.: An Autonomous Trust Construction System Based on Bayesian Method, In: Proceedings of the IEEE/WIC/ACM International Conference on Intelligent Agent Technology (IAT 2006), Hong Kong, China, December 18–22, 357–362 (2006)
13. Peterson, L., Bavier, A., Fiuczynski, M.: el al.: Towards a Comprehensive PlanetLab Architecture. Technical Report PDN–05–030, PlanetLab Consortium, June (2005)
14. Krishna, P., Richard, J., Stefan, S., et al.: Measurement, Modeling, and Analysis of a Peer-to-Peer File-Sharing Workload. SOSP'03, Bolton Landing, New York, USA, Oct (2003)

Managing a Geographic Database from Mobile Devices Through OGC Web Services*

Nieves R. Brisaboa[1], Miguel R. Luaces[1], Jose R. Parama[1], and Jose R. Viqueira[2]

[1] Database Laboratory, University of A Coruña,
Castro de Elviña S/N, 15071 A Coruña, Spain
brisaboa@udc.es, luaces@udc.es, parama@udc.es
[2] Systems Laboratory, University of Santiago de Compostela
Constantino Candeira s/n, 15782 Santiago de Compostela, Spain
joserios@usc.es

Abstract. We present in this paper a system for the management of geographic databases from mobile devices. The architecture of the system is extensible in the sense that it can be adapted to the particular characteristics of a wide variety of heterogenous mobile devices. This is achieved by the incorporation of an extensible middleware between the limited GIS applications running at the Mobile devices and the general purpose web services used at the data tier. The interfaces of these web services are compliant with the well-known Web Map Service(WMS) and Web Feature Service (WFS) specifications proposed by the Open Geospatial Consortium (OGC). A first prototype of the system for a specific PDA hardware with a specific software configuration was also implemented. We believe that the system has many practical applications in wireless environments with connectivity problems.

Keywords: Mobile GIS, Data management, Web Geo-Services, Open Geospatial Consortium, WMS, WFS.

1 Introduction

Traditionally, tools devoted to the management of geographic data use client-server architectures composed of desktop clients and either conventional or spatial Database Management Systems (DBMS) at their servers. Many examples of such desktop clients and DBMSs exist already in the market, both commercially available and derived from open source initiatives. However, the broad bandwidth of the computer networks currently available enables the access to spatial data sources through the web with reasonable delays, even through wireless physical communications. Thus, GIS tools have already appeared that enable the publication on the web of geographic data, both in the form of

* This work was partially supported by Xunta de Galicia (refs. PGIDIT05SIN10502PR and 2006/4) and Ministerio de Educación y Ciencia (refs. TIC2003-06593 and TIN2006-15071-C03-03).

K.C. Chang et al. (Eds.): APWeb/WAIM 2007 Ws, LNCS 4537, pp. 174–179, 2007.

maps for data visualization purposes and in the form of spatial objects and spatial coverages for spatial data editing and analysis purposes. Examples of such tools exist both with proprietary interfaces [6,2] and with well-known open interfaces [1,9,4] based on the standard specifications developed by the Open Geospatial Consortium (OGC) [12,10]. The data published by these tools can now be consumed by both GIS desktop and GIS web applications. The OGC goes beyond these ideas proposing a general architecture for GIS applications based on standard Geo-services. A generic framework for GIS applications based on such an architecture was already proposed in [8].

Regarding mobile environments, GIS vendors already provide tools [3,7] that enable remote access to geographic data sources, generally through proprietary formats and interfaces. The availability of Java Virtual Machines (JVM) for mobile platforms (based on Java ME specification) enables also the use of some Java-based open source GIS initiatives [5,14]. However, these solutions have difficulties to be installed in platforms with limited hardware and software capabilities.

In this paper, we present a system for the management of geographic databases using mobile devices. The architecture of the system uses web services compliant to well-known OGC standard specifications, namely the Web Map Service (WMS) specification [12] and the Web Feature Service (WFS) specification [10]. On top of them, an extensible *Geographic Data Access Middleware* adapts the heterogenous characteristics of a wide variety of mobile devices to the OGC standard interfaces. This is achieved by the incorporation of *Data Access Plug-ins* in the Middleware, one such plug-in for each different type of mobile device, which may range from powerful portable computers and Personal Digital Assistants (PDA) of various types to limited mobile phones. To test the viability of the system, a prototype implementation was also undertaken for a specific type of PDA with a specific software installation.

The rest of the paper is structured as follows. In Sect. 2 the architecture of the system is briefly described. The prototype implementation is shown in Sect. 3. Finally, Sect. 4 provides the conclusions and issues of further work.

2 System Architecture

The proposed architecture for the developed system is next briefly described. As it is shown in Fig.1, the architecture is organized in three tiers. The *Mobile Tier* consists of the client desktop GIS software installed in each of the mobile devices supported by the system. The *Middleware Tier* consists of a *Geographic Data Access Middleware*(GDAM), whose objective is to adapt the heterogeneity of the aforementioned tier to the standard interface of the OGC web services available at the *Data Tier*. This is achieved by extending the GDAM with a *Data Access Plug-in* (DAP) for each different type of mobile configuration. Finally, the *Data Tier* consists (at this stage) of web services implementing the well-known Web Map Service (WMS) and the Transactional Web Feature Service specification (WFS-T) both defined by the OGC. We briefly describe now the functionality of each of the tiers.

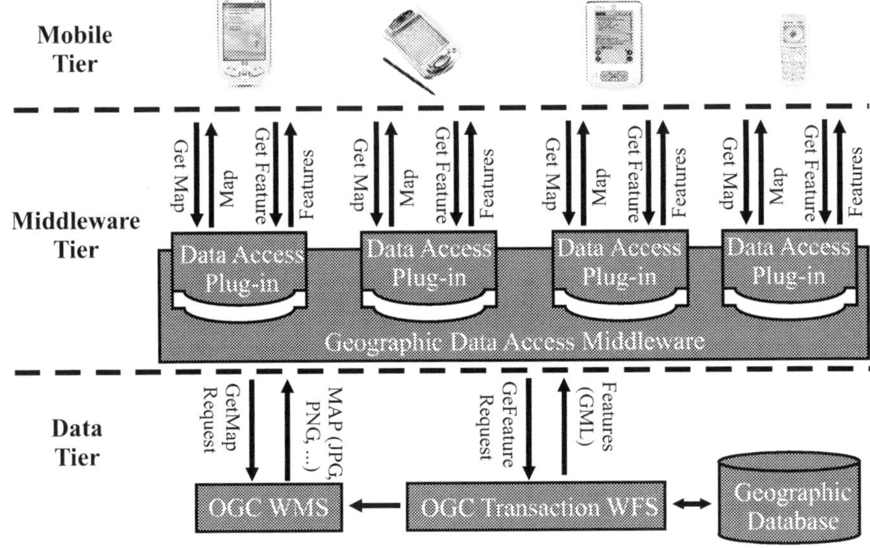

Fig. 1. Architecture of the system

Data Tier: The functionality of this tier is the one provided by the well-known OGC WMS [12] and WFS-T [10] interfaces. These services are accessed through the HTTP protocol, either by a GET or by a POST request. The interfaces define operations for retrieving service metadata and for retrieving a map (in a raster image format) or a collection of features (in GML). In the case of a WFS-T service, an operation for updating the information in the database is also defined.

Middleware Tier: The functionality of the extensible GDAM enables mobile devices with limited capabilities to invoke the operations of well-known OGC web services and assists them in processing the results. This tier supports two tasks: First, downloading *maps* and *features* from the WMS and the WFS services, and second, updating the geographic database through the *Transaction* operation of the OGC WFS-T according to the changes reported by the client PDA GIS application.

The interface between the GDAM and each different type of mobile device is provided by a different Data Access Plug-in. This interface is specifically developed for the specific capabilities of the mobile device. The requests sent by the mobile device are translated to standard requests to the appropriate OGC web services. Similarly, the responses obtained from the OGC services are transformed to formats supported by the specific capabilities of the mobile device. In particular, a *map* obtained from a OGC WMS in some raster format has to be transformed to a raster format supported by the mobile device. Similarly, a collections of *features* encoded in Geography

Markup Language (GML)[11] returned by a OGC WFS service have to be transformed to adapt them to the specific capabilities of the XML parser used by the mobile device.

Mobile Tier: The GIS software installed in the mobile devices supports the rendering and edition of the existing geo-data and the insertion of the new one. In particular, it gives support for the following tasks:

1. Invoking the DAP to download both *maps* and *features* (spatial objects in OGC terminology). The specific DAP accessed is designed and implemented to take into account the particular capabilities and limitations of the mobile device.
2. Rendering both the *maps* and the *features* on the screen of the device, offering the typical GIS navigation functionality (zooms, movements, etc).
3. Selecting *features* and enabling the edition of both their geometries and their conventional properties.
4. Creating new *features*, enabling both the digitation of their geometry and the insertion of the values of their alphanumeric properties.
5. Incorporating the positions retrieved from a GPS device both in the navigation task and also as part of the geometry of new and edited *features.*

3 A Prototype Implementation

A system prototype for a specific mobile platform was implemented based on the architecture proposed in the previous section. The mobile device chosen was a Pocket PC Fujitsu-Siemens Pocket Loox 420 with an Intel XScale PXA 255 processor, 64 MB of RAM and 32 MB of ROM, and a screen of 280x320 pixels and 65536 colors. The operating system of the PDA is Windows CE first edition. Regarding the GPS device, a Fortuna Clip-on BT was chosen, which uses the NMEA protocol and Bluetooth communications. Finally, the GIS client software was developed with the free edition of the SuperWaba [13] virtual machine. The Graphical User Interface (GUI) of the application is shown in Fig.2. A summary of the hardware and software limitations is next given:

- The maximum file size supported is 64 KB. Therefore, both map images and GML documents retrieved from the WMS and the WFS must be broken by the Data Access Plug-in into pieces of an appropriate size.
- Only bitmap images (BMP) of 8 bits per pixel are supported. To the best of these authors knowledge, none of the available tools that implement the WMS specification support such a limited format for the maps.
- An XML parser is not part of the free edition of SuperWaba. An existing parser was adapted to enable the processing of GML documents. However, a couple of limitations had to be assumed. First, the maximum length of an XML element is restricted to 512 Bytes. Second, many special characters cannot be processed by the parser.
- HTTP POST requests cannot be issued from the PDA.

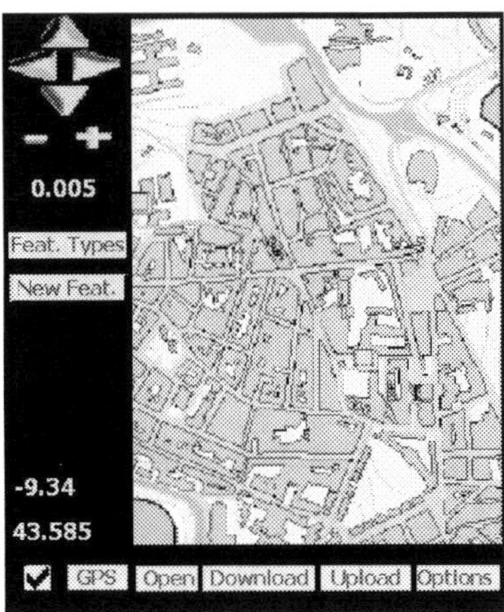

Fig. 2. Graphical User Interface of the GIS Application

The implementation consists of a Middleware Tier and a Data Tier that can be used with many different mobile devices. The limitations of our particular case were solved by a Data Access Plug-in specific to our hardware and software configuration.

Regarding the map downloading funcionality, the map returned by the Data Tier has to be transformed by the DAP before it is sent to the GIS PDA application. For example, due to the file size limitation, the map has to be split (by the DAP) into tiles of the same size and shape and an XML document with metadata describing this transformation has to be sent to the PDA GIS application to enable the reconstruction of the map there. A similar process occurs when the user of the PDA GIS application asks for features to download. Now, the result format of the relevant *GetFeature* operation of the WFS is GML. Such format has to be adapted to the functionality of the XML parser of the PDA GIS application. Thus, for example, the size of the <coordinates> element of GML is likely to exceed the 512 bytes supported by the parser. To solve this problem, the DAP transforms each <coordinates> element to a collection of <coord> elements, also supported in GML.

In the case of the updating functionality, a *Transaction* operation of the OGC Transaction WFS is invoked by the GDAM, according to the GML passed by the PDA GIS application through the DAP. Now the <coord> elements of such GML are transformed back to <coordinates> elements.

4 Conclusions and Future Work

We designed a system for the management of a geographic database from mobile devices and we undertook a prototype implementation. The use of an extensible *Geographic Data Access Middleware* in the architecture of the system enables it to be used from a wide variety of mobile devices with different capabilities. Thus, the incorporation of a new kind of mobile device requires only the development of an appropriate *Data Access Plug-in* to enable its communication with the middleware. At the *Data Tier* of the system, web services compliant with the standard Web Map Service and Web Feature Service specifications of the OGC are used.

Future work include the incorporation of other types of mobile devices to the prototype implementations (mobile phones for example), the improvement of the SuperWaba PDA GIS application and the incorporation of other OGC web service interfaces in the Data Tier (Web Coverage Service, Sensor Observation Service, etc.).

References

1. deegree Home Page. Retrieved (October 2006) from `http://deegree.sourceforge.net/`
2. ESRI. ArcIMS - Publish Maps, Data, and Metadata on the Web
3. ESRI. ArcPad - Mobile GIS Software for Field Mapping Applications
4. GeoServer: Open Gateway for Geospatial Data. Retrieved (December 2007), from `http://docs.codehaus.org/display/GEOS/Home`
5. GeoTools: The Open Sourve Java GIS Toolkit. Retrieved (January 2007), from `http://geotools.codehaus.org/`
6. Intergraph. Geomedia WebMap Professional Online Documentation (2006)
7. Intergraph. Working with IntelliWhere OnDemand (2006)
8. Luaces, M.R., Brisaboa, N.R., Paramá, J.R., Viqueira, J.R.R.: A Generic Framework for GIS Applications. In: Kwon, Y.-J., Bouju, A., Claramunt, C. (eds.) W2GIS 2004. LNCS, vol. 3428, pp. 94–109. Springer, Heidelberg (2005)
9. MapServer Home Page. Retrieved in (January 2007), from `http://mapserver.gis.umn.edu/`
10. Open Geospatial Consortium (OGC). Web Feature Service (WFS). Version 1.0, Retrieved April 2005 from (2002), `http://www.opengeospatial.org`
11. Open Geospatial Consortium (OGC). Geography Markup Language (GML). Version 3.0, Retrieved April 2005 from (2003), `http://www.opengeospatial.org`
12. Open Geospatial Consortium (OGC). Web Map Service (WMS). Version 1.3, Retrieved April 2005 from (2004), `http://www.opengeospatial.org`
13. SuperWaba. Retrieved April (2006), from `http://www.superwaba.com.br`
14. The Jump Project. Retrieved (January 2007), from `http://www.jump-project.org/`

Real-Time Creation Method of Personalized Mobile Web Contents for Ubiquitous Contents Access

SeungHyun Han[1], DongYeop Ryu[1], and YoungHwan Lim[2]

[1] School of Computing, Soongsil University,
[2] School of Media, Soongsil University,
1-1, Sangdo 5-Dong, Dongjak-Gu, 156-743, Seoul, Korea
{power5v1,aceryu}@naver.com, yhlim@computing.ssu.ac.kr

Abstract. Although wireless devices are the most suitable device for ubiquitous environment, they have restrictive capacities when using internet services than desktop environments. Therefore this research proposes a wireless internet service method that uses contents-based personalization. The existing websites can easily and promptly access desired news articles and other data through RSS-linked web contents and by the personalization method. The proposed method will make using wireless internet easier while lowering wireless contents production costs. Moreover, personalized mobile web contents that satisfy the preferences of users can be serviced.

Keywords: Contents, Mobile, Personalization, Ubiquitous, Transcoding.

1 Introduction

The limited browser installed in wireless devices, the low hardware performance, and the inconvenient interface contribute to the inefficiency in terms of time and cost when it comes to accessing the internet with wireless devices.

Various methods that convert original web-pages to mobile web-pages for the OSMU of the digital convergence environment have been studied[1], [2]. Although pages are generated relatively promptly, these methods lack the consideration of the users and the services that are optimized for the different devices as well as too many pages being generated. The following issues must be considered in order to complement such shortcomings. First, a wireless web service that considers the preferences of users will prevent resources from being wasted while enhancing user satisfaction. Secondly, wireless pages optimized to a variety of wireless devices should be serviced. Thirdly, the changes made in an original web contents must be automatically reflected in the wireless devices.

This research proposes a servicing method in which the desired information can be serviced to wireless devices by using concerned areas and concerned keywords. Mobile web contents are generated by classifying the preferred keywords, related keywords, contents suitability and preference of the generated contents. Moreover, by using RSS that can generate active channels, wireless internet can easily be accessed because new contents suitable for mobile environments can be generated at any time.

K.C. Chang et al. (Eds.): APWeb/WAIM 2007 Ws, LNCS 4537, pp. 180–185, 2007.

2 Related Studies

Contents generation and personalization methods for mobile environment, which has more restrictions compared to the PC environment, have actively been researched.

Power Browser[3] provides a function in which existing web contents can be restructured and summarized to a simpler version. Real-time data transfer is difficult when manually converting web contents for mobile[4], [5] use because the overhead consumed in the conversion occupies a large amount. All the methods listed above do not consider the individual preferences of users when it comes to contents conversion.

The basic data used in the personalization method may use explicit data provided by the user and implicit data acquired through the analysis of the flow between the user and the web server and such data analysis is done through database or log files.

Some of the data analyzing methods include reasoning using the association-rule and the CBR(cased-based reasoning)[6]. The latter is based on cases of other users who are not related to a specific user and the former is a method in which it searches for a rule that is identical to a case for reasoning however it is inefficient in that reasoning becomes impossible when a rule is not detected. In addition, the method of simple log analysis has the following shortcomings[7].

First, it is difficult to identify the contents that a user actually wants because only the page access frequency is counted. Secondly, it is troublesome to adopt the large-capacity accumulated log data to the real-time handling system if it has to be analyzed through complex handling process. The following measures can be taken to resolve the above restraints.

3 System Organization

3.1 System's Processing Structure

The structure of the system proposed in this research is as shown in Figure 1. When a wireless device requests for service the device manager analyzes the protocol header information and identifies the connected browser type, device type, and supported image and sound. This information is recorded in the personal profile through the profile manager. The profile manager acquires the requested keyword, related keyword, and preferred genre during the browsing and adds it to the personal profile. These data are used in the individualization through the personalization engine. The contents creator of the contents manager acquires the necessary parts in each item recorded in the RSS channel. Full text feeds are extracted from the linked HTML page.

In addition, media file information and source data are transferred so that the media transcoder can handle the full-text feeds related images. Media transcoder is a set of modules that converts the size, quality, and image format used in webs so that they are suitable for mobile devices. Finally, the XML+XSL that is defined specific to mobile devices through XSLT is serviced in the forms of XHTML and WML2.0.

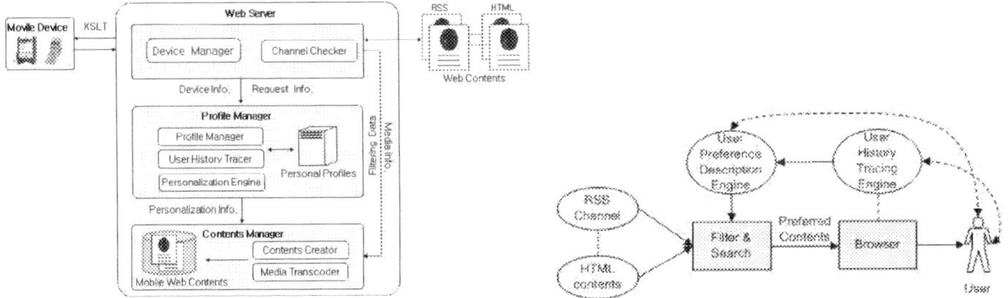

Fig. 1. System Structure

Fig. 2. Flow of user preference reflection

3.2 Applying Personalization Method

In this research, wireless pages specific to individual areas can be created through generating channels according to acquired keywords. Moreover, the preferred genres and keywords are constantly updated by reacquiring user history.

Figure 2 illustrates the reflections made in the personal profile through searching the user preference data and browsing history. Using news article-based RSS in this research, the basic genres are organized as shown in Figure 3. Preferred genres and sub-genres are implemented in the personal profile through such genre classification.

Figure 4 shows the relationship between the keywords that appear in each genre and the names of the genres and sub-genres. This relationship is an important element because when content that lack such connection becomes a concerned content it indicates an error in the personalization method.

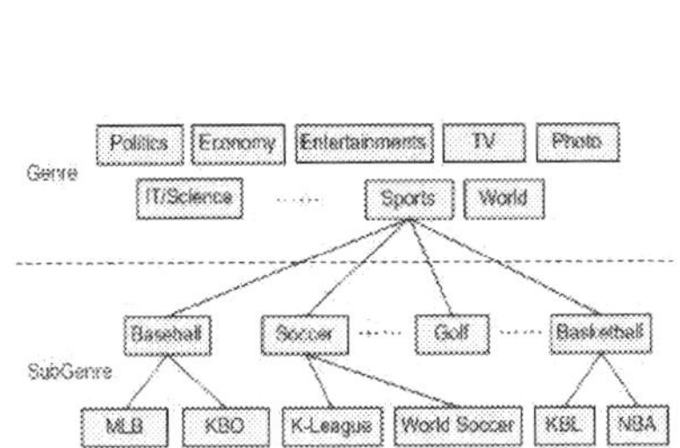

Fig. 3. Genre classification

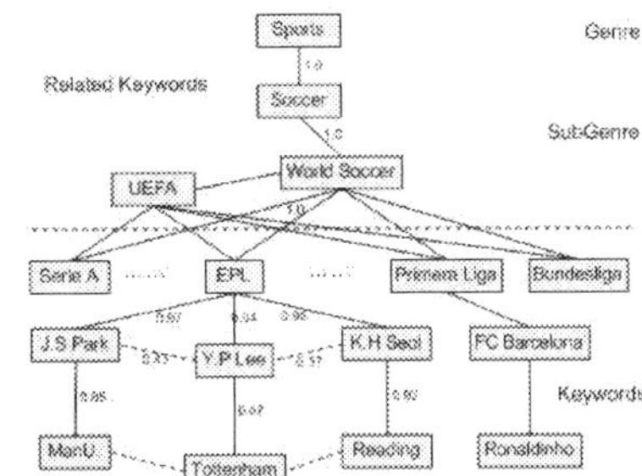

Fig. 4. Keyword classification and Related coefficient between keywords

Therefore the following equations and algorithms were used to examine the preference and the suitability of the generated contents as well as the connection between keywords.

The relational coefficient related to an arbitrary keyword m is obtained from the following equation (1).

$$RC(k_m) = \frac{freq(k_{m.}FTF) + (\sum_j FTF(Rk_j) / freq(k_{m.}FTF))}{\sum_n FTF(Pk_n)} \quad \text{where } (RC(k_m) \leq 1) \tag{1}$$

$RC(k_m)$: Relational Coefficient of Keyword *m*

$freq(k_{m.}FTF)$: Frequency of Keyword *m* in the Full Text Feeds

$FTF(Rk)$: Related Keyword in the Full Text Feeds

$FTF(Pk)$: Preferred Keyword in the Full Text Feeds

The following equation (2) is used for the suitability examination of the contents generated according to a requested keyword.

$$CSC(k_j) = \frac{freq(k_{j.}SJ) + freq(k_{j.}FTF) + \sum_j FTF(Rk_j)}{freq(k_{j.}FTF)} \quad \text{where } (CSC(k_j) > 1) \tag{2}$$

$CSC(k_j)$: Content Suitability Check of Created Channel by Keyword *j*

$freq(k_{j.}SJ)$: Frequency of Keyword *j* in the Subject

$freq(k_{j.}FTF)$: Frequency of Keyword *j* in the Full Text Feeds

$FTF(Rk)$: Related Keyword in the Full Text Feeds

The contents preference of the channel generated by a requested keyword is calculated with the following equation (4). Consequently, whether or not the channel includes the contents that the user is interested in is calculated using all kind of keyword in the channel.

$$FACC(k_i) = freq(k_i, \sum_m FTF_m(k_i)) \tag{3}$$

$$UPC(k_i) = \frac{\sum_g CC(k_g)}{\sum_r RR(k_r)} + FACC(k_i) \tag{4}$$

$FACC(k_i)$: Frequency of Keyword *i* in all Full Text Feeds on a Created Channel by Keyword *i*

$UPC(k_i)$: User Preference of Created Channel by Keyword *i*

$CC(k)$: Channel Created Keyword

$RR(k)$: User's Recently Request Keyword

Algorithm 1 shows the concerned channel configuration process that considers user preferences obtained by using the above (1), (2), (3), (4) equations.

Algorithm 1. Configuring concerned channel of user

(i) Re-enter the request keyword (Rk_n) if it doesn't exist in the keyword classifications in the sub-genre.
(ii) The newly entered keyword is added as a related keyword within the corresponding genre.
(iii) The keywords that other users requested for the same sub-genre is classified as related keywords.

(iv) All full-text-feeds of the created channels are searched and the related coefficients between related keywords are obtained and then applied by equation (1).

(v) All the keywords selected as related keywords use equation (2) for the verification of contents suitability for all the channels created within a genre.

(vi) The user preferences of all the contents within the channel that passed the contents suitability as mentioned in the above (v) are calculated according to equation (4) and the preference of the requested keyword (Rk_n) is recorded in the personal profile.

(vii) The sub-genre that contains the requested keyword (Rk_n) with high user preference is given the highest priority in terms of location and such information is also recorded in the personal profile.

4 Test Results

4.1 Test Environment

The test environment is as follows. Windows2003 Standard Edition O/S, web server IIS 6.0, Pentium-4 3.0GHz CPU, 1GB memory, and Visual Studio 6.0 and JSP were used as the development tools. OpenWaveV.7 Simulator and LG-SD910 mobile phones were used for the client connection test.

4.2 Real-Time Contents Generation and Access

Figure 5 shows mobile browser's average access time for the created contents(*i*~*n*) by the sample keywords(*i*~*n*) versus total service preparation times. A real-time service through wireless internet is possible because total service preparation times are much lower than mobile browser's access times for each channel contents.

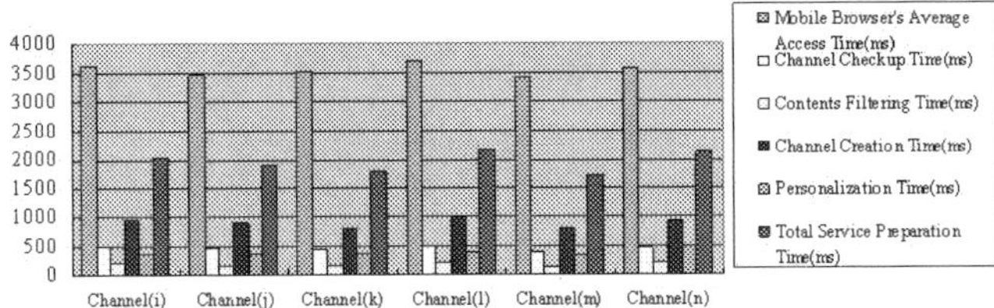

Fig. 5. Mobile browser's average access time for the created contents versus total service preparation time

Figure 6 displays the mobile web contents browsing generated through the method proposed in this research. The application of grouping through personal preference is shown in (c) of Figure 6. (a)~(e) of Figure 6 show the browsing of the personalized mobile web contents, (f)~(g) show advanced option configuration, and (h) shows the entering of the desired keyword if it is not found in the sub-genres.

Fig. 6. (a) genre selection, (b) sub-genre selection, (c) personalized channel keyword selection, (d) contents of the selected channel keyword, (e) browsing the image-included full-text-feeds, (f) image preference option disabled, (g) selection of channel keyword without personalization, (h) additional keyword entering screen for nonexistent channel keyword

5 Conclusion

This research proposes a method in which real-time contents that reflect user preferences can be accessed while diminishing the mobile web access difficulties through a creative personalization method. Successful results were achieved without performing personalization based on complex reasoning and simple frequency calculation through the proposed creation process, channel generation based on keywords, and related keyword analysis. The strength of the proposed system is the fact that services can be optimally personalized for users and wireless devices and that it can be applied to all contents that support RSS servicing.

Acknowledgments. This work was supported by the Korea Research Foundation Grant Funded by the Korea Government (MOEHRD). (KRF-2006-005-J03801).

References

1. Lu, W.W.: Compact multidimensional broadband wireless: the convergence of wireless mobile and access. IEEE Communications Magazine 38(11), 119–123 (2000)
2. Kwan, V., Lau, F., Wang, C.: Functionality Adaptation: A Context Aware Service Code Adaptation for Pervasive Computing Environments, IEEE/WIC International Conference on Web Intelligence, Halifax, Canada, October 13-17 (2003)
3. Buyukkokten, O., Garcia Molina, H., Paepcke, A., Winograd, T.: Power Browser: Efficient Web Browsing for PDAs, In: Proceedings of the Conference on Human Factors in Computing Systems, pp. 430–437 (2000)
4. http://www.avantgo.com
5. http://www.earthlink.net
6. Werner, E., Dietrich, W.: Relational Instance Based Learning, In: Proceedings 13th International Conference on Machine Learning, pp. 122–130 (1996)
7. Domingos, P.: Unifying Instance-Based and Rule-Based Induction. Journal of Machine Learning 24(2), 141–168 (1996)

The Golden Mean Operator Scheduling Strategy in Data Stream Systems

Huafeng Deng, Yunsheng Liu, and Yingyuan Xiao

School of Computer Science and Technology, Huazhong University of Science and Technology, Wuhan, Hubei 430074, China

Abstract. A scheduling strategy, called Golden Mean (GM), is designed to achieve the dynamic balance between the memory usage minimization and latency minimization by taking the status of the future workload, the current memory consumption and QoS requirement of the query into account. In the GM scheduling strategy, the executing order of the operators is decided uniformly according to the Scoring Function. The static parameters in the Scoring Function can be customized to meet the need of various application scenarios, while the dynamical parameters can be adjusted automatically according to the current system status. In addition, the priority levels of queries can be guaranteed by GM. The experimental results indicate that the GM scheduling strategy is very effective in practice.

1 Introduction

In the data stream systems, operator scheduling is very critical [1]. Some simple scheduling strategies such as FIFO and Round Robin do not touch the scheduling objectives of the memory usage minimization and the output latency minimization. Similarly, the preemptive rate-based operator scheduling strategy in [1] aims at the monitoring application to improve the interactive performance of the continuous query. Chain [2] and the scheduling strategy of Aurora [3] consider only one of the performance objectives. By targeting just one performance metric, the gains in this metric may have the side effect of degrading another metric.

Now more and more scheduling strategies consider both scheduling objectives at the same time. The Chain-Flush and Mixed strategies [2] essentially intermediated between Chain and FIFO. The SS scheduling strategy and the TS scheduling strategy [4], which are similar to the Chain-Flush strategy, schedule a time unit for a segment of an Operator Path (OP). The AMoS [5] employs a lightweight learning mechanism to assess the effectiveness of several existing scheduling algorithms and selects the algorithm that probabilistically has the best chance of improving the performance.

However, all the scheduling algorithms mentioned above have not considered the impact of the future workload, the percentage of available memory at runtime and the Quality-of-Service (QoS) requirement of the users towards the application, which directly affect the system performance and should be looked upon

K.C. Chang et al. (Eds.): APWeb/WAIM 2007 Ws, LNCS 4537, pp. 186–191, 2007.

as the important factors to decide the order of the operators. To leverage the strength of existing scheduling algorithms to satisfy both performance objectives, we design the Golden Mean scheduling strategy (simply GM for brevity) which is a dynamical, rate-based and operator-based policy.

2 Scheduling Idea and GM Scheduling Strategy

2.1 The Golden Mean Scheduling Strategy

The model and assumptions in this paper are the same as in [2]. The scheduling strategy in this study is called Golden Mean to reflect the idea of achieving the dynamic balance between minimizing the memory usage and minimizing the output latency according to the current system status.

The Golden Mean scheduling strategy proceeds as follow: at any scheduling time unit, consider all operators that are currently active in the system. Of these, schedule the operator to run, which has the highest score computed according to the Scoring Function in the next subsection. If there are multiple such operators, select the operator whose tuples in the input queues have the earliest arrival time.

2.2 The Scoring Function

The score of each operator op_i is given by Scoring Function (1). The Equation consists of three parts: the first part involving the priority of the query whose plan contains the operator op_i, the second part computed according to the Chain scheduling strategy [2], the third part reflecting the importance of output latency constraint of the queries in the data stream system.

$$SCORE(op_i) = SCP(op_i) + SCC(op_i) + SCL(op_i) \tag{1}$$

The first part is defined in equation (2):

$$SCP(op_i) = 2^k \tag{2}$$

Where k is a positive integer to represent the priority of the query corresponding to the query plan containing operator op_i. In our system, the administrator of system assigns a value between 1 and 128 to k statically according to the importance of the query.

The second part is defined in the equation (3):

$$SCC(op_i) = w.\frac{CMM(op_i)}{\sum CMM(op_j)} \tag{3}$$

Where $CMM(op_i)$ and $CMM(op_j)$ are the Capability of Minimizing the Memory usage which can be computed according to the Chain scheduling strategy [2] and w is the weight of memory minimization and represents the relative importance of minimizing the memory usage. The weight w is expressed as follows:

$$w = (c_l w_t^l + c_u w_t^u).c_{em}/2 \tag{4}$$

We will explain how to compute w in detail in the next subsection.

The third part is expressed as follows:

$$SCL(op_i) = (1 - c_{em}) \frac{UREN(ts_i)}{\sum UREN(ts_j)} \tag{5}$$

Where $UREN$ denotes a function defined in (7) to reflect the urgency of the query, ts denotes the slack time of the query and is defined in (6). In order to calculate the $UREN(ts_i)$, we should maintain a data structure in the form of four-tuple $< q, t_h, r_q, p_q >$ for each queue q in the system as in [2], where t_h is the timestamp of the head tuple of the queue q, r_q is the latency threshold predefined for the query corresponding to the query plan containing q, and p_q is the sum of the average tuple-processing times for gaining the result of the query starting from this operator. Let t_{cur} denote the current time.Then the slack time ts_i for the operator op_i is expressed as follows:

$$ts_i = t_h + r_q - t_{cur} - p_q \tag{6}$$

$$UREN(ts_i) = \begin{cases} 0 & \text{if} \quad ts_i > c_s p_q \\ 1/ts_i & \text{if} \quad 0 < ts_i < c_s p_q \\ ts_i & \text{if} \quad ts_i < 0 \end{cases} \tag{7}$$

Where c_s is the safety coefficient to avoid the occurrence of the output latency. In our system, c_s is set to be $1 - c_{em}$.

2.3 The Weight of Memory Minimization W

In (4), w_t^l is an index to reflect the future workload level of the source streams and w_t^u given by $w_t^u = \frac{memory_{used}}{memory_{total}}$ is an objective index which reflects the percentage of memory consumption. c_l, c_u is the nonnegative coefficient to reflect the relative importance of the w_t^l and w_t^u in the w respectively and c_{em} reflects the user's expectation towards memory usage minimization according to the requirement of the application and the resources of the stream system. The value of c_l, c_u, c_{em}, is less than 1 and more than 0. If the stream arrival rate of the system fluctuates frequently over time, we should set c_l at a higher value. Otherwise, c_l can be set at a lower value. Having abundant memory resources, the impact of memory usage on the scores of operators should be decreased by setting c_u at a lower value. Otherwise, c_l should be set at a higher value. The coefficient c_{em} can be set at a higher value if the users expect to minimize the memory usage. Otherwise, c_{em} should be set at a lower value.The exact value of c_l, c_u, c_{em} should be decided by the experiments.

It is easy to prove that the priorities of the queries can be guaranteed if the GM scheduling strategy is applied in the data stream systems. We also verify this in the first experiment.

3 Experiments

We use our prototype system to study the operator scheduling techniques. The machine that runs the system has a 1.8Ghz Pentium processor and a 512 MB

RAM, running Windows XP and Java 1.4.2 SDK. In our experiments we use synthetic data set as in [2]. The ten input data streams generated are highly bursty streams that have the so-called self-similarity property. We compared the performance of Chain-Flush, Chain and GM for the following three scenarios:

Scenario 1: We run 50 CQs with 281 operators. The value of coefficient c_l, c_u, c_{em} is 0.8, 0.8 and 0.9 respectively for the GM. Fifty CQs are divided into five groups randomly and each group has ten CQs. The priorities of the five groups are set to 64, 65, 66, 67 and 68 respectively.

Scenario 2: We run 50 CQs with 281 operators in the system. The value of coefficient c_l , c_u, c_{em} is 0.8, 0.8 and 0.7 respectively and the levels of priorities of all the queries are set to 64 uniformly.

Scenario 3: We run 40 actual CQs with 205 operators in the system. The value of coefficient c_l, c_u, c_{em} is 0.5, 0.4 and 0.4 respectively. The levels of priorities of all the queries are set to 64 uniformly.

Figure 1 shows the variation in the latency ratio over five query groups of different priority levels. The latency ratio is defined as tn_l/tn_t, where tn_l denotes the number of the tuples missing their deadlines among all the tn_t output tuples. As can be seen from Fig. 1, GM has lower latency ratio in the query groups of higher priority and higher latency ratio in the query groups of the lower priority. However, the latency ratios produced by Chain or Chain-Flush for five groups of queries are almost identical. This can be explained by the fact that the GM scheduling strategy decides the order of operators according to the priority rates assigned to the queries while Chain and Chain-Flush ignore the priority levels of the queries.

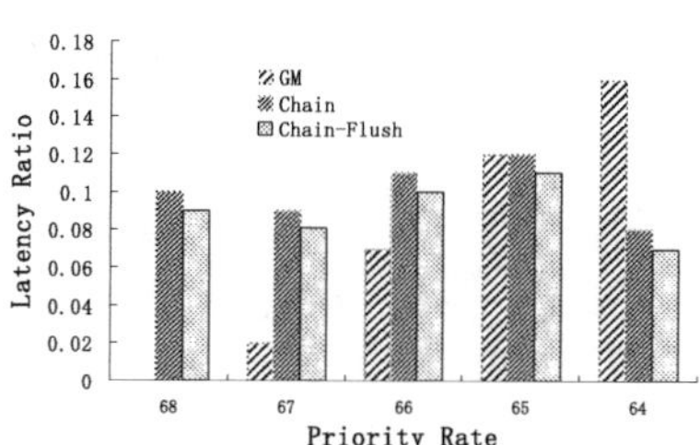

Fig. 1. Latency Ratio vs. Priority

Figure 2 shows the variation in total queue size over time in Scenario 2. We observe that the performance of the GM scheduling strategy lies between that of Chain and Chain-Flush. During the highly bursty periods, GM has a similar memory requirement as Chain, while Chain-Flush requires much more memory. The reason is that GM behaves like Chain, but Chain-Flush needs to invoke flush mode during the highly bursty periods and causes a large number of tuples to be accumulated in the queues. During the less bursty periods, the amount of memory required by Chain-Flush and Chain is similar, while GM consumes much more memory when the memory is surplus. According to the Scoring Function, GM increases the importance of minimizing the output latency by reducing the value of w when the memory is sufficient. It explains GM's performance. The output latency for the different scheduling policies in Scenario 2 is shown in Fig. 3. The output latency here is the average tuple latency of all output tuples within every one second. The tuple latency of an output tuple is computed by taking the difference of its arrival time and its departure time when it leaves the query processing system. The graphs show that the output latency of GM is much lower than that of other scheduling strategies during the less bursty periods because the GM scheduling strategy always tends to minimize the output latency when the memory resource is sufficient. During the highly bursty periods, the performance of GM lies between Chain and Chain-Flush because the flush mode used by Chain-Flush can decrease more output latency than the GM strategy and the Chain strategy. Although the performance of GM is between Chain and Chain-Flush, it is worth noting that the variety of the performance of GM is much less than those of the Chain and Chain-Flush. GM always tends to balance between minimizing the memory usage and minimizing the output latency according to the level of memory usage and future load while Chain-Flush decreases the output latency until the output latency is detected and Chain always focuses on minimizing the memory usage.

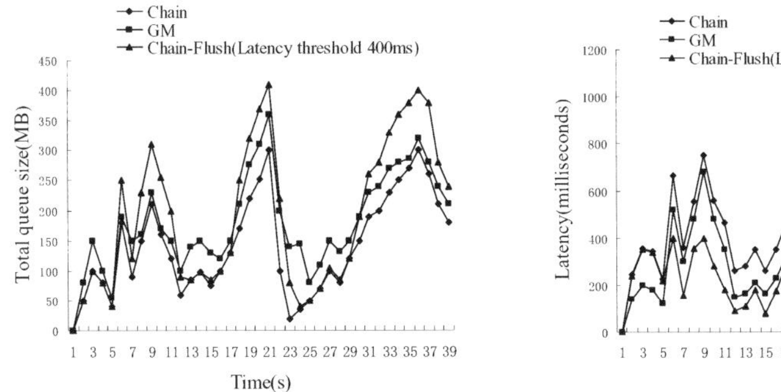

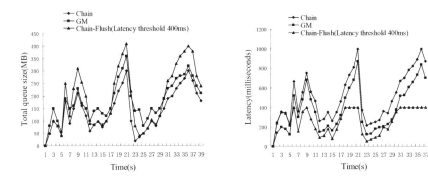

Fig. 2. Queue size vs time (Scenario 2) **Fig. 3.** Output latency vs time (Scenario 2)

Figures 4 and 5 show the variation in total queue size and output latency over time respectively. The performance and the reason in this experiment are similar to the second experiment. However, because the workload in this experiment is lighter than the second experiment, the difference between GM and Chain enlarges. GM performs better in this experiment than in the second one.

4 Conclusion

We designed the GM operator scheduling strategy for minimizing the memory usage and the output latency simultaneously. GM can schedule the operators according to the priority of the queries. Because the executing order of the operators is only decided uniformly by the Scoring Function, the overload of context

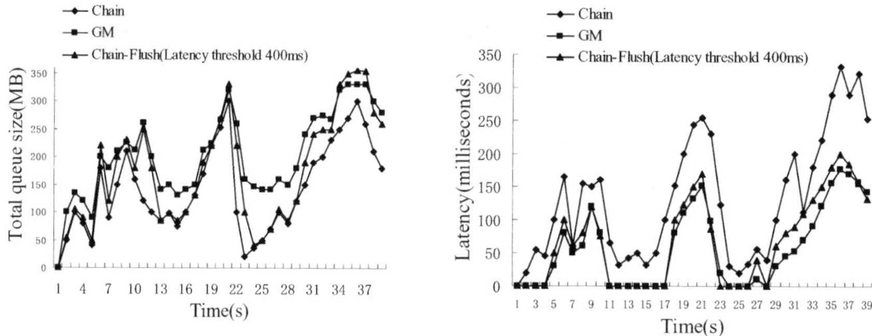

Fig. 4. Queue size vs time (Scenario 3) **Fig. 5.** Output latency vs time (Scenario 3)

switching between different scheduling strategies in the system can be avoided. The parameter w in the Scoring Function can change dynamically to reflect the condition of memory usage and the load variety. GM can perform adaptively in various application scenarios by setting the coefficients in the Scoring Function. Experimental results indicated that GM is very effective in practice.

References

1. sharaf, M.A., Chrysanthis, P.K., Labrinidis, A.: Preemptive rate-based operator scheduling in a data stream management system. In: The 3rd ACS/IEEE international conference on computer systems and applications, pp. 46–55 (2005)
2. Babcock, B., Babu, S., Datar, M., Motwani, R., Thomas, D.: Operator scheduling in data stream systems. VLDB JOURNAL 13, 333–353 (2004)
3. Carney, D., Cetintemel, U., Rasin, A., Zdonik, S., Cherniack, M., Stonebraker, M.: Operator scheduling in a data stream manager. In: Proc. 2003 international conference on very large data bases, pp. 838–849 (2003)
4. Jiang, Q.C., Chakravarthy, S.: Scheduling Strategies for Processing continuous queries over Streams. In: Proc. 2004 21st British National Conference on Databases, pp. 16–30 (2004)
5. Sutherland, T.M., Pielech, B., Zhu, Y.L., Ding, L.P., Rundensteiner, E.A.: An Adaptive multi-objective scheduling selection framework for continuous query processing. In: Proc 2005 the 9th International Database Engineering & Application Symposium, pp. 445–454 (2005)

Continuous Skyline Tracking on Update Data Streams*

Li Tian[1,**], Le Wang[1], AiPing Li[1], Peng Zou[2], and Yan Jia[1]

[1] School of Computer,
[2] Associate Provost of Education Department
National University of Defense Technology, Changsha, Hunan, China 410073
arnotian@163.com

Abstract. We consider the continuous skyline computation in a new scenario named as *update stream*, where the *First-In-First-Out* rule (which is the basic character in sliding window model) does not hold, causing the existing algorithms inapplicable. An algorithm called *BUSM* is developed and analyzed, which can continuous compute the change of skyline caused by update messages; then a progressive algorithm *GUSM* based on a novel grid-indexed data structure is proposed, which makes use of the character that deletion and addition operation appears simultaneously in update streams, and represents the influence region by grids for early elimination. Analytical analysis and experimental evidences show the efficiency of proposed approaches.

Keywords: skyline, update data stream, grid-indexed data structure.

1 Introduction

The skyline operator is important for multi-criteria decision making. For two points X and Y in the d-dimensional space, X *dominates* Y if it is as good as or even better than Y in all dimensions, and better than Y in at least one dimension. Given a set of points P, the skyline operator returns all points which are not dominated by any other point.

Throughout this paper, we will refer to the continuous skyline computation in the following example scenario which represents a typical application of monitoring, and we call it as *update stream* environment. The example application considered here is stock monitoring in a stock market. A stock can be described by several statistic attributes such as *risk*, *volume* and *average price*, etc. Each deal (transaction) may update the attributes of corresponding stock, making it is a real challenge to continuously compute the skyline results. Notice that the above *update stream* environment owns several properties which make it quite different from the sliding window streams. (1) The change of skyline in sliding window stream is caused by the change of points under considering, while in update data stream environment, the skyline is changed because objects move its position in workspace. (2) Points in sliding windows are held in a *First-In-First-Out* manner. Taking full advantage of this property, the existing methods [5,6] expunge tuples from the system as early as possible

* Supported by national 863 high technology development foundation (No. 2004AA112020).
** Corresponding author.

K.C. Chang et al. (Eds.): APWeb/WAIM 2007 Ws, LNCS 4537, pp. 192–197, 2007.

and obtain space and time efficiency. While in the new scenario, the *FIFO* rule is not hold. All these differences make the existing algorithms inefficient or simply useless in this scenario, which is the motivation of our research in the paper.

The problem can be formulized as follows: The *N* objects interested are denoted by $D=\{O_1, O_2, \ldots, O_N\}$, each object is represented as $<id, x_1, x_2, \ldots, x_d>$, where *id* is a unique identifier, x_i $(i=1..d)$ are the *d* attribute values of the object. Update message $tp=<id,\Delta x_1,\Delta x_2,\ldots,\Delta x_d>$ modifies the corresponding object O_{id} according to equation $x_i = x_i + \Delta x_i$. *SK* is used to denote the skyline set, *RE=D-SK*. The goal is to continuous outputs the skyline change stream with form $<t,\pm,O_{id}>$ with the continuous arriving of update messages, which means that at time *t*, the object O_{id} becomes the skyline point (+) or is removed from the current skyline (-).

The rest of this paper includes: Section 2 reviews the related works. In section 3, algorithm *BUSM* is proposed. To cope with shortages of *BUSM*, a grid-indexed structure is presented in section 4, and algorithm *GUSM* is raised based on this structure. Section 5 experimentally verifies the efficiency of the proposed techniques. And lastly, section 6 concludes the paper with directions for future works.

2 Related Works

S. Borzsonyi et al.[1] firstly extend the database systems with a skyline operation and develop two algorithms called *BNL* and *D&C*. A progressive algorithm *bitmap* and an *index*-based method is raised in [2] , both methods can output skyline points without having to scan the entire data input. Kossmann et al.[3] present an algorithm called *NN* due to its reliance on nearest neighbor search. Papadias et al. [4] present a progressive algorithm named *BBS*, which is also based on nearest neighbor search and is proved to be I/O optimal. More recently, system architecture and two algorithmic frameworks are proposed in [5] for continuously monitoring skyline changes over sliding windows. Experimental evaluation illustrates that the indexed implementations have low amortized overhead, and hence, they can support fast streams. X. Lin et al.[6] study the *n-of-N* skyline queries over continuous streams. Properties of sliding windows are fully utilized to minimize the number of elements to be kept. A novel encode schema is developed to map the skyline computation to the *stabbing query* problem. The technique is proved to take effect in supporting on-line skyline computation over very rapid data streams. Several variations of skyline are also researched, such as subspace skyline[7] and k-dominant skyline[8], etc.

Our research differs from all above mainly in considering a quite different scenario and proposes effective algorithms suit for this specific environment.

3 BUSM Algorithm

The *Basic Update-stream Skyline Monitoring algorithm (BUSM)* is a tweak of the algorithm for *Lazy*[5], which streats the update action as the combination of deletion of the corresponding old position and addition of a new position, and processes them independently. The alogrithm is described in Fig. 1.

```
Algorithm: BUSM (tp)
Input: An update tuple tp
Output: Modification of OD, and skyline change stream, if needed.
    r is the old point position corresponding to tp.id
    r' is the new point position corresponding to tp.id
    if r ∈ SK then
        output (t, -, r)
        compute the skyline of data set in RE dominated exclusively by r, denoted by S
        for each point e ∈ S
            output (t, +, e), and move e from RE to SK
    delete point r from OD
    if ∃e ∈ SK , dominate(e,r') then
        add r' to RE
    else
        output(t, +, r') and add r' to SK
        Perform range query on SK and compute the set S' which consists of points dominated by r'
        for each point e ∈ S'
            output (t, -, e), and move e from SK to RE
```

Fig. 1. Pseudo code description of algorithm *BUSM*

4 GUSM Algorithm

Fig. 2 shows an example of *BUSM*'s shortcoming where a skyline point moves from *a* to *b*. *BUSM* firstly computes the skyline of data set exclusively dominated by *a*, assuming is *{c, d, e}*, and adds them into *SK*. Then it continues to process the addition of *b*, removing these three points from *SK* because all of them are dominated by *b*.

The *Grid-based Update-stream Skyline Monitoring algorithm (GUSM)* is present for operation reduction. Similar to those existing approaches[8,9], we use a regular grid to index objects. The data structure in a 2-*d* space is shown in Fig. 3.

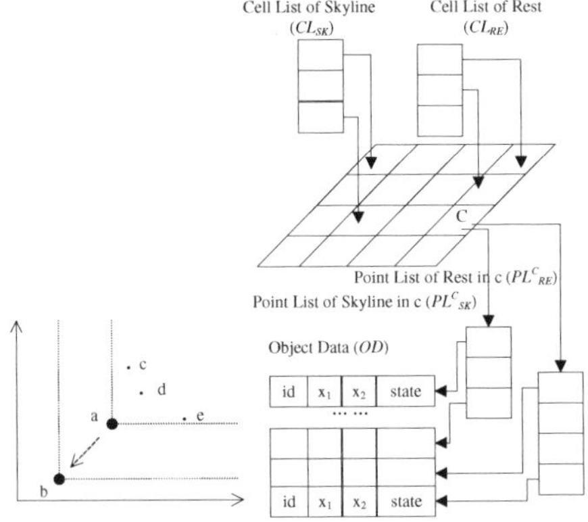

Fig. 2. Shortage of *BUSM* **Fig. 3.** Grid-indexed data structure

Each object is represented as <*id*, x_1, x_2>. The extent of each grid on every dimension is δ, so that grid $C_{i,j}$ at column *i* and row *j* (starting from the low-left corner of the workspace) contains all objects satisfying $i\cdot\delta \le O.x_1 < (i+1)\cdot\delta$ and

$j \cdot \delta \leq O.x_2 < (j+1) \cdot \delta$. Symbol *IN(O,*$C_{i,j}$*)* is used to denote this relationship. There are two point lists in each grid, $C_{i,j}.PL_{SK} = \{O | O \in SK, and, IN(O, C_{i,j})\}$, $C_{i,j}.PL_{RE} = \{O | O \in RE, and, IN(O, C_{i,j})\}$, and two grid lists in the system, $CL_{SK} = \{C_{i,j} | C_{i,j}.PL_{SK} \neq NIL\}$, $CL_{RE} = \{C_{i,j} | C_{i,j}.PL_{RE} \neq NIL\}$. $C_{i,j}.LB$ and $C_{i,j}.RT$ are used to denote the left-bottom and right-top point of grid $C_{i,j}$, respectively. We define that grid $C_{i,j}$ is strictly dominated by point O if *Dominate(O,* $C_{i,j}.LB$*)* is satisfied; $C_{i,j}$ is laxly dominated by O if *Dominate(O,* $C_{i,j}.RT$*)* is hold.

Theorem 1. For a point O, a grid $C_{i,j}$ and arbitrary point e in $C_{i,j}$. (1) If *StrictDominate(O,*$C_{i,j}$*), then Dominate(O,e); (2) If LaxDominate(O,*$C_{i,j}$*) is not hold, then Dominate(O,e) is not satisfied.*

Theorem 2. If an update tuple moves an object from position r to r', we get: (1) Suppose point $e \in RE$ is exclusively dominated by r and IN(e,C), then *LaxDominate(r,C) and NOT StrictDominate(r',C); (2) Suppose point* $e \in SK$ *is dominated by r' and IN(e,C), then LaxDominate(r',C) and NOT StrictDominate(r,C).*

Main idea of *GUSM* is as follows: (1) Grid is used as unit for rapid estimation. Some grids are discarded for further examine with correctness guarantee. (2) The move action is treated holistically, and influence region of update tuple is represented with grids. Pseudo code description of algorithm *GUSM* is shown in Fig. 4. Update operation is considered holistically at line 6 and 16 according to theorem 2, while theorem 1 can be widely used for performance improvement at line 3, 6, 13 and 16.

Algorithm: *GUSM (tp)*

Input: *An update tuple tp*
Output: *Modification of OD, and skyline change stream, if needed.*

1 *r* is the old point position corresponding to *tp.id*, and C_{old} is the cell *r* in.
2 *r'* is the new point position corresponding to *tp.id*, and C_{new} is the cell *r'* in.
3 if $r \in SK$ then
4   Remove r from $C_{old}.PL_{SK}$, and update the data structure
5   *output (t, -, r)*
6     $A = \{ \cup C.PL_{RE} | C \in CL_{RE}$, *LaxDominate(r,C), not StrictDominate(r',C), and is not strict-dominated by any Skyline point}*
7   Compute the skyline of *A* dominated by *r* but not by any other Skyline point, denoted by *S*.
8   for each point $e \in S$ which is in cell C
9   *output (t, +, e)*
10   *move e* from $C.PL_{RE}$ to $C.PL_{SK}$, and update the data structure
11 else
12   Remove r from $C_{old}.PL_{RE}$, and update the data structure
13 if *r'* is a Skyline point then
14   *output(t, +, r')*
15   add *r'* to $C_{new}.PL_{SK}$, and update the data structure
16     $B = \{ \cup C.PL_{SK} | C \in CL_{SK}$, *LaxDominate(r',C), but not StrictDominate(r,C)}*
17   for each point $e \in B$ in cell C, if *Dominate(r',e)* then
18   *output (t, -, e)*
19   remove *e* from $C.PL_{SK}$ to $C.PL_{RE}$, and update the data structure
20 else
21   *add r'* to $C_{new}.PL_{RE}$, and update the data structure

---

Fig. 4. Pseudo code description of algorithm *GUSM*

5 Experimental Evaluation

A simulation system is implemented to evaluate the efficiency of the proposed techniques. All of the experiments have been carried out on a Pentium 4 PC with a 2.8GHz processor, 1GB main memory, and Windows XP OS. The following algorithms are evaluated based on the three most popular synthetic benchmark data, correlated, independent, and anti-correlated[1-6]. (1)*L_BUSM*: instance of *BUSM* which organizes *SK* and *RE* with linked lists. (2)*I_BUSM:* instance of *BUSM* where appropriate indexes and binary search tree are adopted on *RE* and *SK* to improve efficiency. (3) *GUSM*: an implementation of *GUSM* which employs a binary search tree together with the Z-curve[10] technique for grids organization.

Set N=100,000 , and vary d from 2 to 7 to evaluate the effect of dimensions on performance. Fig. 5 (a)-(c) shows that about 10^3 update tuples can be processed even when the data are anti-correlated and d is large. *I_BUSM* performs better than *L_BUSM*, indicating the importance of indexes. *GUSM* stands out only when d is large in independent and anti-correlated data. The reason is: $|SK|$ is small in correlated data set or when d is small, therefore most updates happen in *RE*. In this case the average cost depends mainly on the cost of data structure maintenance. However, when the skyline size is large, the update is skyline-point-related with high probability, thus operations such as *emptiness test* and *range search* are needed , and only in this case *GUSM* shows its predominance.

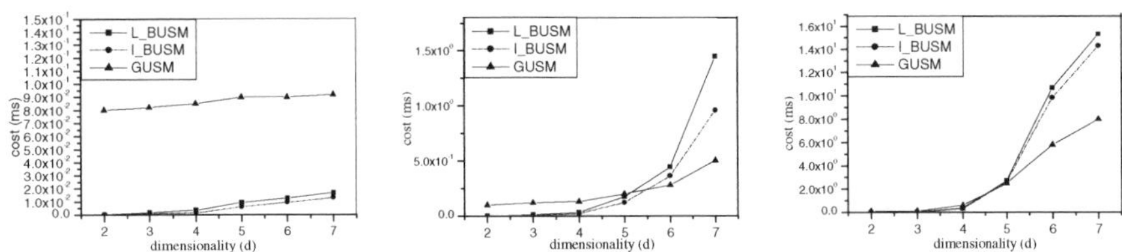

Fig. 5. Average CPU time vs. dimension (N=100,000; δ=30) (a)corr; (b)indep;(c)anti

Take the anti-correlated data as an example, Fig. 6 shows how the performance is affected by N. Comparing with *L_BUSM* and *I_BUSM*, the *GUSM* algorithm is more stable. This is because only parts of grids are needed to make further examination, while the estimation operation on grids are not increased proportionally.

Grid width δ is a key parameter of *GUSM*. Use U to denote the domain of object attribute values in each dimension, then U/δ is the grid number in each dimension.The effect of grid size on *GUSM* performance is show in Fig. 7. The larger a grid is, the shorter CL_{SK} and CL_{RE} are. The estimation operation also reduces. On the other hand, large grid will decrease the effect of estimation operation and cause PL_{SK} and PL_{RE} longer, which will degrade the performance. Therefore, with the decrease of δ, *GUSM* firstly displays an increase trend on performance; however, after some threshold, it performs worse and worse with the decrease of δ.

Taking the independent data as an example, the storage cost of the above algorithms is shown in Fig. 8. The difference in storage cost is caused by different data structure used by algorithm. $2N$ additional memory units are needed for *L_BUSM* to store tuples <id, nextPointer> in *SK* and *RE* list, while *I_BUSM* need 3*N* units because each tuple in binary search tree consists of two pointers. Each none-empty grid in *GUSM* should occupy 3 memory units for Z-Value and binary search tree structure, while PL_{SK} and PL_{RE} in all grids totally need 2N units. Therefore $2N+3(|CL_{SK}|+|CL_{RE}|)$ additional memory units are need for *GUSM*.

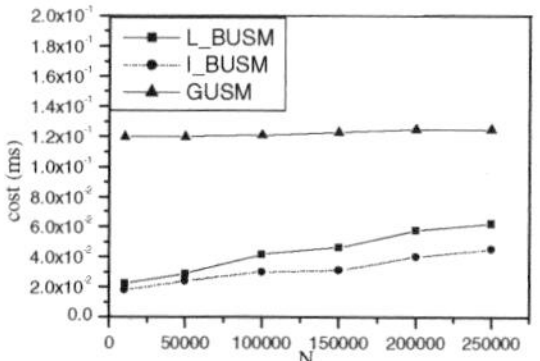

Fig. 6. Average CPU time vs. *N* (anti, d=3, δ=50)

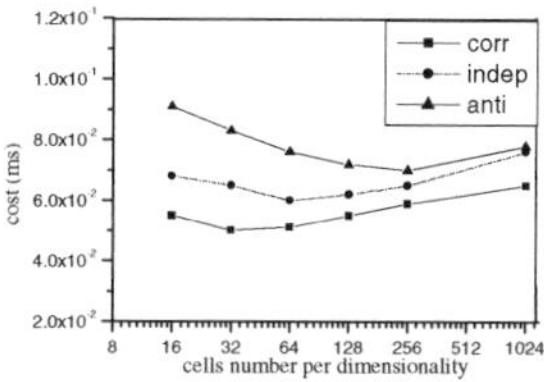

Fig. 7. *GUSM* performance vs. δ (d=2, N=100,000)

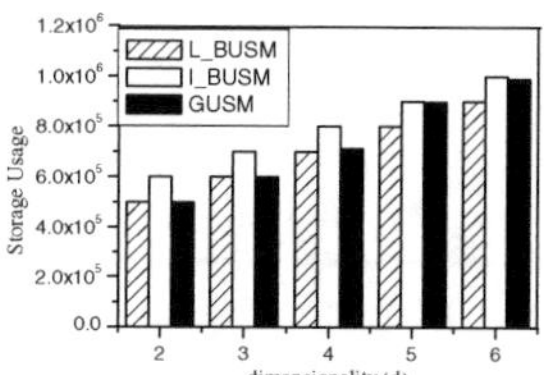

Fig. 8. Storage cost (indep; *N*=100,000; δ=50)

6 Conclusion and Future Works

The continuous skyline computation on update stream environment is researched and two algorithmic frameworks called *BUSM* and *GUSM* are raised. Both are effective on rapid update streams (about 10^3 update tuples can be processed even when the data are anti-correlated and *d* is large). Analytical analysis and experimental evidences show that the former adapts to scenario where skyline size is small, while the latter performs comparatively better when there is a large skyline size or dimensions.

In future steps, we plan to conduct a more extensive analysis and get some theoretic guidance about the influence of grid size on performance. Research on effective algorithm for continuous skyline tracking is an open problem.

References

[1] Borzsonyi, S., Kossmann, D., Stocker, K.: The skyline operator, ICDE (2001)

[2] Tan, K., Eng, P., Ooi, B.: Efficient progressive skyline computation. VLDB (2001)

[3] Kossmann, D., Ramsak, F., Rost, S.: Shooting stars in the sky: an online algorithm for skyline queries, VLDB (2002)

[4] Papadias, D., et al.: Progressive skyline computation in database systems.TODS, 30(1) (2005)

[5] Tao, Y., et. al.: Maintaining sliding window skylines on data streams. TKDE, 18(3) (2006)

[6] Lin, X., et al.: Stabbing the sky: efficient skyline computation over sliding windows. ICDE (2005)

[7] Pei, J., et al.: Catching the best views of skyline: a semantic approach based on decisive subspaces. VLDB (2005)

[8] Koudas, N., et al.: Approximate NN queries on streams with guaranteed error/performance bounds. VLDB (2004)

[9] Mouratidis, K., et al.: Continuous monitoring of top-k queries over sliding windows. SIGMOD (2006)

[10] Orenstein, J., et al.: A class of data structures for associative searching. PODS (1984)

A Hybrid Algorithm for Web Document Clustering Based on Frequent Term Sets and k-Means*

Le Wang, Li Tian, Yan Jia, and Weihong Han

Computer School, National University of Defense Technology, Changsha, China
{wanglelemail,TIAN.L.cn}@163.com
jiayanjy@vip.sina.com
hanweihong@gmail.com

Abstract. In order to conquer the major challenges of current web document clustering, i.e. huge volume of documents, high dimensional process and understandability of the cluster, we propose a simple hybrid algorithm (SHDC) based on top-k frequent term sets and k-means. Top-k frequent term sets are used to produce k initial means, which are regarded as initial clusters and further refined by k-means. The final optimal clustering is returned by k-means while the understandable description of clustering is provided by k frequent term sets. Experimental results on two public datasets indicate that SHDC outperforms other two representative clustering algorithms (the farthest first k-means and random initial k-means) both on efficiency and effectiveness.

1 Introduction

The topic of document clustering has been extensively studied in many literatures [4]. The major challenges for document clustering consist currently in the following three domains [5]: very high dimensionality of the data (~ 10,000 terms), very large size of the databases (particularly the World Wide Web) and understandable description of the clusters.

K-means [7] (or its variants) is a good choice for above challenges, because of its efficiency and effectiveness [8]. However, k-means processes term frequency vectors on full high-dimensional space and can not provide any understandable description of the found clusters. More serious is that k-means is too sensitive to the choice of initial partitions. Frequent term-based text clustering [5][6] is another preferred method for above challenges, since it clusters web documents with understandable description of the clusters and reduces drastically the dimensionality of the data. However, this kind of methods does not outperform other algorithms (variants of k-means) with respect to the cluster quality [1].

It is a nature idea to synthesize these two algorithms into a hybrid clustering method. The hybrid method firstly employs frequent term-based method to produce initial clustering and an understandable description, then uses k-means to refine the

* This project is sponsored by national 863 high technology development foundation (No. 2004AA112020) and the National Grand Fundamental Research 973 Program of China (No. 2005CB321804).

K.C. Chang et al. (Eds.): APWeb/WAIM 2007 Ws, LNCS 4537, pp. 198–203, 2007.

initial result and obtain an optimal clustering. MFI K-means [1] is a good case of this idea. But its clustering effect is unsteady along with documents length. This is discussed detailedly in section 2. We propose a Simple Hybrid Document Clustering algorithm (SHDC) based on top-k frequent terms and k-means algorithm. SHDC eliminates the limit of document length and simplifies the process of generating initial partitions. Experiments performed on real public datasets show that SHDC has outperforming effectiveness and efficiency compared with the same baseline of MFI K-means.

The rest of the paper is organized as follows. The related works are concluded in section 2, and the hybrid algorithm SHDC is present in Section 3. The experimental results are reported in section 4, and finally in section 5, we conclude the paper and discuss future works.

2 Related Works

Frequent term sets are sets of terms (words or phrases) co-occurring in more than a threshold percentage of all documents in a database [3], which can be efficiently discovered by association rule mining algorithms. F. Beil and Benjamin C.M. Fung developed different frequent term based text clustering algorithms [5][6] respectively. Although do not outperform others notably with regard to effectiveness, this kind of algorithms avoid high dimension processing and provide reasonable clustering description, which is extremely remarkable. Besides, the performance of frequent term based clustering algorithms is largely determined by the overlap eliminating function [12].

There are mainly three major families on the initial partitions selection problem of k-means: random sampling methods, distance optimization methods, and density estimation methods [9]. All of these methods are based on high dimension space and computational intensive. When coming to large volume of text corpora, random sampling methods efficiency is suppressed.

The most relevant work is [1], as they also consider the hybrid clustering method and employ maximal frequent item sets to produce the initial partitions for k-means (MFI K-means). If one frequent item set does not have a superset which is also frequent, it is defined as maximal frequent item set. The set of documents which support some maximal frequent item set is an initial cluster for k-means. The clustering result of this method is affected by document length, because the long documents have more opportunity to contain the max frequent term sets than the shorter. So MFI K-means prefers the long documents to produce the initial partitions. In order to fix this problem, that paper generates the document-feature data set by only using the first 300 words of each document. Our work differs from MFI K-means mainly in that we employ a different approach called top-k frequent term sets to produce the initial partitions, which not only avoid the problem of long documents in [1], but simplify the process of generating initial partitions. The simplification of initial partitions generation is beneficial to efficiency and is very important for large volume of corpora.

3 The Hybrid Algorithm

The SHDC algorithm firstly finds top-k frequent term sets using parTFI algorithm [2]. The document set, which contains all terms of one frequent term set, is treated as one cluster candidate and its mean is used as seed to aggregate the similar documents into an initial cluster for k-means. The final optimal clustering is return by k-means on the base of initial clusters.

3.1 Producing Initial Clusters

We use a parallel top-k frequent term sets mining algorithm, called parTFI, to find top-k frequent terms sets. The parTFI is introduced in one of our previous works [2].

Definition 1 (Top-k frequent term set). A term set F is a frequent term set if $sup(F) \geq min\_support$. A frequent term set F is a top-k frequent term set of minimal length min_length if there exist no more than (k-1) term sets of length at least min_length whose support is higher than that of F.

In the process of finding top-k frequent term sets, the $min\_support$ is adaptively selected according to probabilities of document distribution among frequent term sets. min_length is a restriction to length of frequent term set, so that very short term sets, which have extreme high support but very few terms, will not be treated as seeds. Because long term sets will not exist in short documents, the clustering effect of SHDC will not be disturbed by very short documents. Similarly, top-k frequent term sets will not exist only in long documents which are minority. So, long documents could not disturb clustering effect of SHDC too.

All documents that contain some top-k frequent term set are grouped into one initial cluster (candidate). Then means for k initial clusters (candidates) are calculated respectively according formula (1), and used further as seed to aggregate all documents into k clusters (truly initial clusters). Finally, the refined clustering is returned by k-means based on initial clusters above.

$$t_j^{(l)} = 1/n_l \cdot \sum_{d_i \in \mathrm{cov}(F_l)} t_{ij} \tag{1}$$

$t_j^{(l)}$ is the j-th dimension of mean of l-th initial cluster (candidate). $cov(F_l)$ denotes all documents which contain the frequent term set F_l, and n_l denotes the cardinality of $cov(F_l)$. t_{ij} is the j-th dimension of d_i. All vectors of documents have been preprocessed according TF-IDF.

The document in database is grouped into l-th cluster if the similarity between this document and l-th mean $m^{(l)}$ is the biggest one among all clusters (or candidates). The similarity is defined as formula (2).

$$sim(d, m^{(l)}) = \sum \frac{d \cdot m^{(l)}}{|d| \cdot |m^{(l)}|} \tag{2}$$

3.2 Refining the Clustering

K-means would perform the refining process on the initial clusters. The k-means will iteratively operate until there are no changes on document affiliation. SHDC is described as below. We suppose there are m documents and n terms in database.

Algorithm SHDC algorithm
Input: Web document set $D=\{d_1,d_2,\cdots,d_m\}$
Stop word list S
Cluster number k
Frequent term set Length threshold *min_length*
Output: k disjoint clusters of documents
Method:
1: For each document in D, remove the HTML tags and stop words in S and stem all words. Generate the term list T.
2: Calculate the document frequency, df, of terms in T.
For each document in D, generate TF-IDF vector

$$d_i=\left\{tf_1\cdot {}^1\!/_{df_1},tf_2\cdot {}^1\!/_{df_2},\cdots,tf_n\cdot {}^1\!/_{df_n}\right\}$$

3: Employ parTFI to generate the top-k frequent term sets $F=\{F_1,F_2,\cdots,F_k\}$ with the length threshold *min_length*
For each frequent term set, produce the document group in which all documents include this term set
According to formula (1), calculate the means for all document groups $M=\{m^{(1)},m^{(2)},\cdots,m^{(k)}\}$, where $m^{(l)}=(t_1^{(l)},t_2^{(l)},\cdots,t_n^{(l)})$.
4: For each document in D, assign it to the cluster which has the maximal similarity with this document according to formula (2).
5: Employ k-means to produce the clustering result on the base of clustering got on above step. Each cluster can be interpreted by its initial top-k frequent term set.

4 Experimental Evaluations

L. Zhuang [1] compared the validity of MFI K-means with two representative k-means algorithms (the farthest first k-means and random initial k-means). For impartiality consideration, we do not develop the source code self-assertion. We make an indirect comparison between our method and MFI K-means by comparing SHDC with the same two representative algorithms used in [1].

We use two cluster validation methods, i.e. Silhouette Coefficient (SC) [10] and normalized mutual information (NMI) [11], to evaluate the clustering performance on public text datasets. The following two data sets are used: (1) the Reuters-21578 corpus, and (2) 20-newsgroups data. One HP unit with 4 Itanium II 1.4G processors and 48 GB memory is used as hardware platform.

4.1 Experimental Results

For all these three algorithms which are evaluated in the experiments, we use a maximum number of iterations of 10 (to make a fair comparison). Each experiment runs ten times, and at each time, we start from a different random initialization. The averages and standard deviations of the NMI are reported in tables.

Table 1. NMI results on 20-Newsgroup

k	5	15	20	25
Random k-means	.23 ± .03	.28 ± .02	.36 ± .03	.42 ± .02
Farthest first k-means	.37 ± .02	.40 ± .02	.42 ± .01	.53 ± .03
SHDC	**.42 ± .03**	**.48 ± .02**	**.66 ± .01**	**.60 ± .02**

Table 2. NMI results on Reuters-21578

k	40	60	80	100
Random k-means	.22 ± .03	.26 ± .01	.42 ± .02	.39 ± .03
Farthest first k-means	.33 ± .03	.41 ± .02	.53 ± .01	.51 ± .02
SHDC	**.45 ± .03**	**.52 ± .01**	**.64 ± .02**	**.58 ± .02**

The SC and running time results are shown by figures. By the limited space, only main points are displayed.

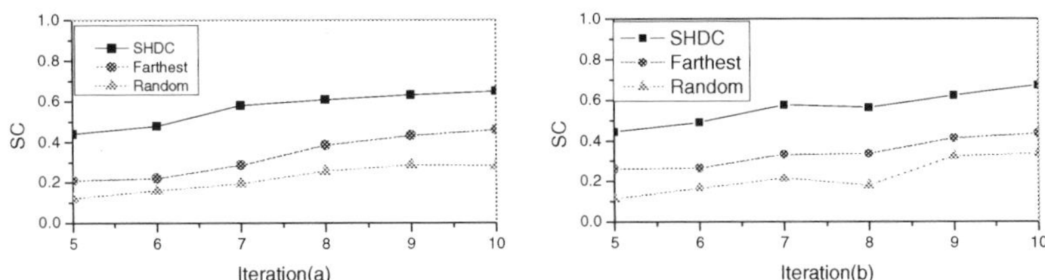

Fig. 1. Comparing the average SC results on NG20(a) and on Reuters-21578(b)

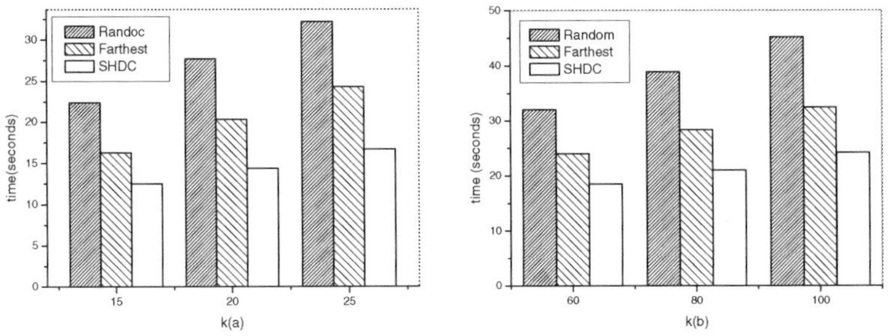

Fig. 2. Comparing the average time performance on NG20(a) and Reuters-21578(b)

From the results we can see that the SHDC get better effectiveness than other two k-means variants, while the random initial partition k-means is the poorest one, which results from the sensitiveness of k-means for initial partitions. Provided optimal starting condition, the k-means could achieve better results than those with random initial partitions.

5 Conclusions and Future Works

In this paper, we propose a hybrid clustering algorithm called SHDC for web document clustering. Top-k frequent term sets are used to produce initial partitions for k-means and provide an understandable description for final clustering. SHDC can process documents fairly with various lengths and has no bias on any length of document. We compared the results of SHDC with two other k-means variants on tow public datasets. SHDC outperforms both on effectiveness and efficiency.

The future work towards SHDC includes enhancing its scalability to accommodate mass text mining applications, e.g. employing parallel implementation.

References

1. Zhuang, L., Dai, H.: A Maximal Frequent Itemset Approach for Web Document Clustering. In: Proceedings of the Fourth International Conference on Computer and Information Technology (CIT'04)
2. Yongheng, W., Yan J., Shuqiang, Y.: Parallel Mining of Top-K Frequent Items in Very Large Text Database. WAIM (2005)
3. Han, J., Kamber, M.: Data Mining. Concepts and Techniques, 2nd edn. Morgan Kaufmann Press, Seattle, Washington, USA (2006)
4. Hotho, A., Nürnberger, A., Paaß, G.: A Brief Survey of Text Mining. LDV FORUM 20, 19–62 (2005)
5. Beil, F., Ester, M., Xu, X.: Frequent Term-Based Text Clustering. In: Proc. 8th Int. Conf. on Knowledge Discovery and Data Mining (KDD '2002), Edmonton, Alberta, Canada (2002)
6. Fung, B.C.M., Wang, K., Ester, M.: Hierarchical Document Clustering using Frequent Itemsets. SDM (2003)
7. MacQueen, J.B.: Some methods for classification and analysis of multivariate observations. In: Proceeding of the 5th Berkeley symposium in mathematics and probability (1967)
8. Steinbach, M., Karypis, G., Kumar, V.: A Comparison of Document Clustering Techniques. In: Proc. TextMining Workshop, KDD 2000 (2000)
9. He, J., Lan, M., et al.: Initialization of cluster refinement algorithms: a review and comparative study. International Joint Conference on Neural Networks (2004)
10. Hotho, A., Maedche, A., Staab, S.: Ontology-based Text Document Clustering. KI 16(4), 48–54 (2002)
11. Strehl, A., Ghosh, J.: Cluster ensembles - a knowledge reuse framework for combining partitions. Journal of Machine Learning Research 3, 583–617 (2002)
12. Shi, Z., Ester, M.: Performance Improvement for Frequent Term-based Text Clustering Algorithm. Technique Report in Computing Science, Simon Fraser University (Apr 2003)

OMSI-Tree: Power-Awareness Query Processing over Sensor Networks by Removing Overlapping Regions*

Wei Zha[1], Sang-Hun Eo[2], Byeong-Seob You[2], Dong-Wook Lee[2], and Hae-Young Bae[3]

Dept. of Computer Science & Information Engineering, Inha University, Yonghyun-dong, Nam-gu, Incheon, 402-751, Korea
[1]zhazhago@hotmail.com
[2]{eosanghun,subi,dwlee}@dblab.inha.ac.kr
[3]hybae@inha.ac.kr

Abstract. Sensor networks have played an important role in our daily life. The most common applications are light and humidity monitoring, environment and habitat monitoring. Window queries over the sensor networks become popular. However, due to the limited power supply, ordinary query methods can not be applied on sensor networks. Queries over sensor networks should be power-aware to guarantee the maximum power savings. In this paper, we concentrate on minimal power consumption by avoiding the expensive communication. A lot of work have been done to reduce the participated nodes, but none of them have considered the overlapping minimum bounded rectangle (MBR) of sensors which make them impossible to reach the optimization solution. The OMSI-tree and OMR algorithm proposed by us can efficiently solve this problem by executing a given query only on the sensors involved. Experiments show that there is an obvious improvement compared with TinyDB and other spatial index, adopting the proposed schema and algorithm.

1 Introduction

Sensor Networks was awarded as one of the 21 most important technologies for the 21st century by Business Week [1]. A sensor network is a computer network that consists of many small, inexpensive and spatially distributed devices using sensors to monitor and detect status in different locations.

While the sensor node processing and monitor capabilities are improving, crucial problems in the application of the sensor networks still remain, power consumption and communication become one of the most urgent problems. All the sensor nodes are equipped with limited power supply. It is almost impossible to recharge or replace these batteries. Once a sensor node is power off, it should be abandoned. In addition, the expense of communication is emphasized to be relatively high. The energy

* This research was supported by the MIC (Ministry of Information and Communication), Korea, under the ITRC (Information Technology Research Center) support program supervised by the IITA (Institute of Information Technology Assessment).

K.C. Chang et al. (Eds.): APWeb/WAIM 2007 Ws, LNCS 4537, pp. 204–210, 2007.

consumption of executing 3 million instructions, using a 100(MIPS)/W processor, is approximately the same as that of transmitting 1Kb data over 100m [2] [3].

Due to the energy scarcity, the centralized index for spatial query is no longer feasible. Many researches have been done to extend the life span of sensor networks by reducing the energy cost during querying [4]. The most common solutions are increasing data processing on single sensor node and limiting the involved sensor nodes.

Madden and Franklin introduced Fjord architecture [6] for multiple queries management over sensors. This architecture focused on maintaining high query throughput with limited sensor resource demands and showed that in sensor environments, communication costs, and power consumption are important issues. Cougar approach [7] discussed the in-network query processing over sensor networks. A central administration that is aware of the location of all the sensor nodes. TiNA schema [11] extends the life of the sensor network by up to 300% by using temporal coherency tolerances to reduce the communication of individual nodes. Soheili et.al, in [8] proposed a distributed spatial index over the sensor networks for processing spatial queries. The research group of TinyDB [4] at UC Berkeley has made a lot of contribution on sensor mote and Tiny OS design. They propose ACQP and prove it is energy-efficiently. A distributed index called Semantic Routing Trees (SRT) is designed to efficiently manage relevant sensor nodes participate in a given query over some constant attribute. FDSI-tree [12] introduced by Eo et.al provided a power-aware range queries based on MBRs which is much the same as our proposed schema. Queries are efficiently limited in relevant sensor nodes to guarantee the maximum power savings. However, the author did not consider the possibility of overlapping MBRs.

In this paper, we propose the design of Overlapping MBR Spatial Index Tree (OMSI-Tree) which efficiently restrains the spatial query to the essential sensors. If the monitoring area MBR of sensor nodes overlaps with each other, their parent node will decide whether both of the sensor nodes are essential to this query. Thus, the query message will only be disseminated to the involved sensor nodes. Although the extra computation of MBR overlapping increases the parameter storage and data processing, the whole performance of the sensor networks improved and energy consumption decreased as we have emphasized before, the communication is much more expensive than processing. The notions of node selection and MBR overlapping are derived from the traditional R-Tree [5] structure. With the strong capability of handling spatial attributes efficiently, R-tree like structure becomes the base for almost all the MBR and spatial index algorithms.

The remainder of this paper is organized as follows: in section 2, we propose a schema called OMSI-tree and OMR algorithm to ensure the minimum energy consumption while querying over sensor networks. In section 3, we make a comparison for our schema and algorithm with previous schemas. The conclusion and future work are given in section 4.

2 OMSI-Tree

In this section, we propose the Overlapping MBR Spatial Index Tree (OMSI-Tree) used for queries in distributed fashion. In addition, Overlapping MBR Remove (OMR) algorithm is introduced.

2.1 Tree Structure

OMSI-Tree is a spatial index like R-tree structure to some extent. However, by storing children node MBRs and relevant overlapping area, OMSI-tree allows each parent node efficiently detect whether any of the child nodes below is necessary to participate in a given query over the queried window.

In order to include the appropriate sensor nodes to a given query, we design the structure of parent node of each sub-branch in the format of <*ChildPts*, *ChildMBRs*, *Overlap-Areas*, *OverallMBR*, *Location-info*>. Here the *ChildPts,* which have the same semantics as the *waiting-list* in Cougar approach, helps the parent node find the children nodes below it. MBR of a specific node confines the corresponding working area. *Location-info* indicates the geographical position of sensor nodes. *Overlap-Areas* stores all the overlapping working area of its children nodes. This parameter is used to remove the redundant nodes when there are two nodes overlapped on the query window. Many approaches have been adopted to detect the efficient MBR of a sensor node, such as quadratic split algorithm [10], detecting by radius and so on, we do not provide any algorithm here in order to leverage the robustness in choosing efficient algorithm for independent implementations of our design.

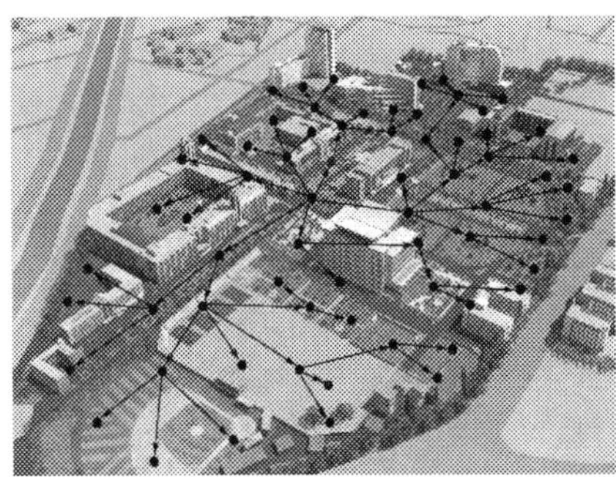

(a) Simulated physical environment area of interest

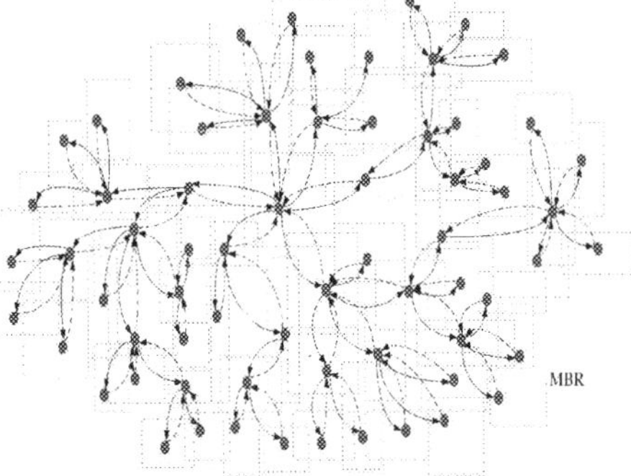

(b) Two steps in building OMSI-tree. The real lines represent the *descending step*, and the dashed lines represent the *Ascending step*. The rectangles illustrate the MBR of each sensor nodes.

Fig. 1. Sensor nodes positions in our test environment

Building OMSI-tree is a two-step process as illustrated in figure 1:

Descending step: the message, which includes the sender location and ID, is disseminated from the gateway nodes. In the descending step, all the sensor nodes will not broadcast themselves to the network in its radio range until they receive a message. All the sensor nodes will keep a list of messages it has received. The descending step is finished when all the sensor nodes heard a message.

Ascending step: if a sensor node has no candidate children nodes, it chooses its parent node. Otherwise, it waits until they select their parent nodes. We call the node, which has no candidate children nodes, as leaf node. When the leaf node confirms its parent node, it updates its MBR to the parent node. When the parent node collects all the children MBR below it, it starts to calculate the *overallMBR* and *overlap-Areas*. Thus, a distributed R-tree like structure is formed among the sensor nodes.

2.2 Overlapping Remove

A lot of previous researches have been done to reduce the communication by limiting the number of involved sensor nodes. The typical strategy is building up an R-tree like spatial index. In [9], they proposed a DSR-tree structure that can efficiently reduce the communication. However, they did not ensure the maximum power savings due to the overlapping nodes. In this section, how to guarantee the minimum number of nodes participated over the evaluation of a query will be illustrated.

To process a range query with OMSI-tree, we propose a straightforward but efficient Overlapping MBR Remove (OMR) algorithm to facilitate the removal of redundant nodes if there are nodes overlapping.

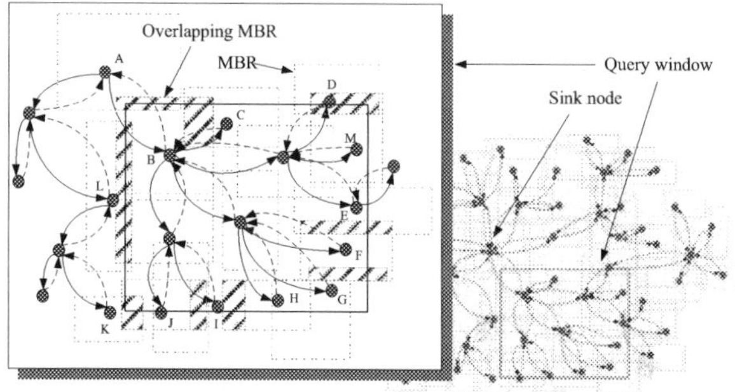

Fig. 2. Part of node position of our test environment

In OMR algorithm, we consider two overlapping situations, namely, overlapping on the vertices of query window and overlapping on the borders of query window. For the first situation, if there is vertices falling into the overlapping region, relevant sensor nodes should be taken into consider. Suppose sensor node A has MBR_A, sensor node B has MBR_B and A, B have overlap area MBR_{AB}, query window is W. The processing step is: first to compare W with MBR_A, if the overlapping area MBR_{WA} of W and MBR_A belongs to MBR_{AB} completely, then it is convinced that node A can not provide any available information in this query. Thus, sensor node A is abandoned and B is selected. Otherwise, we keep sensor A and check sensor B in the same way. There is no specified order for sensor A and B.

In another situation, when the overlapping area appears on the borders of query window, only the sensor node, which is physically inside the query window, will be kept. That means this is decided by their geographical position.

A range query based on OMR algorithm is illustrated in figure 2. An OMSI-Tree has already been built up for testing sensor network. Obviously, there are three vertices placed on the overlapping area. For the first one, sensor node A and B share the overlapping area MBR_{AB}. It is obvious that the shared space between sensor node A and query window W is completely belongs to MBR_{AB}. According to OMR algorithm, sensor node A is abandoned. And sensor node K is abandoned in the same way. However, to sensor node D, more range can be detected in query window W except MBR_{DM}. Based on OMR algorithm, sensor node D is kept and the relevant sensor node M should be checked. Evidently, sensor node M should also be kept.

For the situation where overlapping on the vertices of query window, it is a little CPU costly since we have to use query window *W* to recalculate the overlapping area with two relevant nodes. However, when the overlapping areas appear on the borders of query window, as indicated by OMR algorithm, the calculation is quite cheap because only geographical positions need to be compared. Ultimately, OMR algorithm filters out three redundant sensor nodes: A, L and K. Although this induced excessive computation, comparing with that of communication, the increasing in expense was relatively limited. Consequently the proposed algorithm and schema was able to maintain the minimum count of nodes taking part in the query, and thus ensure the minimum power consumption.

3 Performance Evaluations

To evaluate the performance of our OMSI-tree and OMR algorithm, we set up a testing environment resemble to the one created in DSR-tree due to the similar testing purpose. 400 nodes are placed over a large area in a grid like squares of 20 x 20. Suppose each node could only communicate with that one hop away besides it. In order to make our test result comparable with previous work, we evaluate proposed schema and algorithm against the *best-case* approach and *closest parent* as used by DSR-tree and TinyDB. Figure 3 shows the number of involved sensor nodes changing according to the relevant query size.

It is quite clear that OMSI-tree operated by OMR algorithm can efficiently reduce the number of involved sensor nodes especially when the fraction of query size is between 20% and 70%. That also indicates that when the query size is almost over the whole sensor networks, any of the spatial indexes can not reduce the energy consumption efficiently.

TinyDB concludes that, on average, 41% energy is spent on communication; only 1% goes on ideal consumption. DSR-tree finds their schema can reduce by 20% energy cost in the ideal case compared with TinyDB. In our experiment, OMR algorithm based OMSI-tree performances much better than TinyDB, and requires less sensor nodes to participate in a given query compared with DSR-tree especially when the query size is neither too small nor too big. Approximately, there is 27% decrease in energy consumption on average.

To evaluate the extended lifetime of sensor networks, we make the test based on typical data collection query like TinyDB in much the same way by monitoring the light and humidity. Suppose each sensor node wants to receive results from neighborhood nodes, it listens to its radio for tow seconds per minute. We assume one query consist of following steps: listening, sensing/processing, sending and sleeping. Then, according to the expected power consumption table in TinyDB, we can calculate the energy cost of one sensor per circle is $0.001 \times 58+9.30 \times 0.15+0.50 \times 0.34+0.35 \times 1.3+10.40 \times 0.03+5.00 \times 2=12.39$mAs. That means equipped with a 1000mAh battery, one sensor node can work for approximately 80 hours.

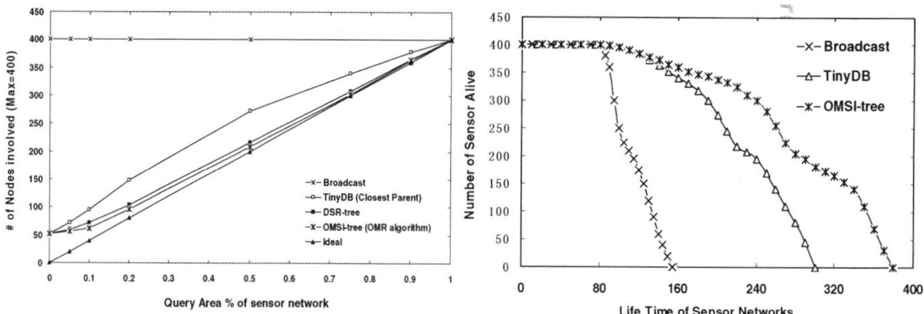

Fig. 3. Query range vs. involved nodes

Fig. 4. Sensor network lifetime comparison

In our scenario, the query area percentage of sensor networks is during 20% to 80% randomly. Figure 4 illustrates the distinct sensor network lifetime by using TinyDB, broadcast and OMSI-tree. From the experiment we can find that after about 50% of sensor nodes were dead, the rest sensors fail much faster because of the extra energy consumption for reconstructing the tree. The experimental results show that there is approximately 245% lifetime improvement by using OMSI-tree.

4 Conclusion and Future Work

In this paper, we propose a spatial index schema OMSI-tree and OMR algorithm in order to reduce the energy consumption while processing query over sensor network. Many strategies have been studied to extend the life time of sensor networks such as TinyDB, SPIX index and DSR-tree which try to execute power-aware query by minimize expensive communication. However, none of these strategies ensured the maximum power savings due to various constrains. By grouping the sensors in a region and removing the redundancy sensors over a given query, our strategy guarantees the minimum power consumption and involved sensors. Thus, longer sensor network lifetime can be obtained.

The overlapping of more than three nodes is not considered in our paper due to the limited sensor processing capability and memory size. In the future work, a proper data structure for *childMBRs* should be studied for large-scale sensor networks. In addition, a schedule strategy for multi-queries is required. Furthermore, we expect a strategy for better grouping sensors that will make the cost of constructing OMSI-tree cheaper.

References

1. 21 ideas for the 21st century, Business Week, pp. 78–167 (August 30, 1999)
2. http://s2k-ftp.cs.berkeley.edu:8000/sequoia/schema/
3. Rappaport, T.: Wireless Communications: Principles and Practice. PH Inc. (1996)
4. Madden, S.R., Franklin, M.J., Hellerstein, J.M., Tiny, D.B.: An Acquisitional Query Processing System for Sensor Networks. ACM Transasctions on Database Systems 30(1), 122–173 (March 2005)

5. Gutman, A.: R-Tree – A dynamic index structure for spatial searching, SIGMOD 1984, Boston, MA (1984)
6. Madden, S., Franklin, M.J.: Fjording the Stream: An Architecture for Queries over Streaming Sensor Data. In: Proc.18th Int. Conference on Data Engineering, pp. 555–566 (2002)
7. Yao, Y., Gehrke, J.: The Cougar approach to in-network query processing in sensor networks. SIGMOD Record 31(3), 9–18 (2002)
8. Soheili, A., Kalogeraki, V., Gunopulos, D.: Spatial queries in sensor networks. In: 13th annual ACM international workshop on Geographic information systems,Bremen, Germany, pp. 61–70, (2005)
9. Eo, S.H., Pandey, S., Park, S.-Y., Bae, H.-Y.: Energy Efficient Design for Window Query Processing in Sensor Networks. APWeb Workshops, pp. 310–314 (2006)
10. Beckmann, N., kriegel, H.-P., Schneider, R., Seeger, B.: The R*-Tree: An Efficient and Robust Access Method for Points and Rectangles. ACM (1990)
11. Sharaf, M.A., Beaver, J., Labrinidis, A., Chrysanthis, P.: TiNA: A Scheme for Temporal Coherency-Aware in-Network Aggregation. In: Proc. of, International Workshop in MobileData Engineering (2003)
12. Eo, S.H., Pandey, S., Kim, M.-K., Oh, Y.-H., Bae, H.-Y.: FDSI-Tree: A Fully Distributed Spatial Index Tree for Efficient & Power-Aware Range Queries in Sensor Network. In: Wiedermann, J., Tel, G., Pokorný, J., Bieliková, M., Štuller, J. (eds.) SOFSEM 2006. LNCS, vol. 3831, pp. 254–261. Springer, Heidelberg (2006)

On Studying Front-Peer Attack-Resistant Trust and Reputation Mechanisms Based on Enhanced Spreading Activation Model in P2P Environments*

Yufeng Wang[1], Yoshiaki Hori[2], and Kouichi Sakurai[2]

[1] College of Telecommunications and Information Engineering, Nanjing University of Posts and Telecommunications, Nanjing 210003, China
[2] Department of Computer Science and Communication Engineering, Kyushu University, Fukuoka 812-0053, Japan

Abstract. In this paper, we proposed two enhanced trust propagation and reputation ranking approaches based on spreading activation model to mitigate the effect of front peer. By front peer it means that these malicious colluding peers always cooperate with others in order to increase their reputation, and then provide misinformation to promote actively malicious peers. One approach is to use adaptive spreading factor to reflect the peer's recommendation ability according to behaviors of the peer's direct/indirect children in trust graph; another way is to investigate the feasibility of propagating distrust to effectively combat front peer. Preliminary simulation results show that those approaches can identify and mitigate the attack of front peer.

1 Introduction

In distributed Peer-to-Peer (P2P) networks, peers should be able to identify reliable peers for communication, which is a challenging task in highly dynamic P2P environments. So, the importance of social control mechanism, that is, reputation and trust management, became more and more crucial in open network and electronic communities. Intuitively, a user in our society is much more likely to believe statements from a trusted acquaintance than from a stranger. And recursively, since a trusted acquaintance will also trust the beliefs of her friends, that is, trusts may propagate (with appropriate discounting) through the relationship network. However, this introduces new challenges, such as how to determine the accuracy of the collected opinions and how to aggregate the conflicting opinions to yield a global reputation. Different propagation schemes for both trust score and distrust score are studied based on a network from a real social community website [1]. The classification of trust metrics is provided in [2], in which trust metrics are distinguished as scalar and group metric. Ref. [3] proposes a local group-based trust metric, Appleseed algorithm, based on spreading activation models to propagate trust value from source peer, and argue

* Research supported by the NSFC Grants 60472067, JiangSu education bureau (5KJB510091) and State Key Laboratory of Networking and Switching Technology, Beijing University of Posts and Telecommunications (BUPT).

K.C. Chang et al. (Eds.): APWeb/WAIM 2007 Ws, LNCS 4537, pp. 211–216, 2007.

this algorithm possesses the property of attack-resistance. EigenTrust algorithm assigns a universal measure of trust (reputation value) to each peer in P2P system (analogous to the PageRank measure for web pages, so-called, global group-based trust metric), which depends on the ranks of referring peers, thus entailing parallel evaluation of relevant nodes thanks to mutual dependencies [4]. So, group-based trust value should be called "reputation rank".

In (semantic) web-based community, there exist great efforts to adopt link structure to demote web spam, which generally means all the deceptive actions that try to increase the ranking of a page in search engines. Ref. [5] shows that propagating trust from a set of highly trusted seed sites helps a great deal in the demotion of web spam. A natural extension of the idea of the conveyance of trust between links is that of the conveyance of distrust. Ref. [6], investigates the possibility of propagating distrust among web pages to demote web spam, in which distrust is a penalty awarded to the source page for linking to an untrustworthy page.

But those above approaches are vulnerable to so-called front peer attack. Front peer attack represents these malicious colluding peers always cooperate with others in order to increase their reputation, and then provide misinformation to promote actively malicious peers. In this paper, we propose two enhanced trust propagation and reputation ranking algorithms based on spreading activation model to mitigate the effect of front peer attack. One way is to use adaptive spreading factor to reflect the peer's recommendation ability; another is to investigate the feasibility of propagating distrust to effectively combat front peer.

The paper is organized as follows: section 2 briefly provides the concept of group trust metrics, the basic spreading activation model and the original Appleseed algorithm. Two enhanced approaches based on spreading activation model, i,e., adaptive spreading factor and reverse distrust propagation, are proposed to combat front-peer attack in section 3. Section 4 briefly introduces the simulation settings and provides preliminary simulation results, which illustrate the effect of our proposals. Finally, we briefly conclude the paper.

2 Basic Models

Generally, there are three inputs to trust metric: direct trust relationship (usually represented as directed trust graph), a designated "seed" peer indicating the root of trust, and a "target" peer, then we wish to determine whether the target node is trustworthy. Each edge from s to t in the graph indicates that probability that s believes that t is trustworthy. According to the classification criteria of link evaluation [3], trust metrics are distinguished as scalar and group metric. Scalar metrics analyze trust assertions independently, while group trust metrics evaluate groups of assertions "in tandem". The entire category of scalar trust metrics fails to resist easily-mounted attacks. While group trust metric is effective in resisting attack, and well suited to evaluating membership in a group, because this evaluation is done over the entire group of nodes, rather than individually for each node. But, one disadvantage of group-based trust metric is that the computed values lack a plausible interpretation on an absolute scale. The scenarios in which they can be used should involve ranking the trust values of many peers and selection of the most trustworthy one(s) among them, that is, group-based trustworthiness should be called "reputation rank".

Ref. [3] proposes local group-based trust metric, Appleseed, which borrows many ideas from spreading activation model in psychology and relates their concepts to trust evaluation in an intuitive fashion. The spreading activation model is briefly given as follows: in a directed graph mode, edge $(x, y) \in E \subseteq V \times V$ connects nodes $x, y \in E$, which is assigned continuous weights $w(x, y) \in [0,1]$. Source node *s* to start the search from is activated through an injection of energy *e*, which is then propagated to other nodes along edges according to some set of simple rules: all energy is fully divided among successor nodes with respect to their normalized local edge weight, i.e., the higher the weight of an edge, the higher the portion of energy that flows along the edge. Furthermore, supposing average outdegree greater than one, the closer node *x* to the injection source *s*, and the more paths leading from s to *x*, the higher the amount of energy flowing into *x* in general. In order to interpret energy ranks as trust ranks (actually, should be called reputation ranks), Appleseed algorithm tailors the above model to trust computation. Specifically, trust rank of peer *x* is updated as follows: $trust(x) \leftarrow trust(x) + (1-d) \cdot in(x)$, where *in*(*x*) represents the amount of incoming trust flowed into peer *x*; $d \cdot in(x)$ the portion of energy divided among peer *x*'s successors. But, there exist several problems in the above Appleseed trust metric algorithm.

① The work in Ref. [2][3], in which Appleseed algorithm is proposed, did not experimentally provides the advantages and disadvantages of Appleseed algorithm when facing various attacks, only illustrated the fast convergence rate of Appleseed algorithm (Appleseed algorithm is inherently recursive).
② In Appleseed algorithm, the spreading factor *d* is regarded as the ratio between trust in the ability of peer to recommend others as trustworthy peers and direct trust. This step collapses functional trust and referral trust into a single trust type, which allows for simple computation, but creates a potential vulnerability but creates a potential vulnerability. A malicious peer can for example behave well during transactions in order to get high normalized trust scores as seen by his local peers, but can report false local trust scores (i.e. too high or too low). By combining good behaviors with reporting false local trust scores, a malicious agent can thus cause significant disturbance in global trust scores.

3 Enhanced Trust Propagation and Reputation Ranking Approaches

In this section, we provide two enhanced trust propagation and reputation ranking approaches based on spreading activation model: adaptive spreading factor approach and reverse distrust propagation approach. Note that our approaches use the passed energy in(*x*) as the reputation rank of peer *x*, and the energy endowed to its children is $d(x) \cdot in(x)$, where *d*(*x*) depends on the behaviors of peer x's children.

① **Adaptive spreading activation factor approach**

In the approach, once source node recognizes the malicious peer previously recommended by other peers (through direction interaction, etc.), then, from the denoted malicious peer, the spreading factor associated with each related peer is updated alone the reverse link in the trust graph according to the following rule:

$$d_u^{new} = \min_{(u,x)\in E}\left\{(1-\alpha)\cdot d_u^{old} + \alpha\cdot\left[(d_x^{new} - d_{init})\cdot w(u,x) + d_{init}\right]\right\} \quad (1)$$

where d_u^{new} (or d_u^{old}) denotes peer u's spreading factor after (before) the update; d_{init} represents the initial spreading value (in Appleseed, $d_{init} = 0.85$). α is the learning rate, a real number in the interval (0,1). Initially, the spreading factor of identified malicious peer is represented as:

$$d_m^{new} = \alpha\cdot d_m^{old} + (1-\alpha)\cdot\rho_m\cdot d_{init} \quad (2)$$

where ρ_m is the source's direct functional trust on malicious peer.

So, the procedure of our trust propagation and reputation ranking algorithm based on adaptive spreading factor is given as follows:

- Whenever, the source peer finds out the malicious peer, then the source updates those related peers' spreading factor according to Eq. (1) and (2);
- Then, based on the updated spreading factor, the modified Appleseed algorithm is used to calculate the reputation rank of related peers. Specifically, the following equations were used to replace the corresponding parts in original Appleseed algorithm:

  $e_{x\to y} = d_x^{new}\cdot in(x)\cdot\frac{w(x,y)^2}{\sum_{(x,i)\in E} w(x,i)^2}$ denotes energy distributed along (x,y) from x to its successor node y, which adopts square weighted methods to splitting the reputation rank of peer x to its child y.

  $reputation\ (x) \leftarrow reputation\ (x) + in(x)$

  and $reputation(x) = in(x) = \sum_{(p,x)\in E} d_p^{new}\cdot in(p)\cdot\left(\frac{w(p,x)^2}{\sum_{(p,i)\in E} w(p,i)^2}\right)$, that is, the pass energy ranks are simply accumulated to represent reputation rank of each peer x.

② **Reverse distrust propagation approach**

Motivated by the idea of trust propagation in group-based trust metrics, we believe that distrust propagation makes sense when malicious peers are used as the distrusted seed set and distrust is reversely propagated from a child to its parent along the indegree direction. In the absence of treatment of distrust in prior work, it is unclear how to model and propagate distrust. We argue that the semantic and propagation of distrust should depend on the specific application scenarios. Specifically, in this paper, distrust semantics is a penalty awarded to the front peer for linking to an untrustworthy peer. In brief, the modified Appleseed algorithm (provided in the subsectoin above) was used to propagate distrust and infer the dis-reputation rank (the accumulation of distrust value in a peer) of related peers.

4 Simulation Settings and Results

Table 1 describes the general simulation setup, including the peer types, behavior patterns.

Table 1. Peer types and behavior patterns in our simulation

	Parameter	Value range
Peer Model	Number of peers in the network Percentage of good peers Percentage of front peers Percentage of malicious peers	300～1000 20%～50% 20%～50% 30%
Behavior patterns	Good Peer: always provide truthful feedback about the transaction party, that is, trust value 0.9 for good peer and front peer; 0.05 for malicious peer; Front peer: like good peer, except providing false feedback about the malicious peer, that is, 1 for interacting malicious peer; Malicious peer: provide bad feedback for good peer (0.05), and provide good feedback for malicious peer and front peer (0.9).	

Since the goal of this paper is to investigate how those proposed reputation ranking algorithms, i.e, adaptive spreading factor and distrust propagation, can help to identify and mitigate the effect of front peers, we will focus on the ranking positions of malicious peers. The percentage of malicious peer in top 20% reputation rank in our proposal and Appleseed algorithm is shown in Fig.1 (total peer number: 1000; learning rate α: 0.7). Obviously, our proposal can recognize more malicious peers recommended by front peers. For the reason of local effect of our proposal and Appleseed, the percentage of malicious peer in top 20% rank only changes slightly with the increase of the percentage of front peer in trust network.

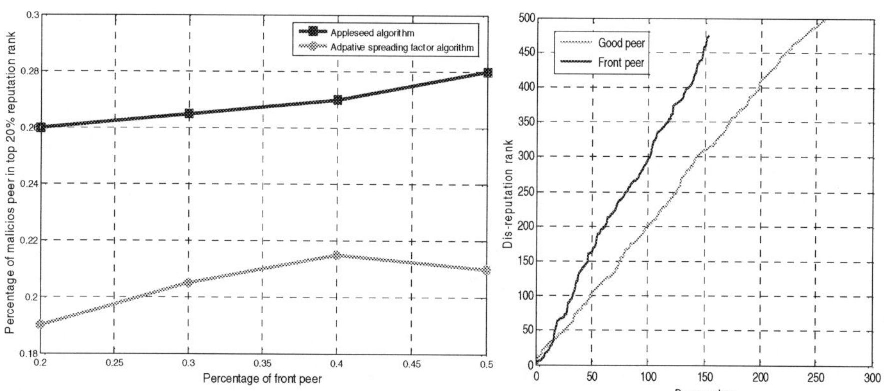

Fig. 1. Our proposal vs. Appleseed (malicious peers' rank)

Fig. 2. Dis-reputation rank

Appleseed algorithm treats the front peers same as the good peer. But, our proposal (reverse distrust propagation algorithm) explicitly uses the dis-reputation to represent the doubt on peer's recommendation ability. Fig. 2 shows the dis-reputation rank of front peer and good peer (simulation environment: 500 peer, 50% good peer). Obviously, the dis-reputation rank of front peer is larger than good peer, and the marginal increase in front peers' dis-reputation rank is also larger than good peer. So, it is reasonable to use dis-reputation rank as the flag of peer's recommendation ability.

5 Conclusion

With the increasing popularity of self-organized communication systems, distributed trust and reputation systems in particular have received increasing attention. Group trust metrics evaluate groups of assertions "in tandem", which have the feature of attack-resistance. But, unfortunately, most group-based trust metrics are vulnerable to front peer attack (a special kind of collusion). In this paper, we argue that group-based trust value should be called as "reputation rank", and propose two enhanced trust reverse propagation and reputation ranking algorithms, namely adaptive spreading factor and distrust propagation, based on spreading activation model, to identify and mitigate the attack of front peer. Specifically, the spreading factor is regarded as the ratio between trust in the ability of peer to recommend others as trustworthy peers and direct trust, which should be adaptively updated according to the behaviors of peer's direct (indirect) children to reflect the current peer's recommendation ability. Thus, front peers can obtain high reputation rank, but can not pass its reputation rank to malicious peers; in distrust propagation algorithm, dis-reputation represents the doubt on peer's recommendation ability, and the natural result is that front peers has relatively higher dis-reputation value than good peers. Preliminary simulation result shows that those algorithms can identify and mitigate the attack of front peer, to which tradition group-based trust metrics and Appleseed are vulnerable.

References

[1] Guha, R., Kumar, R., Raghavan, P., Tomkins, A.: Propagation of trust and distrust. In: Proceedings of the 13th International World Wide Web Conference (WWW2004), New York City, (May 2004)

[2] Ziegler, C.-N., Lausen, G.: Spreading activation models for trust propagation. In: Proceedings of the IEEE International Conference on e-Technology, e-Commerce, and e-Service, Taipei, Taiwan, (March 2004)

[3] Ziegle, C.-N., Lausen, G.: Propagation models for trust and distrust in social networks, Information Systems Frontiers, vol. 7 (2005)

[4] Kamvar, S.D., Schlosser, M.T., Garcia-Molina, H.: The eigentrust algorithm for reputation management in P2P networks. In: Proceedings of the 12th International World Wide Web Conference (WWW2003) (2003)

[5] Gyongyi, Z., Garcia-Molina, H., Pedersen, J.: Combating web spam with TrustRank. In: Proceedings of the 30th International Conference on Very Large Data Bases (VLDB), Toronto, Canada, (September 2004)

[6] Wu, B., Goel, V., Davison, B.: Propagating trust and distrust to demote web spam. In: Proceedings of the 15th International World Wide Web Conference (WWW2006), Edinburgh, Scotland, (May 2006)

A Load Balancing Method Using Ring Network in the Grid Database*

Yong-Il Jang[1], Ho-Seok Kim[1], Sook-Kyung Cho[1], Young-Hwan Oh[2], and Hae-Young Bae[1]

[1] Dept. of Computer Science and Information Engineering, INHA University
[2] Dept. of Information Science, Korea Nazarene University
{yijang,hskim,skyoung}@dblab.inha.ac.kr, yhoh@kornu.ac.kr, hybae@inha.ac.kr

Abstract. In this paper, a load balancing method using ring network in the Grid database is proposed. Past research proposed to solve unbalanced load problem. But, past techniques can not be applied as the Grid database has a number of systems and user's request always changes dynamically. The proposed method connects each node having the same replicated data through ring network. If workload overflows in some node, user's request is transferred to a linked node which has the target data. And, this node stops another request from processing until the workload has significantly decreased. Then, to stop request forwarding from a previous node, it changes the link structure by sending a message to the previous node. Through performance evaluation, this paper shows that the proposed method has increased performance and is more suitable to the Grid database than the existing methods.

1 Introduction

The node that organizes a Grid database contains a huge amount of data. It is replicated among different nodes for performance and availability. This is termed as replica. One replica has several same replicas. These replicas are stored on some nodes for best performance [4].

Processing of user's request requires several data. These requests are distributed to nodes that have the target data. But, some of the nodes can have high request concentration hence the system sufferers from unbalanced request distribution. This reduces overall performance. To avoid this problem, we can choose an optimal node by gathering node information. But, because of huge number of nodes and high message transmission cost, this approach can not be applied in the Grid database [1, 2, 3, 6].

The proposed method connects same replicas in a ring network. In this environment, if some node has overflowed workload then this node transfers user's request to another node via its forward link. This node stops another request from processing until the workload has significantly decreased. Also, it changes the link structure by

* This research was supported by the MIC (Ministry of Information and Communication), Korea, under the ITRC (Information Technology Research Center) support program supervised by the IITA (Institute of Information Technology Assessment).

K.C. Chang et al. (Eds.): APWeb/WAIM 2007 Ws, LNCS 4537, pp. 217–222, 2007.

sending a message to the previous node via the backward link. If the workload has reduced then this node sends a message to the previous node to continue processing requests again. Proposed method presents improved performance and is thus more suitable than existing methods as shown by performance evaluation.

The paper is organized as follows. Chapter 2 consists of related work that explains about load balancing methods. Chapter 3 presents load balancing method with a ring based network, and then in chapter 4, we evaluate the proposed method. The last chapter concludes this paper.

2 Load Balancing Method

The load balancing method can be classified into two types: static method and dynamic method. Static method uses fixed policy that had been setup at the initial phase. Dynamic method changes the scheduling policy continuously to find optimal node [5].

The most basic method is a bidding method. To share the load, this method sends broadcast messages to different nodes and selects optimal node among the nodes that made the response. This method is very simple, but has overhead of network expenses [6]. In load balancing method, which is based on the gradient model, every node changes information with it's neighboring node to distribute the load. In this case the changed message volume will increase. Therefore, this method is not suitable for the Grid database which has workload based on replica [7].

Another research focuses on load balancing in cluster system that uses coordinator or director to control user's request. When coordinator or director gets requests, it will choose the optimal node and transfers requests using algorithms such as Round-Robin(RR), Weighted Round-Robin(WRR), Least-Connection(LC) and so on. But the disadvantage of this method is the performance is only affected by coordinator [8].

3 Load Balancing Method Using Ring Network

In this chapter, the load balancing method, in which a user's request is transferred to a different node based on load information of each node and the ring network that connects all replicas in the Grid database, is explained.

3.1 Ring Network Based Link Structure

The ring network links all replicas in the form of a single ring. Each link used in the ring structure has a forward or backward link. Forward link is used to transfer a request to another node when the workload of a node exceeds its maximum value. Backward link is used to transfer its own status information to the previous node when the workload of a node exceeds its maximum value. The architecture of a database using ring network is presented in Fig. 1.

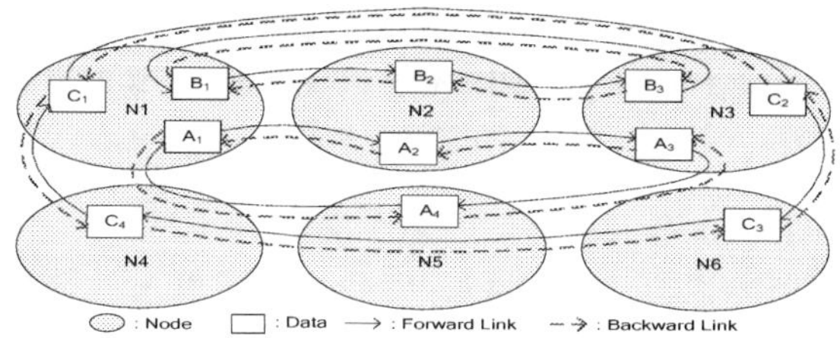

Fig. 1. The Architecture of a Database using Ring Network

Fig. 2 shows the data structure that stores the connection status of the replicas of each node. Forward link(FL) and backward link(BL) are used as explained in Fig. 1 under normal conditions. However, if user's requests are concentrated in the node and additional processing of requests becomes impossible, the structure of FL is changed. From the current structure, this node is removed from the FL of the previous node.

Replica Name (RN)	Forward Link (FL)	Backward Link (BL)

Fig. 2. The Data Structure for Ring Network

When a node is in overload status, this node creates a message that includes its own status information and its FL information. The created massage is transferred to the previous node through its BL. The previous node that receives the message changes its FL to transferred information in the message. If the previous node is also at the state of overload, this message is transferred to its previous node. If only one of the nodes is at the normal condition and all others are overloaded, then forward links of all nodes points to this normal node.

If all nodes have exceeded their maximum value and user requests are processed, the proposed method deletes all forward links of all nodes. Under this condition, there is no need to transfer the requests to other nodes. Because all nodes are processing requests at their maximum capacity, retransmission of requests only increases the load of the network. If a node is changed from an overload state to a normal state, similar to when it becomes overloaded, this node transfers a message containing its status information through its BL. The previous node then receives the message and shifts its FL to point to this node. In addition, if the previous node is at the state of overload, then the message is transferred to its previous node.

3.2 Load Balancing Method Using Ring Network

Load balancing process can be executed after every user's request or at a given time interval. When the state changes from normal to overloaded, the algorithm for handling such change is as follows.

In Fig. 3, as the node changes to the overloaded state, first it reconfigures its own FL. If the FL points to itself, it signifies that all other nodes are in their overloaded state. If the current node also changes to the overloaded state, this means that all nodes are overloaded and the FL is changed to the NULL value. After this, the requests are transferred through the FL (Lines 03-09).

If the value of the FL is already NULL, the request is processed at the current node without transferring it to another node (Lines 10-12). After this, the current node changes its own state to overload and transfers its status information and the information of the FL that has been changed using its BL. Fig. 4 shows the process of a node that changes its state to normal.

```
Input
max_workload: The value of maximum workload
current_workload: The value of current workload
current_node: Current nod ID
current_state: Current state of node (normal or overloaded)
FL: Forward link
BL: Backward link
Output
new_FL: The new value of forward link
Begin
01 : if current_workload is higher than max_workload and current_state is normal
02 :   if FL is not null value
03 :     if FL is current_node
04 :       new_FL ← null value
05 :       FL ← new_FL
06 :       process user's request
07 :     else
08 :       send user's request to FL
09 :     endif
10 :   else
11 :     process user's request
12 :   endif
13 :   set current_state to overloaded
14 :   send message to BL with current node information and new_FL
15 : endif
End
```

Fig. 3. The Processing Algorithm for Changing to overloaded state

When the node changes to a normal state, the user request is processed at the current node and changes its node state to normal. If the value of the FL is NULL, this means that all nodes are in their overloaded state and because the current node is the only node in normal state, it must change its FL to itself. As a result, the value of the FL is changed to reflect itself, and through the BL, a message containing its FL information is transferred.

4 Performance Evaluation

In this chapter, performance of the proposed method and existing methods are compared.

4.1 Test Environment

In this paper, for the evaluation of the proposed method, CSIM, which is a C/C++ language based simulation tool, is used [9]. To implement the proposed method, multiple clients and servers are organized, and the proposed data structure is stored in all servers. The evaluation environment is formed as follows.

Table 1. Evaluation Environment (ut: unit time)

Evaluation Element	Data Range	Evaluation Element	Data Range
Simulation Time	3000 (ut)	Transmission Time	2~4 (ut)
Number of Servers	3 ~ 20	Request Generate Interval	0.1 (ut)
Number of Clients	1 ~ 40	Coordinator Processing Time	0.5~1.5 (ut)
Number of Replica	2 ~ 20	Maximum Workload (Waiting Queue Size)	1~20
Request Processing Time	4~7 (ut)		

- SIMPLE: a method of directly processing the request in the node
- COORD: a method of transmitting the request to the node with the least amount of load by having a controller to catch the current information of all nodes
- RING: the proposed technique is used by materializing the structure that has been proposed in this study

4.2 Test Evaluation

In the evaluation, all results are shown in terms of Throughput and Response Time.

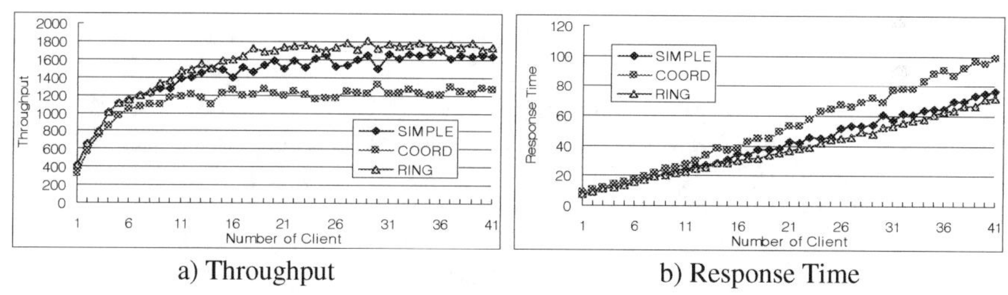

a) Throughput b) Response Time

Fig. 4. Performance Evaluation with Changing the Number of Client

Fig. 4 shows the changes in performance based on the number of clients. Requests are evenly distributed throughout the nodes. The COORD method in Fig. 4. a) does not enhance performance when the number of clients are increased to greater than 11. This seems to occur due to the response limitation of the coordinator. The SIMPLE method and the RING method showed almost identical performance, but the RING method showed a slightly enhanced performance. The RING performs better due to the load balancing. In Fig. 4. b), the COORD method showed differences in its performance based on the number of clients as compared to other methods.

Fig. 5 shows changes in performance based on changes in the number of servers. Through this test, the overall response performance of the COORD method is determined based on the response performance of the coordinator.

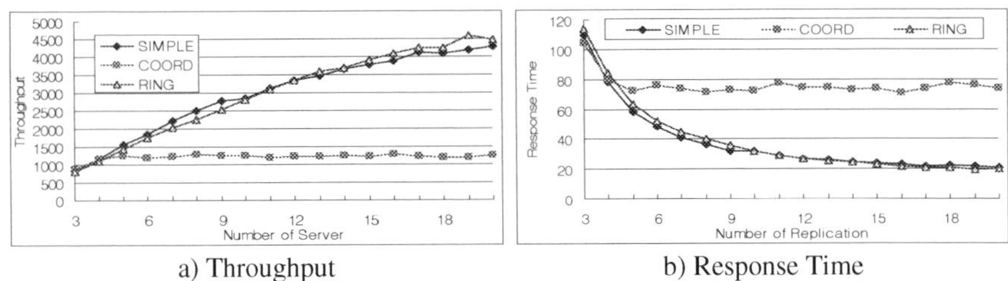

a) Throughput b) Response Time

Fig. 5. Performance Evaluation by Changing the Number of Servers

5 Conclusion

In this paper, we proposed a load balancing method using ring network in the Grid database, and evaluated performance against existing methods. The proposed method organizes connected database through ring network based on replica in the Grid database. And, this ring network is used for load balancing process. In the performance evaluation, proposed method showed increased performance than existing methods.

The proposed method is suitable for the Grid database which has characteristics for unrestricted node creation and dropping, dynamic connection structure, and a number of nodes. This method materialized replica based load balancing. Therefore, it can be used for various applications.

References

1. Anastasiadi, Kapidakis, S., Nikolaou, C., Sairamesh, J.: A computational economy for dynamic load balancing and data replication. In: Proc. of the First International Conference on Information and Computation Economies, pp.166–180, Charleston, SC (October 1998)
2. Farber, D.J.: The Distributed Computer System. In: Proc. 7th Annu. IEEE Comp. Soc. Int.Conf. (February 1973)
3. Gray, J., Helland, P., O'Neil, P., Shasha, D.: The dangers of replication and a solution. ACM SIGMOD International Conference on Management of Data. Published as SIGMOD Record, vol. 25(2), pp. 173–182. ACM, New York (June 1996)
4. Kang, C.: Database Grid for On-Demand and High Availability, KDBC SIGDB, pp.209–224 (2005)
5. Kim, G-S., Eom, Y-I.: Design of the Load Sharing Scheme and Performance Evaluation in Distributed Systems, KIPS Journal, 4(8) (1997)
6. Lan, Z., Taylor, V., Bryan, G.: Dynamic load balancing of samr applications on distributed systems. In: Proceedings of SC'01, IEEE CS Press,Washington (2001)
7. Lin, F.C.H., Keller, R.M.: The Gradient Model Load Balancing Method. IEEE Tr. On Soft. Eng, SE-13 (1) (January 1987)
8. Linux Virtual Server, Virtual server Scheduling Algorithms, (December 2003) http://www.linuxvirtualserver. org/docs/ scheduling.html
9. Mesquite Software. Inc. CSIM19 The Simulation Engine (2005) http://www.mesquite. com

Design and Implementation of a System for Environmental Monitoring Sensor Network

Yang Koo Lee, Young Jin Jung, and Keun Ho Ryu*

Database/Bioinformatics Laboratory, Chungbuk National University, Korea
{leeyangkoo,yjjeong,khryu}@dblab.chungbuk.ac.kr

Abstract. In this paper, we propose a system architecture for handling and storing sensor data stream in real-time to support the spatial and/or temporal queries besides continuous queries. We exploit a segment-based method to store the sensor data stream and reduce the managed tuples without any loss of information, which lead to the improvement of the accuracy of query results. In addition, we offer a method to reduce the cost of join operations in processing spatiotemporal queries by filtering out the list of irrelevant sensors from the query range before making the join operation. We then present a design of the system architecture for processing spatial and/or temporal queries. Finally, we implement a climate monitoring application system.

Keywords: Wireless Sensor Network, Sensor Data Stream, Spatiotemporal Data Stream, Continuous Query, Historical Query.

1 Introduction

Wireless sensor networks (WSN) consist of a large number of sensors that collect and communicate data stream continuously in the physical world [1, 2]. Modern hardware technologies make it possible to gather data by using cheap and small sensor devices (e.g., smart dust and RFIDs) and data are then transferred in the manner of a wireless communication. These sensors collect data about natural phenomenon such as temperature, light, sound, and pressure and then transmit them to a server in real-time.

In this paper, we focus on spatiotemporal query processing and storage of historical data collected from wide area sensor network applications, such as ecology monitoring, environmental monitoring, climate monitoring, and so on. Sensor data can be used to answer not only spatial queries and temporal queries but also spatiotemporal queries continuously. In addition, a sensor query sometimes requires an answer for a long interval. This type of historical query is mainly demanded for periodic analysis or statistics of data stream. Some applications, environmental monitoring for example, have a feature that data measured by sensors are rarely changed for some time interval and all measured data need to be stored.

In this study, we introduce a system architecture using the existing temporal and spatial approaches in order to support spatiotemporal queries and store sensor data

* Corresponding author.

K.C. Chang et al. (Eds.): APWeb/WAIM 2007 Ws, LNCS 4537, pp. 223–228, 2007.

and implement a system for environmental monitoring sensor network. In our architecture, an incoming data stream is stored in the form of a segment that takes some timestamp at which the value of item is changed. And the spatial information about sensors is resident in memory with the assistance of a fixed grid when identifying the sensors' locations.

The remainder of this paper is organized as follows. In Section 2, we introduce how data stream pertaining to spatial and temporal attributes is managed to tackle spatiotemporal queries. The design of system architecture for implementation is introduced in section 3. In section 4, we present the system implementation and run an example illustrating the specific application. Section 5 discusses related work. In Section 6, we conclude the paper and give directions of the future work.

2 Spatiotemporal Data Stream Management

2.1 Temporal Representation to Eliminate Duplicate Data

In this study, before inserting data into the memory, we compare the new value with the previous value of the data and if the value has changed, we update the data that is attached with a timestamp at which the value is changed. If the data has the same value as the previous one, the tuple's timestamp for this data in memory is maintained as "*now*". Let the table be <*S, V, Ts, Te*>, where *S* is sensor identifier, *V* determines the measured value, *Ts* is start time of a new value, *Te* is end time of the value.

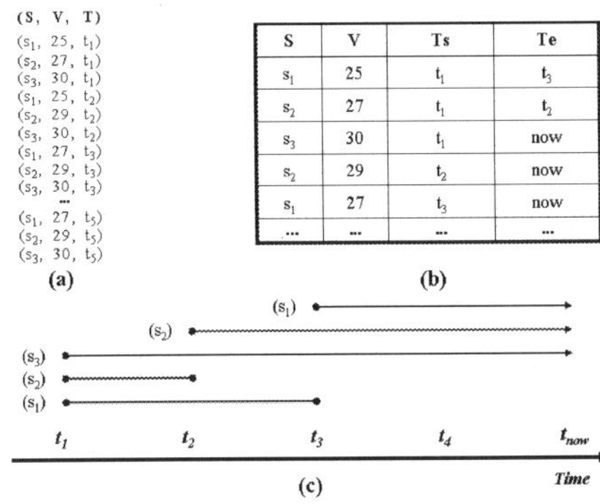

S	V	Ts	Te
s_1	25	t_1	t_3
s_2	27	t_1	t_2
s_3	30	t_1	now
s_2	29	t_2	now
s_1	27	t_3	now
...	...	...	...

Fig. 1. Representation of data stream in memory. (a) incoming data stream. (b) table storing in memory. (c) segments representing the lifetimes of items.

Fig. 1 shows an example of representation of incoming data stream. In Fig. 1, at timestamp t_1, the items s_1, s_2, s_3 are inserted into the tables, each one is associated with a time interval whose *Ts* is t_1 and Te is "*now*". At timestamp t_2, the end time's value "*now*" of s_1 is maintained because the sampled value of s_1 is still the same 25. But *Te* of s_2 at the timestamp t_1 is updated with t_2 since s_2 is sampled and its value has varied from 25 to 27, so the new value 27 has to be stored by inserting a new tuple into the table, the field *Ts* of this tuple is assigned with the timestamp t_2. In the same manner, at timestamp t_3 tuple of s_1 is updated. Finally, when all items of Fig. 1(a) are inserted, the tuples of the table are represented like in Fig. 1(b). Fig. 1(c) displays the timestamps of *Ts* and *Te* over the whole incoming stream.

2.2 Spatial Representation for Spatiotemporal Join

In the case of new incoming data stream, DSMS selects data satisfying query predicate from incoming data stream. If the query pertains to the spatial attribute, query processor executes it along with the spatial attributes in memory. Consider the example query 1.

Query 1: Return the temperature in the State A, every 5 minutes.

In this case, query processor searches the distributed sensors in State A. However, the problem is that query processor searches all the spatial attributes of sensors for the ids satisfying the query's predicates. If query processor can refer to an index scheme for the list of sensors, the cost of query processing is reduced due to searching within the small spatial scope. Therefore, we apply the physical fixed grid [3] in processing continuous queries pertaining to spatial attribute.

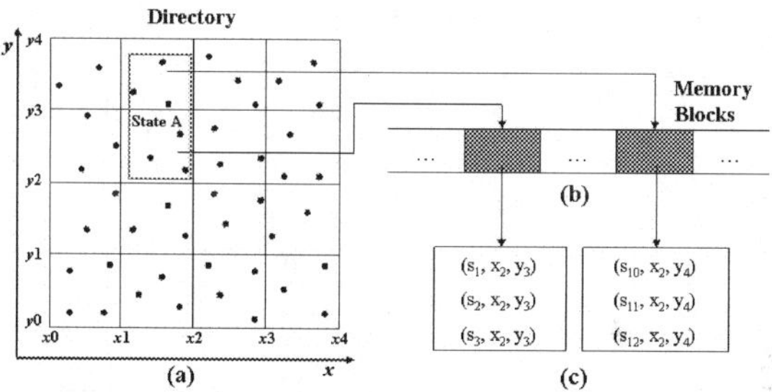

Fig. 2. Spatial mapping for searching location of sensors. (a) directory of fixed grid for location mapping. (b) memory blocks referenced by directory. (c) item set in memory blocks.

Fig. 2 illustrates a spatial mapping for searching locations of sensors. Fixed grid has the advantage that is able to filter the list of sensors out a query predicate in advance. In particular, for a join operation relating to temporal predicate and spatial predicate, this structure can reduce the cost of join operation.

2.3 Spatiotemporal Query Processing

Mostly, queries on sensor data stream involve temporal, spatial, and spatiotemporal attributes. The problem is that a complex query like spatiotemporal query requires a join between spatial and temporal interval. In this case, it is inefficient to join all locations with the time interval. The solution mentioned in the previous section can reduce the cost of query processing using filtering tuples and location before making the join operation. An example query involving spatiotemporal join can be:

Query 2: Return the temperature of the last 5 days in the State B, every 10 minutes.

The join step of query 2 is as follows: First, identify the cells that are limited by the spatial scope in query predicate, and get the list of sensor ids stored in the memory blocks referred by the directory. Then, select all tuples belonging to the last 5 minutes from data stream. Finally, execute join operation with sensor id between two results.

3 System Architecture

DSMS generally presents the system architectures to support the processing of continuous queries. Our system modifies the existing architecture to mange historical data, which are necessary to deal with spatiotemporal queries continuously.

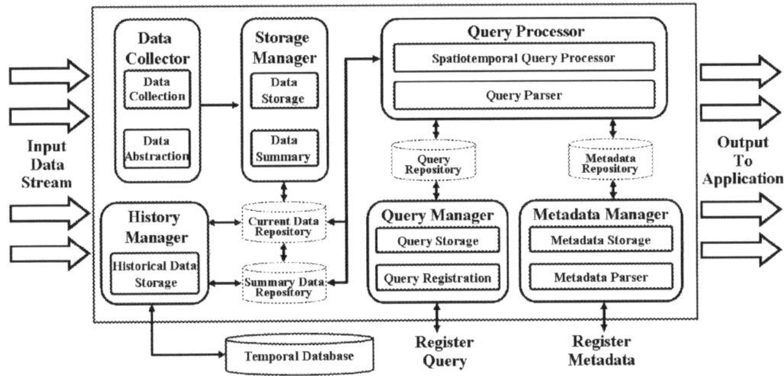

Fig. 3. System architecture for spatiotemporal query processing

Fig. 3 displays the proposed system architecture consisting of Data Collector for gathering incoming data stream, Storage Manager for storing data stream, Query Processor for processing continuous queries, Query Manager for managing user defined queries, Metadata Manager for dealing with the metadata of sensors, and History Manager for storing data stream's history, etc.

4 Implementation and Running Example

We implemented a climate monitoring application system based on our proposed system architecture. In our system, the temperature sensors distributed in a spatial area collect and transmit data stream to a server every 2 seconds. Server receives data from sensors and monitor them in real-time. And users register the continuous queries that they are interested in any or all areas. System performs the registered continuous queries on incoming data stream according to query predicates.

Fig. 4 shows the system interface for climate monitoring. The system displays the data collected from sensors in real-time and then update value of segment or inserts new tuple to represent the changed values after comparing the old values with latest data in memory.

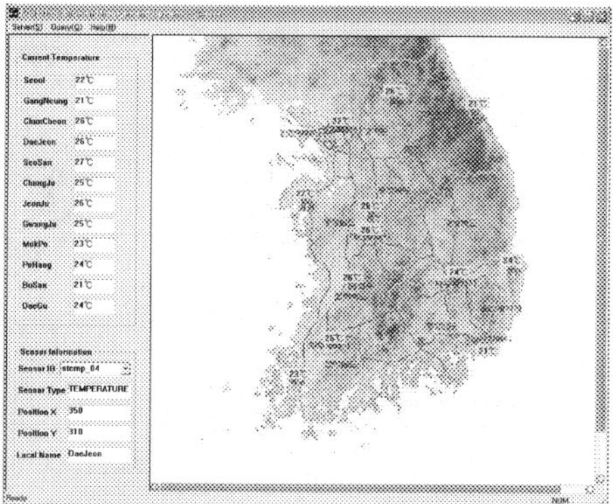

Fig. 4. System interface for climate monitoring

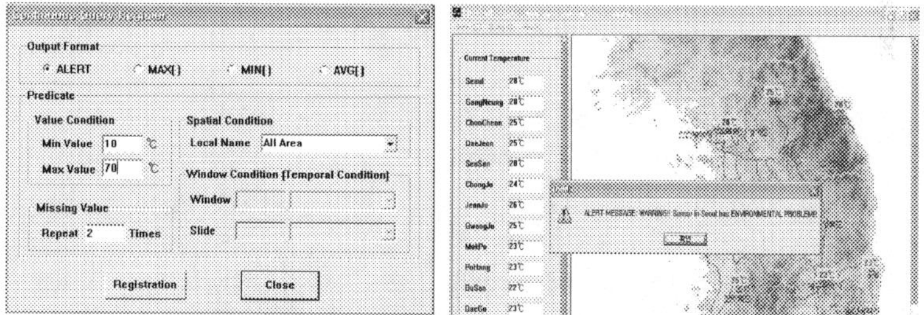

Fig. 5. Continuous query registered to detect environmental event

Fig. 5 shows the view registering a query to detect anomaly data. Query processor continuously generates the results when the temperature sensors measure the values which are less than 10 degree or lager than 70 degree in all areas. Users can register the queries that they want to be informed in any situation.

5 Related Work

Data stream processing has been a very interesting issue over the past few years. Our fundamental motivation is begun by the continuous queries over append-only databases appeared in Tapestry [4] as a means to provide timely responses by utilizing periodic query execution. In [5] a temporal foundation for stream algebra is attempted, which makes a distinction between logical and physical operator levels. Transformation rules are provided between a logical level that refers to query specification and a physical level that covers implementation issues.

The COUGAR project [6] investigates techniques for query processing over sensor data. However, their research focuses on a sensor networks environment where there is a central administration that knows the location of all sensors. Gigascope [7] is a system for managing network flow in large data communication networks, where all stream tuples include an ordering attribute, exactly as in sequence databases.

6 Conclusions

In this study, we proposed the system architecture for handling and storing sensor data stream in real time to support the spatial and/or temporal queries besides continuous queries. We exploited a segment-based method to store the sensor data stream and reduce the stored tuples without any loss of information. Due to this the accuracy of query results is improved. Additionally, we offered the method for reducing the cost of join operations in processing spatiotemporal queries. Finally, we implemented system for climate monitoring application based on our proposed system architecture. We are now further evaluating the performance of the proposed method for query processing in terms of running time and required memory.

Acknowledgments. This work was supported by the RRC program of MOCIE and ITEP, and the MIC (Ministry of Information & Communications), Korea, under the ITRC (Information Technology Research Center) Support Program supervised by the IITA (Institute of Information Technology Assessment).

References

1. Golab, L., Özsu, M.T.: Issues in Data Stream Management. In: SIGMOD Record 32(2), 5–14 (2003)
2. Zhu, Y., StatStream, D. S.: Statistical Monitoring of Thousands of Data Streams in Real Time. In: Proceedings of 28th International Conference on Very Large Data Bases, pp. 358–369 (2002)
3. Samet, H.: Applications of Spatial Data Structures. Addison-Wesley, London, MA (1990)
4. Terry, D. B., Goldberg, D., Nichols, D., Oki, B. M.: Continuous Queries over Append-only Databases. In: Proceedings of ACM SIGMOD, pp. 321–330 (1992)
5. Jensen, C.S., Snodgrass, R.T.: Temporal Data Management. IEEE Transactions on Knowledge and Data. Engineering 11(1), 36–44 (1999)
6. Krämer, J., Seeger, B.: A Temporal Foundation for Continuous Queries over Data Streams. In: Proceedings of COMAD, pp. 70–82 (2005)
7. Johnson, T., Muthukrishnan, S., Shkapenyuk, V., Spatscheck, O.: A Heartbeat Mechanism and its Application in Gigascope. In: VLDB, pp. 1079–1088 (2005)

Policy Based Scheduling for Resource Allocation on Grid*

Sung-hoon Cho[1], Moo-hun Lee[1], Jang-uk In[2], Bong-hoi Kim[3], and Eui-in Choi[1]

[1] Dept. of Computer Engineering, Hannam University
{shcho,mhlee,eichoi}@dblab.hannam.ac.kr
[2] Microsoft.com
jang-uk.in@microsoft.com
[3] UBNC Co., Ltd
kbh@ubnc.net

Abstract. In this paper we discuss a novel framework for policy based scheduling in resource allocation of grid computing. The framework has several features that are associated with the scheduling strategy, efficient resource management, the scheduling method and Quality of Service(QOS). Experimental results are provided to demonstrate the usefulness of the framework.

1 Introduction

Grid computing is becoming a popular way of providing high performance computing for many data intensive, scientific applications. It has the following unique characteristics over the traditional parallel and distributed computing. First, grid resources are geographically distributed and heterogeneous in nature. Research and development organizations, distributed nationwide or worldwide, participate in one or more virtual organizations (VO's). Second, these grid resources have decentralized ownership and different local scheduling policies dependent on their VO. Third, the dynamic load and availability of the resources require mechanisms for discovering and characterizing their status continually.

The dynamic and heterogeneous nature of the grid coupled with complex resource usage policy issues poses interesting challenges for harnessing the resources in an efficient manner[1, 2, 3]. In this paper, we present novel policy-based scheduling techniques. The execution and simulation results show that the proposed algorithm can effectively allocate grid resources to a set of applications under the constraints presented with resource usage policies, perform optimized scheduling on heterogeneous resource using iterative approach and binary integer programming (BIP) and improve the completion time of workflows in integration with job execution tracking modules of SPHINX scheduling middleware.

The rest of the paper is described as follows. In section 2, the optimization model and the proposed algorithm is discussed in detail. Section 3 presents experiment results of our scheduling algorithm. Finally, we conclude this paper, and discuss future works in section 4.

* This work was supported by a grant from Security Engineering Research Center of Korea Ministry of Commerce, Industry and Energy.

K.C. Chang et al. (Eds.): APWeb/WAIM 2007 Ws, LNCS 4537, pp. 229–234, 2007.

2 A Solution Strategy for Policy Based Scheduling

2.1 Notation and Variable Definition

We define the notations and variables that are referred in the proposed scheduling algorithm as the following.

$comp_{ij}$: Computation time of job i on processor j

$comm_{pj}$: Communication time from processor p to j

$prec_i$: A set of the precedent jobs of job i

$succ_i$: A set of the succeeding jobs of job i

$Avail_{ij}$: The available time of processor j for job i

EST_{ij} : Earliest Start Time of job i on processor j

$$EST_{ij} = Max\{Avail_{ij}, Max_{k\in prec_i}\{EFT_{kp} + comm_{pj}\}\}$$

EFT_{ij} : Earliest Finish Time of job i on processor j

$$EFT_{ij} = EST_{ij} + comp_{ij}$$

$compLen_i$: Workflow completion length from job i

$$compLen_i = comp_{ip_i} + Max_{k\in succ_i}(comm_{p_i p_k} + compLen_k)$$

The computation or execution time of a job on a processor ($comp_{ij}$) is not identical among a set of processors in heterogeneous resource environment. The algorithm sets an initial execution time of a job with the mean value of the different time on a set of available processors. The time is updated with an execution time on a specific processor as the algorithm changes the scheduling decision based on the total workflow completion time. The data transfer or communication time between any two processors ($comm_{pj}$) is also various in the environment.

An application or workflow is in the format of directed acyclic graph (DAG)[4]. Each job i has a set of precedent ($prec_i$) and succeeding ($comm_i$) jobs in a dag. The dependency is represented by the input/output file relationship.

The algorithm keeps track of the availability of a processor ($Avail_{ij}$) to execute a job. We assume a processing model in a way that all the jobs in a processor queue should be completed before a new job gets started. We assume a non-preemptive model. The algorithm computes the earliest start time of a job on each processor (EST_{ij}). A job can start its execution on a processor only after satisfying the two conditions; first, a processor should be available ($Avail_{ij}$) to execute the job. Second, all the precedent jobs be completed ($Max_{k\in prec_i}\{EFT_{kp} + comm_{pj}\}$) on the same processor or the others. The earliest finish time of a job on a processor (EFT_{ij}) is defined with the earliest start time (EST_{ij}) and the job completion or execution time on the processor ($comp_{ij}$). The workflow completion time from a job to the end of a DAG ($compLen_i$) is defined recursively from the bottom to the job i.

2.2 BIP Model

We devise a Binary Integer Programming (BIP) model to find an optimal solution to the scheduling problem on the proposed algorithm. We define a scheduling profit function, which utilizes the workflow completion time and the earliest finish time of a job on each heterogeneous processor.

Scheduling profit(p_{ij}): the profit when job i is assigned to processor j

$$p_{ij} = \frac{compLen_i}{EFT_{ij}}, \quad where\ EFT_{ij} > 0$$

The profit function intends to generate higher profit when a job on a critical path is scheduling than another job on a non-critical path. For a job i the profit value is also higher with a processor on which the job finishes early than with another processor. A scheduling algorithm utilizing the function tries to give a higher priority to the job whose completion time is longer than the others' completion time.

$$Max \sum_i \sum_j p_{ij} \cdot x_{ij}$$

$$\begin{aligned}
st.\ & \\
& b_{ij} x_{ij} \le q_{ij} \ \text{for each job } i \text{ and processor } j \quad \text{(Policy)} \\
& \sum_j x_{ij} = 1 \ \text{for each job } i \quad \text{(Assignment)} \\
& \sum_i x_{ij} \le t_j \ \text{for each processor } j \quad \text{(Load)} \\
& x_{ij} = 0\ or\ 1 \quad \text{(Binary)} \\
where\ & \\
& b_{ij}: \text{resource usage requirement of job } i \text{ on processor } j \\
& q_{ij}: \text{resource usage quota of job } i \text{ on processor } j \\
& t_j: \text{the limit of assigned jobs on processor } j
\end{aligned}$$

The objective function of the optimization model utilizes the profit function to select a processor for a job in the schedulable job list. The scheduling algorithm solves the scheduling problem in the optimization model. It tries to assign the jobs in a critical path onto the processors, which provide the jobs with the earliest finish time. The assignment is constrained by a set of constraints.

The BIP optimization model maximizes the profit values of a set of jobs and processors subject to the several constraints. The resource usage constraint is described with the two values, quota (q_{ij}) and requirement (b_{ij}) for a job i and a processor j. The quantitative model is flexible to express the policy of various resource types. The assignment constraint makes it sure that a job is not divided or assigned onto more than two processors. Each processor shouldn't be loaded with a set of assigned jobs over the predefined quota (t_j).

2.3 Policy-Based Scheduling Algorithm

The algorithm uses the iterative scheduling approach to improve a resource allocation decision with a better dag completion time. It also makes an optimal scheduling decision by solving the scheduling problem modeled in Binary Integer Programming (BIP). The algorithm uses the mean value approach to make an initial scheduling decision on heterogeneous. It then modifies the initial scheduling in an iterative way. The execution time of a job is changed with a specific value on a selected processor as the iterative scheduling proceeds. The iteration is terminated when there is no improvement in dag completion time. The algorithm shown in figure 1 is presented with detailed description below.

A workflow formatted in a DAG consists of several jobs. The algorithm generates a list with the jobs in the DAG (1). The algorithm initially sets the job execution time with the mean value on heterogeneous resources (2). In the iterative scheduling approach the best dag completion time in a series of iterations is selected, and the scheduling decision on the corresponding iteration is applied to the resource allocation (3, 4). The workflow completion time from a job to the end of a DAG, $compLen_i$ is computed for each job in the scheduling job list in line 5.

```
Construct a job list, L_unscheduled with the jobs in a workflow                (1)
for each job i in L_unscheduled
        Set exec_i with a mean value of the execution time on processors        (2)
end for
Initialize BestSL and SL with a very large number                               (3)
While SL ≤ BestSL do                                                            (4)
        BestSL = SL
        for each job i in L_unscheduled
          Compute compLen_i                                                     (5)
        end for
        Sort L_unscheduled by the non-increasing order of compLen_i             (6)
        Cluster the sorted L_unscheduled into a set of sub lists                (7)
        for each sub list, L_schedulable
           for each job i in L_schedulable
              for each processor j in the processor list, L_processors
                Compute EFT_ij                                                  (8)
              end for
           end for
           Call optSolver(L_schedulable, L_processors, compLen_i, EFT_ij )      (9)
           Assign each job i in L_schedulable onto a selected processor        (10)
           Update each processor workload and available time                   (11)
        end for
        Set SL with EFT_ep, where EFT_ep is EFT of an exit job e on processor p_e  (12)
        for each job i in L_unscheduled
           Set exec_i with the execution time on the selected processor        (13)
end for end while
```

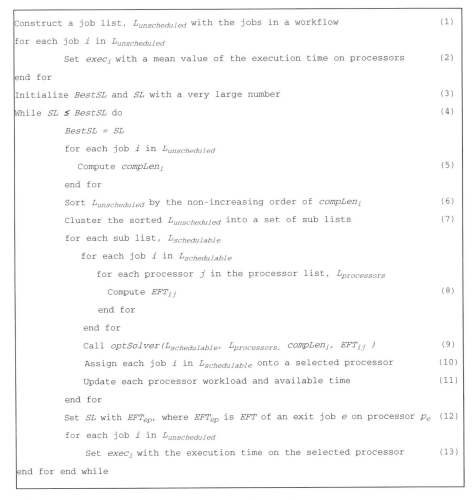

Fig. 1. Policy-based scheduling algorithm on heterogeneous resources

The workflow completion time ($compLen_i$) is used to sort the job list. The non-decreasing order in the sorted list represents the criticality of the jobs to complete the workflow (6). After the sorting the job list is clustered with a set of subsets. The size of subset should be predefined. The algorithm formulates the optimization-scheduling problem with the subset of jobs. Before calling the optimization solver function the algorithm computes the earliest completion time (EFT_{ij}) of each job on each processor (8). The optimization function described in the previous section is called with the parameters, a subset of jobs, a set of processors, *compLen*$_i$ and *EFT*$_{ij}$ (9). The function returns a selected processor for each of the jobs in the subset (10). The workload of each resource is updated based on the job assignment (11). For the workload information the algorithm maintains the number of jobs and the next available time on each processor. The next available time of a processor defines the time when the processor finishes the currently running job, and is ready for executing the next job.

In the iteration of the algorithm a dag completion time resulting from the scheduling is computed (12). The dag completion time is defined with the longest job completion time in a dag. The algorithm updates the job execution time with the time on a selected processor (13). On heterogeneous environment the execution time may be different on each resource.

3 Experiment and Simulation Results

The performance of the proposed policy-based scheduling algorithm is evaluated with the simulation result and the test application execution on the Open Science Grid (OSG). In this section we discuss the simulated and experimental performance to compare the algorithm with the list scheduling which uses the mean value approach to the heterogeneous resource environment.

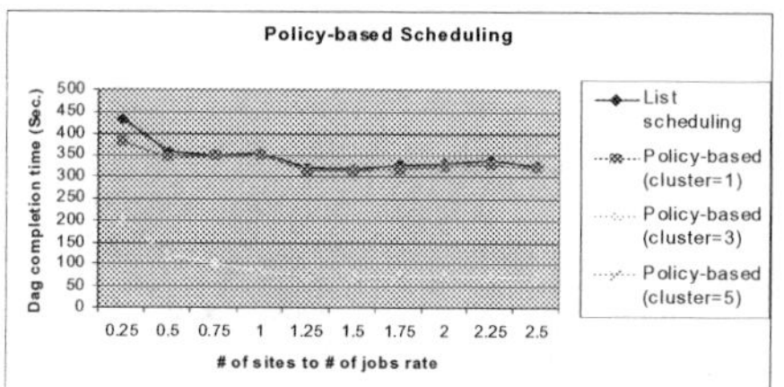

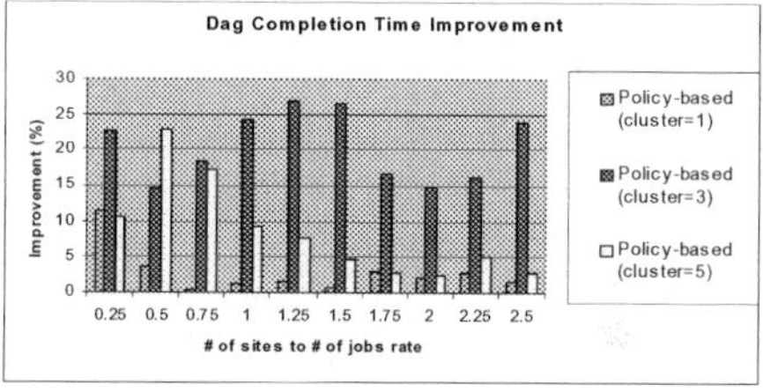

Fig. 2. Dag completion time when the rate of processors to jobs is different

3.1 The Simulated Performance Evaluation

The simulation presents the performance of the algorithms in terms of dag completion time and scheduling improvement when the ratio of the processors to the jobs is different. The graphs in Figure 2 show the performance when the ratio changes from 0.25 to 2.5 increased by 0.25. The policy-based scheduling outperforms the list scheduling by 54% ~ 80% in the first graph. The dag completion time is defined with the longest completion time of a job in the dag. The scheduling improvement represents the ratio of the scheduling time reduction from the initial scheduling time. It is defined with ((Init_Sched_time – Best_Sched_Time) / Init_Sched_Time) * 100.

The graphs in Figure 3 show the performance of the scheduling algorithms when the ratio of the number of jobs to the number of processors is various. The ratio changes from 0.25 to 2.5 increased by 0.25. The policy-based scheduling shows better performance than the list scheduling by 56% ~ 69% in the first graph. The second graph shows the completion time improvement of the policy-based scheduling algorithm with the different cluster size when the ratio is different.

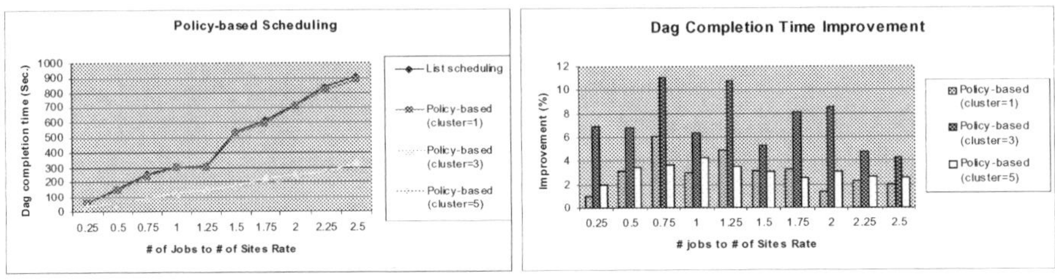

Fig. 3. Dag completion time when the rate of jobs to processors is different

4 Conclusion and Future Works

In this paper we introduce a novel policy-based scheduling algorithm. It allocates grid resources to an application under the constraints presented with resource usage policies. It performs optimized scheduling on heterogeneous resource using iterative approach and binary integer programming (BIP). The algorithm improves the completion time of an application in integration with job execution tracking and history modules of SPHINX scheduling middleware.

We will schedule a set of practical application from high-energy physics experiments such as CMS based on the scheduling decision, and study the execution performance.

References

1. Liu, G.Q., Poh, K.L., Xie, M.: Iterative list scheduling for heterogeneous computing. Journal of Parallel and Distributed Computing 65(5), 654–665 (2005)
2. Kwok, Y.K., Ahmad, I.: Dynamic Critical-Path Scheduling: An Effective Technique for Allocating Task Graphs to Multiprocessors. IEEE Transactions on Parallel and Distributed Systems 7(5), 506–521 (1996)
3. Qin, X., Jiang, H.: Reliability-driven scheduling for real-time tasks with precedence constraints in heterogeneous distributed systems. In: Proceedings of the International Conference Parallel and Distributed Computing and Systems 2000 (PDCS 2000), Las Vegas, USA (November 6-9 2000)
4. Zhao, H., Sakellariou, R.: An Experimental Investigation into the Rank Function of the Heterogeneous Earliest Finish Time Scheduling Algorithm. Euro-Par 2003, LNCS, vol. 2790, Springer, Heidelberg (2003)

Characterizing DSS Workloads from the Processor Perspective

Dawei Liu, Shan Wang, Biao Qin, and Weiwei Gong

School of Information, Renmin University of China, Beijing 100872, P.R. China
{liudawei,swang,qinbiao,wwgong}@ruc.edu.cn
Key Laboratory of Data Engineering and Knowledge Engineering
(Renmin University of China), MOE, Beijing 100872, P.R. China

Abstract. In this paper, we characterized the TPC-H benchmark on an Itanium II processor. Our experiment results clearly demonstrate: (1) On Itanium II processor, the memory stall time is dominanted by first level (L1) instruction cache and third level (L3) data cache misses; (2) Index can reduces L3 data cache misses dramatically but increases a slightly more Condition Branch Instruction misprediction (BMP) rate. These revealed characteristics are expected to benefit database performance optimizations and database architecture design on next-generation processors.

1 Introduction

Previous research calculate the time spent on CPU cache misses by measuring the actual number of misses. This methodology accurately described DBMS's performance on the machines used at that time (Intel Pentium II/Pentium Pro). We characterized the TPC-H benchmark on a DBMS running on an Itanium II workstation. Using an open source database server, PostgreSQL, and TPC-H benchmark, we studied several issues by profiling CPU-related statistics on the Itanium II/Linux platform. Our main focus is on the in-depth analysis of deep memory characteristics. To the best of our knowledge, there is no related research on Itanium II processor reported in the literature. Through these experiments, we obtained a real-time hardware-based characterization from the viewpoint of processor. Analysis of these characterizations provides us more insights into DSS workloads on Itanium II processor.

2 Related Work

There have been several published studies concentrate on characterizing the database workloads. [12] focused on multiprocessor system issues. [2] using TPC-A and TPC-C on another relational DBMS showed that commercial workloads exhibit large instruction footprints with distinctive branch behavior, typically not found in scientific workloads and that they benefit more from large first-level caches. [8] showed that the processor is stalled 50% of the time due to

K.C. Chang et al. (Eds.): APWeb/WAIM 2007 Ws, LNCS 4537, pp. 235–240, 2007.

cache misses when running OLTP workloads. Most of these studies evaluate OLTP workloads[10][5][6], a few evaluate decision support (DSS) workloads [6] and there are some studies that use both [7] [9]. [1] introduces a framework of analyzing query execution time on a DBMS running on a server with a Intel Xeon processor and memory architecture then examines four commercial DBMS running on workstation. In the most recent research [4] characterized the OLTP workloads on a Itanium II based ccNUMA platform. Our research complements the existing works by providing a study of Itanium II deep memory performance running DSS workloads.

3 Query Execution on Modern Processor

3.1 Query Execution Time Breakdown

The pipeline is the basic module of modern processor, that receives an instruction then executes it and stores results into memory. The pipeline works in several sequential stages, and each involves some functional components. An operation at one stage can overlap with operations at other stages. When an operation can not be able to complete immediately, this will cause delay ("stall") in pipeline. Modern processor will cover the stall time by doing other work through using three techniques: non-blocking caches, out-of-order execution and speculative execution with branch prediction.

Even with these techniques, the stalls cannot be fully overlapped with useful computation. Thus, the time to execute a query (T_Q) includes a useful computation time (T_C), a stall time because of memory stalls (T_M), a branch misprediction overhead (T_B), and resource-related stalls (T_R). As discussed above, some of the stall time can be overlapped (T_{OVL}). Thus, the following equation holds [1]:

$$T_Q = T_C + T_M + T_B + T_R - T_{OVL} \tag{1}$$

Table 1 shows the time breakdown into smaller components. The D_{TLB} and I_{TLB} (Data or Instruction Transaction Lookaside Buffer) are page table caches used for translation of data and instruction virtual addresses into physical ones.

4 Experimental Setup

4.1 The Hardware Platform

We used a HP Integrity rx2620-2 server to conduct all of the experiments. The processor of our system is an Itanium II processor running at 1.6 GHz, with 4GB of main memory connected to the processor chip through a 400 MHz system bus, and it uses bundles of six instructions, which is twice the issue width of Intels out-of-order Pentium IV. Experiments were conducted on an open source DBMS, PostgreSQL 8.1.3, running on Linux kernel 2.6.9-34.EL. In our experiment, we preloaded all dataset into the memory, and we configured database with a large size buffer pool to minimize the effects of disk I/O.

Table 1. Execution time components

T_C			**computation time**
T_M			**stall time related to memory hierarchy**
	T_{L1}	T_{L1D}	stall time due to L1 D-cache misses
		T_{L1I}	stall time due to L1 I-cache misses
	T_{L2}	T_{L2D}	stall time due to L2 data misses
		T_{L2I}	stall time due to L2 instruction misses
	T_{L3}	T_{L3D}	stall time due to L3 data misses
		T_{L3I}	stall time due to L3 instruction misses
	T_{DTLB}		stall time due to DTLB misses
	T_{ITLB}		stall time due to ITLB misses
T_B			**branch misprediction penalty**
T_R			**resource stall time**

Table 2. Itanium II cache characteristics

Characteristic	$L_1(split)$	L_2	L_3
Cache size (KB)	16 Data 16 Inst	256	3072
Cache line size (bytes)	64	128	128
Associativity	4-way	8-way	6-way
Miss Penalty (cycles)	7	16	182

4.2 Measurement Tools and Methodology

We use *Calibrator* [11], a cache-memory and TLB calibration tool to get the characteristic values of the following items: each cache level's access and miss latency, main memory access latency, number of TLB levels and each level's TLB miss latency. We use *PerfSuite* [3] to control these counters. *Psrun* can used to measure 60 event types for the results. Using these counters, the number of occurrences for each event type (e.g., number of L1 instruction cache miss during query execution) can be measured. Then we multiplied the number of occurrences by an calibrated penalty. Table 3 shows a detailed list of stall time components and the way they were measured. We were not able to measure T_{FU}, T_{DEP} and T_{ILD} separately, because the event code is not available.

5 Results and Analysis

5.1 Memory Stall

We executed the TPC-H benchmark on 1G dataset. Based on the framework given in section 2.1, Figure 1 and Figure 2 show the breakdown of T_M into the following eight components. In Figure 1, each bar shows the contribution as a portion of the total memory stall time. It can be observed that L1 D-cache,

Table 3. Method of measuring each of the stall time components

Stall time component			Description	Measurement method
T_C			computation time	Estimated minimum μops retired
T_M	T_{L1}	T_{L1D}	$L1$ D-cache stalls	#misses*7cycles
		T_{L1I}	$L1$ I-cache stalls	#misses*7cycles
	T_{L2}	T_{L2D}	$L2$ D-cache stalls	#misses*16cycles
		T_{L2I}	$L2$ I-cache stalls	#misses*16cycles
	T_{L3}	T_{L3D}	$L3$ D-cache stalls	#misses*182cycles
		T_{L3I}	$L3$ I-cache stalls	#misses*182cycles
	T_{DTLB}		DTLB stalls	#misses*16cycles
	T_{ITLB}		ITLB stalls	#misses*8cycles
T_B			branch misprediction penalty	#branch mispredictions retired*17 cycles
T_R	T_{FU}		functional unit stalls	actual stall time
	T_{DEP}		dependency stalls	actual stall time
	T_{ILD}		Instruction-length decoder stalls	actual stall time
T_{OVL}			overlap time	Not measured

L1 I-cache and L3 D-cache stall are significant in all components. Other stalls components are totally insignificant, and we show them in Figure 2 separately. As Figure 2 illustrated, these components contribution is very low, most of which is no more than 7% both in power test and throughput test. These characteristics indicate that when running the DSS workloads on an Itanium II processor, most of the memory stalls are caused by stall of L1 D-cache, L1 I-cache and L3 D-cache misses. To optimize performance under these type workloads, instruction placement should focus on level one, and data placement should focus on level three. The low D_{TLB} and I_{TLB} indicate that the system use few instruction pages, and the I_{TLB} is enough to store the translations for their addresses.

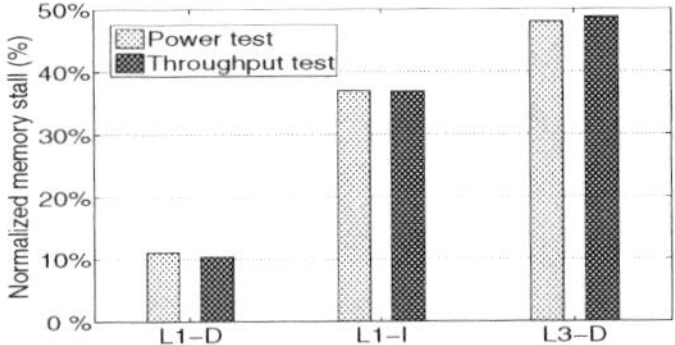

Fig. 1. L1 D-cache, L1 I-cache, L3 D-cache

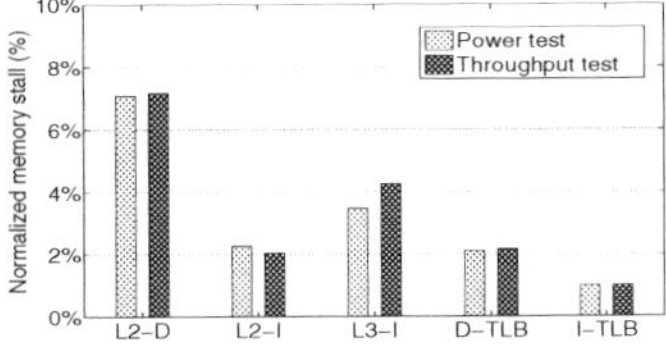

Fig. 2. L2 D-cache, L2 I-cache, L3 I-cache, D-TLB, I-TLB

5.2 Index Influence

From Figure 3 and Figure 4, it can be observed that the system under test with indices has a slightly more L1 D-cache and L1 I-cache stall contribution than that without indices in average. When indices are applied, the percentage of L1 D-cache stall increased by nearly 20%, and L1 I-cache stall increased by about 5% under the power test. While under the throughput test, the L1 D-cache increased by around 5% and L1 I-cache increased by as much as 30%. On the other hand, it can be observed that when indices were applied, L3 D-cache stall contribution was reduced much, it was reduced by as much as 30% under the power test, while a 50% reduction under the throughput test. In addition, indices caused BMP rate increased little more, but caused L3 D-cache miss rate reduced than that without indices, as can be seen in Figure 5 and Fiugre 6. The BMP rate increases by around 5%, and L3 D-cache miss rate reduces by as much as 77%.

Itanium II processor has a feature called branch predication for eliminating branch mispredictions, by allowing executing both the THEN and ELSE blocks in parallel and discard either of the results as soon as the result of the condition becomes known. As illustrated in Figure 5 and Figure 6, DSS workloads on Itanium II processor usually got a lower BMP rate, and both in power test and throughput test it only accounted for around 4%. Even when index is used, the BMP rate is still around 7%. These results indicate that the benefit from index used in Itanium processor exceed that used in Pentium processor.

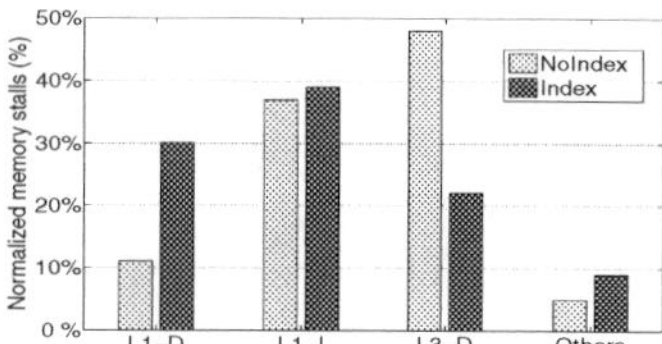

Fig. 3. TPC-H power test

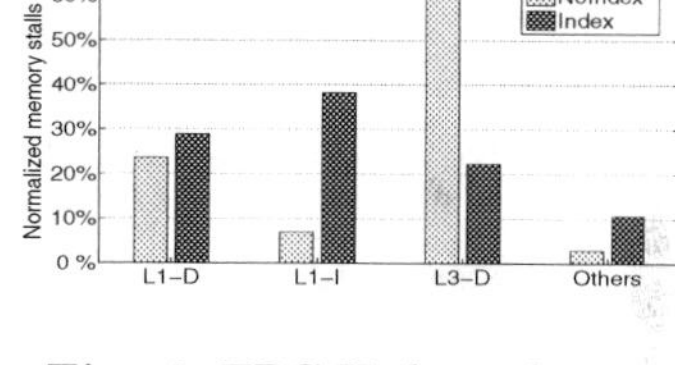

Fig. 4. TPC-H throughput test

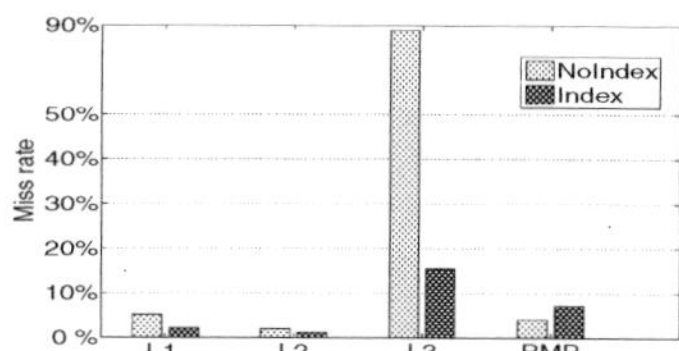

Fig. 5. TPC-H power test

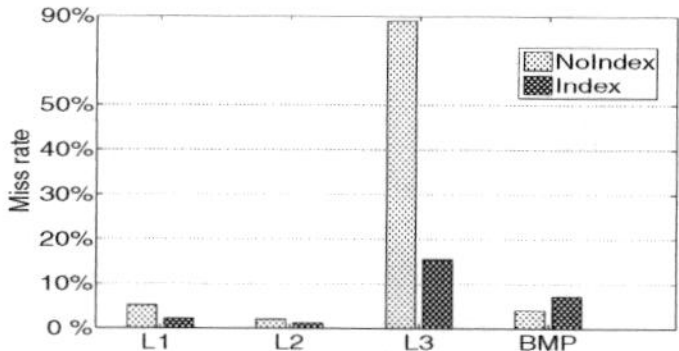

Fig. 6. TPC-H throughput test

6 Conclusions and Future Work

In this paper, we give a processor-level characterization of DSS workloads measured on Itanium II under the TPC-H benchmark. First, the dominant stall time caused by deep memory under the TPC-H benchmark was investigated. Second, how index interacts with deep memory was demonstrated in detail. These revealed characteristics are expect to benefit database performance optimizations and database architecture design on next-generation processors. In the future, we will focus on how to utilize hardware resources efficiently so as to optimize database performance.

Acknowledgement

This work was supported by the National Natural Science Foundation of China (No. 60496325 and No. 60503038), China Grid (No. CNGI-04-15-7A), and by a grant from HP Lab China.

References

1. Ailamaki, A., DeWitt, D.J., Hill, M.D., Wood, D.A.: DBMSs on a Modern Processor: Where Does Time Go? In: Proc. VLDB (1999)
2. Maynard, A.M.G., Donnelly, C.M., Olszewski, B.R.: Contrasting Characteristics and Cache Performance of Technical and Multi-User Commercial Workloads. ASPLOS, pp. 145–156 (1994)
3. Ahn, D., Kufrin, R., Raghuraman, A., Seo, J.H.: `http://perfsuite.ncsa.uiuc.edu`
4. Saylor, G., Khessib, B.: Large scale Itanium 2 processor OLTP workload characterization and optimization. DaMoN vol. 3 (2006)
5. Lo, J.L., Barroso, L.A., Eggers, S.J., Gharachorloo, K., Levy, H.M., Parekh, S.J.: An Analysis of Database Workload Performance on Simultaneous Multithreaded Processors. ISCA, pp. 39–50 (1998)
6. Keeton, K., Patterson, D.A., He, Y.G., Raphael, R.C., Baker, W.E.: Performance Characterization of a Quad Pentium Pro SMP using OLTP Workloads. ISCA, pp. 15–26 (1998)
7. Barroso, L.A., Gharachorloo, K., Bugnion, E.: Memory System Characterization of Commercial Workloads. ISCA, pp. 3–14 (1998)
8. Rosenblum, M., Bugnion, E., Herrod, S.A., Witchel, E., Gupta, A.: The Impact of Architectural Trends on Operating System Performance. SOSP, pp. 285–298 (1995)
9. Ranganathan, P., Gharachorloo, K., Adve, S.V., Barroso, L.A.: Performance of Database Workloads on Shared-Memory Systems with Out-of-Order Processors. ASPLOS, pp. 307–318 (1998)
10. Eickemeyer, R.J., Johnson, R.E., Kunkel, S.R., Squillante, M.S., Liu, S.: Evaluation of Multithreaded Uniprocessors for Commercial Application Environments. ISCA, pp. 203–212 (1996)
11. Manegold, S.: `http://monetdb.cwi.nl/Calibrator`
12. Thakkar, S.S., Sweiger, M.: Performance of an OLTP Application on Symmetry Multiprocessor System. ISCA, pp. 228–238 (1990)

Exploiting Connection Relation to Compress Data Graph

Jun Zhang[1,2,3], Zhaohui Peng[1,2], Shan Wang[1,2], and Jiang Zhan[1,2]

[1] School of Information, Renmin University of China, Beijing 100872, P.R. China
{zhangjun11,pengch,swang,zhanjiang}@ruc.edu.cn
[2] Key Laboratory of Data Engineering and Knowledge Engineering (Renmin University of China), MOE, Beijing 100872, P.R. China
[3] Computer Science and Technology College, Dalian Maritime University, Dalian 116026, P.R. China

Abstract. In this paper, we propose an approach to compress data graph by exploiting connection relations which are some particular entity-relationship relations in relational databses. Our method not only can make a larger data graph fit in memory, but also can improve the performance of existing data graph search algorithms for keyword search over relational databases. Comprehensive experiments are conducted to show the effectiveness and efficiency of our method.

1 Introduction

Keyword Search Over Relational Databases(KSORD) techniques enable casual users or Web users to easily access databases through free-form keyword queries[1]. Many approaches have been proposed to implement KSORD prototypes, including data-graph-based online KSORD(DO-KSORD) systems. Data Graph is a good model for relational databases to support keyword search, in which tuples in a database are modeled as nodes and foreign-key relationships as directed edges. Based on this model, several prototypes of DO-KSORD have been developed, such as BANKS[2], BANKS II[3], DbSurfer[4] and DPBF[5].

However, all of above DO-KSORD prototypes assume that data graph fits in memory. On one hand, data graph for a large database can be too huge to be accommodated in limited main memory. To scale with the size of data graph, disk-based graph algorithms can be considered. Although there is active research in disk-based graph algorithms[9], memory-based graph algorithms tend to be simpler, easier to understand, and easier to implement[6]. On the other hand, the larger data graph becomes, the more poorly DO-KSORD system performs.

Many techniques have been developed to compress Web graph[6,7,8] in Web search engines. These techniques compress Web graph mainly by reducing the edges' storage, and they also require decompression processing when searching the compressed Web graph. To a certain extent, data graph is similar to Web graph, and some compression techniques for Web graph can also be applied to

K.C. Chang et al. (Eds.): APWeb/WAIM 2007 Ws, LNCS 4537, pp. 241–246, 2007.

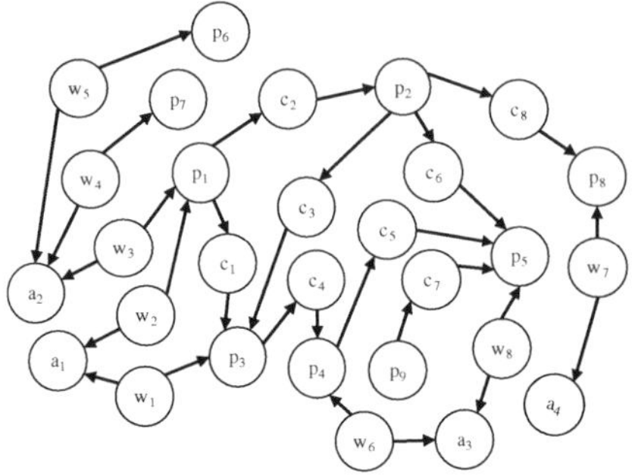

Fig. 1. A sample of DBLP data graph with 29 nodes and 32 edges

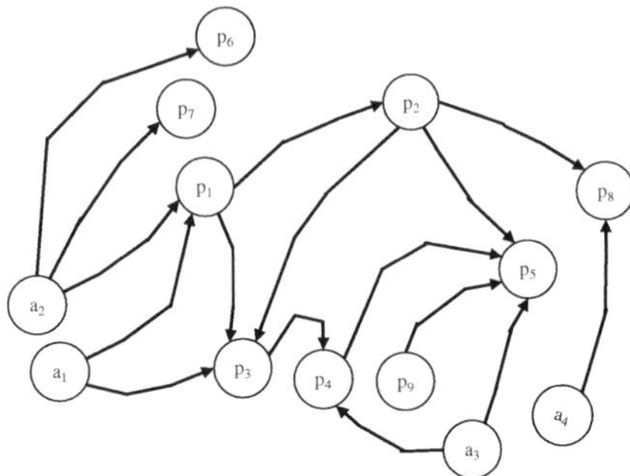

Fig. 2. The compressed sample with 13 nodes and 16 edges

data graph. However, data graph still has some important differences from Web graph. Data graph contains database schema information whereas Web graph doesn't. Unlike Web graph with hyperlinks as directed edges, data graph has Foreign-Key-Primary-Key(FKPF) relationships as directed edges so that the edges have richer semantics than that in Web graph.

By our observations, we find that many database schemas have some particular entity-relationship relations with exactly two foreign keys, which we call connection relations. In the data graph, the nodes coming from connection relations have the degree of exactly two and like bridges connecting their associated primary-key nodes. Intuitively, these nodes can be removed from the data graph, and the edges linked with each removable node can be merged into one edge.

Take DBLP database as an example. There are four relations: Papers (pid, title), Authors(aid, name), Writes(aid, pid) and Cites(citedpid, citingpid). Here relation *Writes* and *Cites* are typical connection relations, and Fig.1 is a sample of DBLP data graph with 29 nodes and 32 edges. Node $w_i(1 < i < 8)$ and $c_j(1 < j < 8)$ in Fig.1, which come from relation *Writes* and *Cites* respectively, are removed in Fig.2, and their associated edges are merged into one edge respectively. The compressed sample in Fig.2 has only 13 nodes and 16 edges.

Therefore, we propose an approach CodCor to compress data graph by exploiting connection relations. CodCor differs from Web graph compression techniques fundamentally. Without loss of information, CodCor compresses the storage of data graph by reducing the number of nodes and edges which come from connection relations. In particular, the compressed data graph doesn't need to be decompressed when searching it. In this way, the data graph will be compressed dramatically, and larger data graph can be accommodated in memory. Moreover, based on compressed data graph, existing memory-based data graph search algorithms(DGSA) can work well with slight modification, and they can also perform more efficiently than before because answers to keyword queries will be provided by matching less nodes. Comprehensive experiments are conducted to show the effectiveness and efficiency of CodCor.

2 Basic Concepts: Data Graph and Connection Relation

Definition 1 Data Graph(G_d). *Denoted as $G_d(V, E)$. Data graph is a directed graph which database can be modeled as. Each tuple in the database is modeled as a node in V, and each FKPK link between the corresponding tuples as a directed edge in E.*

Data graph G_d(V,E) defined in Definition 1 is a sparse graph. This means that the number of edges of G_d is linear to the number of nodes of G_d.The proof is easy, and is omitted due to limited space. Data graph usually is assumed to fit in memory, because the representation of in-memory nodes doesn't need to store any attribute of the corresponding tuples other than the row id of tuples[2,3]. This seems reasonable, even for moderately large databases. However, that's not always the case. For many very large databases, their data graphs will not fit in memory due to limited main memory. It is necessary to compress data graph so that as large as possible data graph can be loaded into memory.

Definition 2 Connection Relation $\Re$. *In a database schema, a relation* R *is called a connection relation $\Re$ if it satisfies the following conditions: (i)* R *has exactly two foreign keys; (ii)* R*'s primary key consists of its foreign keys. (iii)* R *isn't referenced by any relation. Here, (i) and (iii) assure that* R *is only associated with two relations, and (ii) and (iii) assure that R is just an entity-relationship relation, not an entity relation. Furthermore, if R is a connection relation and has only foreign-key attributes, it is called pure connection relation $\Re_p$, otherwise, it is called semi-connection relation $\Re_s$.*

For example, relation Writes and Cites in DBLP database are two pure connection relations $\Re_p$. Take Northwind database which is from Microsoft as another example. There are five relations: Employees(eid, name), Customers(cid, name), Products(pid, name), Orders(oid, cid, eid, orderdate, phone) and OrderDetails(oid, pid, unitPrice, Quantity), where relation OrderDetails is a typical semi-connection relation $\Re_s$ without any text attribute. The third example is the drug-disease database which has three relations: Drugs(druId, title), Diseases(disId, title) and Impact(druId, disId, description). Here relation Impact is a semi-connection relation $\Re_s$ with text attribute.

By our observations and analysis, we find that many database schemas contain some connection relations. In general, a connection relation $\Re$ has the following properties: (i) There are large numbers of tuples in $\Re$ whereas the size of each tuple is relatively small. (ii) The number of tuples in $\Re$ increases more rapidly than that in other relations. (iii) Frequent keyword queries seldom occur in $\Re$, because $\Re$ usually has no text attributes and is used to join tuples from its two referencing relations(Here, its referencing relation means the relation which it references).

It is just the number of tuples in a database that mainly determines the size of the data graph. So, the properties of connection relation state that the size of data graph ought to be dramatically reduced and keyword queries over the data graph will not be affected if all the nodes from connection relations are removed and the edges from the corresponding tuples are reduced.

3 Data Graph Compressing Approach

We propose an approach CodCor to compress data graph by exploiting connection relations, and CodCor is only applicable to those databases with connection relations. Algorithm 1 shows the Compressed Data graph Generation Algorithm(CodGA) in detail. Obviously, if no connection relations are identified(line 1) in Algorithm 1, CodGA will degenerate to be an uncompressed data graph generation algorithm.

Algorithm 1. CodGA: Compressed Data Graph Generation

Input: database schema: S **Output:** data graph: G_d

1: Identify the connection relations $\Re$ from S;
2: **for** each tuple t of each table A in $(S - \Re)$ **do** {
3: Create data graph node n = GraphNode(A.tableid, t.rowid);
4: Add node n to G_d; }
5: Identify all foreign key relationships in S as the set of $G_d.FK$;
6: **for** each f in $G_d.FK$ **do** {
7: **if** (fRel(f) $\in \Re$) **then** Continue;
8: Add f as an edge type e to the set of $G_d.ET$;
9: CREATE VIEW tempv AS SELECT A.rowid, B.rowid
10: FROM pRel(f) as A, fRel(f) as B
11: WHERE pAtt(f) = fAtt(f);
12: **for** each tuple t in tempv **do** {
13: AddAdjLists(G_d, e, A.tableid, A.rowid, B.tableid, B.rowid); }}
14: **for** each r in $\Re$ **do** {
15: CREATE VIEW tempv AS SELECT A.rowid, B.rowid
16: FROM pRel($r.f_1$) as A, r, pRel($r.f_2$) as B
17: WHERE pAtt($r.f_1$) = fAtt($r.f_1$) and pAtt($r.f_2$) = fAtt($r.f_2$);
18: Add composite($r.f_1$,$r.f_2$) as an edge type e to the set of $G_d.ET$;
19: **for** each tuple t in tempv **do** {
20: AddAdjLists(G_d, e, A.tableid, A.rowid, B.tableid, B.rowid); }}
21: Return G_d.

We define three measures to evaluate CodGA, node compressing rate, edge compressing rate and storage compressing rate. First of all, assume that a database D is modeled as the data graph $G_d(V, E)$, each node occupies v bytes and each edge e bytes, and there are $|\Re|$ tuples in one or more than one connection relations $\Re$ in D. After compressed, G_d is called compressed data graph, and denoted as G_{cd}. There will be $|V| - |\Re|$ nodes and $|E| - |\Re|$ edges in G_{cd} according to our approach CodCor. Note that $|X|$ denotes the element number of set X. Because data graph is a sparse graph, so $|E| = k \times |V|$ holds, where k usually is a small number, and reflects the sparse degree to G_d. The smaller k is, the sparser G_d is. Therefore, the node compressing rate α of G_d is defined as the ratio of $\frac{|\Re|}{|V|}$, the edge compressing rate β of G_d is defined as the ratio of $\frac{|\Re|}{|E|} = \frac{|\Re|}{k|V|} = \frac{\alpha}{k}$, and the storage compressing rate γ of G_d is defined as the ratio of $1 - \frac{sizeof(G_{cd})}{sizeof(G_d)} = 1 - \frac{(|V|-|\Re|)\times v+(|E|-|\Re|)\times 2e}{|V|\times v+|E|\times 2e} = \alpha \times \frac{v+2e}{v+2ek}$. Note that each edge is stored twice in the data graph for its two associated nodes. Obviously, α, β and γ vary from 0 to 1, the larger α and β are, the larger γ is and the more effective CodCor is too. By our theoretic analysis, the theoretic upper bound of γ is $\frac{v+2e}{v+4e}$, even though v and e approach zero, the theoretic upper bound of γ can approach 50%. Due to limited space, the proof is omitted.

Then, considering BANKS[2], we discuss how to improve existing DGSAs based on CodCor. Assume that query keywords only occur in text attributes in the database. If $\Re$ is a pure or semi-connection relation without any text attribute, BANKS almost doesn't need to be modified and can work well on G_{cd}. If $\Re$ is a semi-connection relation $\Re_s$ with text attributes, BANKS will be modified in three stages. Firstly, $\Re_s$ will be joined with its two referencing relations to produce a composite tuple set to pick up tuples containing query keywords. If the tuple set is empty, BANKS still doesn't need to be modified, just like $\Re$ is a $\Re_p$. Secondly, when the nodes from $\Re_s$ are searched, they can be viewed as virtual nodes in G_{cd} and can also be searched by their corresponding two primary-key nodes. Finally, the nodes from $\Re_s$ in final answers to keyword queries will be retrieved from the database by their corresponding primary keys.

4 Experimental Evaluation

We ran our experiments using the Oracle 9i RDBMS on the platform of Dawning server with AMD844*4 CPU and 4G RAM memory running Windows 2000 Server. BANKS[2], DPBF[5] and CodCor were implemented in Java and connected to the RDBMS through JDBC. The IR engine was the Oracle9i Text extension. We used two databases, DBLP and Northwind, to test the effectiveness of CodCor, but only DBLP data set was used to test the efficiency of two improved DGSAs: BANKS[2] and DPBF[5].

Table 1 shows the effectiveness of CodCor. This table states that CodCor effectively compressed the data graph for DBLP or Northwind by about half of the uncompressed storage space. In addition, by our experiments, the compressed data graph was generated more quickly than the uncompressed one, for DBLP by 61.3% faster and for Northwind by 45.2% faster, because much less nodes and edges were generated in the compressed data graph.

Table 1. Two Examples of Compressing Data Graph

Data Set	Total number		Coming From $\Re$		Compressing Rate		
	nodes	edges	nodes	edges	α	β	γ
DBLP	1,851,927	2,222,912	1,111,456	2,222,912	0.60	0.50	0.534
Northwind	3,162	5,970	2,155	4,310	0.682	0.361	0.475

We also evaluated the efficiency of DGSAs,BANKS and DPBF. Experiments show that both DPBF and BANKS based on compressed data graph performed two to three times faster than original algorithms respectively. The main reason is that less nodes were matched in the former algorithms than that in the latter ones. For example, if "$a_2 \leftarrow w_3 \rightarrow p_1 \rightarrow c_1 \rightarrow p_3 \rightarrow c_4 \rightarrow p_4$" is one answer to query Q in Fig.1, "$a_2 \rightarrow p_1 \rightarrow p_3 \rightarrow p_4$" will be the answer to Q in Fig.2.

5 Conclusions and Future Work

We presented a novel approach CodCor to compress data graph by exploiting connection relations. CodCor not only can make a larger data graph fit in limited main memory, but also can improve existing DGSAs for keyword search over relational databases. Many databases usually have connection relations, and CodCor can be applied to them. Our experiments show that CodCor is really very effective to compress data graph and can improve the execution efficiency of existing DGSAs by two to three times. In the future work, we consider the relaxation of connection relation's definition so that our approach can benefit more databases, and also try to combine CodCor with other compression techniques to better compress data graph.

Acknowledgements

This work is supported by the National Natural Science Foundation of China (No. 60473069 and 60496325), and China Grid (No. CNGI-04-15-7A).

References

1. Wang, S., Zhang, K.: Searching Databases with Keywords. Journal of Computer Science and Technology 20(1), 55–62 (2005)
2. Bhalotia, G., Hulgeri, A., Nakhe, C., et al.: Keyword Searching and Browsing in Databases using BANKS. ICDE 2002, CA, pp. 431-440 (2002)
3. Kacholia, V., Pandit, S., Chakrabarti, S., et al.: Bidirectional Expansion For Keyword Search on Graph Databases. VLDB 2005, Norway, pp. 505-516 (2005)
4. Wheeldon, R., Levene, M., Keenoy, K.: DbSurfer: A Search and Navigation Took for Relational Databases. In: Proceedings of the 21st Annual British National Conference on Databases, pp. 144-149 (2004)
5. Ding, B., Yu, J., Wang, S., et al.: Finding Top-k Min-Cost Connected Trees in Databases. ICDE 2007, Istanbul, Turkey (2007)
6. Randall, K. H., Stata, R., Wickremesinghe, R., Wiener, J. L.: The link database: Fast access to graphs of the web. In: Proceedings of the Data Compression Conference, Snao Bird, Utah, pp. 122-131 (April 2002)
7. Boldi, P., Vigna, S.: The webgraph framework I: compression techniques. In: Proceedings of the 13th international conference on World Wide Web(WWW 2004), New York, NY, USA, pp. 595-602 (May 2004)
8. Adler, M., Mitzenmacher, M.: Towards Compressing Web Graphs. In: Proceedings of IEEE Data Compression Conference (DCC 2001), pp. 203-212 (March 2001)
9. Chiang, Y., Goodrich, M., Grove, E., Tamassia, R., Vengroff, D., Vitter, J.: External-Memory Graph Algorithms. In: Proc. ACM-SIAM Symposium on Discrete Algorithms, pp. 139-149 (January 1995)

Indexing the Current Positions of Moving Objects on Road Networks

Kyoung Soo Bok[1], Ho Won Yoon[2], Dong Min Seo[2], Su Min Jang[2], Myoung Ho Kim[1], and Jae Soo Yoo[2]

[1] Department of Computer, Korea Advanced Institute of Science and Technology, Korea
{ksbok, mhkim}@dbserver.kaist.ac.kr
[2] Department of Computer and Communication Engineering, Chungbuk National University, Korea
{hwyoon, dmseo, jsm}@netdb.chungbuk.ac.kr, yjs@chungbuk.ac.kr

Abstract. In this paper, we propose an intersection-oriented network model to solve this problem of increasing update costs when objects move to adjacent road segments because of the disconnectivity of adjacent road segments. Our intersection-oriented network model preserves network connectivity by not splitting intersecting nodes that three or more edges meet. We also propose an efficient indexing method based on the intersection-oriented network model for current locations of moving objects on road networks.

Keywords: Spatio-temporal data, moving object, road network, index structure.

1 Introduction

We are currently experiencing rapid technological developments that promise widespread use of on-line mobile personal information appliances. With this proliferation of devices, companies have the opportunity to provide a diverse range of e-services. Among them Location Based Service using positioning systems such as GPS provides location-based information to mobile users. LBS rely on the tracking of the continuously changing positions of entire populations of service users, termed moving objects[1-2].

One effective method to improve the performance of a disk-based index is to increase the memory size. This method can dramatically reduce the I/O cost of queries. But for updates, if one cannot put the entire index into memory, the same strategy is not equally effective. This is because updates are more likely random distributed and can display less locality. Index structures for spatio-temporal data such as the positions of moving objects are classified into two categories; first one deals with object's historical data and second one focuses on current and near future positions of moving objects[3].

In real world applications, movements of moving objects are constrained, that is, objects are moving in a constrained network space such as road networks. Recently, several index structures improve their performance by exploiting the properties of network. FNR-tree separates spatial and temporal components of the trajectories and

K.C. Chang et al. (Eds.): APWeb/WAIM 2007 Ws, LNCS 4537, pp. 247–252, 2007.

indexes the time intervals that each moving object spends on a given network link[4]. The MON-tree further improves the performance of the FNR-tree by representing each edge by multiple line segments (i.e. poly lines) instead of just one line segment[5]. IMORS indices the road's poly line into spatial index structure for processing current positions of objects[6]. Then each object is associated with a poly line. A disadvantage of all these approaches is that their network model doesn't provide network connectivity information. This causes increased update cost when objects move out of current MBR.

In this paper, we develop efficient update techniques for moving objects on road networks using the intersection oriented network model. Intersection oriented network model preserves network connectivity by not splitting intersecting nodes that more than three edges meet always.

2 Problem Description

Each moving object reports very frequently (on a second basis) their identification and current location information (id, x, y) to the central database server. To the best of our knowledge, there are two methods for splitting and indexing road networks. First method splits network according to line segment. Figure 1 (a) shows line-based network model. This model leads to a high number of entries and lots of updates in the index structure, because distinct entries are needed for every line segment the object traverses. Second method divides network into a set of poly lines. Figure 1 (b) shows this method. Poly line based network model has the problem of the high dead space in poly line MBRs.

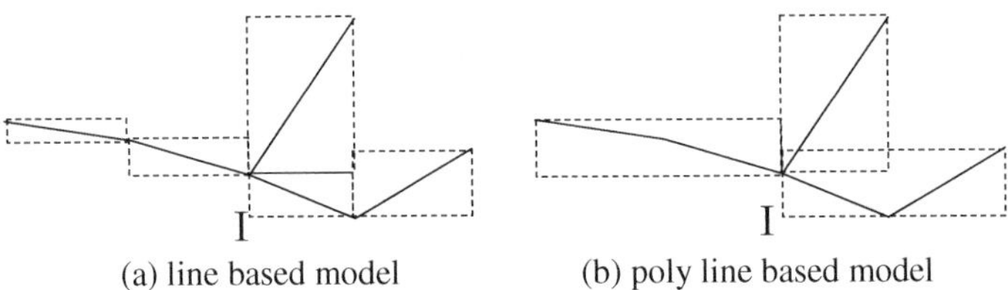

Fig. 1. Network models

Both network models described above have a problem when a moving object traverses from one line(poly line) to adjacent connected line(poly line). To find out the connected line (poly line) the R-Tree must searched from the root node. This can leads serious performance degradation as the number of objects increase. Adjacency component [7] that captures the network connectivity can be used to find out next line or poly line for escaping whole R-Tree search. Using adjacency component can improve the performance but is not a fundamental solution. This also can leads to performance degradation as the number of objects increase because object's update requests are more likely random distributed and can display less locality. If there are millions of objects scattered in networks and each objects send update requests to the server independently we don't know where the next update request occurs in networks. An object going out of a line or poly line means additional disk access to the adjacency component. Figure 2 shows the problem of line or poly line network model based indexing. When object 2 moves to the new line (poly line) in real network, the network R-Tree must searched from the root to insert the object to new line (poly line). Objects going out of MBR cause lots of performance degradation as the number of objects increase.

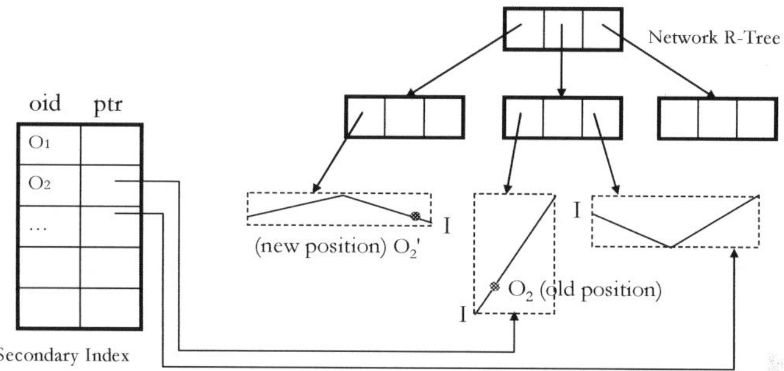

Fig. 2. The problem of existing network model

3 Intersection-Oriented Network R-Tree

3.1 IONR-Tree

We describe the Intersection-Oriented Network model that can be indexed by the IONR-Tree. In line or poly line based network model, intersecting node that more than 3 edges meet always split resulting in connectivity information to be lost. The Intersection-Oriented Network model preserves network connectivity through deliberately split road networks. We preserve network connectivity by not splitting intersecting nodes that more than 3 edges meet. We define IP (Intersection Point) as a node that more than 3 edges meet and CP (Connecting Point) as a point split by MBRs. Figure 3 shows an example of ION model. There are two IPs included in two MBRs and not split.

Figure 4 shows the overall structure of the IONR-Tree. The difference from the existing index structure for current positions on road networks is the structure of the data node because it uses ION model. Figure 5 shows the data node structure.

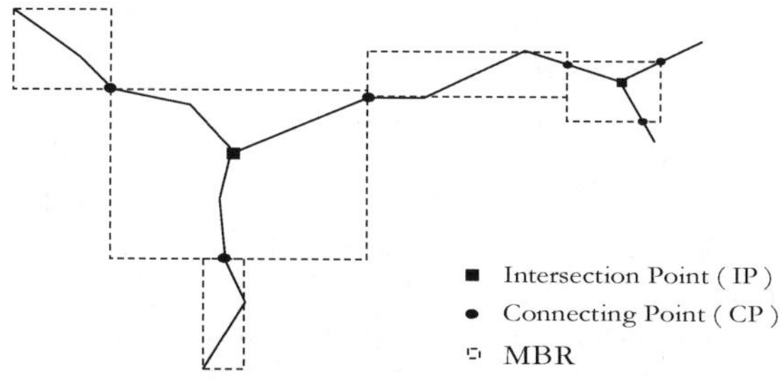

Fig. 3. Example of the Intersection-Oriented Network model

Adjacency list that represents the connectivity information is added additionally. Poly-line part stores the concrete shape between IP and CP or CP and CP. And Objects on the node currently are stored in objects part. There are two kinds of data node according to the number of CP that the node has. Figure 5. (b) represents the adjacency list for a data node that has N (N = 3,4,5…) CP. $Offset_n$ points a point of Poly-line part to represent the real shape between IP and the n-th Connecting Point CP_n. P_n points to a page(data node) has a road segment connected to CP_n. We can directly access the connected road segment by this pointer P_n when objects exit from CP_n.

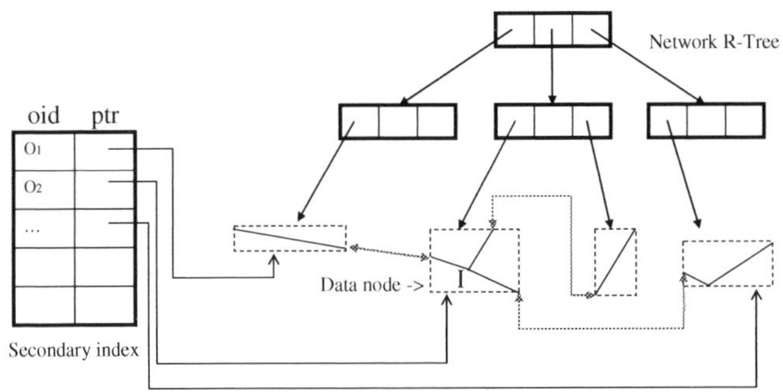

Fig. 4. The IONR-Tree

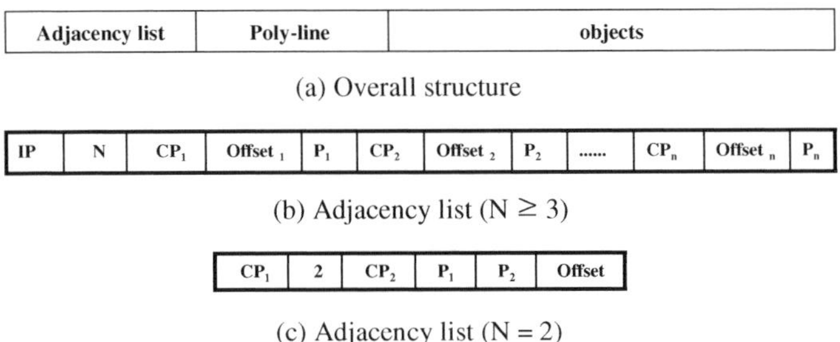

Fig. 5. Data node structure

3.2 Operations of IONR-Tree

A new moving object is registered secondary hash index with object id first. And R-Tree is searched to find out a data block to which the object must included and the object is inserted to the space for objects with id and coordinate. Lastly, the object in secondary index points to this data node to access directly next time. Deletion of an existing object is handled by deleting the object from secondary index and corresponding data node.

Updating an existing object's position is processed by three steps. First, we find a data node storing the object from secondary hash index and check weather the object still lies on the poly lines of the data node. If so, we just overwrite the object's (x, y) position value and update process is terminated. If not so, we delete the object from the data node and calculate the nearest CP to the object's new position in Euclidean distance. Then access the data node this CP points to. Finally, we insert the object in this data node and update secondary index for directly accessing this node next time the object requests update. Figure 6 shows the update algorithm of the IONR-Tree.

```
Update( O, (xn,yn) )

O: moving object, (xn,yn): O's new position

1. Find O in secondary hash index and access a data node N
2. If (xn,yn) is in N then
        Update O's location in N
3. Else
        Delete O from N
        for each connecting point in N do
                Calculate Euclidian distance to (xn,yn)
        Find the nearest CP to (xn, yn)
        Access the page this CP points to
        If O's membership is verified then
                Insert O with (xn,yn) to the new page
        Else
                Search Network R-tree and find corresponding page
                insert O with (xn,yn) to the page
        Update secondary index for object O
```

Fig. 6. Update algorithm of IONR-Tree

In ION model, the main issue is how we determine the size of each MBRs. K.S. Kim [6] has proposed a simple cost model for determining the optimal size l_s of poly line split and indexed. Given a road network, [6] found the optimal number of entries n_{opt} that minimizes the R*-tree leaf node access cost from the root. Assuming the total length of road networks as L, then l_s is determined by $l_s = L/n_{opt}$. We adopt this l_s to be the sum of lengths of road segments in a data node. The size of road segments for a data node that has an IP is obtained by length expansion from the IP to every direction oriented the IP until the sum of lengths expanded reaches ls.

4 Performance Evaluation

We evaluated the IONR-tree scheme by comparing it to the IMORS uses poly line based network model. We implemented both ION R-tree and IMORS in Java and carried out experiments on a Pentium 4, 2.8 GHz PC with 512MB RAM running Windows XP. A network-based data generator[8] is used to create trajectories of moving objects on the real-world road network of Oldenburg comprising 7035 edges and 2238 intersecting nodes that more that 3 edges meet. Figure 7 shows the result of update cost in terms of disk I/O varying the number of moving objects. ION R-tree showed about 2 times better update performance compared to IMORS. Figure 8 shows the region query performance varying query size. Each query test, the number of query was 10k. The result shows that the query performance is degraded very little.

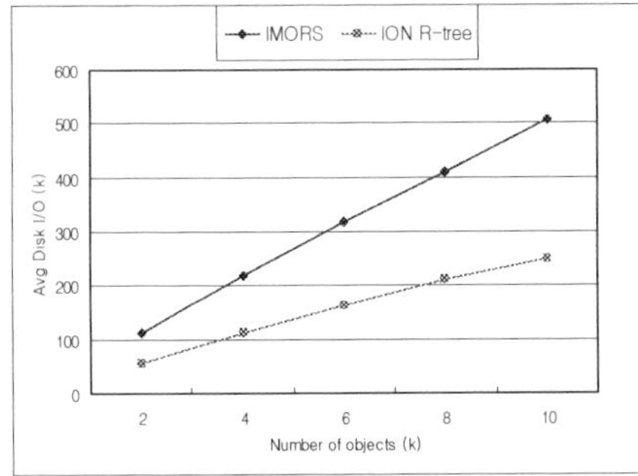

Fig. 7. Update cost

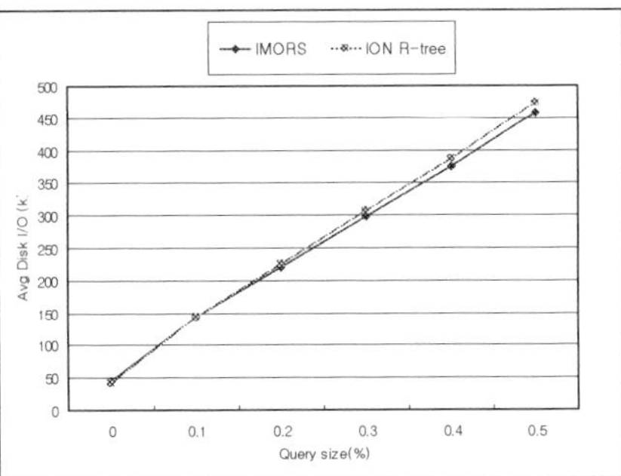

Fig. 8. Query cost

5 Conclusions

In this paper, we proposed the IONR-tree for efficiently updating the current positions of moving objects in road networks. IONR-tree exploits intersection-oriented network model that split networks deliberately to preserve network connectivity. IONR-tree shows about 2 times better update performance and similar query performance compared to IMORS.

Acknowledgements. This work was supported by the Korea Research Foundation Grant funded by the Korean Government (MOEHRD) (The Regional Research Universities Program/Chungbuk BIT Research-Oriented University Consortium) and the Basic Research Program (Grant no. R01-2006-000-1080900) of KOSEF.

References

1. Lee, M. L., Hsu, W., Jensen, C. S., Teo, K. L.: Supporting frequent updates in R-trees: a bottom-up approach. In: Proc. VLDB, pp. 608-619 (2003)
2. Xiong, X., Mokbel, M.F., Aref, W.G.: LUGrid: Update-tolerant Grid-based Indexing for Moving Objects. In: Proc. MDM, p. 13, 8 (2006)
3. Mokbel, M.F., Ghanem, T.M., Aref, W.G.: Spatio-Temporal Access Methods. IEEE Data. Engineering Bulletin 26(2), 40–49 (2003)
4. Frentzos, E.: Indexing objects moving on fixed networks. In: Proc. SSTD, pp. 289–305 (2003)
5. Almeida, V.T., Guting, R.H.: Indexing the Trajectories of Moving Objects in Networks. GeoInformatica 9(1), 33–60 (2005)
6. Kim, K. S., Kim, S., Kim, T., Li, K.: Fast indexing and updating method for moving objects on road networks. In: Proc. WISEW, pp. 34–42 (2003)
7. Papadias, D., Zhang, J., Mamoulis, N., Tao, Y.: Query processing in spatial network databases. In: Proc. VLDB, pp. 802–813 (2003)
8. Brinkhoff, T.: Generating Network-Based Moving Objects. In: Proc. SSDBM, pp. 253–255 (2000)

DBMSs with Native XML Support: Towards Faster, Richer, and Smarter Data Management

Min Wang

IBM T.J. Watson Research Center
Hawthorne, NY 10532, USA

Abstract. XML provides a natural mechanism for representing semi-structured and unstructured data. It becomes the basis for encoding a large variety of information, for example, the ontology. To exploit the full potential of XML in supporting advanced applications, we must solve two issues. First, the integration of structured (relational) data and unstructured or semi-structured data, and on a higher level, the integration of data and knowledge. In this talk, we will address these two issues by introducing a solution that leverages the power of pure XML support in DB2 9.

The semistructured and structured data models represent two seemingly conflicting philosophies: one focuses on being flexible and self-describing, and the other focuses on leveraging the rigid data schema for a wide range of benefits in traditional data management. For many applications such as e-commerce that depend heavily on semistructured data, the relational model, with its rigid schema requirements, fails to support them in an effective way; on the other hand, the flexibility of XML in modeling semistructured data comes with a big cost in terms of storage and query efficiency, which to a large extent has impeded the deployment of pure XML databases to handle such data. We introduce a new approach called eXtricate that taps on the advantages of both philosophies. We argue that semistructured documents, such as data in an E-catalog, often share a considerable amount of information, and by regarding each document as consisting of a shared framework and a small diff script, we can leverage the strengths of relational and XML databases at the same time to handle such data effectively. We also show that our approach can be seamlessly integrated into the emerging support of native XML data in commercial DBMSs (e.g., IBM's recent DB2 9 release with Native XML Support). Our experiments validate the amount of redundancy in real e-catalog data and show the effectiveness of our method.

The database community is on a constant quest for better integration of data management and knowledge management. Recently, with the increasing use of ontology in various applications, the quest has become more concrete and urgent. However, manipulating knowledge along with relational data in DBMSs is not a trivial undertaking. In this paper, we introduce a novel, unified framework for managing data and domain knowledge. We provide the user with a virtual view that unifies the data, the domain knowledge and the knowledge inferable from the data using the domain knowledge. Because the virtual view is in the relational

K.C. Chang et al. (Eds.): APWeb/WAIM 2007 Ws, LNCS 4537, pp. 253–254, 2007.

format, users can query the data and the knowledge in a seamlessly integrated manner. To facilitate knowledge representation and inferencing within the database engine, our approach leverages native XML support in hybrid relational-XML DBMSs. We provide a query rewriting mechanism to bridge the difference between logical and physical data modeling, so that queries on the virtual view can be automatically transformed to components that execute on the hybrid relational-XML engine in a way that is transparent to the user.

A Personalized Re-ranking Algorithm Based on Relevance Feedback*

Bihong Gong, Bo Peng, and Xiaoming Li

Computer Network and Distributed System Laboratory, Peking University, China 100871
gbh@net.pku.edu.cn, pb@net.pku.edu.cn, lxm@pku.edu.cn

Abstract. Relevance feedback is the most popular query reformulation strategy. However, clicking data as user's feedback is not so reliable since the quality of a ranked result will influence the user's feedback. An evaluation method called QR (quality of a ranked result) is proposed in this paper to tell how good a ranked result is. Then use the quality of current ranked result to predict the relevance of different feedbacks. In this way, better feedback document will play a more important role in the process of re-ranking. Experiments show that the QR measure is in direct proportion to DCG measure while QR needs no manual label. And the new re-ranking algorithm (QR-linear) outperforms the other two baseline algorithms especially when the number of feedback is large.

Keywords: Relevance feedback; Re-ranking; Information Retrieval; Personalized.

1 Introduction

Although many search engine systems have been successfully deployed, the current retrieval systems are far from optimal. The main problem is without detailed knowledge of the document collection and of the retrieval environment, most users find it difficult to formulate queries which are well designed for retrieval purposes [1]. That is to say users are unable to express their conceptual idea of what information they want into a suitable query and they may not have a good idea of what information is available for retrieval. But once the system presents them with an initial set of documents, users could indicate those documents that do contain useful information. This is the main idea of relevance feedback.

Relevance feedback is the most popular query reformulation strategy. Users mark documents as relevant to their needs and present this information to the IR system. The retrieval system then uses this information to re-rank the retrieval result list to get documents similar to the relevant ones. Usually a click on a result indicates a relevance assessment: the clicked document is relevant. Many experiments [2][3][4] had shown good improvements in precision when relevance feedback is used.

* This work is supported by the key program of National Natural Science Foundation of China (60435020) and the NSFC Grant (60573166, 60603056).

K.C. Chang et al. (Eds.): APWeb/WAIM 2007 Ws, LNCS 4537, pp. 255–263, 2007.

Although most works would use the clicking data as user relevance feedback, the clicking data are not so reliable or equally relevant to user's interest. Experiments results [5a] indicate that the quality of a ranked result influences the user's clicking behavior. If the relevance of retrieved results decreases, users clicking on abstracts are on average less relevant. This is called *quality bias* [5]. So it's obvious that the interpretation of clicks as relevance feedback should be relative to the quality of current search result.

In this paper, we propose an algorithm to do the re-ranking according to both relevance feedback which in our case is users' clicking on the abstract and the quality of current search result. This is based on the phenomenon that users' feedback is not so reliable or not equally important for the re-ranking process. When the quality of current search result is better, users clicking documents tend to be better feedbacks and more relevant to the users' interest. So we should use the quality of result list to predict the relevance of the feedback and do the re-ranking.

The question is how good a search result list is. This is what we also try to answer in this paper. The *quality* of a ranked result list is defined as how the ranked result satisfies the user's query. Given a user's query indentation, there is a set of documents which contains exactly the relevant documents and no others, all documents are ranked in descending order of relevance. This is referred to as the *ideal result.* Through analyzing the ideal result list, we bring up three features of a result list which can be combined together to evaluate the quality of a ranked result.

Our contributions include:

- An evaluation method to tell how good a ranked result list is.
- A method to calculate the importance of different feedback during the process of re-ranking in one query session.
- An effective algorithm for re-ranking the search result.

The remaining sections are organized as follows. Section 2 discusses the related work. In Section 3, three features of a result list are presented and used to evaluate the quality of a result list. Section 4 describes the QR-linear re-ranking algorithm based on relevance feedback and also the experiment results. Section 5 concludes our work.

2 Related Work

Ranking search results is a fundamental problem in information retrieval. Relevance feedback is the most popular query reformulation strategy. Many works has been done in this area. [1] gave a survey on the relevance feedback technique.[10] provided a brief overview of implicit feedback. It presented a classification of behavior for implicit feedback and classifying the selected key papers in this area accordingly.

In [9], a simulation study of the effectiveness of different implicit feedback algorithm was conducted, and several retrieval models designed for exploiting clickthrough information were proposed and evaluated. [11] presented a decision theoretic framework and develop techniques for implicit user modeling. Several research groups have evaluated the relationship between user behavior and user interest. [12] presented a framework for characterizing observable user behaviors using two dimensions-the underlying purpose of the observed behavior and the scope of the item being acted upon. Joachims et al. [5] presented an empirical evaluation of interpreting clickthrough evidence. Analyzing the users' decision process using eye tracking and comparing implicit feedback against manual relevance judgments, the authors concluded that clicks are informative but biased.

General search engine may do the clickthough log mining which is like treating all users as one special user [6][7]. [8] explored how to capture and exploit the click through information and demonstrated that such implicit feedback information can indeed improve the search accuracy for a group of people.

Our work differs from previous work in several aspects: (1) what we do is a short term relevance feedback, which is to say in one query session, the user' feedback information is collected to improve the current search result immediately while most other work is doing a long-term job. (2) We define the quality of a ranked result which is calculable during the retrieval process while previous work cannot do it without manual work. By using the quality of result, importance of different feedbacks can be calculated.

3 The Quality of a Ranked Result

3.1 Other Measures of a Ranked Result

When the quality of current ranked result is better, users clicking documents tend to be better feedbacks and more relevant to user's interest. So we should use the quality of result to predict the relevance of feedbacks and do the re-ranking. This section describes a method to evaluate the quality of a ranked result. The *quality* is defined as how it satisfies the user's query, more specifically as the difference between it and the ideal result.

Usually a ranked result is evaluated by precision and recall, and other measures derived from precision and recall. Precision is the proportion of retrieval documents that are relevant, while recall is the proportion of relevant documents that are retrieval. But these measures are all set-based measures, which do not take the ranking into account. The *discounted cumulative gain* (DCG) [13][14] measure is used to evaluate a ranked result, the chief advantage of DCG measure is that DCG incorporates multiple relevance levels into a single measure. The discounted cumulative gain vector is defined as Equation (1). Once we get to know the relevance of each result which usually is manually labeled, DCG measure of current ranked result can be calculated.

$$DCG[i] = \begin{cases} G[1], & if\ i = 1 \\ DCG[i-1] + G[i]/\log_b i, & otherwise \end{cases} \tag{1}$$

DCG measure is the real quality of a ranked result. But in real search process, the manual label is impossible to get so that DCG measure is unreachable. In this section we bring up a new measure called *quality of a ranked result* (QR) which is in direct proportion to DCG measure, and can be calculated from three features of a ranked result without manual label.

Section 3.2 will discuss the three features which are all in direct proportion to DCG measure and are derived from the observation on ideal result. Combined three features together, we get the new measure QR. Experiments about the features and new measure are discussed in section 3.3.

3.2 Three Features and QR Measure

Given a user's query indentation q and initial retrieval results set $< d_1, d_2, \cdots, d_n >$, the ideal ranked list should be: $< d_1, \cdots d_i, d_{i+1} \cdots, d_n >$. In the ideal list, all documents are ranked in descending order of relevance and $< d_1, \cdots d_i >$ is the relevant document sets while $< d_{i+1} \cdots, d_n >$ is the irrelevant. Through the observation on ideal result, following features came up:

1. All the documents are in descending order of relevance.
2. There is no outlier in $< d_1, \cdots d_i >$, and the relevant documents are similar.
3. There is no relevant document in $< d_{i+1} \cdots, d_n >$. None of the irrelevant document is similar to the relevant.

These three features are derived from the study on ideal result and are intuitively right. But they are still not calculable. So let's make it more specific. Here are three new features that are just a more detailed version of the above.

1. Rank: $< d_1, d_2, \cdots, d_n >$ is in descending order of the relevance. This is a basic requirement which can be easily satisfied but impossibly done well. Since the actually relevance is no way to know, the value calculated by current ranking algorithm is used to rank the result. In this way, this requirement can be easily satisfied, so the measure QR (quality of a ranked result) will not consider this requirement.
2. Radius: the radius of relevant document set $< d_1, \cdots d_i >$.The smaller the radius is, the higher the quality of ranking result is. This is derived from the feature "relevant documents are similar". Here *radius* means the average distance between each result in $< d_1, \cdots d_i >$ to the center and is calculated by Equation 2.When it comes to an ideal result, all relevant documents are ranked in top of the result list and they are more similar to each other than to the irrelevant ones. It'll make the radius of relevant document set smaller when it's a better ranked result.

$$radius = \frac{1}{rel\,num} \sum_{i=1}^{rel\,num} dist(d_i, d_{center}) \tag{2}$$

3. Ratio: the ratio of documents in irrelevant set $< d_{i+1} \cdots, d_n >$ whose distance to the center of relevant set is smaller than radius. The smaller the ratio is, the higher the quality of ranked list is. This is derived from the feature "There is no relevant document in $< d_{i+1} \cdots, d_n >$. None of the irrelevant document is similar to the relevant". It's calculated by Equation 3 where "radius" means the radius of relevant documents set, d_{center} is the centroid of relevant set. If a document is similar to those relevant documents, there is a very high probability that it also is a relevant

one. Hence the ratio of documents in irrelevant set which is likely to be a relevant one should be smaller when it comes to a better ranked result.

$$ratio = \frac{|\{d_j \mid dist(d_j, d_{center}) < radius \ \wedge d_j \in irelset|}{|irelset|} \tag{3}$$

Combined the features described above, we got QR measure presented by Equation 4. QR value is in direct proportion to DCG measure so it can evaluate the quality of a ranked result. The adjustment factor (log function) is to do smoothing.

$$QR(D_n) = \frac{1}{radius} + \frac{1}{\log(ratio)} \tag{4}$$

There is a note on three features and QR measure. The features cannot describe all the aspects of ideal result. These three are just the most important feature of the ideal; they are just necessary but not sufficient conditions. With more good features, new measure QR should be better.

3.3 The Experiments of New Measure QR

We tested our three features and QR measure on TREC [15] collection: GOV2 dataset with topics 751-800. In all cases, the titles of topic description are used as initial query, since they are closer to the actual queries used in real world.

Given a query, use TianWang [16] search engine to get an initial retrieval result. To test the second and third feature, increase the quality of result list which means increasing DCG value since manual labeled relevant results are available, then check the changing of radius and ratio feature. Fig. 1 and Fig.2 shows that radius and ratio features are relative to DCG value. When DCG value goes up which means the quality of result list increases, the radius and ratio value go down.

Fig 3 shows QR measure is in direct proportion to DCG value, which means QR can describe the quality of a ranked list and can be used to evaluate the quality of a result list. All figures are the average of 30 queries.

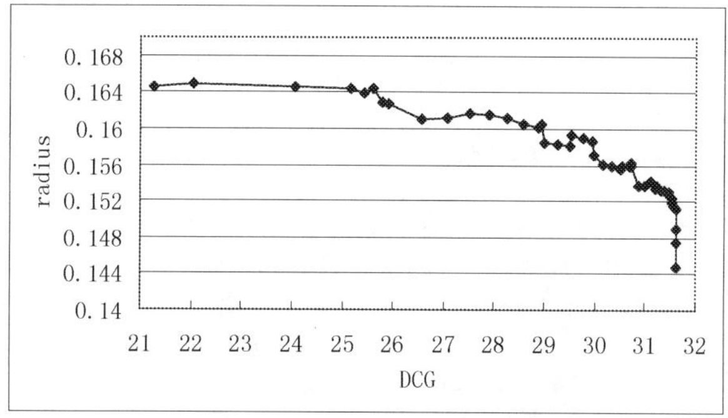

Fig. 1. It shows when the quality of result list increase which means the DCG values goes up, the radius goes down. It validates the second feature.

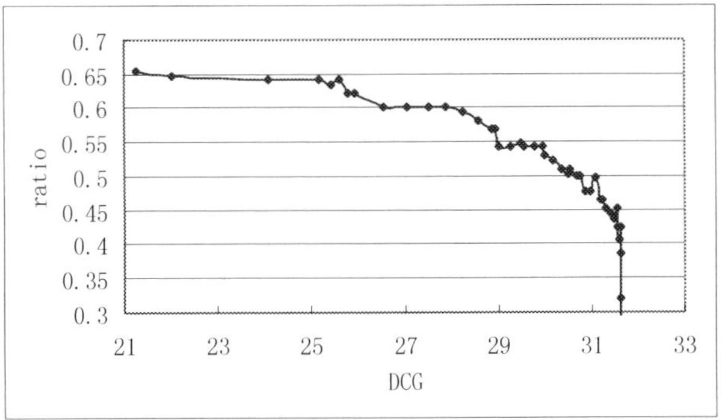

Fig. 2. It shows when the quality of result list increase which means the DCG values goes up, the ratio goes down. It validates the third feature.

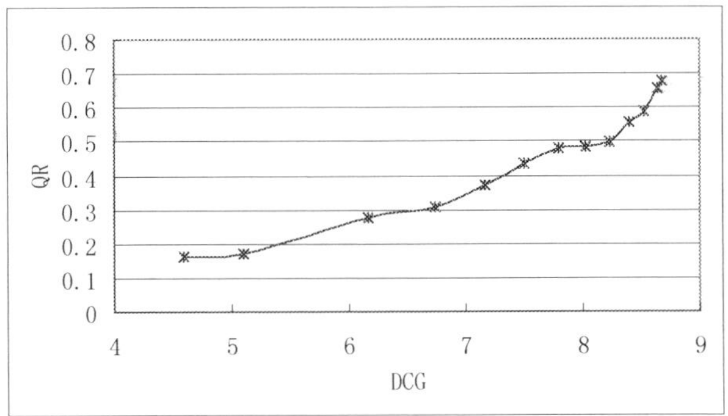

Fig. 3. QR measure is in direct proportion to DCG value, which means QR can describe the quality of a ranked list and can be used to evaluate the quality of a result list

4 The QR Weighted Re-ranking Algorithm and Experiments

As mentioned above, although most works would use the clicking data as user relevance feedback, the clicking data are not so reliable or equally relevant to user's interest. If the relevance of retrieved results decreases, users' clicking on abstracts is on average less relevant. So it's obvious that the interpretation of clicks as relevance feedback should be relative to the quality of current search result.

With QR measure described in section 3, we proposed the new weighted re-ranking algorithm. A feedback document tends to be a more relevant feedback if current QR is higher. So QR value is used to re-weight different feedback.

Three re-ranking algorithms are tested here:

- Ide-Regular algorithm [17] as Equation 5. It's a classical query expansion and term reweighting algorithm for the Vector Space Model. Here it's used as a baseline algorithm on relevance feedback based on query refinement.

$$Ide-\text{Regular}: \vec{q}_m = \alpha\vec{q} + \beta \sum_{\forall \vec{d}_j \in D_r} \vec{d}_j - \gamma \sum_{\forall \vec{d}_j \in D_n} \vec{d}_j \quad (5)$$

- Linear-Combination algorithm as Equation 6. Given n feedbacks $\{f_1, f_2, ..., f_n\}$, the relevance of a document is calculated by a linear combination of every w_{d,f_i} which is the similarity between the document and each feedback. Here it's used as a baseline algorithm on relevance feedback based document similarity.

$$relevance\ \ of(d_i \mid f_1, f_2, \ldots f_n) = \sum_{j=1}^{n} w_{f_j,d_i} \quad (6)$$

- Our QR weighted re-ranking algorithm as Equation 7. Each feedback is weighted by the quality of result list when it is clicked.

$$relevance\ \ of(d_i \mid f_1, f_2, \ldots f_n) = \sum_{j=1}^{n} QR_{current} w_{f_j,d_i} \quad (7)$$

The experiment also used TREC collection GOV2 and we used Tianwang seach engine to index and retrieve this dataset. The dataset has 50 topics while we chose around 40 topics. Each topic, first 500 abstracts in the initial retrieval list are taken out to do the pseudo feedback and re-rank. The experiment result is shown in Fig. 4. It's clear that QR-linear algorithm outperforms the other two baseline algorithm and is effective for re-ranking especially when the number of feedback is large.

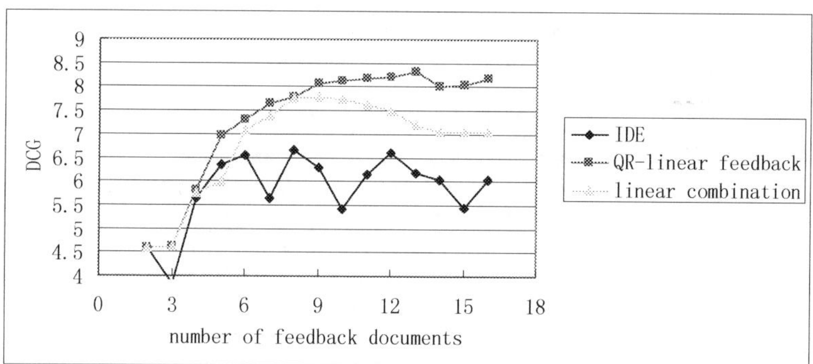

Fig. 4. It shows the performance of three feedback algorithm. It's clear that QR-linear algorithm outperforms the other two baseline algorithm especially when the number of feedback is large.

5 Conclusion

In this paper, first we brought up a new measure called *quality of a ranked result* (QR) to answer the question: how good is a search result list. This measure QR can be calculated from three features of a ranked result without manual label. And experiments show that QR is in direct proportion to DCG measure which means QR can be used to evaluate the quality of a result list. Three features are:

1. Rank: $<d_1, d_2, \cdots, d_n>$ is in descending order of the relevance.
2. Radius: the radius of the relevant document set $<d_1, \cdots d_i>$. The smaller the radius is, the higher the quality of the ranking list is.
3. Ratio: the ratio of documents in irrelevant set $<d_{i+1} \cdots, d_n>$ whose distance to the center of the relevant set is smaller than the radius. The smaller the ratio is, the higher the quality of ranking list is.

After discussion about QR, the quality of current ranked result is used to predict the relevance of different feedbacks and do the re-ranking, that is the QR-weighted re-ranking algorithm. In this way, good feedback document will play a more important role in the process of re-ranking. Experiments show that new re-ranking algorithm (QR-linear) outperforms the other two baseline algorithm. And it is effective for re-ranking especially when the number of feedback is large.

References

1. Ruthven, I., Lalmas, M.: A survey on the use of relevance feedback for information access systems. Knowledge Engineering Review 18(2), 95–145 (2003)
2. Salton, G., Buckley, C.: Improving retrieval performance by relevance feedback. Journal of American Society of Information System 41(4), 288–297 (1990)
3. Robertson, S.E., Sparck Jones, K.: Relevance weighting of search terms. Journal of the American Society of Information Sciences 27(3), 129–146 (1976)
4. White, R., Jose, J., Ruthven, I.: Comparing explicit and implicit feedback techniques for web retrieval: Trec-10 interactive track report. In: Text Retrieval Conference(TREC) (2001)
5. Joachims, T., Granka, L., Pan, B.: Accurately interpreting clickthrough data as implicit feedback. In: Annual ACM SIGIR Conf. on Research and Development in Information Retrieval(SIGIR), pp. 154–162 (2005)
6. Agichtein, E., Brill, E., Dumais, S., Ragno, R.: Learning User Interaction Models for Predicting Web Search Result Preferences. In: annual ACM SIGIR Conf. on Research and Development in Information Retrieval(SIGIR), pp. 3–11 (2006)
7. Beitzel, S. M., Jensen, E. C., Chowdhury, A., Grossman, D., Frieder, O.: Hourly analysis of a very large topically categorized query log. In: Proceedings of SIGIR2004, pp. 321–328 (2004)
8. Joachims, T.: Optimizing search engines using clickthrough data. In: Proceedings of SIGKDD 2002 (2002)
9. White, R.W., Ruthven, I., Jose, J.M., Van, Rijsbergen, C.J.: Evaluating implicit feedback models using searcher simulations. ACM Transactions on Information Systems (TOIS) (2005)
10. Kelly, D., Teevan, J.: Implicit Feedback for Inferring User Preference: A Bibliography. sigir forum (2003)
11. Shen, X., Tan, B., Zhai, C.: Implicit user modeling for personalized search, CIKM (2005)
12. Oard, D., Jim, J.: Modeling information content using observable behavior. In: proceedings of the 64th Annual Meeting of the American Society for Information Science and Technology (2001)

13. Voorhees, E.M.: Evaluation by highly relevant documents. In: 24th Annual International ACM SIGIR Conference on Research and Development in Information Retrieval, New Orleans, LA (2001)
14. Jarvelin, K., Kekalainen, J.: IR evaluation methods for retrieving highly relevant documents. In: proceedings of the 23rd Annual International ACM SIGIR conference on research and development in Information Retrieval, pp. 41–48 (2000)
15. TREC. http://trec.nist.gov
16. TianWang search engine. http://e.pku.edu.cn
17. Ide, E.: New experiments in relevance feedback. In: Salton, G. (ed.) The Smart Retrieval System, pp. 337–354. Prentice-Hall, Englewood Cliffs (1971)

An Investigation and Conceptual Models of Podcast Marketing

Shuchih Ernest Chang* and Muharrem Cevher

Institute of Electronic Commerce, National Chung Hsing University,
250 Kuo Kuang Road, Taichung City 402, Taiwan
Tel.: +886 4 22859465; Fax: +886 4 22859497
eschang@dragon.nchu.edu.tw, m_cevher@mynet.com

Abstract. While podcastig is growing rapidly and gains huge popularity, advertising has shown up as an emerging topic throughout the podcasting world. The objective of this study is to shed light on the potentials of podcasting as a new media technology for marketing, by investigating its applicability and effectiveness for targeted advertising and its value as a new way of communicating with captive audiences. Our study also tried to propose conceptual models for taking advantage of the unique strength of podcast technology to overcome the limitations of traditional channels and to enhance the effectiveness and efficiency of traditional marketing practice. For example, considered as a pull mechanism, podcast might be used to enhance marketing strategy for attracting a niche audience, particularly in the event that when traditional marketing approach becomes outdated or inconvenient for customers. A qualitative case study approach was adopted in our research to explore subject matter experts' experiences and feelings concerning adaptability of podcast to advertising, and moreover their expectations of podcasting as an advertising medium. Our research findings may be referenced by business executives and decision makers for the purpose of making favorable marketing tactics and catching the revolutionary opportunity and benefit of podcast advertising.

1 Introduction

Just two years ago, the term 'podcast' was considered for inclusion in the New Oxford American Dictionary of English (NOAD). Due to its rapid growth in popularity, 'podcast' was thereafter not only being added into NOAD in early 2006, but declared as the Word of the Year for 2005 by the editors of NOAD. NOAD defines the term 'podcast' as "*a digital recording of a radio broadcast or similar program, made available on the Internet for downloading to a personal audio player*" [1]. Wikipedia defines that a podcast is a media file that is distributed by subscription (paid or unpaid) over the Internet using syndication feeds, for playback on mobile devices and personal computers, and states that the publish/subscribe model of podcasting is a version of push technology, in that the information provider chooses which files to offer in a feed and the subscriber chooses among available feed

* Corresponding author.

K.C. Chang et al. (Eds.): APWeb/WAIM 2007 Ws, LNCS 4537, pp. 264–275, 2007.

channels [2]. A podcast is a web feed of multimedia files placed on the Internet for audiences' subscriptions, and podcasters' websites provide both direct download of their files and the subscription feed of automatically delivered new content is what distinguishes a podcast from a simple download or real-time streaming [3].

Podcastig is growing rapidly and gains huge popularity. According to Nielsen//NetRatings, a global company in Internet media and market research, 6.6 percent of the U.S. adult online population (about 9.2 million Web users) have recently downloaded an audio podcast, and 4.0 percent (about 5.6 million Web users) have recently downloaded a video podcast [4]. A report from eMarketer, a market research and trend analysis firm, forecasted that the number of podcasting audiences in U.S. will reach 25 million in 2008 and 50 million by 2010, and the same report also mentioned that podcast advertising spending would increase from an estimated $80 million in 2006 to $300 million by 2010 [5]. It is suggested by the aforementioned forecasts that podcasting holds considerable potentials for marketing, and therefore provides a good deal of opportunities for marketers.

Referring to distribution of ads to mobile devices such as cellular phones, PDA's, and other handheld devices, mobile advertising not only offers sending unique, personalized and customized ads, but also enables engaging consumers to discussions and transactions with the advertiser [6]. As a valuable and desirable approach of target options of mobile advertising, the ads that match with the users' personal interests and current needs can be sent, making sure that the customer will only receive ads that they are willing to. Hence it can make the advertiser achieve high view-through rates by targeting the ads appropriately and effectively [7]. Such specialization of advertising and targeted advertising can help increase the awareness and response from audience, and the effect of conducting targeted advertising is useful for optimizes ad budgets as well.

The objective of this study is to shed light on the potentials of podcasting as an advertising tool, by investigating its applicability and effectiveness for targeted advertising and its value as a new way of communicating with captive audiences. Specialization of ads and targeted advertising are the main concepts our study would like to focus on, and therefore our discussions would aim to probe the ways of utilizing this new technology to reach determined target groups with suitable one-to-few or one-to-many contents customized for various targeted groups. In addition, this study would like to find out the advantages of podcasting by comparing it with other existing advertising media, and then develop suitable conceptual models of using podcasting as an effective advertising medium.

2 Research Backgrounds

2.1 Podcast for Advertising

A podcast, which contains audio or video files, can be downloaded automatically. When a new podcast show is available, there is no need for interested audiences to visit website again and again, since the new show will be automatically delivered. This feature positions podcasting as a unique marketing and communication tool [8]. As podcasting is growing rapidly and puts large markets forward to parts of value chain, and opportunities for business applications of podcasting are showing up at each step in the value chain, the requirement of new businesses and business models for full commercialization of podcasting started to emerge [9]. Podcasting provides listeners the convenience of time-shifting and space independence to listen to media files, i.e., after downloading podcast programs onto their handheld devices, listeners can listen to or view the content anytime and anyplace at their convenience. That is why it has got fast growth. Nevertheless, podcasting has also been pointed out as a shift for personalization of media from mass broadcasting [10].

Nowadays, it is really getting difficult to reach people by e-mail due to various barriers and issues such as junk mails and spam filters, and similarly, it is hard to get or sustain people's loyalty on a website since we cannot be sure that people will visit the website again. However, podcasting takes the advantage of offering the capability and opportunity to listeners for subscribing to their interested podcast programs, and consequently we can be sure that the subscribers will receive the new shows/contents. Thus, podcasting helps businesses to increase their marketing reach and online visibility, and also promises to get regular line of communication with subscribers and to obtain their loyalty [8].

Podcasting is quite a virgin medium which promises to open up new avenues in corporate marketing, especially in a sense that podcasting has been posited to get a captive audience group. The corporate marketers should use podcasting to complement traditional marketing channels. With audio, the promotional inserts can be conveyed. Audio ads are rarely ignored completely by audiences if they are aimed at the correct target. This is because listeners are seldom bothered to skip the ads, if the duration of ads is adjusted well or the ads are interspersed in the program. Podcasting is considered as a suitable medium to achieve the goals of audio advertising, but determining the true numbers and demographics of listeners is considered as not only critical information but serious challenge for advertisers. In addition, podcast advertising has some problems. For example, podcast ads may have already been over, but outdated programs/shows might be available on the net for long time and the programs still can be shared among the listners indefinetely. Another problem with advertising with podcasts is measurement. Getting a true idea of how much response there was to the advertised product or brand as the result of a particular podcast advertisement is very difficult, beyond increased hits of the podcasts on the net.

The easiest way of advertising in podcasting is sponsorship, and it is beneficial to implement and reinforce brand profiles as opposed to specific products. Sponsorship in podcasting could be inexpensive and highly related to the content of podcasting programs. While podcast can be easily produced without any technical background, its production costs depend on how simple or elaborate the desired podcast is. However, the costs of radio (audio) production are generally low and manageable compared to other mass mediums. For example, it is known that quality radio commercials can be produced for a fraction of the cost of a quality television commercial [11].

2.2 Comparison of Relevant Advertising Mediums

Since podcasting is capable of distributing audio, video, and text contents to handheld devices, advertising via podcast shows shares some similar characteristics with some existing advertising mediums, including radio, televesion, cable television, short messaging service (SMS), and e-mail. For the purpose of comparing these various advertising approaches, more descriptions of these advertising mediums are summarized in the rest of this subsection.

2.2.1 Radio Advertising

Often described as an "intimate" medium, radio offers a form of entertainment and information such as: music, news, weather reports, traffic conditions that attracts listeners while they are doing almost anything in daily life. In additoin to the advantage that reaching people by radio is relatively inexpensive and cost-effective, radio has some other advantages including the ability to easily change and update scripts, and its capability to support the printed advertising. On the other hand, radio has some disadvantages as well. For example, radio ads are not suitable for products that must be seen by the listeners, radio commercials can not be replayed by the listeners, and radio listeners usually do not pay attention to advertisements. Furthermore, there are a lot of radio stations, and therefore, the total listening audience for any one station is just one small piece of a much larger population.

2.2.2 Broadcast TV Advertising

Television is the "king" of the advertising media, but it is also the "king" of advertising costs. There are some advantages of advertising a product or service via TV - instant validity and prominence can be given and it provides visibility effect. It is easy to reach the target audiences by TV, and creative advertising opportunities are also existing via TV. On the other hand, TV advertising contains disadvantages such as high advertising costs. In addition, costs of TV advertising vary significantly based on the time of show, and the advertising of some kind of goods is not permitted on TV by law. TV advertisements can be skipped by watchers and most of them may be ignored and becomes meaningless.

2.2.3 Cable TV Advertising

Cable advertising is a lower cost alternative to advertising on broadcast television. It has many of the same qualities (both advantages and disadvantages) as broadcast television and, in fact, it is even easier to reach a designated audience since it offers more programming. There are also some limitations with cable TV. The trouble with cable is that it does not reach everyone in the market area, since the signal is wired rather than broadcast and not everyone subscribes to cable. However, In US, Cable TV was legalized in 1993 and the penetration rate rapidly increased to 80 percents by the end of 2001.

2.2.4 Short Messaging Service (SMS) Advertising

The increasing popularity and adoption rate of mobile devices (such as cell phones) in the general population has created challenges facing businesses for how to use cell phones for doing advertisement. According to the GSM Association, mobile users send more than 10 billion SMS messages each month. Based on this fact, corporations may be able to take advantage of SMS in conducting mobile marketing. Actually, using SMS to promote goods or services is popular and it makes sales more effective. The response rate of customers is relatively high in terms of using SMS in marketing.

In contract, traditional marking skill was outdated to promote in mobile environment because it is not convenient or not visible for mobile customers. However, SMS in marketing only props up text contents, and therefore, talking about the visual activities is out of SMS marketing's concern. SMS in marketing presents only limited occasions for advertiser.

2.2.5 E-Mail Advertising

Sending e-mails to potential customers can be viewed as a marketing tool to promote products or services, and it is one of the cheapest ways to transmit ads. However, advertisers are in a jam with spam filters, and therefore, may not be sure that the customer received the e-mail or not. On the other hand, customers do not show respect to e-mail ads, and often delete e-mails containing ads without having a look. Thus, e-mail advertising might become wasting time for advertiser. E-mail advertising is still in use mainly because of its fee free feature, though it may be difficult to stay in touch with target customers through e-mail.

2.3 RSS as an Important Ingredient of Podcasting

One of the most important ingredients of podcasting is in its offering direct download or streaming of content automatically, by using software capable of reading data feed in standard formats such as Really Simple Syndication (RSS). According to Wikipedia, RSS is a simple XML-based system that allows users to subscribe to their favorite websites. Podcasters, defined as the hosts or authors of podcast programs, can put their content into RSS format, which can be viewed and organized through RSS-aware software or automatically conveyed as new content to another website. Such capability of RSS for delivering content to end-users is unprecedented because it is unique in a sense that it forces marketers to become more relevant and sensitive to the needs of their target audiences. RSS offers marketers to easily, inexpensively and quickly get their content delivered to their customers, business partners, the media, employees, and others throughout the World Wide Web. Nevertheless, RSS content can be delivered to other websites, search engines, and specialized RSS directories. RSS is a content delivery channel that allows marketers to easily deliver Internet content to the target audiences, while eliminating a large portion of unwanted noises and shortcomings of other delivery channels.

By using RSS, podcasting presents the business opportunities valuable for general marketing communications, direct marketing, PR, advertising, customer relationship management, online publishing, e-commerce to internal enterprise communications, and internal knowledge management, especially in terms of getting a niche but attentive audience and gaining the repeat consumer exposure [12]. In addition to the adoption of RSS as the blog-inspired subscription format, podcasting also inherits advantages from other aspects including: high-speed Internet connections suitable for downloading large files, availability of digital music player software and weblog software, increasing popularity of digital audio and video devices such as iPods, MP3 and MP4 players, and the ubiquitous MP3 audio format.

3 Conceptual Models

Based on the characteristics of podcast technology and the identified potentials described in the previous section, a model was proposed by this study for emphasizing and capturing the concept of specialization of podcast content. As shown in Fig.1, the model was proposed to illustrate the specialized advertising, which takes advantage of the benefit of customer segmentation for conducting marketing activities. Indeed, our proposed podcast model for specialized advertising stresses the following characteristics:

- Consumers can interact with the medium
- Firms and customers can interact in a way that is actually a radical departure from traditional marketing environment
- Firms can provide specialized contents to the medium and interact with each target group

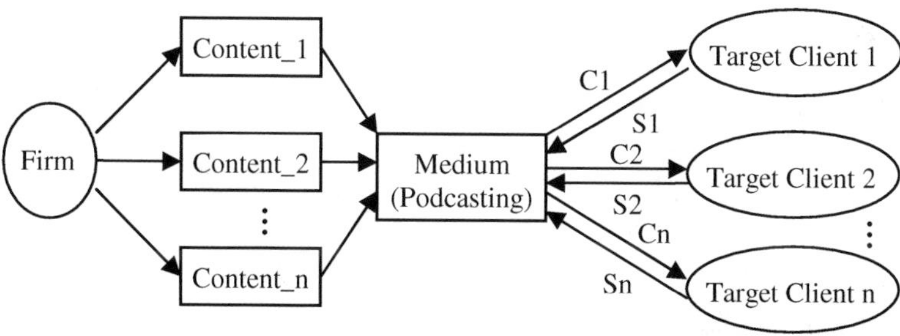

Fig. 1. Conceptual Model Posited for Specialized Advertising (where C1, C2, Cn represent the delivery of Content_1, Content_2, and Content_n respectively, and S1, S2, Sn represent the subscription of Content_1, Content_2, and Content_n respectively)

Since podcasting is by itself very selective in a sense that it allows marketers to target the right segment of potential customers, the production process of podcast commercials could customize the content to suit various client segments. For example, podcast ads for promoting sports related products may be delivered via sports shows. The core concept of this model is to produce specific content (ads) for each determined target groups, and to achieve the goal of offering the opportunity to respond to the needs of each client segment properly. It is also possible to make different brand image and get different product positing for each client segment.

Marketers may reference this proposed model to reconstruct their existing advertising models for incorporating podcasting into their marketing mix as a new interactive medium. However, such restructured models must account for the fact that consumers may not only actively choose whether or not to approach the firm's websites [13], but also exercise unprecedented control over the management of the content they interact with, i.e., decide whether or not to subscribe to any particular podcast show together with targeted podcast ads attached to the show.

In this era of digital economy, most companies more or less involve in collaborative activities with other organizations, in order to become more efficient and cost-effective in terms of operating various business functions including marketing. Our second conceptual model (see Fig. 2) for conducting podcast advertising addresses the need of collaborative marketing, which allow companies to share resources (people, teams, hardware and software, knowledge and expertise, talent, and idea) with business partners. In particular, this model emphasizes the synergy of integrating various marketing teams from different companies into a more competent consortium, in which many teams can collaboratively work together for achieving the goals of promoting their products, servicing podcasting customers, and developing new marketing strategies. It is suggested that a more comprehensive analysis on customer survey with regards to market sensing, marketing plans, and business practices is needed to close the knowledge gaps between the service providers and customers [14].

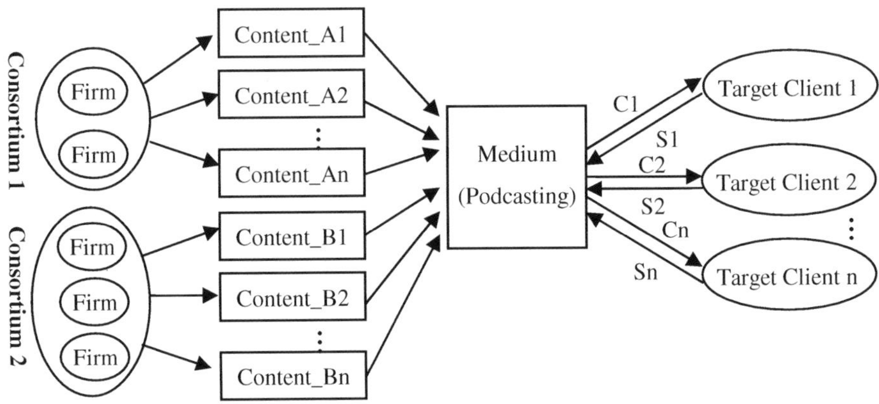

Fig. 2. Conceptual Model for Collaborationist Advertising (where C1, C2, and Cn represent the delivery of Content_A1/Content_B1, Content_A2/Content_B2, and Content_An/ Content_Bn respectively, and S1, S2, and Sn represent the subscription of Content_A1/ Content_B1, Content_A2/Content_B2, and Content_An/Content_Bn respectively)

As shown in Fig. 2, the conceptual model of "Collaborationist Podcast for Advertising" is talking about sharing a cooperative strategy for producing specific advertising contents, distributing ads, and going ahead with marketing projects. Companies can do collaboration to improve the efficiency and quality of the advertising activities, and producing more professional advertisements and perpetuating long-term projects which needs to spend big money can get easy by this collaborative strategy. However, the marketing concept must be broadened to include new views such as appreciating the customer as an active individual in an interactive process, and developing an effective business strategy in utilizing an emerging medium for a cooperative effort that includes the customer.

4 Evaluation of the Proposed Models by Qualitative Research

4.1 A Qualitative Research Approach

A qualitative research approach, which uses subjective and experiential knowledge for collecting and analyzing data, was used in our study to facilitate the development of practical and theoretical understanding of podcast advertising and the validation of the proposed concepts/models in regard to podcast advertising initiatives. Originally developed in the social sciences to enable researchers to study social and cultural phenomena, qualitative research takes the researcher's interaction with the field as an explicit part of knowledge production and includes the subjectivities of the researcher and participants as part of the research process [15]. Understanding endeavor of how humans construct meanings in their contextual settings, and deep understanding of human behavior are the aims of qualitative research, and therefore, qualitative research is useful in the early phases of research (for description derivation or concept development) where there exist no sufficient constructs or prior work available for guidance.

As a matter of fact, qualitative research is often applied to studying innovations. An innovation is considered as the process of developing and fulfillment a new idea. Diffusion of an innovation is a social process of communication whereby potential adopters become aware of the innovation and are impressed to adopt the innovation [16]. Since the application of podcast technology for marketing is considered innovative, and the applicability of podcast advertising is essentially exploratory in nature, using a qualitative research approach in this study is therefore deemed more appropriate than quantitative research approaches (such as linear regression) which are more appropriate for evaluating well established constructs related to the topic of research. Nevertheless, this study is motivated by a desire to explore subject matter experts' experiences and feelings concerning adaptability of podcasting to advertising, and moreover their expectations of podcasting as an advertising medium, by taking up the qualitative research approach to find scientific support.

4.2 Data Collection

Interview is an appropriate method when there is a need to collect in-depth information on people's opinions, thoughts, experiences, and feelings. Interviews are useful when the topic of inquiry relates to issues that require complex questioning and considerable probing. Specifically, face-to-face interviews are suitable when target population can communicate through face-to-face conversations better than communicating through writing or phone conversations. There are benefits of collecting data by interview, such as: (1) Level of interview structure can vary; (2) It is flexible and adaptable; and (3) It provides the researcher with more freedom than a survey or questionnaire.

Face-to-face interviews were chosen to collect data for this study. Open-ended questions were directed to the interviewees to ask them for their opinions about the proposed podcast advertising models. Each of the interviews took nearly one and half hour. We used recorders at two interviews, and wrote down the answers directly for the third interview without using a recorder. All three companies which cooperated with us for the interviews located in Taiwan. One of the interviewees is a marketing manager of an integrated marketing and advertising company. The other one is a marketing manager in the wealth management department of a financial company. The last one is an executive of a system company providing software and services. The data was collected by the following questions asked at each interview:

- Do you think if the podcasting is useful for advertising?
- Why you think companies should add this new technology into their marketing mix?
- Do companies need to do specialization of their ads?
- How you think podcasting can help in advertising/marketing?
- What are your opinions about the two-way interaction function of the models?

The titles below were discussed with the interviewees mainly to get their opinions about our project titled "An Investigation and Conceptual Models of Podcast Marketing".

- Current Situation of Advertising Sector
- Advertising Needs of Companies
- The Adoptability of the Project
- The Usefulness of the Project
- Advantages and Disadvantages of the Project
- The Threats on the Project
- Two Way Interaction Feature of the Project

Although the interview method possesses the advantage of offering the rich details, it reaches thoughts and feelings rather than direct observations. As mentioned earlier, the qualitative research approach (interview) used in this study may include the subjectivities of the researcher and participants as part of the research process. The inclusion of subjectivity is due to the ratification that there may be several different perspectives which should be explored. Indeed, retaining subjectivity may be advantageous to promote multiplicity of viewpoints in trying to understand the phenomena deeply [15].

5 Results

It is mentioned by the interviewees that although there were disadvantages of using traditional broadcasting channels as the mediums of advertising, those traditional mediums were still useful for advertising companies. According to one interviewee's experience, podcasting may provide an alternative advertisement channel particularly valuable for integrated marketing, and an effective integrated marketing strategy should always consider the adoption of innovative and useful channel, such as podcasting, into its marketing mix. It is also suggested by the interviewee that podcasting may be easily accepted as an advertising tool by companies focused on consumer goods, but at the same time, it would be difficult to clearly measure the performance of this kind of advertising activities because of the integrated marketing concept. For example, while podcasting web site can allow users to discuss on the

Internet, it may cause some unwanted side effect as well. Thus, the interviewees worry about the proposed model regarding that it lets everyone know the comprehensive product information including consumers' perceptions and comments to the product, good or bad. If the product is not good enough in meeting customers' expectation, there might be negative impacts, generated from web-based discussions and interactions which are expected to be part of the integrated marketing practice, on product images.

Podcasting can categorize advertising contents more clearly than other mediums, such as television, cable TV, SMS, and e-mail. The key point of podcasting is categorizing the contents to establish effective connections with the target groups. Since podcasting allows transmitting music and video programs (i.e., it is both audio and visual), it is better than SMS advertising via cell phones. However, for advertisement companies to adopt and apply this concept of podcast advertising, they demand more successful and convincing cases. Since traditional mediums may not be cost-effective enough or not suitable for some emerging application domains (such as in the ubiquitous computing environment) to satisfy the advertisers expectation of innovation, all interviewees agree that companies should consider adding podcast advertising into their marketing mix. In terms of using podcasting as an advertising medium, low cost, two-way interaction, visual content, and opportunity of producing specific content are the main favorable features.

In terms of customer segmentation, podcasting is particularly adaptable to promote products or services for young generation by helping businesses distributing messages about new products and services to young people, due to the fact that the majority of podcast users are relatively young. All interviewees stated that podcasting is convenient to distribute advertisements, especially when both the popularity of podcasting and the quantity of various categories of podcast programs are increasing. While the demand of the devices supporting podcasting is rising up, this trend undoubtedly presents innovative opportunities to podcast advertising. Cost of efficiency also attracts advertisers to use this medium, but at the same time, how to effectively identify appropriate target groups for podcast advertising remains as an interesting and challenging area subject to future research.

6 Conclusions

Through literature survey and interviews with subject matter experts, this study has identified the perspectives of podcast advertising listed below:

1. Getting better communication with audience, clients and target market
2. Reaching more customers, clients and prospects in a short time
3. Setting up strong relationships between company and customers
4. Getting more information about market, customers, or rivals for marketing research
5. Distributing latest news about company or products to prospects or clients quickly
6. Improving customer service and client satisfaction levels
7. Making positive impact on the brand and brand extension

Our interview results suggested that podcasting may become a useful tool for advertising and marketing, and this new technology can be used by companies to target in particular the market segment of young people since podcast consumers are typically younger, more affluent, and more likely to be influencers. Indeed, younger consumers are more attentive, giving brands the opportunity to target their advertisements more specifically by podcasting. Furthermore, the two-way interaction feature was applauded by the subject matter experts, and the opinions derived from them were generally positive upon the proposed conceptual models (see Fig. 1 and Fig. 2) for conducting podcast advertising. As a consequence, podcasting technology has the potentials to create new marketing impact and generate valuable result, alter the competitive landscape of business, and change existing societal and market structures. Podcasting enables interactions between customer and advertiser that become increasingly rapid and easy. However, still little is known about how this new technology can be used successfully in terms of being incorporated into the integrated marketing activities.

In summary, podcast advertising refers to the transmission of advertising messages via RSS which provides subscription opportunity, and specialization of ads is possible by podcast, although conceptually the use of podcasting for conducting mass media advertising is non-personal. Podcasting containing specialization of ads and interactivity effects may make it as important as we point out.

References

1. Oxford University Press: Podcast is the Word of the Year. Oxford University Press, USA (December 2005), http://www.oup.com/us/brochure/NOAD_podcast/?view=usa/
2. Wikipedia: Podcast. Wikipedia, The Free Encyclopedia, (accessed January 23, 2007) http://en.wikipedia.org/wiki/Podcast
3. Bruen Productions: Is Your Company Podcasting? 7th Annual Circulation Summit, Arizona, USA (January 26-27, 2006) http://bruen.com/pdf/PodcastingForNPs.pdf
4. Bausch, S., Han, L.: Podcasting Gains an Important Foothold among U.S. Adult Online Population, According to Nielsen/NetRatings. Nielsen/NetRatings, New York, (July 2006)
5. Chapman, M.: Podcasting: Who Is Tuning In? eMarketer, New York, (March 2006)
6. Dickinger, A., Haghirian, P., Murphy, J., Scharl, A.: An Investigation and Conceptual Model of SMS Marketing. In: Proceedings of the 37th Hawaii International Conference on System Sciences, pp. 31-41 (2004)
7. Balasubraman, S., Peterson, R.A., Jarvenpaa, S.L.: Exploring the Implications of M-commerce for Markets and Marketing. Journal of the Academy of Marketing Science 30(4), 348–361 (2002)
8. Rumford, R.L.: What You Don't Know About Podcasting Could Hurt Your Business: How to Leverage & Benefit from This New Media Technology. Podblaze, California (2005) http://www.podblaze.com/podcasting-business-whitepaper.php
9. Necat, B.: One Application to Rule Them All. ITNOW (Focus: The. Future of the Web) 48(5), 6–7 (2006), http://itnow.oxfordjournals.org/cgi/reprint/48/5/6
10. Crofts, S., Dilley, J., Fox, M., Retsema, A., Williams, B.: Podcasting: A New Technology in Search of Viable Business Models. First Monday vol. 10(9) (2005) http://www.firstmonday.org/issues/issue10_9/crofts/index.html

11. Radio Advertising Bureau: Media Facts - a Guide to Competitive Media. Radio Advertising Bureau, Texas (accessed July 10, 2006) http://www.rab.com/public/media/
12. Nesbitt, A.: The Podcast Value Chain Report: An Overview of the Emerging Podcasting Marketplace. Digital Podcast, California (2005) http:// www.digitalpodcast.com/ podcastvaluechain.pdf
13. Hoffman, D.L., Novak, T.P., Chatterjee, P.: Commercial Scenarios for the Web: Opportunities and Challenges. Journal of Computer-Mediated Communication 1(3) (1995) http://sloan.ucr.edu/blog/uploads/papers/hoffman,%20novak,%20and%20chatterjee%20(1995)%20JCMC.pdf
14. Ulaga, W., Sharma, A., Krishnan, R.: Plant Location and Place Marketing: Understanding the Process from the Business Customer's Perspective. Industrial Marketing Management 31(5), 393–401 (2002)
15. Flick, U.: An Introduction to Qualitative Research, 3rd edn. Sage Publications, California (2006)
16. Rogers, E.M.: Diffusion of Innovations, 4th edn. Free Press, New York (1995)

A User Study on the Adoption of Location Based Services

Shuchih Ernest Chang[1], Ying-Jiun Hsieh[2,*], Tzong-Ru Lee[3], Chun-Kuei Liao[4], and Shiau-Ting Wang[4]

[1] Institute of Electronic Commerce, National Chung Hsing University, 250 Kuo Kuang Road, Taichung City 402, Taiwan
eschang@dragon.nchu.edu.tw

[2] Graduate Institute of Technology and Innovation Management, National Chung Hsing University, 250 Kuo Kuang Road, Taichung City 402, Taiwan
Tel.:+886 4 22840547; fax: +886 4 22859480
arborfish@dragon.nchu.edu.tw

[3] Department of Marketing, National Chung Hsing University, 250 Kuo Kuang Road, Taichung City 402, Taiwan
trlee@dragon.nchu.edu.tw

[4] Institute of Electronic Commerce, National Chung Hsing University, 250 Kuo Kuang Road, Taichung City 402, Taiwan
ec@mail.nchu.edu.tw

Abstract. Based on the end user's exact location, providing useful information and location based services (LBS) through wireless pervasive devices at right place and right time could be beneficial to both businesses and their customers. However, the adoption rates of these location-aware pervasive services from the consumption side are still low, implying that there might be some reasons keeping the potential users away from using LBS. This research attempted to find out such reasons by investigating what factors would negatively influence users' adoption of LBS. A hybrid approach, integrating a qualitative method, ZMET, with the quantitative data analysis of the samples collected from a subsequent questionnaire survey, was designed and implemented in this study to elicit and validate potential LBS users' in-depth feelings. Our study results show that cost, worry of security & privacy Issues, worry of quality of LBS information, and lack of cognition of LBS are the barriers impeding mobile service users' adoption of LBS applications. Our findings can be referenced by service providers for the purpose of the design and development of successful business applications to catch the revolutionary opportunity and benefit of LBS.

1 Introduction

To survive the highly competitive environment, businesses are adopting the strategy of providing more abundant and desirable services to their clients. Based on the end user's exact location, providing useful information and location based services (LBS)

[*] Corresponding author.

K.C. Chang et al. (Eds.): APWeb/WAIM 2007 Ws, LNCS 4537, pp. 276–286, 2007.

through wireless pervasive devices at right place and right time can be beneficial to both businesses and their customers. Consequently, LBS applications (such as local traffic condition announcements, automatic route guidance based on onboard electronic maps and local traffic announcements, fleet management, etc.) have been deployed to the market for years. However, the adoption rates of these location-aware pervasive services from the consumption side are still far from satisfactory [1], implying that there might be some reasons keeping the potential users away from using LBS. This research attempted to find out such reasons by investigating what factors would negatively affect users' attitude toward adopting LBS. Our findings can be referenced by service providers for the purpose of the design and development of successful business applications to catch the revolutionary opportunity and benefit of LBS.

The remaining sections of this article are structured as follows. Section 2 provides more information about LBS, and describes a qualitative research approach, ZMET, which was used in this study. Afterwards, Section 3 describes the applied methodology, including the hypotheses development, questionnaire design, and data collection. Section 4 presents our study result, and Section 5 concludes this paper after the discussions. The limitations of this study are described in Section 6.

2 Research Backgrounds

2.1 Location-Based Service

LBS can be defined as services that integrate a mobile device's location with other information so as to provide added value to users [2]. For mobile LBS users, the awareness of the current, past, or future location forms an integral part of the services for providing end users with right information at right time and right place by locating their mobile devices [3]. The mobility is an added value to this new technology, which needs to service subscribers through multiple networks. Initially, tracking of mobile location information was used in the 1980s for trucking and freight industries [4]. Nowadays, LBS can be viewed as an application of mobile commerce (M-commerce), utilizing technologies such as wireless sensor network technology to locate the position of mobile device, and then according to its position provide the relevant information such as maps, routes, points of interest, emerging call, roadside assistance and so on through service provider's pervasive computing environment.

LBS are already introduced in global mobile commerce for various application domains including intelligent navigation support, transport logistics, mobile working environment, rescue/emergency support, location/context-based events, and spatial information systems [3]. However, the penetration rate of LBS is still low, failing to form the basis for wider application scope [2]. A research, which surveyed 25 cellular operators and 412 mobile users in UK, France and Germany, reported that 'Because I didn't know they existed' and 'Because the services are not useful' were two major reasons why mobile users did not use LBS more than they do [1]. That survey also found that only 3% of the respondents claimed to have used LBS, and around half of those LBS experienced respondents were dissatisfied with the services they used. In Taiwan, we found that although four major cellular operators offer various LBS applications to subscribers, the adopting rate is far from expectation. Thus, we would like to investigate the barriers for Taiwanese users to adopt LBS applications, and provide valuable insights to LBS service providers and business decision makers.

2.2 Zaltman Metaphor Elicitation Technique (ZMET)

In consumer and marketing researches, it might be more difficult to help consumers express their real demands, feelings, and thoughts than to make them understand product and service offering, because most of such underlying meaning is exchanged nonverbally [5, 6]. It is recognized that eliciting and compiling these nonverbal communications, such as facial expression, physical gesture, attire, scent, and so on, are crucial in understanding customers' true meanings. When customers are able to represent their thought in nonverbal terms, they are closer to the state in which thoughts occur and, thus, we can understand them better [5]. Based on these findings, the Zaltman Metaphor Elicitation Technique (ZMET) was developed by Zaltman in early 1990s for eliciting interconnected constructs that influence thought and behavior [6]. ZMET is a qualitative approach that integrates a variety of behavioral/social research methods including the visual projection technique, in-depth personal interview, and a series of qualitative data-processing techniques such as categorization, abstraction of categories, comparison with each respondent' data, and extraction of key issue from these data.

The main concepts of ZMET are image-based and metaphor-focused, and the typical application of ZMET usually includes the activities of interviewing, constructs and the consensus mapping, and results presenting. In addition to explicit knowledge, ZMET can draw out implicit imagery that represents the respondent's deepest thoughts and feelings related to the research topic, by assisting respondents to express their in-deep, latent and undisclosed perceptions and recognitions via verbal and non-verbal metaphor elicitation and storytelling [7, 8]. Generally speaking, ZMET is a good choice when the researcher wants to investigate some consumer behaviors but just has little prior research as reference. However, there is no standard procedure for ZMET, and the specific steps involved in implementing ZMET vary according to the project focus [5, 6]. More details of ZMET can be found in [5, 6, 7, 8].

3 Research Methodology

3.1 An Approach Integrating ZMET and Questionnaire Survey

A hybrid approach integrating ZMET with a questionnaire based survey was used in this study. To take advantage of the ability of obtaining a deep and rich understanding of LBS users' perceptions, ZMET interviews were conducted in the first stage of our study to capture respondents' requirements, opinions, and objective comments from LBS users' point of view. The captured information about users' needs and comments were subsequently analyzed by using qualitative data-processing techniques to extract factors reflecting the reasons why mobile service users did not adopt LBS as expected. In the second stage, the extracted factors were used to form a research framework and postulate research hypotheses, and the hypotheses were further used to design and conduct a questionnaire survey. Afterwards, the survey results were used to re-confirm that the factors extracted by ZMET were valid barriers affecting the adoption of LBS. This integrated approach was able to eliminate the shortage caused by insufficient sample size in the ZMET stage by conducting the second stage of questionnaire based survey, since the research model, hypotheses, questionnaire items, and collected data would be checked to assure their corresponding reliability and validity statistically.

In our study, ZMET processes were separated into two parts: data collection via interviews, and data analysis. Convenience sampling was used to select interview participants from mobile service subscribers, and in-depth personal interviews were conducted on those selected. Some pictures were shown to the participants for guiding them to think of any disadvantages or barriers that might be associated with using LBS. Although original ZMET require interviewees to select some pictures by themselves and bring these pictures to the personal interviews, the concept of LBS is somehow too new to request these interviewees to do so. For this reason, we altered this step a little bit by preparing pictures for interviewees in advance. As a matter of fact, we went through the following five steps to dig out major important factors and utilize these factors to set the hypotheses and design the questionnaire:

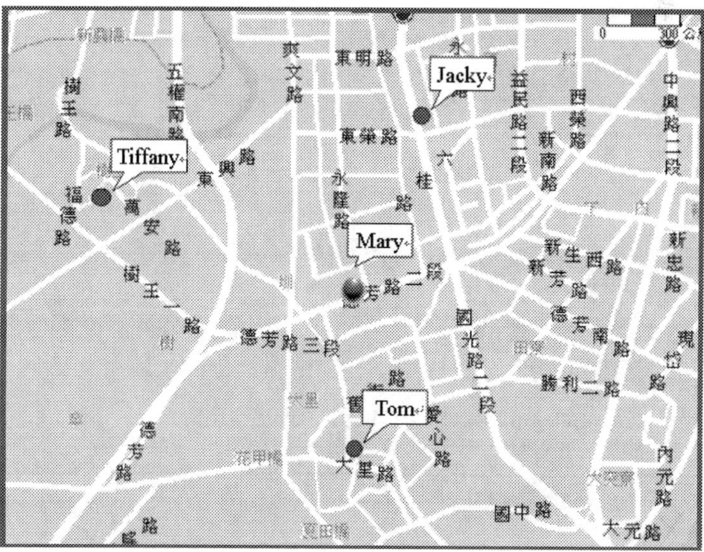

Fig. 1. An example of the pictures shown to the interviewees of ZMET processes

1. Telling stories - Participants were asked to describe the content of each picture provided by the interviewer. Fig. 1 shows one example of such pictures.
2. Summarizing images - Participants were asked to describe any images or opinions they got from these pictures.
3. Sorting issues - We sorted participants' issues into meaningful factors and generalized major factors that resulting in users' unwillingness to adopt LBS. We arranged the interview results of nine respondents, sorted the issues mentioned in every respondent's descriptions, counted the number of times of each item mentioned, and classified these items into five measuring indicators including cost, complexity of adoption process, worry of security & privacy issues, worry of quality of LBS information, lack of cognition of LBS.
4. Setting hypotheses - We then utilized those major factors abstracted from prior step to set the hypotheses.
5. Designing questionnaire - The questionnaire was designed based on the hypotheses set in the previous step.

3.2 Research Framework

From the aforementioned ZMET processes, a model for investigating the barrier to using LBS applications was developed (see Fig. 2), and the following five hypotheses were postulated accordingly:

H1: *Cost* has a negative effect on consumer's adoption of LBS.
H2: *Complexity of adoption process* has a negative effect on consumer's adoption of LBS.
H3: *Worry of security & privacy issues* has a negative effect on consumer's adoption of LBS.
H4: *Worry of quality of LBS information* has a negative effect on consumer's adoption of LBS.
H5: *Lack of cognition of LBS* has a negative effect on consumer's adoption of LBS.

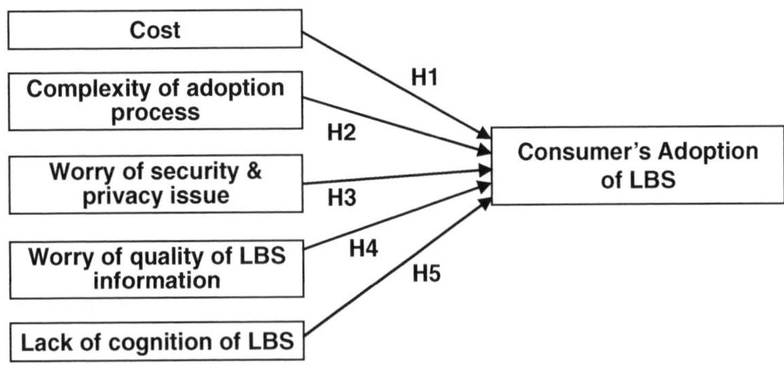

Fig. 2. Research framework

3.3 Questionnaire Design

Based on the derived hypotheses, the questionnaire was then designed as an instrument of data collection. The questionnaire included three major parts: (1) demographic information, (2) barriers to adopting LBS, and (3) willingness to adopt LBS. The demographic characteristics included gender, age, education level, internet experience, and mobile service experience. Part 2 covered the five major barriers extracted from the previous ZMET processes. Part 3 surveyed on users' actual usage of LBS. In total, the questionnaire consists of 26 items collecting demographic information and measuring the six variables listed in Table 1. Except for the demographic data in Part 1, questionnaire items are measured through five-point Likert scale, ranging "strongly disagree" (extremely unimportant) to "strongly agree" (extremely important).

Table 1. The variables used in the framework

Construct (Variable)	Concept
Cost	Users spend extra expenses including service fee, Web access fee, cost of devices supporting LBS, etc.
Complexity of adoption process	Users believe that using LBS is impractical, and operating LBS devices is complex.
Worry of security & privacy issues	Users concern with data security, personal security, and privacy protection when they used LBS.
Worry of quality of LBS information	Users concern with the accuracy, comprehensiveness, and explicitness of location based information provided by LBS.
Lack of cognition of LBS	Users concern with lack of understanding of LBS product
Adoption of LBS	Behavior of adopting LBS applications

3.4 Data Collection

Empirical data was collected by conducting a field survey of using online questionnaire. The survey subjects were supposed to have experience using mobile services. To ensure that the question items could be understood and measured validly, pre-test was conducted with a small group 40 respondents. From the feedback derived from pre-test and the subsequent discussion with experts, the questionnaire was modified and refined. In the formal survey conducted during the period from April 18, 2006 to May 7, 2006, 136 copies of questionnaire were gathered.

4 Results and Analysis

4.1 Descriptive Statistics

SPSS for Windows 11.0 was used in our study to analyze the sample. The formal questionnaire is used by confirmatory factor analysis to analyze collected data. Survey data were evaluated for their adequacy and construct validity, and the hypotheses were tested using correlation and regression analyses.

The characteristics of respondents were described using descriptive statistics. After a total of 136 responses were gathered, invalid survey results were identified by techniques such as the use of reverse questions. Overall, 129 valid questionnaires were collected and used for analysis. Among them, 49.6% were male, and 50.4% were female. The majorities were from two age groups: 20 to 25 years old (31.8%) and 26 to 30 years old (29.4%). 38% had monthly incomes of less than NT$20,000 (US$625), and 34% had monthly incomes between NT$20,000 (US$625) and NT$40,000 (US$1,300). Most respondents (77.5%) had education level of college degree or above, and 78 out of the 129 respondents (60.5%) had more than 6 year experience in using mobile services.

4.2 Data Analysis and Findings

Reliability analysis was used to ensure the consistency of measurement by checking Cronbach's Alpha, a measure of internal consistency. Our analysis results revealed that all constructs had good Cronbach's Alpha values ranging from 0.7194 to 0.8985, exceeding the acceptable reliability coefficient of 0.7 indicated by Nunnally and Bernstein [9]. Convergent validity, a type of construct validity, was used to assure that the items used for every construct had at least moderately correlated. It is considered that a factor loading reaching or exceeding 0.7 could be an evidence of convergent validity [10]. We evaluated convergent validity via factor analysis, and derived the standardized factor loadings of the observed variables. As presented in Table 2, all factor loadings of the measurement items ranged from 0.700 to 0.849, passing the acceptable standard of 0.7, and therefore, the convergent validity was demonstrated.

Subsequently, regression analysis was used to investigate the structural relationships among the research variables and derive the corresponding standardized coefficients. As shown in Fig. 3, the results obtained from this analysis confirm our hypotheses (H1, H3, H4, and H5) that *cost*, *worry of security & privacy issues*, *worry of quality of LBS information*, and *lack of cognition of LBS* have negative impacts on *customer's adoption of LBS*. The results indicate that *cost* significantly and negatively influences *customer's adoption of LBS* (β=-0.248, P<0.001). It means that *cost* is an important determinant of LBS adoption. When the cost of using LBS applications is high, the tendency of consumers to adopting LBS becomes low. *Worry of security & privacy issues* has a significant and negative effect on *customer's adoption of LBS* (β=-0.229, P<0.01), *Worry of quality of LBS information* has a significant and negative effect on *customer's adoption of LBS* (β=-0.155, P<0.05), and *lack of cognition of LBS* also has a significant and negative effect on *customer's adoption of LBS* (β=-0.221, P<0.01). These results offer suggestions to the service providers that they should put more efforts on the security and privacy protection, the improvement of information quality (such as the timely update, data abundance, and information accuracy), and, last but not least, the promotion of their products/services.

Table 2. Standardized factor loadings and composite reliability estimates (Cronbach's alpha)

Item	Concept	Factor loading	Cronbach's Alpha
Cost1	Service fee	0.793	
Cost2	Web access fee	0.821	
Cost3	Cost of devices which support LBS	0.762	0.8213
Complexity1	Impracticality of service	0.798	
Complexity2	Complexity of operation	0.739	
Complexity3	Concern with getting need information automatically	0.700	0.7423
Security1	Worry of personal security	0.841	
Security2	Worry of data security	0.849	
Security3	Worry of privacy protection	0.837	0.8986
Quality1	Worry of accuracy of information	0.774	
Quality2	Worry of explicitness of information	0.793	
Quality3	Worry of comprehensiveness of information	0.735	
Quality4	Worry of timeliness of information update	0.724	
Quality5	Worry of Inaccuracy of location detection	0.729	0.8863
Cognition1	Lack of promotion activities of service providers	0.800	
Cognition2	Lack of understanding of product	0.769	0.7194

As shown in Table 3, while H1, H3, H4, and H5 were supported, one hypothesis, H2, was not supported by our empirical results, i.e., *complexity of adoption process* did not have a significant effect on *customer's adoption of LBS*. The penetration rate of mobile phone (100.31%) in Taiwan is much higher than the ones of local telephone (59.63%) and Internet (40.96%), and mobile phone/device is actually connected tightly with people's everyday life [11]. Since most mobile service users in Taiwan are seasoned mobile device/service users familiar with the usage of mobile device and the operating procedures of accessing mobile services (e.g., the Short Message Service), the complexity of adoption process of LBS (which can be viewed as yet another mobile service) is not a significant concern/barrier toward users' adoption of LBS.

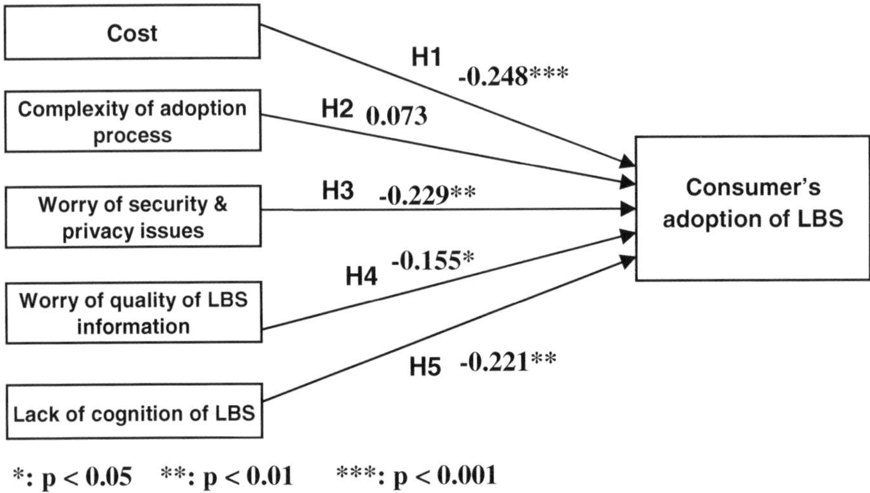

Fig. 3. The result of this empirical study

Table 3. The result of hypothesis test

Hypothesis	β	p	Result
H1	-0.248	0.000	Supported
H2	0.073	0.303	Not supported
H3	-0.229	0.002	Supported
H4	-0.155	0.031	Supported
H5	-0.221	0.002	Supported

5 Discussions and Conclusions

LBS have already been introduced to various application domains, such as local traffic condition announcements, automatic route guidance, fleet management, etc. It is speculated that LBS applications may have great potentials for service providers to improve their service quality, enhance their product/service values, and increase their business profits. However, recent researches show that LBS applications are not able to form the basis for wider application scenarios nor are they adequate to support a common infrastructure platform [1, 2]. Our research was targeted to elicit and sort out the barriers impacting users' adoption of LBS applications,

A qualitative research approach, ZMET, was adopted in this study to extract rich insights instead of general concepts of the respondents through interviews. The ZMET part of our research approaches revealed that there were five major barriers influencing consumers' adoption of LBS services. A research model was then formulated to include such five major factors/barriers. The research model together with five hypotheses was empirically tested using data collected from a questionnaire survey against mobile service users in Taiwan. This hybrid approach, integrating the qualitative ZMET method with the quantitative data analysis of the samples collected from a subsequent questionnaire survey, was designed and implemented in this study to elicit and validate potential LBS users' true meanings (in terms of demands, feelings, and thoughts).

Our study results show that *cost*, *worry of security & privacy Issues*, *worry of quality of LBS information*, and *lack of cognition of LBS* negatively and significantly impact *customer's adoption of LBS*. However, *complexity of adoption process* was not significant in affecting the adoption of LBS. The validated model (see Fig. 3) and its corresponding study results (see Table 3) can be referenced by enterprise executives and decision makers to make favorable tactics for taking advantage of the opportunity available through LBS. Consistent with the result from the prior study conducted in Europe [1], one highly important finding obtained from our samples is that about 80% of the respondents have never or seldom used LBS. This demonstrates that frustrating adoption rate of LBS is not just a hunch; it is indeed a fact and serious issue facing service providers. LBS will not be widely accepted if the customers have no idea about LBS or do not know how to obtain such services.

6 Limitations and Future Studies

There exist limitations in our study of investigating the barriers to users' adoption of LBS. Firstly, the participants of ZMET interviews were selected by convenience sampling. Actually, the interviewees were from the group with undergraduate degree or above, and they tended to be more familiar with information technology issues than the general consumer population. Future researches may modify the sampling procedure for comparative studies. Secondly, since the 3G mobile phone is still in its infancy in Taiwan, the insufficient understanding of LBS applications might be an overlooked factor impacting consumers' adoption behavior. Future studies are recommended to take it into considerations. In this paper, we assume LBS will provide values to customers, but is it always true? The excess of information and the usefulness of information provided by LBS may need to be researched further. Besides, applying LBS concept to specific industries and developing the framework for choosing the target customers of LBS applications are also worth investigating.

References

1. BWCS Research: Wireless Location on Target? (2004) http://www.trueposition.com/WirelessLocOnTarget.pdf
2. Schiller, J., Voisard, A.: Location-Based Services. In: Morgan Kaufmann Series in Data Management Systems, Morgan Kaufmann, Washington (2004)
3. Kühn, P.J.: Location-Based Services in Mobile Communication Infrastructures. AEU - International Journal of Electronics and Communications 58(3), 159–164 (2004)
4. Pura, M.: Linking Perceived Value and Loyalty in Location-Based Mobile Services. Managing Service Quality 15(6), 509–538 (2005)
5. Zaltman, G.: Metaphorically Speaking: New Technique Uses Multidisciplinary Ideas to Improve Qualitative Research. Marketing Research 8(2), 13–20 (1996)
6. Catchings-Castello, G.: The ZMET Alternative. Marketing Research 12(2), 6–12 (2000)

7. Coulter, R.A., Zaltman, G., Coulter, K.S.: Interpreting Consumer Perceptions of Advertising: An Application of the Zaltman Metaphor Elicitation Technique. Journal of Advertising 30(4), 1–21 (2001)
8. Zaltman, G.: How Customers Think: Essential Insights into the Mind of the Market. Harvard Business School Press, Boston, MA (2003)
9. Nunnally, J.C., Bernstein, I.H.: Psychometric Theory. McGraw-Hill, New York (1994)
10. Bagozzi, R.P., Yi, Y.: On the Evaluation of Structural Equation Model. Journal of Academy of Marketing Science 16(1), 74–94 (1998)
11. Directorate General of Telecommunications: Penetration Rates of Major Telecom Services, Annual Report 2004. Ministry of Transportation and Communications, Taiwan, ROC (2005) http://www.dgt.gov.tw/chinese/About-dgt/Publication/94/images/pic-jpg/06.jpg

Email Community Detection Using Artificial Ant Colony Clustering

Yan Liu, QingXian Wang, Qiang Wang, Qing Yao, and Yao Liu

Information Engineering Institute, Information Engineering University
Zhengzhou, 450002, P.R. China
liu_yan_hello@yahoo.com.cn

Abstract. The investigation of community structures in networks is an important issue in many domains and disciplines. Several types of algorithms exist for revealing the community structure in networks. However, Most of these algorithms consider only structure of the network, ignoring some additional conditions such as direction, weight, semantic, etc. In this paper we consider the behaviors of each individuals and describe an ant colony clustering algorithm for automatically identifying social communities from email network. This algorithm is successfully tested and evaluated on the Enron email dataset of 517,431 emails from 151 users, and shows that the method is effective at identifying true communities, both formal and informal.

1 Introduction

Email has become the predominant means of communication in the information society. It pervades business, social and technical exchanges and as such it is a highly relevant area for research on communities and social networks. Email is also a tantalizing medium for research because it provides plentiful data on personal communication in an electronic form. Given its ubiquity, it is a promising resource for tapping into the dynamics of information within organizations, and for extracting the hidden patterns of collaboration and leadership that are at the heart of communities of practice.

Community structure, which is a property of complex networks, can be described as the gathering of vertices into groups such that there is a higher density of edges within groups than between them[1] as shown in Fig.1. The ability to find and analyze such groups can provide invaluable help in understanding and visualizing the structure of networks.

The rest of the paper is organized as follows. In Section 2, related work is described. In section 3, the email community and ant colony are compared and show some common characters in their self-organization and social nature. An email community detection model is built up in Section 4 and Section 5 outlines some features in the email network which are used to calculus the distance between two email users. Email Social Community Detection Algorithm is discussed in Section 6 using a variant of ant colony clustering algorithm, and experimental evaluation is presented in this section. Finally, we conclude our work.

K.C. Chang et al. (Eds.): APWeb/WAIM 2007 Ws, LNCS 4537, pp. 287–298, 2007.

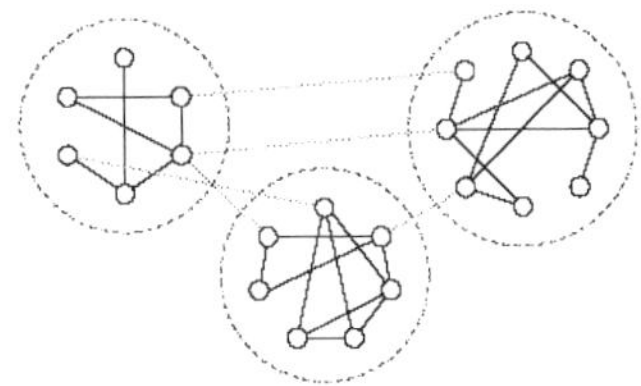

Fig. 1. A small network with community structure of the type considered in this paper. In this case there are three communities, denoted by the dashed circles, which have dense internal links but between which there are only a lower density of external links.

2 Related Works

There have been many approaches and algorithms to analyze the community structure in complex networks. The algorithms use methods and principles of physics, artificial intelligence, graph theory and even electrical circuits. One of the most known algorithms proposed so far, Girvan-Newman (GN) algorithm, is based on betweenness centrality. The algorithm is a divisive method and has $O(e^3)$ time-complexity. Radicchi proposed a similar methodology with GN[3], however used a new metric, edge-clustering coefficient whose computation time is less than GN's betweenness centrality, which decreases Radicchi's time-complexity to $O(e^2)$.

Hierarchical agglomerative method based on network modularity, Q, is a fast algorithm for detecting communities[4]. The algorithm has $O(n^2)$ time-complexity, which is better than GN algorithm and produces very accurate results as well. After the proposal of this algorithm, Newman, Clauset and Moore[1] proposed an improved version of the algorithm that is suitable for large networks.

Community detection using extremal optimization[5] also uses the network modularity. The algorithm tries to optimize the network modularity, Q, using artificial intelligence method in a recursive divisive manner. It starts with one community, representing the whole network and continues until the point from which the modularity Q cannot be improved further.

The current algorithms are successful approaches in community detection. However there are some drawbacks of current algorithms.

1. Most of these algorithms have time complexities that make them unsuitable for very large networks.
2. some algorithms have data structures like matrices etc., which are hard to implement and use in very large networks.
3. Most of the algorithms also need some priori knowledge about the community structure like number of communities etc., where it is impossible to know these values in real-life networks.
4. Some algorithms need threshold values as well, which is another problem for an algorithm, because of variant nature of different complex networks.

3 Email Community and Ant Colony

In several species of ants, workers have been reported to sort their larvae or form piles of corpses, literally cemeteries, to clean up their nests. Denebourg et al. [6] have proposed one model (BM -basic model) to account for the phenomenon of corpse clustering in ants. The general idea is that isolated items should be picked up and dropped at some other location where more items of that type are present.

Email has been established as an indicator of collaboration and knowledge exchange and the widely usage of email make it tightly related to the people's lives. As a result, the complexity of real society influences the characters of email network. So the email network is a complex open dynamic network, exhibiting a self-organizing adaptive behavior similar to ant society.

Thus, if the email addresses contained in the data set are viewed as the objects (such as corpses or larvae) scattered in the ant society, the community detection problem can be transformed into the progress of cleaning up the ants' nest. The main idea of our algorithm is to construct a virtual space for the community detection problem, to produce some virtual ants which deal with the single sub-problem, including picking up the object, drop down the object and move randomly to the other places. And at last, depending on the intelligence of ant colony, virtual ants can ultimate achieve clustering the tightly contacted email address.

4 Community Detection Model

In the community detection model, the agent represents the ant, and its purpose is to compare the similarity of the object (email address) to its surrounding environment and to decide whether move it from the present place or keep it staying at the position. Its behavior is simple and repetitive. We can formally define the major factors of the Community Detection Model(CDM) as follows.

4.1 Problem Description

Definition 1. Let the given email address set: $X = \{x_1, x_2, \cdots, x_n\}$, $\forall i \in \{1, 2, \cdots, n\}$, $x_i = \{x_{i1}, x_{i2}, \cdots, x_{ip}\}$ is the ith object of X, $\forall l \in \{1, 2, \cdots, p\}$, x_{il} is the lth attribute of x_i.

The purpose of email community detection is to break the set X into $C = \{C_1, C_2, \cdots, C_k\}$ according to the communication characters of the email address pairs, where $\cup_{i=1}^{k} C_i = X, \forall i, j \in \{1, 2, \cdots, k\}, C_i \neq \Phi, C_j \neq \Phi, and(C_i \wedge C_j = \Phi), i \neq j$.

$K = \{X, C\}$, named clustering space of email society network, and C_i is named as the ith community of the space.

4.2 Problem Space

The moving space of ants in the nature is a 3-dimentional continuous space. In order to simplify the problem, we convert the 3-dimention continuous space into 2-dimentional discrete grid.

Definition 2. Let G represents a 2-dimentional reference frame, the coordinates x, y is integer, $x \in [0, Len]$,$y \in [0, Len]$, where Len is the scale of the problem space. Fig.2(a) shows a example of the problem space. $G(x, y) \in \{0, 1\}$ represent

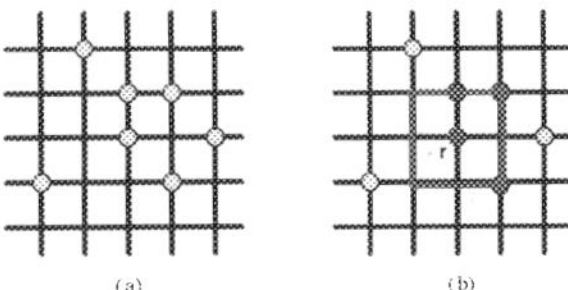

Fig. 2. (a) The example of problem space. (b) The local environment of the object located in r.

whether the position (x, y) is null or not.

$$G(x, y) = \begin{cases} 1, \text{some object is located in the position (x,y)} \\ 0, \ else \end{cases}$$

Definition 3. Let an object represents an email address by $object_i$, to represent the ith object, $i \in [1, N]$,N is the number of the email address. The position of $object_i$ is represented by (x_i, y_i), namely $G(object_i) = G(x_i, y_i)$.

We use Len to limit the moving scope of the ants. Bigger of Len means incompact objects and more times of recursion. And smaller of Len means crowded space and as a result lower quality. In our experimentation $Len = 2\left\lceil\sqrt{N}\right\rceil$.

4.3 Virtual Ants

Virtual ants are the simulation of the ants in the real nature that can accomplish the similar task as corpses forming. Virtual ants can be viewed as simple intelligent agents dealing with the single solution.

Definition 4. Let the virtual ant $k(k = 1, 2, \cdots, M)$, where M is the amount of ants, the location of the kth ant is represented by (x^k, y^k),namely $G(k) = (x^k, y^k)$. The state of k is $Loaded(k)$.

$$Loaded(k) = \begin{cases} 1, \text{the ant is carrying objects} \\ 0, \ else \end{cases}$$

Let the number of the objects that each ant carries is ω In the problem space there are N objects, satisfying

$$N = \sum_{(x,y)\in[0,Len-1]^2} G(x, y) + \sum_{k\in[1,M]} \omega \times Loaded(k)$$

4.4 Local Environment

The virtual ants repetitively decide whether to pick up or to carry or to drop down the object in the period of the processing. The key is to calculate the similarity of the object and its local environment. Local environment represents the ant's eyesight, which is a sub-class of the problem space. Fig.2(b) describes the local environment of the object in location r.

Definition 5

$$N(object_i) = \{(x, y) | \, |x - x_i| \leq S_x, |y - y_i| \leq S_y\}$$

$$N(x_i, y_i) = N(object_i)$$

$$L(object_i) = L(x_i, y_i) = \{(x, y) | (x, y) \in N(object_i), G(x, y) = 0\}$$

Let S_x and S_y are the eyesight limits of $object_i$ in the horizontal and vertical directions. $N(object_i)$ is used to denote $object_i$'s neighbor whose size is $(2S_x + 1) \times (2S_y + 1)$, $L(object_i)$ is used to denote a set of empty position in $N(object_i)$.

Definition 6 (Swarm Similarity). The swarm similarity of a data object is the integrated similarity of a data object with other data objects within its local enviroment, namely

$$f(object_i) = \max \left\{ 0, \frac{\sum\limits_{object_j \in N(object_i)} 1 - \frac{1}{\alpha}(1 - distance(object_i, object_j)/num)}{(2S_x + 1) \times (2S_y + 1)} \right\}$$

Where $distance(object_i, object_j)$ is determined by the distance between $object_i$ and $object_j$, representing the communication closeness between $object_i$ and $object_j$. In the email communication network, the measurement of $distance$ $(object_i, object_j)$ should consider many factors of the communication as discussed in section 5. num is the number of the factors considered in the section 5 and α is a parameter for adjusting the swarm similarity.

4.5 Probability Conversion Function

Probability conversion function is a function which converts the swarm similarity of a data object into picking-up or dropping probability for a simple agent. Wu Bin et al. [7] simplified the basic model and gave a simple function and successfully used to the customer behavior analysis and Web document clustering.

Definition 7 (Probability of "Pick up"). The probability of an unloaded ant picking up a object is

$$P_{pickup} = \begin{cases} 1 - \varepsilon, \; f(object_i) \leq 0 \\ 1 - k \times f(object_i), 0 < f(object_i) \leq 1/k \\ 0, \; f(object_i) > 1/k \end{cases}$$

Definition 8 (Probability of "Drop down"). The probability of a loaded ant dropping down a object is

$$p_{putdown} = \begin{cases} 1-\varepsilon, & f(object_i) \geq 1/k \\ k \times f(object_i), & 0 < f(object_i) < 1/k \\ 0, & f(object_i) \leq 0 \end{cases}$$

Where ε is a small positive number, used to help convergence of the algorithm.

5 Features Extraction from Email Dataset

Email communication is a special network. This kind of network has large scale of participants and large numbers of messages. If we view the network as a graph, in which nodes represent the email owners and edges denote the relationship among the email owners, the number of edges will be greatly more than that of nodes. Thus, the number of edges has more influence on the calculus complexity. To improve the efficiency of algorithm should reduce the computation related to edges.

Traditional method is to omit some edges by means of statistic and further more reduce the calculate complexity. However, this technique will loose some underlying factors. Here we use the other method that is measuring the distance between nodes from a macroscopical viewpoint by some features.

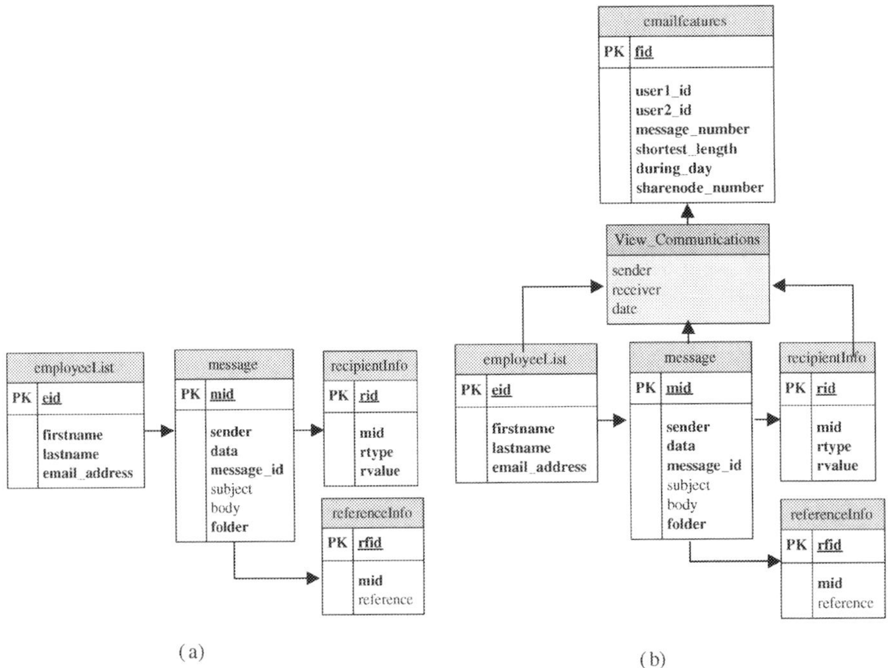

Fig. 3. Enron database schema

5.1 Enron Email Dataset

The Enron email dataset contains all kind of emails, personal and official. William Cohen from CMU has open the dataset on the web for researchers[8]. This version of the dataset contains around 517,431 emails from 151 users distributed in 3500 folders. The dataset was also stored in the MySql database for the dataset to catalyze the statistical analysis of the data.

Fig.3(a) shows the schema of the database. The Enron database contains four tables: "employeelist","message","recipientInfo" and "referenceInfo".

The original database pays attention to only the message and the user, not considering more about the relationship. In order to analyze the relationship between the communication pairs and extract the email features from the dataset, we adopt View_Communications as the analyzing basic and add some description of relationships in the email network. New database schema is shown in Fig.3(b) which is added a view(View_Communications) and a table(emailfeatures).

5.2 Email Features

In the email communication network, the measurement of $distance(object_i, object_j)$ should consider many factors of the communication. By analyzing the characters of the email network, we bring forward four features to calculate the distance of the message owners.

Messages Number. More frequently communication implies more tightly relationship. The messages number between the email owners denotes their social action in some perspective.

The messages number is achieved by both directions of communication (send and receive):

$$comm\_num_{ij} = send_{ij} + receive_{ij}$$

Communication Duration. Longer period of communication implies more tightly relationship. The communication duration between the email owners denotes their social action in some perspective.

The communication duration is achieved by the difference of the latest day and the earliest day achieved from the messages of a pair:

$$dur\_day_{ij} = latest\_day_{ij} - earliest\_day_{ij}$$

Shortest Length. Assuming that relationship between the pair of email users is symmetrical, the graph of the email network can be viewed as an undirected graph. The shorter the shortest length in the undirected graph, the closer relationship they hold.

For each pair of email owners, we count the shortest length in the graph by Breadth-First Search (BFS), namely $shortest\_len_{ij}$, In Enron dataset, the shortest length from one node to itself is assigned to 0, and length to the unreachable node is assigned to 10.

Share Neighbors. In the real society, if the persons share more friends they are often more familiar with each other.

The share neighbors are achieved by the number of joint set of two users' direct neighbors in the undirected graph:

Let

$$neighbor_i = \{u|u \in G\ andshortest\_len_{iu} = 1\}$$
$$neighbor_j = \{v|v \in G\ andshortest\_len_{jv} = 1\}$$

The number of share neighbors of i, j is:

$$sharenode\_num_{ij} = |share\_node_{ij}| = |share\_node_{ij} = neighbor_i \cap neighbor_j|$$

5.3 Distance Measurement

The distance between two objects is important in the calculation of swarm similarity. Whether a user is close to the other could be reflected in many perspectives, such as the messages number, communication duration, share neighbors, and shortest length. By taking all the four factors into account, we use the formula below to achieve the distance:

$$distance(object_i, object_j) = \frac{comm\_num_{ij}}{Max\_num} + \frac{dur\_day_{ij}}{Max\_day} + \frac{sharenode\_num_{ij}}{Max\_node} + \left(1 - \frac{shortest\_len_{ij}}{Max\_len}\right)$$

where Max_num is the maximum messages number among the user pairs, Max_day is the longest period of communication among the user pairs, Max_node is the maximum number of common friends among the user pairs, and Max_len is the longest path length among the user pairs.

6 Email Social Community Detection Algorithm(ESCDA)

6.1 Algorithm Description

Based on the features analysis and extraction in email network, we design the social community detection algorithm ESCDA as shown in Fig.4.

6.2 Algorithm Complexity

Time Complexity. In the process of Algorithm ESCDA, step 1~3 are initial steps and the main step is step 4. Step 4 includes the main work of M ants and each ant repeats its work n times. Based on the email features being achieved and stored in the database, the time complexity is mainly related to the calculus of swarm similarity, including:

(1) Time complexity of wandering in the local environment is $O(S_x \times S_y)$.
(2) Swarm similarity computation: $O(Aver)$, where $Aver$ is the average number of data objects in the local environment.

Therefore, if the ant number is M and the times of recurrence is n, the time complexity of step 4 is: $T = O(n \times M \times (Aver + S_x \times S_y))$.

Algorithm ESCDA:
1. Initialize the parameters, M, N, α, S_x, S_y, k, n, clusterno etc;
2. Take the email addresses as data objects and scatter the objects onto the grid randomly;
3. for each ant do
   Place ant at a randomly selected object in the grid. The state of ant is loaded.
   End for
4. for i=1,2,....,n;
   For j=1,2,..., M ; // M is ant number
   (x ,y): current coordinates, CrtAnt: current ant;
   (1) If (loaded(CrtAnt)==0 && G(x,y)!=0)
   ◇ Compute the swarm similarity of the data object with its local region according to definition 6;
   ◇ Compute a random probability Pr;
   ◇ Compute picking-up probability Pp according to definition 7, if Pp>Pr, CrtAnt picks up this object, loaded(CrtAnt)=1, else the ant does not pick up it;
   (2) If(loaded(CrtAnt)==1 & G(x,y)==0)
   ◇ Compute the swarm similarity of the data object carried by CrtAnt with its local region according to definition 6;
   ◇ Compute a random probability Pr;
   ◇ Compute dropping-down probability Pd according to definition 8, if Pd>Pr, CrtAnt puts down this object, loaded(CrtAnt)=0, else the ant keeps carrying it;
   (3) assign a new random pair of coordinates to the ant.
5. for i= 1, 2, ..., N;
   label the object to its correlative *clusterno*;

Fig. 4. Algorithm Description

Space Complexity. Space complexity concerns about the storing space for problem description and data structures used in the program running, including three main parts listed below:

(1) The graph to describe the email network: $O(N^2)$

(2) The 2-dimention grid: if we limit the problem space scale to $Len = 2\left\lceil\sqrt{N}\right\rceil$, the space complexity is $O(2\left\lceil\sqrt{N}\right\rceil^2) \approx O(N)$

(3) The ants list: $O(M)$

Therefore, the space complexity in the algorithm is $S = O(N^2 + N + M)$.

6.3 Experiment

We use the Algorithm ESCDA to the real email dataset of Enron. The experiment is designed in four steps:

(1) Analyze the Enron dataset and prepare the dataset for the next step of features extraction;

(2) Extract the email features and store into the database;

(3) Run the algorithm ESCDA in the Enron dataset to adjust the parameters;

(4) Email community detection using ESCDA.

Data Analysis. In this step, we use the statistical method to analysis some outstanding characters of email network, a special network and draw some conclusions.

(1) Messages related to per user

Fig.5 shows the distribution of emails per user. The users in the corpus are sorted by ascending number of messages along the x-axis. The number of messages is represented in log scale on the y-axis. The horizontal line represents the average number of messages per user (647). As can be seen from the graph, the messages are distributed basically exponentially, with a small number of users having a large number of messages. And there are users distributed along the entire graph from one message to 100,000 messages.

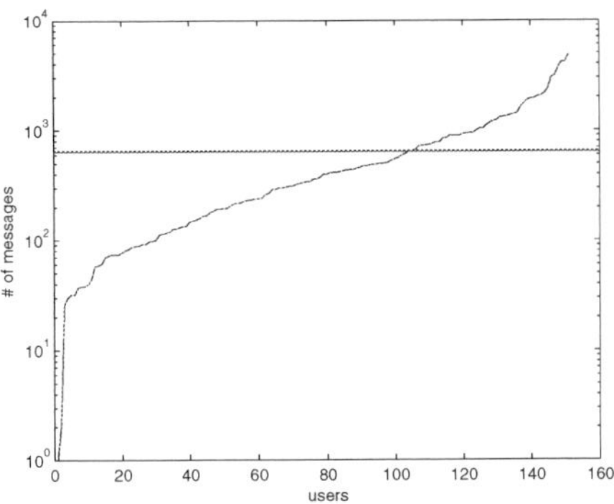

Fig. 5. Messages Distribution per user

(2) Friends related to per user

We then concentrate on single users' email associates. Fig.6 shows the distribution of neighbors per user. The users in the corpus are sorted by ascending number of neighbors along the x-axis. The number of neighbors is represented in log scale on the y-axis. The horizontal line represents the average number of friends per user (21). As can be seen from the graph, the neighbors are distributed basically exponentially. And there are users distributed along the entire graph from no friend to nearly 100 friends.

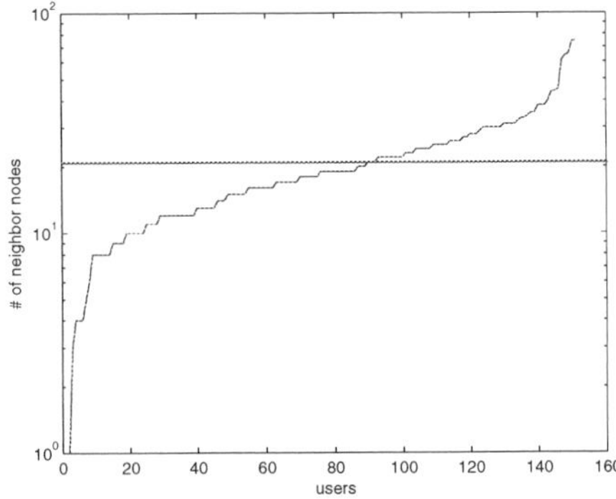

Fig. 6. Neighbors distribution per user

Community Detection Result. In the experiment, a 26*26 grid was used and the initial distribution of the email address objects is shown in Fig.7(a), email addresses are scattered in the grid randomly. Fig.7(b) shows the data distribution after 5000 iterations respectively. The parameters used in the dataset are $\alpha = 0.5, k = 1, S_x = S_y = 3, n = 5000$.

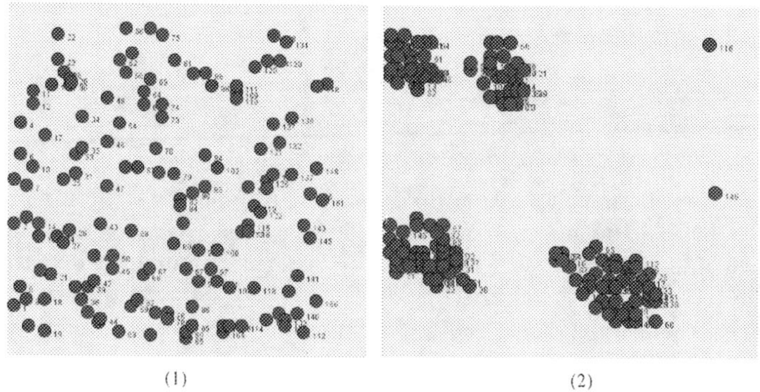

Fig. 7. the Process of clustering of ESCDA

From the result shown in Fig.7 we can see that the algorithm successfully find out four clusters in the grid. It is worth noting that because few real-world networks have community metadata associated with them to which we may compare the inferred communities, some type of manual checks of the veracity and coherence of the algorithm's output is often necessary.

As the analysis of time complexity above, the number of iterations is the main factors of time complexity. Compared with the amount of edges, it is notable that our algorithm can reduce the complexity of calculation in such a large scale and dense edges network as Enron email network.

There are some isolated nodes in the result in our experiment. By analyzing the isolated nodes, we find that node 116 has no communication with the other email address, and the result is agreed to the real society. However, node 149 should be moved into one cluster. The node was untouched in the whole process for no ants having touched the object. In order to avoid the isolated node as node 149, we should modify the strategy of ants' movement in the next research to furthermore improve the precision of the result.

7 Conclusion

Email has become the predominant means of communication in the information society. It pervades business, social and technical exchanges and as such it is a highly relevant area for research on communities and social networks.

The investigation of community structures in networks is an important issue in many domains and disciplines. Several types of algorithms exist for revealing the community structure in networks. However, Most of these algorithms consider only structure of the network, ignoring some additional conditions such as direction, weight, semantic, etc.

We consider the behaviors of each individuals and describe an ant colony clustering algorithm ESCDA for automatically identifying social communities from email network. This algorithm is successfully tested and evaluated on the Enron email dataset of 517,431 emails from 151 users, and shows that the method is effective at identifying true communities.

References

1. Clauset, A., Newman, M.E.J., Moore, C.: Finding community structure in very large networks, Physical Review E, 70:066111 (2004)
2. Newman, M.E.J., Girvan, M.: Finding and evaluating community structure in networks, Physical Review E, 69:026113 (2004)
3. Radicchi, F., Castellano, C., Cecconi, F., Loreto, V., Parisi, D.: Defining and identifying communities in networks. In: Proceedings of National Academy of Science in USA, vol. 101, pp. 2658-2663 (2004)
4. Newman, M.E.J.: Fast algorithm for detecting community structure in networks, Physical Review E, 69:066133 (2004)
5. Duch, J., Arenas, A.: Community detection in complex networks using extremal optimization, Pre-print condmat/0501368 (2005)
6. Deneubourg, J.-L., Goss, S., Franks, N., Sendova- Franks, A., Detrain, C., Chretien, L.: The Dynamic of Collective Sorting Robot-like Ants and Ant-like Robots. In: Meyer, J.A., Wilson, S.W. (eds.) SAB'90 - 1st Conf. On Simulation of Adaptive Behavior: From Animals to Animats, pp. 356–365. MIT Press, Cambridge (1991)
7. Wu, B.,Shi, Z.: A clustering algorithm based on swarm intelligence[A]. In: Proceedings IEEE international conferences on info-tech and info-net proceeding[C]. Beijing, pp. 58-66 (2001)
8. Enron Email Dataset (2005) `http://www.cs.cmu.edu/~enron`

EviRank: An Evidence Based Content Trust Model for Web Spam Detection

Wei Wang, Guosun Zeng, Mingjun Sun, Huanan Gu, and Quan Zhang

Department of Computer Science and Technology, Tongji University, Shanghai 201804, China
Tongji Branch, National Engineering & Technology Center of High Performance Computer, Shanghai 201804, China
The Key Laboratory of Embedded System and Service Computing, Ministry of Education
willtongji@gmail.com

Abstract. Creating an effective spam detection method is a challenging task. Traditional works usually regard this kind of work as a problem of binary classification. In this paper, however, we argue that it is more property to use the notion of content trust for it, and regard it as a ranking or ordinal regression problem. Evidence is utilized to define the feature of spam web pages, and machine learning techniques are employed to combine the evidence to create a highly efficient and reasonably-accurate detection algorithm. Experiments on real web data are carried out, which improve the proposed method performs very well in practice.

Keywords: web spam; evidence; content trust; ranking; SVM, learning.

1 Introduction

Creating an effective spam detection method is a challenging task [1]. In the context of search engines, spam can be a great nuisance for several reasons. First, since there are financial advantages to be gained from search engine referrals, web site operators generating spam deprive legitimate sites of the revenue that they might earn in the absence of spam. Second, if a search engine returns spam web pages to its users, and they will be confronted with irrelevant results, which may lead to frustration and disappointment in the search engine's services. Finally, a search engine may waste significant resources on spam pages [2].

There are pages on the Web that do not try to deceive search engines at all and provide useful and reliably contents to Web users; there are pages on the Web that include many artificial aspects that can only be interpreted as attempts to deceive search engines, while not providing useful information at all and of course can be regarded as distrusted information; finally, there are pages that do not clearly belong to any of these two categories. So, web spam detection can not be simply considered as a problem of binary classification which most of the traditional work do [1] [2] [4]. In fact, it can be regarded as a ranking problem which arises frequently in the social science and in information retrieval where human preferences play a major role [11] [12].

K.C. Chang et al. (Eds.): APWeb/WAIM 2007 Ws, LNCS 4537, pp. 299–307, 2007.

In this paper, we explore a novel content trust model, EviRank, based on evidence for detecting spam. The notion of content trust was first introduced by Gil et al. to solve the problem of reliability of the web resource [18]. But they only proposed the preliminary notion of content trust, and considered only exploring it through the transfer of trust using a resource's associations which not took the information content into account actually. In our opinion, spam web pages are a salient kind of distrusted web resource which can utilize content trust to model it. So, we developed a content trust model with ranking algorithms to detect web spam. Experiments show that our method performs very well in finding spam web pages.

The rest of this paper is organized as follows. We review some related work in Section 2. Section 3 introduces the proposed content trust model for detecting web spam. We first describe the key evidence for the model, and then a rank learning algorithm is proposed to detect the spam. We evaluate our approach and analyze the experiments results in Section 4. Section 5 concludes the paper.

2 Related Work

Web search has become very important in the information age. Increased exposure of pages on the Web can result in significant financial gains and/or fames for organizations and individuals. Unfortunately, this also results in spamming, which refers to human activities that deliberately mislead search engines to rank some pages higher than they deserve. The following description of web spam taxonomy is based on [2] and [4].

Content-based spamming methods basically tailor the contents of the text fields in HTML pages to make spam pages more relevant to some queries. This kind of spamming can also be called *term spamming*, and there are two main term spam techniques: repeating some important terms and dumping of many unrelated terms.

Link spam is the practice of adding extraneous and misleading links to web pages, or adding extraneous pages just to contain links. An early paper investigating link spam is Davison [5], which considered nepotistic links. Baeza-Yates et al. [6] present a study of collusion topologies designed tot boost PageRank [7] while Adali et al. [8] show that generating pages with links targeting a single page is the most effective means of link spam. Gyongyi et al. [9] introduce TrustRank which finds non-spam pages by following links from an initial seed set of trusted pages. In [1] Fetterly et al. showed ways of identifying link spam based on divergence of sites from power laws. Finally, Mishne et al. [10] present a probabilistic method operating on word frequencies, which identifies the special case of link spam within blog comments.

Hiding techniques is also used by spammers who want to conceal or to hide the spamming sentences, terms and links so that Web users do not see them. *Content hiding* is used to make spam items invisible. One simple method is to make the spam terms the same color as the background color. In *cloaking*, Spam Web servers return a HTML document to the user and a different document to a Web crawler. In this way, the spammer can present the Web user with the intended content and send a spam pages to the search engine for indexing.

On the other hand, trust is an integral component in many kinds of human interaction, allowing people to act under uncertainty and with the risk of negative consequences. Human users, software agents, and increasingly, the machines that provide services all need to be trusted in various applications or situations. Trust can be used to protect data, to find accurate information, to get the best quality service, and even to bootstrap other trust evaluations. In order to evaluate the reliability of the web resource, content trust was proposed as a promising way to solve the problem [18], such as spam filtering. So, it is promising to use content trust to model the reliability of the information, and solve the problem of web spam detection. Content trust was first introduced by Gil et al. on the International World Wide Web Conference in 2006. They discussed content trust as an aggregate of other trust measure, such as reputation, in the context of Semantic Web, and introduced several factors that users consider in detecting whether to trust the content provided by a web resource. The authors also described a simulation environment to study the models of content trust. In fact, the real value of their work is to provide a starting point for further exploration of how to acquire and use content trust on the web.

3 Content Trust Model Based on Evidence

In this paper, based on previous research [1-4] [10] [13], we explore a set of evidences, most of them based on the content of web pages. And some of these evidences are independent of the language a page is written in, others use language-dependent statistical properties.

The overview of the proposed content trust model can be descried in Figure 1.

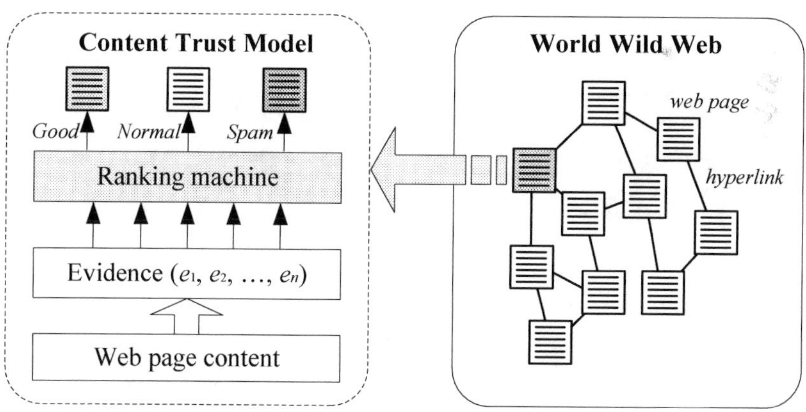

Fig. 1. Overview of the content trust model

We first analysis the content of the web page, and extract some salient evidence which can be used to evaluate the reliability of the content. Then, we train a ranking machine using the evidence as the feature to predict the trustworthy of future web pages. It is obvious that the evidence extraction and the rank machine training is the key. We descript them in more detail in the following section.

3.1 Evidence for Detecting Web Spam

There are many salient factors that affect how users determine trust in content provided by Web pages. So we extract the following evidence for detecting web spam based on previous research [1] [2] [4]. Table 1 describes the major evidence used in this paper.

Table 1. Evidence for spam detection

	Name	How to calculate
1	Number of words in the page	the number of words in the page
2	Number of words in the page title	the number of words in title
3	Average length of words	$\frac{\sum \text{the length (in characters) of each non-markup words}}{\text{the number of the words}}$
4	Amount of anchor text	$\frac{\text{all words (excluding markup) contained in anchor text}}{\text{all words (excluding markup) contained in the page}}$
5	Fraction of visible content	$\frac{\text{the aggregate length of all non-markup words on a page}}{\text{the total size of the page}}$
6	Compressibility	$\frac{\text{the size of the compressed page}}{\text{the size of the uncompressed page}}$
7	Fraction of page drawn from globally popular words	$\frac{\sum \text{the number of each words among the N most common words}}{\text{the number of all the words}}$
8	Fraction of globally popular words	$\frac{\text{the number of the words among the N most common words}}{N}$
11	Various features of the host component of a URL	
12	IP addresses referred to by an excessive number of symbolic host names	
13	Outliers in the distribution of in-degrees and out-degrees of the graph induced by web pages and the hyperlinks between them	
14	The rate of evolution of web pages on a given site	
15	Excessive replication of content	

3.2 Evidence Based Ranking Machine

As we have discussed above. One way of combining our evidence methods is to view the spam detection as a ranking problem. In this case, we want to create a ranking model which, given a web page, will use the page's features jointly in order to (correctly, we hope) rank it in one of several ordered classes, such as good, normal and spam. We follow a standard machine learning process to build out ranking model. In general, constructing a ranking machine involves a training phase during which the parameters of the classifier are determined, and a testing phase during which the performance of the ranking machine is evaluated. The whole process can be described in Figure 2.

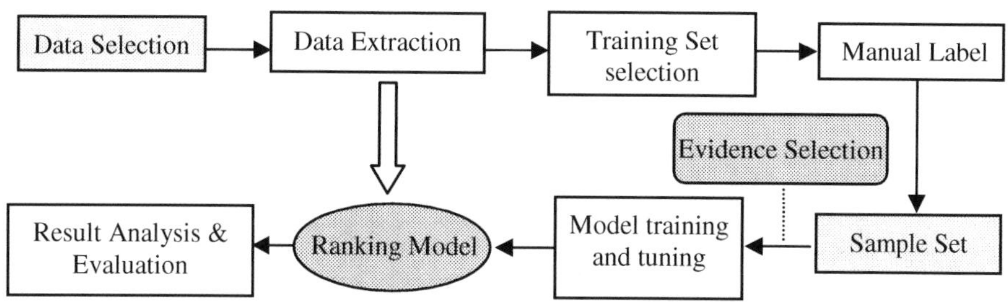

Fig. 2. Process of factoid/definition mining from content

The most important process in Figure 2 is evidence selection which forms the features of the proposed ranking model. Besides evidence described above, we also use some normal text features. The total number of the feature is 24 in our implementation of the model. For every web page in the data set, we calculated the value for each of the features, and we subsequently used these values along with the class label for the training of our ranking machine.

In learning to rank a number of candidates are given and a total order is assumed to exist over the categories. Labeled instances are provided. Each instance is represented by a feature vector, and each label denotes a rank. Ranking SVM [12] is a method which formalizes learning to rank as learning for classification on pairs of instances and tackles the classification issue by using SVM. The reason why we use ranking SVM is because it performs best compare to the other method, such as Naïve Bayesian [14] and decision tree [15] for ranking problem. The experiments result is described in section 4 lately. Here, we only introduce our method of adapting ranking SVM to the problem of spam detection.

Ranking SVM is a method which formalizes learning to rank as learning for classification on pairs of instances and tackles the classification issue by using SVM.

Assume that there exists an input space $X \in R^n$, where n denotes number of features. There exists an output space of ranks, or categories, represented by labels $Y = \{c_1, c_2, \cdots, c_q\}$ where q denotes number of ranks. Further assume that there exists a total order between the ranks $c_q \succ c_{q-1} \succ \cdots \succ c_1$, where $\succ$ denotes a preference relationship. A set of ranking functions $f \in F$ exists and each of them can determine the preference relations between instances:

$$\bar{x}_i \succ \bar{x}_j \Leftrightarrow f(\bar{x}_i) \succ f(\bar{x}_j) \tag{1}$$

Assume that we are given a set of ranked instances $T = \{(\bar{x}_i, y_i)\}_{i=1}^{m}$ from the space $X \times Y$. The task here is to select the best function f' from F that minimizes a given loss function with respect to the given ranked instances.

Herbrich et al. [12] propose formalizing the above learning problem as that of learning for classification on pairs of instances, and constructing the Ranking SVM model is equivalent to solving a novel Quadratic Optimization problem. More detail can be found in reference [12].

The relation $\vec{x}_i \succ \vec{x}_j$ between instance pairs $\vec{x}_i$ and $\vec{x}_j$ is expressed by a new vector $\vec{x}_i - \vec{x}_j$. Next, Herbrich takes any instance pair and their relation to create a new vector and a new label. Let $\vec{x}^{(1)}$ and $\vec{x}^{(2)}$ denote the first and second instances, and let $y^{(1)}$ and $y^{(2)}$ denote their ranks, then we have

$$\left(\vec{x}^{(1)} - \vec{x}^{(2)}, z\right), z = \begin{cases} +1, y^{(1)} \succ y^{(2)} \\ -1, y^{(1)} \prec y^{(2)} \end{cases} \tag{2}$$

From the given training data set *S*, we create a new training data set *S'* containing *m* labeled vectors.

$$S' = \{\vec{x}_i^{(1)} - \vec{x}_i^{(2)}, z_i\}_{i=1}^{m} \tag{3}$$

S' can be taken as classification data and construct a SVM model that can assign either positive label $z = +1$ or negative label $z = -1$ to any vector $\vec{x}^{(1)} - \vec{x}^{(2)}$.

Constructing the SVM model is equivalent to solving the following Quadratic Optimization problem [12]:

$$\min_{\vec{w}} \sum_{i-1}^{m} \left[1 - z_i \left\langle \vec{w}, \vec{x}_i^{(1)} - \vec{x}_i^{(2)} \right\rangle\right] + \lambda \|\vec{w}\|^2 \tag{4}$$

When Ranking SVM is applied to spam detection, an instance (feature vector) is created from the evidence we described in Section 3.1. Each feature is defined as a function of the document content.

4 Simulation Results and Performance Evaluation

4.1 Data Configuration

The data set in the following experiments is collected through the following process showed in Figure 3. The process of assembling this collection consists of the following two phases: web crawling and then labeling, which are described in the rest of this section.

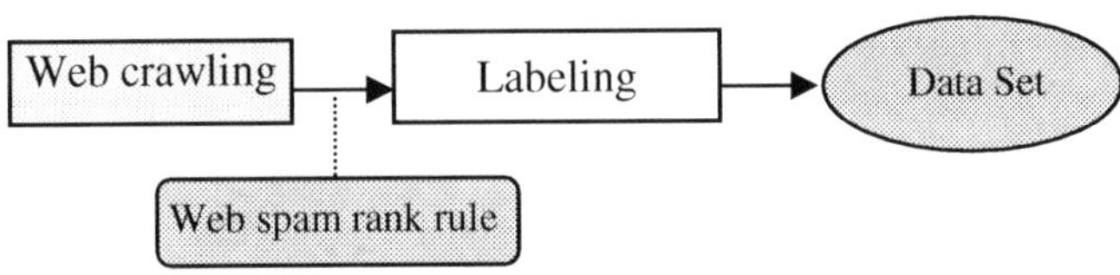

Fig. 3. Process of web spam data collection

We started in October 2006 by collecting a large set of web pages. The crawl was done using the *TrustCrawler* which developed by us in breadth-first-search mode for cross-host links. The crawler was limited to the .cn and .com domain and to 8 levels of depth, with no more than 50,000 pages per host. The obtained collection includes 0.5 million pages, and includes pages from 1000 hosts. The collection was stored in the WARC/0.9 format which is a data format proposed by the Internet Archive, the non-profit organization that has carried the most extensive crawls of the Web. WARC is a data format in which each page occupies a record. A record includes a plain text header with the page URL, length and other meta-information, and a body with the verbatim response from the Web servers, including the HTTP header. A total of ten volunteer students were involved in the task of spam labeling. The volunteers were provided with the rules of spam web pages, and they were asked to rank a minimum of 200 hosts. Further, we divide out data set in two groups according to the language used in the page. The first data set is composed with English web pages (DS1), and the other is Chinese web pages (DS2).

In order to train our rank machine, we used the pages in the manually classified data set to serve as our training data set. For our feature set, we used all the metrics described in Section 3. But for Chinese data set, some of the evidence is not suitable, such as evidence 3 in Table 1, and we ignore such features. For every web page in the data set, we calculated the value for each of the features, and we subsequently used these values along with the class label for the training of our ranking model.

4.2 Ranking Techniques Comparison

We experimented with a variety of ranking techniques: decision- tree based ranking techniques (R-DT) [15], Naïve Bayesian based ranker (R-NB) [14] and ranking support vector machine (R-SVM) [11], which modified by us in section 3.2 to suit the problem of spam detection. All algorithms are implemented within the Weka framework [17].

The metric we used to compare the different algorithm here is the ROC (Receiver Operating Characteristics) curve [16], or simply AUC. If a ranking is desired and only a dataset with class labels is given, the area under AUC can be used to evaluate the quality of rankings generated by an algorithm. AUC is a good "summary" for comparing two classifiers across the entire range of class distributions and error costs. It has been shown that, for binary classification, AUC is equivalent to the probability that a randomly chosen example of class - will have a smaller estimated probability of belonging to class + than a randomly chosen example of class + [9]. Thus, AUC is actually a measure of the quality of ranking. The AUC of a ranking is 1 (the maximum AUC value) if no positive example precedes any negative example.

In order to evaluate the accuracy of our method, we employed a technique known as ten-fold cross validation. Ten-fold cross validation involves dividing the judged data set randomly into 10 equally-sized partitions, and performing 10 training/testing steps, where each step uses nine partitions to train the classifier and the remaining partition to test its effectiveness.

Using the metric of AUC, we found that R-SVM based techniques performed best both on DS1 and DS2, but that the other techniques were not far behind. The result is showed in Table 2.

Table 2. Comparison of varies ranking algorithms on AUC

	AUC (%)		
Data Set	R-DT	R-NB	R-SVM
DS1	90.12	89.23	93.56
DS2	88.34	86.45	90.78

5 Conclusions

In this paper, we explore a novel content trust model for spam detection method based on content of the pages. This method takes the web spam detection task as a ranking problem. And we present how to employ machine learning techniques that combine our evidence to create a highly efficient and reasonably-accurate spam detection algorithm. Experiments show that our method performs very well on the crawling data sets.

Acknowledgements

This research was partially supported by the National Natural Science Foundation of China under grant of 60673157, the Ministry of Education key project under grant of 105071 and SEC E-Institute: Shanghai High Institutions Grid under grant of 200301.

References

1. Fetterly, D., Manasse, M., Najork, M.: Spam, Damn Spam, and Statistics: Using statistical analysis to locate spam web pages. In: 7th International Workshop on the Web and Databases (2004)
2. Ntoulas, A., Najork, M., Manasse, M., et al.: Detecting Spam Web Pages through Content Analysis. In: proceedings of WWW 2006, May 23–26, Edinburgh, Scotland (2006)
3. Wang, W., Zeng, G. S., Liu, T.: An Autonomous Trust Construction System Based on Bayesian Method, In: Proceedings of the IEEE/WIC/ACM International Conference on Intelligent Agent Technology (IAT 2006), Hong Kong, China, pp. 357-362 (December 18-22 2006)
4. Gyongyi, Z., Garcia-Molina, H.: Web Spam Taxonomy. In: 1st International Workshop on Adversarial Information Retrieval on the Web (May 2005)
5. Davison, B.: Recognizing Nepotistic Links on the Web. In AAAI-2000 Workshop on Artificial Intelligence for Web Search (July 2000)
6. Baeza-Yates, R., Castillo, C., Liopez, V.: PageRank Increase under Different Collusion Topologies. In: 1st International Workshop on Adversarial Information Retrieval on the Web (May 2005)

7. Page, L., Brin, S., et al.: The PageRank Citation Ranking: Bringing Order to the Web. Stanford Digital Library Technologies Project (1998)
8. Adali, S., Liu, T., Magdon-Ismail, M.: Optimal Link Bombs are Uncoordinated. In: 1st International Workshop on Adversarial Information Retrieval on the Web (May 2005)
9. Gyiongyi, Z., Garcia-Molina, H., Pedersen, J.: Combating Web Spam with TrustRank. In: 30th International Conference on Very Large Data Bases (August 2004)
10. Mishne, G., Carmel, D., Lempel, R.: Blocking Blog Spam with Language Model Disagreement. In: 1st International Workshop on Adversarial Information Retrieval on the Web (May 2005)
11. Cao, Y. B., Xu, J., Liu, T. Y., et al.: Adapting Ranking SVM to Document Retrieval. In: Proceedings of the 29th Annual International ACM SIGIR Conference On Research and Development in Information Retrieval, pp. 186-193 (2006)
12. Herbrich, R., Graepel, T., Obermayer, K.: Large Margin Rank Boundaries for Ordinal Regression. Advances in Large Margin Classifiers, pp. 115-132 (2000)
13. Wang, W., Zeng, G. S., Yuan, L. L.: A Semantic Reputation Mechanism in P2P Semantic Web. In: Proceedings of the 1st Asian Semantic Web Conference (ASWC), LNCS, vol. 4185, pp. 682-688 (2006)
14. Zhang, H., Su, J.: Naive Bayesian classifiers for ranking. In: Proceedings of the 15th European Conference on Machine Learning (ECML2004), Springer, Heidelberg (2004)
15. Provost, F.J., Domingos, P.: Tree Induction for Probability-Based Ranking. Ma.-chine Learning 52(3), 199–215 (2003)
16. Provost, F., Fawcett, T.: Analysis and visualization of classifier performance: comparison under imprecise class and cost distribution. In: Proceedings of the Third International Conference on Knowledge Discovery and Data Mining, AAAI Press, California, pp. 43-48 (1997)
17. Witten, I.H., Frank, E.: Data Mining–Practical Machine Learning Tools and Techniques with Java Implementation. Morgan Kaufmann, Washington (2000)
18. Gil, Y., Artz, D.: Towards content trust of web resources. In: Proceedings of the 15th International World Wide Web Conference (2006)

A Novel Factoid Ranking Model for Information Retrieval

Youcong Ni and Wei Wang

[1] Department of Mathematics and Physics, Institute of Architecture & Industry of Anhui, Hefei 230601, China
nyc@hubu.net
[2] Department of Computer Science and Technology, Tongji University, Shanghai 201804, China
willtongji@gmail.com

Abstract. How can we distinguish accurate information from inaccurate or untrustworthy information is a big challenge in the field of information retrieval. This paper discusses trust as a factoid learning problem, which extracts factoid from the content and then rank them according to their likehood as trustworthy ones. Learning methods for performing factoid ranking are proposed in this paper, and then we combine our method with the famous PageRank algorithm to form a more powerful method for retrieval reliable information. Evaluating of the model and the experimental results were presented.

Keywords: information retrieval; SVM; ranking, web mining.

1 Introduction

Information retrieval (IR) is finding material of an unstructured nature (usually text) that satisfies an information need from within large collections. However, how can we distinguish accurate information from inaccurate or untrustworthy information is a big challenge. As it becomes easier for people to add information to the Web, it is more difficult, and also more important, to distinguish reliable information and sources from those that are not.

In this paper we consider the acquisition of quality of information as a problem of 'factoid mining' on the web. Factoid here refers to something which can reflect the truth of the content, such as the definition of one thing which describe the problem of "what" in information retrieval, and the other kinds of factoid may include the information about "when", "where", "who", "why", and "how" about one thing. We identify key factors of factoid learning, and implement it by using ranking support vector machine algorithm, and then we combine our method with the famous PageRank algorithm to form a more powerful method for retrieval reliable information.

The rest of this paper is organized as follows. Section 2 introduces the proposed factoid ranking model. We first describe the basic concept of this model, and then describe the key factors of the model. We evaluate our approach and analyze the

K.C. Chang et al. (Eds.): APWeb/WAIM 2007 Ws, LNCS 4537, pp. 308–316, 2007.

experiments results in Section 3. We review some related work in Section 4. Section 5 concludes the paper.

2 Ranking Model Based on Factoid Learning

2.1 Basic Concept

Trust is often subjective, and there are many factors that determine whether content could or should be trusted, and in what context [5]. Some sources are preferred to others depending on the specific context of use of the information. Some sources are considered very accurate, but they are not necessarily up to date. This kind of properties can give users some useful information and knowledge which is very valuable in the context of information retrieval. Information may be considered sufficient and trusted for more general purposes. Information may be considered insufficient and distrusted when more fidelity or accuracy is required. In addition, specific statements by traditionally authoritative sources can be proven wrong in light of other information. [5]

In this paper, we discuss factoid mining as a ranking problem. Learning methods for performing factoid ranking are proposed, which formalize the problem as ordinal regression. The overview of the proposed factoid ranking model can be descried in Figure 1.

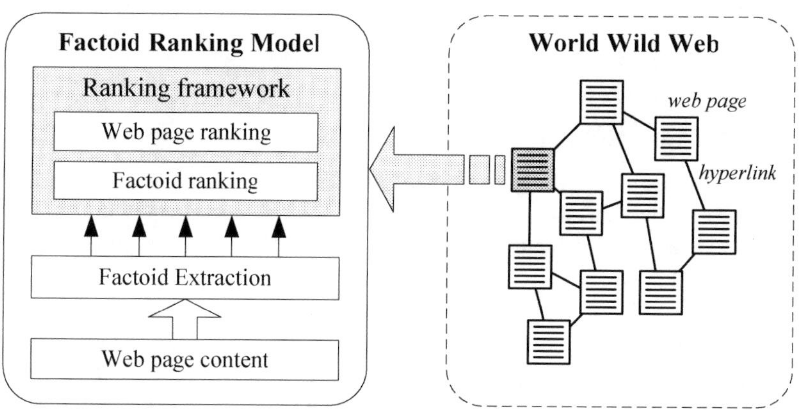

Fig. 1. Overview of the content trust model

We first analysis the web page content, and extract some factoid about a term which can be used to evaluate the reliability of them. Then, we propose a two-level ranking framework to evaluate the trustworthy of the web pages. In the first level, a ranking SVM is trained to learn good factoid from bad ones, and in the second level, we combine our factoid learning method with PageRank to form a powerful algorithm to evaluate the trustworthy of the web pages.

2.2 Trustworthiness of Factoid

When we decide whether to trust information on the web, the factoid in the content can help us to make decision. Normally, it is hard to judge whether a factoid is trustworthy or not in an objective way. However, we can still provide relatively objective guidelines for the judgment. In our opinion, the better factoid in the content, the more trustworthy the content may be regarded. In fact, our method is largely based on the idea from Microsoft, where Xu et al. [10] developed a supervised learning approach to search of definitions.

Here, without loss of generation, we first consider a kind of simple factoid, say definitions, because definitions describe the meanings of terms and thus belong to the type of frequently accessed factoid. Definition can be regarded as one kind of factoid which related to "what" [10]. The other types of factoid may be related with when, where and how about something. In light of this, we develop the following rules to definitions as well as factoid evaluation.

We then create three categories for factoids which represent their trustworthiness as factoid: 'good factoid/definition', 'indifferent factoid/definition' and 'bad factoid/definition'.

A *good factoid/definition* must contain the general notion of the term (i.e., we can describe the term with the expression "是一种" (is a kind of)) and several important properties of the term. From a trusted factoid, one can understand the basic meaning of the term.

An *indifferent factoid/definition* is one that between trusted and distrusted factoid.

A *bad factoid/definition* neither describes the general notion nor the properties of the term. It can be an opinion, impression, or feeling of people about the term. One cannot get the meaning of the term by reading it.

2.3 Level One: Factoid Ranking

First, we collect factoid/definition candidates (paragraphs) using heuristic rules. That means that we filter out all unlikely candidates. Second, we calculate the score of each candidate as factoid using a Ranking SVM. As a result, we obtain triples of <Term, Candidate, Score>.

In learning to rank a number of candidates are given and a total order is assumed to exist over the categories. Labeled instances are provided. Each instance is represented by a feature vector, and each label denotes a rank. Ranking SVM [11] is a method which formalizes learning to rank as learning for classification on pairs of instances and tackles the classification issue by using SVM. We adapt ranking SVM to the problem of factoid learning as follows.

Ranking SVM is a method which formalizes learning to rank as learning for classification on pairs of instances and tackles the classification issue by using SVM.

Assume that there exists an input space $X \in R^n$, where n denotes number of features. There exists an output space of ranks, or categories, represented by labels $Y = \{c_1, c_2, \cdots, c_q\}$ where q denotes number of ranks. Further assume that there exists a total order between the ranks $c_q \succ c_{q-1} \succ \cdots \succ c_1$, where $\succ$ denotes a preference relationship. A set of ranking functions $f \in F$ exists and each of them can determine the preference relations between instances:

$$\vec{x}_i \succ \vec{x}_j \Leftrightarrow f(\vec{x}_i) \succ f(\vec{x}_j) \tag{1}$$

Assume that we are given a set of ranked instances $T = \{(\vec{x}_i, y_i)\}_{i=1}^{m}$ from the space $X \times Y$. The task here is to select the best function f' from F that minimizes a given loss function with respect to the given ranked instances.

Herbrich et al. [12] propose formalizing the above learning problem as that of learning for classification on pairs of instances, and constructing the Ranking SVM model is equivalent to solving a novel Quadratic Optimization problem. More detail can be found in reference [12].

The relation $\vec{x}_i \succ \vec{x}_j$ between instance pairs $\vec{x}_i$ and $\vec{x}_j$ is expressed by a new vector $\vec{x}_i - \vec{x}_j$. Next, Herbrich takes any instance pair and their relation to create a new vector and a new label. Let $\vec{x}^{(1)}$ and $\vec{x}^{(2)}$ denote the first and second instances, and let $y^{(1)}$ and $y^{(2)}$ denote their ranks, then we have

$$\left(\vec{x}^{(1)} - \vec{x}^{(2)}, z\right), z = \begin{cases} +1, y^{(1)} \succ y^{(2)} \\ -1, y^{(1)} \prec y^{(2)} \end{cases} \tag{2}$$

From the given training data set *S*, we create a new training data set *S'* containing *m* labeled vectors.

$$S' = \{\vec{x}_i^{(1)} - \vec{x}_i^{(2)}, z_i\}_{i=1}^{m} \tag{3}$$

S' can be taken as classification data and construct a SVM model that can assign either positive label $z = +1$ or negative label $z = -1$ to any vector $\vec{x}^{(1)} - \vec{x}^{(2)}$.

Constructing the SVM model is equivalent to solving the following Quadratic Optimization problem [12]:

$$\min_{\vec{w}} \sum_{i-1}^{m} \left[1 - z_i \left\langle \vec{w}, \vec{x}_i^{(1)} - \vec{x}_i^{(2)} \right\rangle\right] + \lambda \|\vec{w}\|^2 \tag{4}$$

We next collect from the document collection all the paragraphs that are matched with heuristic rules and output them as factoid candidates.

First, we parse all the sentences in the paragraph with a Base NP (base noun phrase) parser and identify <Term> using the following rules.

- <Term> is the first Base NP of the first sentence.
- Two Base NPs separated by 'of' or 'for' are considered as <Term>.

In this way, we can identify not only single word <Term>s, but also more complex multi-word <term>s.

Next, we extract factoid candidates with the following patterns,

- <Term>是* (<Term> is a|an|the *)
- <Term>是指* (<Term>, *, a|an|the *)

- <Term>是一个/一种* (<Term> is one of *)
- 所谓<Term>是* (The notion of <Term> is *)
- ...

Here, '*' denotes a word string containing one or more words and '|'denotes 'or'.

The uses of other sets of rules for candidate selection are also possible. However, they are not essential for conducting factoid ranking, and we can skip this step or reinforce it by using more sophisticated rules.

In search of factoids/definitions, given a query term, we retrieve all the triples matched against the query term and present the corresponding factoids in descending order of the scores. Our data set was collected from both Google and MSN search.

All the data necessary for factoids search is stored in a database table in advance. The data is in the form of <Term, Candidate, Score> triples. For each term, the corresponding factoid candidates and scores are grouped together and the factoid candidates are sorted in descending order of the scores.

During search, we retrieve the sorted factoid candidates with regard to the search term by table lookup. We retrieved both Chinese and English web pages, and here we only give the results of English ones. Given the query term 'Data mining', we retrieve the ranked list of the factoid candidates as those in Table 1. The results indicate the usefulness of this approach.

Table 1. Ranked list of factoids for 'Data Mining'

1. **Data mining** (sometimes called data or knowledge discovery) is the process of analyzing data from different perspectives and summarizing it into useful information - information that can be used to increase revenue, cuts costs, or both.	0.846
2. **Data mining** is a rapidly growing field that is concerned with developing techniques to assist managers to make intelligent use of these repositories.	0.781
3. **Data mining** allows you to find the needles hidden in your haystacks of data.	0.628
4. **Data mining** uncovers patterns in data using predictive techniques	0.520
5. **Data mining** software is one of a number of analytical tools for analyzing data.	0.496
6. **Data mining** is primarily used today by companies with a strong consumer focus	0.448
7. **Data Mining** and Knowledge Discovery is one of the fast growing computer science fields.	0.355
8. **Data Mining** is a prospective science that focuses on the discovery of previously unknown relationships among existing data.	0.321
9. **Data mining** has a promising future.	0.171
10. **Data mining** is difficult to study for me.	0.121

2.4 Level Two: Factoid Ranking with PageRank

PageRank is a well know algorithm that uses link information to assign global importance scores to all pages on the web, which can be regarded as the reputation of the pages.

The intuition behind PageRank is that a web page is important if several other important web pages point to it. Correspondingly, PageRank is based on a mutual reinforcement between pages: the importance of a certain page influences and is being influenced by the importance of some other pages. The PageRank score $r(p)$ of a page p is defined as:

$$r(p) = \alpha \cdot \sum_{q:(q,p)\in \varepsilon} \frac{r(q)}{\omega(q)} + (1-\alpha)\cdot\frac{1}{N} \tag{5}$$

where α is a decay factor. $\omega(\cdot)$ is the number of outdegree of a web page, ε is the set of directed links(edges) that connect pages, and N is the number if the all web pages.

Here, we combine our factoid ranking method with the PageRank algorithm in the following way to evaluate the trustworthy of a given page:

$$ConTrust(p) = \lambda \cdot FacRank(p) + (1-\lambda)\cdot r(p),$$

$$\text{and } FacRank = \sum_{c\in p} \text{score}(c)/|c| \tag{6}$$

where score $(\cdot)$ is the factoid ranking score described above, and $\lambda \in (0, 1)$ donate the weight of each term. If the content itself is more important, or the page is a new one and can not find any other pages to link to it, when evaluate the reliability of a page, one can choose $\lambda > 0.5$.

3 Experiments and Results Analysis

We prepare the data sets for our following experiments as follows. The data set is collected through Google search. The process of assembling this collection consists of the following two phases: web crawling and then labeling (trusted, normal, distrusted). The obtained collection includes 5,000 pages with respect to 100 terms. The collection was stored in the WARC/0.9 format [13] which is a data format proposed by the Internet Archive, the non-profit organization that has carried the most extensive crawls of the Web. WARC is a data format in which each page occupies a record. A total of ten volunteer students were involved in the task of labeling. The volunteers were provided with the rules of trusted web pages, and they were asked to rank a minimum of 200 web pages. Further, we divide out data set in two groups according to the language used in the page. The first data set is composed with English web pages (DS1), and the other is Chinese web pages (DS2). In all the experiments, λ in equation (10) is set to 0.5.

Based on the previous work in information retrieval [10], we made use of three measures for evaluation of factoid ranking. They are: error rate of preference pairs and R-precision. The definition of them can be given as follows:

$$\textit{Error rate} = \frac{|\textit{mistakenly predicted preference pairs}|}{|\textit{all preference pairs}|} \tag{7}$$

$$\textit{R-precision}(\textit{term}_i) = \frac{|\textit{good term at R highest ranked candidates}|}{R} \tag{8}$$

where R is the number of good factoid form term_i.

$$R\text{-}precision = \frac{\sum_{i=1}^{T} \textit{R-precision}(\textit{term}_i)}{T} \tag{9}$$

The baseline used in the experiments is Okapi [14] and random ranking of factoid candidates. In Okapi, given a query term, it returns a list of paragraphs or sentences ranked only on the basis of relevance to the query term, not consider the reliability of the candidate. Random ranking can be viewed as an approximation of the existing method of information retrieval method.

Similar to most information retrieval method, we also use recall and precision measurement. In order to evaluate the accuracy of our method, we employed a technique known as ten-fold cross validation. Ten-fold cross validation involves dividing the judged data set randomly into 10 equally-sized partitions, and performing 10 training/testing steps, where each step uses nine partitions to train the classifier and the remaining partition to test its effectiveness.

3.1 Result of the Experiments

The result of the experiments can be described in Table 2 and 3. From Table 2, we can see that our method perform best on both error rate and R-precision. The ranking accuracy after the ten-fold cross validation process is encouraging as showed in the tables. We can also summarize the performance of our factoid ranking method using a precision- recall matrix. The precision-recall matrix shows the recall (the true-positive and true-negative rates), as well as the precision showed in Table 3 also improve that our method performs best.

Table 2. Evaluation of error rate and R-precision

Baseline	DS1		DS2	
	Error rate	R-Precision	Error rate	R-Precision
Okapi	0.539	0.291	0.574	0.274
Random ranking	0.410	0.372	0.459	0.332
Factoid ranking	0.213	0.561	0.254	0.517

Table 3. Evaluation of recall and precision

Baseline	DS1		DS2	
	Recall	Precision	Recall	Precision
Okapi	0.856	0.837	0.816	0.813
Random ranking	0.763	0.734	0.736	0.711
Factoid ranking	0.916	0.882	0.898	0.855

4 Related Work

Trust is an integral component in many kinds of human interaction, allowing people to act under uncertainty and with the risk of negative consequences. The need for trust spans all aspects of computer science, and each situation places different requirements on trust. Human users, software agents, and increasingly, the machines that provide services all need to be trusted in various applications or situations. Recent work in trust is motivated by applications in security, electronic commerce, peer to peer (P2P) networks, and the Semantic Web, which all may use trust in differently. Traditional trust mechanisms were envisioned to address authentication, identification, reputation and proof checking [1-4]. To trust that an entity is who it says it is, authentication mechanisms have been developed to check identity [6], typically using public and private keys. To trust that an entity can access specific resources (information, hosts, etc) or perform certain operations, a variety of access control mechanisms generally based on policies and rules have been developed.

On the other hand, current search engines, such as Google or MSN search, do not capture any information about whether or not a user accepts the information provided by a given web resource when they visit it, nor is a click on a resource an indicator of acceptance, much less trust by the users that have visited it. Popularity is often correlated with trust. One measure of popularity in the Web is the number of links to a Web site, and is the basis for the widely used PageRank algorithm [7]. Authority is an important factor in content trust. Authoritative sources on the Web can be detected automatically based on identifying bipartite graphs of 'hub' sites that point to lots of authorities and 'authority' sites that are pointed to by lots of hubs [8]. Reputation of an entity can result from direct experience or recommendations from others. Varieties of trust metrics have been studied, as well as algorithms for transmission of trust across individual webs of trust, including ours previous research [4].

5 Conclusions

In this paper, we propose a content trust model based on factoid learning to solve the problem of evaluation trustworthiness through web content. We identify key factors of factoid learning in modeling content trust in open sources. We then describe a model that integrates a set of trust features to model content trust by using ranking support vector machine. We hope that this method will help move the content trust closer to fulfilling its promise.

Acknowledgements

This research was partially supported by project of sustentation fund for master & doctor scientific research in Institute of Architecture & Industry of Anhui (No: 2005110124).

References

[1] Berners-Lee, T., Hendler, J., Lassila, O.: The Semantic Web. Scientific American 284(5), 34–43 (2001)

[2] Resnick, P., Zeckhauser, R., Friedman, R., et al.: Reputation systems. Communications of the ACM 43(12), 45–48 (2000)

[3] Golbeck, J., Hendler, J.: Inferring reputation on the semantic web. In: Proceedings of the 13th International World Wide Web Conference (2004)

[4] Wang, W., Zeng, G.S., Yuan, L.L.: A Semantic Reputation Mechanism in P2P Semantic Web. In: Mizoguchi, R., Shi, Z., Giunchiglia, F. (eds.) ASWC 2006. LNCS, vol. 4185, pp. 682–688. Springer, Heidelberg (2006)

[5] Gil, Y., Artz, D.: Towards content trust of web resources. In: Proceedings of the 15th International World Wide Web Conference (2006)

[6] Miller, S.P., Neuman, B.C., et al.: Kerberos authentication and authorization system. Tech. rep. MIT, Cambridge, MA (1987)

[7] Brin, S., Page, L.: The anatomy of a large-scale hypertextual web search engine. In: Proceedings of the 7th International World Wide Web Conference (1998)

[8] Kleinberg, J.M.: Authoritative sources in a hyperlinked environment. Journal of the ACM 46(5), 604–632 (1999)

[9] Wang, W., Zeng, G.S., Yuan, L.L.: A Reputation Multi-Agent System in Semantic Web. In: Shi, Z.-Z., Sadananda, R. (eds.) PRIMA 2006. LNCS (LNAI), vol. 4088, pp. 211–219. Springer, Heidelberg (2006)

[10] Xu, J., Cao, Y.B., Li, H. et al.: Ranking Definitions with Supervised Learning Methods. In: Proceedings of the 14th International World Wide Web Conference (2005)

[11] Cao, Y.B., Xu, J., Liu, T.Y. et al.: Adapting Ranking SVM to Document Retrieval. In: Proceedings of the 29th Annual International ACM SIGIR Conference On Research and Development in Information Retrieval, pp. 186–193 ((2006)

[12] Herbrich, R., Graepel, T., Obermayer, K.: Large Margin Rank Boundaries for Ordinal Regression. Advances in Large Margin Classifiers, pp. 115–132 (2000)

[13] Robertson, S., Hull, D.A.: The TREC-9 Filtering Track Final Report. In: Proceedings of the 9th Text Retrieval Conference, pp. 25–40 (2000)

[14] Robertson, S.E., Walker, S. et al.: Okapi at TREC-4. In: Proc. of the 4th Annual Text Retrieval Conference, National Institute of Standards and Technology, Special Publication, pp. 500–236 (1995)

Dynamic Composition of Web Service Based on Coordination Model*

Limin Shen[1], Feng Li[1], Shangping Ren[2], and Yunfeng Mu[1]

[1] Department of Computer Science, Information College, Yanshan University
Qinhuangdao, Hebei 066004, China
shenllmm@sina.com, ysu_lifeng@126.com, yfm@ysu.edu.cn
[2] Department of Computer Science, Illinois Institute of Technology
Chicago, IL 60616, USA
ren@iit.edu

Abstract. To deal with Web service dynamicity and changes of application constraints in open distributed environment, a coordination model for dynamic service composition is presented. Based on the separation of concerns, there are three different categories of entities in a Web service-based application: Web service, role and coordinator. The Web service is only responsible for performing pure functional service, and carrying out the task assigned by the role; the role is an abstraction for certain properties and functionalities, responsible for binding Web service according to constraints, and actively coordinating Web service to achieve coordination requirements; the coordinator is responsible for the coordination among roles by imposing coordination policies and binding constraints. The logic separation of Web services, roles, and coordinators in the model, decouples the dependencies between the coordinators and Web services. Thus, the model shields the coordinator layer from the dynamicity of Web services. Finally, a vehicle navigation application including traffic control Web services, GPS Web services and a navigator is illustrated how the model can be used to achieve the interaction adaptation by means of dynamic composition of Web service.

Keywords: dynamic service composition; separation of concern; Web service; coordination; role.

1 Introduction

Building an application system based on Web-service composition is a current trend. There are two approaches of Web-service composition, static composition and dynamic composition [1]. Static composition takes place during design-time when the architecture and the design of the software system are planned. The components to be used are chosen, linked together, and finally compiled and deployed. This may work well as long as the web service environment, business partners and service components does not (or only rarely) change. Microsoft Biztalk and Bea WebLogic are examples of static composition engines [2].

* Supported by the State Scholarship Foundation, China (Grant No. 2003813003).

K.C. Chang et al. (Eds.): APWeb/WAIM 2007 Ws, LNCS 4537, pp. 317–327, 2007.

Web services are in open distributed environment. An application may demand to add or reduce a web service meanwhile web services may be not available or newer web services appear [3]. At a time of service, the application may need different Web services because of different user's constraints or environment conditions. The interaction topology among Web services may change. Furthermore, Web service is a black box that allows neither code inspection nor change, and most Web-service's QoS is uncontrollable, and Web services do not know about any time deadline. All of these above have made distributed computation system based on web service more dynamic, uncertain, and real-time requirements more difficult satisfy. In open environment, static composition may be too restrictive. A static Web service composition cannot handle a change that occurs while it is under execution. In fact, it happens that a change has not been considered during the development stage of the composite service. In other words, the static composition that software designers adopted—carefully elicit change requests, prioritize them, specify them, design changes, implement and test, then re-deploy the software—doesn't work anymore.

Dynamic composition takes place during runtime, i.e. systems can discover and bind Web service dynamically to the application while it's executing. Dynamic composition is considered as a good way to deal with and even take advantage of the frequent changes in Web services with minimal user intervention. However, the current technology doesn't support expression, at design time, of the requirements and constraints to be fulfilled at runtime in the discovery and selection phase to identify the services to be bound [1]. This drawback makes development of systems exploiting this runtime binding capability complex, difficult and almost impossible in practice.

In order to handle the challenges of dynamic environments and complexity of dynamic composition, based on the principle of "separation of concern" [4] [11], we present a coordination model, the Web Service, Role and Coordinator (WSRC) model. The focus of the WSRC model is to better address the dynamicity and scalability issues inherent in open Web service environment. The WSRC model has the following characteristics:

- There are three different categories of entity in a Web service-based application, Web services, roles, and coordinators.
- The Web services are only responsible for performing pure functional services and carrying out the task assigned by a role.
- The role is an abstraction for certain properties and functionalities, responsible for binding a Web service according to constraints, and actively coordinating the Web service to achieve coordination requirements.
- The coordinators are responsible for the coordination among roles by imposing coordination policies and binding constraints.
- The logic separation of Web services, roles, and coordinators in the model, decouples the dependencies between the coordinators and Web services. Thus the model shields the coordinator layer from the dynamicity of Web services.

The rest of the paper is organized as follows: Section 2 discusses related work. Section 3 discusses the model in detail. Section 4 gives a dynamic composition approach. Section 5 illustrated an application of this model through a vehicle navigation example. Section 6 concludes the paper and presents future research work.

2 Related Works

Web service composition, especially dynamic service composition, is a very complex and challenging task. Many different standards have been proposed and various approaches have been taken to create a widely accepted and usable service composition platform for developing composite web services. Generally speaking, industrial area focuses on developing composition description language, manipulative tools and executable engine. While academic area focuses on semantic description and artificial intelligence for automatic composition, then validates QoS of composition systems using formalizable approaches.

Because the first-generation composition languages—IBM's Web Service Flow Language (WSFL) and BEA Systems' Web Services Choreography Interface (WSCI)—were incompatible, researchers developed second-generation languages, such as the Business Process Execution Language for Web Services (BPEL4WS [5], or BPEL), which combines WSFL and WSCI with Microsoft's XLANG specification. Nonetheless, the Web Services Architecture Stack still lacks a process-layer standard for aggregation, choreography, and composition. Sun Microsystems, for example, has proposed standard called WS-CAF [6] (Web Services Composite Application Framework), in which coordination, transaction, context and conversation modeling is discussed to support various transaction processing models and architectures. The literature [7] introduces a framework that builds on currently existing standards to support developers in defining service models and richer web service abstractions. Ontology-based composition and semantic web services composition [8][9] as counter-part to manual composition techniques such as provided by BPEL have been provided. The method builds an optimized graph for service composition based on domain ontology and its reasoning capability. The above arguments lack a mechanism for discovering and binding changed Web services.

In our model, the concept of roles that represent abstractions for system behaviors and functionalities is introduced as a remedy to conceal the dynamicity. As a semantic concept, a role comes with its own properties and behaviors. Features of a service can be role-specific. With this semantic concept, a role-based interaction model not only facilitates access control, but also offers flexible interaction mechanism for adapting service-oriented applications.

3 Web Service, Role and Coordinator (WSRC) Model

3.1 A Layer Model

Open Web service-based systems have three main characteristics: (1) dynamicity, the interaction topology among Web service entities changes dynamically; (2) constraints, there are some QoS constraints and functional constraints imposed on Web services; (3) functionality, the systems are designed to execute tasks and achieve goals. Based on the "separation of concern" principle, the concerns in these three aspects can be designed to be orthogonal to each other, and they are specified and modeled independently as Web services, roles, and coordinators. A Web Service, Role and Coordinator (WSRC) Model is shown as Fig.1. The WSRC model may be conceptualized as the composition of three layers, Web Service layer, Role layer and Coordinator layer. The separation of concerns is apparent in the relationships involving the layers. Coordinators deliver interaction policies to roles; roles impose functional constraints and QoS constraints on Web services; Web service layer is dedicated to functional behavior enacted in the role.

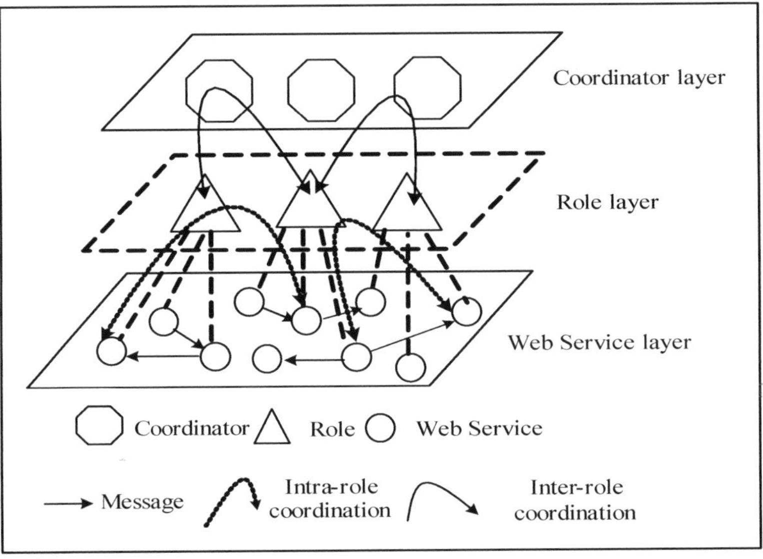

Fig. 1. Model of WSRC

The coordinator layer is oblivious to the Web service layer and is reserved to inter-role coordination interaction. A coordinator can specify constraint strategies and binding strategies for role, and can enact these strategies on roles according to the user's requirements. The role layer bridges the Web service layer and the coordinator layer and may therefore be viewed from two perspectives. From the perspective of a coordinator, a role enables the coordination policies for a set of Web services without requiring the coordinator to have fine-grained knowledge of the individual Web service because the role owns the static description of abstract behavior of the Web services that play the role. From the perspective of Web service, a role is an active coordinator that manipulates the messages sent and received by the Web service. The roles and coordinators form a coordination layer responsible for imposing coordination policies and application constraints among Web services. QoS requirements can be mapped to coordination constraints and imposed on Web services by roles.

3.2 The Model Composition

Web Service's Responsibilities. A Web service is an accessible application that roles can automatically discover and bind. Web services are defined as self-contained, modular units of application logic that provide business functionality to other applications via an Internet connection. Web services support the interaction of business partners and their processes by providing a stateless model of "atomic" synchronous or asynchronous message exchanges [10]. These "atomic" message exchanges can be composed into longer business interactions by providing message exchange protocols that show the mutually visible message exchange behavior of each of the partners involved. The issue of how web services are to be described can be resolved in various ways.

Role's Responsibilities. Because of the intrinsic dynamicity and a large scale of distribute environment, the underlying Web services could be both very dynamic and large in number. The stability and scalability of coordination policies are difficult to maintain if coordination is based on these large numbers of highly dynamic Web services. However, in an application, the set of functional tasks are limited, stable and less dynamic. Hence, the concept of roles that represent abstractions for system tasks and functionalities is introduced as a remedy to conceal the dynamicity.

The role in our model not only is an abstraction for certain properties and functionalities, but also it is responsible for actively coordinating its players to achieve coordination requirements. Hence, the roles serve two purposes in the WSRC model. Firstly, the roles provide static abstractions (declarative properties) for functional behaviors that must be realized by Web services. Secondly, a role is an active entity that is composed of states and behaviors, and it can actively coordinate Web services performing abstract behavior to satisfy coordination requirements.

The abstract behaviors are composed of many tasks, which depict functional logic of system. The formal definition of abstract behaviors is given as follows:

Abstract Behavior: {

$$task1:(I_1,I_2,\cdots,I_n;O_1,O_2,\cdots,O_n);$$
$$task2:(I_1,I_2,\cdots,I_n;O_1,O_2,\cdots,O_n);$$
$$\vdots$$
$$taskn:(I_1,I_2,\cdots,I_n;O_1,O_2,\cdots,O_n);$$

}

Each task contains lots of input values I and output values O. The parameter I denotes input and O denotes output expected by user.

Web services are dynamic and autonomy, which often change frequently in their lifetime. At the same time, application constraints, including QoS constraints and functional constraints, can change dynamically too. To maintain Web services accessibility, roles should bind dynamically the needed Web services that must be appropriate for users. According to abstract behavior, binding strategies and application constraint strategies, dynamic binding means that the role can find dynamically the needed Web services from service registry centers. The roles can save these services interface and related technology specification in the service manager after it finds appropriate Web services. Hence, roles can invoke services to achieve tasks.

The purpose of introducing role into model is as follows: (1) roles can describe abstractly functional behaviors of system, while coordinators control interaction relationships among roles and direct Web service composition; (2) in open distributed environment, Web services and their composition relationships change dynamically but the interaction relationships among roles are relatively stable and persistent, so that Web service composition based on roles is stable and effective; (3) because roles and coordinators are independent of Web services, the programmer can design them independently.

Coordinator's Responsibilities. Coordinators manage interaction and coordination among roles. As an active entity, a coordinator can enact coordination policy, constraint strategies and binding strategies for Web services, and impose these strategies on corresponding roles according to runtime environment and users' requirements. Coordination policy is an abstract behavior composition, which is a workflow process composed by many tasks. Binding strategies are high-level instructions that define binding policies by specifying the management tasks to perform rather than how to implement them. The binding strategy helps in separating management strategies from task implementation, thus increasing flexibility and adaptability. Constraint strategies can ensure Web services availability. These strategies specify some basic conditions for web services depending on users' requirements, application security, scalability and reliability. The format of Constraint strategies is defined as follows:

Constraint strategies: {
$Rule_1$:
$Rule_2$:
$Rule_3$:
⋮
$Rule_n$:
}

Users can also add constraint rules into constraint strategies according to their requirements. Administrator can browse, edit, update, and delete binding strategies and constraint strategies to adapt to changes of environment and users' requirements. The coordinator contains detail information about roles, such as role identity, abstract tasks, etc. The binding strategies and constraint strategies are stored respectively in the binding strategy repertory and the constraint repertory.

A coordinator doesn't bind Web services but transfers constraint to roles that take charge of binding and composing Web services. Because of the static abstract functionalities described by roles, it is simpler for coordinator to specify and enact binding strategy and constraint policy.

4 Dynamic Composition of Web Services

4.1 Dynamic Binding of Web Service

The Web service environment is a highly flexible and dynamic environment. New services become available on a daily basis and the number of service providers is constantly growing. In this case, static binding may be too restrictive and cannot transparently adapt to environment changes, and cannot adapt to users' requirements with minimal user intervention. We describe a dynamic binding engine in this section, which can bind, discover a Web service dynamically.

The dynamic binding engine includes several modules: a service registry (UDDI) to provide a web service repository; a binder manager to find proper services that meet user's requirements in the service registry; a Web service manager to store trace information of web services execution and mapped them with abstract behaviors; an event server and a monitor to monitor events and notify the binder manager, as illustrated in Fig.2. The event server and the monitor are the infrastructure provided by environment. We will mainly introduce the functionalities of binder manager and Web service manager as follows.

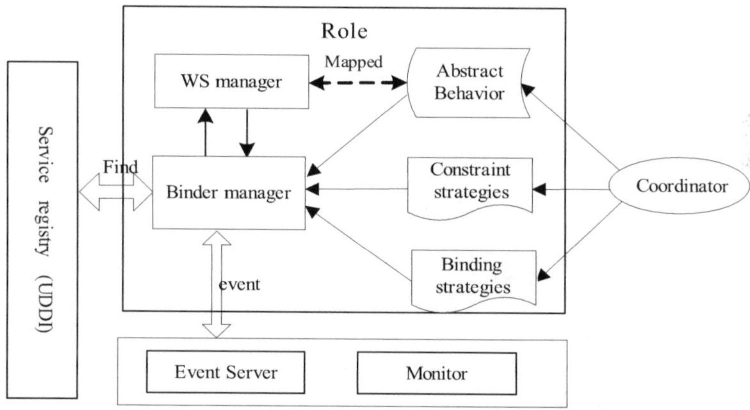

Fig. 2. Dynamic binding engine

The binder manager can find proper services that meet user requirements from the service registry. The Web service that was found by binder manager must satisfy to abstract behaviors, constraint strategies and bind strategies specification that specify user requirement and environment conditions. Binder manager can receive an event that tells the change of state of a web service from the event server and the monitor. The process of dynamic binding consists of four steps as follows:

- Web service providers publish their services to service registry (UDDI).
- Binder manager decomposes role's abstract behaviors into an abstract service by semantic translation function, and then composes it with binding strategies and constraint strategies and sends a SOAP request to the service registry to find the proper services.
- The Web service registry returns a set of concrete services information to binder manager; the binder manager sends a SOAP request to the concrete services and binds them to roles. At the same time, the binder manager sends Web services' information to Web service manager to store these information and notify event server and monitor to monitor this web service.
- When the binder manager receives the event of Web service failure, it will rebind a new Web service.

The Web service manager stores some trace information of web service execution and builds a mapped relationship between web services and tasks described by role's abstract behavior, as illustrated in Fig.3. There is a mapped relationship table that records all Web services execution descriptor, task descriptor and mapped relationship information in Web service manager. Each mapped relationship table entry includes the task identifier (TID), Web service identifier (WSID) and the mapped relationship (MR). When the role executes a task, it firstly finds corresponding Web services that can perform this task in mapped relationship table. Then, the role invokes these Web services bound by binder manager.

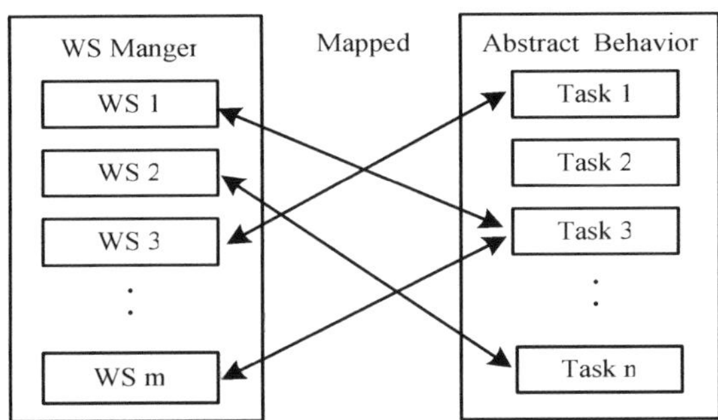

Fig. 3. Mapped between Web services and tasks

4.2 Dynamic Composition Based on Abstract Behavior

In order to adapt to unpredictable changes, we describe a dynamic composition approach based on role's abstract behaviors. Our approach separates abstract behaviors from their implementation. The program designer only concerns abstract functional behaviors that are performed by Web services. The designers can composite abstract functional behaviors dynamically according to user's requirements and application constraints, for example, $role_1.task_j \to role_3.task_i \to \cdots \to role_m.task_n$, which can perform a complex functionality by composition of many tasks. Finally, roles perform these tasks by binding proper Web services. Thus, when a new Web service is provided, or is replaced with others, the abstract functional behavior composition is yet stable.

The implementation of dynamic composition consists of three steps as follows:

- The coordinator specifies coordination strategies and enacts them to corresponding roles according to users' requirement $(I_1, I_2, \cdots, I_n; O_1, O_2, \cdots, O_n)$ and role's tasks; the format of coordination policy is specified as follows:

  Coordination Policy: $\{role_1.task_j \to role_3.task_i \to \cdots \to role_n.task_j\}$.
- When the role receives coordination policy from coordinator, it firstly finds web services that can perform the task in a mapped relationship table, and then the role invokes the Web services to implement the tasks. Finally, the role receives the message from Web services and reroutes the message to other roles to activate other tasks.
- The role sends the results to coordinator after all tasks are performed.

5 Case Study

In this section, we will present a vehicle navigation system example to illustrate Web services dynamic composition ability of the WSRC model.

A vehicle navigation system is aided by a GPS (Global Position System) Service and traffic control services. Given a destination, the GPS system is able to navigate the vehicle to destination. There may be different optimization goals for the vehicle in deciding the path to the destination, such as shortest or fastest path, etc. Taking road condition, traffic condition or environment condition in general into account, the shortest path or the regular highway path may not be the quickest path. Therefore, the vehicle needs frequently communicate with the traffic control services to get the current road and traffic information in detail. Different traffic control service controls different regions. As the vehicle moves on, it needs to communicate with different traffic control services to get accurate and updated traffic information. Hence, the vehicle needs all information in detail from the GPS Service and the traffic control service to arrive at a destination. This example unveils two important characteristics of this type of applications. Firstly, the web services might change unexpectedly, because the vehicle at different location might need different traffic control service. Secondly, the GPS Service and traffic control Services are composed dynamically.

With the WSRC model, the navigation system is separated of five different parts: a navigator, a coordinator, two roles, GPS S (GPS Service) and Tcc S (traffic control Service), as illustrated in Fig.4.

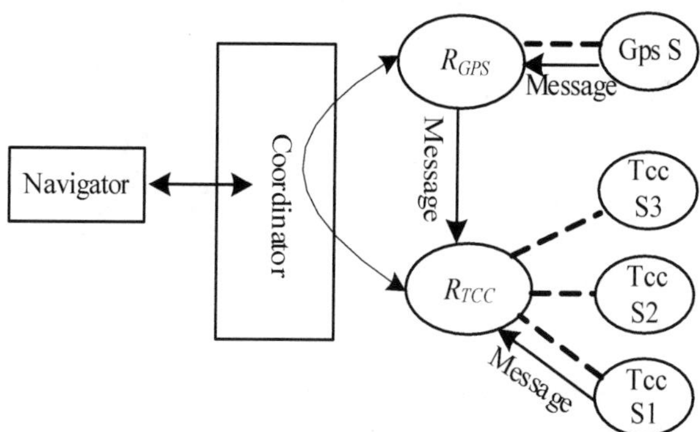

Fig. 4. The model of vehicle navigation system

R_{GPS} is a role that describes abstract functional behaviors of GPS service, and can bind dynamically a GPS service according to binding strategies and constrains strategies. The abstract behavior is defined as follows:

Abstract behavior: {task (destination^{i}, current-location^{o}, path^{o})}

R_{Tcc} is also a role that describes abstract functional behaviors of traffic control service and binds dynamically an appropriate Tcc service according to binding strategies and constrains strategies. The abstract behavior is defined as follows:

Abstract behavior: {task (path^i, current-location^i, traffic-condition^o, road-condition^o)}

The GPS service provides path and current location information for navigator, and implements the task described by R_{GPS}. GPS service is described as follows:

GPS S (destination^i, path^o, current-location^o)

Tcc S service provides traffic and road information for navigator, and implements the task described by R_{Tcc}. Traffic control service is described as follows:

Tcc S (path^i, current-location^i, traffic-condition^o, road-condition^o)

The navigator is a decision-maker, which only makes decisions based on information delivered to it. It sends a request message to coordinator to get some traffic information, road information and path information. The format of message is defined as follows:

Message (destination^i, path^o, current-location^o, traffic-condition^o, road-condition^o)

The coordinator enacts coordination policy, binding strategies and constraint strategies to R_{GPS} and R_{Tcc} according to navigator's requirements. Binding strategies and constraint strategies trigger and control R_{GPS} and R_{Tcc} to bind a proper GPS services and different Tcc services. The definition of coordination policy is as follows:

Coordination policy: {GPS.task $\rightarrow$ T_{Tcc}.task}

The implementation process of navigation system is described as follows:

- The roles of R_{GPS} and R_{Tcc} bind the most proper GPS service and Tcc services respectively according to abstract tasks, binding strategies and constraint strategies that they have described.
- The navigator sends a request message to coordinator for acquiring path information, including current position, traffic condition and the states of road.
- The coordinator specifies coordination policies according to navigator's message contents, and enacts them to R_{GPS} and R_{Tcc}.
- According to coordination policies, R_{GPS} executes its task by invoking the bound GPS service, and then it will send results of execution to R_{Tcc}.
- R_{Tcc} begins to execute its task after it receives a message from R_{GPS}, and then it will send results of execution to the coordinator.
- Finally, the coordinator returns information that the navigator needs.

The case shows the ability of the WSRC Model to adapt for dynamic changes of Web services and application constraints. In the case, the task composition is independent on concrete Web services, and is established by the abstract functional behavior described by roles. Hence, roles could rebind new Web services to complete tasks when Web services change.

6 Conclusion

We have described a WSRC model for composing web services in open distributed environments. The model and the vehicle navigation application show the following benefits. Firstly, coordinators, roles and Web services are defined as orthogonal dimensions, so that they can be modeled and designed independently. Secondly, the model simplifies the design and implementation of applications based on dynamic composition of Web services, and increases software reusability and flexibility. Finally, coordinators can dynamically impose or release coordination strategies and application constraints, and can dynamically reconfigure the interaction topology among roles, which leads to the dynamic adaptability of interaction topology among Web services.

This is our initial investigation on design of the model. There is still a lot of work to be done in the future as follows: the performance analysis, the formalized description, and the exploration of implementation approaches.

References

1. Baresi, L., Nitto, E.D., Ghezzi, C.: Toward Open-World Software. Issues and Challenges, IEEE Computer 39(10), 36–44 (2006)
2. Sun, H., Wang, X., Zhou, B., Zou, P.: Research and Implementation of Dynamic Web Services Composition. In: Zhou, X., Xu, M., Jähnichen, S., Cao, J. (eds.) APPT 2003. LNCS, vol. 2834, pp. 457–466. Springer, Heidelberg (2003)
3. Schreiner, W.: A survey on Web Services Composition. Int. J. Web and Grid Services 1(1), 1–30 (2005)
4. Ren, S., Shen, L., Tsai,: Reconfigurable coordination model for dynamic autonomous real-time systems. In: The IEEE International Conference on Sensor Networks, Ubiquitous and Trustworthy Computing, pp. 60–67. IEEE, New York (2006)
5. Doulkeridis, C., Valavanis, E., Vazirgiannis, M., Benatallah, B., Shan, M-C. (eds.): TES 2003. LNCS, vol. 2819, pp. 54–65. Springer, Heidelberg (2003)
6. Bunting, D., Hurley, M.C.O., Little, M., Mischkinsky, J., Newcomer, E., Webber, J., Swenson, K.: Web Services Composite Application Framework (WS-CAF) Ver1.0 (2003)
7. Benatallah, B., Casati, F.: Web Service Conversation Modeling. A cornerstone for E-business Automation, IEEE Internet Computing 8(1), 46–54 (2004)
8. Rao, J., Su, X., Li, M., et al.: Toward the Composition of Semantic Web Services. In: Li, M., Sun, X.-H., Deng, Q.-n., Ni, J. (eds.) GCC 2003. LNCS, vol. 3033, pp. 760–767. Springer, Heidelberg (2004)
9. Agarwal, S., Handschuh, S., Staab, S., Fensel, D., et al.: ISWC 2003. LNCS, vol. 2870, pp. 211–226. Springer, Heidelberg (2003)
10. Koehler, J., Srivastava, B.: Web Service Composition: Current Solutions and Open Problems ICAPS 2003, Workshop on Planning for Web Services. pp. 28–35 (2003)
11. Ren, S., Yu, Y., Chen, N., Marth, K., Poirot, P., Shen, L.: Actors, Roles and Coordinators – a Coordination Model for Open Distributed and Embedded Systems. In: Ciancarini, P., Wiklicky, H. (eds.) COORDINATION 2006. LNCS, vol. 4808, pp. 247–265. Springer, Heidelberg (2006)

An Information Retrieval Method Based on Knowledge Reasoning

Mei Xiang, Chen Junliang, Meng Xiangwu, and Xu Meng

State Key Laboratory of Networking and Switching Technology,
Beijing University of Posts and Telecommunications,
Beijing, China
meixiang98@gmail.com, chjl@bupt.edu.cn

Abstract. As the competition of Web Search market increases, there is a high demand for accurately judging the relations between the web pages and the user's requirement. In this paper, we propose an information retrieval method that tightly integrates description logic reasoning and traditional information retrieval technique. The method expresses the user's search intention by description logic to infer the user's search object, and selects high-quality keywords according to the semantic context of the search object. Further, fuzzy describing logic is introduced to confirm the relations between the web pages and the user's search requirement, and the method to calculate the membership degree of web pages w.r.t the search requirement is presented. A prototype is implemented and evaluated, and the results show large improvements over existing methods.

1 Introduction

With the advent of Internet and WWW, the amount of information available on the web grows daily. Search engines assist users in finding information, but most of the results are irrelevant to the question. There are two main reasons: one is the keyword based search approach is not able to capture the user's intention exactly. The other is a "semantic gap" existing between the meanings of terms used by the user and those recognized by the search engines.

Semantic web makes the machines understand the meanings of web pages, and has been accepted as the next generation network. Ontology [1] which works as the core of semantic web, defines the conceptions and the relations among conceptions to describe the sharing knowledge. Semantic search [2] can capture the user's intention and search exactly and effectively. But semantic search requires two major prerequisites. First, the entire collection of web pages must be transformed to the ontological form. Second, there is a common agreement on the presentation of the ontology, the query and reasoning mechanisms. These two prerequisites are not satisfied now.

In this paper, we propose an information retrieval method based on knowledge reasoning, named IRKR. This method tightly integrates description logic (*DL*) [3] reasoning and classical information retrieval (*IR*) technology, and describes the user's

K.C. Chang et al. (Eds.): APWeb/WAIM 2007 Ws, LNCS 4537, pp. 328–339, 2007.

intention in *DL* and finds relevant web pages by *IR* search engines. It analyzes the relevancies of web pages w.r.t the user's intention by fuzzy description logic [4].

This paper intends to make three main contributes. First, we propose a way based on *DL* to express user's intentions, and build *IR* query string exactly. Second, we propose an approach to compute the relevancies of web pages w.r.t instances in knowledge base (*KB*), and in the approach we take semantic association [5] into account. Third, we introduce fuzzy description logic to *IR*, and establish the corresponding *KB*, further propose the way to compute the membership degrees of web pages w.r.t the user's intention conceptions. A prototype is implemented and the experimental results show that IRKR yields significant improvements over the origin results.

The rest of the paper is organized as follows. Section 2 presents related research. The way to describe the user's intention and build *IR* query string is discussed in Section 3. After that, we describe the establishment of fuzzy description logic *KB* in Section 4 and the way to judge the relevancies of web pages w.r.t the user's question in Section 5. Section 6 describes the implementation of a prototype and shows the evaluation. Section 7 concludes the paper.

2 Related Work

There are many ongoing research efforts in integration logic reasoning with information retrieval.

A formal *DL* query method for the semantic web is proposed in [2]. However, [2] is a pure logical approach and uses only semantic information. Our method extends it to use and search also textual information.

Our work closely relates to the problem of question answering on the web. Singh [6] extends fuzzy description logic for web question answering system, and queries in it are simple conceptions with property restrictions. The method uses heuristics to find web pages that matches a query, and relies more on natural language processing techniques. Our model uses explicit semantic in a *KB* for query answering and the query can be any *DL* conception expression.

In the semantic search and retrieval area, [7, 8, 9, 10] use semantic information to improve search on the web. In TAP [7], search requests are primarily answered by a traditional web search engine, when the search request matches some individuals in the backend *KB*, information about the individual from the *KB* is also presented to the end user. This method uses only keywords as search request and does not provide formal query capability. It therefore lacks a tight integration of the two methods. SHOE [8] formulates an ontology-based query on the *KB* to find web pages. When returning very few or no results, SHOE automatically converts the query into an *IR* query string for a web search engine. *IR* thus is used only as a complementary method. There is no tight integration of the two methods also. In OWLIR [9] search requests can contain both a formal query for semantic information and a keyword search for textual information. Web pages that satisfy both of them are returned as results. From the paper, it is hard to tell whether it provides other types of integration beyond this conjunction.

WSTT [10] specifies the user's search intention by conceptions or instances in *KB*, then transforms the semantic query into *IR* query strings for existing search engines, and ranks the resulting page according to semantic relevance, syntactic relevance, et al. The method synthesizes semantic search and *IR* in some extent, but it can only express very simple requests, and lacks the formal reasoning capability as provided in our method.

3 User's Intention

3.1 User's Intention Description

User's intention is the objects or the properties of objects, which the user is interested in. The information that the users provide to express their intentions can be categorized to two classes:

① Object Constraint, describes the conditions and restrictions on target objects;

② Property Constraint, restricts which properties of target objects are in the user's interests.

ALC [3] is a description language of *DL*. IRKR uses ALC conception to model object constraint, and uses the set of ALC roles to describe property constraint. The conception standing for the object constraint in a search question is expressed as *OC*:

$$OC \equiv D_1 \vee D_2 \vee \ldots \vee D_n$$
$$D_i \equiv A_{i1} \wedge A_{i2} \wedge \ldots \wedge A_{im} \quad 1 \leq i \leq n$$

D_i is called sub-object constraint. A_{ij} $(1 \leq j \leq m)$, named atomic object constraint, is the basic component in object constraint. Atomic object constraints can be categorized to two classes: ①atomic conception, $\perp$, $\top$; ② conception expressions $\neg$C, $\forall$R.C, $\exists$R.C (C is a conception, and no "$\vee$" in C). In term of the features of ALC, any conception expression can be transformed to the form of *OC*.

Property constraint in a search question is a set *P*. p_j $(p_j \in P)$, named atomic property constraint, is a certain property of search objects.

Let object constraint in search question *s* be $D_1 \vee D_2 \vee \ldots \vee D_n$, $n \geq 1$, Property constraint in *s* be *P*, *k* is the element number of *P*. When *k*>0, the combination of $D_i(1 \leq i \leq n)$ and $p_j(p_j \in P, 1 \leq j \leq k)$ is called a Sub-Search Question of *s*; When *k*=0, each $D_i(1 \leq i \leq n)$ is a Sub-Search Question of *s*. The pages related to any Sub-Search Question of *s* are also related to *s*.

When the atomic property constraint or role in object constraint goes beyond the scope of *KB*, IRKR automatically adds it as a property called expansion role to *KB*. When the conception in object constraint goes beyond the scope of *KB*, IRKR automatically adds it as a conception called expansion conception to *KB*. Expansion roles and conceptions have no associations with any instances in *KB*.

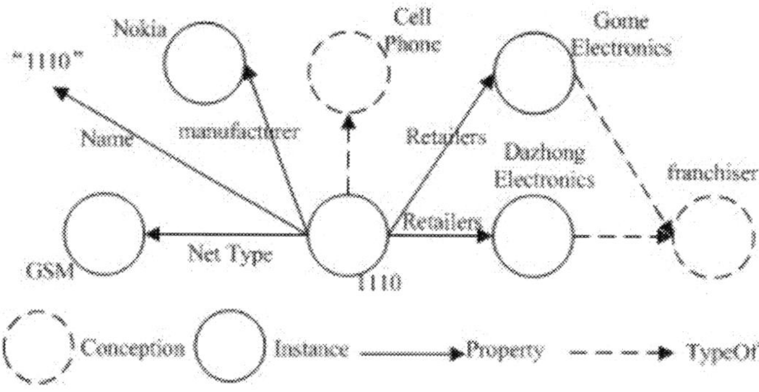

Fig. 1. Snippet of Knowledge base

Take the search question s "the retailers and user comments about Nokia cell phone 1110" for example. "Nokia cell phone 1110" describing the object which the user is searching for, belongs to object constraint; "retailers and user comments" restricting the search object's property which the user is interested in, belongs to property constraint. Based on the *KB* in figure 1, s's object constraint has one sub-object constraint, and OC = Cell Phone ∧ ∃manufacturer.Nokia ∧ ∃Name.1110, s's property constraint P={ Retailers, User comment } ; s has two Sub-Search Questions:

① s_1={Cell Phone ∧ ∃manufacturer.Nokia ∧ ∃Name.1110; Retailers }
② s_2={Cell Phone ∧ ∃manufacturer.Nokia ∧ ∃Name.1110;User comment}

"User comment" is out of the *KB*, and it is an expansion role.

3.2 Query Keywords

By the reasoning of ALC, we can find the instances which the user is interested in. For the information of instances in *KB* is very limited, IRKR converts the query to *IR* query string, and searches the web pages which contain the information more detailed. To solve the problem of polysemy and synonymy, IRKR expands the keywords based on the instances' semantic context (the conceptions and instances which are related to the given instances).

Suppose that search request s contains sub-search questions s_1, s_2, …, s_w; $s_i(1 \leq i \leq w)$ contains sub-object constraint D_i and atomic property constraint p_i(p_i may be null). The set of instances which satisfy D_i is $A_i=\{a_{i1}, a_{i2},\ldots,a_{ik}\}$. $R_i=\{r_{i1}, r_{i2},\ldots,r_{in}\}$, where each r_{im} is property p_i value of an instance in A_i, and $r_{im} \neq$ null $(1 \leq m \leq n)$. When p_i= null, $R_i=\phi$. Let K_i be the keyword set of s_i, we discuss the method to compute K_i in three cases:

① case1: $A_i=\phi$. No instance in *KB* satisfies the constraints of D_i, IRKR connects the literal descriptions of D_i and p_i, and uses it as the one and only element in K_i;

② case2: $A_i \neq \phi$ and $R_i \neq \phi$, or $A_i \neq \phi$ and p_i=null. Let T be the set of objects which the user is searching for. When $A_i \neq \phi$ and $R_i \neq \phi$, the instances in R_i are the targets which the user is searching for, $T=R_i$; when $A_i \neq \phi$ and p_i=null, the instances in A_i are the targets which the user is searching for, $T= A_i$. For the keyword kc_j of instance

$c_j \in T$, IRKR takes the literal description of c_j as basic keyword, draws 3 elements from c_j's semantic context and uses the literal descriptions of them as expansion terms. For information contents of many conceptions in *KB* are very small (information content of conception C: I(C) = -logPr[C], Pr[C] is the probability which any instance in KB belongs to C), using them to expand keyword does not work on solving the problem of polysemy and synonymy. IRKR only chooses the conceptions whose information contents are bigger than ε (ε is the min-value of information content). The keyword set of s_i is $K_i=\{k_{i1}, k_{i2},......,k_{im}\}$, $k_{ij}=kc_j$, m is the number of the instances in T;

③ case3: $A_i \neq \phi$, $p_i \neq$ null, $R_i=\phi$. The property p_i values of instances in A_i are targets that the user is searching for. For property p_i of instance a_{ij}, IRKR takes the literal descriptions of a_{ij} and p_i as the basic keyword; draws 2 elements from a_{ij}'s context, and uses the literal descriptions of them as expansion terms, then get the keyword kc_{ij}. The keyword set of s_i is $K_i=\{k_{i1}, k_{i2},..., k_{in}\}$, $k_{ij}=kc_{ij}$, n is the number of instances in R_i.

Let K be the keyword set of search request s, θ specified by users be the threshold of K's element number. K is composed of the elements drawn from the keyword sets of s's sub-search questions. The probabilities of choosing the keyword from the different keyword sets will be discussed in 6.3.

4 FALC Knowledge Base

The relevancies of instances and conceptions in ALC are based on a binary judgment (either 0 or 1), but the relevancies of web pages and the user's intention are uncertain. We can't tell the relevancies of web pages and the user's intention, and justly can analyze the probabilities of web pages related to the user's intention. IRKR introduces fuzzy description logic FALC [4] to search process, and uses FALC to describe the relevancies of web pages and the user's intention: user's intentions and web pages are modeled as FALC conceptions and instances respectively. In this way, the problem of judging the relevancies between the web pages and the user's intention is then reduced to compute the membership degrees of web instances w.r.t the user's intention conception.

Assume that $\sum_{ALC}$ is the standard ALC *KB* which is used to describe user's intention, and s is a search request. We build FALC *KB* $\sum_F$ based on them:

① create a new FALC *KB* $\sum$, set ABox and TBox to null;

② for every sub-object constraint D_i in s, add a new atomic conception DC_i to $\sum$TBox. Let the set of instances in $\sum_{ALC}$ which satisfy D_i be $A_i=\{a_{i1}, a_{i2},...,a_{ik}\}$, for a_{ij} ($1 \leq j \leq k$), add an assertion $<a_{ij}: DC_i=1>$ to $\sum$ABox;

③ for every atomic property constraint p_i in s, add a new atomic conception PC_i to $\sum$TBox;

④ let s_i be a sub-search question of s, and contain sub-object constraint D_i and atomic property constraint p_i. DC_i and PC_i are the conceptions in $\sum$ responding to D_i and p_i respectively. The set of instances in $\sum_{ALC}$ which satisfy D_i is A_i. $R_i=\{r_{i1}, r_{i2},...,r_{in}\}$, where each r_{im} is property p_i value of an instance in A_i, and $r_{im} \neq$ null ($1 \leq m \leq n$) (When p_i= null, $R_i=\phi$). If $R_i \neq \phi$, add a new atomic conception SC_i to

$\sum$TBox, for each r_{im} in R_i, add an assertion $<r_{im}: SC_i=1>$ to $\sum$ABox; if $p_i \neq$ null and $R_i=\phi$, add a complex conception SC_i, $SC_i \equiv DC_i \wedge PC_i$ to $\sum$TBox; if p_i=null, add a new atomic conception SC_i to $\sum$TBox, for each instance a_{ij} in A_i, add an assertion $<a_{ij}: SC_i=1>$ to $\sum$ABox;

⑤ let the search questions s contain sub-search questions: $s_1, s_2, \ldots, s_w$. Add a complex conception S, $S \equiv SC_1 \vee SC_2 \vee \ldots \vee SC_w$ to $\sum$TBox, where $SC_i (1 \leq i \leq w)$ is the conception in $\sum$ corresponding to s_i;

⑥ let $\sum'$ be the KB that we get after ①②③④⑤. Add a new atomic conception Doc to $\sum'$TBox, and define all web pages are instances of Doc. The set of web instances is defined as $D_{\sum}=\{d | \sum' \models d{:}Doc\}$. Then we enrich $\sum'$ with a set of new fuzzy assertions about the membership degrees of web instances w.r.t conceptions, and define the enriched $\sum'$ as $\sum_F$:

$$\sum_F = \sum' \vee \{<i{:}D_q = RSV(i, D_q)> | \forall i \in D_{\sum}, \forall D_q \in C\}$$

C is the set of fuzzy conception in $\sum'$, RSV(i, D_q) calculates the membership degree of web page i w.r.t conception D_q, and returns a degree in [0,1], we will discuss it in Section 5 in detail.

5 The Relevance of Web Page and User's Intention

In $\sum_{ALC}$, the membership degrees of instances w.r.t conceptions are certain, after we extend $\sum_{ALC}$ to $\sum_F$, these membership degrees are certain too (0 or 1). By analyzing the relevancies of web pages and instances in $\sum_{ALC}$, IRKR infers the membership degrees of web pages and user's intention conceptions in $\sum_F$.

5.1 Semantic Association

Semantic association describes the intensities of the relevancies among entities (conception, property and instance). In ontology KB, two entities e_1 and e_2 are semantically associated [5] if one or more property sequences $e_1, p_1, e_2, p_2, e_3, \ldots, e_{n-1}, p_{n-1}, e_n$ exist. Taking semantic association into account can analyze the relevancies of web pages and instances more effectively. For example, in figure 1 entities "Nokia", "GSM", "Cell Phone" and "1110" are strongly semantically associated. When we analyze the relevance of a page and the instance "1110", for the ambiguity of keyword we can't tell the key word "1110" related to the instance "1110" or not. In this case, if there are other keywords of instances which are semantically associated to the instance "1110" in the same page, for example "Nokia", "GSM", "Cell Phone", the page is more likely to be related to the instance "1110". Boanerges [11] has proposed the way of ranking semantic association, we don't discuss it here, just use $AS(a,b)$ to describe the semantic association between entities a and b.

5.2 The Relevance of Web Page and Instances

Take semantic association into account, we can analyze the relevancies of web pages and instances more exactly. For reducing the complicacy of computing, IRKR justly considers the entities whose semantic associations with the given instance are bigger than γ (γ is specified by user). Let T(a)={a_1, a_2,..., a_n} be the set of entities whose semantic association with instance a are bigger than γ, and MAX_SA is the max semantic association between two entities. The relevance degree of web page p and instance a is

$$\text{relationDegree}(a, p) = \begin{cases} 1, \sum_{i=1}^{n} \text{relation}(p, a_i) * \dfrac{\text{AS}(a,a_i)}{\text{MAX\_SA}} > 1 \\ \sum_{i=1}^{n} \text{relation}(p, a_i) * \dfrac{\text{AS}(a,a_i)}{\text{MAX\_SA}}, \text{ else} \end{cases}$$

relation(p,a_i) is the vector space similarity degree of web page p and entity a_i, and the bigger semantic association of a_i and a is, the more effect *relation(p,a_i)* will have on the relevance degree of page p and a. The max value of *relationDegree(a,p)* is 1.

$$\text{relation}(p,a_i) = \frac{\sum_{j=1}^{m} td_j \times \mu tc_{ij}}{\sqrt{\sum_{j=1}^{m} td_j^2 * \sum_{j=1}^{m} \mu tc_{ij}^2}}, \qquad 1 \le i \le n \qquad (1)$$

m is the total number of keywords of page p and instance a_i. When a_i is an instance, we take the literal description of a_i as its basic keyword, and draw 3 elements from a_i's context as the keyword expansion; when a_i is a conception, we take the literal description of a_i as its keyword. td_j and tc_{ij} are the TF*IDF results of page a and a_i's keywords respectively; μ is the weight of basic key words and expansion terms, when tc_{ij} is basic key word, μ=2, otherwise μ=1.

5.3 Membership Degree

IRKR analyzes the relevancies of the web pages and instances in Σ_{ALC} to deduce the membership degrees of web pages and conceptions in FALC *KB* Σ_F.

① conception responding to sub-search question: *SC*

When *SC* is an atomic conception, let $B=\{b_1, b_2,\ldots,b_x\} \subset \Delta^I$ (Δ^I is the instance set in the domain)be the set of instances which satisfy $SC^I(b_i)=1$. The membership degree of page p w.r.t conception *SC* is

$$SC^I(p) = \begin{cases} 1, \sum_{i=1}^{x} \text{relationDegree}(b_i, p) > 1 \\ \sum_{i=1}^{x} \text{relationDegree}(b_i, p), \text{else} \end{cases}$$

When *SC* is a complex conception, and $SC \equiv DC_i \wedge PC_i$, by the properties of FALC.

$$SC^I(p) = (DC_i \wedge PC_i)^I(p) = \min\{DC_i^I(p), PC_i^I(p)\}.$$

② conception responding to sub-object constraint: *DC*

When $\exists\, a_i \in \Delta^I$, $DC^I(a_i)=1$. Let $A=\{a_1, a_2,\dots,a_t\} \subset \Delta^I$ be the set of instances which satisfy $DC^I(a_i)=1$, the membership degree of page p w.r.t conception DC is

$$DC^I(p)=\begin{cases} 1, \sum_{i=1}^{t} \text{relationDegree}(a_i, p) > 1 \\ \sum_{i=1}^{t} \text{relationDegree}(a_i, p), \text{else} \end{cases}$$

When $\forall\, a_i \in \Delta^I$, $DC^I(a_i) \neq 1$. Let D_i be the sub-object constraint responding to DC, take the literal descriptions of the conceptions and properties in D_i as the keywords of DC. $DC^I(p)$ is equal to the vector space similarity degree of web page p and the keywords of DC, computed by formula (1).

③ conception responding to property constraint: *PC*

Let p_i be the atomic property constraint responding to PC, take the literal description of property p_i as the keywords of PC. $PC^I(p)$ is equal to the vector space similarity degree of web page p and the keywords of PC, computed by formula (1).

④ conception responding to user's search question: *S*

Let $S \equiv SC_1 \vee SC_2 \vee \dots \vee SC_n$. By the properties of FALC, $S^I(p)= (SC_1 \vee SC_2 \vee \dots \vee SC_n)^I(p) = \max\{ SC_1^I(p), SC_2^I(p),\dots, SC_n^I(p)\}$.

6 Implementation and Evaluation

6.1 Implementation

We implement a prototype system SSKR (Search System based on Knowledge Reasoning), figure 2 shows the structure of the system.

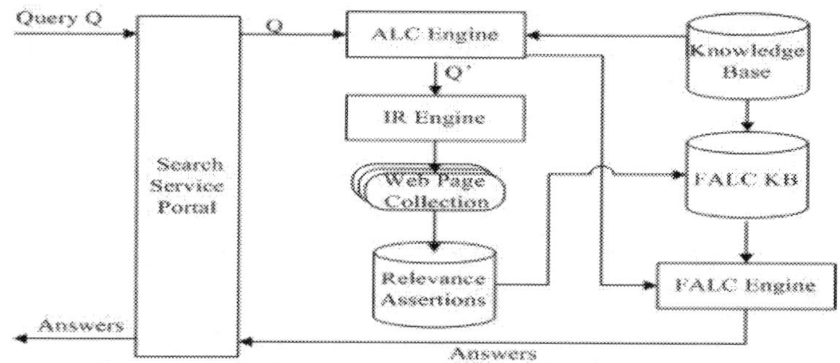

Fig. 2. Architecture of the SSKR system

Search Service Portal accepts a query, returns a list of pages in the descent order of the membership degree. ALC Engine is the reasoning engine of ALC, which is constructed on Jena API [12]. IR Engine transfers the query to classical search engine and gets the relevant Web Page Collection, which is realized by calling Google Web APIs [13]. Relevance Assertions computes the membership degree of web pages w.r.t fuzzy conception. Knowledge Base can be any OWL *DL KB*, the following experiments are based on SWETO (medium) [14], which includes 115 conceptions, 69 properties, 55876 instances, 243567 assertions; SSKR takes the "label" property values of instances, properties and conceptions as their literal descriptions. FALC *KB* is the fuzzy extending edition of Knowledge Base. FALC Engine is the FALC reasoning engine based on alc-F [15].

The workflow of the system consists of the following stages: ① the portal accepts an query Q; ② ALC Engine transforms Q to the keyword set Q' by the method in Section 3.2; ③ IR Engine finds the web pages related to the keywords in Q'; ④ Relevance Assertions computes the membership degree of web pages w.r.t fuzzy conception, and establishes the FALC *KB*; ⑤ FALC Engine computes the membership degree of pages w.r.t the user's intention, and reorders the web pages.

In stage ③, Google may return millions of web pages for every keywords. For reducing the complicacy of computing, IR Engine just returns the 50 top-ranked pages for one keyword, if Google may return too many web pages.

6.2 Impact of Number of Expansion Terms

When searching instances or properties of instances, IRKR takes the instance contexts as query expansion. In general, the number of expansion terms should be within a reasonable range in order to produce the best performance. This chapter analyzes the impacts that the number of expansion terms has on the precision.

Firstly, we analyze the case in querying instances. We draw 50 instances from the SWETO *KB*. For each instance, we take its literal description as the basic keyword, the literal description of an elements which drawn from the instance context as an expansion term (the information content of chosen conception should be bigger than ε), then search by Google. Judging the relevancies of pages and the instance is done by manual work. Considering the 20 top-ranked pages, the experiment result is showed by "instance query" curve in figure 3. When using basic keywords to search only, the precision is 33.7%; as the increase of the number of expansion terms, the precision increases, when the number is 3, the precision arrives at the max value, after that, the precision begins to descend.

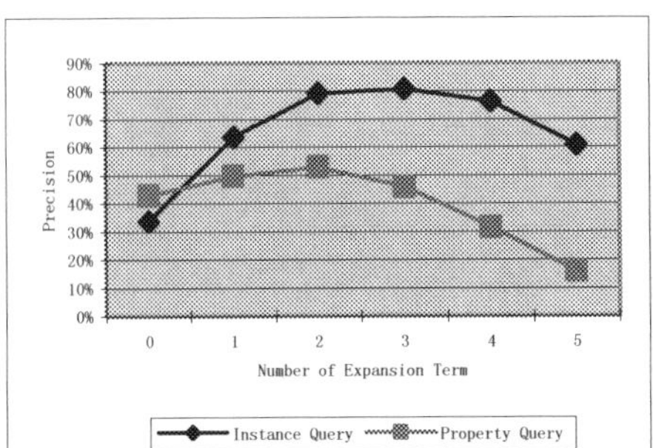

Fig. 3. Impact of various number of expansion terms

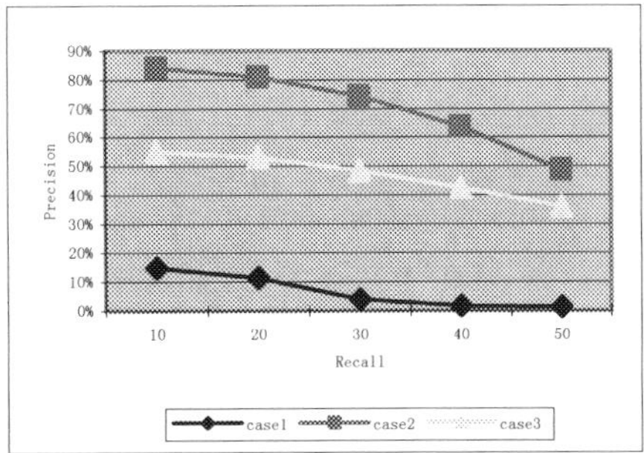

Fig. 4. Retrieval Effectiveness in Three Cases

Secondly, we analyze the case in querying the properties of instances. We draw 50 instances from the SWETO *KB*. For each instance, we choose one of its properties as target to query. Take the literal description of chosen instance and property as the basic keyword, the literal description of an element which drawn from the instance context as an expansion term (the information content of chosen conception should be bigger than ε), then search by Google. Judging the relevancies of pages and the instance is done by manual work. Considering the 20 top-ranked pages, the experiment result is showed by "property query" curve in figure 3. When the number of expanding terms is 2, the precision arrives at the max value, after that the precision begins to descend.

The experiment shows that when searching instances or properties of instances, taking the instance context as query expansion can improve precision effectively, but the precision will not always increase along with the increasing of the number of expansion terms, when the number of expansion terms exceeds a certain value, precision will descend instead.

6.3 Quality of Keywords

In this chapter, we compare the qualities of keywords generated in the 3 cases mentioned in Section 3.2. Bring forward 30 queries for each case, and one query only contains a sub-search question. Draw an element from the keyword set of every query, and query by Google. Judging the relevancies of pages and the instance is done by manual work; figure 4 shows the experiment results. Considering the 20 top-ranked pages, the descent orders of the precision are case2, case3, case1, and 81%, 53%, 11% respectively. The precision of case1 is much lower than case2 and case3. It is because the keywords in case2 and case3 contain the direct description about the search object, and the KB in case1 does not contain the search object, so the keywords in case1 are more likely to be added some irrelevant terms. And we find in the experiment, when the search question is complex, keywords generated in case1 often exceed the length limit that search engine can accept, which may also lead to performance deterioration.

Let the probabilities of choosing the keywords from the keyword set in case1, case2, case3 be p1, p2, p3, then p2> p3>> p1 should be satisfied.

6.4 Retrieval Effectiveness

In this chapter, we compare the retrieval performance of SSKR with the method in WSTT. SSKR and WSTT use Google as classical search engine both. 5 volunteers in our lab are recruited, each volunteer proposes 20 questions (for satisfying the condition which WSTT executes in, the instance set responding to each question should not be null), and queries by WSTT and SSKR respectively(WSTT executes after ALC reasoning).

Figure 5 shows the experiment results. WSTT1 queries using the keywords generated by WSTT(if there exist several lists of returned web pages, which are responding to different keywords, then WSTT1 shows the average result); WSTT2 queries using the keywords generated by WSTT, and reorders by WSTT; SSKR1 queries using the keywords generated by SSKR(if there exist several lists of returned web pages, which are responding to different keywords, then SSKR1 shows the average result); SSKR2 queries using the keywords generated by SSKR, and reorders by SSKR; WSTT3 queries using the keywords generated by SSKR, and reorders by WSTT. Considering the 20 top-ranked pages, the descent orders of the precision are SSKR2, WSTT3, SSKR1, WSTT2, WSTT1, and 69.5%, 64.6%, 60.3%, 45.7%, 38.2% respectively.

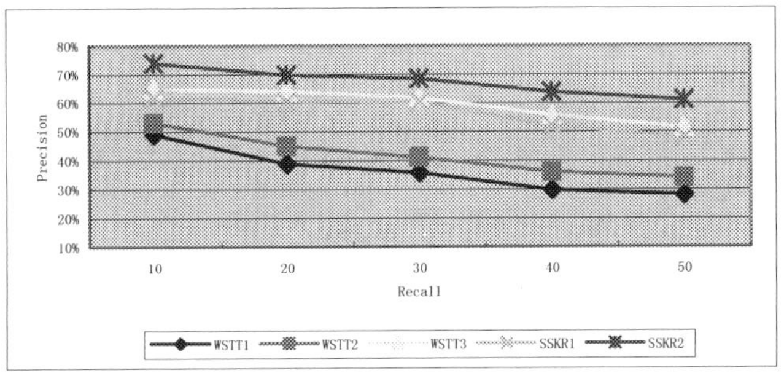

Fig. 5. Average precision for WSTT, SSKR

We can see that the precision of SSKR1 is much higher than that of WSTT1. WSTT uses the descriptions of the searched instances and the searched instances' parent classes as the keyword, which often involves some useless words. When the searched instances have too many parent classes, keywords may exceed the length limit that search engine can accept, which may also lead to performance deterioration. SSKR takes the information content into account and constrains the length of keywords reasonably, the keywords generated by SSKR can express user's intentions more exactly; SSKR2 and WSTT3 is the result that SSKR and WSTT analyze the same page set, so the higher precision of SSKR2 than that of WSTT3 shows that semantic association and FALC reasoning based SSKR can judge the relevance degree of web page and user's intention more effectively.

Because the SSKR needs some reasoning, the time cost of SSKR is about 1.5 times more than WSTT on the average in the process of re-ranking web pages. But taking the precision improvement which SSKR produces into account, we think that the cost is valuable.

7 Conclusion

To improve the expression and reasoning ability of search engines is an important topic in information retrieval area. In this paper, we have presented an information retrieval method based on knowledge reasoning, which tightly integrates the ability of expression and reasoning of *DL* and classical *IR* technology. This method expresses user's intentions based on *DL*, which can express all kinds of complex user's intentions. It solves the problem of polysemy and synonymy by taking the semantic context of the search target into account, and introduces FALC for describing the uncertain relations of web pages and user's intentions. The way to compute the membership degree of web page w.r.t conception standing for user's intention is proposed, and a series of experiments showed that the approach mentioned in this article is able to effectively improve the search capability of the search engines.

As a future work, we plan to integrate visual query formulation into our system. We are also going to optimize algorithm efficiency.

Acknowledgments. The National Natural Science Foundation of China under grant No. 60432010 and National Basic Research Priorities Program (973) under grant No. 2007CB307100 support this work. We gratefully acknowledge invaluable feedbacks from related research communities.

References

1. Gruber, T.: Towards principles for the design of ontologies used for knowledge sharing. International Journal of Human-Computer Studies 43(5/6), 907–928 (1995)
2. Horrocks, I., Tessaris, S.: Querying the Semantic Web: a Formal Approach[A]. In: Proc. of the 13th Int. Semantic Web Conf [C], pp. 177–191. Springer-Verlag, Heidelberg (2002)
3. Baader, F., Calvanese, D., McGuinness, D., et al.: The description logic handbook. Cambridge University Press, Cambridge (2003)
4. Straccia, U.: Reasoning within fuzzy description logics. Journal of Artificial Intelligence Research, 14 (2001)
5. Anyanwu, K., Sheth, A.: ρ -Queries:enabling querying for semantic associations on the semantic web[A]. In: Proceeding of the WWW2003[C], pp. 690–699. ACM Press, New York (2003)
6. Singh, S., Dey, L., Abulaish, M.: A framework for extending fuzzy description logic to ontology based document processing. In: Proc. 2nd Intl Atlantic Web Intelligence Conf. (2004)
7. Guba, R., McCool, R.: Semantic search[A]. In: Proceeding of the WWW2003[C], pp. 700–709. ACM Press, New York (2003)
8. Heflin, J., Hendler, J.: Searching the web with SHOE. In: Proc. of AAAI-2000 Workshop on AI for Web Search (2000)
9. Shah, U., Finin, T.: Information retrieval on the semantic web. In: Proc. Of the 11th Intl. Conf. on Information and Knowledge Management, pp. 461–468 (2002)
10. Kerschberg, Larry, Kim, Wooju, Scime, Anthony: A personalizable agent for semantic taxonomy-based web search. In: Truszkowski, W., Hinchey, M., Rouff, C.A. (eds.) WRAC 2002. Lecture Notes in Artificial Intelligence (Subseries of Lecture Notes in Computer Science), vol. 2564, pp. 3–31. Springer, Heidelberg (2003)
11. Aleman-Meza, B., Halaschek, C.: Context-Aware semantic association ranking[A]. In: Proceeding of Semantic Web and Databases Workshop[C] (2003)
12. McBride, B.: Jena: A Semantic Web Toolkit[J]. IEEE Internet Computing 6(6), 55–59 (2002)
13. Google Web APIs., http://www.google.com/apis/
14. Aleman-Meza, B.: SWETO: large-scale semantic web test bed[A]. In: Proceedings of the 16th international Conference on Software Eng. & Knowledge Eng(SEKE,: Workshop on Ontology in Action[C].Banff, Canada: Knowledge Systems Inst. 2004. pp. 490–493 (2004)
15. Straccia, U., Lopreiato, A.: alc-F: A fuzzy ALC reasoning engine (2004), http://faure.iei.pi.cnr.it/ straccia/software/alc-F/

Research on Personalized Recommendation Algorithm in E-Supermarket System

Xiong Feng[1,2] and Qi Luo[3,4]

[1] School of Management, Wuhan University of Technology, Wuhan 430070, China
[2] School of Business, Ningbo University, Ningbo, 315211, China
[3] College of Engineering and Technology, Southwest University, Chongqing 400715, China
[4] Information Engineering school, Wuhan University of Science and Technology and Zhongnan Branch, Wuhan 430223, China
xiongfeng@nbu.edu.cn

Abstract. The rapid growth of e-commerce had caused product overload where customers were no longer able to effectively choose the products that they were needed.To meet the personalized needs of customers in E- supermarket, the technologies of web usage mining, collaborative filtering and decision tree were applied in the paper, while a new personalized recommendation algorithm were proposed. Personalized recommendation algorithm were also used in personalized recommendation service system based on E-supermarket (PRSSES). The results manifest that it could support E-commerce better.

Keywords: Personalized recommendation, Association mining, Web usage mining, Decision tree.

1 Introduction

With the development of E-commerce and network, more and more enterprises have transferred to the management pattern of E-commerce [1]. The management pattern of E-commerce may greatly save the cost in the physical environment and bring conveniences to customers. People pay more and more attention to E-commerce day by day. Therefore, more and more enterprises have set up their own E-supermarket websites to sell commodities or issue information service. But the application of these websites is difficult to attract customer' initiative participation. Only 2%-4% visitors purchase the commodities on E-supermarket websites [2]. The investigation indicates that personalized recommendation system that selecting and purchasing commodities is imperfect. The validity and accuracy of providing commodities are low. If E-supermarket websites want to attract more visitors to customers, improve the loyalty degree of customers and strengthen the cross sale ability of websites, the idea of personalized design should be needed. It means that commodities and information service should be provided according to customers' needs. The key of personalized design is how to recommend commodities based on their interests.

At present, many scholars have carried on a great deal of researches on personalized recommendation algorithms, such as collaborative recommendation

K.C. Chang et al. (Eds.): APWeb/WAIM 2007 Ws, LNCS 4537, pp. 340–347, 2007.

algorithm and content-based recommendation algorithm [3]. Although collaborative recommendation algorithm is able to mine out potential interests for users, it has some disadvantages such as sparseness, cold start and special users. Similarly, the content-based recommendation algorithm also has some problems such as incomplete information that mined out, limited content of the recommendation, and lack of user feedback [4].

According to the reasons mentioned above, the technologies of web usage mining, collaborative filtering and decision tree are applied in the paper, while a new personalized recommendation algorithm is proposed, personalized recommendation algorithm is also used in personalized recommendation service system based on E-supermarket (PRSSES). The results manifest that it can support E-commerce better.

2 Personalized Recommendation Algorithm Structure

Personalized recommendation algorithm consists of four steps, as shown in Fig. 1.

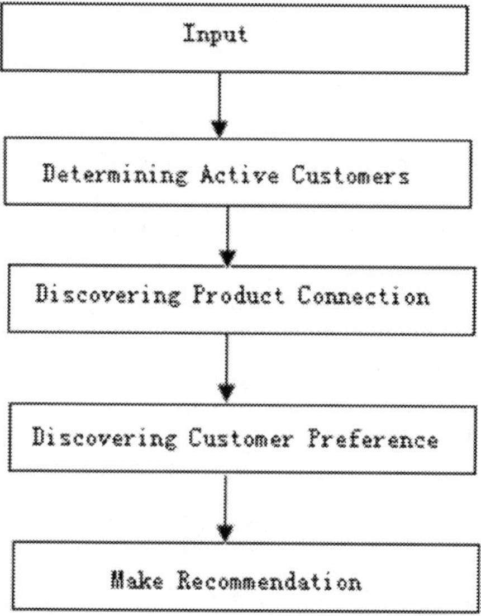

Fig. 1. Personalized recommendation algorithm Structure

Step 1: Active customers are selected using the decision tree induction technique.

Step 2: Two association rule sets are generated from basket placement and purchase data set, and used for discovering connections between products.

Step 3: Tracking individual customer's previous shopping behavior in an E-supermarket site is used to make preference analysis. Active customer's preferences across products are analyzed. Active customer's preferences across products are analyzed.

Step 4: A personalized recommendation list for a given active customer is produced by integrating product connections and customer preferences discovered in previous phases. A more detailed description of each phase is provided in the following subsections.

2.1 Determining Active Customers

Making a recommendation only for customers who are likely to buy recommended products could be a solution to avoid the false positives of a poor recommendation. This phase performs the tasks of selecting such customers based on decision tree induction. The decision tree induction uses both the model set and the score set generated from customer records. To generate the model set and the score set of our recommendation problem $\mathrm{Re}c(l,n,p,t)$, we also needs two more sets; one is the model candidate set, which is a set of customers who constitute the model set, and the other is the score candidate set, which is a set of customers who form the score set.

To build an effective model, the data in the model set must mimic the time frame when the model will be applied. The time frame has three important components: past, current and future [5]. The past consists of what has already happened and data that has already been collected and processed. The present is the time period when the model is being built. The future is the time period for prediction. Since we can predict the future through the past, the past is also divided into three time periods: the distant past used on the input side of the data, the recent past used to determine the output, and a period of latency used to represent the present. Given such a model set, a decision tree can be induced which will make it possible to assign a class to the dependent variable of a new case in the score set based on the values of independent variables.

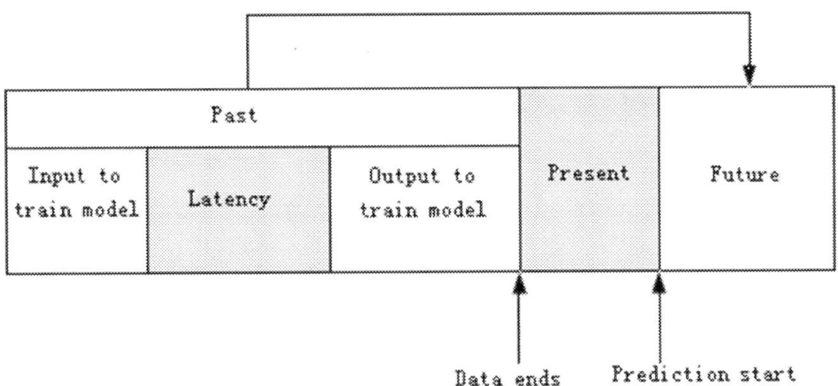

Fig. 2. The modeling time frame

Let msst, pd, pl and pr be the start time of the model set, the time period for the distant past, the time period of latency, and the time period for the recent past, respectively. Then, a model candidate set is defined as a set of customers who have purchased p or more products at the level-l product class between msst and msst +pd. Table 1 illustrates an example of determining the model candidate set from customer purchase records in the case that taxonomy, Rec(1,2,1,2005-12-1), msst =<May1,2005>, pd = <four months>, pl= <one month>, and pr = <one month>. Here, we obtain <101, 103,104> as the model candidate set since 101, 103 and 104 follow the definition of the model candidate set.

2.2 Discovering Product Connection

In this phase, we first search for meaningful relationships or affinities among product classes through mining association rules from large transactions. As defined in the problem statement section, association rule mining is performed at the level-l of product taxonomy. Unlike the traditional usage of association rule mining, we look for association rules from the basket placement transaction set as well as the purchase transaction set in order to capture more accurately the e-shopper's preference. The steps for mining level-l association rules from two different transaction sets are as follows:

Step 1: Set the given time period as the time interval between *msst* and *t- 1*. For each of the purchase transaction set and basket placement transaction set.

Step 2: Transform all the transactions made in the given time period into a single transaction of the form < customer ID, {a set of products}>.

Step 3: Find association rules at level-l by the following sub-steps:

(1) Set up minimum support and minimum confidence. Note that the minimum support is higher in the case of the basket placement transaction set.

(2) Replace each product in the transaction set with its corresponding level-l product class.

(3) Calculate frequent 2-itemsetsusing Apriori or its variants [6].

Generate association rules consisting of a single body and a single head from the set of all frequent 2-itemsets.

(4) In the case of mining association rules from the purchase transaction set, denote the set of generated association rules as Rule set. In the case of the basket placement transaction set, denote as Ruleset_b .

Step 4: Set Ruleset $\text{Ruleset}_{all} = \text{Ruleset}_p \cup \text{Ruleset}_b$ Here, if there are common rules in both Ruleset_p and Ruleset_b , then use the common rule with b the highest confidence.

Next, we compute the extent to which each product class appeals to each customer by applying the discovered rules to him /her. This work results in building a model called the product appeal matrix. Let PurSet(i) be a product class set which the active customer i has already purchased, and let AssoSet(i) be a product class set which is inferred by applying PurSet(i) to Ruleset_{all} .Then, a product appeal matrix A=(a_{ij}), i =1, . . . ,M (total number of active customers), j= 1, . . . ,N (total number of level-l product classes) is defined as follows:

$$a_{ij} = \begin{cases} conf(* \Rightarrow j), if & j \in AssoSet(i) - PurSet(i) \\ 0, & otherwise \end{cases} \tag{1}$$

Note that the notation $conf(* \Rightarrow j)$ means the maximum confidence value among the confidences of rules having head j in Ruleset all.

2.3 Discovering Customer Preference

As mentioned in previous sections, this research intends to apply the results of analyzing the preference inclination of each customer. For this purpose, we propose a customer preference model represented by a matrix. The customer preference model is constructed based on two shopping steps of buying products in an Internet shopping mall: click-through and basket placement. Let p_{ij}^c and p_{ij}^b be the total number of occurrences of click and the total number of occurrences of basket placements of customer i for a given *level-l* product class *j*, respectively. p_{ij}^c And p_{ij}^b are calculated from the raw click-stream data as the sum over the given time period, and so reflect an individual customer's behavior. From the above definition, we define the customer preference matrix P = (p_{ij}), i =1, . . . , M, j= 1, . . . ,N,as follows

$$p_{ij} = \left(\frac{p_{ij}^c - \min_{1\le j\le N}(p_{ij}^c)}{\max_{1\le j\le N}(p_{ij}^c) - \min_{1\le j\le N}(p_{ij}^c)} + \frac{p_{ij}^b - \min_{1\le j\le N}(p_{ij}^b)}{\max_{1\le j\le N}(p_{ij}^b) - \min_{1\le j\le N}(p_{ij}^b)}\right) \times \frac{1}{2} \tag{2}$$

p_{ij} Implies a simple average of the normalized value j of p_{ij}^b and p_{ij}^c .

2.4 Making Recommendations

In the preceding phases, we have built a product appeal model and a customer preference model. Based on the product appeal model and the personal preference inclination of customer preference model, this phase determines what product classes will be recommended to the customer .We propose a matching score s_{ij} between customer i and level-l product class j as follows:

$$s_{ij} = \frac{2 \times a_{ij} \times p_{ij}}{a_{ij} + p_{ij}} \tag{3}$$

3 PRSSES

Personalized recommendation service system based on E-supermarket model (PRSSES) is Fig.3.There are three layer structures of PRSSES, browse, Web server, and database server. The Web server includes WWW server and application server.

Client: Based on browse. User logs in E- supermarket system website, browses websites or purchases commodity after inputting register and password, All information is stored in database server though Web server collection.

Web server: collecting user personalized information and storing in user interest model. Simultaneously the commodity recommendation pages are presented to user.

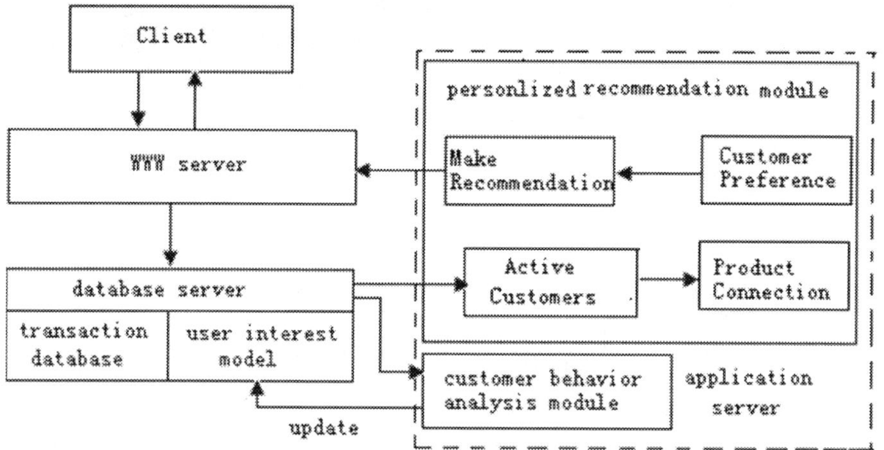

Fig. 3. Personalized recommendation service system based on E-supermarket model

Database server: Including user transaction database and user interest model, user purchasing detailed records are stored in transaction database The user interest model stores user personality characteristic, such as name, age, occupation, purchasing interest, hobby and so on.

Personalized recommendation module: It is the core module of personalized recommendation service system based on E-supermarket. Personalized recommendation algorithm consists of four steps, such as determining active customers, discovering product connection, and discovering customer preference and makes recommendation.

Customer behavior analysis module: Users' behavior pattern information is traced according to user browsing, purchasing history, questionnaire investigation and feedback opinions of some after-sale service. User interest model is updated though analyzing users' behavior pattern information.

4 Performance Evaluation

On the foundation of research, we combine with the cooperation item of personalized service system in community. The author constructs a personalized recommendation system website based on E-supermarket. In order to obtain the contrast experimental result, the SVM classification algorithm, content-based recommendation algorithm based on VSM and an adaptive recommendation algorithm are separately used in the module of personalized recommendation. X axis represents recall, Y axis represents precision. The experimental results are Fig.4.

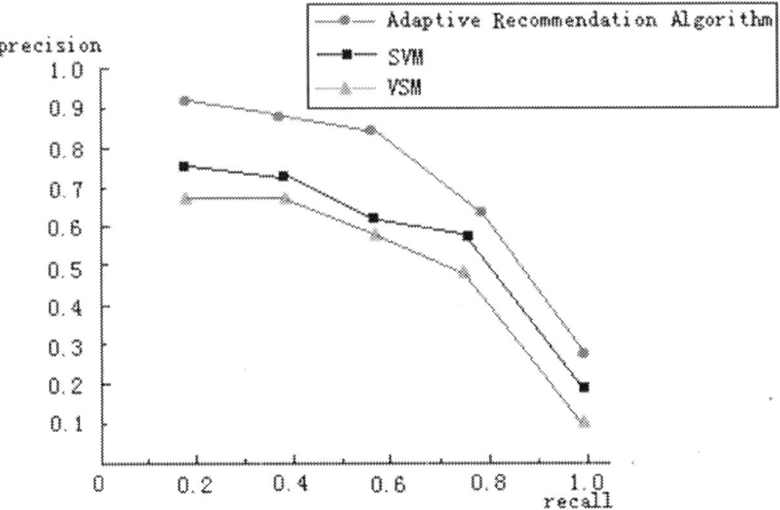

Fig. 4. The performance of adaptive recommendation algorithm, SVM and VSM

We also compare adaptive recommendation algorithm with other algorithms such as k-NN, Rocchio [7]. The experimental results are Fig.5.

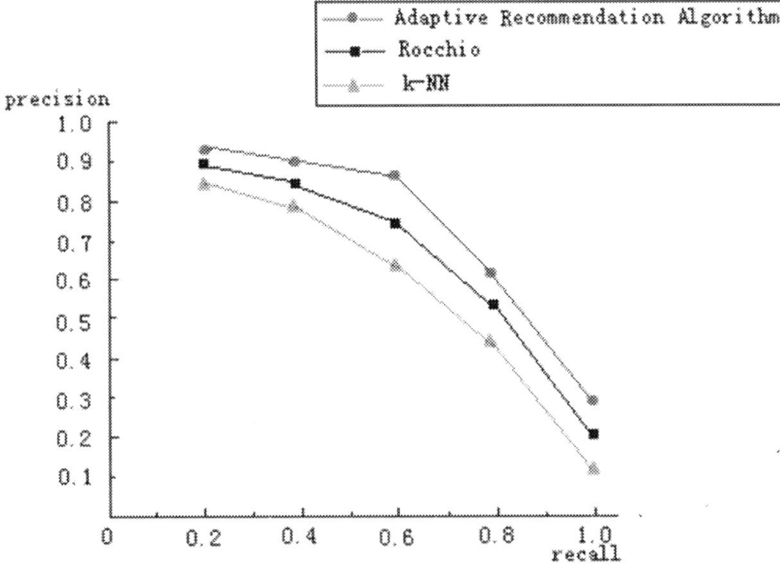

Fig. 5. Comparison with other algorithms

From Fig.4, 5, the precision of adaptive recommendation algorithm is higher than separate classification algorithm. The recall of adaptive recommendation algorithm is more effective than separate content-based recommendation algorithm.

5 Conclusion

In summary, the technologies of web usage mining, collaborative filtering and decision tree are applied in the paper, while a new personalized recommendation algorithm is proposed. The algorithm is also used in personalized recommendation service system based on E-supermarket. Finally, we recommend highly efficient products by integrating product connections and customer preferences .The system can support E-commence better. The results manifest that the algorithm is effective through testing in personalized recommendation service system based on E-supermarket. I wish that this article's work could give references to certain people.

References

1. Yanwen, W.: Commercial flexibility service of community based on SOA. The fourth Wuhan International Conference on E-Business, Wuhan, pp. 467–471 (2005)
2. Yu, L., Liu, L.: Comparison and Analysis on E-Commence Recommendation Method in china. System Engineering Theory and Application, pp. 96–98 (2004)
3. Bingqi, Z.: A Collaborative Filtering Recommendation Algorithm Based on Domain Knowledge. Computer Engineering, Beijing 31, 29–33 (2005)
4. Ailin, D.: Collaborative Filtering Recommendation Algorithm Based on Item Clustering. Mini-Micro System, pp. 1665–1668 (2005)
5. Berry, J.A., Linoff, G.: Mastering Data Mining. In: The Art and Science of Customer Relationship Management, Wiley, New York (2000)
6. Qi, L., Qiang, X.: Research on Application of Association Rule Mining Algorithm in Learning Community. CAAI-11, Wuhan, pp. 1458–1462 (2005)
7. Joachims, T.: Text categorization with support vector machines. In: Proceedings of the European Conference on machine learning, pp. 1234–1235. Springer, Heidelberg (2002)

XML Normal Forms Based on Constraint-Tree-Based Functional Dependencies

Teng Lv[1,2] and Ping Yan[1,*]

[1] College of Mathematics and System Science, Xinjiang University, Urumqi 830046, China
yanping@xju.edu.cn

[2] Teaching and Research Section of Computer, Artillery Academy, Hefei 230031, China
lt0410@163.com

Abstract. This paper studies the normalization problem of XML datasets with DTDs as their schemas. XML documents may contain redundant information due to some anomaly functional dependencies among elements and attributes just as those in relational database schema. The concepts of XML partial functional dependency and transitive functional dependency based on constraint tree model and three XML normal forms: 1xNF, 2xNF and 3xNF, are defined, respectively.

Keywords: XML; Functional dependency; Normal form.

1 Introduction

XML (eXtensible Markup Language)[1] has become de facto standard of data exchange on the World Wide Web and is widely used in many fields. XML datasets may contain redundant information due to bad designed DTDs, which cause not only waste of storage space but also operation anomalies in XML datasets. So it is necessary to study normalization problem in XML research field, which is fundamental to other XML research fields, such as Querying XML documents [2], mapping between XML documents and other data forms [3-5], etc. Some schemas for XML documents are proposed, such as XML Schema [6], DTD (Document Type Definition)[7], etc. DTDs are widely used in many XML documents and supported by many applications and product providers. Although normalization theory of relational database has matured, there is no such mature and systematic theory for XML world because XML is new comparing to relational databases, and there are so many differences between relational schemas and XML schemas in structure and other aspects.

Related work. Normalization theory such as functional dependencies and normal forms for relational databases [8-10] can not be directly applied in XML documents as there are significant differences in their structures: relational model are flat and structured while XML schemas are nested and unstructured or semi-structured.

For XML functional dependencies, there are two major approaches in XML research community. The first approach is based on paths in XML datasets, such as

* Corresponding author.

K.C. Chang et al. (Eds.): APWeb/WAIM 2007 Ws, LNCS 4537, pp. 348–357, 2007.

Refs. [11~16]. Unfortunately, they do not deal with tree-structured situation proposed in this paper. The second approach is based on sub-graph or sub-tree in XML documents, such as Ref.[17], but it does not deal with tree-structured situation with some constraint conditions proposed in the paper. Ref. [18] deals with XML functional dependencies with constraint condition, but without specifying what kind of constraint they allowed for. More discussion can be found in Sub-section 3.2.

For XML keys, Refs. [19-21] propose the concept of XML keys based on paths in XML datasets.

For XML normal forms, Refs. [22, 23] propose some XML normal forms based on paths in XML datasets. A normal form called NF-SS [24] for semi-structured data can reduce data redundancies in the level of semantic relationship between entities in real world. A schema called S3-Graph [25] for semi-structured data is also proposed, which can only reduce transitive dependencies partly between elements. The three XML normal forms proposed in this paper are based on constraint-tree-based functional dependencies, which can deal with more general semantic information than previous related XML normal forms. More discussions can be found in Sub-section 3.2.

In this paper, we first give a definition of XML functional dependency based on constraint tree model with formal specified constraint condition. XML keys and other three special functional dependencies such as full functional dependencies, partial functional dependencies, and transitive functional dependencies are also given. Then concepts of three XML normal forms including first XML normal form, second XML normal and third XML normal form are defined based on partial functional dependencies, transitive functional dependencies, and keys. The three XML normal forms proposed in our paper can remove more data redundancies and operation anomalies.

Organization. The rest of the paper is organized as follows. Some notations are given in section 2 as a preliminary work. In Section 3, we first give definitions of XML functional dependencies, full XML functional dependencies, partial XML functional dependencies, transitive XML functional dependencies, and XML keys based on constraint tree model. Then three XML normal forms are proposed to normalize XML documents. Section 4 concludes the paper and points out the directions of future work.

2 Notations

In this section, we give some preliminary notations.

Definition 1[26]. A DTD (Document Type Definition) is defined to be *D*=(*E*, *A*, *P*, *R*, *r*), where (1) *E* is a finite set of element types; (2) *A* is a finite set of attributes; (3) *P* is a mapping from *E* to element type definitions. For each $\tau \in E$, $P(\tau)$ is a regular expression α defined as $\alpha ::= S \mid \varepsilon \mid \tau' \mid \alpha \mid \alpha \mid \alpha, \alpha \mid \alpha^*$, where S denotes string types, ε is the empty sequence, $\tau' \in E$, "|", " , " and " * " denote union (or choice), concatenation and Kleene closure, respectively; (4) *R* is a mapping from *E* to the power set $\mathcal{P}(A)$; (5) $r \in E$ is called the element type of the root.

Example 1. Consider the following DTD D_I which describes the information of some courses, including the name of a course, a pair (a male and a female) taking the course, and an element *community* which indicates if the course is in a course community. We suppose that two courses are in the same course community if the two courses are taken by a same pair, i.e., the two courses have some similarity in aspect of having the same students. Moreover, all courses have this similarity construct a course community.

```
<!ELEMENT courses (course*)>
<!ELEMENT course (pair*,community)>
  <!ATTLIST course name CDATA #REQUIRED>
<!ELEMENT pair (he,she)>
<!ELEMENT he (#PCDATA)>
<!ELEMENT she (#PCDATA)>
<!ELEMENT community (#PCDATA)>
```

We illustrate the structure of DTD D_I in Fig. 1, which just shows the necessary information of DTD for clarity.

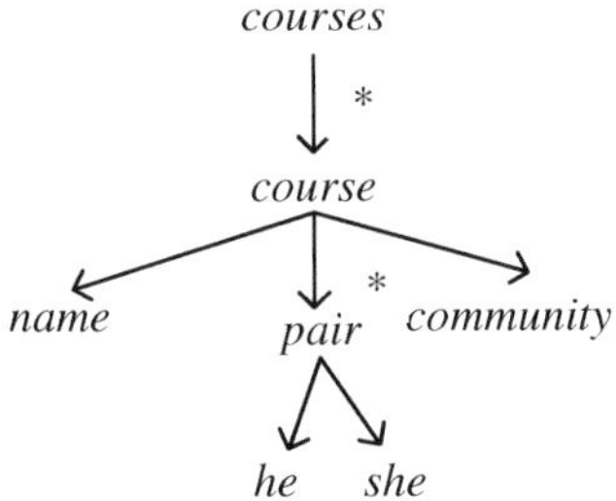

Fig. 1. A tree-structured DTD D_I

A path p in $D=(E, A, P, R, r)$ is defined to be $p=\omega_1.....\omega_n$, where (1) $\omega_1=r$; (2) $\omega_i \in P(\omega_{i-1})$, $i \in [2,\ldots,n-1]$; (3) $\omega_n \in P(\omega_{n-1})$ if $\omega_n \in E$ and $P(\omega_n) \neq \varnothing$, or $\omega_n=S$ if $\omega_n \in E$ and $P(\omega_n)= \varnothing$, or $\omega_n \in R(\omega_{n-1})$ if $\omega_n \in A$. Let $paths(D)=\{p \mid p$ is a path in D $\}$. Similarly, we can define a path in a part of DTD.

Definition 2. Given a DTD D, suppose v is a vertex of D. A v-subtree is a tree rooted on v in D. Similarly, we can define the path in the v-subtree as a part of path started from vertex v. If a v-subtree contains all vertexes which can be reached from root v through all the paths in v-subtree, then it is called a full v-subtree in D.

Example 2. Fig. 2 is a *course*–subtree of DTD D_I. As it contains all paths reached from vertex *course*, it is a full *course*–subtree. Fig. 3 is also a *course*–subtree of DTD D_I. As it does not contains all vertexes reached from vertex *course* (such as vertex *she* through path *course.pair.she*), it is not a full *course*–subtree.

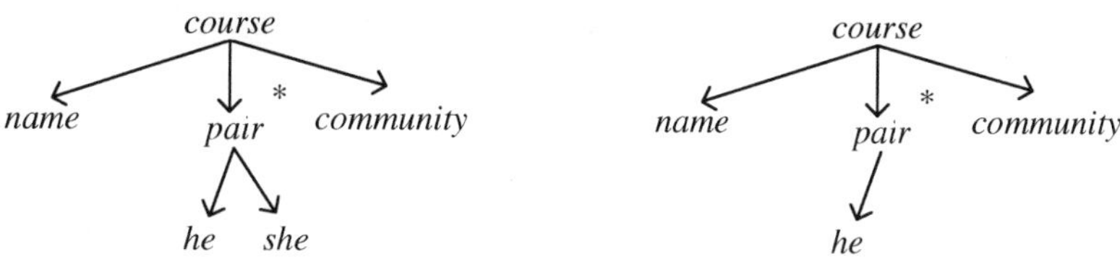

Fig. 2. A full *course*–subtree of DTD D_1 **Fig. 3.** A *course*–subtree of DTD D_1

Definition 3[26]**.** Let D=(E, A, P, R, r). An XML tree T conforming to D (denoted by $T|=D$) is defined to be T=(V, *lab*, *ele*, *att*, *val*, *root*), where (1) V is a finite set of vertexes; (2) *lab* is a mapping from V to $E \cup A$; (3) *ele* is a partial function from V to V^* such that for any $v \in V$, $ele(v)=[v_1, \ldots, v_n]$ if $lab(v_1), \ldots, lab(v_n)$ is defined in $P(lab(v))$; (4) *att* is a partial function from V to A such that for any $v \in V$, $att(v)=R(lab(v))$ if $lab(v) \in E$ and $R(lab(v))$ is defined in D; (5) *val* is a partial function from V to S such that for any $v \in V$, $val(v)$ is defined if $P(lab(v))=S$ or $lab(v) \in A$; (6) $lab(root)=r$ is called the root of T.

Example 3. Fig. 4 is an XML tree T_1 conforming to DTD D_1, which says that there are 4 courses (“c1”, “c2”, “c3” and “c4”).

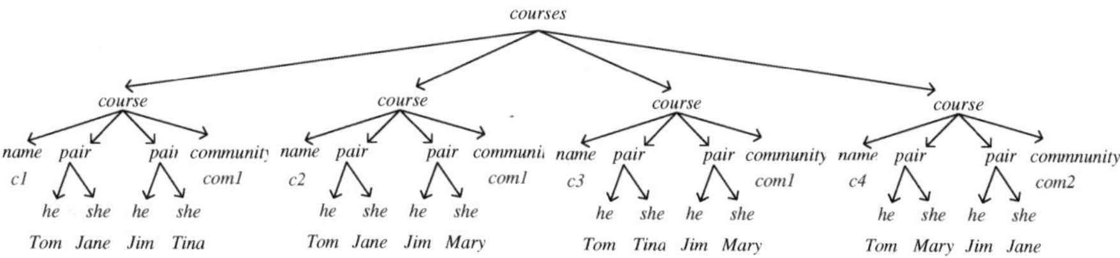

Fig. 4. An XML tree T_1 conforming to DTD D_1

3 Normal Forms for XML Datasets

3.1 XML Constraint-Tree-Based Functional Dependency (xCTFD)

We give the following XML functional dependencies definition based on Ref. [18] with specified constraint condition.

Definition 4. An XML constraint-tree-based functional dependency (xCTFD) has the form $\{v : X \mid_C \rightarrow Y\}$, where v is a vertex of DTD D, X and Y are v-subtrees of D, and C is a constraint condition of X. An XML tree T conforming to DTD D satisfies xCTFD $\{v : X \mid_C \rightarrow Y\}$ if for any two pre-images W_1 and W_2 of a full v-subtree of D in T, the projections $W_1(\Pi_Y) = W_2(\Pi_Y)$ whenever condition C is satisfied, which is defined as the following form:

$\{\exists v'\text{-subtree}_1 = v'\text{-subtree}_2 \mid v'$ is a vertex of DTD, $v'\text{-subtree}_1 \subseteq W_1(\Pi_X)$, and v'-subtree$_2 \subseteq W_2(\Pi_X)\}$.

Of course, more complicated conditions can be defined, but we only consider a specific type of condition *C* in xCTFDs in this paper, which is the most common and useful constraint in real XML applications.

Example 4. There is a xCTFD $\{course : X\ |_C \rightarrow Y\}$ in XML tree T_1 (Fig. 4), where *X* (Fig. 5) is the *course*-subtrees with leaves "he" and "she" of DTD D_1, *Y* (Fig. 6) is the *course*-subtrees with leaves "community" of DTD D_1, and *C* is the condition that there exists a *pair* is equal, i.e.,

$\{\exists\ pair\text{-subtree}_1 = pair\text{-subtree}_2 \mid pair\text{-subtree}_1 \subseteq W_1(\Pi_X)$ and $pair\text{-subtree}_2 \subseteq W_2(\Pi_X)\}$.

Fig. 5. A *course*-subtree *X* of DTD D_1 **Fig. 6.** A *course*-subtree *Y* of DTD D_1

The semantic meaning of the above xCTFD is that the two courses belong to the same course community if there exists a pair are equal in the two courses. The intuitive meaning implied by this xCTFD is that if two courses are taken by a same pair, then the two courses have some similarity in aspect of having the same students. Moreover, all courses have this similarity construct a course community. Let's examine why XML tree T_1 satisfies xCTFD $\{course : X\ |_C \rightarrow Y\}$:

(1) Fig. 7 is the four pre-images of full *course*-subtree of DTD D_1 in XML tree T_1;

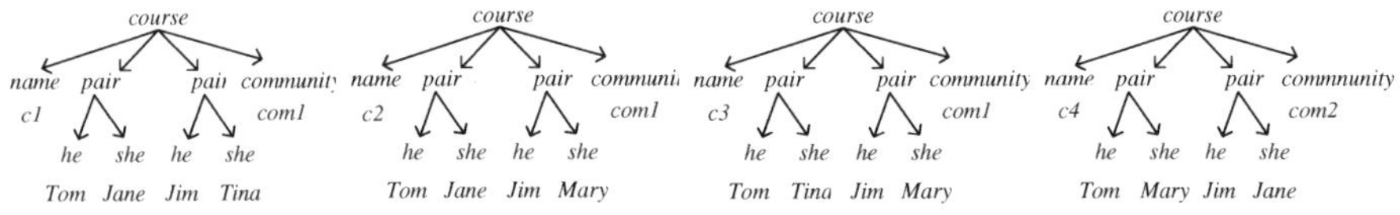

Fig. 7. Four pre-images of full *course*-subtree of DTD D_1 in XML tree T_1

(2) Fig. 8 is the four projections of the four full *course*-subtrees of Fig. 7 on *course*-subtree *X* (Fig. 5.);

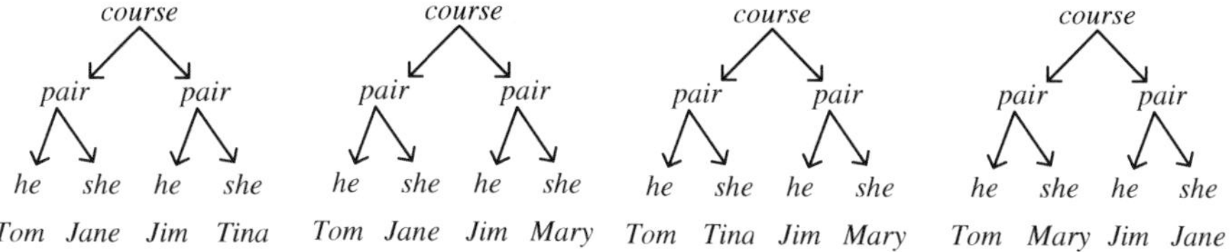

Fig. 8. Four projections of Fig. 7. on *course*-subtree *X* (Fig. 5.)

(3) Fig. 9 is the four projections the four full *course*-subtrees of Fig. 7 on *course*-subtree *Y* (Fig. 6.).

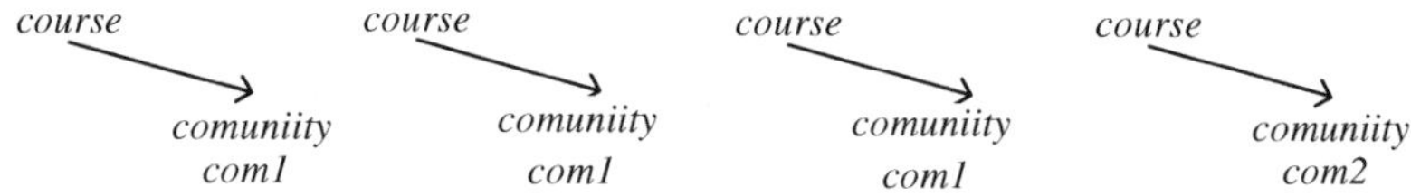

Fig. 9. Four projections of Fig. 7. on *course*-subtree *Y* (Fig. 6.)

We can see that for any two pre-images in Fig. 7, if they are equal on the projections on *X* under condition *C* (note: the former 3 projections on *X* are equal under condition *C* in Fig. 8: for the first and second projections, the first pair of each are equal, and for the second and third projections, the second pair of each are equal), then they are equal on the projections on *Y* (the former 3 projections on *Y* are equal in Fig. 9). By definition of xCTFD, we have xCTFD $\{course: X\,|_C\!\rightarrow Y\}$ in XML tree T_1.

3.2 Discussions

The above xCTFD can not be expressed by earlier related XML functional dependencies. For example, path-based XML FDs [11~16,22,23] can only express the above xCTFD in the following FD form: {courses.course.pair→courses.course.community}, which just says that a pair determines a community. Ref. [17] can only express the above xCTFD in the form {*course*: *X*→*Y*} without condition *C*, which says that all the set of pairs determines a community. Ref. [18] does not specify the formal form of constraint condition *C*. So they do not capture the exact semantic of xCTFD $\{course: X\,|_C\!\rightarrow Y\}$ defined here, which says that a pair in a set of pairs determines a community.

There is a relationship between xCTFD and XFD_2 proposed in Ref. [17], i.e., xCTFD $\{course: X\,|_C\!\rightarrow Y\}$ is equal to XFD_2 if condition *C* is null. We can see that XFD_2 is a more strong XML functional dependency than xCTFD because XFD_2 can not express this kind of functional dependencies in real XML documents. As proposed in Ref. [17], XFD_2 has more expressive power than other related XML functional dependencies proposed in earlier work, the major contribution of our definition xCTFD is the expressive power of XML functional dependencies.

3.3 Normal Forms for XML Datasets

In this section, we will give three normal forms for XML datasets, i.e., the First XML Normal Form (1xNF), the Second XML Normal Form (2xNF), and the Third XML Normal Form (3xNF).

First, we give some related definitions including Full xCTFD, Partial xCTFD, Transitive xCTFD, and XML Key.

Definition 5. xCTFD $\{v : X \mid_C \rightarrow Y\}$ is a Full xCTFD (FxCTFD) if xCTFD $\{v : X' \mid_C \rightarrow Y\}$ does not exist for each $X' \subset X$.

Example 5. xCTFD $\{course : X \mid_C \rightarrow Y\}$ in Example 4 is a FxCTFD as there is no *course*-subtree $X' \subset X$ such that $\{course : X \mid_C \rightarrow Y\}$ is satisfied. Otherwise, there are data redundancies and operation anomalies in the XML tree considering redundant information stored in $X - X'$. This kind of anomaly xCTFD is defined as following:

Definition 6. xCTFD $\{v : X \mid_C \rightarrow Y\}$ is a Partial xCTFD (PxCTFD) if there exist xCTFD $\{v : X' \mid_C \rightarrow Y\}$ where $X' \subset X$.

Another kind of xCTFD can cause data redundancies and operation anomalies by introducing transitive functional dependencies between xCTFDs. This kind of anomaly xCTFD is called Transitive xCTFD (TxCTFD) and defined as following:

Definition 7. xCTFD $\{v : X \mid_C \rightarrow Z\}$ is a TxCTFD if there exist xCTFDs $\{v : X \mid_C \rightarrow Y\}$ and $\{v : Y \mid_C \rightarrow Z\}$, but xCTFD $\{v : Y \mid_C \rightarrow X\}$ does not exist.

The intuitive meaning behind TxCTFD is that xCTFD $\{v : X \mid_C \rightarrow Z\}$ is formed by a transitive relation from $\{v : X \mid_C \rightarrow Y\}$ to $\{v : Y \mid_C \rightarrow Z\}$, but there are no direct relation in $\{v : X \mid_C \rightarrow Z\}$, which is confirmed by the fact that xCTFD $\{v : Y \mid_C \rightarrow X\}$ does not exist in Definition 7.

Definition 8. The key of XML has the form $\{v : Y(X)\}$ where *v* is a vertex of DTD *D*, *X* and *Y* are *v*-subtrees of *D*, and $X \cup Y$ is the full tree rooted on vertex *v*. An XML tree *T* conforming to DTD *D* satisfies key $\{v : Y(X)\}$ if for any two pre-images W_1 and W_2 of a full *v*-subtree of *D* in *T*, the projections $W_1(\Pi_Y) = W_2(\Pi_Y)$ whenever $W_1(\Pi_X) = W_2(\Pi_X)$. If *v* is the root of an XML document or is null, then key $\{v : Y(X)\}$ is simplified as $\{Y(X)\}$, and is called a global XML key which means that the key is satisfied in the whole XML tree ; otherwise, it is called a local XML key which means that the key is satisfied in a sub-tree rooted on the vertex *v*.

From definitions of XML key and xCTFD, it is easy to obtain the relationship between XML key and xCTFD as the following theorem shows:

Theorem 1. An XML tree satisfies an XML key $\{v: Y(X)\}$ iff it satisfies xCTFD $\{v: X\mid_C \rightarrow Y\}$ where condition *C* is null.

As those normal forms in relational database field, we give the definitions of three XML normal forms:

Definition 9. For a given DTD *D* and an XML tree $T \models D$, if the value for each leaf vertex v (i.e. *ele*(v)= $\varnothing$ or $v \in A$) is an atomic value, then *D* is in the First XML Normal Form (1xNF).

1xNF requires that only one value for each element or attribute is stored in an XML document; otherwise some new elements or attributes are needed to store them. For any DTD that is not in 1XNF, it is easy to be converted into 1xNF just by adding new attributes or elements to store additional values. 1xNF is the basic requirement for a DTD to be normalized to obtain the clear semantics, but other more advanced XML normal forms are needed to avoid data redundancy and operation anomalies.

Definition 10. DTD *D* is in the Second XML Normal Form (2xNF) if *D* is in 1xNF, and each xCTFD $\{v: X\mid_C \rightarrow Y\}$ is a FxCTFD if $\{v: Y(X)\}$ is a key and $Y \not\subset X'$ for all keys $\{v: Z(X')\}$.

2xNF removes data redundancies and operation anomalies of XML documents caused by anomaly PxCTFDs.

Definition 11. DTD *D* is in the Third XML Normal Form (3xNF) if *D* is in 2xNF, and each xCTFD $\{v: X\mid_C \rightarrow Y\}$ is not a TxCTFD if $\{v: Y(X)\}$ is a key and $Y \not\subset X'$ for all keys $\{v: Z(X')\}$.

3xNF removes data redundancies and operation anomalies of XML documents caused by anomaly TxCTFDs.

As xCTFD defined in this paper has more general expressive power in XML documents, the three normal forms based on xCTFD defined here have more general significance then earlier related work [22, 23].

4 Conclusions and future work

Normalization theories are important for XML research field, which are fundamental to other related XML research topics such as designing native XML databases. Although we choose DTD rather than other XML schemas as a start point to research XML normal forms, the concepts and methods used in the paper can be generalized to the field of XML Schema just with some changes in formal definitions as there is similarity between XML Schema and DTD in structure.

Although the normal forms of the paper can solve redundancies caused by PxCTFDs and TxCTFDs, there are still other redundancies in XML documents such as those caused by multi-valued dependencies. We plan to research this topic in our future work.

Acknowledgements

This work is supported by Natural Science Foundation of Anhui Province (No. 070412057), Natural Science Foundation of China (NSFC No. 60563001), and College Science & Research Plan Project of Xinjiang (No. XJEDU2004S04).

References

1. Bray, T., Paoli, J., et al.: Extensible Markup Language (XML) 3rd edn. http://www.w3.org/TR/REC-xml
2. Deutsch, A., Tannen, V.: Querying XML with mixed and redundant storage. Technical Report MS-CIS-02-01 (2002)
3. Lv, T., Yan, P.: Mapping DTDs to relational schemas with semantic constraints. Information and Software Technology 48(4), 245–252 (2006)
4. Lee, D., Mani, M., Chu, W.W.: Schema conversion methods between XML and relational models. In: Knowledge Transformation for the Semantic Web, Frontiers in Artificial Intelligence and Applications, vol. 95, pp. 1–17. IOS Press, Amsterdam (2003)
5. Lu, S., Sun, Y., Atay, M., Fotouhi, F.: A new inlining algorithm for mapping XML DTDs to relational schemas. In: ER workshops 2003. LNCS, vol. 2814, pp. 366–377. Springer, Heidelberg (2003)
6. XML Schema Part 0: Primer 2nd edn. W3C Recommendation, http://www.w3.org/TR/, /REC-xmlschema-0-20041028/
7. W3C XML Specification, DTD. (Jun 1998), http://www.w3.org/XML/1998/06/xmlspec-report-19980910.htm
8. Abiteboul, S., Hull, R., Vianu, V.: Foundations of Databases. Reading, MA (1995)
9. Hara, C.S., Davidson, S.B.: Reasoning about nested functional dependencies. In: Proc of ACM Symp on principles of Database Systems(PODS), pp. 91–100. ACM Press, Philadelphia (1999)
10. Mok, W.Y., Ng, Y.K., Embley, D.W.: A normal form for precisely characterizing redundancy in nested relations. ACM Trans. Database Syst 21(1), 77–106 (1996)
11. Vincent, M., Liu, J., Liu, C.: Strong functional dependencies and their application to normal forms in XML. ACM Transactions on Database Systems 29(3), 445–462 (2004)
12. Lee, M.L., Ling, T.W., Low, W.L.: Designing Functional Dependencies for XML. In: Jensen, C.S., Jeffery, K.G., Pokorný, J., Šaltenis, S., Bertino, E., Böhm, K., Jarke, M. (eds.) EDBT 2002. LNCS, vol. 2287, pp. 124–141. Springer, Heidelberg (2002)
13. Vincent, M., Liu, J.: Functional dependencies for XML. In: Zhou, X., Zhang, Y., Orlowska, M.E. (eds.) APWeb 2003. LNCS, vol. 2642, pp. 22–34. Springer, Heidelberg (2003)
14. Liu, J., Vincent, M., Liu, C.: Local XML functional dependencies. In: Proc. of WIDM'03, pp. 23–28
15. Liu, J., Vincent, M., Liu, C.: Functional dependencies from relational to XML, Ershov Memorial Conference, pp. 531–538 (2003)

16. Yan, P., Lv, T.: Functional Dependencies in XML Documents. In: Shen, H.T., Li, J., Li, M., Ni, J., Wang, W. (eds.) Advanced Web and Network Technologies, and Applications. LNCS, vol. 3842, pp. 29–37. Springer, Heidelberg (2006)
17. Hartmann, S., Link, S.: More functional dependencies for XML. In: Kalinichenko, L.A., Manthey, R., Thalheim, B., Wloka, U. (eds.) ADBIS 2003. LNCS, vol. 2798, pp. 355–369. Springer, Heidelberg (2003)
18. Lv, T., Yan, P.: XML Constraint-tree-based functional dependencies. In: (ICEBE2006). Proc. of 2006 IEEE Conference on e-Business Engineering, pp. 224–228. IEEE Computer Society Press, Los Alamitos (2006)
19. Buneman, P., Davidson, S., Fan, W., Hara, C., Tan, W.: Keys for XML. Computer Networks 39(5), 473–487 (2002)
20. Buneman, P., Fan, W., Simeon, J., Weistein, S.: Constraints for semistructured data and XML. ACM SIGMOD Record 30(1), 47–54 (2001)
21. Buneman, P., Davidson, S., Fan, W., Hara, C., Tan, W.-c.: Reasoning about keys for XML. In: Ghelli, G., Grahne, G. (eds.) DBPL 2001. LNCS, vol. 2397, pp. 133–148. Springer, Heidelberg (2001)
22. Lv, T., Gu, N., Yan, P.: Normal forms for XML documents. Information and Software Technology 46(12), 839–846 (2004)
23. Arenas, M., Libkin, L.: A normal form for XML documents. In: Symposium on Principles of Database Systems (PODS'02), Madison, Wisconsin, U.S.A, pp. 85–96. ACM press, New York (2002)
24. Wu, X., et al.: NF-SS: A normal form for semistructured schema. In: International Workshop on Data Semantics in Web Information Systems (DASWIS'2001), Germany, pp. 292–305. Springer, Heidelberg (2002)
25. Lee, S.Y., Lee, M.L., Ling, T.W., Kalinichenko, L.A.: Designing good semi-structured databases. In: Akoka, J., Bouzeghoub, M., Comyn-Wattiau, I., Métais, E. (eds.) ER 1999. LNCS, vol. 1728, pp. 131–145. Springer, Heidelberg (1999)
26. Fan, W., Libkin, L.: On XML Integrity Constraints in the Presence of DTDs. Journal of the ACM (JACM) 49(3), 368–406 (2002)

Untyped XQuery Canonization

Nicolas Travers[1], Tuyêt Trâm Dang Ngoc[2], and Tianxiao Liu[3]

[1] PRiSM Laboratory-University of Versailles, France
Nicolas.Travers@prism.uvsq.fr
[2] ETIS Laboratory - University of Cergy-Pontoise, France
Tuyet-Tram.Dang-Ngoc@u-cergy.fr
[3] ETIS Laboratory - University of Cergy-Pontoise, France
Tianxiao.Liu@u-cergy.fr

Abstract. XQuery is a powerful language defined by the W3C to query XML documents. Its query functionalities and its expressiveness satisfy the major needs of both the database community and the text and documents community. As an inconvenient, the grammar used to define XQuery is thus very complex and leads to several equivalent query expressions for one same query. This complexity often discourages XQuery-based software developers and designers and leads to incomplete XQuery handling.

Works have been done in [DPX04] and especially in [Che04] to reduce equivalent forms of XQuery expressions into identified "canonical forms". However, these works do not cover the whole XQuery specification.

We propose in this paper to extend these works in order to canonize the whole untyped XQuery specification.

Keywords: XQuery evaluation, canonization of XQuery, XQuery processing.

1 Introduction

The XQuery [W3C05] query language defined by the W3C has proved to be an expressive and powerful query language to query XML data both on structure and content, and to make transformation on the data. In addition, its query functionalities come from both the database community, and the text community. From the database languages, XQuery has inherited from all data manipulation functionalities such as selection, join, ordering, set manipulation, aggregation, nesting, unnesting, ordering and navigation in tree structure. From the document community, functions as text search, document reconstruction, structure and data queries have been added.

The XQuery query language is expressed using the famous FLWOR (FOR ...*exp*... LET ...*exp*... WHERE ...*exp*... ORDER ...*exp*... RETURN...*exp*...) expression form. But this simple form is not so simple: thus, any expression *exp* can also be recursively a FLWOR expression but also a full XPath expression.

In Table 1, Query A is a complex XQuery expression that defines a function that selects `books` with constraints on `price`, `keywords` and `comments` and that

K.C. Chang et al. (Eds.): APWeb/WAIM 2007 Ws, LNCS 4537, pp. 358–371, 2007.

returns `price` and `isbn` depending on the number of returned `titles`. This query contains XPath Constraint, Filter, Quantifier, Document construction, Nesting, Aggregate, Conditional and Set operation, Ordering, Sequence and Function.

However, by using XQuery specifications, some expressions are found to be equivalents (ie. give the same result independently of the set of input documents). Thus, the Query B in Table 1 is an equivalent form of the previous Query A.

Table 1. Two equivalent XQuery queries

Query A

```
declare function local:f($doc as xs:string) as element()
{
   for $x in
      (doc("rev.xml")/review|doc("$doc")/catalog) [.
      contains("Robin Hobb")]/book/[.//price > 15]
   where some $y in $x/comments
      satisfies contains ($y, "Excellent")
   order by $x/@isbn
   return
      <book>
         {$x/@isbn}
         <price>{$x//price/text()}</price>
         {
            if (count($x/title) > 2)
            then {for $z in doc("books.xml")/book
               where $z/@isbn = $x/@isbn
               return
                     <title>{($z/title)[3]}</title>}
            else <title/>
         }
      </book>
}
```

Query B

```
declare function local:f($doc as xs:string) as element()
{
   let $l1 := for $f1 in doc("rev.xml")/review
      for $f2 in doc("$doc")/catalog
      return ($f1 | $f2)
   for $f3 in $l1
   for $x in $f3/book
   let $l2 := for $y in $x/comments
      where contains ($y, "Excellent")
      return $y
   let $l3 := orderby ($x, $x/@isbn)
   for $ordered in $l3
   let $l4 := count ($ordered/title)
   let $l5 := for $z in doc("books.xml")/book
      let $l6 := $z/title
      where
         $z/@isbn = $ordered/@isbn
         and $z/position () == 3
      return <title>{$l6}</title>
   where
      contains($f3, "Robin Hobb")
      and $x//price > 15
      and count ($l2) > 0
   return
      <book>
         {$ordered/@isbn}
         <price>{$ordered//price/text()}</price>
         {
            if ($l4 > 2)
            then {$l5}
            else <title/>
         }
      </book>
}
```

XQuery can generate a large set of equivalent queries. In order to simplify XQuery queries studies, it is useful to identify sets of equivalent queries and associate them with a unique XQuery query called : *Canonical query*. This decomposition is used in our evaluation model called TGV [TDL06, TDL07] in which each canonized expression generates a unique pattern tree. This paper aims at allowing all XQuery representation by adding missing canonization rules (not studied in [Che04] and [OMFB02]).

The rest of this paper is organized as follows. The next section describes related works, especially canonical XQuery introduced by [Che04]. Section 3 focuses on our extension of [Che04]'s work to the canonization of the full untyped XQuery. Section 4 reports on validation of our canonization rules and finally, section 5 concludes.

2 Related Work

2.1 GALAX

GALAX [FSC+03] is a navigation-based XQuery processing system. It has first propose a full-XQuery support by rewriting XQuery expression in the XQuery core using explicit operation. The major issue of the navigational approach is to evaluate a query as a series of nested loops, whereas a more efficient evaluation plan is frequently possible. Moreover, the nested loop form is not suitable in a system using distributed sources and for identifying dependencies between the sources.

2.2 XPath

[OMFB02] proposes some equivalence between XPath axes. Those equivalences define XPaths in a single form with child and descendant expressions. Each "*or-self*" axis is bound to a union operator. A "*Parent*" or "*Ancestor*" axis is bound to a new variable with an "*exist()*" function a child/descendant. Table 2 illustrates some canonization of XPath axis.

Table 2. XPath canonization

XPath with specific axis	**Canonized XPath**
for $i in //a/parent::b	for $i in **//b** where **exists ($i/a)**
for $i in //a/ancestor::b	for $i in **//b** where **exists ($i//a)**
for $i in //a/descendant-or-self::b	for $i in //a(**//b | /.**)
for $i in //a/ancestor-or-self::b	for $k1 in **//b** **for $k2 in $k1//a** for $i in **($k1 | $k2)**

2.3 NEXT

Transformation rules suggested by [DPX04] are based on queries minimization of [AYCLS01] and [Ram02] in NEXT. They take as a starting point the *group-by* used in the OQL language, named *OptXQuery*. In order to eliminate redundancies while scanning elements, NEXT restructures the requests more efficiently to process nested queries. We do not take into account those transformation rules since [Che04] proposes transformation rules that create "*let*" clauses (and not a *group by* from OQL).

2.4 GTP

Works on GTP [Che04] propose transformation rules for XQuery queries. Aiming at structuring queries, XQuery queries are transformed in a canonical form of XQuery. The grammar of canonical queries is presented in table 3. This form is more restricted than XQuery specifications, but it allows us to cover a consequent subset of XQuery.

Table 3. Canonical XQuery in GTPs

```
expr ::= ( for $fv1 in range1, ... , $fvm in rangem )?
( let $lv1 := "(" expr1 ")", ... , $lvn := "(" exprn ")" )?
( where φ )?
return <result>
            < tag1 >{arg1}< /tag1 > ... < tagn >{argn}< /tagn >
        < /result>
```

Thus, we obtain a specific syntax that enables us identifying XQuery main properties. These canonized queries must match the following requirements:

- XPath expressions should not contain building filters.
- *expr* expressions are XPaths or canonical XQuery queries.
- Expression φ is a Boolean formula created from a set of atomic conditions with XPaths and constants values.
- Each *range* expression must match the definition of a field of value.
- Each *range* expression is an XPath or an aggregate function.
- Each aggregate function can be only associated to a `let` clause.

In [Che04], it is shown that XQuery queries can always be translated into a canonical form. Lemmas enumerated below show canonical transformation rules.

1. XPath expressions can contain restrictions included in filters (between "[ ]"). With XQuery specifications, those filters can be replaced by defining new variables that are associated with predicate(s) (within the filter) into the *where* clause. Table 4 illustrates a transformation of a filter.

Table 4. Query with filters

XQuery query	**Canonized form**
for $i in doc("cat.xml")/catalog/book [@isbn="12351234"]/title return {$i}	for $j in doc("cat.xml")/catalog/book **for $i in $j/title** where **$i/@isbn = "12351234"** return {$i}

2. A FLWR expression with nested queries can be rewritten into an equivalent expression in which FLWR expressions are declared in *let* clauses. The new declared variable is used instead of the nested query. An example given in table 5 redefined a nested query in the *let* clause: "*let $l: = (...)*", and the return value becomes *$t*.

Table 5. Nested queries transformation

XQuery query	**Canonized form**
for $i in doc("cat.xml")/catalog/book return <book> {for $j in $i/title return {$j}} </book>	for $i in doc("cat.xml")/catalog/book **let $l** := (for $j in $i/title return {$j}) return <book>**{$l}**</book>

3. A FLWR expression with a quantifier "every" can be transformed into an equivalent one using an expression of quantity. XQuery syntax defines quantifiers `every` as a predicate associated to the Boolean formula φ. The quantifier checks if each selected tree verifies the predicate. Table 6 returns all books for which all prices which are strictly higher than 15 euros. In order to simplify and to canonize this query, the "let" clause is created, containing books whose prices are lower or equal than 15 euros. If the number of results is higher than 0, then the selected tree ($i) does not satisfy the quantifier "every" and is not returned.

Table 6. Transformation of a quantifier "every"

XQuery query	Canonized form
for $i in doc("cat.xml")/catalog/book where every $s in $i/price satisfies $s > 15 return {$i}	for $i in doc("cat.xml")/catalog/book **let $l** :=(for $j in $i/price where **$j** <= **15** return {$j}) where **count($l)** = **0** return {$i}

4. In the same way, a FLWR expression, containing a quantifier "some", can be transformed. It is the same transformation, but the tree is selected if there is at least a tree that checks the condition (in the "let" clause).
5. Aggregates functions defined in FLWR expressions can be rewritten in "*let*" clauses, associated to a new variable. This variable replaces the aggregate function at the previous location.

Table 7 shows transformation of a nested query, an aggregate and a filter.

As we can see, rules minimization [DPX04] and canonization [OMFB02] [Che04] helps at transforming XQuery queries into a canonical form. The [Che04] approach is more likely to deal with our needs, but it does not handle: **Ordering**

Table 7. Canonization of a nested query, an aggregate Function and a filter

XQuery query	Canonized form
for $y in doc("rev.xml")/review [. contains ("daulphin")]/book where $y/price > 15 return <result> {$y/@isbn} {$y/price} <nb_titles>{ for $z in collection ("books")/book where $z/@isbn = $y/@isbn return {count ($z/title)} }</nb_titles> </result>	for $x in doc("rev.xml")/review, $y in $x/book let $l₁ := (for $z in collection ("books")/book let $l₂ := count ($z/title) where $z/@isbn = $y/@isbn return {$l₂}) where $x contains ("dauphin") and $y/price > 15 return <result> {$x/@isbn} {$y/price} <nb_titles>{$l1}</nb_titles> </result>

operators, Set operators, Conditional operators, Sequences and **Functions declaration**.

Thus, we propose some more canonization rules in order to handle those XQuery requirements, making it possible to cover a more consequent set of the XQuery queries. Those new canonization rules will allow us to integrate those expressions in our XQuery representation model: TGV [TDL07] (Tree Graph View).

3 Canonisation

As said in the previous section, transformation rules transform a query into a canonical form. Since, it covers a subset of XQuery; we propose to cover much more XQuery queries. Thus, we add new canonization rules that handle all untyped XQuery queries.

In [Che04], five categories of expression are missing: ordering operators, set operators, conditional operators, sequences and function declaration. We thus propose to add canonization rules for each of those expressions.

3.1 Ordering (Order by)

Ordering classifies XML trees according to one or more given XPaths. The order of the trees is given by nodes ordering on values, coming from XPaths. This operation takes a set of trees and produces a new ordered set.

Lemma 3.1 : Ordering

An *XQuery* query containing an `Order By` clause can be transformed into an equivalent query without this clause. It is declared in a `let` clause with an aggregate function `orderby()` whose parameters are ordering fields with XPaths, and the ascending/descending sorting information. The `orderby` function results a set of sorted trees. The new linked variable replaces original used variables into the `return` clause. To keep the XML trees flow, a `for` clause is added on the given variable.

To obtain a canonical query, the `order by` clause must be transformed into a `let` clause. In fact, ordering is applied after `for`, `let` and `where` clauses, and before the `return` clause. Thus, results of preceding operations can be processed by the aggregate function: `orderby()`. This function orders each XML trees with a given XPath. Then, this aggregate function is put into a `let` clause, as specified in the canonical form. The new variable replaces all variables contained into the `return` clause.

Proof: Take a query Q. If Q does not contain an *orderby* clause, it is then canonical (for the order criteria).

Let us suppose that Q has n *orderby* clauses: *order by* $\$var_1/path_1$, $\$var_n/path_n$. Using the transformations lemmas on XPaths, $path_x$ are in a canonical form. The query Q is said to be canonical if the *orderby* clause is replaced by a *let* clause with an aggregate function *orderby*, and each transformed corresponding variable.

It is then necessary to study 3 cases of *orderby* clause:

1. If a variable is declared: *order by* $\$var_1/path_1$ *return* $\$var_1/path_2$, then: *let* $\$t := orderby\ (\$var_1, \$var_1/path_1)$ *return* $\$t/path_2$;
2. If two variables (or more) are declared, but identical: *order by* $\$var_1/path_1$, $\$var_1/path_2$ *return* $\$var_1/path_3$, then: *let* $\$t := orderby\ (\$var_1, \$var_1/path_1, \$var_1/path_2)$ *return* $\$t/path_3$;
3. If two variables (or more) are declared, but different: *order by* $\$var_1/path_1$, $\$var_2/path_2$ *return* $\{\$var_1/path_3, \$var_2/path_4\}$, then: *let* $\$t_1 := orderby\ (\$var_1, \$var_1/path_1)$, $\$t_2 := orderby\ (\$var_2, \$var_2/path_2)$ *return* $\{\$t_1/path_3, \$t_2/path_4\}$.

Then, the $(n+1)^{th}$ *orderby* expressions in query Q can be written with n *orderby* expression, since a query with no *orderby* expression is canonical, then recursively, Q can be written without *orderby* clause.

Here is a example of an *orderby* clause canonization:

Table 8. *Orderby* canonization example

XQuery query	Canonized form
for \$i in /catalog/book order by \$i/title return \$i/title	for \$i in /catalog/book let \$j := orderby (\$i, \$i/title) for \$k in \$j return \$k/title

In table 8, the *for* clause selects a set of *book* elements contained in *catalog*. Then, it is sorted by values of the *title* element, and linked to the \$j variable. The *orderby* clause canonization gives a *let* clause: \$j, whose ordering function *orderby*() takes the variable \$i for the input set, and \$i/title to sort. The result set is then defined into the *for* clause (\$k), in order to build a flow of XML trees. This new variable is used in the *return* clause by modifying XPaths (\$k/title instead of \$i/title).

Then, we obtain a canonized query without *orderby* clauses. This let clause creates a step of evaluation that would be easily identified in the evaluation process.

3.2 Set Operators

Set operators express unions, differences or intersections on sets of trees. It takes two or more sets of trees to produce a single set. A *union* operator gathers all sets of trees, a *difference* operator removes trees of the second set from the first one and an *intersection* operator keeps only trees that exist in the two sets.

Lemma 3.2 : Set Operator

An XQuery query containing a set operator can be transformed into an equivalent query where the expression is decomposed and contains a *let* clause with two canonized expressions. The *return* clause contains the set operator between the two expressions.

Proof: Let's take a query Q. If the query Q does not contain a set operator between two FLWR expressions, then it is known as canonical.

When a query Q contains $n+1$ set operators between two expressions (other than variables), using canonization lemmas, we can say that this expressions are canonical. Let's take ξ, the set operator defined as $\{union, intersect, except\}$ (union, intersection, difference), then the table 9 illustrates the four possibilities of transformation:

Table 9. Transformation of different set expressions

Set expression	Canonized expression	Comments
$(expr_1\ \xi\ expr_2)$	let $\$t_3$:= for $\$t_1$ in $expr_1$ for $\$t_2$ in $expr_2$ return ($\$t_1\ \xi\ \$t_2$)	each expression is defined by a new variable. Those are linked by the operator.
$(expr_1\ \xi\ expr_2)/P$	let $\$t_3$:= for $\$t_1$ in $expr_1$ for $\$t_3$ in $expr_2$ return ($\$t_1\ \xi\ \$t_2$) ... $\$t_3/P$	The expression is broken up. 1) the set operator 2) the expression is replaced by the variable.
$\$XP(P_1\ \xi\ P_2)$	for $\$t_x$ in XP let $\$t_3$:= for $\$t_1$ in $\$t_x/P_1$ for $\$t_2$ in $\$t_x/P_2$ return ($\$t_1\ \xi\ \$t_2$)	A new variable is created. Apply the set operator (rule 1) on the new variable
$\$XP(P_1\ \xi\ P_2)/P_3$	for $\$t_x$ in XP let $\$t_3$:= for $\$t_1$ in $\$t_x/P_1$ for $\$t_2$ in $\$t_2/P_2$ return ($\$t_1\ \xi\ \$t_2$) ... $\$t_3/P$	Use the second and third decomposition rule on set expressions between XP et P_3

Thus, a query Q that contains $n+1$ set operators between two expressions can be rewritten with n set operators. If there are no set operators, it is canonical. Then, recursively, any query Q can be canonized without set operators.

Here a canonization example of a set expression:

Table 10. Canonization of a set expression

XQuery query	Canonized form
for \$i in (/catalog | /review)/book return \$i/title	let $\$i_3$:= for $\$i_1$ in /catalog for $\$i_2$ in /review return ($\$i_1$ | $\$i_2$) for \$i in $\$i_3$/book return \$i/title

In table 10, the *for* clause contains a union "|" between two sets. The first set is */catalog* and the second one */review*. On each one, the *book* element is selected. The title is then projected for each book. The canonization of the union operator (shortened "|") gives a *let* clause ($\$i_3$) containing two expressions $\$i_1$ and $\$i_2$. Each one is defined by a *for* clause on expected paths. The *let* clause $\$i_3$ returns the union of the two variables. Then, the XML trees flow is rebuilt by the *for* clause i_3 on the *book* element. We then obtain a canonized query where set operators are decomposed to detail each step of the procedure.

3.3 Conditional Operators

Conditional operators bring operational processing on XML documents. Indeed, results of conditional operators depend on a given predicate. Then, the first result is returned if the constraint is *true*, the second one else. In the possible results, we can find XPath expressions, nested queries, tags or strings. In the case of nested queries, it is then necessary to canonize them to create a single canonized form.

Lemma 3.3 : Conditional Operators

An XQuery query containing a conditional operator (if/then/else) and a nested query, this one can be transformed into an equivalent query where the nested query will be declared in a clause *let*.

This lemma can be demonstrated in the same way of unnested queries [Che04] (section 2.4). Thus, recursively, we are being able to show that any query containing a nested query in a conditional operator can be canonized.

Here is a canonization example of a query with a conditional operator:

Table 11. Canonization example of conditional operators

XQuery query	Canonized form
for $i in /catalog/book return {if contains ($i/author, "Hobb") then (for $j in $i//title return $j) else ($i/author)}	for $i in /catalog/book let $l := for $j in $i//title return $j return {if contains ($i/author, "Hobb") then ($l) else ($i/author)}

In table 11, a conditional operator is declared in the *return* clause with a constraint on the author's name that must contain the word *Hobb*. If the word is contained, the nested query $j returns the *title*(s) of *book* else the *author* is returned. We obtain a canonized query where nested queries in conditional operators are set in a *let* clause.

3.4 Sequences

Sequences are sets of elements on which operations are applied. Indeed, when a constraint is applied on a sequence using brackets (*XPath*), the constraint is applied on the set of the trees defined by XPath (and not on each one). This operation gathers sets of trees in order to produce a unique set one which we apply the given constraint.

Lemma 3.4 : Sequences

An XQuery query containing a sequence can be rewritten in an equivalent query without sequences. Each sequence is translated in a *let* clause on which operations are put.

Sequences' filters behave like on current XPaths. They applied on results of the sequence. So, the proof is similar to the filter's one in lemma (2.3.1) of [Che04]. Sequences are built by grouping information. Thus any sequence expression is declared in a *let* clause, generating a new variable that could be used in the remaining query.

Table 12. Example of sequences canonization

XQuery query	Canonized form
for $i in (/catalog/book)[2] return $i/title	let $i_1 := for $x in /catalog/book return $x for $i in $i_1 where $i/position() == 2 return $i/title

In table 12, a sequence is defined in the *for* clause. The catalog's book set is aggregated. Then the second book element is selected (and not the second element of each set). Then, its title is projected. The canonization step produces a *let* clause in which the *for* clause is declared on required elements. Then, the new variable is used in the *for* clause $i with a constraint on position. Finally, the title is returned.

3.5 Functions

Function definition is useful to define a query that could be re-used many times, or to define queries with parameters. In XQuery, functions take parameters in input and a single set in output. Inputs and output are typed.

Lemma 3.5 : Functions

An XQuery function containing an XQuery expression can be rewritten in an equivalent function containing a canonical expression.

Table 13. Function transformation

XQuery query	Canonized form

declare function local:section
($i as element()) as element ()* {
for $j in $i/book
return
<book>
{$j/title}
{for $s in $i/section/title
return <section>
{$s/text()}
</section>}
</book>
}
for $f in doc("catalog.xml")/catalog
return local:section($f)
	declare function local:section
	($i as element()) as element ()*
	{
	for $j in $i/book
	let $l := (for $s in $i/section/title
	return <section>
	{$s/text()}
	</section>)
	return
	<book> {$j/title} {$l} </book>
	}
	for $f in doc("catalog.xml")/catalog
	return local:section($f)

In Table 13, a function is defined (*local: section*) with a parameter in input. This input is defined by the *for* clause: *for \$f in doc("catalog.xml")/catalog*, which set of trees will be used in the called function: *local:section (\$f)*. In the function, each *book* element returns its title, and the set of all the titles contained in the sections (*\$/section/title*). As we can see, the function contains a nested query. The unnesting canonization step transforms the query into a canonized form inside the function.

3.6 Canonical XQuery

Thus, using the previous lemmas and those proposed by [Che04], we can cover a broad set of expressions over XQuery. We can now cover: ① XPath expressions with filters ② *for*, *let* and *return* clauses ③ Predicates in the *where* clause ④ Nested queries ⑤ Aggregate functions ⑥ Quantifiers ⑦ Ordering operators ⑧ Set operators ⑨ Conditional operators ⑩ Sequences ⑪ Definition of functions. The only part of XQuery we do not consider yet is typing. Adding typing to the canonized form needs some works using XQuery/XPath typing consideration [GKPS05] on validated XML document.

Table 14 summarizes the additional canonization rules we propose. Those rules allow us to cover all untyped XQuery queries.

Table 14. Proposed canonization rules

	Expressions	Canonical Form
R1	*order by var/xp*	⇒ *let* $\$l_1$:= *orderby*(*var*, *var/xp*)
R2	($expr_1$ union $expr_2$)	⇒ *let* $\$i_3$:= for $\$i_1$ in $expr_1$, $\$i_2$ in $expr_2$ return ($\$i_1$ union $\$i_2$)
	($expr_1$ intersect $expr_2$)	⇒ *let* $\$i_3$:= for $\$i_1$ in $expr_1$, $\$i_2$ in $expr_2$ return ($\$i_1$ intersect $\$i_2$)
	($expr_1$ except $expr_2$)	⇒ *let* $\$i_3$:= for $\$i_1$ in $expr_1$, $\$i_2$ in $expr_2$ return ($\$i_1$ except $\$i_2$)
R3	*if* $expr_1$ *then* $expr_2$ *else* $expr_3$	*let* $\$l_1 := expr_2, \$l_2 := expr_3$ ⇒ *if* $expr_1$ *then* $\$l_1$ *else* $\$l_2$ (if each $expr_2$ and $expr_3$ are nested queries)
R4	$(expr_1)/expr_2$	⇒ *let* $\$l_1 := expr_1$ ... $\$l_1/expr_2$

Using all these rules, we can now deduce that the canonized form of Query A of Table 1 is the Query B of Table 1.

Theorem 3.1 : Canonization

All untyped XQuery queries can be canonized.

With all previous lemmas, we can infer theorem 3.1 that defines a grammar for canonical XQuery queries (Table 15). We can see that canonical queries start with a FLWR expression *Expr* and zero or more functions. The canonical form of *Expr* is composed of nested queries, aggregate functions, XPaths and non-aggregate functions. Moreover, set operators are integrated in these expressions, while the conditional operations are integrated into *ReturnClause*. The *Declaration* has also a canonical form that prevents any nested expressions. XPaths do not contained anymore filters, sequences, nor set operators, since those are canonized.

Table 15. Untyped Canonical XQuery

```
XQuery ::=          (Function)* FLWR;
FLWR ::=            ( "for" "$" STRING " in " Declaration
                       (, "$" STRING " in " Declaration)*
                      | "let" "$" STRING "::=" "(" Expr ")"
                       (, "$" STRING "::=" "(" Expr ")")* )+
                    ("where" Predicate ( ( "and" | "or" ) Predicate )*)?
                    "return " ReturnClause ;
ReturnClause ::=    "{" CanonicExpr "}"
                    | "{" "if" Predicate "then" "(" Expr ")" "else" "(" Expr ")" "}"
                    | "<" STRING ">" ( ReturnClause )* "</" STRING ">" ;
Expr ::=            FLWR | "(" Path SetOperator Path ")" | CanonicExpr | aggregate_function ;
CanonicExpr ::=     Path | non_aggregate_function;
Declaration ::=     "collection" "(' " STRING " ')" (XPath)? | CanonicExpr;
Path ::=            "$" STRING XPath (EndXPath)?;
Predicate ::=       Val Comp Val | QName "(" ( ( Val "," )* Val)? ")";
Comp ::=            ">" | "<" | "=" | "<=" | ">=" | "! =" ;
Val ::=             ' STRING ' | Number | Path;
XPath ::=           ("/" Element | "//" Element)+;
SetOperator ::=     "|" | "-" | "/" ;
EndXPath ::=        "/" ( Attribut | Element | "text()");
Element ::=         ( QName | "." | "..");
Attribute ::=       "@" QName;
QName ::=           (STRING ":")? STRING;
Function ::=        "declare function" QName "(" "$" STRING "as" "element"
                    ("," "$" STRING "as" "element")* ")" "as" "element" "{" FLWR "}";
```

4 Validation

The use cases listed in Table 16 were created by the XML Query Working Group to illustrate important applications for an XML query language. Each use case is focused on a specific application area. Each use case specifies a set of queries that might be applied to the input data, and the expected results for each query. They are designed to cover the most part of XQuery specification.

Table 16. Use cases, number of specified queries, number of supported queries

Use case	Description	Specified	Recognized
"XMP"	Experiences and Exemplars	12	12
"TREE"	Queries that preserve hierarchy	6	6
"SEQ"	Queries based on Sequence	5	5
"R"	Access to Relational Data	18	18
"SGML"	Standard Generalized Markup Language	10	10
"STRING"	String Search	5	5
"NS"	Queries Using Namespaces	8	8
"PARTS"	Recursive Parts Explosion	1	1
"STRONG"	queries that exploit strongly typed data	12	0

Our canonization rules cover 8 of 9 use cases [DPD+05] of the W3C: XMP, TREE, SEQ, R, SGML, STRING, NS, PARTS. The use-cases category not covered by our canonization rules is the STRONG category that queries type information.

We have already implemented the XQuery canonization in our XML-based mediation system: XLive[TD07]. With the canonization, the XQuery processor has been easier to design and implement, and any untyped XQuery can be evaluated with XLive.

5 Conclusion

In this paper, we have extended the works of [OMFB02] and [Che04] in order to recognize the full untyped XQuery specification.

We claim that thanks to our canonization rules, all works that aim to manipulate XQuery could handle the full untyped XQuery specification with only a minimal XQuery subset to recognize. Especially for our TGV model [TDL06, TDL07] which is a simple translation from canonized XQuery queries [TDN07].

Adding typing to the canonized form needs some works using typing consideration [GKPS05] on validated XML document. We are currently working on this issue.

This work has been implemented as a module in the XLive mediation system that evaluates any XQuery query on distributed heterogeneous sources.

Acknowledgement. This work is supported by the ACI Semweb and the ANR PADAWAN projects. The research prototype XLive system is an Open Source software and can be downloaded on: *http://www.prism.uvsq.fr/users/ntravers/xlive*

References

[AYCLS01] Amer-Yahia, S., Cho, S., Lakshmanan, L.V.S., Srivastava, D.: Minimization of Tree Pattern Queries. In: SIGMOD Conf (2001)

[Che04] Chen, Z.: From Tree Patterns to Generalized Tree Patterns: On Efficient Evaluation of XQuery (2004)

[DPD+05] Chamberlin, D., Fankhauser, P., Florescu, D., Marchiori, M., Robie, J.: XML Query Use Cases, september (2005)

[DPX04] Deutsch, A., Papakonstantinou, Y., Xu, Y.: The NEXT Framework for Logical XQuery Optimization. In: VLDB, pp. 168–179 (2004)

[FSC+03] Fernández, M.F., Siméon, J., Choi, B., Marian, A., Sur, G.: Implementing xquery 1.0: The galax experience. In: VLDB (2003)

[GKPS05] Gottlob, G., Koch, C., Pichler, R., Segoufin, L.: The complexity of XPath query evaluation and XML typing. ACM (JACM), 52 (2005)

[OMFB02] Olteanu, D., Meuss, H., Furche, T., Bry, F.: XPath: Looking Forward. In: Chaudhri, A.B., Unland, R., Djeraba, C., Lindner, W. (eds.) EDBT 2002. LNCS, vol. 2490, Springer, Heidelberg (2002)

[Ram02] Ramanan, P.: Efficient Algorithms for Minimizing Tree Pattern Queries. In: ACM SIGMOD, pp. 299–309 (June 2002)

[TD07] Travers, N., Dang, T.-T.: Ngoc. XLive: Integrating Source With XQuery. WebIST (March 2007)

[TDL06] Travers, N., Dang-Ngoc, T.-T., Liu, T.: TGV: An Efficient Model for XQuery Evaluation within an Interoperable System. Interoperability in Business Information Systems (IBIS) (December 2006)

[TDL07] Travers, N., Dang Ngoc, T.-T., Liu, T.: TGV: a Tree Graph View for Modelling Untyped XQuery. Database Systems for Advanced Applications (DASFAA international conference) (April 2007)

[TDN07] Travers, N., Dang-Ngoc, T.-T.: An extensible rule transformation model for xquery optimization. In: The 9th International Conference on Enterprise Information Systems (ICEIS), Madeira, Portugal (2007)

[W3C05] W3C. An XML Query Language (XQuery 1.0) (2005)

Web Search Tailored Ontology Evaluation Framework

Darijus Strasunskas and Stein L. Tomassen

Department of Computer and Information Science,
Norwegian University of Science and Technology, NO-7491 Trondheim, Norway
{dstrasun, steint}@idi.ntnu.no

Abstract. Ontologies are increasingly used in various applications (e.g., semantic interoperability, data integration). In particular, there is a trend in applying ontologies to improve results of search. Quality of ontology plays an important role in these applications. An important body of work exists in ontology quality assessment area. However, there is a lack of task- and scenario-based quality assessment methods. In this paper we propose a framework to assess fitness of ontology for use in ontology-based search. We define metrics for ontology fitness to particular search tasks and metrics for ontology capability to enhance recall and precision.

1 Introduction

In this article we investigate the application of ontology to enhance search tasks. Since information quality is critical for organizations, ontologies are being applied in a number of ontology-based information retrieval systems [1], [2], [3], [4] in order to improve the performance of these systems.

The ontology's ability to capture the content of the universe of discourse at the appropriate level of granularity and precision and offer the application understandable correct information are important features that are addressed in many ontology quality frameworks (e.g., [1], [5], [6], [7]). Both, qualitative and quantitative evaluation approaches are used to evaluate ontologies either as a final ontology specification or during its development. However, ontologies are not fixed specifications but always depend on the context of use. There are many different criteria proposed for ontology evaluation (e.g. [6]), but in order "*to be meaningful and relevant, criteria need to be connected to scenarios of use, and these scenarios to be explained and further analyzed need to be connected to activity models*" [8]. Therefore, the evaluation of the ontology also needs to take into account usage scenarios as well as the behavior of the application.

The literature reports on improvement of web search using ontology-based information retrieval tools (e.g., [4], [9]), as well indicates that inexperienced users find ontology helpful in comprehending domain, familiarizing themselves with the terminology and formulating queries [4], [10]. However, there is no investigation on what ontology features enhance or impair IR performance. Therefore, the objective of this paper is to analyze ontology quality role in information

K.C. Chang et al. (Eds.): APWeb/WAIM 2007 Ws, LNCS 4537, pp. 372–383, 2007.

retrieval (IR). Here we propose a framework for Evaluation of Ontology Quality for Search - the EvOQS framework. The proposed framework is defined by a set of metrics used to stepwise selection of relevant ontologies.

Related work comes from two main areas, information retrieval and ontology quality. Consequently, in Section 2 we discuss ontology-based information retrieval approaches, and in Section 3 - ontology quality evaluation approaches. In Section 4 we define the EvOQS framework and metrics. Finally, in Section 5 we conclude the paper and outline future work.

2 Information Retrieval

An increasing number of recent information retrieval systems make use of ontologies to help the users clarify their information needs and expand users' queries. First, we analyze information needs and typical search scenarios. Second, we summarize ontology based information retrieval (ObIR) methods, taking a closer look what role ontology plays in the methods proposed.

2.1 Information Needs and Search Strategies

There are many studies of users' information needs, their search strategies and behavior (e.g. [11], [12], [13], [14]) resulting in different classification of search strategies. For instance, Guha *et al.* [15] distinguish two different kinds of search, namely, *navigational search* and *research search*. Navigational search is defined as the one where user provides a phrase or keywords and expects to find them in the documents, i.e. the user is using a search engine to navigate to a particular document. While in the research search the user provides a phrase or keywords which are intended to denote object or phenomena about which the user wants to gather information, i.e. the user is trying to locate a collection of documents which will provide required information [15].

Similarly, Rose & Levinson [16] report on three top-level categories of search goals, namely, *navigational*, *informational* and *resource*. Where navigational and informational search goals correspond to the ones identified in [15]. While the last search goal categorizes searches dealing with finding and obtaining a resource, not information, available on Web. Informational and resource search goals are further subdivided into sub-categories [16], such as *locate*, *advice*, *download*, *interact*, etc.

With emergent Semantic Web there is envisioned a shift in IR from retrieval of appropriate Web pages to answering questions without extraneous information [17]. This being separate and important area in information retrieval and knowledge management that requires robust ontology quality, reasoning and fine-grained annotation of documents. However, a precise question answering is the most ambitious information retrieval task, but still inevitable and required feature of web search. Therefore, we consider a fact-finding searches being able to partially substitute question answering on the Web. For this reason, here we adopt classification of search tasks into three categories, such as *fact-finding*, *exploratory*, and *comprehensive* search tasks [11]. In fact-finding, a precise set of results is important, while the amount of retrieved documents is less important. In exploratory search task, the user wants to obtain a general understanding about the search topic, consequently, high precision of the result set is not necessarily the most important thing, nor is high level of recall [11]. Finally, a concern of comprehensive search task is to find as many documents as possible on a given topic, therefore the recall and precision should be as high as possible.

2.2 Ontology-Based Information Retrieval

The basic assumption of ontology-based information retrieval (ObIR) systems is as follows. If a person is interested in information about `B`, its likely that she will find information about `A` interesting, with a condition that `A` and `B` are closely related terms/concepts in an ontology. I.e. ObIR exploits semantic relationships. In the simplest way, user's query is expanded by hypernyms (superclasses)- i.e. generalization [18] or hyponyms (subclasses) - i.e. focalization (more detailed knowledge) [18] or other related concepts (e.g., sibling concept and other neighborhood concepts).

However, there are more sophisticated approaches to ObIR. We classify them into three categories as follows.

Ontology based query expansion (`QE`). These approaches are focusing on using ontologies in the process of enriching queries [3], [19]. There ontology typically serves as thesaurus containing synonyms, hypernyms/hyponyms (as discussed above), and do not consider the context of each term, i.e. every term is equally weighted.

Knowledge Base based ObIR (`KB`). These approaches use reasoning mechanism and ontology querying languages to retrieve instances form Knowledge Base. There, documents are treated either as instances or are annotated using ontology instances [20], [21], [22], [23], i.e. there focus is on retrieving instances rather than documents. Main disadvantages of these approaches are as follows, use of formal ontology querying languages straitens their adoption by inexperienced users; requires annotation of web resources - the process is tedious and results may be misused by the content providers for the purpose of giving the documents a misleading higher ranking by the search engines. These characteristics make the KB based approaches problematic for large-scale web search.

Integrated with vector space model (`VS`). These approaches combine ObIR with already traditional vector space model. Some start with semantic querying using ontology query languages (e.g. SPARQL, RDQL, OWL-QL) and use resulting instances to retrieve relevant documents using vector space model [2], [20], [24], [25]. Where Castells *et al.* [2] use weighted annotation when associating documents with ontology instances. The weights are based on the frequency of occurrence of the instances in each document, i.e. term frequency. Nagypal [24] combines ontology usage with vector-space model by extending a non-ontological query. There ontology is used to disambiguate queries. Simple text search is run on the concepts' labels and users are asked to choose the proper term interpretation.

3 Evaluation of Ontology Quality

An important body of work exists in ontology quality assessment area (e.g., [1], [6], [7]). Most of them aim at defining a generic quality evaluation framework and, therefore, do not take into account specific application of ontologies. For instance, the Ontometric [7] methodology defines Reference Ontology that consists of metrics to evaluate ontology, methodology, language and tool (used to develop ontology) - 117 metrics in total. The OntoQA framework [26] is proposed to evaluate ontologies and knowledge bases. There metrics are divided into two categories: schema metrics and instance metrics. The first category of metrics evaluates ontology design and its potential for rich knowledge representation. The second category evaluates the effective usage of the ontology to represent the knowledge modeled in ontology.

Analysis of the literature shows that ontologies are typically examined according to five aspects: syntax, vocabulary, structure, population of classes and usage statistics. Where *evaluation of syntax* checks whether an ontology is syntactically correct. This quality aspect is most important in any ontology-based application, since syntactic correctness is a prerequisite to be able to process an ontology. Syntactic quality is a central quality aspect in most quality frameworks (e.g., [1], [7]).

Cohesion to domain and vocabulary. Congruence between an ontology and a domain is another important aspect in ontology quality evaluation. There ontology concepts (including taxonomical relations and properties) are checked against terminology used in the domain. In the OntoKhoj approach [27] ontologies are classified into a directory of topics by extracting textual data from the ontology (i.e. names of concepts and relations). Similarly, Brewster *et al.* [28] extracted a set of relevant domain-specific terms from documents. The amount of overlap between the domain-specific terms and the terms appearing in the ontology is then used to measure the fit between the ontology and the corpus. Similar lexical approach is taken in EvaLexon [29] where recall/precision type metrics are used to evaluate how well ontology triples were extracted from a corpus. Burton-Jones *et al.* [1] define a metric called *accuracy* that is measured as a percentage of false statements in the ontology.

Structural evaluation. Structural evaluation deals with assessment of taxonomical relations vs. other semantic relations, i.e. the ratio of `IsA` relationships and other semantic relationships in ontology is evaluated. Presence of various semantic relationships would identify the richness of ontology. In OntoSelect [30] a metric, called *structure*, is used. The value of the structure measure is simply the number of properties relative to the number of classes in the ontology. Similarly, *Density Measure* defined in [31] indicates how well a given concept is defined in the ontology. While *relationship richness* [26] reflects the diversity of relations and placement of relations in the ontology.

Population of classes. This quality aspect is based on instance related metrics. Tartir *et al.* [26] define *class richness* that measures how instances are distributed across classes. The amount of classes having instances is compared with the overall number of classes. *Average population* [26] indicates the number of instances compared to the number of classes. It is used to determine how well the knowledge base has been populated.

Usage statistics and metadata. Evaluation of this aspect focuses on the level of annotation of ontologies, i.e. the metadata about an ontology and its elements. There are defined three basic levels of usability profiling in [6] as follows. *Recognition annotations* take care of user-satisfaction, provenance and versioning information; *efficiency annotations* deal with application-history information; and the last level is about *organizational-design information.* Burton-Jones [1] define similar metrics, namely, *relevance* assesses the amount of statements that involve syntactic features marked as useful or acceptable to the user/agent; *history* accounts for how many times a particular ontology has been accessed relatively to other ontologies. Futhermore, Supekar [32] proposes a framework and a tool for peer-review based ontology evaluation, while Swoogle approach [33] ranks retrieved ontologies based on references between them. Analogical metric to Swoogle's is defined in [1] and is called *authority* - i.e. how many other ontologies use concepts from this ontology.

Table 1. Summary of existing approaches to ontology evaluation

Quality framework	Syntax evaluation	Domain cohesion	Structural evaluation	Population of classes	Usage statistics
AKTiveRank [5]		X	X		
OntoClean [34]			X		
OntoKhoj [27]		X			X
Ontometric [7]	X				
OntoQA [26]			X	X	
OntoSelect [30]			X		
oqval [6]		X			X
Semiotic metrics [1]	X	X			X
Swoogle [33]					X
Other	[35]	[28],[35],[29]	[35]		[32]

Table 1 summarizes ontology evaluation approaches with respect to the five aspects discussed above. In summary, manual evaluation (e.g., [7], [34]) by domain experts or ontology users is difficult to apply in ontology-based information retrieval, since ontologies for a particular search session will be most likely retrieved on-flow, except KB based approaches. In evaluation of a cohesion to domain terminology, a direct match of the vocabulary used to denote concepts in the ontology with a terminology used in text corpora has positive impact on overall ObIR performance. Lexical fit allows better adoption of an ontology, both from user and document collection perspectives. However, that is not vital for every single approach to ObIR. For instance, an approach [25] aligns terminologies by the help of a feature vector constructed for each of the concepts in the ontology based on terms collocation in a document collection. Evaluation of a structural aspect determines richness of ontology, therefore it is important for KB and vector-space model based ObIR.

Consequently, some of the above discussed metrics and criteria are applicable and feasible to assess capability of ontologies to enhance information retrieval. However, there is a lack of a systematic framework to assess fitness of ontologies for a particular search strategy and/or ObIR approach. Adequate optimality criteria should be selected to enable quality estimation of ObIR. These measures should be related to the users' information needs.

4 The EvOQS Framework

There are conducted evaluations of the ontology based IR systems which indicate that ontology based IR systems perform better for more generic queries, i.e. helps to specify queries where users are not familiar with the domain [4], [10]. In such cases, visualization of the ontology is a certain quality of an ObIR system, but that concerns graphical user interface, not ontology itself. In addition, it was found that linguistic enhancements (inclusion of synonyms) close the gap between ontology concepts and document text [9], [10], and enable the ontology to perform better for queries that are required to find only a small number of documents.

The EvOQS (Evaluation of Ontology Quality for Searching) framework including functional steps and assessment criteria is defined in Figure 1. It consists of three steps as follows.

① *Generic quality evaluation*	
Syntactical correctness	Domain fitness

② *Search task fitness*		
Fact-finding	Exploratory	Comprehensive

③ *Search enhancement capability*	
Recall Enhancement	Precision Enhancement

Fig. 1. The EvOQS framework for ontology fitness in information retrieval

Step 1. ***Generic quality evaluation.*** This initial step concerns filtering out poor quality (i.e. syntactically incorrect) and irrelevant ontologies. More detail account on this step is provided in subsection 4.1.

Step 2. ***Search task fitness.*** This step concerns evaluation of ontology fitness for a particular search task. Typical search tasks were discussed in section 2.1. For instance, ratio of taxonomic vs. non-taxonomic relationships is important when selecting an appropriate ontology for exploratory and comprehensive search tasks. For more detail the reader is referred to subsection 4.2.

Step 3. ***Search enhancement capability.*** This final step in our framework concerns evaluating vocabulary of ontologies. Here we account for availability of internal lexical resources in ontologies, i.e. presence of specified synonyms, alternative labels that might potentially be used for a query expansion [36]. More detail account on this step is given in subsection 4.3.

4.1 Generic Quality Evaluation

This step evaluates syntactic correctness and domain fitness. For the syntactic correctness we define a straightforward measure (Eq. 1).

$$SC = \lambda \frac{1}{|E|}. \tag{1}$$

Where, E is the number of error messages generated by a parser, and $\lambda \in \Lambda$ and Λ is a set of OWL sub-language[1] preference weights, i.e. $\Lambda = \{0.0; 0.5; 1.0\}$. For instance, based on a particular implementation of ObIR, OWL DL might be a preferable ontology language, though OWL Lite would be second choice. Correspondingly, an ontology in OWL DL is given a preference weight λ=1.0; OWL Lite, λ=0.5; and OWL FULL, λ=0.0. Furthermore, these coefficients can be related to a particular search task. For instance, an ontology specification in a form of subject hierarchy/taxonomy is enough to support an exploratory search task (for more details, see next subsection), therefore, an ontology specified in OWL Lite is an appropriate for this task.

For the domain fitness sub-step we adopt the AKTiveRank algorithm [5]. Currently, we are experimenting with the algorithm with the purpose to tune *Semantic Similarity Measure* and *Betweenness Measure* [5]. Since these two measures are based on query terms that are supposingly used to retrieve an ontology. In the discussed case [5] only two terms were used, while including more terms it may require substantially more computational resources. Furthermore, it is not clear how these measures will perform when computed for more terms/concepts.

4.2 Search Task Fitness

We have identified three typical search tasks in section 2.1. Here we discuss what ontology features are needed to support these tasks.

Fact-finding. Here, high precision can be achieved by using precise terms or phrases in the query, and typically, by formulating a query consisting of several terms. In order to enhance results in fact-finding search task, provided concepts need to be extended by their instances, phrases matched to triples. Consequently, concepts, their instances and properties are essential here. Information need example: *lens type of fujitsu cameras.*

Exploratory search. Here, the user may find topic-related documents by extending simple keyword-based search with parent- and child-concepts. For instance, extending search *camera* with hypernyms (e.g., *photographic equipment*)

[1] http://www.w3.org/TR/2004/REC-owl-guide-20040210/#OwlVarieties

and hyponyms (e.g., *digital camera*, *SLR camera*, etc.) in order to explore the topic of interest.

Comprehensive search. In order to cover broader-topic in addition to hypernyms and hyponyms, sibling concepts and semantic relationships are included in the query, to cover the most important aspects of the search topic. Continuing the *camera* example, *flash*, *lens*, etc. can be added.

In Table 2 we summarize ontology support necessary to support search tasks as discussed.

Table 2. Search tasks and ontology support

Search tasks		**Ontology support**
Navigational search	Fact-finding	Concepts, their instances, object and datatype properties
Research search	Exploratory	Super- and sub-concepts (taxonomic hierarchy)
	Comprehensive	Super- and sub-concepts, sibling- concepts, object properties

Corresponding to Table 2, we define a coefficient for ontology's Fact Finding Fitness (FFF):

$$FFF = \alpha\frac{|I|}{|C|} + \beta\frac{|OP| + |DP|}{|C|}. \tag{2}$$

Where, I is the number of instances specified in an ontology, OP and DP are OWL constructs `owl:ObjectProperty` and `owl:DatatypeProperty`, correspondingly. Here α, β are adjustment weights. Their purpose is discussed later.

Fitness of ontology for exploratory search task is defined by the Exploratory search task Fitness (EXF) coefficient in Eq. 3.

$$EXF = \frac{|SubC|}{|OP|}. \tag{3}$$

Where, $SubC$ is the number of subclasses defined in an ontology and OP is the number of object properties (i.e. `owl:ObjectProperty`).

Finally, ontology fitness for comprehensive search task is defined by the Comprehensive search task Fitness (COF) coefficient in Eq. 4.

$$COF = \frac{1}{|C|}\sum_{i=1}^{|C|}(\alpha|OP_i| + \beta(|SupC_i| + |SubC_i| + |SibC_i|)). \tag{4}$$

Where, C is the number of concepts, as above. OP_i is the number of object properties for the i concept, and $SupC_i$, $SubC_i$ and $SibC_i$ are amount of super-, sub- and sibling concepts for the i concept, respectively.

4.3 Search Enhancement Capability

In order to improve the result of search, query expansion is typically used, where a query is refined to improve both, recall and precision. Table 3 summarizes a role of main ontology elements (and corresponding OWL constructs) in query expansion. Our aim is to define metrics to assess capability of ontologies to provide lexical resources for enhancement of precision and recall. As it was mentioned above, ontology lexicon could improve recall. Ontology lexicon is a set of lexical entries for the concepts of ontology (synonyms). Each concept is represented by one or more lexical entries that are extracted from the concept name, and synonyms specified by the `rdfs:label` construct.

Table 3. OWL language constructs relevance for IR performance

Search enhancement	**Ontology elements**	**OWL constructs**
Precision	sub-concepts and related concepts	`rdfs:subClassOf`, `owl:intersectionOf`, `owl:unionOf`
	disjoint concepts (*to be used with boolean operator NOT*)	`owl:complementOf`, `owl:disjointWith`
	properties	`owl:ObjectProperty`, `owl:DatatypeProperty`, `rdfs:subPropertyOf`
	instances (*w/ boolean operator NOT*)	`owl:differentFrom`
Recall	super- and sub-concepts, siblings	`rdfs:subClassOf`
	instances	`owl:sameAs`
	synonyms	`owl:equivalentClass`, `rdfs:label`
	related concepts	`owl:intersectionOf`, `owl:unionOf`

Therefore, we define a recall enhancement capability (REC) that shows average amount of synonyms and related terms specified in the ontology (see Eq. 5). The use of this coefficient should be cautious, since certain concepts may have specified several synonyms, and that would wrest the coefficient. Therefore, it might be useful to calculate REC coefficient for every concept in ontology and then calculate the fraction of concepts having specified synonyms.

$$REC = \alpha \frac{|L| + |eC|}{|C|} + \beta \frac{|uO| + |iO|}{|C|}. \tag{5}$$

Where, L=`rdfs:label`, eC=`owl:equivalentClass`, iO=`owl:intersectionOf`, C=`owl:Class`, uO=`owl:unionOf`, and α, β are adjustment weights.

Supplementally, the criterion proposed by [1] can be used for the concepts having no specified synonyms, namely, *interpretability* (whether term appears in WordNet) [1], if concept name exists in WordNet - its synset can be used in query expansion.

A precision enhancement capability (PEC) is defined based on OWL constructs provided in Table 3 as follows (see Eq. 6).

$$PEC = \alpha \frac{|cO| + |dW|}{|C|} + \beta \frac{|uO| + |iO|}{|C|}. \quad (6)$$

Where, iO=`owl:intersectionOf`, cO=`owl:complementOf`, uO=`owl:unionOf`, C=`owl:Class`, dW=`owl:disjointWith`, and α, β are adjustment weights.

However, applicability of the above defined metrics depends a lot on a particular implementation of ObIR. Therefore, we include adjustment weights[2] ($\alpha + \beta$=1) to tailor metrics (by specifying preferable OWL constructs) to a particular implementation. Furthermore, all coefficients are normalized to fall into range [0..1].

5 Conclusions and Future Work

There is a trend to improve information retrieval by the help of ontologies. However, the outputs of the application and its performance in a given task, might be better or worse partly depending on the ontology used. Therefore, evaluation criteria need to be connected to scenarios of use with a purpose to enhance particular search tasks. In this article we have proposed the EvOQS framework to assess ontology fitness and capability to improve ontology-based search. The framework consists of three functional steps that guide in selecting appropriate ontology for a particular search task. In summary, first step filters out syntactically incorrect and irrelevant ontologies. Second step classifies ontologies according their fitness for a particular search task. Whereas the last step classifies ontologies based on their characteristics to enhance recall and precision.

A value of specific ontology properties is very dependent on the actual use of the ontology in ontology-based information retrieval systems. Therefore, a major future work is to conduct a series of experiments that would test utility of the EvOQS framework.

Acknowledgments. This work is funded by the Integrated Information Platform for reservoir and subsea production systems (IIP) project, which is supported by the Norwegian Research Council (NFR), the project number 163457/S30.

References

1. Burton-Jones, A., Storey, V., Sugumaran, V., Ahluwalia, P.: A semiotic metrics suite for assessing the quality of ontologies. Data and Knowledge Engineering 55(1), 84–102 (2005)
2. Castells, P., Fernandez, M., Vallet, D.: An adaptation of the vector-space model for ontology-based information retrieval. IEEE Transactions on Knowledge and Data Engineering 19(2), 261–272 (2007)

[2] See, equations 2, 4, 5 and 6.

3. Ciorascu, C., Ciorascu, I., Stoffel, K.: knOWLer - ontological support for information retrieval systems. In: SIGIR 2003 Conference, Workshop on Semantic Web, Toronto, Canada (2003)
4. Suomela, S., Kekalainen, J.: Ontology as a search-tool: A study of real user's query formulation with and without conceptual support. In: Losada, D.E., Fernández-Luna, J.M. (eds.) ECIR 2005. LNCS, vol. 3408, pp. 315–329. Springer, Heidelberg (2005)
5. Alani, H., Brewster, C., Shadbolt, N.: Ranking ontologies with AKTiveRank. In: Cruz, I., Decker, S., Allemang, D., Preist, C., Schwabe, D., Mika, P., Uschold, M., Aroyo, L. (eds.) ISWC 2006. LNCS, vol. 4273, pp. 1–15. Springer, Heidelberg (2006)
6. Gangemi, A., Catenacci, C., Ciaramita, M., Lehmann, J.: Ontology evaluation and validation. An integrated formal model for the quality diagnostic task. Technical report, ISTC-CNR, Trento, Italy (2005)
7. Lozano-Tello, A., Gomez-Perez, A.: Ontometric: A method to choose appropriate ontology. Journal of Database Management 15(2), 1–18 (2004)
8. Giboin, A., Gandon, F., Corby, O., Dieng, R.: Assessment of ontology-based tools: a step towards systemizing the scenario approach. In: EON2002 workshop. (2002)
9. Aitken, S., Reid, S.: Evaluation of an ontology-based information retrieval tool. In: Gomez-Perez, A., Benjamins, V., Guarino, N., Uschold, M. (eds.) Workshop on the Applications of Ontologies and Problem-Solving Methods, ECAI 2000, Berlin (2000)
10. Brasethvik, T.: Conceptual modelling for domain specific document description and retrieval: An approach to semantic document modelling. PhD thesis, NTNU, Trondheim, Norway (2004)
11. Aula, A.: Query formulation in web information search. In: IADIS International Conference WWW/Internet (ICWI 2003), Algarve, Portugal, IADIS, pp. 403–410 (2003)
12. Fox, S., Karnawat, K., Mydland, M., Dumais, S., White, T.: Evaluating implicit measures to improve web search. ACM Transactions on Information Systems 23(2), 147–168 (2005)
13. Jansen, B., Spink, A., Saracevic, T.: Real life, real users, and real needs: A study and analysis of user queries on the web. Information Processing and Management 36(2), 207–227 (2000)
14. Spink, A., Wolfram, D., Jansen, B., Saracevic, T.: Searching the web: the public and their queries. Journal of the American Society for Information Science 52(3), 226–234 (2001)
15. Guha, R., McCool, R., Miller, E.: Semantic search. In: (WWW2003). 12th International Conference on World Wide Web, pp. 700–709. ACM Press, New York (2003)
16. Rose, D., Levinson, D.: Understanding user goals in web search. In: Proceedings of WWW2004, pp. 13–19. ACM Press, New York (2004)
17. McGuinness, D.: Question answering on the semantic web. IEEE Intelligent Systems 19(1), 82–85 (2004)
18. Bonino, D., Corno, F., Farinetti, L., Bosca, A.: Ontology driven semantic search. WSEAS Transaction on Information Science and Application 1(6), 1597–1605 (2004)
19. Braga, R., Werner, C., Mattoso, M.: Using ontologies for domain information retrieval. In: Proceedings of the 11th International Workshop on Database and Expert Systems Applications, pp. 836–840. IEEE Computer Society, Los Alamitos (2000)

20. Kiryakov, A., Popov, B., Terziev, I., Manov, D.D.: Semantic annotation, indexing, and retrieval. Journal of Web Semantics 2(1), 49–79 (2004)
21. Paralic, J., Kostial, I.: Ontology-based information retrieval. In: Proceedings of the 14th International Conference on Information and Intelligent systems (IIS 2003), Varazdin, Croatia, pp. 23–28 (2003)
22. Rocha, C., Schwabe, D., de Aragao, M.: A hybrid approach for searching in the semantic web. In: Proceedings of WWW 2004, ACM Press, pp. 374–383. ACM Press, New York (2004)
23. Song, J.F., Zhang, W.M., Xiao, W., Li, G.H., Xu, Z.N.: Ontology-based information retrieval model for the semantic web. In: EEE 2005, pp. 152–155. IEEE Computer Society, Los Alamitos (2005)
24. Nagypal, G.: Improving information retrieval effectiveness by using domain knowledge stored in ontologies. In: Meersman, R., Tari, Z., Herrero, P. (eds.) OTM Workshops. LNCS, vol. 3762, pp. 780–789. Springer, Heidelberg (2005)
25. Tomassen, S., Gulla, J., Strasunskas, D.: Document space adapted ontology: Application in query enrichment. In: Kop, C., Fliedl, G., Mayr, H.C., Métais, E. (eds.) NLDB 2006. LNCS, vol. 3999, pp. 46–57. Springer, Heidelberg (2006)
26. Tartir, S., Arpinar, I., Moore, M., Sheth, A., Aleman-Meza, B.: OntoQA: Metric-based ontology quality analysis. In: IEEE Workshop on Knowledge Acquisition from Distributed, Autonomous, Semantically Heterogeneous Data and Knowledge Sources, Houston, TX, USA, pp. 45–53. IEEE Computer Society, Los Alamitos (2005)
27. Patel, C., Supekar, K., Lee, Y., Park, E.: OntoKhoj: A semantic web portal for ontology searching, ranking and classification. In: Proceedings of the Workshop on Web Information and Data Management, pp. 58–61. ACM Press, New York (2003)
28. Brewster, C., Alani, H., Dasmahapatra, S., Wilks, Y.: Data driven ontology evaluation. In: International Conference on Language Resources and Evaluation, Lisbon, Portugal (2004)
29. Spyns, P., Reinberger, M.L.: Lexically evaluating ontology triples generated automatically from texts. In: Gómez-Pérez, A., Euzenat, J. (eds.) ESWC 2005. LNCS, vol. 3532, pp. 563–577. Springer, Heidelberg (2005)
30. Buitelaar, P., Eigner, T., Declerck, T.: OntoSelect: A dynamic ontology library with support for ontology selection. In: Proceedings of the Demo Session at the International Semantic Web Conference, Hiroshima, Japan (2004)
31. Alani, H., Brewster, C.: Ontology ranking based on the analysis of concept structures. In: 3rd Intl. Conf. on Knowledge Capture (K-CAP 05), Banff, Canada, pp. 51–58. ACM Press, New York (2005)
32. Supekar, K.: A peer-review approach for ontology evaluation. In: Proceedings of 8th International Protege Conference, Madrid, Spain (2005)
33. Ding, L., Finin, T., Joshi, A., Pan, R., Cost, R.S., Peng, Y., Reddivari, P., Doshi, V., Sachs, J.: Swoogle: A search and metadata engine for the semantic web. In: Proceedings of the 13th ACM Conference on Information and Knowledge Management (CIKM), pp. 652–659. ACM Press, New York (2004)
34. Guarino, N., Welty, C.: 8. An Overview of OntoClean. In: Handbook on Ontologies, pp. 151–172. Springer, Heidelberg (2004)
35. Supekar, K., Patel, C., Lee, Y.: Characterizing quality of knowledge on semantic web. In: Proceedings of AAAI Florida AI Research Symposium (FLAIRS-2004) (2004)
36. Tomassen, S., Strasunskas, D.: Query terms abstraction layers. In: Meersman, R., Tari, Z., et al. (eds.) OTM 2006 Workshops. LNCS, vol. 4278, pp. 1786–1795. Springer, Heidelberg (2006)

An Overview of the Business Process Maturity Model (BPMM)

Jihyun Lee, Danhyung Lee, and Sungwon Kang

Software Technology Institute, Information and Communications University, Seoul, Korea
{puduli, danlee, kangsw}@icu.ac.kr

Abstract. This paper presents a Business Process Maturity Model (BPMM) for measuring and improving business process competence. The BPMM comprises maturity levels that are associated with the scope of influence of process areas, the capability of monitoring and controlling processes and the influence on process improvement It is based on the principle that any business process essentially consists of activities belonging to four categories; Input, Mechanism, Control, and Output. While constructing our BPMM, we aligned it with the terms, maturity levels, and some elements of Key Process Areas (KPAs) of CMM/CMMI, IS12207, and IS15288. We incorporated the results of the existing researches on Process Maturity Model (PMM) and Process Management Maturity Model (PMMM) and conducted a survey on a group of companies that are actively pursuing Business Process Management (BPM).

Keywords: Business Process, Maturity Model, Process Improvement, Software Engineering.

1 Introduction

Companies today are increasingly trying to manage their processes, customers, suppliers, products, and services as an integrated whole. Additionally, they hope to improve and innovate their business processes in near real-time to cope with market dynamics. The Business Process Management System (BPMS) is considered to be a technical enabler that helps realize such needs of enterprises [6, 11, 15, 21].

Processes help an organization's workforce meet business objectives by helping them work not harder, but more efficiently and with improved consistency. Effective processes also provide a vehicle for introducing and using new technology in a way that best meets the business objectives of the organization [3]. However, a business process is quite different from software or systems in the respect that a business process does not adapt only a project or process focus; it assumes an organizational focus because the business process cannot create business values without aligning with the business strategy of the organization. A Business Process Maturity Model (BPMM) also provides a roadmap for the company and must adapt to the process areas (PA) for different business processes [4, 5]. Therefore, it is essential for a BPMM to adopt an organizational focus, not merely a process focus, and provide a common improvement framework for the company.

K.C. Chang et al. (Eds.): APWeb/WAIM 2007 Ws, LNCS 4537, pp. 384–395, 2007.

The BPMM is a conceptual model that compares the maturity of an organization's current practices against an industry standard. It helps the organization set priorities for improving its product/service (P/S) operations using a proven strategy and developing the capability required to execute its business strategy. Through a BPMM, an organization can efficiently and effectively manage their business process in while trying to achieve and realize its business objectives and values. We can also analyze whether the process meets the needs and expectations of the related stakeholders by reviewing the "as-is" process and performing a BPMM-based gap analysis.

There are several researches on business process maturity [4, 7, 10, 20]; however, business process maturity has not been defined and standardized well enough to be applied to an organization's business process in order to improve its performance [23]. The existing Process Maturity Model (PMM) and Project Management Maturity Model (PMMM) are not business specific or too abstract to apply in practice.

Our BPMM is based on certain generic principles and the terms, the maturity levels, and many elements of the KPAs of CMM/CMMI, IS12207, and IS15288, which are all widely accepted and implemented in various industry sectors [3, 13, 14]. Additionally, our BPMM reflects the existing researches of the PMM and the PMMM, so as to incorporate the fundamental elements of the framework of business process specific maturity. Another key characteristic of the presented maturity model is that it identifies KPAs based on the Input, Mechanism, Control, and Output (IMCO) viewpoint of business processes.

This paper is organized as follows. In Section 2, we describe the existing researches on business processes and business process maturity and explain their limitations. Section 3 presents the approaches to process maturity, such as maturity concepts for business process and validation, in order to represent the maturity model. In Section 4, we discuss the results of the survey that we conducted in order to collect the practitioners' opinions, and thus represent BPMM. In Section 5, we discuss our contributions, conclusions, and future works.

2 Related Works

Among the business process models that were proposed in the past, the PMM presented by Curtis [4, 5] is the only comprehensive model. Curtis coined well-defined and verified CMMI concepts and introduced them into the field of business process studies. In his study, KPAs are categorized into service operations support, service operations work performance, service operations management, organizational process improvement, and organizational management according to role responsibility. However, unfortunately, Curtis's PMM leaves room for improvement because it relies excessively on experts' intuition, and unclear definitions caused due to this could lead to a misunderstanding among practitioners. In addition, there are no details of activities and tasks for each field of practice, which makes the model difficult to use practically.

On the other hand, Fisher considered business process as a process having multi-dimensional and non-linear characteristics, unlike the software project/system life cycle [7]. Fisher defined actions on the grounds that PMM is represented as five levers of change and five states of process maturity. Due to its high level of abstraction on actions, Fisher's model only provides ends, with no means to these ends. The model could be improved by incorporating suggested actions and achieving capability in a progressive manner.

The PMM described by Harmon regards all the core and support processes as a value chain, starting from the resource right up to the final product [10]. It also provides a checklist for accessing organization/process maturity. Although Harmon's approach shows the need for including values in PMM, his maturity model does not provide the means to achieve these values. In addition, it is a heuristic and informal approach that assesses the maturity level based on just a few checklists.

Smith introduces the PMMM [20]. He insists that the PMMM should be taken into consideration because process management maturity has an orthogonal relationship with process maturity. Smith also assumes that CMM can be used as a BPMM. However, CMM/CMMI is not suitable as a BPMM because of the differences between the context of software and business process.

Rosemann describes BPMM as a three dimensional structure that consists of Factor (IT/IS, Methodology, Performance, Accountability, Culture and Alignment), Perspective (Align, Design, Execute, Control and Improve), and Organization Scope (includes time and area, the entity to which the model is applied, one dimension location, a division, a business unit or a subsidiary) [18,19]. Rosemann's model has an advantage over other models in that it is supported by surveys and case studies. However, Rosemann's model is an unorganized and complex three dimensional structure.

The 8 Omega Framework of BPMG encompasses the four dimensions of Strategy, People, Process, and Systems; the framework applies DADVIICI (Discovery, Analysis, Design, Validate, Integrate, Implement, Control, and Improve) to all four dimensions [1]. The framework is simple and intuitive but it does not have any principles or guidelines for its application.

Further, the possibility of applying CMM to PMM is currently being studied and tested by experts from industry and academia [12]. In this section on related works, it becomes evident that many researchers attempted to construct a PMM based on CMM/CMMI; else, they based the construction on the principles of software engineering.

3 Our Approach to Constructing the Business Process Maturity Model

In this section, we describe our approach to the construction of our BPMM. The architectural construction of the BPMM has two stages. In the first stage, we structure the BPMM so that it has five layers of maturity levels where each level is associated with the scope of the influence of PAs, the capability of monitoring and controlling processes, and the influence on process improvement.

In the second stage, we identify key business process areas with the viewpoint that any business process essentially consists of four kinds of generic IMCO activities—Input, Mechanism, Control, and Output. In order to produce a product or provide a service, a company consumes various resources. A company also requires a mechanism that turns the resources into products and services that will then be provisioned to the customer. In order to ensure that these three activities are performed effectively and efficiently, they should be monitored and controlled. Thus, according to this viewpoint, a particular KPA would either belong to one of the four generic activities or it would be a cross-activity process that has aspects of more than one activity.

While the architecture of BPMM is designed as stated above, we align the components and contents of BPMM to the greatest extent possible with the terms, maturity levels, and many elements of KPAs of CMM/CMMI, IS12207, and IS15288, which are all widely accepted and implemented in various industry sectors. We also incorporate the results of the existing researches on PMM and PMMM into the BPMM.

3.1 Five Process Maturity Levels

The BPMM has a five-level structure like the CMMI and the existing PMM. The five-level structure is widely used in many reference models and has an advantage in that it is comprehensible and practical. Some CMMI books define maturity based on predictability, control, and effectiveness [10]. Certainly, these are important aspects that should be reflected when defining maturity levels. We generalize and extend the characteristics of the maturity levels to include concepts such as the scope of influence of PAs measurement & analysis, monitoring & control, and organizational process improvement activities. Table 1 shows the result of defining maturity levels of BPMM accordingly based on the following elements: "Focus of KPA," "Measurement & Analysis," "Monitoring & Control," and "Organizational Process Improvement." As the maturity level increases, capabilities in these aspects are escalated, thereby making them useful in determining business process maturity levels. Organizational process[1] in process capability hierarchy for maturity levels is a defined process[2] that is executed in an organization for producing P/S.

Table 1. Characteristics of Business Process Maturity Levels

	Level 2	Level 3	Level 4	Level 5
Focus of KPAs	Work unit (product focus)	Organization-wide (product focus)	Organization-wide (product & process focus)	Organization-wide (competitive advantage focus)
Measurement & Analysis	Black-box with control points	Gray-box (all process areas)	White-box (statistically analyzed)	White-box (statistical predictability)
Control	Reactive	Reactive/Adaptive	Adaptive/Proactive	Proactive
Influence on Process Improvement	Partially controlled	Controlled	Partially systematic	Systematic

[1] In CMMI, an organizational process is an organizational standard process.

[2] Defined process is a managed process that is tailored from the organization's standard process, based on the organization's tailoring guidelines [3].

Table 2 shows the process capability hierarchy that is defined according to the characterization of process maturity levels by dividing the organizational process into IMCO aspects, which will be introduced shortly.

Table 2. Concepts for Business Process Maturity

Level	Characteristics	Concepts
Optimizing	• Monitoring and controlling process performance in a proactive way • Systematically using process performance data to improve and optimize process	input <<org. process>> output mechanism
Quantitatively Managed	• Measuring process performance quantitatively • Systematically controlling process performance • Using performance data in an ad hoc manner for process improvement	input <<org. process>> output mechanism
Defined	• Defining process • Measuring process and mechanism performance for overall organization • Monitoring and controlling process performance for overall organization • Using the partial performance data only in an ad hoc manner for process improvement	input <<org. process>> output mechanism
Managed	• Not defining or partially defining process • Measuring process performance partially • Monitoring and controlling process performance for a work unit • Unable to use performance data for process improvement	input <<org. process>> output mechanism
Initial	• Ad hoc manner	

⇧ process is monitored in an ad hoc manner ⬆ process is monitored in a systematic manner.

⇩ process is controlled in an ad hoc manner ⬇ process is controlled in a systematic manner.

process is systematically observed and the results are used to control process in a proactive manner.

input/process/output/mechanism standards are defined and managed for each work unit.

organization-wide standards for input/process/output/mechanism are defined and managed.

control degree (black/gray/white box).

Standards, process descriptions, and procedures are defined according to process instance in CMMI Level 2, while the set of standard processes of an organization can be tailored to fit the purpose of the organizational unit within the allowance of the tailoring guideline in CMMI Level 3 [3]. On the other hand, because business process usually comprises cross multiple work units, the distinction between Level 2 and Level 3 depends on whether the focus of the KPAs is on business work units or is organization-wide. Specifically, Level 3 focuses on the P/S in a manner similar to Level 2 but it differs from Level 2 in that it manages the P/S life cycle. Level 4 implements in-depth process measurement for improving P/S quality, not just control for the sake of process. Moreover, Level 4 defines PAs for managing P/S quality quantitatively, considering specificities of the business sector as well as the quantitative management on process performance. Similarly, in BPMM, we separately define KPAs for process and P/S in order to give equal focus to each of them. In the case of Level 5, KPAs for proactive improvement are added because the prompt response to the market environment change is important in the business sector [2, 8].

3.2 Generic PAs Based on IMCO

The previous PMM/PMMM does not provide clear rationale for the derivation of process areas (PAs). However, in this research, we explicitly discuss the rationale behind the BPMM structure by showing the procedure for deriving the basic activities of the business process in the maturity levels from the perspective of IMCO.

Integration Definition for Function Modeling provides a means for completely and consistently modeling the functions (activities, actions, processes, operations) required by a system or enterprise. It can be used when a modeling technique for the analysis or development of a system is required [16]. We use the ICOM concept of IDEF0 to model the necessary activities for executing business process. IMCO[3] framework maps contain the essential PAs of conducting business processes with regard to the four perspectives of IDEF0. However, the IMCO concept is used for deriving necessary activities according to the maturity concepts shown in Tables 1 and 2, and not for decomposing functions of the IDEF0. Throughout these mapping, we attempted to develop a more complete BPMM that includes all the necessary PAs of the business processes. For this purpose, each quadrant of the IMCO framework is redefined as follows:

[3] We term the ICOM of IDEF0 as "IMCO framework" in that the Mechanism transforms the Input into Output.

- **Input Quadrant:** PAs for providing/managing inputs (money, material, acquisition, baseline, etc.) and business processes that are performed early in the P/S production, which is a thorough preparation for successful businesses
- **Mechanism Quadrant:** the means used to produce P/S (PAs for providing means (tools, man) to transform inputs to P/S)
- **Control Quadrant:** the PAs required to monitor and conduct statistical analyses for producing correct P/S
- **Output Quadrant:** the PAs used to deliver and maintain produced P/S

In this study, we derive KPAs in the same way that we analyzed the general IMCO of activities for producing and map these KPAs to IMCO quadrants. All the companies perform the functions of IMCO to produce P/S. Companies produce P/S from man, money, and material (3M) via organizational processes and control for improving productivity and P/S quality. It is possible that there are more activities than these, but most of the activities are performed in an ad hoc manner. Further, most outcomes are achieved as a result of the ad hoc activities of individuals. P/S requirements gathering, P/S development, and P/S provisioning are the main activities of the Initial Level (Level 1); however, these are not planned and standardized processes.

PAs are not included at the Initial Level since they are basic production mechanisms that are common to all business processes. In contrast, at the Managed Level (Level 2), generic PAs related to inputs are derived as a result of analyzing IMCO functionalities, according to the maturity concepts presented by Tables 1 and 2. This is because we should manage Inputs preferentially for producing P/S that is compliant with the organizational business goal [Fig. 1].

Various management activities for IMCO appear at the Managed Level. Essentially, the initial I/M/O activities should be managed. At the Managed Level, activities of the Mechanism quadrant are not defined and managed on an

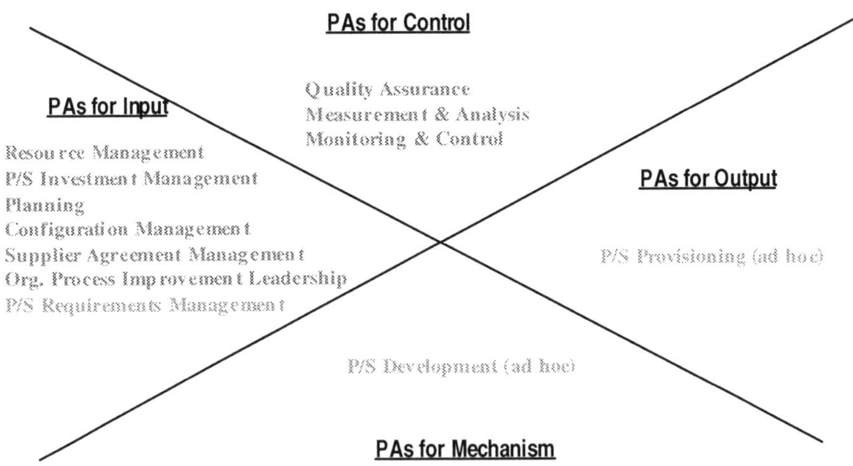

Fig. 1. PAs for the Managed Level

organization-wide level. In addition, the Control quadrant contains Quality Assurance, Measurement & Analysis, and Monitoring & Control. In the Input quadrant, P/S Requirements Management, Supplier Agreement Management, Planning, and Resource Management are obvious. Configuration Management is assigned to the Input quadrant because the focus is on producing and releasing the baselines of configuration items rather than changing management activity. Investment management for P/S is also allocated to the Input quadrant because here, the focus is on assessing business opportunities and making decisions and commitments rather than concluding whether the investment on the organizational process should be continued or not.

At the Defined Level (2nd Layer in Fig. 2), organizational processes are defined and standardized. In other words, P/S lifecycles such as input, mechanisms that transform inputs into outputs, output management, and control activities are defined. At Level 3, the organizational process becomes defined and organizational support for the process is provided. Additionally, reuse is emphasized and common assets are explicitly managed.

At the Quantitatively Managed Level (3rd Layer in Fig. 2), mechanism and output activities are managed quantitatively, and the controls on them are enacted by integrating organizational process and P/S.

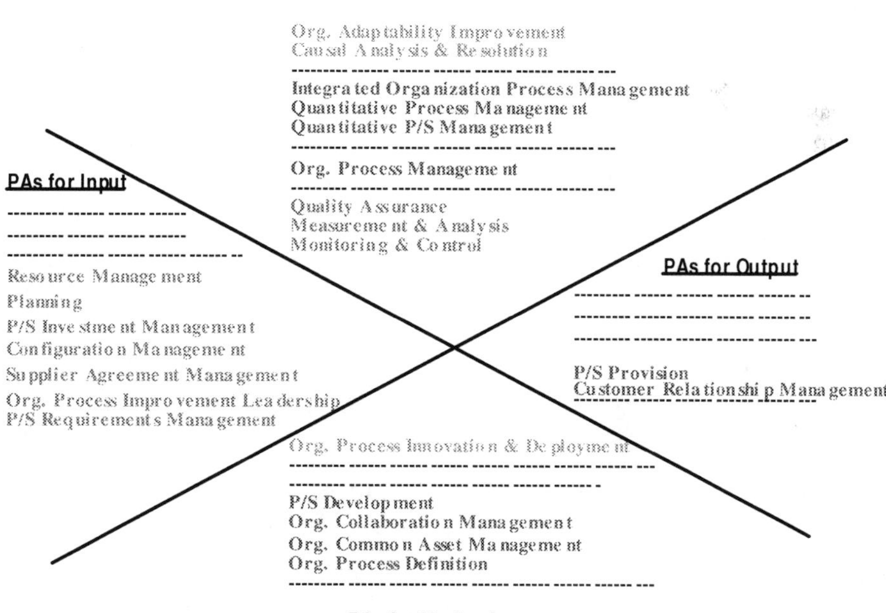

Fig. 2. PAs for All Maturity Level

At the Optimizing Level (4th Layer in Fig. 2), the control is conducted proactively, and organization- wide and process innovation and improvement are implemented.

We also reorganized a referenced model based on P/S and analyzed survey results in business process or P/S operations to help an organization. Moreover, the BPMM structure is evaluated via survey and the IMCO framework in order to ensure that it reflects the principles of generic processes and business process characteristics.

Initially, we ensured that our BPMM complied with models proposed by past researchers [5, 10, 12, 13, 14, 18, 20]. Subsequently, in order to confirm that business characteristics are reflected in the BPMM, we mapped the business PAs in correspondence with the value chain [17] of a manufacturing company [9]. By doing this, we attempted to check whether the BPMM structure was complete, which would imply that the model included all the necessary activities.

4 The Business Process Maturity Model

4.1 Analysis of Survey

We conducted a survey to extract those characteristics of the maturity model that are suitable for the business sector. While doing the survey, it was difficult to frame appropriate questions to obtain relevant and useful information in order to conduct a meaningful analysis. Since many questions may discourage respondents from answering a question, it is imperative that the number of questions be organized in an easy-to-answer structure that facilitates the retrieval of quantitative and qualitative data. The survey was developed to define the organization's needs regarding the implementation and operation of the BPMM system. Survey questionnaires were completed by 12 individuals in four different business sectors, including the manufacturing industry, financial businesses, service industry, and government agencies. Since participation in the survey was voluntary, it was difficult for us to gather a complete data set. However, a significant amount of relevant and useful information was obtained.

We used certain results of the survey as input for constructing the BPMM. The survey revealed important market needs, such as the necessity for improved leadership, customer relations, and separation of P/S from process management. In addition, the ability to monitor and control the business process and calibrate business goals, and Key Performance Indicators (KPIs) were considered as important. These survey results revealed the need to develop capabilities, such as the construction of a Business Process Management Group (BPMG), process management, P/S life cycle management, customer relationship management, common asset management, and adaptability management.

4.2 KPAs of the BPMM

Table 3 exhibits the resulting structure of the BPMM based on the principles and approaches presented in Section 3.

Table 3. KPAs of the BPMM

Level	Focus	KPAs (Key Process Areas)
Level 5: Optimizing	Proactive Process Improvement (Product & Process Focus)	Causal analysis & Resolution Org. Process Innovation & Deployment Org. Adaptability Improvement
Level 4: Quantitatively Managed	Quantitative Management (Product & Process Focus)	Integrated Organizational Process Management Quantitative Process Management Quantitative Product & Service Management
Level 3: Defined	Organizational Process Standardization (Product Focus)	Org. Process Definition Org. Process Management Org. Collaboration Management Product & Service Management Product & Service Development Product & Service Provision Org. Common Asset Management Customer Relationship Management
Level 2: Managed	Work Unit Process Management (Product Focus)	Org. Process Improvement Leadership P/S Requirement Management Planning Monitoring & Control Configuration Management Quality Assurance Measurement & Analysis Resource Management P/S Investment Management Supplier Agreement Management
Level 1: Initial	Ad-hoc	Ad-hoc

5 Conclusions and Future Works

In this research, we presented BPMM that has characteristics of practices of the business sector and is in compliance with CMM/CMMI. It was designed such that it could overcome the limitations of the existing software-oriented maturity models by reflecting the voice of the market and the engineers. We derived and checked generic PAs based on the IMCO framework and the maturity concept principles in order to guarantee that all business activities are included within the defined KPAs. In particular, we presented a BPMM reflecting the business characteristics based on terms and some concepts of CMM/CMMI that have already been verified in various fields (IS15288 for system life cycle process and the existing PMM/PMMM). In a similar manner, we mapped the value chain, applied practically in a manufacturing organization to check whether or not the BPMM is business process-oriented.

Consequently, we assure that our BPMM is designed to reflect practical business activities.

With the help of the BPMM, companies should be able to analyze the strengths and the weaknesses of their current business processes and develop "to-be" models to achieve the organization's business objectives. Moreover, BPMM should guide companies to achieve business objectives by executing the practices of the model. The BPMM presented in this study is distinct in the following way:

- As a result of conducting a survey on the execution of BPMM in organizations, we grasped the organizations' current business processes, the purpose of executing BPM, and the direction of business process improvement. Additionally, we included business processes being valued in a real organization.
- KPAs that include flavored software and systems processes with business process specific characteristics and covering the existing PMM/PMMM.
- Defining KPAs derived from IMCO framework and not from empirical intuition.
- Reflecting the opinions of experts such as practicing BPMS engineers and consultants.
- Clarifying the difference between maturity levels clearly, unlike the case with established maturity models.

As Spanyi insists, a business process maturity model has to focus on competence [22]. In order to apply our BPMM practically, we need to improve the model in order to provide assurance for equivalent competence to maturity levels. The presented BPMM is currently a prototype and has not yet been applied. However, we conducted a survey for the organizations executing BPM to validate our BPMM, and we established a plan to apply the model to the selected organizations. In the future, we are going to refine BPMM as best practices based on the feedback.

References

1. The 8 Omega Framework, http://www.bmpg.org
2. Burlton, R. In: search of BPM excellence, BPM: From Common Sense to Common Practice, Meghan-Kiffer Press (2005)
3. Chrissis, M.B., Konrad, M., Shrum, S.: CMMI: Guidelines for Process Integration and Product Improvement. Addison Wesley, London, UK (2003)
4. Curtis, B.: Overview of the business process maturity model, San Antonio SPIN (2004)
5. Curtis, B., Alden, J.: The business process maturity model: An overview for OMG members, (2006)
6. Elzinga, D.J., Horak, T., Lee, C., Bruner, C.: Business process management: Survey and methodology. IEEE Transactions on Engineering Management 42(2), 119–128 (1995)
7. Fisher, D.M.: The business process maturity model: A practical approach for identifying opportunities for optimization, BP Trends (2004)
8. Fisher, D.M.: Getting started on the path to process-driven enterprise optimization, BP Trends (2005)
9. Garretson, P.: How Boeing A&T manages business processes, BP Trends (2005)
10. Harmon, P.: Evaluating an organization's business process maturity, BP Trends (2004)

11. Harrington, H.J.: Business Process Improvement: The Breakthrough Strategy for Total Quality, Productivity, and Competitiveness. McGraw-Hill, NY (1991)
12. Ho, D.: Business process maturity model–A CMM-based business process reengineering research, Center of Excellence in Business Process Innovation, Stevens Institute of Technology
13. ISO/IEC 12207: 1995 (ISO/IEC 12207) Standard for Information Technology Software Life Cycle Processes (1995)
14. ISO/IEC 15288: 2002 (ISO/IEC 15288) Standard for Information Technology System Life Cycle Processes (2002)
15. Khan, R.N.: Business Process Management: A Practical Guide, Meghan-Kiffer Press (2004)
16. NIST, Integration definition for function modeling (IDEF0), Draft Federal Information Processing Standards Publication 183, National Institute for Standards and Technology, Washington D.C
17. Porter, M.E.: Competitive Advantage: Creating and Sustaining Superior Performance. Free Press, NY (1985)
18. Rosemann, M., de Bruin, T.: Application of a holistic model for determining BPM maturity, BP Trends (2005)
19. Rosemann, M., de Bruin, T.: Towards a business process management maturity model, Proceedings of the 13th European Conference on Information Systems (2005)
20. Smith, H., Fingar, P.: Process management maturity models, BP Trends (2004)
21. Smith, H., Neal, D., Ferrara, L., Hayden, F.: Business process management, Business Process Management Summit (2001)
22. Spanyi, A.: Beyond process maturity to process competence, BP Trends (2004)
23. Watson, G.H.: The benchmarking workbook: Adapting best practices for performance improvement. Productivity Press, NY (1992)

Process Mining: Extending α-Algorithm to Mine Duplicate Tasks in Process Logs

Jiafei Li, Dayou Liu, and Bo Yang

College of Computer Science & Technology JiLin University,
Changchun, 130012, China
Key Laboratory of Symbolic Computation and Knowledge Engineering of Ministry of Education, JiLin University, Changchun, 130012, China
{jiafei, liudy, ybo}@jlu.edu.cn

Abstract. Process mining is a new technology which can distill workflow models from a set of real executions. However, the present research in process mining still meets many challenges. The problem of duplicate tasks is one of them, which refers to the situation that the same task can appear multiple times in one workflow model. The "α-algorithm" is proved to mine sound Structured Workflow nets without task duplication. In this paper, basing on the "α-algorithm", a new algorithm (the "α*-algorithm") is presented to deal with duplicate tasks and has been implemented in a research prototype. In eight scenarios, the "α*-algorithm" is evaluated experimentally to show its validity.

Keywords: Process mining, workflow mining, duplicate tasks, Petri nets, workflow nets.

1 Introduction

During the last decade workflow management concepts and technology have been applied in many enterprise information systems [1,7,10]. These systems usually require formal models of business processes to start the application. However, it is a time-consuming and error-prone task to acquire workflow models and adapt them to changing requirements, because knowledge about the whole process is usually distributed among employees and paper procedures. As indicated by many researchers, this requirement makes the workflow management systems inflexible and difficult to deal with change [1,10]. Process mining can be seen as a technology to contribute to this which aims at extracting a structured process description from a set of real executions recorded in a workflow log. And it can be viewed as a three-phase process: pre-processing, processing and post-processing [8].

Data mining is the name given to the task of discovering information in data, which provide a stable foundation for process mining [10]. Different data mining methods can target different kind of data, such as relation database, images, time series and sequence data. Process mining handles the data which is the information recorded in the event logs and belongs to sequence data. Information systems using transactions (such as ERP, CRM and SCM) can provide such kind of data. The goal

K.C. Chang et al. (Eds.): APWeb/WAIM 2007 Ws, LNCS 4537, pp. 396–407, 2007.

of process mining is to distill information about processes from event logs which record every event that occurred during workflow process execution. The event here refers to a task in a workflow instance and all events are totally ordered. The framework of process mining is depicted in Figure 1.

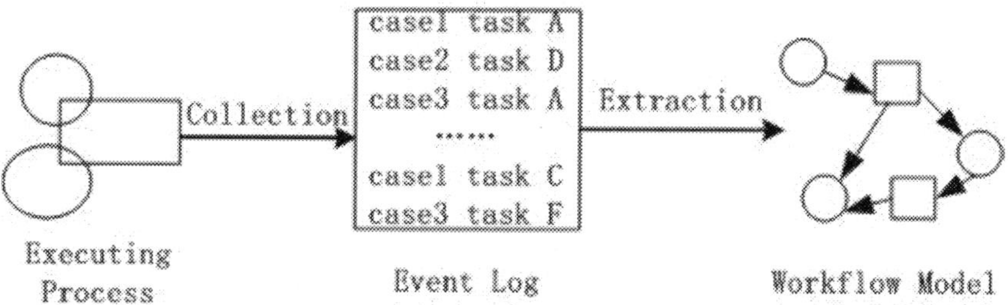

Fig. 1. Framework of Process Mining

Most research in process mining focuses on mining heuristics primarily based on binary ordering relations of the events in a workflow log. A lot of work has been done on utilizing heuristics to distill a process model from event logs and many valuable progresses are made in the domain. However, all the existing heuristic-based mining algorithms have their limitations [8,9]. There are still many challenging problems that the existing mining algorithms cannot handle. Duplicate tasks are one of them. It refers to the situation that one process model (e.g., a Petri net) has two or more nodes referring to the same task. Figure 2 shows a workflow model with three duplicate tasks (i.e. task X, task D and task E) represented in Petri nets. However, it is very difficult to automatically construct a process model from the event log of this model, because it is impossible to distinguish the task in one case from the task owning the same name in the other cases.

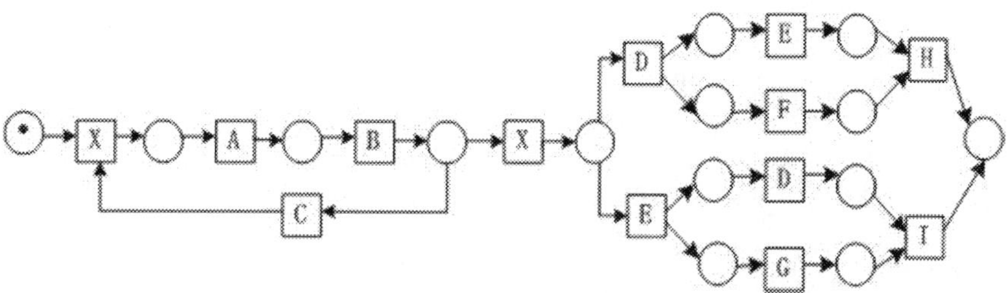

Fig. 2. A workflow model with duplicate tasks

The "α-algorithm" [8] is proved to correctly distill sound Structured Workflow nets (SWF-nets, [8]) which have no task duplication [9]. The main idea of our method to handle the duplicate tasks is as follows. First, in the pre-processing phase, those tasks with same label are identified by our heuristic rules and marked with different labels in the log. Then, the "α-algorithm" is adopted to discover a workflow model from the identified log. Finally, during post-processing, the distilled model (in our case a Petri-net) is fine-tuned by recovering the marked task to their original label and a workflow model with duplicate tasks is obtained. The new mining algorithm based on the "α-algorithm" is name as "α*".

The remainder of this paper is organized as follows. Section 2 discusses related work. Section 3 presents the new approach to tackle task duplication using the "α-algorithm". Section 4 concludes the paper and points out future work.

2 Related work

The idea of process mining is accepted widely for several years [3,4,5,8,10]. In the beginning, the research results are limited to sequential behavior. To extend to concurrent processes, Cook and Wolf propose several metrics (entropy, event type counts, periodicity, and causality) and apply them to distill models from event streams in [4]. However, they do not give any method to generate explicit process models. In [5, 6] Herbst and Karagiannis are also use an inductive approach to perform process mining in the context of workflow management. Two different workflow induction algorithms which are based on hidden Markov models are provided in [5]. The first method is a bottom-up, specific-to-general method and the other applies a top-down, general-to-specific strategy. These two strategies are limited to sequential models. The approach described in [6] is extended to tackle concurrency. Their approach is divided into two steps: induction step and transformation step. In the induction step task nodes are merged and split in order to extract the underlying process which is represented by stochastic task graphs. The stochastic task graph is transformed into an ADONIS workflow model in the transformation step. A notable difference with other approaches is that the approach allows for task duplication. The work of Aalst and his team members is characterized by the focus on workflow processes with concurrent behavior. In [10] a heuristic approach is provided to construct so-called "dependency/frequency tables" and "dependency/frequency graphs". The approach is practical for being able to deal with noise. Another formal algorithm called "α-algorithm" is provided and proved to correctly distill workflow models represented in Petri-net from event logs and an extended version of the "α-algorithm" to incorporate short loops (i.e. length-one loops and length-two loops) is also presented in [8]. However, these algorithms are restricted to process models without duplicate tasks.

Compared with existing work, our work is characterized by the focus on concurrent workflow processes with task duplication behavior. Therefore, we want to distinguish duplicate tasks in the workflow log explicitly. To achieve this goal, the machine learning techniques are combined with Workflow nets (WF-nets, [2]) in this paper. Actually, WF-nets are a subset of Petri nets that provide a graphical but formal language to represent the workflow model. Our approach results in a workflow model of Petri-net directly without additional transformation step.

3 Solution to Tackle Duplicate Tasks

In this section the details of the new algorithm that can handle duplicate tasks are presented. First, the predecessor/successor table (P/S-table) of task which helps us to find duplicate tasks is constructed. Then, according to the P/S-table, several heuristic rules are given to identify the duplicate tasks. Last, an algorithm (called "α*") that correctly mines sound WF-nets with duplicate tasks is provided.

3.1 Construction of the Predecessor/Successor Table

The starting point of our algorithm is to construct P/S-table of each task. For each task *A* that occurs in every workflow trace, the following information is abstracted out of the workflow log: (i) the name of the task that directly precedes task *A* (notation T_P), (ii) the name of the task that directly follows task *A* (notation T_S). The distilled information of task *A* is reserved in P/S-table.

Table 1. An event log of the model of Fig. 2

case id	event trace
"δ_1"	*X A B X D E F H*
"δ_2"	*X A B X E D G I*
"δ_3"	*X A B X D F E H*
"δ_4"	*X A B X E G D I*
"δ_5"	*X A B C A B X D E F H*
"δ_6"	*X A B C A B X E D G I*
"δ_7"	*X A B C A B C A B X D F E H*
"δ_8"	*X A B C A B C A B X E G D I*

According to the process model of Figure 2, a random workflow log with 1000 event sequences (10550 event tokens) is generated. As an example, Table 1 shows the distinctive workflow traces which represent all the possible occurrences of every task in the log. The preceding task and the following task of every task *X* in each representative trace are listed in Table 2. The P/S-table seems clear without extra explanation except the notation of task identifier. The meaning of "$\delta_i(t,N)$" is the Nth occurrence of task named by *t* in the workflow trace called "δ_i". For example, "$\delta_1(X,1)$" is the first occurrence of task *X* in "δ_1" and "$\delta_5(X,2)$" is the second occurrence of task *X* in "δ_5". Notice that two nodes of task *X* both belong to the sequential event stream, while one node of task *D* is included in a concurrent event stream (the AND-split in *E*). To support the analysis of the information in P/S Table, we give the definition of cross-equivalent.

Definition 1 (cross-equivalent). Let "$\delta_i(t,N)$" and "$\delta_j(t,N')$" be two occurrences of task t in traces "δ_i" and "δ_j". T_P and T_S are predecessor and successor of "$\delta_i(t,N)$" in "δ_i", T_P' and T_S' are predecessor and successor of "$\delta_j(t,N')$" in "δ_j". If there is the situation that $T_P = T_S'$ or $T_P' = T_S$, then the situation is called cross-equivalent.

Table 2. An example P/S-table for task *X*

task identifier	T_P	T_S
"$\delta_1(X,1)$"	~	A
"$\delta_1(X,2)$"	B	D
"$\delta_2(X,1)$"	~	A
"$\delta_2(X,2)$"	B	E
"$\delta_3(X,1)$"	~	A
"$\delta_3(X,2)$"	B	E
"$\delta_4(X,1)$"	~	A
"$\delta_4(X,2)$"	B	D
"$\delta_5(X,1)$"	~	A
"$\delta_5(X,2)$"	B	D
"$\delta_6(X,1)$"	~	A
"$\delta_6(X,2)$"	B	E
"$\delta_7(X,1)$"	~	A
"$\delta_7(X,2)$"	B	D
"$\delta_8(X,1)$"	~	A
"$\delta_8(X,2)$"	B	E

Table 2 indicates that (i) the predecessors of "$\delta_1(X,1)$" and "$\delta_1(X,2)$" are different, (ii) the successors of "$\delta_1(X,1)$" and "$\delta_1(X,2)$" are also distinct, (iii) the predecessors and successors of "$\delta_1(X,1)$" and "$\delta_2(X,1)$" are identical, (iv) the predecessors of "$\delta_1(X,2)$" and "$\delta_2(X,2)$" are same while their successors are unlike. Finally, (v) if *X* is preceded by *B*, sometimes it is followed by *D* and sometimes by *E*. Table 3 shows the P/S table of task *D*.

Table 3. An example P/S-table for task *D*

task identifier	T_P	T_S
"$\delta_1(D,1)$"	X	E
"$\delta_2(D,1)$"	E	G
"$\delta_3(D,1)$"	X	F
"$\delta_4(D,1)$"	G	I
"$\delta_5(D,1)$"	X	E
"$\delta_6(D,1)$"	E	G
"$\delta_7(D,1)$"	X	F
"$\delta_8(D,1)$"	G	I

Table 3 depicts the predecessor and successor of task *D*. It can be concluded from Table 3 that (i) the predecessor and successor of "$\delta_1(D,1)$" are quite different with those of "$\delta_2(D,1)$" and "$\delta_4(D,1)$", (ii) the predecessors of "$\delta_1(D,1)$" and "$\delta_3(D,1)$" are same while their successors are unlike, (iii) the predecessors and successors of "$\delta_2(D,1)$" are cross-equivalent with those in "$\delta_1(D,1)$" and "$\delta_4(D,1)$". It is remarkable that the other occurrences of *X* and *D* in the left traces is similar with the above situations. In the next section we will use the P/S-table in combination with several relatively simple heuristics to identify the duplicate tasks.

3.2 Identification of Duplicate Tasks

The identification of duplicate tasks in a sequential workflow model is relatively easy. If it always the case that, the predecessors and successors of the tasks with same name are different, then it is plausible that they are two tasks owing same name. On the other hand, if the tasks sharing same name also have same predecessors and successors, it is no doubt that they refer to unique task. Nevertheless, the situations in a concurrent workflow model or choice workflow model are more complicated. In many cases, although the tasks owning the same name in two workflow traces have distinct predecessors and successors, we can not decide whether the two tasks are duplicate tasks or not, because the predecessors and successors may be cross-equivalent. This occurs not only when the unique task belongs to a concurrent event stream but also when there are duplicate tasks. Another possible case is that one task is both preceded and succeeded by a choice structure although there is no cross-equivalence. In this case, though the occurrences of task have different predecessors and successors in different traces, they are still corresponding to one unique task instead of duplicate tasks. Task E in Figure 3 shows the above situation. To avoid the situation that the unique task is determined as a duplicate task, we give the definition of selection relation.

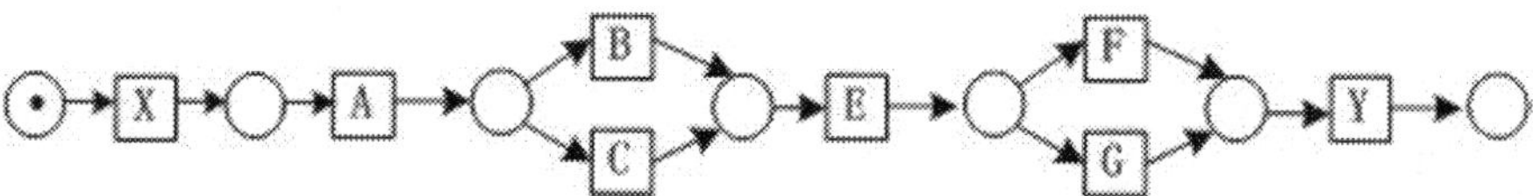

Fig. 3. Example Model of Selection Relation

Definition 2 (selection relation). Let T be the task set, W be a loop-complete workflow log over T. And let "$a,b \in T$", Task a and b have the selection relation if and only if there are traces "$\delta_1=t_1t_2t_3...t_n$", "$\delta_2=t_1't_2't_3'...t_n'$" and "$i \in \{1,...,n-2\}$" such that "$t_i = a \wedge t_i' = b \wedge a \# b \wedge t_{i+1} = t_{i+1}' \wedge t_{i+2} = t_{i+2}'$". The notation of selection relation is "$\sum_W$".

Task B and C in Figure 3 have the selection relation, because in traces "δ_1=$XABEFY$" and "δ_2=$XACEFY$", we can find an integer $i=3$ which makes "$t_3 = B \wedge t_3' = C \wedge B \# C \wedge t_4 = t_4' \wedge t_5 = t_5'$".

In the previous section we observed that the information in the X-P/S-table strongly suggests that "$\delta_1(X,1)$" and "$\delta_1(X,2)$" are duplicate tasks because their predecessors and successors are quite different, and are not cross-equivalent also. Furthermore, their predecessors haven't the selection relation. Basing on the information in the in the D-P/S-table, the similar conclusion can be drawn on "$\delta_1(D,1)$"and "$\delta_4(D,1)$". In line with these observations, rule (1), the first heuristic rule to identify duplicate tasks is given below, let U be the set of duplicate tasks:

$$\text{IF } ((T_P \neq T_P') \text{ AND } (T_S \neq T_S') \text{ AND } (T_P \neq T_S') \text{ AND } (T_S \neq T_P') \text{ AND } (\text{not } T_P \textstyle\sum_W T_P')) \quad (1)$$
$$\text{THEN } <\delta_i(t,N_1),\ \delta_j(t,N_2)> \in U$$

In rule (1), the first condition $(T_P \neq T_P')$ is used to judge that the predecessors of two occurrences of task t are different. The second condition determines the difference of their successors. And the third condition and the fourth one state the requirement that there are not cross-equivalence between the preceding tasks and the following tasks. Finally, the fifth condition is to judge there is no selection relation between their predecessors. If five conditions are all satisfied, we can conclude that the tuple consisting of two occurrences of task t belongs to U, Applying this heuristic rule on the P/S-tables extracted from the log in Table 1, we obtain the result workflow log in Table 4. Comparing the workflow log of Table 4 and the process model of Figure 2, it can be seen that some of the duplicate tasks such as X is identified correctly. However, rule (1) can not identify all the duplicate tasks correctly. For instance, "$\delta_1(D,1)$" and "$\delta_2(D,1)$ " correspond to different task D, but they are not be marked separately in Table 4.

In fact, in the case of cross-equivalence, if we can determine the task belongs to a concurrent event stream, the occurrences in two workflow traces can be confirmed to be a unique task, otherwise the two occurrences are corresponding to duplicate tasks. The property of the task in the concurrent case is illustrated by the following representative example. First, two functions of "$pred(\delta, t)$"and "$succ(\delta, t)$" are defined to get the predecessor and successors of task t in trace δ respectively.

Table 4. An identified event log of the log in Table 1

case id	event trace
"δ_1"	$X\ A\ B\ X_1\ D\ E\ F\ H$
"δ_2"	$X\ A\ B\ X_1\ E\ D\ G\ I$
"δ_3"	$X\ A\ B\ X_1\ D\ F\ E\ H$
"δ_4"	$X\ A\ B\ X_1\ E\ G\ D_1\ I$
"δ_5"	$X\ A\ B\ C\ A\ B\ X_1\ D\ E\ F\ H$
"δ_6"	$X\ A\ B\ C\ A\ B\ X_1\ E\ D\ G\ I$
"δ_7"	$X\ A\ B\ C\ A\ B\ C\ A\ B\ X_1\ D\ F\ E\ H$
"δ_8"	$X\ A\ B\ C\ A\ B\ C\ A\ B\ X_1\ E\ G\ D_1\ I$

In event trace"δ_3" of Table 1, the predecessor of "$\delta_4(D,1)$" is G and "$pred(\delta_4,G)$" is E which is just the predecessor of "$\delta_2(D,1)$ ". In"δ_2", the successor of "$\delta_2(D,1)$" is G and "$succ(\delta_2,G)$" is I which is just the successor of "$\delta_4(D,1)$". The "$\delta_1(E,1)$" and "$\delta_3(E,1)$" also have the similar property. In line with the observations, the first heuristic rule (1) is extended with rule (2) and (3):

$$\text{IF } ((T_P = T_S') \text{ AND } ((T_P' \neq pred(\delta_i, T_P)) \text{ OR } (T_S \neq succ(\delta_j, T_S')))) \quad (2)$$
$$\text{THEN } <\delta_i(t,N_1),\ \delta_j(t,N_2)> \in U$$

$$\text{IF } ((T_S = T_P') \text{ AND } ((T_P' \neq pred(\delta_j, T_P')) \text{ OR } (T_S' \neq succ(\delta_i, T_S))) \quad (3)$$

$$\text{THEN } <\delta_i(t,N_1), \delta_j(t,N_2)> \in \text{U}$$

Rule (2) and (3) specify the situation without concurrency. In rule (2), the first condition $(T_P = T_S')$ is used to judge that the predecessor of "$\delta_i(t,N_1)$" and the successor of "$\delta_j(t,N_2)$" are cross-equivalent. The second condition determines the difference of the predecessor of "$\delta_j(t,N_2)$" and the predecessor of T_P. And the third condition is similar with the second one which states the requirement that the successor of "$\delta_i(t,N_1)$" and the successor of T_S' are not equal. If these three conditions are all be met, we can determine that the tuple consisting of two occurrences of task t belongs to U. Rule 3 prescribes another similar situation. In the next section, the three rules are applied to identify the duplicate tasks in a workflow log. Table 5 gives the result after applying the three heuristics rules on the log in Table 4. It can be seen from Table 5 that all the duplicate tasks are identified correctly.

Table 5. An identified event log of the log in Table 4

case id	workflow trace
"δ_1"	$X A B X_1 D E F H$
"δ_2"	$X A B X_1 E D_2 G I$
"δ_3"	$X A B X_1 D F E H$
"δ_4"	$X A B X_1 E G D_2 I$
"δ_5"	$X A B C A B X_1 D E F H$
"δ_6"	$X A B C A B X_1 E D_2 G I$
"δ_7"	$X A B C A B C A B X_1 D F E H$
"δ_8"	$X A B C A B C A B X_1 E G D_2 I$

3.3 Generating WF-nets from the Identified Workflow Log

The solution to tackle duplicate tasks in sound WF-nets focuses on the pre- and post-processing phases of process mining [8]. The assumption about the completeness and noise free of a log is continued to use. The main idea is to identify the duplicate tasks and give them different identifiers. Any duplicate tasks can be identified by searching and checking the P/S-table of the task with the three heuristic rules above. The inspection of P/S-table follows the sequence of the occurrence of task in every event trace. If the task is determined to belong to duplicate tasks, it is renamed at the same time. The method of renaming is to append a serial number to the original task name.e.g.B1, B2, 1A, 1B. It is convenient that the original task name and the serial number are taken from distinct character sets. In fact, if the task to check has been renamed already, it is unnecessary to compare it with its forwards tasks owning the same name because the same task before it has compared with them already.

The algorithm called "α*" based on these heuristics is presented in Figure 4. Let T be a set of tasks and W be a workflow log over T, the "α-algorithm" as in Definition 2.16 and the ordering relations as in Definition 2.14 in [8].

```
Algorithm  α*(W, N)
/*the extended α-algorithm to
   tackle duplicate tasks*/
1. T_log ← {t ∈ T | ∃_σ∈W [t ∈ σ]}.
2. W^-DT ← W.
3. isDup ← false.
4. isIdentify ← false.
5. FOR ∀t ∈ T_log DO
      (λ ← buildPSTable(t, W^-DT).
       FOR ∀θ ∈ λ DO
        (λ' ← {θ' ∈ λ | θ' ≠ θ}.
         FOR ∀θ' ∈ λ' DO
          (isDup ← judgeDuplicate(θ, θ').
           IF isDup THEN
              (// t' is the renamed task of
                 task t
                 t' ← renameTask(t, θ', λ).
                 T_log ← T_log ∪ {t'}.
                 isIdentify ← true.))).
       IF isIdentify THEN
          W^-DT ← IdentifyLog(W^-DT, λ).).
6. (P_W-DT, T_W-DT, F_W-DT) ← α(W^-DT).
7. P_W ← P_W-DT
8. T_W ← eliminateTMark(T_W-DT).
9. F_W ← eliminateFMark(F_W-DT).
10. N ← (P_W, T_W, F_W) ▮
```

Fig. 4. The "α*-algorithm" to mine duplicate tasks

The algorithm of "α*" works as follows. First, it examines the log traces (Step 1). Then the input log W^{-DT} to be processed by the "α-algorithm", the flag *isDup* to describe whether there are duplicate tasks and the flag *isIdentify* to depict whether to identify the original input log *W* are initialized in steps 2 to 4. Then, in Step 5, the P/S- table "λ" of each task is generated, each table is checked to find the duplicate tasks based on the previous three heuristic rules in function *judgeDuplicate*, the found duplicate tasks are identified in tuple "θ' " of "λ", the renamed task *t'* is added in T_{log} for further inspection and accordingly the previous log W^{-DT} is also marked and the result is still reserved in W^{-DT}. In Step 6, the "α-algorithm" discovers a workflow net based on the identified workflow log W^{-DT} and the ordering relations as defined in Definition 2.14 in [8]. The identifiers of the duplicated tasks are recovered to the original task name and their respective input and output arcs are adjusted accordingly in steps 7 to 9. Finally, the workflow net with the duplicate tasks is returned in Step 10. In the next section our experimental results of applying the above-defined approach on other workflow logs are reported.

3.4 Experiments

To evaluate the above described heuristics we have developed a research prototype which includes the "α*-algorithm". The prototype can read a text file containing workflow traces produced by a WFMS of Staffware. By using Staffware, a wide variety of workflow traces of workflow models with different sizes and structural complexities can be generated. Eight different workflow traces are used to test our approach. One of these examples is taken from Herbst [5] to simply compare our method with model splitting. For each model a random workflow logs with 1000 event sequences is generated. In each log, the execution of tasks completely follows the sequence which is pre-defined in the WF-nets model. Then, we study the data in these logs under the experimental environment of P4(2.0GHz), 512 Ram and Windows 2000. The results of experiment are listed in Table 6 which is expressed in the metrics of event tokens number, duplicate tasks number, resultant model data (original model data) and mining time.

Table 6. Experiment Results

Model Name	Event Trace Number	Event Token Number	Duplicate Tasks Number	Resultant Model Data Transition Number/Place Number/Arc Number (Original Model Data)	Mining Time
M1	1000	5975	2	5/6/12 (5/6/12)	600ms
M2	1000	8750	1	5/6/11 (5/6/11)	670ms
M3	1000	2000	1	2/3/4 (1/2/2)	176ms
M4	1000	5000	1	5/7/12 (5/7/12)	687ms
M5	1000	6000	1	10/12/24 (10/12/24)	603ms
M6	1000	4000	1	5/5/10 (5/5/10)	562ms
M7	1000	8240	2	12/13/30 (12/13/30)	816ms
M8	1000	10550	3	12/14/33 (12/14/33)	1107ms

It can be learned from the experiment results in Table 6 that with the increase of the event token number in log, the time to mine model becomes longer gradually. And the more the duplicate tasks number in the WF-nets is, the longer the mining time becomes. The "α*-algorithm" can discover the example model (M1) adopted by Herbst correctly in less time and with higher efficiency. It is also shown in Table 6 that "α*-algorithm" can cost less time to recover most of the WF-nets to the original models and have stable performance. The reason that M3 can not be discovered precisely is the information included in the P/S-tables is few which makes our algorithm is unable to study correct knowledge from these limited data. In a word, the experiment results indicate that the "α*-algorithm" which identifies the duplicate tasks in logs and distills the identified logs later is valid and effective.

4 Conclusion and Future Work

In this paper, we focus on the extension of the "α-algorithm" so that it can mine WF-nets with duplicate tasks. The learning algorithm is named as "α*". The "α-algorithm" is proven to correctly discover sound SWF-nets without task duplication. Changes in the pre- and post-processing phases are mainly involved in the extension. The details of "α*" is presented in three steps: Step (i) the construction of the P/S-table, step (ii) the identification of duplicate tasks based on P/S-table, and step (iii) the generation of the WF-net out of the identified workflow log using "α-algorithm".

In the experimental section, we applied our algorithm on eight different workflow models with duplicate tasks. Sequential process, concurrent process and loops are included in these different models. For each model, we generated a random workflow log with 1000 event sequences. The experimental results show that our approach is valid to induce WF-nets with duplicate tasks.

In spite of the reported results, a lot of future work needs to be done. Firstly, the number of our test logs are limit, more experiments must to be done. Secondly, the theoretical basis of our learning algorithm needs to be improved. Finally, the above results are obtained by presupposing that the logs are complete and noise free. However, this situation appears rarely in real logs. Thus, to make our approach more practical, the heuristic mining techniques and tools which are less sensitive to noise and the incompleteness of log must to be developed.

Acknowledgments. This work is supported by NSFC Major Research Program, Basic Theory and Core Techniques of Non Canonical Knowledge (No. 60496321); National Natural Science Foundation of China (No. 60373098, 60573073, 60503016), the National High-Tech Research and Development Plan of China (No. 2003AA118020), the Major Program of Science and Technology Development Plan of Jilin Province (No. 20020303), the Science and Technology Development Plan of Jilin Province (No. 20030523).

References

1. van der Aalst, W.M.P., Desel, J., Oberweis, A. (ed.): Business Process Management: Models, Techniques, and Empirical Studies. LNCS, vol. 1806. Springer, Heidelberg (2000)
2. van der Aalst, W.M.P.: The Application of Petri Nets to Workflow Management. The Journal of Circuits, Systems and Computers 8(1), 21–66 (1998)
3. Agrawal, R., Gunopulos, D., Leymann, F.: Mining process models from workflow logs. In: Proceedings of the Sixth International Conference on Extending Database Technology, pp. 469–483 (1998)
4. Cook, J.E., Wolf, A.L.: Event-Based Detection of Concurrency. In: Proceedings of the Sixth International Symposium on the Foundations of Software Engineering (FSE-6), pp. 35–45 (1998)
5. Herbst, J., Karagiannis, D.: Integrating Machine Learning and Workflow Management to Support Acquisition and Adaptation of Workflow Models. International Journal of Intelligent Systems in Accounting, Finance and Management 9, 67–92 (2000)

6. Herbst, J.: Dealing with Concurrency in Workflow Induction. European Concurrent Engineering Conference, SCS Europe (2000)
7. Jablonski, S., Bussler, C.: Workflow Management: Modeling Concepts, Architecture, and Implementation. International Thomson Computer Press (1996)
8. de Medeiros, A.K.A., van Dongen, B.F., van der Aalst, W.M.P., Weijters, A.J.M.M.: Process Mining: Extending the "α-algorithm" to Mine Short Loops. BETA Working Paper Series, WP 113, Eindhoven University of Technology, Eindhoven (2004)
9. de Medeiros, A.K.A., van der Aalst, W.M.P., Weijters, A.J.M.M.: Workflow Mining: Current Status and Future Directions. In: Meersman, R., Tari, Z., Schmidt, D.C. (eds.) CoopIS 2003, DOA 2003, and ODBASE 2003. LNCS, vol. 2888, pp. 389–406. Springer, Heidelberg (2003)
10. Weijters, A.J.M.M., van der Aalst, W.M.P.: Process Mining: Discovering Workflow Models from Event-Based Data. In: Proceedings of the 13th Belgium-Netherlands Conference on Artificial Intelligence, pp. 283–290. Springer, Heidelberg (2001)

A Distributed Genetic Algorithm for Optimizing the Quality of Grid Workflow*

Hongxia Tong, Jian Cao, and Shensheng Zhang

Collaborative Computing and Intelligence Technology Lab,
Shanghai Jiaotong University, Shanghai, 200240, China
thx781212@sjtu.edu.cn

Abstract. The advancement of Grid and Web service technologies greatly facilitates the aggregation of distributed applications. As the grid workflow generally involves long lasting execution tasks, the quality optimization for grid workflow has much significant importance. In order to accurately define the quality of grid workflow, an extended quality model is proposed which takes account of message compositionality and composition rationality between Web services. Based on the extended quality model, a distributed multi-objective genetic algorithm to optimize the quality of grid workflow is proposed. This approach focuses on the distributed nature of grid environment that consists of autonomous domains and can deal with the global multiple objectives and constrains. The experimental results show that the distributed multi-objective genetic algorithm proposed in this paper can effectively optimize the services selection for grid workflow and has ideal performance.

1 Introduction

Grid workflow can be seen as a collection of tasks that are processed on distributed resources in a well-defined order to accomplish a specific goal, which enables the scientists to explore research issues in their domain at greater and finer scales than ever before[1]. The tasks in a grid workflow are also called abstract services[2]. The abstract service describes the functions of each task in grid workflow and is specified without referring to specific Web service for task execution, which provides a flexible way for users to define workflows without concerning details of low-level implementation. As a single Web service is most likely inadequate to satisfy the functional requirements of a grid workflow, a group of Web services are composed, which is called composite Web service. Since the number of Web services in grid environment that can perform a certain abstract service may be very large[3], there can be more than one combinations

* This work is supported by National Science Foundation of China under grant 60503041, Shanghai Commission of Science and Technology/International Cooperation Project under grand 05SN07114, Scientific Research Project for Shanghai 2010 World Expo under grand 2005BA908B09 and National High-Tech Research and Development Plan of China under grand 2006AA04Z152.

K.C. Chang et al. (Eds.): APWeb/WAIM 2007 Ws, LNCS 4537, pp. 408–419, 2007.

of Web services to accomplish the functions of a grid workflow. Therefore, there is need to distinguish them by using a set of well-defined quality criteria. Meanwhile, since the grid workflow generally involves long lasting execution tasks with large data flow and utilizes heterogeneous and dynamic resources from multiple organizations[1], the quality optimization for grid workflow has significant importance. In this paper, an extended quality model to define the quality of grid workflow is proposed which includes the general quality criteria, message compositionality and composition rationality between Web services. General quality criteria include price, time, reliability, reputation and security. Message compositionality reflects matching degree of the input parameters and output parameters between Web services, and the composition rationality gives an approximation of the composition soundness from the aspect of domain knowledge.

Given a grid workflow that consists of a sequence of abstract services, the Web services selection for each abstract service to optimize the quality of grid workflow is a problem of NP-hard[4]. Motivated by the powerful computation ability of genetic algorithm[5], a distributed multi-objective genetic algorithm is proposed in this paper. This approach focuses on the distributed nature of grid environment where there is no centralized manager and can deal with the global multiple objectives and constrains provided by service requester. The experimental results show that the distributed multi-objective genetic algorithm proposed in this paper can effectively optimize the services selection for grid workflow.

The remainder of this paper is organized as follows. Section 2 introduces some related work. Section 3 presents an extended QoS model for grid workflow. Section 4 describes the problem of optimizing the services selection for grid workflow. Section 5 presents the distributed multi-objective genetic solution for QoS optimization. Section 6 reports the experimental results. Finally some conclusions and future work are drawn.

2 Related Work

Quality of Web service has been highlighted in [6,7], and some related work has been conducted. Jorge Cardoso et al. [8] presented a predictive QoS model which can automatically compute the quality of service for workflows based on QoS attributes of elementary Web services. Chintan Patel et al. [9] proposed a QoS oriented framework that can not only conduct the adaptive selection process but also simultaneously provide binding and execution of Web Services for the underlying workflow. However, only general quality criteria were concerned and they didn't take into account the problem of global quality optimization.

In order to optimize services selection for the composite Web service, the linear integer programming approach was proposed in [10], but only general quality criteria were concerned. Liu et al. [11] model the quality optimization problem as the Multiple Choice Knapsack Problem. This solution is usually too complex for run-time decisions.

The most similar work to ours was done by Geeardo Canforda[12], where a simple genetic algorithm was employed to optimize the Web services selection.

The distinct difference between their work and ours is that here a distributed multi-objective genetic algorithm is applied. One the one hand, this approach can deal with the global multiple objectives and constraints. One the other hand, distributed genetic algorithm can effectively reduce computation time for its applicability to parallel computer, which is most important in dynamic grid environment. Meanwhile, the application of distributed genetic algorithm is appropriate as the number of isolated management domains in grid environment is large and each autonomous domain has incomplete information about others.

3 Web Services Quality Model

Quality of service is a generic term used to denote the non-functional aspects of a service. In the scope of Web services, QoS concerns the non-functional aspects of the service being provided to the Web service requesters. In order to accurately define the quality of a composite Web service, an extended quality model is proposed. Besides the general quality criteria, the extended quality model for the composite Web service also includes message compositionality and composition rationality between Web services, which will be detailed in the following.

3.1 QoS Model for Elementary Service

In this paper, five important quality criteria for elementary services are considered. They are price, time, reliability, reputation and security. More quality requirements can refer to [7] and some conceptions discussed in the following are partly based on [7,10].

Price. For a Web service s , the price criterion $Q_{price}(s)$ measures the money that the service consumer has to pay to the owner of Web service for executing the service s.

Time. This quality criterion is defined as the total time needed for a Web service instance to transform a set of inputs into outputs. For a Web service s, the time quality is calculated by the expression $Q_{time}(s) = T_{process}(s) + T_{delay}(s)$, where $T_{process}(s)$ indicates the time that an instance of a Web service s takes while being processed and $T_{delay}(s)$ represents non-value-added time such as queuing time needed in order for an instance of a Web service s to be processed.

Reliability. Reliability represents the ability of a service to function correctly and consistently and provide the same service quality despite system or network failures, which is calculated by $Q_{rel}(s) = N_c(s)/k$, where $N_c(s)$ is the number of times that the service s has been successfully delivered within the maximum expected time frame, and k is the total number of invocations.

Reputation. The reputation criterion measures the trustworthiness of a service s, which can be calculated by the expression $Q_{rep}(s) = \sum_1^n R_i/n$, where R_i is the end user's rank on a service's reputation, and n is the number of times that the service has been graded.

Security. Security is the quality aspect of the Web service of providing confidentiality and non-repudiation by authenticating the parties involved, encrypting messages and providing access control. Since Web service invocation occurs over the public Internet, security has added importance. Given a Web service s, the security level $Q_{sec}(s)$ is evaluated by trusted third parties.

According to the discussion above, the quality vector of an elementary Web service s is defined as $QoS(s) = \{Q_{price}(s), Q_{time}(s), Q_{rel}(s), Q_{rep}(s), Q_{sec}(s)\}$.

3.2 QoS Model for Composite Service

For a composite Web service S which is composed of a sequence of Web services $\{s_1, s_2, \ldots, s_n\}$, the aggregation quality of general criteria is defined as:

(1) $Q_{price}(S) = \sum_{i=1}^{n} Q_{price}(s_i)$
(2) $Q_{time}(S) = \sum_{i=1}^{n} Q_{time}(s_i)$
(3) $Q_{rel}(S) = \sqrt[n]{\prod_1^n Q_{rel}(s_i)}$
(4) $Q_{rep}(S) = \frac{1}{n}\sum_{i=1}^{n} Q_{rep}(s_i)$
(5) $Q_{sec}(S) = \min\{Q_{sec}(s_1), \ldots, Q_{sec}(s_n)\}$

In order to correctly evaluate the quality of a composite Web service, the quality model for the composite service should not only include the above general quality criteria, but also take account of message compositionality and composition rationality between Web services.

For any pair of Web services (s_i, s_{i+1}) in the composite Web service $S = \{s_1, s_2, \ldots, s_n\}$, message compositionality $MC(s_i, s_{i+1})$ measures the matching degree of the outputs of s_i and the inputs of s_{i+1}, where s_i is the direct precursor of s_{i+1} and the output parameters of s_i are given to s_{i+1} as the input parameters. The output parameters of the service s_i and the input parameters of the service s_{i+1} are represented by $O_i = \{o_1, o_2, \ldots, o_l\}$ and $I_{i+1} = \{i_1, i_2, \ldots, \imath_m\}$ respectively, where all parameters are conceptualized within the same ontology. Given any pair of parameters (o_h, i_k), the function $Sem(o_h, i_k)$ evaluates the similarity of two concept classes o_h and i_k, which is defined as follows[13]:

$$Sem(o_h, i_k) = \begin{cases} 1 & o_h = i_k \\ 1 & o_h \subseteq i_k \\ \frac{|pr(o_h)|}{|pr(i_k)|} & o_h \supseteq i_k \\ Sim(o_h, i_k) & otherwise \end{cases}$$

Where $o_h = i_k$ represents the two concepts are the same, $o_h \subseteq i_k$ represents the concept o_h is subsumed by the concept i_k, $o_h \supseteq i_k$ represents the concept o_h subsumes the concept i_k, $|pr(o_h)|$ and $|pr(i_k)|$ represent the number of properties of the concepts o_h and that of i_k respectively, and the value of $Sim(o_h, i_k)$ is calculated as follows[13]:

$$Sim(o_h, i_k) = \sqrt{\frac{|pr(o_h) \cap pr(i_k)|}{|pr(o_h) \cup pr(i_k)|} * \frac{|pr(o_h) \cap pr(i_k)|}{|pr(i_k)|}}$$

Based on the above discussion, the concept similarity of all pairs of parameters (o_h, i_k), O_i and I_{i+1} are constructed into a complete bipartite graph $G = (O_i, I_{i+1}, E)$, where $O_i = \{o_1, o_2, \ldots, o_l\}$, $I_{i+1} = \{i_1, i_2, \ldots, i_m\}$, E is the set of edges which all go between O_i and I_{i+1}, and the weight of the edge (o_h, i_k) is equal to $Sem(o_h, i_k)$. The maximum weight bipartite match of the graph G is to find the maximum match $M \subseteq E$ which has the maximum weight sum denoted by $W(M) = \sum_{e \in E} w(e)$, where $w(e)$ is the weight of the edge e.

According to the above discussion, the message compositionality $MC(s_i, s_{i+1})$ is computed by the expression $MC(s_i, s_{i+1}) = W(M)/m$, where m is the number of input parameters of the Web service s_{i+1}. Then the value of message compositionality for a composite Web service S composed of a sequence of Web services $\{s_1, s_2, \ldots, s_n\}$ is defined as:

$$Q_{mc}(S) = \frac{1}{n-1} \sum_{i=1}^{n-1} MC(s_i, s_{i+1})$$

For any pair of Web services (s_i, s_{i+1}) in S, the composition rationality denoted by $CR(s_i, s_{i+1})$ gives an approximation of the composition soundness, which is defined by domain experts from the aspect of domain knowledge. $CR(s_i, s_{i+1}) = 0$ indicates that the two Web services can not be composed, and $CR(s_i, s_{i+1}) = 1$ means that the two Web services can be composed very well. For a composite Web service S which is composed of a sequence of Web services $\{s_i, s_2, \ldots, s_n\}$, the value of composition rationality is defined as:

$$Q_{cr}(S) = \frac{1}{n-1} \sum_{1}^{n-1} CR(s_i, s_{i+1})$$

Based on the above discussion, the quality vector of a composite Web service S is defined as:

$$QoS(S) = \{Q_{price}(S), Q_{time}(S), Q_{rel}(S), Q_{rep}(S), Q_{sec}(S), Q_{mc}(S), Q_{cr}(S)\}$$

4 Solution of Distributed Multi-objective Genetic Algorithm

4.1 Description of Problem

Given a grid workflow that has a sequence of abstract services denoted by $T = \{t_1, t_2, \ldots, t_n\}$, and the Web services that can be assigned to abstract service t_i is represented as $sc(t_i) = \{s_{i1}, s_{i2}, \ldots, s_{im}\}$. Let x_i be the decision variable to determine which Web service to be chosen for the abstract service t_i and $v_i(x_i)$ indicates the selected Web service in $sc(t_i)$. So the problem is to dynamically select a group of Web services $\{v_1(x_1), v_2(x_2), \ldots, v_i(x_i), \ldots, v_n(x_n)\}$ to compose a composite Web service S whose quality satisfies the global multiple objectives and constraints imposed by the service requesters. In this paper, two global objective functions and five global constraints are formulated as follows:

Global objective functions:

(1) $min(Q_{price}(S)) = \sum_{i=1}^{n} Q_{price}(v_i(x_i))$
(2) $min(Q_{time}(S)) = \sum_{i=1}^{n} Q_{time}(v_i(x_i))$

Global constraints:

(1) $Q_{rel}(S) = \sqrt[n]{\prod_1^n Q_{rel}(v_i(x_i))} \geq b_1$
(2) $Q_{rep}(S) = \frac{1}{n}\sum_{i=1}^{n} Q_{rep}(v_i(x_i)) \geq b_2$
(3) $Q_{sec}(S) = \min(Q_{sec}(v_1(x_1)), Q_{sec}(v_2(x_2)), \ldots, Q_{sec}(v_n(x_n))) \geq b_3$
(4) $Q_{mc}(S) = \frac{1}{n-1}\sum_{i=1}^{n-1} MC(v_i(x_i), v_{i+1}(x_{i+1})) \geq b_4$
(5) $Q_{cr}(S) = \frac{1}{n-1}\sum_{i=1}^{n-1} CR(v_i(x_i), v_{i+1}(x_{i+1})) \geq b_5$

It is not easy to solve the above multi-objective function optimization problem, especially in grid environment. Grid environment consists of a number of autonomous domains and there is no centralized manager. Each domain owns a Web service registry where service providers publish their Web services and each autonomous domain only exchanges information with its neighboring domains. Focusing on the distributed nature of grid environment, a distributed multi-objective genetic algorithm is proposed to solve the above optimization problem. The distributed services selection model in grid environment is illustrated as Fig. 1.

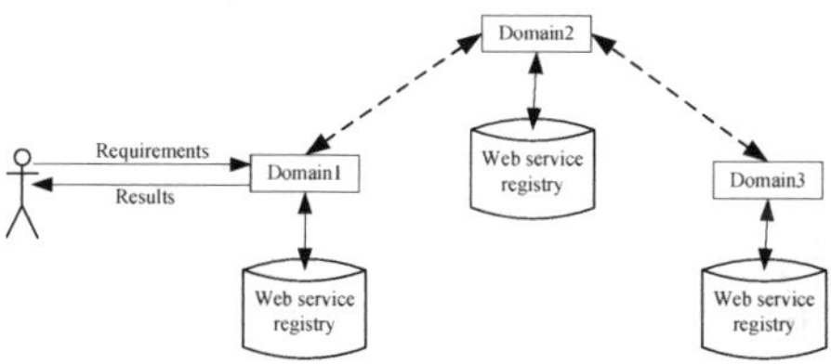

Fig. 1. Distributed services selection in grid environment

As Fig. 1 shown, the user submits the service requirements to a autonomous domain. Upon receiving the service requirements from the user, the autonomous domain will forward service requirements to its neighboring domains and run the genetic algorithm model to optimize the services selection according to service requirements and the Web services published in its service registry. According to the principle of distributed genetic algorithm, migration operation is carried between neighboring domains after the predefined migratory interval. The algorithm terminates after a certain number of generations or the time limit has been exceeded. Each autonomous domain returns its optimal solutions to the neighboring domain from which it receives the service requirements. After all the results from its neighbor domains are received, the autonomous domain that receives the service requirements from user combines these results and returns the optimal solutions to the user.

4.2 Distributed Multi-objective Genetic Algorithm

Distributed genetic algorithm has been one of the active topics over the last two decades. The basic idea of distributed genetic algorithm is that the global population is divided into several subpopulations, and each subpopulation evolves independently. The communication between subpopulations occurs during the migration phase, which takes place at a regular interval: A fixed proportion of each subpopulation is selected and sent to another subpopulation. In return, the same number of individuals is received from some other subpopulation and replaces the individuals selected according to some criteria. Distributed genetic algorithm has several advantages. One is it can effectively reduce computation time for optimal solution, which comes from its applicability to parallel computer. The other is the improvement of its searching ability, which stems from its avoidance of the premature convergence of a population by applying genetic operations locally. The application of distributed genetic algorithm is appropriate as the number of isolated domains in grid marketplace is large and each of the domains has incomplete information about others. The basic steps of distributed multi-objective genetic algorithm are illustrated briefly as follows:

(1) Generate randomly the initial population.

(2) Divide the initial population into subpopulations $P_1, P_2, \ldots, P_n$.

(3) Evaluate the individuals according to the defined fitness functions. If the termination condition is satisfied, terminate the algorithm. Otherwise, each subpopulation executes in paralleled the next steps.

(4) Perform the selection operation.

(5) Carry out crossover and mutation operations.

(6) Evaluate the individuals according to the defined fitness functions.

(7) If the condition of migration is true, perform the migration operation, and then perform the evaluation operation.

(8) If the termination criterion is not fulfilled, return to 4, else terminate algorithm.

The detail of our approach is illustrated in the following sections.

Chromosome Encoding. One of the primary decisions to be made concerning any genetic algorithm is the encoding. According to the above discussion, as there is only one Web service in $sc(t_i)$ that will be selected to accomplish the abstract service t_i, the integer encoding scheme is employed. Each chromosome is encoded as an array of integers, where the ith element in the array is equal to k which indicates the selected service assigned to the abstract service t_i is the kth Web service in $sc(t_i)$. For a grid workflow with a sequence of abstract services $\{t_1, t_2, \ldots, t_n\}$, an example of chromosome is encoded as Fig .2.

t_1	t_2	···	t_i	···	t_{n-1}	t_n
4	1	···	2	···	3	4

Fig. 2. Chromosome encoding

Compared with the traditional binary string encoding, the integer encoding provides several advantages. One is it provides a straightforward and natural way to express the mapping from representation to the solution domain. The other is it facilitates the implementation of genetic operations such as crossover and mutation, which improves the computation efficiency.

Evaluation. For the constrained multi-objective optimization problem in this paper, the solution space contains two parts: feasible solutions and infeasible solutions. A solution x that satisfies all constraints and variable bounds is called feasible (as opposed to infeasible). In order to evaluate feasible solutions in the population, domination relation is employed.

Definition 1. A feasible solution x is said to dominate another feasible solution y, if the solution x is not worse than y in all objectives and x is strictly better than y in at least one objective.

As far as the problem in this paper is concerned, for any two feasible solutions x and y, x dominates y if any of the following conditions are true:

(1) $Q_{price}(x) < Q_{price}(y)$ and $Q_{time}(x) \leq Q_{time}(y)$
(2) $Q_{price}(x) \leq Q_{price}(y)$ and $Q_{time}(x) < Q_{time}(y)$

A feasible solution x in current generation is said to be pareto optimal if there are no other feasible solutions in current population that can dominate individual x[14]. The pareto optimal set includes all pareto optimal solutions in the current generation. For every infeasible solution x in current generation, the penalty function[15] is defined as:

$$p(x) = \sum_{i=1}^{5} \frac{\triangle b_i(x)}{\triangle b_i(x)^{max}}$$

Where $\triangle b_i(x)$ refers the value that the solution x violates for constraint i, and the $\triangle b_i(x)^{max}$ is the maximum value that the solutions in current population violate for constraint i. To compare dominance relation of any two individuals x and y in the current population, the principle of constrained-domination[16] is applied, which is detailed in definition 2.

Definition 2. An individual x is said to constrained-dominate another individual y, if any of the following conditions are true:

(1) Individual x is the feasible solution and the individual y is not.
(2) Both x and y are infeasible solutions, but $p(x) < p(y)$.
(3) Both x and y are feasible solutions and x dominates y according to the definition 1.

As the definition 2 shown, feasible solutions constrained-dominate any infeasible solutions and two infeasible solutions are compared only based on their constraint violation. However, when two feasible solutions are compared, they are evaluated according to the definition 1.

Selection Operation. Selection operation is carried out according to the fitness of the individuals. Based on the definition of constrained-domination, the binary tournament selection policy is adopted in this paper, where for the randomly selected two individuals in the current population, if one individual x

constrained-dominates another individual y, then the individual x wins the tournament, otherwise randomly select one from x and y.

Mutation Operation. In order to improve the fitness of the whole population, a set of biological inspired operators such as crossover and mutation are employed. Crossover is a generic operator that combines two chromosomes to produce a new chromosome. The idea behind crossover is that the new chromosome may be better than both of the parents if it takes the best characteristics from each of the parents. Mutation is the genetic operator that randomly changes one or more of the chromosome's gene. The purpose of the mutation operator is to prevent the population from converging to a local minimum and to introduce to the population new possible solutions. In this study, only mutation operation is adopted. For the integer encoding scheme, the mutation operation facilitates to hybrid the neighborhood search techniques to produce an improved offspring. Fig. 3 illustrates an example of this mutation with neighborhood search technique.

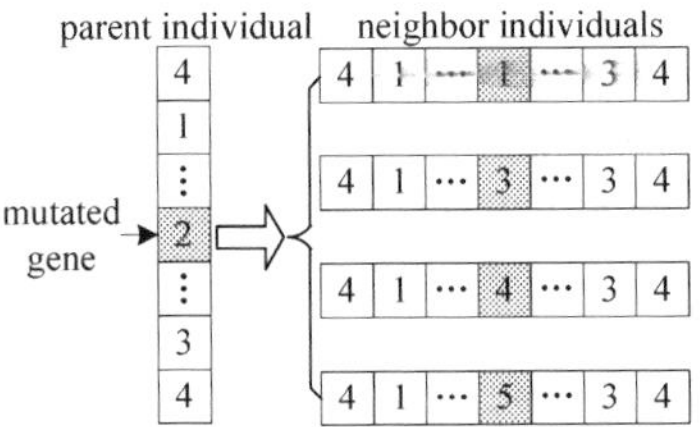

Fig. 3. Mutation with neighborhood search

Where the mutated gene is at i^{th} element in the encoding array and the number of Web services that can fulfill the functional requirements of abstract service t_i is 5.

Migration Operation. Migration operation occurs at a regular interval. The interval generation between the migrations is called migration interval and the number of individuals migrated between two neighboring subpopulations is called migration rate. During the process of migration operation, a certain number of pareto solutions are chosen randomly and sent to its neighboring domains which are determined by network topology. Meanwhile, each subpopulation replaces the worst solutions in its population by the solutions received from its neighboring domains. Migration operation between autonomous domains can avoid the premature convergence efficiently and obtain optimal global solutions, which will be validated by the experimental results of the next section.

5 Experiments

As an initial attempt to investigate the feasibility of the above approach to optimize Web services selection for grid workflow, a group of experiments are conducted. In our experiments, 10 autonomous domains are considered and each subpopulation owns 50 individuals. In order to facilitate the neighborhood search, crossover operation is descarded and only mutation operation is adopted, which is described in section 4.2. The migration interval and the migration rate are 10 and 2 respectively. During the process of migration operation, each subpopulation randomly selects two individuals from its pareto set and migrates them to exactly one of the neighboring subpopulations. Meanwhile, each subpopulation replaces the two worst solutions in its population by the solutions received from its neighboring domain.

Given a grid workflow with 16 abstract services, and the number of alternatives for each abstract service is 32, $b_1 = 0.7, b_2 = 70, b_3 = 7, b_4 = 0.7, b_5 = 0.7$, The pareto set of some generations is illustrated in Fig. 4.

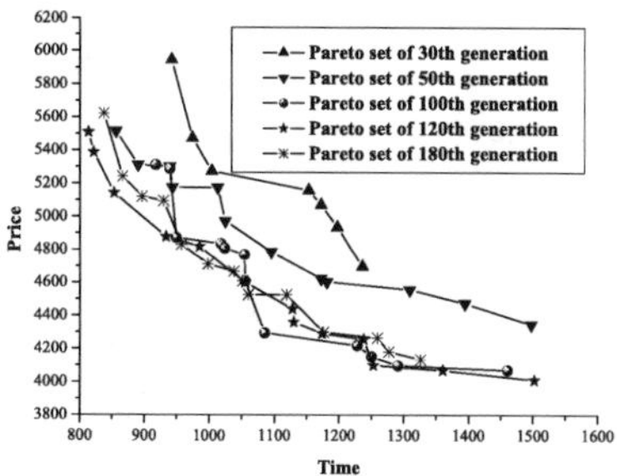

Fig. 4. Pareto set of some generations

As Fig .4 shown, in the early stage of the algorithm, the pareto solutions become more and more better. But after a certain number of generations, the pareto solutions of different generations arrive at a relative stable status and further evolution has no obvious effect on both the convergence and diversity of the pareto set. Therefore, an appropriate number of evolution generations should be chosen in order to achieve the tradeoff between solution optimization and the computation time of the distributed multi-objective genetic algorithm. For the objective criteria, the average values of pareto solutions for different generations are showed in Fig. 5 and Fig. 6.

Fig. 5 and Fig. 6 show that the average values of pareto solutions for both price and time criteria decrease markedly in the early stage of the algorithm. However, with the further evolution of population, these two values fluctuate between a small range and do not converge at a certain value, which indicates that these two objective criteria are independent and any criterion can not further optimize without increasing the value of another criterion. Therefore, comprehensive consideration of all objective criteria is needed in order to achieve the global optimization.

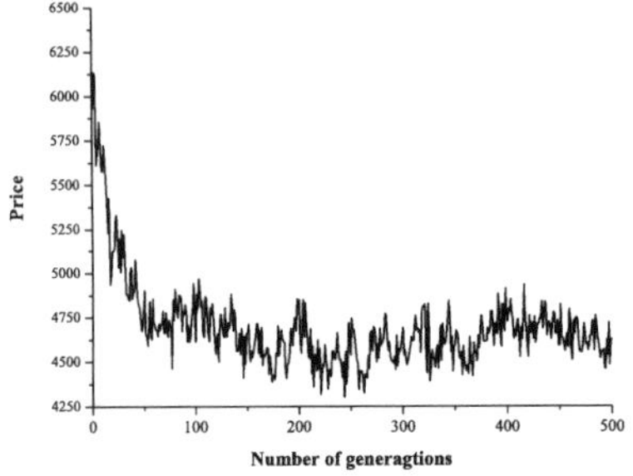

Fig. 5. Average value of price criterion

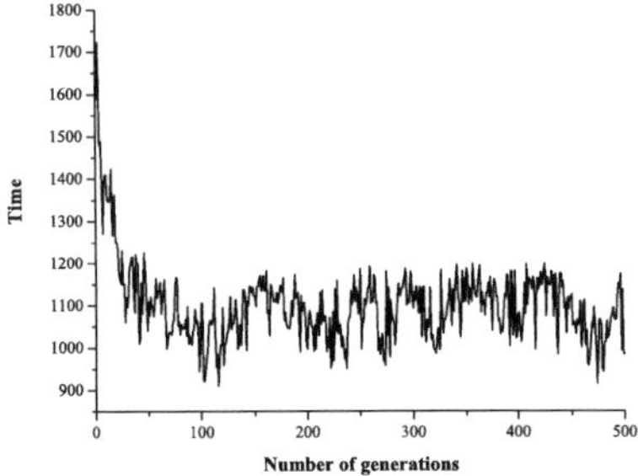

Fig. 6. Average value of time criterion

6 Conclusion

To accurately define the quality of the grid workflow, an extended quality model is proposed. Based on the extended quality model, an approach of distributed genetic algorithm to optimize quality of grid workflow is proposed. The application of distributed genetic algorithm is appropriate as the number of isolated management domains in grid environment is large and each of the domains has incomplete information about others. The distributed nature of the algorithm scales well in grid environment where there are a large number of service providers and their behaviors vary dynamically, and the experimental results show that this approach has ideal performance.

It should be pointed out that the quality of service in this paper is static preset, but in the open grid environment, the quality of service is continuously changing and the quality of service that a given service instance will deliver can not be easily obtained. How to dynamically obtain the quality of service will be our future work. Meanwhile, the quality of service in this paper is mainly provided by service providers, which will raise the problem of cheating, thus how to detect the cheating behaviour of service providers is another work needed to be studied in the future.

References

1. Yu, J., Buyya, R.: A Taxonomy of Workflow Management Systems for Grid Computing. In: ACM SIGMOD Record, 34(3), 44–49 (2005)
2. Deelman, E., Blythe, J., Gil, Y.: Mapping Abstract Complex Workflows onto Grid Environments. In: Journal of Grid. Computing 1(1), 25–39 (2003)
3. Bubak, M., Gubala, T., Kapalka, M.: Workflow Composer and Service Registry for Grid Applications. In: Future Generation Computer Systems 21(1), 79–86 (2005)
4. Bonatti, P.A., Festa, P.: On Optimal Service Selection. In: Proceedings of the 14th international conference on World Wide Web, Chiba, Japan, pp. 530-538 (2005)
5. Fonseca, C.M., Fleming, P.J.: An Overview of Evolutionary Algorithms in Multiobjective Optimization. In: Journal of Evolution Computation 3(1), 1–16 (1995)

6. Menasc, D.A.: QoS Issues in Web Services. In: IEEE Internet Computing, 6(6), 72–75 (2002)
7. Papazoglou, M.P., Dubray, J.-j.: A Survey of Web service technologies. In: Technical Report, DIT-04-058, University of Trento (June 2004), http://eprints.biblio.unitn.it/archive/00000586/01/mike.pdf
8. Cardoso, J., Sheth, A., Miller, J.: uality of Service for Workflows and Web Service Processe. In: web Semantics: Science, Services and Agents on the World Wide. Web. 1(3), 281–308 (2004)
9. Patel, C., Supekar, K., Lee, Y.: A QoS Oriented Framework for Adaptive Management of Web Service based Workflows. In: Proceeding of 14th Database and Expert Systems Applications Conference, pp. 826-835 (2003)
10. Zeng, L., Benatallah, B.: QoS-Aware Middleware for Web Services Composition. In: IEEE Transactions on Software Engineering, 30(5), 311–327 (2004)
11. Liu, Y., Ngu, A., Zeng, L.: QoS Computation and Policing in Dynamic Web Service Selection. In: Proceedings of the 13th International Conference on World Wide Web, pp. 66-73 (May 2004)
12. Canfora, G., Di Penta, M., Esposito, R.: An Approach for QoS-aware Service Composition Based on Genetic Algorithms. In: Proceedings of the 2005 conference on Genetic and evolutionary computation, Washington DC, pp. 1069-1075 (2005)
13. Cardoso, J., Sheth, A.: Semantic E-Workflow Composition. In: Intelligent Information Systems, 21(3), 191–225 (2003)
14. Deb, K.: Multi-Objective Genetic Algorithms: Problem Difficulties and Construction of Test Problems. In: Journal of Evolutionary Computation, 7(3), 205–230 (1999)
15. Gen, M., Cheng, R.: A Survey of Penalty Techniques in Genetic Algorithms. In: Proceedings of, IEEE International Conference on Evolutionary Computation, Nayoya University, Japan, 1996, pp. 804-809 (1996)
16. Deb, K., Pratap, A., Agarwal, S.: A Fast and Elitist Multiobjective Genetic Algorithm:NSGA-II. In: IEEE Transactions on Evolutionary Computation 6(2), 182–197 (2002)

Safety Analysis and Performance Evaluation of Time WF-nets

Wei Song[1,2], Wanchun Dou[1,2], Jinjun Chen[3], and Shaokun Fan[1,2]

[1] State Key Laboratory for Novel Software Technology
[2] Department of Computer Science and Technology, Nanjing University, Nanjing, China, Post Code 210093
[3] Faculty of Information and communication technology, Swinburne University of Technology, Melbourne, Australia
douwc@nju.edu.cn

Abstract. Time management in workflow systems is crucial in determining and controlling the life cycle of a workflow process. Research on time modeling and analysis is important to guarantee workflow plans to be efficiently implemented and to make enterprises more competitive. Time WF-nets derived from time Petri nets and WF-nets are an effective model for workflow time management. In this paper, considering multiple instances of one workflow model run concurrently, we concern how to determine the workflow instance arrival cycle in order to make the waiting time of each instance as short as possible. Meanwhile the safety property of Time WF-nets is preserved. The key contribution of our work is twofold. First, instance arrival cycle is calculated in order to satisfy the safety property of Time WF-nets. Second, performance evaluation for Time WF-nets is proposed, which is based on instance arrival cycle and instance throughput cycle derived from safety analysis.

1 Introduction

Workflow management now is a key technique that aims to help business goals to be achieved with high quality and efficiency. Correctness verification and performance evaluation for business processes are of great importance in workflow modeling phase. Hence, the first step for workflow management is to build an appropriate and analyzable workflow model to represent the business process. There are several popular workflow modeling techniques [1-4], directed acyclic graph (DAG), Petri nets (PN), unified modeling language (UML), State-Entity-Activity-Model (SEAM), to name a few. Among these models, Petri nets are used comprehensively, due to their formal semantics, simple representation of concurrency and synchronization. Using Petri nets to model workflow processes results in a class of Petri nets which is called workflow nets (WF-nets) [2].

Workflow model analysis is conducted mainly at three levels, i.e., logical, temporal, and performance levels [5]. In order to facilitate the time management and time performance analysis of workflow processes, Petri nets with time extensions are utilized for workflow modeling [5-7]. In this paper, we will discuss the safety property of Time WF-nets, in view of instance arrival cycle that is an important factor

K.C. Chang et al. (Eds.): APWeb/WAIM 2007 Ws, LNCS 4537, pp. 420–432, 2007.

for performance evaluation. Please note that the temporal analysis should be conducted after the soundness verification, which means that the WF-net considered here is free of logical errors.

Some works have been done for workflow correctness verification and performance evaluation [1-2,5-7]. However, they assume that two workflow instances of one process can run concurrently without any conflict between them, namely, one workflow instance does not impact on the other one. However, this assumption is not always true in real-life situations. Yet, little work has been concerned about the concurrency control of several workflow instances of the same workflow process. In this paper, we will focus on this.

The rest of this paper is organized as follows. Section 2 describes a motivating example of our work. Section 3 introduces critical task and critical section in Time WF-nets. Section 4 presents the methodologies for safety analysis and performance evaluation of Time WF-nets. Section 5 illustrates the proposal by an example based on a real-life complaint handling process. Section 6 briefly discusses a comparison with related work. Finally, the conclusion and future work are given in Section 7.

2 A Motivating Example

In this part, a motivating example is used to introduce the safety problem in workflow processes. Fig.1 draws the outline of the process that a person who catches a cold and goes to see a doctor in a hospital. t_1, t_2, t_3, t_4 denote registration, consultation, taking medicine, taking an injection, respectively. When the patient comes into the hospital, before consulting the doctor, she/he should register first. The doctor will make a suggestion according to the situations of the patient while the patient can determine to take an injection or take medicine instead. If two patients come into the hospital together, as there is only one doctor, the doctor can only serve one patient at a time. This means that task t_2 should be well organized to avoid conflict. The situation is the same to t_3, t_4, as there are limited nurses in the hospital. Service time is an important factor of QoS (Quality of Service) for this workflow process. Thus, the workflow tasks are assigned with time information in order to enhance the treatment rate of each task and organize the overall process properly.

Although this is a simple workflow process, two interesting topics are summarized below to highlight the motivations of our research work presented in this paper.

(1) How to determine the time delay between two workflow instances in order to make the waiting time of each workflow instance as short as possible, meanwhile to be able to avoid the conflict between two patients? Safety analysis is the general way to verify whether conflicts exist in Petri nets. Hence, in this paper, the least time delay between two workflow instances will be calculated so as to preserve the safety property of Time WF-nets.

(2) How to evaluate performance of the workflow process when considering multiple instances running concurrently? Given the amount of resource, the more workflow instances running in the process at the same time, the better of the performance. However, in view of the time delay between two workflow instances, performance evaluation should be conducted based on the delay derived from safety analysis.

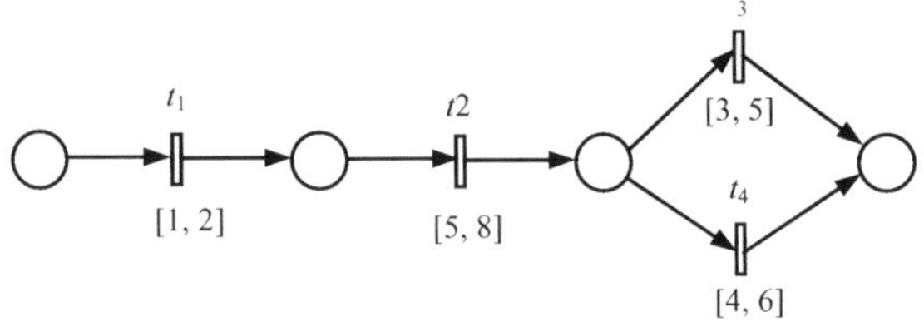

Fig. 1. Using WF-nets to model the process of seeing a doctor

These two problems will be discussed later in this paper. To achieve this goal, we first introduce the preliminaries of Time WF-nets and definitions of critical task and critical section, and then discuss safety problems and performance issues for workflow processes based on Time WF-nets.

3 Critical Task and Critical Section in Time WF-nets

3.1 Time WF-Nets Preliminaries

In this section, the basic notations of the Time WF-nets and relevant content are stated. We assume that readers have some knowledge of Petri nets.

Definition 1 (WF-net). A Petri net $N=(P, T, F)$ is a WF-net if and only if [2]:

(1) N has two special places: source place i and sink place o where ${}^{\bullet}i = \varphi$, $o^{\bullet}=\varphi$.

(2) If a transition t^* is added to N, connecting place o with i so that ${}^{\bullet}t^*= o$ and $t^{*\bullet}=i$, the Petri net N^* obtained is strongly connected.

Generally speaking, tasks in a workflow process are modeled by transitions, and causal dependencies are modeled by places or arcs. A place corresponds to a condition that can be used as pre and/or post condition for the task [what ref.?]. While tokens, represent the workflow cases. Roughly speaking, we can call a case as the intermediate state of a workflow instance or a workflow instance that is running [2].

Definition 2 (Time WF-net). A WF-net $N=(P, T, F)$ is a Time WF-net if we put a static time interval $[\alpha_1, \beta_1]$ to every transition, i.e., there is a mapping FI: $T\rightarrow INT$ [7]. Where $INT=\{[y, z] | y\leq z$ and $y, z\in \text{REAL}\}$, $FI(t)=[\alpha_1, \beta_1]$, α_1 and β_1 are real numbers and relative to the moment when t is enabled.

Facilitating further discussion, new time semantics is presented by extending Definition 2 where α_1 and β_1 can be regarded as the enabling duration and the firing duration of task t. This new time semantics will be illustrated in Section 3.2.

3.2 Critical Task and Critical Section

There are three basic ways to representing time in Petri nets: firing durations, holding durations, and enabling durations [8]. In general, timed Petri nets are augmented with firing and holding durations, and time Petri nets are with enabling durations. In this paper, in view of business processes, we assume that transitions are associated with enabling and firing durations. This endows Time WF-nets with new time semantics. Considering the example in Fig.2 which depicts a simple Time WF-net with enabling duration (α_1) and maximum firing duration (β_1) associated with transition t_1, the concrete running procedure of t_1 is shown in Fig.2.

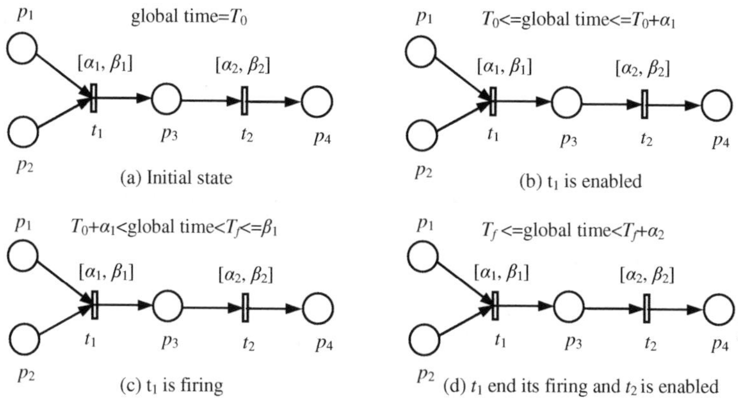

Fig. 2. A Time WF-net with enabling and firing durations

Definition 3 (Preparation time interval and maximum execution time interval). Consider the static time interval $[\alpha_1, \beta_1]$ associated with a transition t. $[0, \alpha_1]$ and $(\alpha_1, \beta_1]$ are the corresponding preparation time interval and maximum execution time interval of task t, respectively.

See Fig.3. Note that if a case comes into place p_1, the case should wait for α_1 time units and prepare for enough resources in order to fulfill the pre conditions of transition t_1. And after α_1 time units, the transition begins to execute, while (α_1, β_1) is the maximum execution time interval of the task. The task can end its execution at an arbitrary time point T_f in $(\alpha_1, \beta_1]$. Before firing, the case is contained in t_1.

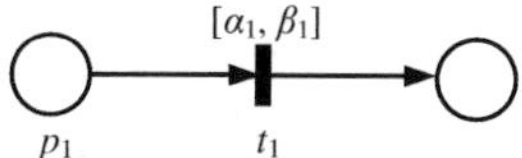

Fig. 3. A fragment of time WF-nets

Soundness reflects the correctness when only one case runs in the workflow process [2]. However, in productive and administrative workflows that emphasize on the repeated executions of workflow processes, concurrent execution of several workflow cases is very important in order to maximize the throughput. In the rest of our paper, we assume that the WF-net considered here is already sound. This enables us to concentrate on the safety property analysis of WF-nets when multiple instances run simultaneously.

Given a concrete workflow process, if several cases run concurrently, we should make sure that all the cases can run properly without any conflict among them. The notion is called time safety, which allows us to control multiple cases of workflow competing for limited resources. In analogy to operating system (OS) terminology, critical task and critical section are defined below.

Definition 4 (Critical task). A task in a Time WF-net is called a critical task, if the task cannot serve more than one case at a time due to the limited resources.

Definition 5 (Critical section). A critical section [9] in a Time WF-net is a non empty set of several related tasks that cannot serve more than one case at a time due to the limited resources. A critical section is a continuous fragment of places and transitions in the Time WF-net.

4 Workflow Safety Analysis and Performance Evaluation

4.1 Safety Analysis for Sequential Routings

After discussing the concepts of critical task and critical section, in this section, time safety property is particularly emphasized in a Time WF-net. Time safety means two aspects: place safety and transition safety. If a task is critical, the transition that represents this critical task should be safe, as well as the input places of the corresponding transition. Namely, not only the input places [7] but also the corresponding transition should be safe.

Definition 6 (Place safety). A place p is said to be safe if it contains at most one token at a time.

Definition 6 means that when several workflow instances execute concurrently, a place cannot contain more than one workflow case in order to avoid contradiction.

Definition 7 (Transition safety). A transition t is said to be safe if it contains at most one token when in running.

Definition 7 means that when several workflow cases execute concurrently, a transition can handle only one case at a time.

Place safety in Time WF-nets has been introduced in [7], while transition safety is a conception related to the new time semantics of Time WF-nets. Both situations have their sense in the workflow management. However, for a non-critical task, i.e., the task that can serve several cases at a time, there is no need to restrict place or transition safety in order to support the high degree for concurrency of workflow instances. If two cases arrive at a place in tandem, like the multiple enabledness issues of time Petri nets [10], we would like to adopt the age semantics [10-11] and follow the FCFS (First come, fist served) rule.

Definition 8 (Time WF-nets Safety). A Time WF-net is safe if and only if all critical tasks are safe, i.e., the corresponding places and transitions are safe.

Definition 9 (Place safety cycle). Place safety cycle is defined as the least time delay between two workflow cases that satisfies the safety of place *p*.

Definition 10 (Transition safety cycle). Transition safety cycle is **defined as** the least time delay between two workflow cases that satisfies the safety of transition *t*.

Definition 11 (Place safety cycle of a Time WF-net). Place safety cycle of a Time WF-net is defined as the least time delay T_p between two workflow cases that satisfies place safety of all critical tasks in the Time WF-net.

Definition 12 (Transition safety cycle of a Time WF-net). Transition safety cycle of a Time WF-net is defined as the least time delay T_t between two workflow cases that satisfies transition safety of all critical tasks in the Time WF-net.

Definition 13 (Instance arrival cycle). Instance arrival cycle T_a is defined as the time delay between the executions of two workflow instances in order to satisfy place safety and transition safety of all critical tasks in a Time WF-net.

Definition 14 (Instance throughput cycle). Instance throughput cycle is defined as the time duration for a case to run from the source place to the sink place.

Obviously, for a sequential Time WF-net with n transitions, the instance throughput cycle is within interval $[\sum_{j=0}^{n}\alpha_j, \sum_{j=0}^{n}\beta_j]$, where $T_l=\sum_{j=0}^{n}\alpha_j$ and $T_u=\sum_{j=0}^{n}\beta_j$ are the lower and upper bounds of the instance throughput cycle, respectively.

Theorem 1. In Fig.4, given a sequential Time WF-net composed of n critical tasks, then

(1) $T_t(k)=\sum_{j=0}^{k}(\beta_j-\alpha_j)$ and $T_p(k)=\sum_{j=0}^{k-1}(\beta_j-\alpha_j)+\alpha_k(1\leqslant k\leqslant n)$, are the transition safety cycle of t_k and place safety cycle of p_k, respectively.

(2) $T_t=T_t(n)$, $T_p=\max\{T_p(i), 1\leqslant i\leqslant n\}$ are the transition safety cycle and place safety cycle of this sequential Time WF-net, respectively.

Proof. It can be verified by induction on i.

1) For the basis case (i=1) (see Fig.3), suppose that the second case comes into place p_1 $T_p(1)$ time units later than the first one. In order to avoid contact in place p_1, when the second case comes in to p_1, the first one should begin to run. This means that $T_t(1)$ should not be less than α_1, i.e., $T_p(1)\geq\alpha_1$. Meanwhile, the second case can not run until the first one ends up its firing, i.e., $T_t(1)+\alpha_1\geq\beta_1$. So, $T_t=\beta_1-\alpha_1$ and $T_p=\alpha_1$ are the transition safety cycle and the place safety cycle of this Time WF-net in Fig.3.

2) Assume that the assertion holds for $i\leq k$, then $T_t(k)=\sum_{j=0}^{k}(\beta_j-\alpha_j)$, $T_p(k)=\sum_{j=0}^{k-1}(\beta_j-\alpha_j)+\alpha_k$. And $T_t=T_t(k)$, $T_p=\max\{T_p(i), 1\leqslant i\leqslant k\}$ are the transition safety cycle and place safety cycle of the workflow task sequence from t_1 to t_k, respectively. Now, consider $i=k+1$. See Fig.4. Let $T_t(k+1)$ and $T_p(k+1)$ be the transition safety cycle of t_{k+1} and place safety cycle of p_{k+1}. Note that $T_t(k)$ also represents that after $T_t(k)$ time units when case two finishes the firing of t_k, case two may have the chance to catch up with case one.(Case one ends its firing of t_k on upper bound of the instance throughput cycle from t_1 to t_k, while case two ends its firing of t_k on the lower bound.)

So, in order to ensure the safety of the place p_{k+1} and transition t_{k+1}, following two equations should be satisfied:

$$T_t(k+1)\text{-}\ T_t(k)+\ \alpha_{k+1}=\beta_{k+1} \tag{1}$$

$$T_p(k+1)\text{-}\ T_t(k)=\alpha_{k+1} \tag{2}$$

From equations (1)(2),we can get $T_t(k+1)=\sum_{j=0}^{k+1}(\beta_j-\alpha_j)$, $T_p(k+1)=\sum_{j=0}^{k}(\beta_j-\alpha_j)+\alpha_{k+1}$, respectively. And because $T_t(k+1)=T_t(k)+(\ \beta_{k+1}\text{-}\alpha_{k+1})>T_t(k)$, so $T_t=\max\{T_t(k+1), T_t(k)\}=T_t(k+1)$, while $T_p=\max\{T_p, T_p(k+1)\}=\max\{T_p(i), 1\leqslant i\leqslant k+1\}$.

Thus, in overall terms, the theorem holds.

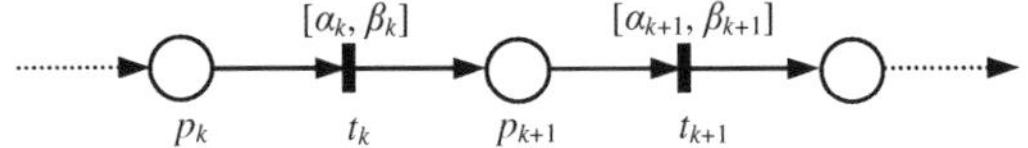

Fig. 4. A sequential routing of time WF-nets

Theorem 1 can be generalized as follows:

Corollary 1. Given a sequential Time WF-net composed with n tasks, there are m critical tasks in it. Let C be the set of numbers that represent the critical tasks' position in the sequential routing, and b is the maximal number of C. Then, $T_t=T_t(b)$, $T_p=\max\{T_p(j), j\in C\}$ are the transition safety cycle and place safety cycle of this Time WF-net, respectively.

Proof. This corollary can be derived from Theorem 1 and the above discussion.

Corollary 2. The instance arrival cycle for the sequential Time WF-net is $T_a=\max\{T_t, T_p\}$, where $T_t=T_t(b)$, $T_p=\max\{T_p(j), j\in C\}$.

Proof. Proof of this corollary follows directly from Corollary 1 and Definitions 13.

Theorem 2. In Fig.4, given a sequential Time WF-net composed of n tasks, if the first k tasks construct a critical section of the overall workflow process, then $T_t(k)=\sum_{j=0}^{k}\beta_j$ - α_1, $T_p(k)=\sum_{j=0}^{k-1}\beta_j+\alpha_k$ are the transition safety cycle and place safety cycle of the sequential Time WF-net from task t_1 to t_k, respectively.

Proof. we can verify it by induction on i.

1) For the basis case (i=1), see Fig.3. As there is only one task in the critical section, this problem in this situation degenerates to one critical task. According to Theorem 1, $T_t=\beta_1\text{-}\alpha_1$ and $T_p=\alpha_1$ are the transition safety cycle and place safety cycle for this peculiar critical section.
2) Assume that the assertion holds for $i\leq k$, then $T_t(k)=\sum_{j=0}^{k}\beta_j$ -α_1 and $T_p(k)=\sum_{j=0}^{k-1}\beta_j+\alpha_k$.

   Now, consider i=k+1. See Fig.4. Let $T_t(k+1)$ and $T_p(k+1)$ be the transition safety cycle and place safety cycle for the sequential critical section from task t_1 to t_k, respectively. From Fig.4, It can be verified that the following equations should be satisfied to avoid conflict in the critical section.

$$T_t(k+1) - T_t(k) = \beta_{k+1} \quad (3)$$

$$T_p(k+1) - T_p(k) = \beta_k - \alpha_k + \alpha_{k+1} \quad (4)$$

As $T_t(k)=\sum_{j=0}^{k}\beta_j - \alpha_1$ and $T_p(k)=\sum_{j=0}^{k-1}\beta_j + \alpha_k$, we could get from equation (3) and (4) that $T_t(k+1)=\sum_{j=0}^{k+1}\beta_j - \alpha_1$ and $T_p(k+1)=\sum_{j=0}^{k}\beta_j + \alpha_{k+1}$, respectively.

Thus, in overall terms, the theorem holds.

Although all these results are obtained from sequential Time WF-nets, we can generalize them to generic Time WF-nets. Further details are discussed in section 4.2.

4.2 Performance Evaluation Based on Instance Throughput Cycle and Instance Arrival Cycle Derived from Safety Analysis

In this section, we will discuss how to use instance arrival cycle and instance throughput cycle to conduct performance evaluation for a generic Time WF-nets. First, the transformation rules from parallel and conditional routings to sequential ones are given below in order to use the results in Section 4.1. For generic Time WF-nets, four routings have been identified: sequential, parallel, conditional and iteration. Iteration is often considered to be an undesirable form of routing [2]. So, this paper does not concern iteration routings.

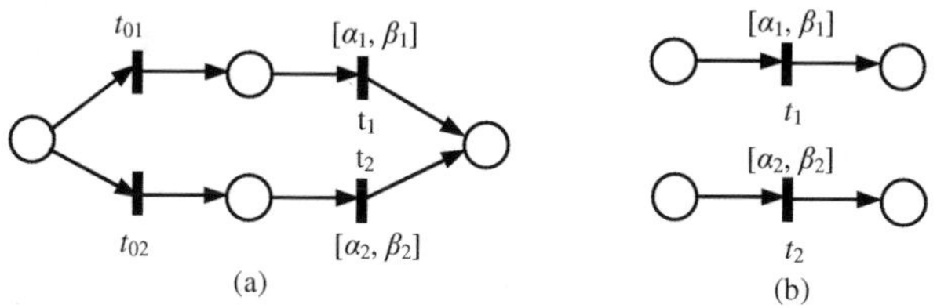

Fig. 5. Transform the conditional routing to sequential one of Time WF-nets

Conditional routing is used to allow for a routing that may vary between cases. It is pointed out that time Petri nets are not appropriate to model conditional structures [12]. However, instantaneous transitions can used to conquer this drawback. An instantaneous transition is a transition that can be fired at once after it is enabled. Using instantaneous transitions, a concrete conditional routing is illustrated in Fig.5 (a), which preserves the fairness.

Time WF-nets we used here are free choice Petri nets, suggested by Aalst in [2]. This is the reason why the conditional routings in Fig.5 (a) can be decomposed into two sequential routings as illustrated in Fig.5 (b).

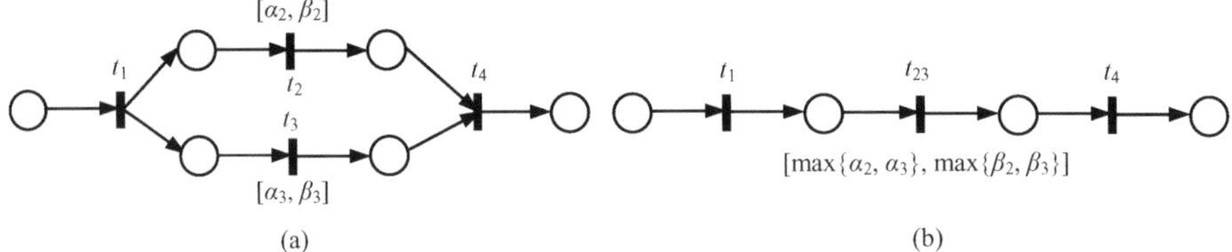

Fig. 6. Transform the parallel routing to sequential one of Time WF-nets

Parallel routing is used in situations where the order of execution is less strict. For example, in Fig.6 (a), both transitions t_2 and t_3 need to be executed but the order of execution is arbitrary. In other words, they can be executed concurrently. Fig.6 (b) depicts the time equivalent form of the parallel routing in Fig.6 (a) by combining t_2 and t_3 into t_{23}. This transformation rule can be verified using component-level reduction rules discussed in [13]. Due to the page limit, the proof is omitted here.

After discussing how to transform conditional, parallel, iteration routings into sequential ones, based on the results of Section 4.1, we can calculate the instance arrival cycle, as well as the lower and upper bounds of instance throughput cycle for any generic Time WF-net.

As known to all, instance throughput cycle is of great importance for evaluating performance of workflow processes. However, instance arrival cycle is also one of the most important performance indicators in industry, which reflects the instance disposal rate while multiple instances run concurrently. Now, we would like to discuss the performance of the WF-net based on these two factors.

Definition 15 (Degree of concurrency). Degree of concurrency is defined as the maximum number of instances a Time WF-net can serve at a time, i.e., $D=\lceil \max T_u / T_a \rceil$ where $maxT_u$ is the upper bound of the instance throughput time.

The degree of concurrency can reflect the performance character of a workflow process to some extent. We can make a verdict that the larger of D, the better of performance. However, we cannot draw the conclusion that the larger of $maxT_u$, the better of performance. In fact, this problem cannot be analyzed only from the granularity of instance throughput cycle. A wider time range is needed. If we assume that a workflow process works t time units continuously one day, then we may ask how many workflow instances can be handled? By the simple calculation, we can get the minimum and maximum number of instances the workflow process handled. They are $(t\text{-}minT_l)/T_a$ and $(t\text{-}maxT_u)/T_a$, respectively.

5 A Case Study

To illustrate the notion of safety in Time WF-nets, and performance evaluation based both on the instance throughput cycle and instance arrival cycle, we use a workflow example of a complaint handling process in an insurance company. The general process is represented as a WF-net depicted in Fig. 7.

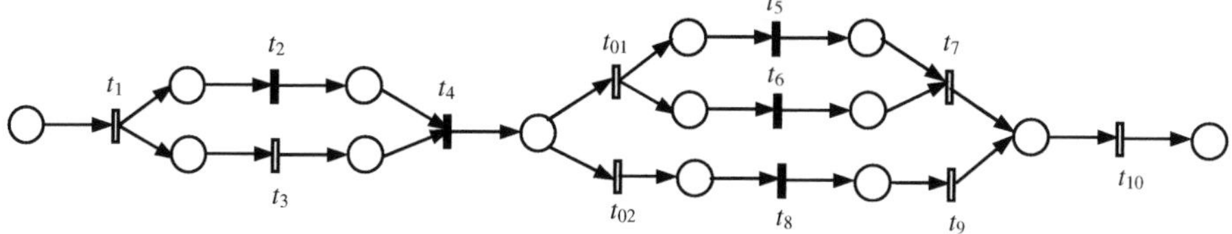

Fig. 7. A Time WF-net modeling the process of complaint handling

Descriptions and some other important information of the workflow tasks are presented in Table 1. Task t_{2c} and task t_4 constitute a critical section that can serve two workflow cases at a time. Note that the hollow transitions in Fig. 7 are non-critical, while the solid ones are critical. Now, the instance throughput cycle and instance arrival cycle of this Time WF-net should be calculated in order to facilitate the safety analysis and performance evaluation.

Table 1. Legend for Fig.7

Transition	Description	Type	Timing interval
t_1	Register	Non-critical	[1, 3]
t_2	Contact client	Critical	[3, 5]
t_3	Contact department	Non-critical	[2, 4]
t_4	Collect	Critical	[3, 4]
t_{01}	Agree to compensate	Non-critical	[0, 0]
t_{02}	Disagree to compensate	Non-critical	[0, 0]
t_5	Pay	Critical	[4, 6]
t_6	Send letter	Critical	[3, 4]
t_7	record	Non-critical	[1, 3]
t_8	Send letter	Critical	[3, 4]
t_9	record	Non-critical	[1, 3]
t_{10}	Archive	Non-critical	[0, 3]

Based on Section 4.2, the Time WF-net in Fig. 7 can be transformed into two sequential routings. Note that if a parallel routing has at least one critical task, then the corresponding task in the sequential routing after transformation is critical. Based on Theorem 1 the calculation results are summarized in Table 2. Particularly, the place safety cycle and transition safety cycle for the critical section from task t_{23} to task t_4 are calculated according to Theorem 2:

$$T_p=((3-1)+(5+3))/2=5,\ T_t=((3-1)+(5+4-3))/2=4.$$

Thus, the safety cycle for this critical section is max{5, 4}=5.

From Table 2, the instance throughput cycle can be easily obtained, i.e., $minT_l$=11, $maxT_u$=25. As the safety cycle for the critical section is 5, the and instance arrival

Table 2. Calculation results of Fig. 7

Transition	Type	Timing interval	$T_p(i)$	$T_t(i)$
t_1	Non-critical	[1, 3]	1	2
t_{23}	Critical	[3, 5]	5	4
t_4	Critical	[3, 4]	7	5
t_{01}	Non-critical	[0, 0]	7	5
t_{02}	Non-critical	[0, 0]	7	5
t_{56}	Critical	[4, 6]	9	7
t_7	Non-critical	[1, 3]	7	8
t_8	Critical	[3, 4]	8	6
t_9	Non-critical	[1, 3]	7	8
t_{10}(a)	Non-critical	[0, 3]	9	12
t_{10}(b)	Non-critical	[0, 3]	8	11
		[11, 25]	9	7

cycle is T_a= max{9,7}=9. We can get the degree of concurrency D of this Time WF-net. $D=\lceil \max T_u / T_a \rceil = \lceil 25/9 \rceil = 3$. If the insurance company works for 200 time units one day, the minimum and maximum number of instances the workflow process handled one day can be obtained: $(t\text{-}minT_l)/T_a$=(200-25)/9=19, $(t\text{-}maxT_u)/T_a$=(200-11)/9=21.

To automate the safety analysis and performance evaluation of Time WF-nets, we have implemented a software tool which is coded in Java. This tool can also support workflow modeling and time Petri nets simulation. Fig. 8 and Fig. 9 show two screenshots of this software tool.

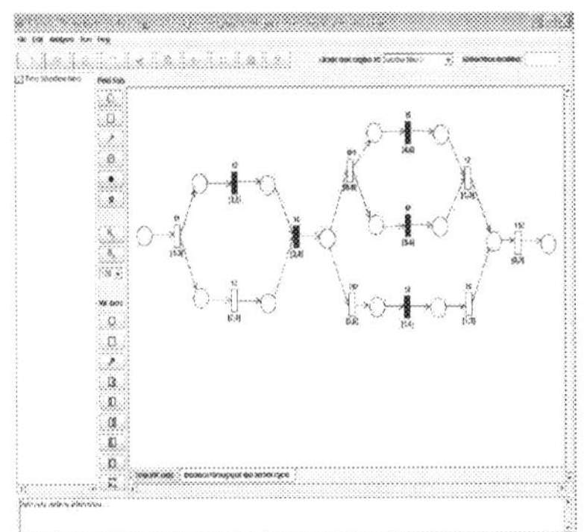

Fig. 8. Screenshot of workflow modeling using Time WF-nets software tool

Fig. 9. Screenshot of safety analysis and performance evaluation using Time WF-nets software tool

6 Comparisons with Related Work

In the field of time management for workflow processes, to the best of our knowledge, very few projects have taken time safety into account. Moreover, there is little work discussing performance evaluation of workflow processes using instance arrival cycle and instance throughput cycle.

Sea Ling and Heinz Schmidt used time Petri nets to model workflow processes [7]. They presented the soundness verification method for Time WF-nets, as well as the primary notion of safety in workflow processes. This inspires us for this work. Our work extends the notion of safety in workflow by presenting two kinds of safety: place safety and transition safety of critical tasks. Moreover, tokens are not used to model resources, i.e., they mean cases only. [5] discussed performance modeling based on multidimensional workflow net (MWF-net) and [6] used TCWF-net to study the boundedness of workflow models. However, they did not consider critical task (critical section), and ignored the safety problem in workflow systems. [9] utilized workflow control net to synchronize activity flows in the critical sections of the process. They addressed control flow logic between two workflow cases. However, we concern this problem from the time perspective in workflow processes.

7 Conclusions and Future Work

Time WF-nets allow multiple cases to run simultaneously in the system. This feature is more useful in practice and provides us with richer semantics for the workflow model. Due to the limited resources in the workflow process, critical task and critical section are introduced in order to define the safety property in Time WF-nets. Several definitions and theorems are presented to facilitate the safety analysis of Time WF-nets. Through safety analysis, the instance arrival cycle and the instance throughput cycle can be obtained. These are two important performance indicators by which we can conduct performance evaluation of a workflow process. A complaint handling process is used as an example to demonstrate the feasibility of our work.

In future, we plan to work on safety analysis and performance evaluation on inter-organizational workflows. This helps to pilot collaborations between distributed and virtual enterprises.

Acknowledgements. This paper is partly supported by the NSFC Project under Grant No.60303025 and No.60673017, National Grand Fundamental Research 973 Program of China under Grant No.2006CB303000, and Jiangsu Provincial NSF Project under Grant No. BK2004411 and No. BK2005208.

References

1. Sadiq, W., Orlowska, M.E.: Analyzing Process Models using Graph Reduction Techniques. Information systems 25(2), 117–134 (2000)
2. van der Aalst, W.M.P.: The Application of Petri nets to Workflow Management. The Journal of Circuits, Systems and Computers (1998)
3. Bastos, R.M., Ruiz, D.D.A.: Extending UML Activity Diagram for Workflow Modeling in Production Systems. Proceedings of the 35th Hawaii International Conference on System Sciences (2000).

4. Bajaj, A., Ram, S.: SEAM: A State-Entity-Activity-Model for a Well-Defined Workflow Development Methodology. IEEE Transactions on Knowledge and Data. Engineering 14(2), 415–431 (2002)
5. Li, J.Q., Fan, Y.S., Zhou, M.C.: Performance Modeling and Analysis of Workflow. IEEE Transactions on Systems, Man, and Cybernetics-Part. A: Systems and Humans, 34(2), 229–242 (2004)
6. Li, J.Q., Fan, Y.S., Zhou, M.C.: Timing Constraint Workflow Nets for Workflow Analysis. IEEE Transactions on Systems, Man, and Cybernetics-Part. A: System and Humans, 33(2), 179–193 (2003)
7. Ling, S., Schmidt, H.: Time Petri nets for Workflow Modeling and Analysis. In: Proceedings of the IEEE International Conference on Systems, Man, and Cybernetics (2000)
8. Bowden, F.D.J.: Survey and Synthesis of the Roles of Time in Petri Nets. Mathematical and Computer Modeling, 31, 55–68 (2000)
9. Kotb, Y.T., Baumgart, A.S.: An Extended Petri net for Modeling Workflow with Critical Sections. In: Proceedings of the 2005 IEEE International Conference on e-Business Engineering (2005)
10. Berthomieu, B., Diaz, M.: Modeling and Verification of Time Dependent Systems using Time Petri Nets. IEEE Transactions on Software Engineering 17(3), 259–273 (1991)
11. Boyer, M., Diaz, M.: Multiple Enabledness of Transition in Petri Nets with Time. In: Proceedings of the 9th International Workshop on Petri Nets and Performance Models (2001)
12. Tsai, J.J.P., Yang, S.J., Chang, Y.H.: Timing Constraint Petri Nets and Their Applications to Schedulability Analysis of Real-Time System Specifications. IEEE Transactions on Software Engineering 21(1), 32–49 (1995)
13. Wang, J.C., Deng, Y., Zhou, M.C.: Compositional Time Petri Nets and Reduction Rules. IEEE Transactions on Systems, Man, and Cybernetics-Part. A: Cybernetics 30(4), 562–572 (2000)

Dual Workflow Nets: Mixed Control/Data-Flow Representation for Workflow Modeling and Verification

Shaokun Fan[1,2], Wanchun Dou[1,2], and Jinjun Chen[3]

[1] State Key Laboratory for Novel Software Technology
[2] Department of Computer Science and Technology, Nanjing University, Nanjing, China, Post Code 210093
[3] Faculty of Information and communication technology, Swinburne University of Technology, Melbourne, Australia
douwc@nju.edu.cn

Abstract. A WFMS(workflow management system) contains two basic elements: the workflow model and the workflow engine. It is important to verify workflow models before they are put to execution. Traditional workflow models mainly describe workflows either from the control perspective or from the data perspective. In fact, the control flow and the data flow are two important aspects for workflow modeling and they are not independent from each other. A new workflow modeling technique, named Dual Workflow Nets (DWF-nets), is proposed to explicitly model the control flow and data flow of workflow processes. Besides, the control/data flow interactions can be captured in DWF-nets. Moreover, the control/data inconsistency, which is neglected by traditional modeling techniques, can be detected by verification of DWF-nets.

1 Introduction

These years have witnessed many workflow management systems implemented on a wide range of organizations. From business areas to scientific areas, the workflow technology is playing a more and more important role [1], [2]. Workflow management systems are being used to model, execute and monitor business/scientific processes in real or virtual enterprises.

A WFMS consists of two basic elements: the workflow model and the workflow engine [3]. Recently, various researchers have stressed the need for formal foundations of process specifications techniques, and many workflow modeling techniques and methodologies were promoted. Among these techniques, there are several popular ones: Petri nets [4], directed acyclic graph [5], unified modeling language [6], information control net [7], and so on.

Based on these model methods, some researchers use them to model the control flow, while others devoted themselves to the data flow. The control flow and the data flow are two important aspects of workflows [8]. The control flow determines the order and selection of activities that need to be performed to satisfy underlying business objectives of the workflow model. The data flow determines the flow of data from provider activities to receiver activities [9].

K.C. Chang et al. (Eds.): APWeb/WAIM 2007 Ws, LNCS 4537, pp. 433–444, 2007.

The control flow and the data flow of workflow processes are not completely independent. On the contrary, a control flow can determine the execution path of a data flow, while the result of a data flow operation can influence a control flow. These control/data flow interactions are not a trivial matter and have been long neglected. Traditional workflow modeling approaches are either circumscribed to the control domain or constitute the basic structure of the data domain.

We believe that a workflow model should differentiate between control and data components. Besides, mechanisms for modeling the control/data interactions and still being able to reach an analytical conclusion from them should be provided in all cases. By extending Dual Flow Net (DFN) [10], which was first introduced by Varea in the embedded system modeling area, we propose a new net structure—Extended Dual Flow Nets (EDFN). Then, based on EDFN, the concept of Dual Workflow Nets (DWF-nets) is presented for workflow modeling.

The main contributions of this work are twofold. First, we propose a new workflow modeling technique—DWF-nets, which can explicitly model the control flow, the data flow, and the interactions between the two flows. Second, a formal verification method is presented to detect control/data flow inconsistencies of a workflow model. The rest of this paper is organized as follows. In section 2, the underlying motivation for this work is outlined. Section 3 defines our model from a structural and behavioral perspective. The verification method of such a model is carried out in Section 4, while Section 5 shows a case study of the DWF-nets model. Section 6 is related works. Finally, conclusions are given in Section 7.

2 Motivations

In this section, let's elaborate our discussion by an example. To ease readers to follow, we refer to a simple *crane design process*, which is less complicated than real life applications. But it still outlines the problem that we want to tackle in this paper.

The research and development department of company X designs new cranes for the company to meet varied demands of customers. The design process can be divided into four subtasks: *analyze customers' requirements, design the boom, design the power system, and combined test*. First, the department should communicate with customers to analyze customers' requirements. At this step, customers' requirement specification will be produced for the following steps. Second, the boom will be planed to design based on the customers' requirement specification. Third, the power system will be designed based on the customers' requirement specification and some designed results of the boom, because the power system must guarantee that the boom can work safely. Finally, according to the data of the previous three steps, combined test should be carried out.

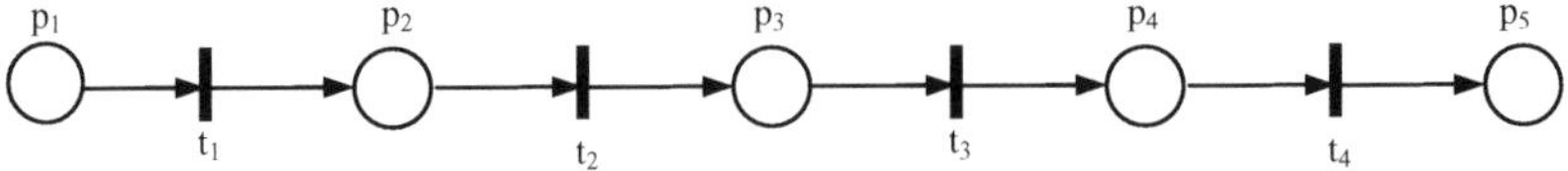

Fig. 1(a). The control flow model of *the crane design process*

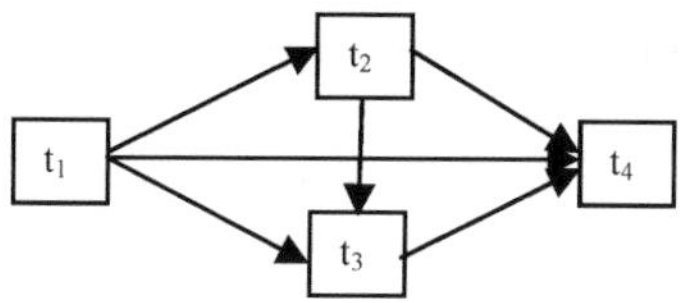

Fig. 1(b). The data flow model of *the crane design process*

Fig. 1(a) and Fig.1(b) depict the *crane design process* from the control flow aspect and the data flow aspect, respectively. In Fig.1(a), we use the WF-net[11] to model the control flow of this process. According to the correctness criterion in [11], it is apparently that Fig.1(a) shows a correct process model. In Fig.1(b), we use Directed Graph to model the data flow. It is also clear that this data passing logic itself is correct. Thus, we can claim that this process model is correct according to the correctness criterion.

However, if the execution sequence of $\mathbf{t_2}$ and $\mathbf{t_3}$ is changed and the modified control logic as is shown in Fig.2, it is also correct according to the traditional correctness criterion. Note that, in this case, we still use the data flow in Fig.1(b), which itself has been proved to be correct, too. Unfortunately, when this workflow model executes, there will be a confliction between $\mathbf{t_2}$ and $\mathbf{t_3}$, because $\mathbf{t_2}$ should satisfy the conditions simultaneously: (1) it should execute after $\mathbf{t_3}$ according to the new control flow; (2) it should execute before $\mathbf{t_3}$ according to the data flow. This execution logic is named as the "***modified*** *crane design process*" in the rest of this paper.

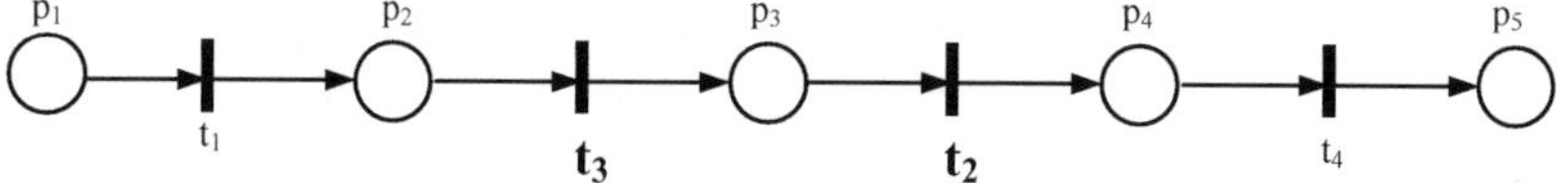

Fig. 2. The control flow of the ***modified*** *crane design process*

From the above example, we can draw the conclusion that logic conflictions may still exist when both the control logic and the data logic are correct. In this paper, we define this kind of confliction between control flows and data flows as **control/data flow inconsistency**. In this simple example, it is not so difficult for us to find the control/data flow inconsistency. But for complicated real-life applications, which might have hundreds of sub-tasks, manual verification for such process models is complex and hard to achieve with an adequate dose of accuracy.

In the rest of this paper, in order to detect the **control/data flow inconsistency,** the DWF-nets model is proposed, which can model the data flow and control flow in a single graph. Then, the formal verification method is presented for detecting the **control/data flow inconsistency**.

3 Formalization of DWF-nets

3.1 Preliminary Knowledge of DFN

The DFN model [10] is first introduced in the embedded system modeling area. It assumes a structure where three types of elements (rather than two, as in standard

P/T nets) compose the entire system: storage, reactive, and transformational units. Given a weighted, directed, tripartite graph, its vertices $V = P \cup T \cup Q$ can be used to map each of these elements as follows: Storage elements ($p \in P$) relate to memory components in the system (e.g., registers, memory cells, latches, variables, etc.), reactive elements ($t \in T$) allude to components in the control part and transformational elements ($q \in Q$) refer to arithmetic operations performed among storage elements.

Definition 1. A Dual Flow Net is a tuple $S = (P, T, Q, F, W, G, H, \mu_0)$, where:

(1) P is a finite, non-empty set of places;

(2) Q, T are a finite set of transitions and hulls respectively, such that $P \cap T = \emptyset$, $P \cap Q = \emptyset$, and $T \cap Q = \emptyset$ holds;

(3) $F \subseteq (P \times T) \cup (T \times P) \cup (P \times Q) \cup (Q \times P) \cup (T \times Q) \cup (Q \times T)$, is a binary relation, called the flow relation;

(4) $W : F \rightarrow Z^+ \cup Z^-$ is a weight function;

(5) $G : T \rightarrow \# \cup \{T\}$ is a guard function, where $\# = \{=, \neq, >, <, \geqslant, \leqslant\}$, and T is a symbol of truth value;

(6) $H : Q \rightarrow Z$ is an offset function;

(7) μ_0 is an initial marking function.

Each transition $t \in T$ is either labeled with a symbol from the set # or T, according to the guard function $G(t)$. Hulls $q \in Q$ are labeled with integers, corresponding to the offset function $H(q)$. The guard function $G(t)$ provides a mechanism for data flow to interfere in the control flow, i.e., to have conditions taken from the data flow. The offset function $H(q)$, on the other hand, enhances the functionality of the hull. The basic operation of a hull is to sum over the data domain, so the $H(q)$ function is provided in order to cover those situations where a constant is needed. For the length being, the details of DFN are omitted here. Interested readers can refer to [10].

3.2 Formalization of EDFN and DWF-nets

There are some essential differences between embedded systems and workflow processes. Straightforward using of DFN can not model workflow processes exactly. In this section, some extensions are made to DFN for workflow modeling.

In contrast to DFN, EDFN extends it from the following aspects: (1) Some unnecessary parts are omitted, i.e., the guard function $G(t)$ and the offset function $H(q)$. This makes our model more readable and concise. (2) Some concepts are changed, i.e., the flow relation F and the Marking function. These changes are made according to the semantics of workflow processes. (3) The behavioral model is changed due to the modification of the net structure.

Def.2 introduces the concept of EDFN. Please note that the EDFN is just a mathematical model. When it is used to model workflow process, some extraordinary conditions should be satisfied. In this paper, we name this special kind of EDFN as DWF-nets.

Definition 2. An Extended Dual Flow Net is a six-tuple $S =(P, T, Q, F, W, M_0)$, where (1) P is a finite set of places, T is a finite set of transitions, Q is a finite set of hulls;

(2) $F \subseteq (P \times T) \cup (T \times P) \cup (P \times Q) \cup (Q \times P) \cup (T \times Q)$, is a binary relation, called the flow relation;

(3) $W \rightarrow Z^{+} \cup Z^{-}$ is a weight function;

(4) M_0 is an initial marking function, and the marking function is defined by Def.5.

P, T and Q are disjoint, i.e., $P \cap T = \emptyset$, $P \cap Q = \emptyset$ and $T \cap Q = \emptyset$, and both P and $T \cup Q$ are not empty sets. The cardinality of each set P, T and Q is represented by n, m and h respectively, where $n > 0$, $m > 0$, $h > 0$ and $m+h > 0$ holds. The elements $p \in P$, $t \in T$, and $q \in Q$ are referred as places, transitions, and hulls of the EDFN net, and are graphically represented by a circle, a bar, and a box, respectively.

The control domain and the data domain of this model are formalized in Def.3 and Def.4, where the concepts of pre- and post- sets, from the classical Petri nets, are extended to support any element $p \in P$, $t \in T$, and $q \in Q$ of the new EDFN model.

Definition 3. Given a certain $p \in P$, $t \in T$, $q \in Q$, the following subsets are defined:

(1) The control preset of a place, $\bullet p = \{t \in T | (t, p) \in F\}$;

(2) The control postset of a place, $p \bullet = \{t \in T | (p, t) \in F\}$;

(3) The control preset of a transition, $\bullet t = \{p \in P | (p, t) \in F\}$;

(4) The control postset of a transition, $t \bullet = \{p \in P | (t, p) \in F\}$;

(5) The control preset of a hull, $\bullet q = \{t \in T | (t, q) \in F\}$;

Definition 4. Given a certain $p \in P$, $t \in T$, $q \in Q$, the following subsets are defined:

(1) The data preset of a place, $\circ p = \{q \in Q | (q, p) \in F\}$;

(2) The data postset of a place, $p \circ = \{q \in Q | (p, q) \in F\}$;

(3) The data postset of a transition, $t \circ = \{q \in Q | (t, q) \in F\}$;

(4) The data preset of a hull, $\circ q = \{p \in P | (p, q) \in F\}$;

(5) The data postset of a hull, $q \circ = \{p \in P | (q, p) \in F\}$.

Note that $\circ t = \emptyset$and $q \bullet = \emptyset$ because the flow relation F excludes any pair in $(Q \times T)$. A transition $t \in T$ has two different post-sets $t \bullet$ and $t \circ$, since a control action would normally perform changes in the state of the process, i.e. places $p \in t \bullet$, and the active data path, i.e. hulls $q \in t \circ$. However, A hull $q \in Q$ can only have one post-set $q \circ$ since the data transformation would only influence the data value in places $p \in q \circ$.

The EDFN should have places which are capable to capture not only the control domain, but also the data domain. Because of its two-dimensional nature, the complex number(C) is used to represent tokens in places. Please note that although DFN also marks places as complex numbers and it uses the modules and angles of complex numbers to represent the control domain and the data domain. However, in EDFN, real parts and imaginary parts of complex numbers are adopted.This makes the marking function more intuitive and still has the ability to describe both control and data domain in one place. The marking function is defined in Def.5.

Definition 5. The marking function M: $P \rightarrow$ C, where C is the set of all complex numbers. The marking $M(p)$ of a place $p \in P$, is written as a complex number in Cartesian coordinates, e.g., $X + Yi$. The following notation is used hereafter: $M = (M(p_1), M(p_2), \ldots, M(p_n))$.

For further discussions, two operators on complex numbers are defined:

(1) REAL(X+Y**i**) = X, as the control part of the complex-value token;
(2) IMAG(X+Y**i**) = Y, as the data part of the complex-value token.

Def.5 declares the marking as an assignment of complex numbers (C) into places of the net, as opposed to the classical marking function defined in the natural (N) domain. This marking function is capable of considering the effects of both control and data domains when analyzing the dynamics of a workflow model, because of its extended structure that allows two independent but related sets of quanta, i.e., X and Y, to share a place at any time. The control domain captured as the real part X of a complex number and the data domain as its imaginary part Y.

An EDFN which models a workflow process definition (i.e. the life-cycle of one case in isolation) is called a Dual Workflow-net (DWF-net). The concept of DWF-nets is formally defined by Def.6.

Definition 6. An EDFN is a DWF-net if and only if:

(1) $i \subseteq P \wedge o \subseteq P$, where $\bullet i = \emptyset \wedge \circ i = \emptyset$; $o\bullet = \emptyset \wedge o\circ = \emptyset$;
(2) $\forall x \in P \cup T \cup Q \wedge x \neq i \wedge x \neq o$, x is on the path from i to o.

The item(i) is used to describe that a workflow process should have one start point and at least one terminal point. The item(ii) ensures that there are no dangling tasks. Every task should contribute to the processing of cases.

3.3 Behavioral Model of DWF-nets

The behavior of the DWF-nets model is described in terms of enabling and firing transitions, as in classic Petri nets, in addition to a synchronized data flow operation scheme. Please note that the behavioral model is different from that in [10] because of the modifications of the net structure.

Definition 7. A transition t is said to be enabled, for a given marking M_k, if the following condition holds:

(1) All places in preset $p_i \in \bullet t$ contain at least $W(p_i, t)$ tokens, which is:
$\forall p_i\ ((p_i \in \bullet t) \rightarrow (\text{REAL}(M_{k+1}(p_i)) \geq W(p_i, t)))$
(2) All hulls in post set $q_i \in t\circ$ can be executed, which is:
$\forall q_i\ ((q_i \in t\circ) \rightarrow (\forall p_m(p_m \in \circ q_i) \rightarrow (\text{IMAG}(p_m) \geq W(p_m, q_i))))$

Def.7 states whether a transition is enabled or not in the DWF-nets model. On the control perspective, the enabling condition of a transition depends on not only the data tokens, but also the control tokens. In fact, this transition enabling condition puts the data influence on the control domain. When a transition is enabled, it will be fired according to the following definition.

Definition 8. The firing of an enabled transition t_j changes a marking M_k into M_{k+1} by means of the following rules:

(1) A finite number of control tokens are removed from $p_i \in \bullet t_j$:

$$\text{REAL}(M_{k+1}(p_i)) = \text{REAL}(M_k(p_i)) - W(p_i, t_j),\ \forall p_i \in \bullet t_j$$

(2) A finite number of control tokens are added to $p_i \in t_j\bullet$:

$\text{REAL}(M_{k+1}(p_i)) = \text{REAL}(M_k(p_i)) + \text{W}(t_j, p_i), \forall p_i \in t_j\bullet$

(3) Each hull $q \in t_j\circ$ is executed (c.f. Def.9).

From the behavioral point of view, the execution of hulls $q \in Q$ are synchronized with transitions $t \in T$ in the net, i.e., no hull q can fire non-deterministically. This is where the control flow influences the data flow in DWF-nets. By Def. 9, the execution process of q is described.

Definition 9. The firing of any transition $t \in \bullet q$ produces the execution of the hull q, which changes a marking M_k into M_{k+1} as follows:

(1) A finite number of data tokens are removed from $p_i \in \circ q$

$\text{IMAG}(M_{k+1}(p_i)) = \text{IMAG}(M_k(p_i)) - \text{W}(p_i, t_j), \forall p_i \in \circ q$

(2) A finite number of data tokens are added to $p_i \in q\circ$

$\text{IMAG}(M_{k+1}(p_i)) = \text{IMAG}(M_k(p_i)) + \text{W}(t_j, p_i), \forall p_i \in q\circ$

Thus far, the definitions introduced in this section are based on classic Petri nets. Actually, the control part of DWF-nets is semantically equivalent to Petri nets, while the data part is modeled based on the complex number. By combining the two aspects, we obtain a model that not only can describe both the data flow and the control flow of a workflow, but also can capture the interactions between the two flows.

4 Verification of Workflows Based on DWF-nets

The correctness of the workflow model supported by WFMSs is very important. A workflow process definition which contains errors may lead to failures at running time. These running time failures will cost much more than expected. Hence, it is vital to verify a workflow process definition and make it correct before it is put into execution. But what is the definition of "correct"? Based on the correctness criterion of WF-nets, the correctness criterion of DWF-nets we use in this paper is a set of minimal requirements any workflow process definition should satisfy:

(1) For any case, the procedure will terminate eventually and the moment the procedure terminates there is a control token in place o and all the other places are empty. Please note that, there is no constrain on data tokens.

(2) There should be no dead tasks, i.e., it should be possible to execute an arbitrary task by following the appropriate route though the WF-net.

These two properties are called soundness property and the formal definition of soundness in this paper is similar to that in [11]:

Definition 10. A procedure modeled by a DWF-net is sound if and only if:

(1) For every state M reachable from state i, there exists a firing sequence leading from state M to state. Formally: $\forall M (i \xrightarrow{*} M) \Rightarrow (M \xrightarrow{*} o)$;

(2) State o is the only state reachable from state i with at least one token in place o. Formally: $\forall M (i \xrightarrow{*} M \wedge M \geq o) \Rightarrow (M = o)$, where $M \geq o$ iff for all $p \in P$: $\text{REAL}(M(p)) \geq \text{REAL}(o(p))$;

(3) There are no dead transitions. Formally: $\forall t \in T \exists M, M' (i \xrightarrow{*} M \xrightarrow{t} M')$.

In order to check soundness of a DWF-net, we first introduce the concept of occurrence graph (OG), which is actually a directed graph. The nodes (M_n) of this graph are all the reachable markings and there is a directed arc labeled (M_1, t, M_2) from M_1 to M_2 when $M_1 \xrightarrow{t} M_2$. The following algorithm shows how to construct an occurrence graph for a DWF-net. The enabling condition (Def.7) and the firing transitions (Def.8 and Def.9) are used in this algorithm.

```
Algorithm 1. GenerateOG
Input:A DWF-net
Output:An OG corresponding to the DWF-net
W := ∅; OG := ∅;
Add M0 into W;
Add a node valued M0 into OG;
Repeat
Select a node M1∈W;
X := ∅;
For all t∈DWF-net do
  If t is enabled by the Marking M1(Def.7) then
    M2:= the result of firing t in M1;(Def.8 and Def.9)
    Put (t, M2) into X;
End for
For all elements in X do
  If M2 does not exist in G then
    Add M2 into W and add a node valued M2 into OG;
  If arc (M1, b, M2) does not exist in G then
    Add an arc valued (M1, b, M2) into OG;
End for
Remove M1 from W;
Until W := ∅;
Return OG;
```

Here, the behavioral model of DWF-nets is utilized to generate the OG. According to Def.10, algorithm 2 is presented to check whether a DWF-net is sound.

```
Algorithm 2. Soundness Checking
Input:A DWF-net and its OG
Output:A Boolean value to show whether DWF-net is sound.
W: = ∅;
For each node M in OG do
  BreadFirstSearch(OG, M) and put searched nodes in W;
  If o∉W then return false;
  W := ∅;
End for
For each node M in OG do
  If Real(M)≥1&& M≠o then return false;
End for
For each t in DWF-net do
  If t is not in at least one OG arc then return false;
End for
Return true;
```

Proof. Nodes in the OG are all reachable states M, so this algorithm is a straightforward consequence of Def. 10.

Note that we do not explicitly consider the data part in the verification algorithm. Since the control/data flow interaction is considered in the transition conditions which are used to generate OG, actually data parts can also impact the soundness property.

Further analysis of the two algorithms will reveal that this formal verification method is not efficient enough. When a DWF-net has many places, transitions and hulls, the number of nodes in OG of the DWF-nets model will be tremendous large. Hence, part of our future work will be devoted to more efficient verification methods.

5 Case Study

In this section, we illustrate our approach by using the examples mentioned in Section 2. In the first part of this section, we use the DWF-nets to model the *crane design process* and the ***modified** crane design process,* respectively. In order to present the behavioral model of DWF-nets, the execution sequence of the *crane design process*'s DWF-net model is also analyzed. In the second part of this section, the formal verification method is carried out to check these two DFW-nets models.

5.1 Workflow Modeling

Fig.3 is the DWF-nets model of the *crane design process*. In this model, *P* captures both control states and data states of this process; *T* captures the changes in states, i.e., the control flow, and exposes its influence over the third set *Q*; *Q* captures all transformations that are relevant to the data flow of the workflow.

Please note that, as in Petri nets, in order to reduce notational clutter, the weight functions are only written for nontrivial cases, i.e., numbers 1 are not explicitly written down across the net. All weight values of the arcs in this example are 1, so there is no number attached to arcs shown in Fig. 3. Besides, three kinds of arcs are used in Fig.3 to represent different flow semantics. However, this is just to make the graph more readable and there is no different constrains on these different arcs.

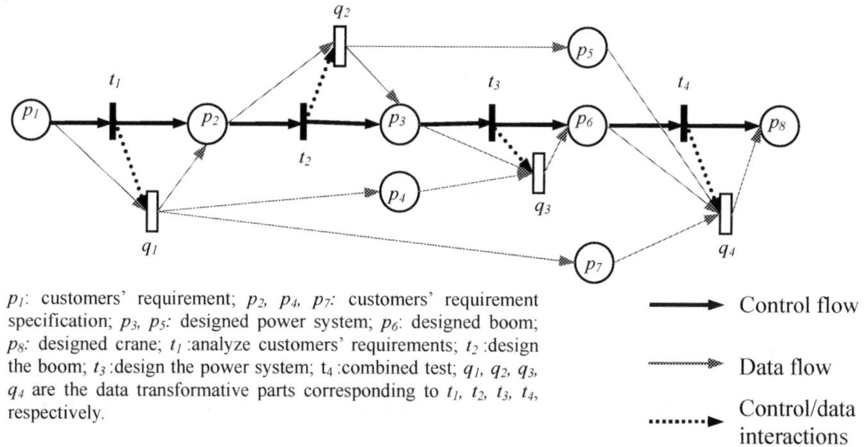

Fig. 3. Using DWF-nets to model the crane design process

The DWF-nets model of the ***modified*** *crane design process* is shown in Fig.4. The semantics of places, hulls, transitions, and flow arcs are the same with those in Fig. 3. The key difference between Fig.4 and Fig.3 is the control logic of $\mathbf{t_2}$ and $\mathbf{t_3}$.

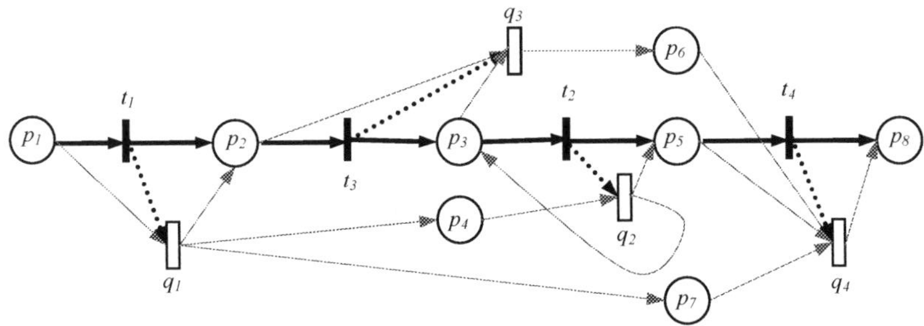

Fig. 4. Using DWF-net to model the ***modified*** *crane design process*

In order to show the behavioral model of DWF-nets, the execution sequence of Fig.3 is analyzed as bellow:

1. At the beginning of this process, M_0 is ((1+1**i**), (0+0**i**), (0+0**i**), (0+0**i**), (0+0**i**), (0+0**i**), (0+0**i**), (0+0**i**)). According to Def.7, $t_1$is fired.
2. Therefore, according to Def.8:
   a) One control token is removed from p_1;
   b) One control token is added to p_2;
   c) $t_1\circ = \{q_1\}$, so q_1 is executed according to Def.9.
3. The execution of q_1 have the following effects: Each place of p_2, p_4 and p_7 will be added one data token.
4. The marking function M_1 is ((0+0**i**), (1+1**i**), (0+0**i**), (0+1**i**), (0+0**i**), (0+0**i**), (0+1**i**), (0+0**i**)). M_1 satisfies the following two conditions:
   d) $\forall p_i\ ((p_i \in \bullet t_2) \rightarrow (\text{ REAL}(M_{k+1}(p_i)) \geq \text{W}(p_i\ , t_2)))$
   e) $\forall q_i\ ((q_i \in t_2\circ) \rightarrow (\forall p_m(p_m \in \circ q_i) \rightarrow (\text{IMAG}(p_m) \geq \text{W}(p_m\ , q_i))))$
   Then t_2 is fired.
5. After the execution of t_2 and q_2, M_2 is ((0+0**i**), (0+0**i**), (1+1**i**), (0+1**i**), (0+1**i**), (0+0**i**), (0+1**i**), (0+0**i**)).
6. Then t_3 is fired according to Def.7. And after the execution of t_3 and q_3, M_3 is ((0+0**i**), (0+0**i**), (0+0**i**), (0+0**i**), (0+1**i**), (1+1**i**), (0+1**i**), (0+0**i**)).
7. At last, t_4 is fired. And M_4 is ((0+0**i**), (0+0**i**), (0+0**i**), (0+0**i**), (0+0**i**), (0+0**i**), (0+0**i**), (1+1**i**))

5.2 Workflow Verification

According to Algorithm 1, the OG of the DWF-net of the *crane design process* is shown in Fig.5, where values of M_0-M_4 are shown in section 5.1. Based on Algorithm 2, we can claim that DWF-nets model of the *crane design process* is correct according to our correctness criterion.

Fig. 5. The OG of the DWF-net of the *crane design process*

For the ***modified*** *crane design process*, which has a control/data inconsistency, its OG is shown Fig.6. But in this case, Algorithm 2 will show that it is not a sound DWF-net. So, such kind of control/data inconsistency can be detected before execution in the DWF-nets model.

M_0 —t_1→ M_1

$M_0 = ((1+1\mathbf{i}), (0+0\mathbf{i}), (0+0\mathbf{i}), (0+0\mathbf{i}), (0+0\mathbf{i}), (0+0\mathbf{i}), (0+0\mathbf{i}), (0+0\mathbf{i}))$
$M_1 = ((0+0\mathbf{i}), (1+1\mathbf{i}), (0+0\mathbf{i}), (0+1\mathbf{i}), (0+0\mathbf{i}), (0+0\mathbf{i}), (0+1\mathbf{i}), (0+0\mathbf{i}))$

Fig. 6. The OG of the DWF-net of the ***modified*** *crane design process*

6 Related Works and Comparison Analysis

Aalst [4] identifies three reasons for using Petri Nets in workflow modeling. Firstly, Petri Nets possess formal semantics despite their graphical nature. Secondly, instead of being purely event-based, Petri Nets can explicitly model states, and lastly it is a theoretical proven analysis technique. But Petri nets are only mentioned to model the control flow of workflow processes. Other researches use high level Petri nets to model workflow processes from different perspectives. S. Ling [12] uses time Petri nets to model the control flow of workflow processes with time constrains. Aalst [13] discusses Object Petri nets and uses it to model control, data, resources, tasks, and operations. But the control/data inconsistency is not mentioned in this paper. Dongsheng Liu [3] adopts colored Petri nets to model a family of workflow processes instead of individual processes.

There are also some approaches focused on data flows. Shazia Sadiq [9] proposes a number of properties of data that should constitute an essential part of a data model and data validation problems are concerned in this work. Also some data-control flow inconsistencies are presented. But formal verification methods are not presented. Jintae Lee [14] uses dataflow diagram, which consists of a collection of processes, stores, and terminators linked by flows, to model the dataflow of processes.

In contrast to the above workflow modeling techniques, the DWF-nets model technique has the following distinct characteristics:

(1) It explicitly describes both the control flow ($P \times T$ and $T \times P$) and the data flow ($P \times Q$ and $Q \times P$), which facilitates the runtime interpreting for workflow engines.

(2) It has the ability to model the control/data flow interactions ($T \times Q$ and $P \times T$), and consequently be able to detect control/data flow inconsistencies before execution.

(3) It uses complex numbers to depict data and control in places.

(4) Separating control flow ($P \times T$ and $T \times P$) and data flow ($P \times Q$ and $Q \times P$) also makes the DWF-nets model easier for modelers and readers to understand.

7 Conclusions

The control flow and the data flow are two important aspects of workflow processes. Based on DFN, this paper defines a new workflow model (DWF-nets), which can describe both the data flow and the control flow. The control/data flow interactions are also concerned in DWF-nets, which makes it conveniently fitted into a formal verification framework, allowing the inconsistency detection of workflow models.

Our future work mainly includes two parts. First, we are going to work on more efficient methods for workflow verification. Second, a workflow engine based on DWF-nets will be developed for workflow modeling, verification and execution.

Acknowledgements. This paper is partly supported by the NSFC Project under Grant No.60303025 and No.60673017, National Grand Fundamental Research 973 Program of China under Grant No.2006CB303000, Jiangsu Provincial NSF Project under Grant No.BK2005208, and Program for New Century Excellent Talents in University. We also appreciate Dr. Varea for his helpful discussion.

References

1. van der Aalst, W.M.P., ter Hofstede, A.H.M., Weske, M.: Business Process Management: A Survey. International Conference on Business Process Management (BPM 2003) (2003)
2. Yu, j., Buyya,: A taxonomy of scientific workflow systems for grid computing. In: Special Issue on Scientific Workflows, SIGMOD Record. 34(3), pp. 44–49. ACM Press, New York (2005)
3. Liu, D., Wang, j., Chan, S.C.F., Sun, J., et al.: Modeling workflow processes with colored Petri nets. Computer in Industry 49, 267–281 (2002)
4. van der Aalst, W.M.P.: Three Good Reasons for Using a Petri Net-Based Workflow Management System. In: Proceedings of the International Working Conference on Information and Process Integration in Enterprises (IPIC'96) (1996)
5. Sadiq, W., Orlowska, M.E.: Analyzing Process Models using Graph Reduction Techniques. Information systems 25(2), 117–134 (2000)
6. Bastos, R.M., Ruiz, D.D.A.: Extending UML Activity Diagram for Workflow Modeling in Production Systems. Proceedings of the 35th Hawaii International Conference on System Sciences (2000)
7. Ellis, C.A., Nutt, G.J.: Modelling and Enactment of Workflow Systems. Application and Theory of Petri Nets 1993, Lecture Notes in Computer Science 691 (1993)
8. Russell, N., ter Hofstede, A.H.M., Edmond, D., van der Aalst, W.M.P.: Workflow data patterns. QUT Technical report, Queensland University of Technology (2004)
9. Sadiq, S., Orlowska, M., Sadiq, W., Foulger, C.: Data Flow and Validation in Workflow Modelling. In: Proceedings of the 15th Australasian database conference (2004)
10. Varea, M., Al-Hashimi, B.M., Corte´s, L.A., Eles, P., Peng, Z.: Dual Flow Nets: Modeling the Control/Data-Flow Relation in Embedded Systems. ACM Transactions on Embedded Computing Systems, 5(1), 54–81 (2006)
11. van der Aalst, W.M.P.: The Application of Petri nets to Workflow Management. The Journal of Circuits, Systems and Computers (1998)
12. Ling, S., Schmidt, H.: Time Petri nets for Workflow Modeling and Analysis. In: Proceedings of IEEE International Conference on Systems, Man, and Cybernetics (2000)
13. van der Aalst, W.M.P., Moldt, D., Wienberg, F.: Enacting Interorganizational Work using nets in nets. In: Proceedings of the 1999 Workflow Management Conference (1999)
14. Lee, J., Wyner, G.M.: Defining specialization for dataflow diagrams. Information Systems 28(6), 651–671 (2003)

Bridging Organizational Structure and Information System Architecture Through Process

Vladimír Modrák

Technical University of Košice, Faculty of Manufacturing Technologies, Bayerova 1, 08001 Prešov, Slovakia
modrak.vladimir@fvt.sk.

Abstract. Because the architecture of an enterprise information system determines relations of its individual parts with its surrounding environment, an effective use of an information system (IS) requires, apart from other things, its painless integration into the enterprise management system. The IS architecture can be seen by different angles. The focus of this paper is in particular on the application architecture of an IS. The reason for analyzing organizational structure modeling approaches and IS architecture is to eliminate common semantic gaps between the languages used by IS users and designers. The architectural design of enterprise IS might be tightly interconnected with enterprise organizational architecture. However, in reality, architectural models of enterprise IS are not fairly understandable for users. For both of these architectural models the orientation on processes often is a joint base. Thus, process approach can prospectively bridge existing gaps between enterprise organizational models and IS architectures. In coherence with that goal the paper presents methodological aspects of how cross-organizational process support can be effectively used for designing the first stage of information system.

Keywords: IS architecture, Diagram, Business process, Organization.

1 Introduction

Statistics on information systems evaluation in generally indicate that less number of IS projects were successful than those, which were abandoned or failed. Rueylin [19] explains the failure of enterprise IS projects as an implication of complex contextual patterns including organizational context issues. The present research dwells about organizational context in the first stages of IS design and analysis. The object of interest is the transformation of an object-oriented organization into process-oriented organization. Contrary to latter ones, the traditional organizational structure models are organized based on objects [2]. Because a modern organization must be adjustable to the major or minor changing situation, it is practically constantly transformed. While it is possible to plan retaliations for major change, minor changes may be unnoticed even thought they are implicate important functional changes. Therefore minor organizational changes may also pose a serious problem for IS project management. Accordingly, organizational structure models might be basically

K.C. Chang et al. (Eds.): APWeb/WAIM 2007 Ws, LNCS 4537, pp. 445–455, 2007.

integral to application architecture of an IS. The practical problem seems to be lying in an inconsistency of organization transformation and IS analysis methods. Hess and Oesterle [9] compared 12 different business process redesign methodologies showing how they fail to address the issue of company culture or subculture. Company culture or subculture issues in this sense are closely related with familiarity of management to applied methodologies. Solving methodological problems requires cooperation between the IS designers that develop the tools and the users of information system. As a common base for organization structure model and architectural model of IS might be orientation on process with the aim to bridge existing semantic gaps. The paper is structured in the following way. After research background on a common base of organization models and IS architecture are presented methodological aspects of the proposed process modeling technique for designing of business process models. Finally findings and research conclusions are discussed.

2 A Common Base of Organization Models and IS Architecture

The role of information systems in business processes is tangible. In opposite way the position of business processes from viewpoint of their reengineering is equally obvious. Despite of that there are evident differences between the underlying theories of business processes and information systems. These differences result from different subject matters and their theoretical grounds [8]. Business process management provides the foundation for the handling changes in the organizational structure to the underlying information systems that might be in place to power process-managed enterprises. On the other hand, theories on information systems are often oriented towards technological and informational aspects [12]. By Österle [17], the promising way to the understanding contextual character of information systems leads through business processes. By Marash [14], the term of the business process means the totality of all of the individual activities that the business performs. In this connection, integrated management system modeling can be considered as pertinent approach to designing of business process models. According to Flinchbaugh [6], the optimal design of a management system should bridge company functions in new and integrated ways. According to Lee and Walden [11], the journey of learning how the various management systems fit together and how to design integrated management systems for an organization is at its beginning stages. As significant component of integrated management system are considered business process maps. Based on a such map one can begin to define, which related processes are involved in analyzed workflow and how one can flow chart them in detail [14]. From another viewpoint, the process mapping produces workflow diagrams that bring forth a clearer understanding of a process or of a series of parallel processes. Workflow diagrams subsequently can be used in a designing of process-oriented IS architectures.

The term IS architecture from different aspects can be viewed. The term *information system architecture* is rather complicated. Its indeterminateness came from the term itself. For the purpose of present research we will come out from recognition of information, application and technological architecture of IS. The focus of this paper is in particular on the application architecture of an IS. Its objective is a detailed model of IT application areas with the aim to make the organization's activities more effective. The model of the application architecture of an enterprise IS basically identifies the areas of information flow automation inside business processes, their interfaces and integration requirements as well as relations to external processes. As it is known, the term 'architecture' does not origin in informatics. Zachman [21] compares directly the procedure for the architectural design of a building with the procedure for designing an IS. The architecture of a system in both areas may have a very different content depending on complexity and other characteristics of the system. A topical issue of IS architecture is its function. In the case of a complex system, there will be more functions. The functions are basically represented by models of processes in terms of flows and information processing. The solution therefore requires gradual decomposition of complex processes into partial ones and the subsequent decomposition of complex functions into partial ones and the possible decomposition of functions into activities or/and another smaller elements. The overall philosophy of such attitude implies that the decomposition takes place gradually in the top-down direction.

According to Vasconcelos et al. [20], application architecture of enterprise IS has to be a part of an enterprise organizational model. By them an architectural design of IS has to be closely interconnected with designing of enterprise architecture (EA) model and these two architectural models require being fully compatible and comprehensible for both involved parties users and system engineers. For this reason, especially EA model should be as transparent and unequivocal as possible. These requirements on EA model can be effectively accommodate via business process maps represented by workflow diagrams. In that connection, it is often necessary to introduce changes in the organizational structure, which is the platform for enterprise reengineering (see for instance [4], [5], [7], [13], [18]). However, the effective adoption and use of process models in practice has been slow [1].

3 Bridging Semantic Gaps Between IS Users and Designers

Information system development methodologies (object-oriented, structured or other) have traditionally been inspired by programming concepts, not organizational ones, leading to a semantic gap between the system and its environment [3]. Contrariwisely, semantic gap between IS users and designers is one of the potential realm in the deployment of the enterprise information systems (EIS) applications.

Basically, this phenomenon results from different views of users and designers to IS. The users of IS understand it in simplified way as an operation automation tool, mainly used in data processing and accessing. Designers of IS or its part understand it principally as HW/SW subsystems and their interaction. These contradictory notions are shown in Fig. 1, in which chosen components are used as an example.

It is obvious that outlined semantic gap can not be replaced completely. The goal should be to reduce it to a level allowing to relieve the shortage made by it, when creating and deploying IS. The way of minimizing this semantic gap is possible through process modeling approach leading up to merge of process models to process maps in the phase before IS design and development. A precondition for such process

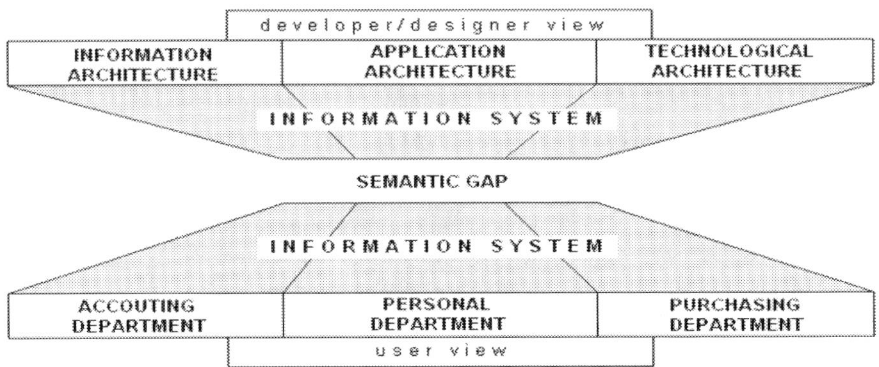

Fig. 1. Different views of users and designers to IS

maps might be understandability for users and simultaneously utilizabibility for IS developers (consultants). In opposite way, even though the EIS is sophistically designed, possible deficiencies are practically anticipated. In other words, to ensure better operatibility of enterprise activities through IS without detailed knowledge of process relations in various levels of decomposition of enterprise activities is rather complicated. For designers of IS this potential risk is not a crucial issue, moreover if they know that client/user is responsible for specifications of requirements for software system. But companies are not fully known of this reality. It confirms the fact that companies mostly use object-oriented model of organization structure without formal determination of work and information flows instead of process-oriented model of organization structure. The goal of next section is to outline possible way of process mapping, which makes provision for requirements on narrowing semantic gap mentioned above and transformation of object-oriented organization structure to process-oriented process structure.

4 Business Process Mapping

4.1 Framework of Business Processes Classification

One of the important roles of business process reengineering should be building a logistical concept of organization, which should involve co-ordination and management of all material and information flows. Especially higher-level modelling of business process structures substantially support the company's successful running. Such process models structures normally start by establishing a framework for the systematic classification of company processes. A classification framework for the systematic rebuilding of processes can be built from three hierarchical levels, which are (bottom to top) [15]:

- *Elementary process (EP)* represented by a set of complex tasks, consisting of the smallest elements-activities;

- *Integrated process (IP)* which represents a set of two or more Elementary processes with the purpose to create the autonomic organizational unit at the second hierarchical level;
- *Unified enterprise process (UEP),* which consists of one or several integrated processes to the extent that is conditioned by its capability to flexibly and effectively secure customers' requirements.

This classification framework is rigidly applied in the business processes modelling approach, which is described in the following sub-section.

4.2 Business Process Mapping Technique

Further described process mapping technique is based on process decomposition that is resulting in a set of business structure models, which are represented by diagrams in the order given [16]:

- *System diagram,*
- *Context diagrams,*
- *Commodity flow diagrams,*
- *State transition diagrams.*

An example of the first three simplified diagrams is illustrated on Fig. 2.

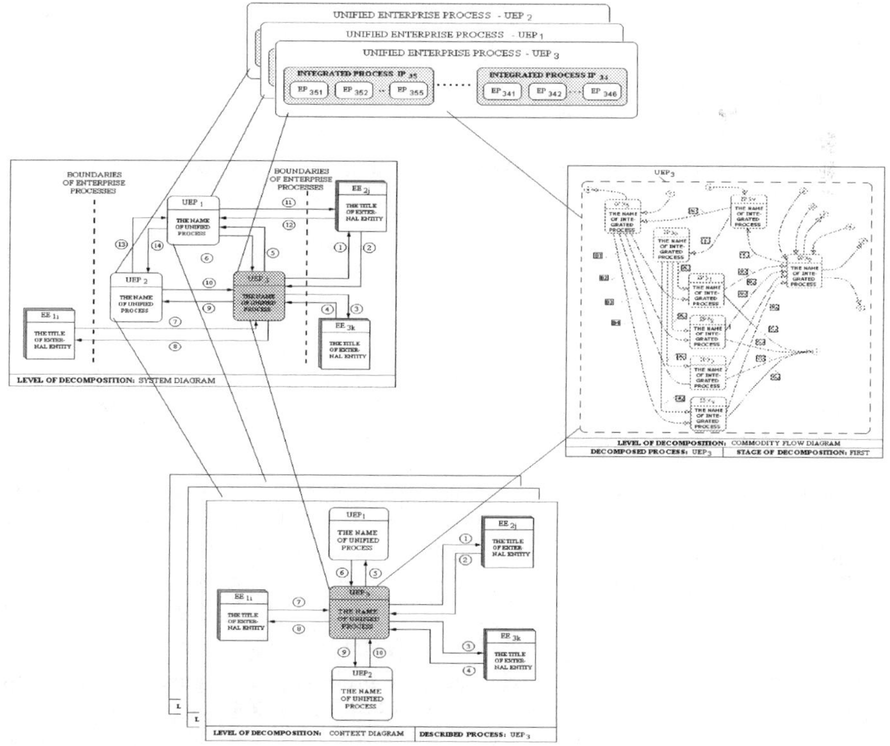

Fig. 2. A fragment of the structure of process diagrams

According to the procedure outlined for redesigning enterprise processes, the first step of this method is the creation of a System Diagram. Its purpose is to separate so-called Unified Enterprise Processes (UEP) from the original arrangement of processes. Subsequently, relations between them and the environment of the enterprise are specified. The environment is represented in a System Diagram by External Entities, with which the system communicates, while their content is not a subject of analysis in the following steps. They usually represent the initial source of commodity flows, or their end consumer. In the fact it represents the starting base of modelling processes, from which other diagrams are derived using the principle of process decomposition. Fig. 3 shows an example of System diagram.

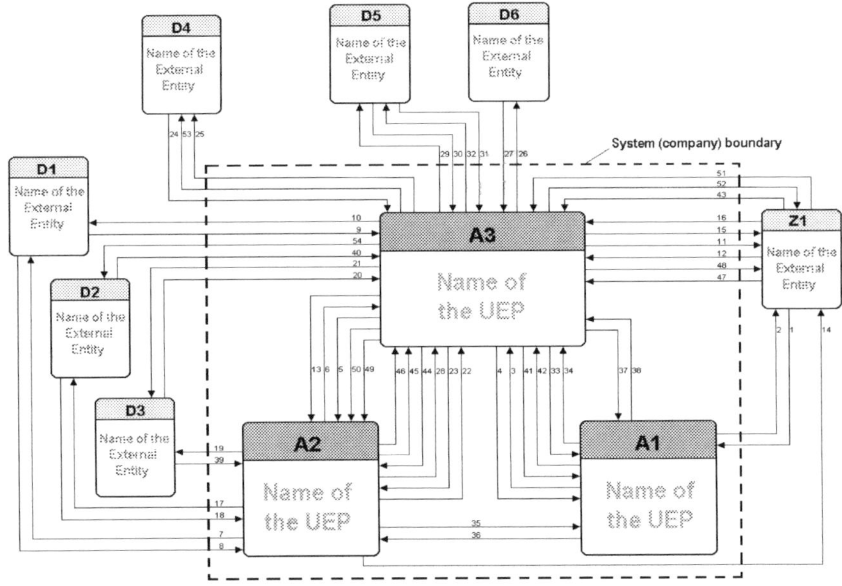

Fig. 3. System Diagram that reflects the starting situation of EA modelling, where A1, A2, A3 represent level of Unified enterprise processes and D1-D6, Z1 depicts the external entities

In terms of the reengineering objectives it is recommended to establish an optimal number of Unified enterprise processes. The number of types of external objects depends on the real or essential structure of external suppliers and customers.

The subsequently created context diagrams were pointed to improve the readability during the gradual modelling of processes. Context Diagrams are created for each Unified enterprise process drawn in a System Diagram. Individual Context Diagrams express relations of the given Unified enterprise process with its environment. These surrounding elements, irrespective of whether they represent objects outside the enterprise or internal processes, are approached as External Entities. It means that External Entities are considered in the same way as internal processes of the System Diagram. System decomposition at the given level emphasizes the need for the creation of an equal customer approach, in which there should not be differences between internal and external customers. One of the Context Diagrams derived from a previous System Diagram is shown on Fig. 4.

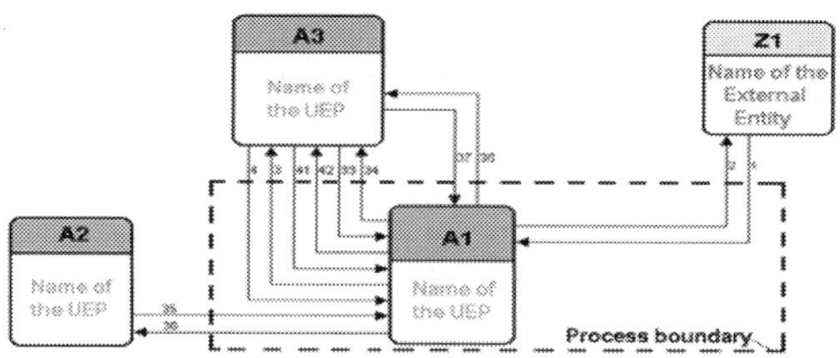

Fig. 4. The Context diagram for the process A1

Consecutively, Commodity Flow Diagrams are designed for A1, A2 and A3, which describe relations at the level of Integrated Processes. Two of them for the A2 and A3 processes are shown on Fig. 5 and Fig. 6.

The essence of the Commodity Flow Diagrams is gradual decomposition of UEP, up to the level of so-called elementary or primitive processes. The diagrams at the first stage of decomposition start from Context Diagrams. In the sense of the proposed classification, the mutual links of Integrated Processes (IP) are described in them.

Despite that, the Commodity Flow Diagrams do not provide any details about modelled processes; their purpose is to provide a general overview of the follow-up of processes, which allows their owners at different levels to see the boundaries of their

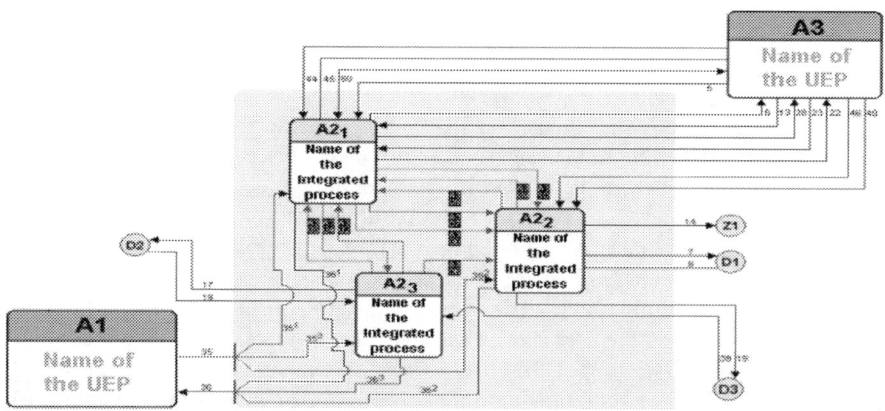

Fig. 5. Commodity flow diagram for the process A2. The processes A21, A22, A23 represent a level of integrated processes.

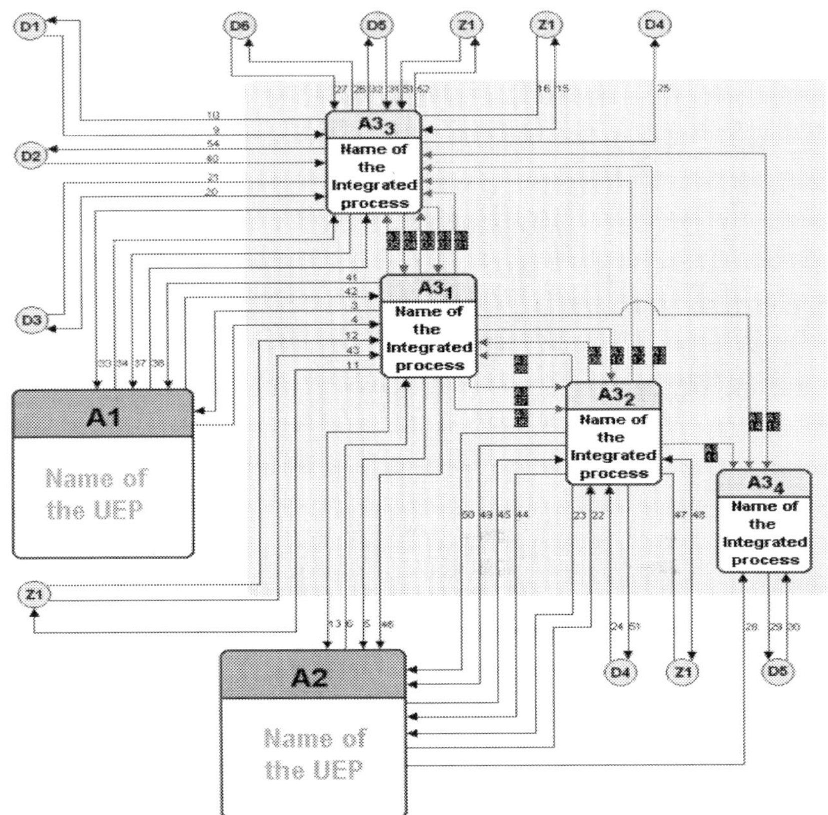

Fig. 6. Commodity Flow Diagram for the process A3

own as well as subsequent processes. Commodity Flow Diagrams of the second stage are constructed in an analogous way as Commodity Flow Diagrams of the first stage. It is the last stage of commodity flow diagrams because the Elementary processes, which present the objects of modelling, are considered to be the primitive processes.

The objective of the State Transition Diagram is the description of the dynamics of Elementary processes by modelling states, in which objects can be presented and transition periods between these states. These diagrams also describe events that initiate transitions between states and conditions for the realization of these transitions. In analogical way, State Transition Diagrams are sequentially created up to the level of State Transition Diagrams at the second stage.

An important condition in the designing and subsequent modelling of process diagrams is to maintain the consistency of inputs and outputs, so that it is possible to create process maps, starting at the level of Commodity Flow Diagrams at the first stage, up to the level of state transition diagrams. An example of the creation of the process map from two Commodity Flow Diagrams at first stage specified in previous figures 4 and 5 is shown in Fig. 7.

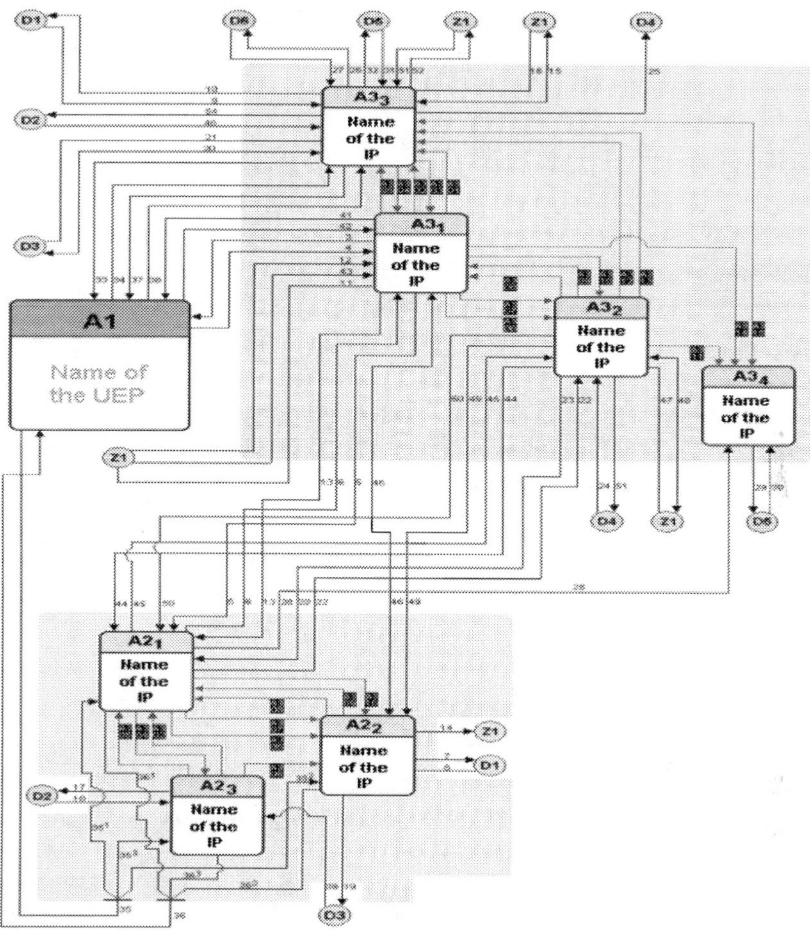

Fig. 7. Example of a process map by the merging of two Commodity Flow Diagrams at the first level of decomposition

5 Discussion and Conclusion

The sense of the process approach to organizational structure and information system architecture design lies in a increasing of the effectiveness of the organisation management and a creation of preconditions for the effective EIS development. As outlined above, the process approach is the common feature of enterprise architecture and of application architecture of information system.

The formation of a process-oriented organisation cannot be narrowed to the redefining of processes in the form of their new description and redesign on the basis of the abstract models creation. The transition to the process-oriented organisation envisages a noticeable change in its very existence. That includes the use of potent management tools, such as information systems, which automate business processes by controlling the sequence of activities and by the activation of necessary resources.

Awareness of organizational context of information system in the scope of outlined approach helps to design and maintenance IS. The classification of business processes, as a form of clustering, moreover serves for logical processes coding. Accordingly, it enables simple identification of individual parts of the process model in cases of updates and how these updates are related to other processes. An application of this approach also helps overcoming difficulties in interactions between users of various professions and specialists on information systems.

References

1. Austin, S., Baldwin, A., Li, B., Waskett, P.: Analytical Design Planning Technique (ADePT). A Dependency Structure Matrix Tool to Schedule the Building Design Process. Construction Management and Economics 2, 173–182 (2000)
2. Brickley, J.A., Smith, C.W., Zimmerman, J.: Managerial Economics & Organizational architecture. Mc Graw-Hill/ Irwin, New York, USA (2003)
3. Castro, J., Kolp, M., Mylopoulos, J.: Towards requirements-driven information systems engineering. the Tropos project. Information Systems 6, 365–389 (2002)
4. Crowe, T.J., Fong, P.M., Bauman, T.A., Zayas-Castro, J.L: Quantitative Risk Level Eestimation of Business Process Reengineering Efforts. Business Process Management Journal 5, 490–511 (2002)
5. Davenport, T.H., Stoddard, D.B.: Reengineering: business change of mythic proportions? MIS Quarterly. June, 121-127 (1994)
6. Flinchbaugh, J.: Using Integrated Management Systems to Design a Lean Factor. Center for Quality of Management Journal 2, 23–30 (1998)
7. Grant, D.: A wider view of business process reengineering. Communications of the ACM 2, 85–90 (2002)
8. Goldkuhl, G., Röstlinger, A.: Towards an integral understanding of organizations and information systems: Convergence of three theories. In: Proceedings of the 5th International Workshop on Organizational Semiotics, Delft (2002)
9. Hess, T., Oesterle, H.: Methods for Business Process Redesign: Current State and Development Perspectives. Business Change and Reengineering 2, 73–83 (1996)
10. Kosanke, K., Zelm, M.: CIMOSA modeling processes. Computers in Industry 2(3), 141–153 (1999)
11. Lee, T.H., Walden, D.: Designing Integrated Management Systems. Center for Quality of Management Journal 1, 3–18 (1998)
12. Lind, M.: Contextual Understanding of Information Systems – Characteristics of Process Oriented Information Systems. In: Campbell B., Underwood J., Bunker, D. (eds.) Proceeding of the 16th Australasian Conference on Information Systems (ACIS 2005), University of Technology, Sydney (2005)
13. Maull, R.S., Tranfield, D.R., Maull, W.: Factors Characterizing the Maturity of BPR Programmes. International Journal of Operations and Production Management 6, 59 (2003)
14. Marash, S.A.: A Process Approach to ISO 9001: Quality Digest, Retrieved January 20, (2000), from http://www.qualitydigest.com/Oct00/html/marash.html

15. Modrák, V.: Evaluation of Structural Properties for Business Processes. In: Sixt International Conference of Enterprise Information Systems: ICEIS, - Proceedings, Universidade Portucalense, Portugal. Porto, 2004, pp. 619–622 (2004)
16. Modrák, V.: Business Process Improvement through Optimization of its Structural Properties. In: Fischer, L. (ed.) Workflow Handbook, Future Strategies, Book Division Lighthouse Point, FL, USA, 2005, pp. 75- 90 (2005)
17. Österle, H.: Business in the Information Age – Heading for New Processes. Springer, Berlin Heidelberg New York (1995)
18. O'Neill, P., Sohal, A.S.: Business Process Reengineering: Application and Success-an Australian study. International Journal of Operations & Production Management 9(10), 832–864 (1998)
19. Hsiao R.: Why IT-enabled Change Fail. Working Paper of the National University of Singapore. Retrieved (January 2007), from http/school.nus.edu.sg/Research/ paperseries.htm
20. Vasconcelos, A., Sousa, P., Tribolet, J.: Information System Architecture Evaluation: From Software to Enterprise Level Approaches. In: 12th European Conference on Information Technology Evaluation (ECITE 2005), Turku, Finland September (2005)
21. Zachman, J.A.: A framework for information systems architecture. IBM Systems Journal 3, 276–292 (1987)

Relation-Driven Business Process-Oriented Service Discovery

Yu Dai, Lei Yang, and Bin Zhang

Institute of Computer Application and Technology, Northeastern Univeristy, Shenyang, Liaoning
Neu_daiyu@126.com

Abstract. In order to discover services in business-driven Web service composition conveniently, accurately, and efficiently, this paper proposes a business process-oriented service discovery approach. In order to link business property with services, this paper proposes a business related description model of operations. Based on the business related relation model proposed in this paper, a relation-driven discovering algorithm for business process-oriented service discovery is presented. Compared with other discovering approach, the proposed discovering approach considers the business relation between operations and can solve the problem of discovering services with meaningful business relation. The experimentation shows the better performance of this algorithm.

Keywords: business process, web service composition, business-oriented service discovery, web service.

1 Introduction

Web services are loosely coupled software components, published, located and invoked across the Web. People can use Web services to access software for use as components to build their own applications and services. Lots of business-driven approaches for service-oriented development have been proposed [1-3] which aims at linking business and IT in order to minimize the complexity of construction of business application. The business-driven approach is a top-down composition. In this approach, a business process is firstly designed which contains a set of business operations. Through matching business operations with operation of services, the business process can be bound with a set of services to form an instance of the business process. As the growing number of Web services available within an organization and on the Web, service discovery becomes a challenging problem in the business-driven composition.

The current UDDI [4] specification is a solution for describing and discovering services. However, the current UDDI repositories only enable users to retrieve services based on the keywords (IBM, 2001). In particular, if users are not familiar with the pre-specified service categories, they usually cannot get the satisfied retrieval results. Additionally, semantic-based service discovery [5-8] is proposed. This approach adds semantics to services for automated Web service discovery. However,

K.C. Chang et al. (Eds.): APWeb/WAIM 2007 Ws, LNCS 4537, pp. 456–466, 2007.

most of these approaches focus on discovery for single operations. And the discovery for the whole business process is different from the discovery for single operation.

Firstly, the business process-oriented discovery needs to consider the business dependent relationships between business operations in the business process. For example, in a business process, there are two business operations: one is business operation of "Plane-Ticket-Booking" and the other is the business operation of "Plane-Ticket-Querying". If the two operations are provided by "Plane company A" and "Plane company B" respectively, for Plane company B" having no ticket of "Plane company A", the business process will not be run properly. Since the invocation of the latter one depending on the outcome of the former one, these two business operations should be implemented by the operations with the business dependent relationship with each other. However, the traditional semantic approaches of Web service discovery only focus on discovery for single operations while neglect the business dependent relationships between operations in the business process.

Moreover, the business process-oriented discovery needs to consider the communicational protocol between operations. For that the operation is invoked by message, the message format which is based on the communicational protocol (e.g. HTTP, SOAP) is essential in the interoperation between operations. The communicational dependent relationship between operations is used to describe how two operations can be interoperated from the syntactic view.

Finally, the business process-oriented discovery needs to consider the semantic interoperation between operations. Since the syntactic interoperation deals with the message format, the semantic interoperation deals with the message meaning. For example, although concept "Address" and "Postal Code" can be matched to some extent, the first concept cannot be the input parameter to an operation whose output parameter is the latter one. Thus, the discovery needs to consider the semantic interoperation between operations but not the semantics of single operation.

In order to discover services for business process conveniently, accurately, and efficiently, this paper proposes an approach for business process-oriented service discovery. More precisely, this paper's contribution focuses on the following:

- Business-related service description. A major issue in the business process-oriented service discovery is how to express the business property of the operation. The business property of the operation is a key problem to express the dependent business relationship between operations. A business-related service description model is proposed in this paper.
- Business-related relation model between operations. We propose a business-related relation model for comparing business relation, syntactic relation and semantic relation between operations.
- Business process-oriented service discovering algorithm. We propose an algorithm for discovery operations for business process while preserving the aforementioned relation rules.
- Prototype implementation and experiments. We provide a prototype implementation in our service-design oriented web service composition platform. We also conduct a set of experiments to evaluate the performance and scalability of our approach.

The remainder of the paper is organized as follows. In section 2, related works of service discovery is discussed. Section 3, we present our approach for business-related description of operations. Section 4 shows the proposed relation model for operations. In section 5, we present the proposed discovering algorithm. Section 6 describes our experimental evaluation and section 7 concludes.

2 Related Works

UDDI is a universal description and discovery standard for Web services. However, UDDI is limited to the use of keywords-based techniques for matching Web services and limited to the discovery of single operation, which leads to low precision results and cannot be used in the business process-oriented discovery.

OWL-S [5] supports the use of a domain ontology to describe the elements of a service profile. OWL-S describes the service process model which describes the relationships between operations. For OWL_S having no indication of business relations between operations, such business-relatedness model does not perform well in the business process-oriented discovery.

[6] proposes a multilevel model for checking composabliliy of operations. However, such approach has no business relation rule of operations and thus, cannot support the business process-oriented discovery.

The reason why current approach cannot support the discovery for business process is that the current service description model cannot express the business relation between operations. For such reason, this paper firstly proposes a business-related description model for operations. And then based on such model, the business relation rule is presented. This rule is used to discover operations with business dependent relation.

3 Business Related Description of Operations

The business process-oriented service discovery requires the description of each operation. In order to support the business process-oriented service discovery, the description of operations should be expressed the business features of operations. Current description languages, such as WSDL[8], DAML_S[9], support the operational features of operations. However, such languages do not describe the business features of operations. Thus, the discovering approaches based on such description languages cannot support the business process-oriented discovery effectively. For such reasons, this paper firstly proposes a business-related description model for operations.

3.1 Business Related Description of Operations

The example of business dependent relation between operations is that the ticket of "Plane company A" can be booked only by the operation having business relation with "Plane company A". In order to express such relation, this paper proposes a use of constraints on input and output parameters.

Definition 1. Web Services. A Web service can be defined as a 4-vector: $WS=(ID, Fc, OPs, BE)$, where ID is the identification of the service, Fc is the functional category, OPs is the set of operations and BE is the provider of the service.

Definition 2. Operation. An operation can be defined as a 7-vector: $OP=(ID, Fc, I, O, CI, CO, Q, B)$, where
(1) ID is the identification of OP;
(2) Fc is the functional category, which is defined as a 2-vector: <Verb, Target>. Verb is the action and Target is the subject of the action. For example, if Fc of a service is <Booking, Plane-Ticket>, it means that the function of the service is to book plane-ticket;
(3) $I=\{p_1, p_2... p_n\}$ is a set of input parameters and for each parameter p_i, $p_i=(name, Role, DataType)$。 $name$ is the name of p_i, $Role$ is the semantic of p_i and $DataType$ is the data type of p_i;
(4) $O=\{p_1, p_2... p_n\}$is a set of output parameters and the meaning of each parameter p_i is as same as meaning of p_i in I;
(5) $CI=\{c_1, c_2... c_m\}$is a set of constraints towards input parameters and for each c_i, $c_i=(name, Role, DataType, D)$。 $name$ is the name of c_i which is also related with a parameter in I, $DataType$ is the constraint of data type of c_i and D is the constraint of domain value of c_i;
(6) $CO=\{c_1, c_2... c_m\}$is a set of constraints towards output parameters and the meaning of each constraint c_i is as same as meaning of c_i in Pre;
(7) Q is a set of non-functional properties. For each $Q_i \in Q$, $Q_i=(Na, Role, Unit, D)$, where Na is the name of Q_i, $Role$ is the semantics of Q_i, $Unit$ is the unit of Q_i, D is the domain value of Q_i.
(8) B is a set of banding modes of the operation.

Example 1. Operation. Take an example of "Plane Ticket Booking" operation, where:

(1) ID =PTB0001;
(2) Fc = <Booking, Plane-Ticket>
(3) $I=\{p_1, p_2\}$, $p_1=(No, PlaneNo, String)$, $p_2=(DateT, Take\text{-}off\text{-}DateTime, DateTime)$;
(4) $O=\{p_1\}$, $p_1=(No, TicketNo, String)$;
(5) $CI = \{c_1, c_2\}$, $c_1=(No, PlaneNo, String$, ({PlaneCompanyA's PlaneNo,PlaneCompanyB's PlaneNo}));
(6) $CO = \{c_1\}$, $c_1=(No, TicketNo, String$, ({PlaneCompanyA's TicketNo, PlaneCompanyB's PlaneNo}));
(7) $Q=\{Q_1, Q_2\}$, $Q_1=(Time, ResponseTime, seconds$, ((0,20))), $Q_1=(Cost, Cost, dollar$, ((10,10))).

3.2 Description of Business Process

Definition 3. Business Operation. Business operation is a 7-vector: $BOP=(ID, Fc, I, O, Pre, Post, Q)$, where:

(1) ID is the identification of BP;
(2) Fc is the functional category;

(3) $I=\{p_1, p_2... p_n\}$ is a set of input parameters and for each parameter p_i, p_i=(*name*, *Role*, *DataType*, *Type*)。 *name* is the name of p_i, *Role* is the semantic of p_i and *DataType* is the data type of p_i. Compared with parameters defined in operation, p_i is a variable which can be obtained from input of users, or from input or output parameters of other business operations. If pi is from the input of users, type=1. Else, type=other business operatons.parameter.
(4) $O=\{p_1, p_2... p_n\}$is a set of output parameters and the meaning of each parameter p_i is as same as meaning of p_i in *I*;
(5) $CI=\{c_1, c_2... c_m\}$is a set of constraints towards input parameters and for each c_i, c_i=(*name*, *Role*, *DataType*, *D*)。 *name* is the name of c_i which is also related with a parameter in *I*, *DataType* is the constraint of data type of c_i and *D* is the constraint of domain value of c_i. The constraints show the business relations between business operations;
(6) $CO=\{c_1, c_2... c_m\}$is a set of constraints towards output parameters and the meaning of each constraint c_i is as same as meaning of c_i in *Pre*;
(7) *Q* is a set of non-functional properties. For each $Q_i \in Q$, Q_i=(*Na*, *Role*, *Unit*, *D*), where *Na* is the name of Q_i, *Role* is the semantics of Q_i, *Unit* is the unit of Q_i, *D* is the required domain value of Q_i.

Definition 4. Business Process. A business process can be defined as a 2-vector: *BP*=(*BOPs*, *Rs*), where:

(1) *BOPs* is a set of business operations;
(2) *Rs* is a set of control logic relations between business operations. For each $r_i \in Rs$, r_i is a relation between business operations as [10] introduced.

Definition 5. Instance of Business Process. An instance of business process can be defined as 4-vector: *IBP*=(*BOPs*, *Rs*, *OPs*, *fs*), where *BOPs* and *Rs* have the same meaning as defined in business process, *OPs* is a set of operations, *fs* is a set of mapping relationship between *BOPs* and *OPs*, and $\forall\ bop_i \in BOPs$, $\exists\ f_k \in fs$, $f_k(bop_i) \in OPs$.

Example 2. Business Process. Take an example of "Query Plane Ticket and Book it" business process. Where:

(1) $BOPs=\{BP_1, BP_2\}$, where:
BP_1.*Fc*=<Plane-Ticket, Query>, BP_2.Fc=<Plane-Ticket, Book>,
$BP_1.I=\{p_1, p_2\}$, p_1=(*No*, *PlaneNo*, *String*, 1), p_2=(*DateT*, *Take-off-DateTime*, *DateTime*, 1)
$BP_2.I=\{p_1, p_2\}$, p_1=(*No*, *TicketNo*, *String*, $BP_1.O.p_1$)
$BP_1.O=\{p_1\}$, p_1=(*No*, *TicketNo*, *String*)
$BP_2.CI=\{c_1\}$, c_1=(*No*, *TicketNo*, *String*, ($BP_1.CO$));
(2) $Rs=\{BP_1$Sequence $BP_2\}$

The aim of business process-oriented service discovery is to find operations for each business operation in business process in order to form instances of business process.

4 Business-Related Relation Model

4.1 Business-Related Similar Relation Model

Definition 6. Equal Functional Relation between Business Operation and Operation (*ERBOP*). *ERBOP* $\subset$ *BOP* $\times$ *OP*, $\forall$ $bop_i \in BOP$, $\forall$ $op_j \in OP$, $<bop_i, op_j> \in ERBOP$, if and only if the following conditions are satisfied. If $<bop_i, op_j> \in ERBOP$, bop_i is equal to op_j, signified as $bop_i = op_j$ or $op_j = bop_i$:

(1) $bop_i .Fc.Verb=op_j.Fc.Verb \wedge bop_i .Fc.Subject=op_j.Fc.Subject$;
(2) let $I_i=bop_i.I$, $I_j=op_j.I$, $\forall\ i_{jk} \in I_j$, $i_{jk} \in I_i \wedge \forall\ i_{il} \in I_i$, $i_{il} \in I_j$
(3) let $O_i=bop_i.O$, $O_j=op_j.O$, $\forall\ o_{jk} \in O_j$, $O_{jk} \in O_i \wedge \forall\ o_{il} \in O_i$, $O_{il} \in O_j$
(4) let $CI_i=bop_i.CI$, $CI_j=op_j.CI$, $\forall\ c_{jk} \in CI_j$, $\exists\ c_{il} \in CI_i$, $c_{il} \Leftrightarrow c_{jk} \wedge \forall\ c_{il} \in CI_i$, $\exists\ c_{jk} \in CI_j$, $c_{il} \Leftrightarrow c_{jk}$
(5) let $CO_i=bop_i.CO$, $CO_j=op_j.CO$, $\forall\ c_{jk} \in CO_j$, $\exists\ c_{il} \in CO_i$, $c_{il} \Leftrightarrow c_{jk}$. $\wedge$ $\forall\ c_{il} \in CO_i$, $\exists\ c_{jk} \in CO_j$, $c_{il} \Leftrightarrow c_{jk}$.

Definition 7. Equal Functional Relation between Operations (*EROP*). *EROP* $\subset$ *OP* $\times$ *OP*, $\forall$ $op_i, op_j \in OP$, $<op_i, op_j> \in ERBOP$, if and only if the following conditions are satisfied. If $<op_i, op_j> \in ERBOP$, op_i is equal to op_j, signified as $op_i = op_j$ or $op_j = op_i$:

(1) $op_i .Fc.Verb=op_j.Fc.Verb \wedge op_i .Fc.Subject=op_j.Fc.Subject$;
(2) let $I_i=op_i.I$, $I_j=op_j.I$, $\forall\ i_{jk} \in I_j$, $i_{jk} \in I_i \wedge \forall\ i_{il} \in I_i$, $i_{il} \in I_j$
(3) let $O_i=op_i.O$, $O_j=op_j.O$, $\forall\ o_{jk} \in O_j$, $O_{jk} \in O_i \wedge \forall\ o_{il} \in O_i$, $O_{il} \in O_j$
(4) let $CI_i=op_i.CI$, $CI_j=op_j.CI$, $\forall\ c_{jk} \in CI_j$, $\exists\ c_{il} \in CI_i$, $c_{il} \Leftrightarrow c_{jk} \wedge \forall\ c_{il} \in CI_i$, $\exists\ c_{jk} \in CI_j$, $c_{il} \Leftrightarrow c_{jk}$
(5) let $CO_i=op_i.CO$, $CO_j=op_j.CO$, $\forall\ c_{jk} \in CO_j$, $\exists\ c_{il} \in CO_i$, $c_{il} \Leftrightarrow c_{jk}$. $\wedge$ $\forall\ c_{il} \in CO_i$, $\exists\ c_{jk} \in CO_j$, $c_{il} \Leftrightarrow c_{jk}$.

Definition 8. Compatible Functional Relation between Operations (*CROP*). *CROP* $\subset$ *OP* $\times$ *OP*, $\forall$ $op_i, op_j \in OP$, $<op_i, op_j> \in CROP$, if and only if the following conditions are satisfied. If $<op_i, op_j> \in CROP$, op_i is compatible with op_j, signified as $op_i \lozenge op_j$:

(1) $op_i .Fc.Verb=op_j.Fc.Verb \wedge op_i .Fc.Subject \supseteq op_j.Fc.Subject$;
(2) let $I_i=op_i.I$, $I_j=op_j.I$, $\forall\ i_{jk} \in I_j$, $i_{jk} \in I_i$;
(3) let $O_i=op_i.O$, $O_j=op_j.O$, $\forall\ o_{jk} \in O_j$, $O_{jk} \in O_i$;
(4) let $CI_i=op_i.CI$, $CI_j=op_j.CI$, $\forall\ c_{jk} \in CI_j$, $\exists\ c_{il} \in CI_i$, $c_{il} \Rightarrow c_{jk}$;
(5) let $CO_i=op_i.CO$, $CO_j=op_j.CO$, $\forall\ c_{jk} \in CO_j$, $\exists\ c_{il} \in CO_i$, $c_{il} \Rightarrow c_{jk}$.

Definition 9. Compatible Functional Relation between Operation and Business Operation (*CRBOP*). *CROP* $\subset$ *OP* $\times$ *BOP*, $\forall$ $op_i \in OP$, $\forall$ $bop_j \in BOP$, $<op_i, bop_j> \in CRBOP$, if and only if the following conditions are satisfied. If $<op_i, bop_j> \in CRBOP$, op_i is compatible with bop_j, signified as $op_i \lozenge bop_j$:

(1) op_i $.Fc.Verb=bop_j.Fc.Verb \wedge op_i .Fc.Subject \supseteq bop_j.Fc.Subject$;
(2) let $I_i=op_i.I$, $I_j=op_j.I$, $\forall i_{jk}\in I_j$, $i_{jk}\in I_i$;
(3) let $O_i=op_i.O$, $O_j=op_j.O$, $\forall o_{jk}\in O_j$, $O_{jk}\in O_i$;
(4) let $CI_i=op_i.CI$, $CI_j=op_j.CI$, $\forall c_{jk}\in CI_j$, $\exists c_{il}\in CI_i$, $c_{il}\Rightarrow c_{jk}$;
(5) let $CO_i=op_i.CO$, $CO_j=op_j.CO$, $\forall c_{jk}\in CO_j$, $\exists c_{il}\in CO_i$, $c_{il}\Rightarrow c_{jk}$.

4.2 Business-Related Composable Relation Model

A major issue of the business process-oriented service discovery is to find services and ensure the dependent relations of them. In this section, we identify three sets of composable relation.

Since operations may suppor different communicational protocols, it is important to insure that they understand each other at the message format and protocol level.

Definition 10. Syntactic Composable Relation between Operations (*SyCRO*). $SyCRO \subset OP \times OP$, $\forall$ op_i, $op_j \in OP$, $<op_i, op_j> \in SyCRO$, if and only if $op_i.B \cap op_i.B \neq \Phi$. If $<op_i, op_j> \in SyCRO$, op_i is syntactic composable with op_j, signified as $op_i \propto^{syn} op_j$.

Interoperations between operations involve the exchange of input and output parameters. The semantic composable relation is to ensure that two operations are composed.

Definition 11. Semantic Composable Relation between Operations (*SeCRO*). $SeCRO \subset OP \times OP$, $\forall$ op_i, $op_j\in OP$, $<op_i, op_j>\in SeCRO$, if and only if the following conditions are satisfied. If $<op_i, op_j>\in SeCRO$, op_i is semantic composable with op_j, signified as $op_i \propto^{sem} op_j$.

(1) $\exists bp\in BP$, bop_k, $bop_l\in BP$, and $\exists p_{lu}\in bop_l.I$, p_{lu} is from the p_{im} of $bop_i.O$ and $op_i=bop_k$, $op_j=bop_l$;
(2) $op_i.p_{im}.Role=op_j.p_{ju}.Role$;
(3) $op_i.p_{im}.Unit= op_j.p_{ju}.Unit$;
(4) $op_i.p_{im}.DataType= op_j.p_{ju}.DataType$.

Business dependent relation between operations means that two operation can be composed from business aspect.

Definition 12. Business Composable Relation between Operations (*BuCRO*). $BuCRO \subset OP \times OP$, $\forall$ op_i, $op_j\in OP$, $<op_i, op_j>\in BuCRO$, if and only if the following conditions are satisfied. If $<op_i, op_j> \in BuCRO$, op_i is semantic composable with op_j, signified as $op_i \propto^{bus} op_j$.

(1) $\exists bp\in BP$, bop_k, $bop_l\in BP$, and $\exists p_{lu}\in bop_l.I$, p_{lu} is from the p_{im} of $bop_i.O$ and $op_i=bop_k$, $op_j=bop_l$ and domain constraint on p_{lu} is domain constraint on p_{im};
(2) $op_i.p_{im}.Role=op_j.p_{ju}.Role$;
(3) $op_i.p_{im}.Unit= op_j.p_{ju}.Unit$;

(4) $op_i.p_{im}.DataType= op_j.p_{ju}.DataType$;
(5) $op_j.p_{ju}.D \supseteq op_i.p_{im}.D$.

5 Relation-Driven Business Process-Oriented Service Discovery

Based on the proposed relation model, we propose an approach for the business process-oriented service discovery. This approach consists of 2 separate phases: discovery for each business operation in the required business process, and discovery for the whole business process.

5.1 Business-Related Similar Relation Driven Discovery for Single Business Operations

The discovering algorithm for single business operation is used for discovering operations for single business operation. The discovering algorithm contains 2 parts: the algorithm firstly finds operations which have equal relation with required business operation; then the algorithm finds operations with compatible relation with the operations which have equal relation with the required business operation. The algorithm of this generic process of business operation discovery is showed as following.

Algorithm 1. Discovering Algorithm for Single Business Operation (DASBO)

```
void DASBO(bop: BusinessOperation)
{
   For each op∈ Registry
   {
    If op=bop then
    {
     Put op to Set_of_Result;
    }
    If op◊bop then
    {
     Put op to Result_Set_of_bop;
    }
   }
}
```

5.2 Business-Related Composable Relation Driven Discovery for Business Process

For every pair of business operations, it also needs to check the composable relation between operations which can implement these two business operations. Through this checking phase, the instance of business process can be obtained. The algorithm can be seen in Algorithm 2.

Algorithm 2. Discovering Algorithm for Business Process

```
Void DABP(bp: BusinessProcess, Result_set_of[bop_i][op_ik])
{
    for each bop_i ∈ bp
    {
     for each op_ik ∈ Result_set_of[bop_i][op_ik]
     {
       For each op_{i+1,l} ∈ Result_set_of[bop_{i+1}][op_{i+1,l}]
       {
        If (op_ik ∝^syn op_{i+1,l}) and (op_ik ∝^sem op_{i+1,l}) and (op_ik ∝^bus op_{i+1,l})
          i=i+1;
        else
          l=l+1;
       }
       k=k+1;
     }
    }
}
```

6 Experimentation

We have developed a prototype SDOWSCP [11] for web services composition. In this system, we conduct our approach for business process-oriented service discovery. The aim of the experimentation is to show the scalability of our approach.

The experiments are run with a PC of Intel Pentium IV 2.4G CPU and 1GB RAM. The operation system is Windows Server 2000. And the algorithms involved in the selection part are written by Java. Because there is limited number of Web Services in hand of us, then in this paper, we randomly simulated data of Web Services to be samples.

We do the experimentation on the situations that the business process contains 5, 10, and 20 business operations. And the number of operations is 50. We evaluate the time for discovering services. The experimentation result is shown as Fig.1. From Fig.1, with the increase of the number of business operations, the time of discovering will be increased.

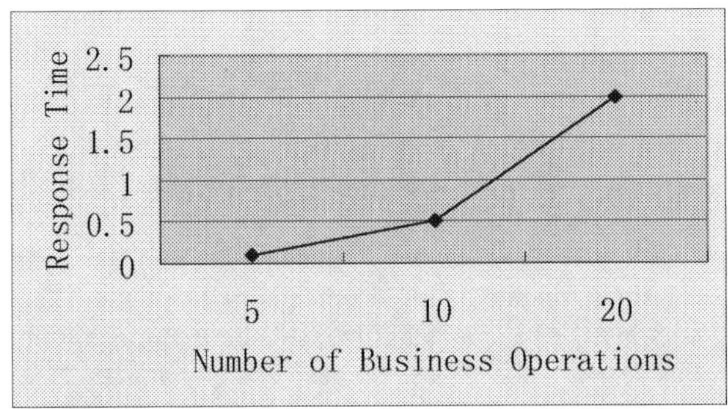

Fig. 1. Response Time of Discovering When Number of Business Operations is Varied

We do the experimentation on the situations that the business process contains 20 business operations. And the number of operations is 50, 100, 200 and 400. We evaluate the time for discovering services. The experimentation result is shown as Fig.2. From Fig.2, with the increase of the number of operations, the time of discovering will be increased.

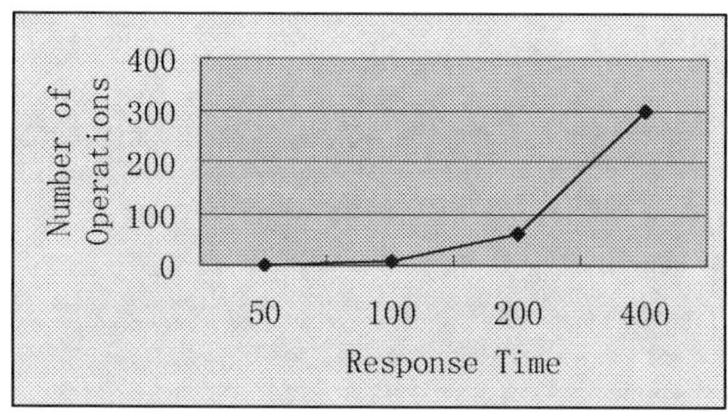

Fig. 2. Response Time of Discovering When Number of Operations is Varied

7 Conclusions

For the problem of business process-oriented service discovery, this paper firstly proposes a business-related description model for operations. This description model provides a new way to express the business relation between operations and can discover business-related operations for the business process. Moreover, in order to discover business-related operations for business process, this paper also provides a business-related relation model of operations. Based on such relation model, a relation-driven discovering algorithm for business process is presented. Compared with other discovering algorithm, this algorithm considers the relation between operations and especially the business relation. And thus, this algorithm can ensure the business property of the business process.

References

1. Allen, P.: Realizing e-Business with Components. Addison Wesley, London, UK (2001)
2. Scheben, U.: Hierarchical composition fro component-based modeling. Science of Computer Programming 55, 161–183 (2005)
3. Han, Y., Geng, H., Li, H., et al.: VINCA-A visual and personalized business-level composition language for chaining Web-based services. In: Proceedings of the 1st International Conference on Service-Oriented Computing, Trento, Italy, pp. 165-177 (2003)
4. UDDI Project. UDDI Technical White Paper, September, (2000): http://www.uddi.org
5. OWL Services Coalition. OWL_S: Semantic Markup for Web Services. OWL_S V.1.1, White Paper (November 2004): http://www.daml.org/services/owl-s/1.1/
6. Medjahed, B., Bouguettaya, A.: A multilevel composablility model for semantic web service. IEEE transactions on knowledge and data engineering 17, 954–968 (2006)

7. MCIlraith, A., Tran, C., Honglei, Z.: Semantic Web services. EEE intelligence systems. 16(2), 46–53 (2001)
8. Christensen, E., Curbera, F., Meredith, G.: Web Services Description Language (WSDL) 1.1[EB/OL], (2001) http://www.w3.org/TR/2001/NOTE-wsdl-20010315
9. DAML Joint Committee. Daml+oil language. (2001) http:// www.daml.org/ 2001/03/ daml+oil-index.htm
10. Hollingsworth, D.: The workflow reference model, Workflow management coalition, (1995): http://www.wfmc.org/standards/docs/tc003v11.pdf
11. Yang, L., Dai, Y., Zhang, B.: A Service Design Oriented Web Services Composite Platform. Journal of Wuhan University. 11(1), 160–164 (2006)

Workflow Message Queue's Performance Effect Measurements on an EJB-Based Workflow Management System

Hyungjin Ahn[1], Kiwon Lee[1], Taewook Kim[1], Haksung Kim[2], and Ilkyeun Ra[3]

[1] Collaboration Technology Research Lab.
Department of Computer Science
KYONGGI UNIVERSITY, South Korea
{hjahn, kwlee, twkim, kwang}@kyonggi.ac.kr
Http://ctrl.kyonggi.ac.kr
[2] Dept. of Web Contents Development
DONGNAM HEALTH COLLEGE, South Korea
hskim@dongnam.ac.kr
[3] Distributed Computing and Networking Lab
Department of Computer Science and Engineering
UNIVERSITY OF COLORADO AT DENVER, USA
ikra@carbon.cudenver.edu

Abstract. The scalability and performance issues about large-scale workflow are the most impeccable topics, in recent, because the size of workflow system and its applications are becoming larger and larger and requiring, nevertheless, much more reliable and faster services. These technological pressures push us to develop an EJB-based workflow management system targeting at covering these two issues, which is named e-Chautauqua. In this paper,[1] we present an asynchronous communication mechanism—workflow message queue mechanism—providing a reliable communication channel among the constituent objects organizing the enactment engine of e-Chautauqua, and try also to measure how much the mechanism effects on accomplishing the scalable and performable e-Chautauqua. From a series of test scenarios based on the message queue properties such as message alive time and maximum pool size, we have observed that the message alive time property sensitively effects on the scalability, and while on the other the maximum pool size property has a great influence on the performance of the workflow management system.

Keywords: Very Large-Scale Workflow Management Systems, Message Queue Mechanism, Performance Measurement.

1 Introduction

In recent, workflow is becoming one of the hottest technologies in the enterprize information technology arena; according as workflow systems have been rapidly

[1] The research was supported by the BEIT special fund of Kyonggi University.

K.C. Chang et al. (Eds.): APWeb/WAIM 2007 Ws, LNCS 4537, pp. 467–478, 2007.

spreading their applications over the various industry domains[4, 6, 8]. Nevertheless, many of the workflow products have been experienced being ineffective in some serious workflow applications due to the lack of necessity in modeling capability, scripting tools, dynamic change supports, and many other advanced features[6, 7]. The recent requirements on the workflow systems have been set up and characterized by the challenge - Scalability and Performance[2, 8, 12].

As a feasible solution to the scalability and performance challenge, we have developed a large-scale workflow architecture and its system (e-Chautauqua) that runs on an enterprize java beans (EJB) framework. The workflow enactment architecture of e-Chautauqua, which is called workcase-oriented workflow enactment architecture[9, 11], exerts quite acceptable performances predicted through the mathematical analysis, as shown in [9]. The goal of e-Chautauqua is to get used to dealing with those very large scale workflow applications without any serious performance degradation, where we assume that each workflow model is able to spawn up to 500,000 or 1 million workcases.

In terms of the performance requirement for the workflow enactment components of e-Chautauqua, we have got to reach at the final decision after a couple of long technical discussions; it must be no doubt that the communication channels connecting all active components comprising the enactment engine become a bottleneck of the system's performance as well as the system's recoverability. The communication channel used often to be realized by either synchronous mechanism or asynchronous mechanism. The typical synchronous communication mechanism is the remote procedure call mechanism; however, we won't agree on that this mechanism fits very well to the very large scale workflow enactment engine like e-Chautauqua, which is comprised of a huge number of active objects instantiated from the very large scale workflow models. While on the other, the message queue mechanism is the most representative asynchronous communication mechanism, and it also provides a very reliable communication channel too. There are several benefits to the message queue mechanism. First, requesting objects can proceed with its processing independently of their replying objects. Second, it allows batch and parallel processing of messages. Third, there is less demand on keeping connections active, which makes the system less exposed to crashes. From these benefits, we can expect that the message queue mechanism is able to exert very positive influences on realizing a scalable and highly performable workflow enactment engine.

In this paper, we try to measure the performance effect of the message queue mechanism through a series of test scenarios based on the message queue properties—message alive time and maximum pool size—set their values up through the JMS's parameters. Through these measurement results, we strongly expect to get some practical conclusions about how much these message queue properties effect on the overall system-level performance of a very large scale workflow enactment engine, specifically e-Chautauqua's. As a consequence of this, now the architecture and system, which is developed by the EJB framework approach, is able to provide acceptable performance for much larger and much more complex workflow applications, and now it is fully deployable and operable.

2 Related Works

Nowadays, there exist hundreds of commercial workflow management products. The genesis of workflow management software was in automating document-driven workflow procedures such as officeTalk-D[5], etc.. Some of the early products were extensions to the document imaging and management software. The emphases of those workflow products have been on routing, sharing, application launching, and cooperating activities of workflows. Meanwhile, the current products are pursuing the issues of transactional workflows and large scale workflows. Especially, the feature of supporting the '*Large-scale Workflows*' is still hot-issued in the literature[1, 2, 9, 10, 11, 12, 13], because the architectures of almost all workflow products are belonging to a family of the client-server architecture that suffers from performance and scalability problems in the situation where it is faced with large scale workflow applications. More scalable architectures[14, 15, 16] need to be supported for the large scale workflows.

As mentioned before, many workflow management systems have been developed and described in the workflow literature; we cannot describe them all here. Almost all commercialized products, such as BizFlow, InConcert, Staffware, FloWare, Notes, ActionWKF, FlowWorks and FlowMark workflow management systems, have adopted a certain type of advanced workflow architectures employing a sort of the EJB enterprize middleware platform. Particularly, JBoss jBPM is a typical example of the new paradigm recently happening in the literature. It is pursuing a flexible, extensible workflow management system as the open source enterprize middleware platform. JBoss jBPM can be used in the simplest environment like an ant task and scale up to a clustered J2EE application. However, the performance and scalability issues with the message queue mechanism that we are concerning about in this paper have never addressed in the literature, yet, after being firstly addressed by [1]. So, we hope that this paper may ignite the research issues, again.

3 An EJB-Based Workflow Management System

We have developed a workflow management system that aims for very large scale workflows applications, and it is named e-Chautauqua workflow management system[10]. e-Chautauqua is based on the workcase-oriented workflow architecture[11], and especially we have implemented it by the Enterprize Java Beans framework approach. In this section, we describe e-Chautauqua's overall system architecture and details of the components comprising the workflow enactment engine.

3.1 The Overall System Architecture of e-Chautauqua

The functional descriptions of e-Chautauqua's architectural components and the ways they cooperate with each other are briefly explained in this section. Basically, e-Chautauqua is fully implemented so as to provide almost all services and

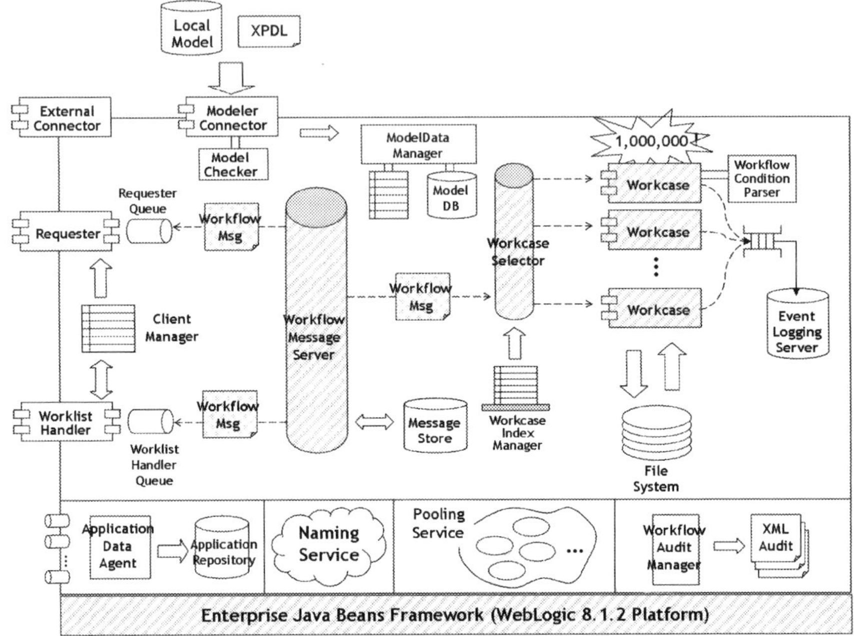

Fig. 1. The Overall System Architecture of e-Chautauqua

facilities specified in the international workflow standards from the interfaces 1 through 5 released by WfMC.

The overall system architecture of e-Chautauqua as shown in Fig. 1. All components are deployed on an Enterprize Jave Beans middleware platform[2] and they asynchronously communicate with each other through a JMS-based message queue mechanism named the workflow message server. The details of this message queue mechanism will be described in the next section. The core component that distinguishes e-Chautauqua from other workflow engines is the workcase object component. That is, almost all conventional workflow engines are based on the activity-oriented workflow architectural style[6], but on the other hand e-Chautuaqua's engine is based on the workcase-oriented workflow architectural style[6]. In the activity-oriented workflow engines, the control flow management functionality is performed through interactions among activity objects, because the core component is the activity object component. In contrast to this, the control flow management is done by the workcase objects in the workcase-oriented workflow engine (e-Chautauqua), and so the activity precedence information is stored to the inside of each workcase object as data. Conclusively, the most valuable benefit that we expect from the workcase object component is that the control flow management functionality can be done by the much simpler

[2] Note that we used the Weblogic 8.1.2 package, but it won't be dependent on a specific EJB package. Recently we have successfully deployed our system on JBoss enterprize middleware platform, too.

mechanism, and also it can be efficiently done with much smaller amount of computing resources, because the workcase-oriented workflow engine is able to dramatically decrease the number of objects resided and managed in the system. These facts mean also that the workcase-oriented workflow engine like e-Chautuaqua must be more appropriate and durable for the very large scale workflows, and it must give satisfaction in the higher degree of workflow instantiation complexity.

3.2 Details of e-Chautauqua's Engine Components

In this section, we describe the details of a set of component types that is a *sine qua non* for successful implementation of the workcase-oriented workflow enactment architecture.

Table 1. Types of Components and Their Layouts on EJB

Component Type	Component Layout	Description
PDLAgent	Stateless Session Bean	Handling Import/Export workflow models in XML-based Process Definition Language (XPDL)
OMAgent	Stateless Session Bean	Manipulating Organizational Information such as departments, roles, and actors
Requester	Stateless Session Bean	Handling WAPIs such as Connection, Process Control, Activity Control, Process Status, and Activity Status
Worklist Handler	Stateless Session Bean	Handling WAPIs manipulating the runtime clients' worklists
Workcase	Stateful Session Bean	Executing objects of Workflow instances
Workflow Queue	Message-Driven Bean	Message queue objects
Script Interpreter	Class Module	Handling conditional scripts on control flow transitions of OR-split
Tool Agent	Stateless Session Bean	Handling WAPIs bridging the engine/the runtime clients and invoked applications
Logger	Using Open Source Library (Log4j)	Handling WAPIs logging the engine's runtime events

Based on the component-based software development methodology using the spiral model[3], we have developed a set of component types reflecting the workcase-oriented workflow architecture. The core of the component types ought to be the workcase component as mentioned in the previous section, and its related component types are PDLAgent Component, OMAgent, Worklist Handler, Requester, Workflow Queue (Workflow Message Server), Logger, Script Interpreter, and ToolAgent Component, which are additionally needed for performing the workflow enactment functionality. Table 1 gives the component types and their layout properties to be deployed on an enterprize java beans framework.

These all components of e-Chautauqua communicate with each other through the workflow message server controlling workflow message queues. The workflow message queue mechanism of the server uses a message-driven bean for realizing the reliable message transmission, and it is realized by JMS (Java Messaging Service) queue of an EJB framework. The detailed description of the mechanism and its performance effect measurement results will be given in the next section.

4 Performance Effect Measurements on the Workflow Queue Component

As stated in the previous sections, our primary goal was to develop a robust workflow enactment engine not only that is endurable for a huge number of workflow instances, but also that is able to give the acceptable performance in enacting those workflow instances, (from now we call them workcases). In other words, the engine has got to be implemented so as to cope with the scalability issue as well as the performance issue. In terms of the scalability issue, we are going to show you how many workcases are able to simultaneously reside on the engine under the given hardware configuration of the computer. While on the other, in the performance aspect we will show you how long the engine take to create those workcase objects, each of which fulfills a workflow instance. A series of test scenarios presented in this section is based on the strong belief that there is a very close connection between these issues and the workflow message queue mechanism.

4.1 e-Chautauqua's Message Queue Mechanism

The workflow message queue mechanism adopted in e-Chautauqua is based upon a workflow queue component embodied as a message-driven bean of JMS[17] embedded in the enterprize java beans platform of e-Chautuaqua. The operations procedure (labeled by sequential numbers) of the workflow message server implemented through the workflow queue component is illustrated in Fig. 2. That is, if one of the engine components tries to call a method of the target object, then the message sender composes a message representing the callee's method and its parameters' values, and transmits the sending-message event to the workflow message server after enqueueing the composed message to a JMS message queue[10, 17], which is specified as the destination of the message. As soon as the workflow message server receives the sending-message event, it transmits the event to the message receiver, and then the receiver, after recognizing the arrival of the event through its event listener, dequeues an entry message from the JMS message queue and looks up the message map to find out the target object and its method with parameters' values by interpreting the received message. Finally the message receiver invokes the method of the target object with passing its parameters' values.

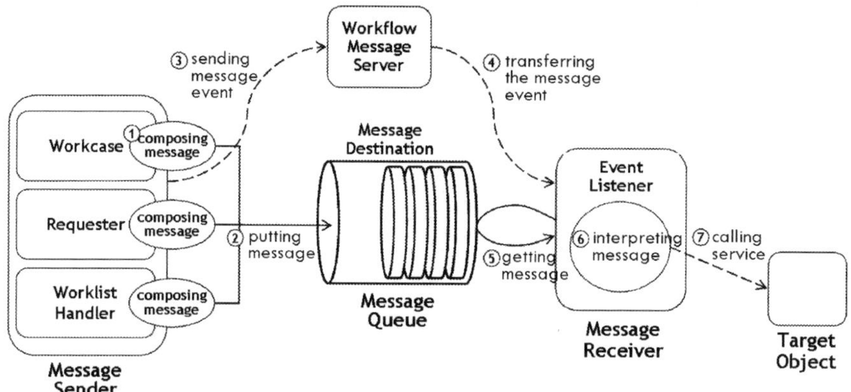

Fig. 2. e-Chautauqua's Workflow Message Queue Mechanism

4.2 Performance Effect Measurements

In this section, we try to observe e-Chautauqua's runtime executions through the message queue's performance for practically measuring its performance and evaluating whether e-Chautauqua is affordable for the higher degree of the workflow instantiation complexity as a very large scale workflow management system. Especially for the observation, which measures how much the workflow message queue mechanism effects on the engine's performance, we have developed a series of test scenarios based on the workflow message queue properties—message alive time and maximum queue size—the values of which are specified into the parameters of JMS message queue[10, 17].

Test Scenario. In general, it is very hard to measure the elapsed time or the processing time of a workcase in a workflow system, because it's not easy to define or predict the processing time of each activity in the underlined workflow model. Conclusively speaking, we decide to observe e-Chautauqua's runtime executions through measuring the response time for creating a workcase object rather than measuring the elapsed time for processing all activities of a workcase fulfilling its corresponding workflow instance, because it is enough to measure how much the workflow message queue mechanism effects on the engine's performance.

So, as depicted in Fig. 3, we develop a test scenario supported by an automatic method calls generator that is able to produce a huge number of creating-workcase method calls and measure their response times. Also, Table 2 gives the software and hardware computing environment for the test scenario. The response time implies the elapsed time, from the start time when e-Chautauqua's requester component submits a creating-workcase message into the workflow queue, to the end time when the corresponding workcase object is created and resided in the workcase pool after dequeueing the message from the workflow queue. More minutely analyzing the response time, it is to sum up the following detailed processing times:

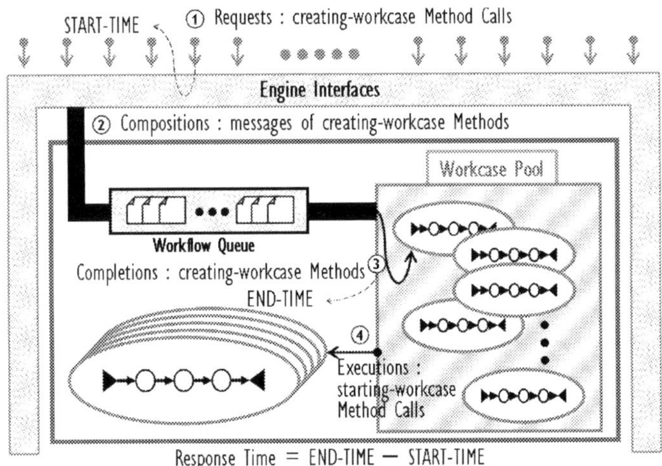

Fig. 3. The Test Scenario Structure

Table 2. SW/HW Computing Environments for the Test Scenario

Type	Classification	Model
Software	Operating System	Windows 2000 Server
	EJB Framework	JBoss 3.x
	DBMS	MySQL 4.x
Hardware	Computer Model	COMPAQ Proliant ML 530
	CPU	Intel Xeon Pentium 1.0 GHz (* 2 = Dual)
	HDD	SCSI 90G (30G * 3)
	RAM	1024 Mb (133MHz ECC SDRAM)

- *The Enqueuing Time : from requesting a creating-workcase method call to enqueuing its corresponding message entry into the workflow queue*
- *The Waiting Time : the duration of that the message entry is dwelling in the workflow queue*
- *The Dequeuing Time : from dequeuing the message entry to creating its corresponding workcase object into the workcase pool*

Test Cases and Measurements on MAT Property. Based on the test scenario, we develop three test cases with respect to the message alive time property as shown in Table 3. By setting up three different message alive times on the workflow queue of JMS message queue, we try to measure the response times to create workcase objects and the maximum number of workcase objects that e-Chautauqua's engine can handle under the computing environment given in Table 2.

Fig. 4 depicts a graph showing the measurement results for the test cases of Table 3. Additionally, Table 4 is to summarily present the measurement results according to the total numbers of created workcases and their average response

Table 3. Test Cases of the Message Alive Time Property

Test Case	Queue Minimum, Maximum Size	Message Alive Time (msec)
1st Case	1, 15	30,000
2nd Case	1, 15	10,000
3rd Case	1, 15	5,000

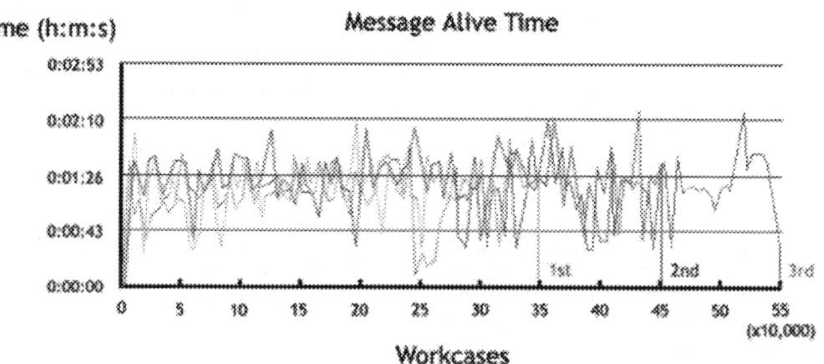

Fig. 4. e-Chautauqua's Measurement Results on the Message Alive Time Property

Table 4. The Response Times and the No. of Workcases on the MAT Test Cases

Test Case	The Average Response Time(h:m:s)	The No. of Workcases
1st Case	00:01:12	348,944
2nd Case	00:01:20	450,649
3rd Case	00:01:16	543,766

Table 5. Test Cases of the Queue Maximum Size Property

Test Case	Queue Minimum, Maximum Size	Message Alive Time (msec)
1st Case	1, 100	30,000
2nd Case	1, 1,000	30,000
3rd Case	1, 10,000	30,000
4th Case	1, 20,000	30,000

times for the three test cases. From the interpretation of the measurement results, we can observe that the number of created workcase objects is dramatically increasing according as the message alive time is decreasing. The fact that the messages's alive time dwelling in the workflow queue is longer will cause to make the workflow queue full always, and after all it will happen that a lot of requester threads is never forced to be out of the workflow queue' service. On this situation, the requester threads are holding the computing resources, and swiftly occupying the whole resources, and finally the engine will be crashed. The 1st test case causes the system to crash when the number of workcase objects reaches at 348,944, on the other hand, the 3rd test case makes it at 543,766 workcase objects. Therefore, we can conclude from this observation that

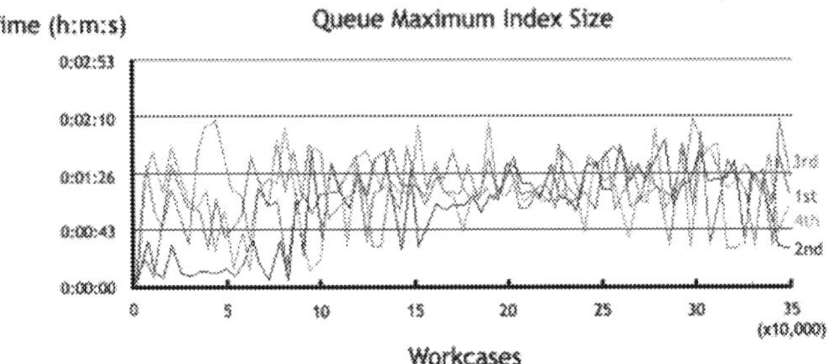

Fig. 5. e-Chautauqua's Measurement Results on the Queue Maximum Size Property

the message alive time property of the workflow queue effects on how much the engine can be scalable for the very large scale workflow applications, and it has a close connection to the scalability issue.

Test Cases and Measurements on QMS Property. Additionally, we develop four test cases with respect to the queue maximum size property and execute them on the same test scenario described in the previous section. Table 5 shows the test cases, each of which has a different maximum queue size. And Fig. 5 is present a graph reflecting the measurement results for the test cases. Through Table 6, we summarize the measurement results according to the average response time and the total response time that sums up all response times being required for creating all of the created workcase objects. Based upon the measurements, we can interpret that the queue maximum pool size property of the workflow queue gives a meaningful effect on realizing a highly performable workflow engine for the very large scale workflow applications. As a consequence, the QMS property has something to do with the performance issue.

Table 6. The Response Times on the QMS Test Cases

Test Case	The Average Response Time(h:m:s)	The Total Response Time
1st Case	00:01:10	68:00:00
2nd Case	00:00:59	57:21:40
3rd Case	00:01:14	71:01:00
4th Case	00:01:11	69:00:00

5 Conclusions

In order to conduct the message queue's performance effect measurement works, in this paper we built a test scenario with an automatic method calls generator of the start workcase request messages, and give the measurement results gained from the test cases. Based upon the performance effect measurement results, we made a decision that e-Chautauqua is able to guarantee the acceptable performance and scalable extention for the very large scale workflow applications.

We would say that the maximum number of workcases in the system should be closely related with the message alive time property of the workflow message queue, while on the other the response time of workcase creation depends on the queue maximum size property of the workflow message queue. If we deploy the system on a more powerful hardware configuration like a clustering or grid platform, then we can surely expect that the performance measurement result will be much more reasonable for the very large scale workflow environment. Based upon the results of this paper, we got the basis for further digging out our research into the world of Grid-based workflow architectures or/and P2P-based workflow architectures for exploring more advanced workflow message queue mechanisms and their systems targeting at the very large scale workflow applications.

Acknowledgement. The research was supported by the BEIT special fund of Kyonggi University.

References

1. Alonso, D., et.al.: "Exotica/FMQM: A Persistent Message Based Architecture for Distributed Workflow Management". In: Proceedings of the IFIPS Working Conference on Information Systems for Decentralized Organizations (1995)
2. Alonso, G., Schek, H.-J.: "Research Issues in Large Workflow Management Systems". In: Proceedings of NSF Workshop on Workflow and Process Automation in Information Systems: State-of-the-Art and Future Directions (1996)
3. Boehm, B.W.: "A Spiral Model of Software Development and Enhancement," IEEE Computer May 1998, pp. 61-72 (1988)
4. Ellis, C.: "Team Automata". In: Proceedings of the ACM Group'97 Conference (1997)
5. Ellis, C., Maltzahn, C.: "Chautauqua: A Flexible Workflow System". In: Proceedings of the 30th HICSS Conference (1997)
6. Ellis, C., Kim, K.: "A Framework and Taxonomy for Workflow Architectures". In: The Proceedings of ACM GROUP2000: the Fouth International Conference on Design for Cooperative Systems (2000)
7. Jablonski, S., Bussler, C.: "MOBILE: A Modular Workflow Model and Architecture," University of Erlangen Internal Report (1995)
8. Kim, K.: Practical Experiences and Requirements on Workflow. In: Conen, W. (ed.) Collaborative Technology for Collaboration Applications. LNCS, vol. 1364, pp. 145–160. Springer, Heidelberg (1998)
9. Kim, K., Ellis, C.: Performance Analytic Models and Analyses for Workflow Architectures. Information Systems Frontiers: A Journal 3(3), 339–355 (2001)
10. Kim, K.-H., Ahn, H.-J.: "An EJB-based Very Large Scale Workflow System and Its Performance Measurement", LNCS, vol. 3739, pp. 526–537 (2005)
11. Ahn, H.-J., et al.: "Workcase-oriented Workflow Enactment Components for Very Large Scale Workflows", LNCS, To be published (2007)
12. Kim, K.-H.: "A Layered Workflow Knowledge Grid/P2P Architecture and Its Models for Future Generation Workflow Systems", Future Generation Computing Systems, To be published (2007)
13. Miller, A., et al.: "CORBA Based Run Time Architectures for Workflow Management Systems," LSDIS Internal Lab. Report, University of Georgia (1996)

14. Perry, D., Wolf, A.: "Foundations for the Study of Software Architectures" ACM SIGSOFT Software Engineering Notes, vol. 17(4) (1992)
15. Wallnau, K., Long, F., Earl, A.: "Toward a Distributed Mediated Architecture for Enterprise-wide Workflow Management" Proceedings of the NSF Workshop on Workflow and Process Automation (1996)
16. Weissenfels, J., et al.: "An Overview of the Mentor Architecture for Enterprise Wide Workflow Management" Proceedings of the NSF Workshop on Workflow and Process Automation (1996)
17. Curry, E., Chambers, D., Lyons, G.: "A JMS Message Transport Protocol for the JADE platform" IEEE/WIC International Conference on Volume, pp. 596–600 (2003)

RFID Application Model and Performance for Postal Logistics

Jeong-Hyun Park[1] and Boo-Hyung Lee[2]

[1] Postal Technology Research Center,
Electronics and Telecommunications Research Institute (ETRI)
161 Kajong-Dong, Yusong-Ku, Daejeon, 305-700, Korea
jh-park@etri.re.kr
[2] Department of Computer Science and Engineering, Kongju University
275, BooDae-Dong, CheonAnn, ChungNam, 330-717, Korea
bhl1998@kongju.ac.kr

Abstract. In this paper, we suggest a postal RFID application model that can be used for real time trace and tracking system implementation of parcel processing and pallet management. This paper also shows the tag recognition performance of parcels by speed and mounting tag material such as thin cans, water bottles, and paper using an implemented postal RFID system to find the best solution for RFID adaptation in postal logistics.

Keywords: RFID Application Model, Postal Logistics, Parcel & Pallet.

1 Introduction

RFID technology has broad applicability to the Automatic Identification and Data Capture (AIDC) industry in item management. But the performance characteristics of devices (tags and interrogation equipment) may vary drastically due to application factors as well as the particular RF air interface (frequency, modulation, protocol, etc...) being supported. Of key concern is the matching of the various performance characteristics to user applications [1]. This paper details how RFID technology adapts parcel processing such as registration, collection, sorting, distribution, sending, arriving, and delivery, and real time trace and tracking of pallet management in postal logistics environments. This paper suggests a postal RFID application model that can be referred to for adaptation of RFID technology in postal logistics environments in section 2. We also show the recognition performance of parcel tag by speed and tag mounting material such as water bottles, paper, and thin cans using implemented RFID application system for postal logistics in section 3. We conclude with summary and further study in section 4.

2 RFID Application Model for Postal Logistics

To minimize trial and error for the adaptation of RFID on postal environments, to check the problem of an RFID application system on postal logistics environments, to

K.C. Chang et al. (Eds.): APWeb/WAIM 2007 Ws, LNCS 4537, pp. 479–484, 2007.

find an RFID adaptation model, and to find the application requirement profile of RFID on postal logistics environments before establishing a real RFID system on a postal field, we need to design a postal RFID process and application model, RFID tag data structure and code, and postal RFID application system. Fig. 1 shows the postal RFID application model that was proposed for the adaptation of RFID technology in postal logistics environments. The postal RFID application model can be done using a real time trace and tracking of parcel processing and pallet management based on RFID. There are mail offices which can deal with the RFID tag issue and label printing for parcel registration. Fig. 1 shows a mail center that can do RFID tag recognition for the sending and arrival processing of parcel sand pallets at docks and for auto-sorting at sorting machines, Fig. 1 also shows a delivery office that can do RFID tag recognition of delivery parcels, and data reset for the preparation of tag reuse after parcel delivery. The following are the detailed functions and interactions of each component for a real time trace and tracking parcel process and pallet management in a postal RFID application model.

■ Mail Office

A mail office, which can conduct parcel registration through issuing an RFID tag and label printing, sends parcel issue data such as a postal code and address automatically using an RFID reader with internet to a postal logistics system via a local server. The mail office also gets mail trace data such as parcel processing and delivery data from a local server after the registered parcel moved to the mail center.

■ Mail Collection & Distribution Center

A mail center that can do RFID tag recognition at docks and has a sorting machine for the auto-sorting, sending, and arrival processing of parcels and pallets sends RFID tag recognition data of the parcels and pallets, such as mail center code and parcel processing data, automatically using an RFID reader with internet to a postal logistics system via a local server. The mail center also gets mail trace data such as parcel processing and delivery data from the local server after the parcels and pallets have moved to the delivery office.

■ Delivery Office

A delivery office that can do RFID tag recognition for the delivery of parcels and a tag data reset for reuse preparation of parcel and pallet tags sends RFID tag recognition data of the parcels and pallets, such as delivery office code, delivery man, and delivery data automatically using a hand held RFID reader with wireless LAN (Local Area Network) and CDMA (Code Division Multiple Access) to a postal logistics system via a local server. The delivery office also gets processing data such as parcel registration and sorting data from the local server after the sorted parcels have moved to the delivery office.

■ Monitoring Center

The monitoring center can do a real time trace and tracking of parcel processing and pallet management via a local server and postal logistics system. The monitoring center can also check how many parcels are registered in real time, how many pallets with parcels have moved among mail centers, and how many empty pallets are stored in each mail center. This is also important for the effective use of parcel and pallet management, customer service parcel upgrades, and for knowing which mail center has a problem with parcel processing and pallet management.

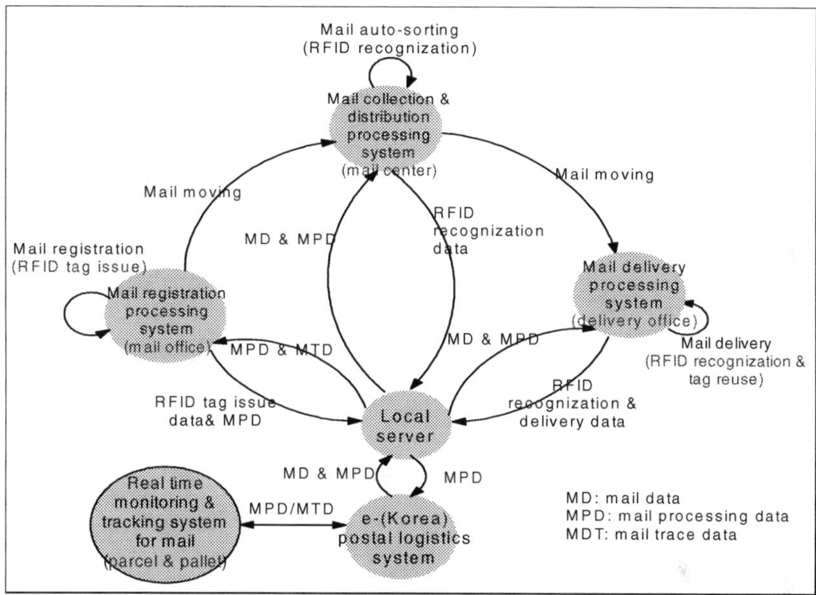

Fig. 1. RFID Application Model for Postal Logistics

3 Performance

In order to prove the tag recognition performance based on the implemented postal RFID system [2][3] and find the best installation solution of the system in postal logistics environments, we tested the tag recognition rate of parcels by tag mounting material, tag orientation, and moving speed using a postal RFID application system. And the recognition rate of pallets and boxes by antenna type and tag position was also done in postal logistics environments. We use only a UHF linear and circular antenna (Intermac) with RFID tag (ISO 18000-A, SAMSys) for the performance test based on our implemented and installed system in the field.

3.1 Recognition Rate of Parcel on Pallet by Tag Position and Speed and Tag Mounting Material

Fig. 2 shows the recognition rate of parcel by tag mounting material and speed. The recognition rates of paper and aluminum are about 90 % and 80 %, respectively.

The recognition rates of cans and water are about 50 % and 35 %, respectively. There is also a small difference of recognition rate of paper by moving speed such as slow, medium, and fast, and by tag orientation. Medium means that the walking speed for passing the installed system & RFID antenna from the 5 m front area to 5 m backend area is 8 seconds, slow is 14 seconds, and fast is 4 seconds. We also installed the antenna 130 m high in an overhead type RFID system for testing the reading rate of water, 170 m high for paper, 150 m high for aluminum, and 73 m high for a can.

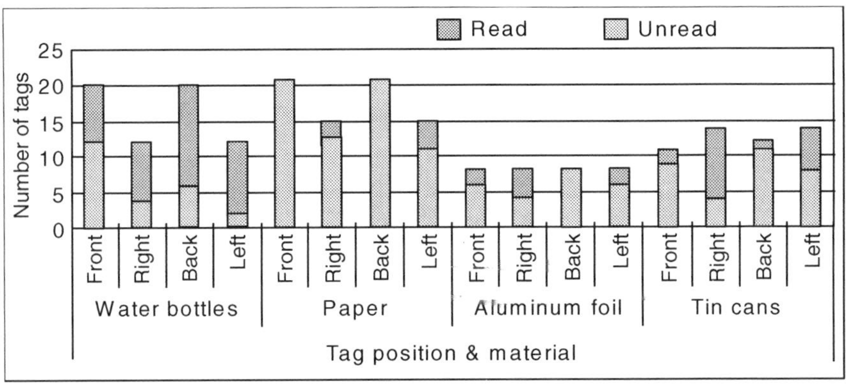

Fig. 2. Recognition Rate of Parcel on Pallet by Tag Position and Speed and Tag Mounting Material

3.2 Recognition Rate of Box on Pallet by Antenna Type and Tag Position

To find a postal RFID application requirement profile such as antenna type and direction for registration, takeover, and sending and arrival of parcels, we performed a field test by box tag orientation on pallet and antenna type using an RFID test bed. Table 1 shows the recognition rate of box [plastic] tags on a pallet by antenna type and tag position. A circular and ceiling type antenna with maximum 98 % reading rate is better than a circular and stand type antenna with maximum 44.3 % reading rate, and linear and stand type antenna with maximum 62 % reading rate. Table 2 also shows that the best recognition rate of box tags on pallets occurs when we attached the tag under the box. And it shows that the maximum recognition rate of multiple boxes on pallets is 20 box tags. Based on this preliminary field test, we consider antenna type, tag orientation, optimal number for multiple parcels, and pallet recognition in postal logistics environments as necessary parts of an application requirement profile. For parcel and pallet management using RFID, the best situation is a flat type antenna for parcel registration, circular-overhead type antenna for parcel takeover, circular-ceiling type antenna for sending and arrival of multiple parcels, and a pallet with 20 tagged boxes.

Table 1. Recognition Rate of Box Tag in Pallet by Antenna Type and Tag Position (Reading Tag/Total Tag)

Antenna Type	Tag Position	Number of Test					Average Reading
		1	2	3	4	5	
Circular/ Stand Type	Box Outside	1/20	3/20	3/20	0/20	2/20	1.8/20
		3/40	5/40	1/40	3/40	4/40	3.2/40
		5/60	3/60	2/60	2/60	1/60	2.6/60
	Box Inside /Under	2/20	1/20	0/20	0/20	1/20	0.8/20
		4/40	6/40	2/40	4/40	4/40	4/40
		2/60	6/60	4/60	4/60	3/60	3.8/60
Circular/ Ceiling Type	Box Out-side	5/20	5/20	6/20	4/20	6/20	5.2/20
		8/40	10/40	8/40	8/40	10/40	6.8/40
		9/60	9/60	8/60	10/60	11/60	9.4/60
	Box Inside /Under	19/20	17/20	19/20	18/20	19/20	19/20
		34/40	28/40	29/40	22/40	30/40	28.6/40
		28/60	37/60	29/60	38/60	39/60	34.2/60
Linear/ Stand Type	Box Out-side	0/20	1/20	3/20	1/20	1/20	1.2/20
		3/40	2/40	2/40	2/40	2/40	2.2/40
		3/60	1/60	3/60	3/60	2/60	2.4/60
	Box Inside /Under	2/20	1/20	0/20	1/20	3/20	1.4/20
		4/40	3/40	1/40	4/40	2/40	3.2/40
		4/60	3/60	4/60	3/60	5/60	3.8/60

3.3 Recognition Rate by Power and Distance

The reading distance of a tag depends on the field conditions, electric conditions between tag and reader, the reader's power, and so on. The minimum value for normal operation of a tag that has to get signal power from an RFID reader is 35 μW. So we can calculate the maximum reading distance of a tag from the RFID reader signal using the following formula.

$$R_{max} = \sqrt{\frac{P_{EIRP} \times G_{tagantenna} \times \lambda^2}{(4 \times \pi)^2 \times P_{chip}} \times Loss}$$

- P_{EIRP} = Reader Power (EIRP), R_{max} = Reading Distance, $G_{tagantenna}$= Tag Antenn a Gain, f (Operation Frequency), P_{chip} = Received Power of Tag, Loss (Mismat ching and Tag Antenna Loss)
- P_{EIRP} = 4 W, $G_{tagantenna}$= 1.64 dB, f = 912.5 MHz, P_{chip} = 35 μW, Loss = 3 dB
- R_{max} = 7.56m

We also tested the reading distance by RFID reader power using a postal RFID system in postal logistics environments. Fig. 3 shows the received power at a tag by distance from the RFID reader. It shows that the received power of a tag at 6 m is

56 μW, and 35 μW at 7.56 m as the maximum reading distance. We can see that the measured value is almost the same with the calculated value using the above formula.

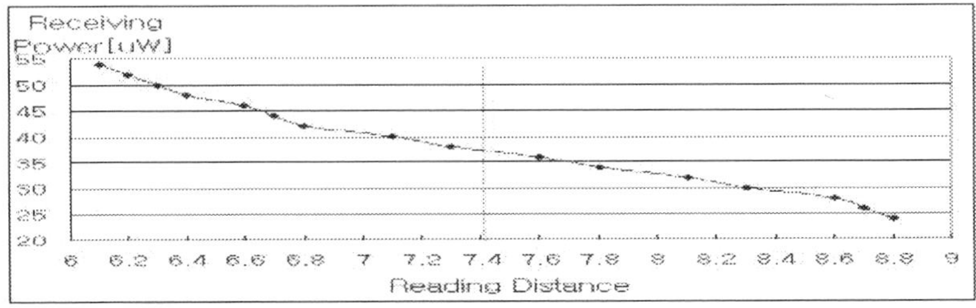

Fig. 3. Recognition Rate by Power & Distance

4 Concluding Remark

In this paper, we suggested a postal RFID application model that can be used for real time trace and tracking of parcel processing and pallet management. This application model and architecture can enable a developer to compose a postal RFID application system and service. This paper also showed the recognition performance of parcels by speed and tag mounting material such water bottles, thin cans, and paper in postal logistics environment. In order to identify the postal RFID application requirement profile such as antenna type and direction, we performed field tests according to box tag orientation in the pallet and antenna type using the RFID testbed, and tested the reading distance according by the RFID reader power. However, there are still some considerations for adaptation of RFID technology in postal logistics environments such as multiple tag recognition rate by speed, distance, content, and environment; interference problem of antennas; visibility problem of recognized multiple parcels; visibility of non-recognized parcels; visibility of RFID tag data; reuse of RFID tags & labels; specific label type design problem for parcels and pallets; postal code development for postal logistics; and a new standard postal logistics process & platform development based on RFID.

We now keep going field test to find the best solution and installation guide for adaptation of RFID technology in postal logistics environments for further study.

References

1. DoC, Radio Frequency Identification - Opportunities and challenges in Implementation, April (2005)
2. Park, J.: Postal RFID System in Korea as RFID Pilot Project, ISO/IEC SC31/WG4/SG5, SG5n0010, USA (September 22, 2005)
3. Park, J.: RFID Testbed Construction and Development for Parcel Processing and Pallet Management in Postal Logistics Environments, KPF 2005 Proceeding, Seoul (June 1, 2005)

An Organization and Task Based Access Control Model for Workflow System

Baoyi Wang and Shaomin Zhang

School of Computer, North China Electric Power University, Baoding 071003, Hebei, China
wangbaoyi@126.com

Abstract. The application of traditional Role-Based Access Control in workflow system can reduce the complexity of privilege management. However, merely using role concept can hardly reflect enterprise' organization structure and can not specify organization unit for task; moreover, permission inheritance relation among roles isn't consistent with the fact of workflow system. This paper proposed an organization and task based access control model to amend the above shortcomings. Combined with the proposed model, through adding users' blacklist to each task, a dynamic access control algorithm is given. In practice, the model OTBAC with the access control algorithm is flexible and feasible.

Keywords: workflow system, organization modeling, access control, authorization, constraint.

1 Introduction

Access control is the crucial technology to prevent security breaches in workflow system. RBAC model[1] associates permissions with roles and assigns users to roles, and users get permissions by their assigned roles. Because roles' quantities in an enterprise is much smaller then users', RBAC facilitate the management of permissions. Several application example of RBAC model in workflow have been proposed in recent years, e.g., WAM[2] and SRWM[4], and RWAM[5].

However, in workflow system, usually there exists same roles in different organization units, when we assign task to roles, we also want to point out the organization unit the role belongs. To reflect enterprise's organization structure, the concept of organization unit is necessary. Although role hierarchy avoids redundant assignment of permission, in fact, the senior role may not inherit all permissions of the junior one. For example, department director is senior to department accountant in role hierarchy, but department director cannot have permissions belong to department accountant. Of course, we can define private role [1] to avoid unsuitable permissions inheritance, but it is obscure and uneasy to be accepted. Moreover, static authorization constraints in RBAC cannot meet the dynamic requirement of workflow. This paper proposed an organization and task based access control model for workflow to amend these shortcomings.

K.C. Chang et al. (Eds.): APWeb/WAIM 2007 Ws, LNCS 4537, pp. 485–490, 2007.

2 Organization and Task Based Access Control Model: OTBAC

2.1 OTBAC Model

OTBAC model is shown in Fig.1. We take task as group of permissions and divided task to two classes, common task and professional task, the former is assigned to organization unit to which all member belong can execute the task, and it can be inherited by the lower organization unit; the latter is assigned to a certain role in a certain organization unit and cannot be inherited. We remove the permission inheritance relation among roles to avoid unsuitable permission inheritance. Users acquire task permissions by assigned organization units or roles. The formalized model is described as follow.

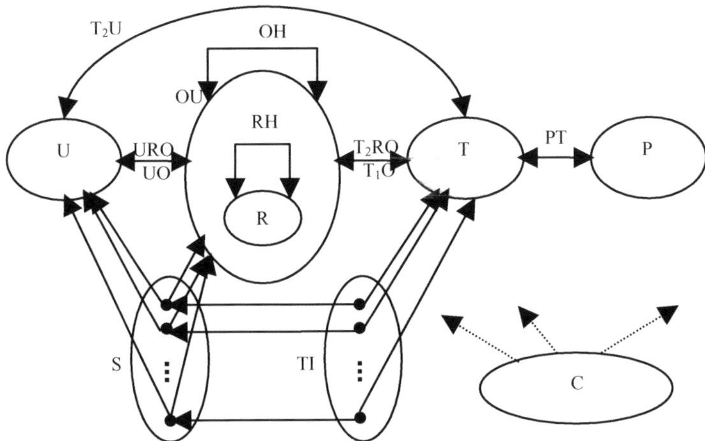

Fig. 1. Organization and task based access control model

2.2 Formalized OTBAC Model

2.2.1 Definitions of Sets in OTBAC

$U = \{u_i \mid i = 1,2,\cdots n\}$ is set of users.

$R = \{r_i \mid i = 1,2,\cdots n\}$ is set of roles, each role belongs to an organization unit.

$OU = \{ou_i \mid i = 1,2,\cdots n\}$ is set of organization unit.

$T = T_1 \cup T_2 = \{t_i \mid i = 1,2,\cdots n\}$, T_1 is set of common tasks in workflow, and T_2 is set of professional tasks in workflow, and $T_1 \cup T_2 = T, T_1 \cap T_2 = \varnothing$.

$P = \{p_i \mid i = 1,2,\cdots n\}$ is set of permissions.

$S = \{s_i \mid i = 1,2,\cdots n\}$ is set of sessions initiated by users.

$TI = \{ti_i \mid i = 1,2,\cdots n\}$ is the current task instance set in workflow system.

$C = \{c_i \mid i = 1,2,\cdots n\}$ is set of authorization constraints.

2.2.2 Definitions of Map Relations in OTBAC

$RH = \{(r_i, r_j) \mid r_i \le_r r_j\} \subseteq R \times R$ is a local partial order set between roles, denoted as $\le_r$. It denotes roles' orders when assigning tasks to roles.

$OH = \{(ou_i, ou_j) \mid ou_i \le_o ou_j\} \subseteq OU \times OU$ is partial order set between organization units, if $ou_i \le_o ou_j$, then the members of ou_i is also implicit members of ou_j, and ou_i inherits common tasks assigned to ou_j.

$UO = \{(u, ou) \mid u$ is a member of $ou\} \subseteq U \times OU$ is an assignment relation mapping users to the organization units they belong to.

$RO = \{(r, ou) \mid r$ belongs to $ou\} \subseteq R \times OU$ is an assignment relation mapping roles to the organization units they belong to.

$URO = \{(u, (r, ou)) \mid u$ is assigned r in $ou\} \subseteq U \times RO$ is an assignment relation mapping users to the roles in some organization unit.

$T_2RO = \{(t, (r, ou)) \mid$ professional task t is assigned to r in $ou\} \subseteq T_2 \times RO$ denotes an assignment relation mapping roles in some organization to professional tasks they are assigned.

$T_1O = \{(t, ou) \mid$ common task t is assigned to organization unit $ou\} \subseteq T_1 \times OU$ is an assignment relation mapping organization units to common tasks they are assigned.

$T_2U = \{(t, u) \mid$ professional task t is assigned to user $u\} \subseteq T_2 \times U$ is an assignment relation mapping users to professional tasks they are assigned.

A class of task only can be assigned to one among users, roles and organization units, if we use $domR$ denoting the domain of relation R, we get the following,

$$dom(T_2U) \cup dom(T_2RO) \cup dom(T_1O) = T$$
$$dom(T_2U) \cap dom(T_2RO) \cap dom(T_1O) = \varnothing.$$

$PT = \{(p, t) \mid$ permission p is assigned to task $t\} \subseteq P \times T$ is an assignment relation mapping permissions to tasks they are assigned to.

2.2.3 Description of Constraints in OTBAC

Bertino etc have done much work[3] in definition and enforcement constraints in workflow system. They mainly discuss dynamic and static constraints of separation of duty (SOD). In workflow system, constraints of binding of duty (BOD), that means that two or more tasks must be executed by the same user, is also necessary. Here we discuss SOD and BOD constraints. We define exclusive relation between tasks for SOD, and define binding relation between tasks for BOD as following.

$TT_{SOD} = \{(t_i, t_j) \mid t_i$ is exclusive with $t_j\} \subseteq T \times T$,

$TT_{BOD} = \{(t_i, t_j) \mid t_i$ is binding with $t_j\} \subseteq T \times T$.

If $(t_1, t_2) \in TT_{SOD}$, then t_1 and t_2 can not be assigned to the same organization unit, role or user; if $(t_1, t_2) \in TT_{BOD}$, then t_1 and t_2 must be assigned to the same organization unit, role or user, and corresponding task instances must be executed by the same user. Similarly, we can define exclusive relation and binding relation between roles, permissions or organization units to implement related SOD and BOD constraints. To enforce constraints, we can adopt the method of blacklist[6].

2.2.4 Definitions of Functions in OTBAC

$user\_s : S \rightarrow U$ is a function mapping each session to a single user.

$role\_ou\_s : S \rightarrow 2^{RO \cup OU}$ is a function mapping each session to activated organization units or activated roles contained in organization units,

$$role\_ou\_s(s) \subseteq \{(r, ou), ou_i \mid ((user\_s(s), ou_j) \in UO \wedge (ou_j \leq ou_i)) \vee (user\_s(s), (r, ou)) \in URO\}.$$

$task(s)$ denotes tasks set that session s can execute,

$$task(s) = \cup_{(r,ou), ou_i \in role\_ou\_s(s)} \{t \mid \quad (t, (r, ou)) \in T_2RO, \quad or \quad \exists ou_j \in OU, ou_i \leq ou_j \wedge (t, ou_j) \in T_1O, or\ (t, user\_s(s)) \in T_2U\}.$$

$task : RO \cup OU \rightarrow 2^T$ is a function mapping each role in some organization unit or organization unit to task set they can execute,

$$task((r, ou)) = \{t \mid (t, (r, ou)) \in T_2RO\},$$

$$task(ou_i) = \{t \mid \exists ou_j \in OU, ou_i \leq_o ou_j \wedge (t, ou_j) \in T_1O\}.$$

$user\_t : T \rightarrow 2^U$ is a function mapping each task to user set who can execute it,

$$user\_t(t) = \{u \mid (t, u) \in T_2U, or \exists ou \in OU, t \in task(ou) \wedge (u, ou) \in UO, or \exists r \in R, \exists ou \in OU, t \in task((r, ou)) \wedge (u, (r, ou)) \in URO\}.$$

$session : TI \rightarrow S$ is a function mapping each task instance ti to a single session $session(ti)$.

3 The Algorithm and the Application Example

This section proposes an access control algorithm. Assuming the beginning time of each task in workflow is sequential [7], we give the access control algorithm according to the OTBAC model.

Algorithm name: access control algorithm
Input: (1) task t_i being asked for executing; (2)the constraint sets TT_{SOD} and TT_{BOD}.
Output: Return true if approve the application; Return false, otherwise.

We use sod_blacklist(t_i) and bod_blacklist(t_i) denoting SOD blacklist and BOD blacklist of task t_i respectively. First, get user_t(t_i) for task t_i using related interfaces of access control model. When a certain session s asks for executing instance of task t_i, see whether t_i belongs to task(s) and whether user_s(t_i) belongs to user_t(t_i); if yes then check sod_blacklist(t_i) and bod_blacklist(t_i); if sod_blacklist(t_i) is not null and u belongs to sod_blacklist(t_i), then refuse the requirement, else approve the requirement; if bod_blacklist(t_i) is not null and u belongs to bod_blacklist(t_i), then approve the requirement, else refuse the requirement. Finally, modify blacklist of task related to task t_i according to constraint set C.

Let us discuss a purchase workflow that consists of six tasks: the creation of purchase order (crtPO), the check of purchase order (checkPO), signing a goods received note (signGRN), issuing check (issueCHK), the approval of the check (apprCHK) and signing check (signCHK), as shown in Fig.2. There exists constrains $TT_{BOD}(crtPO, signGRN)$ and $TT_{SOD}(crtPO, checkPO)$. The organization structure consists of a root organization units of enterprise (E), and sub organization units including manufacture department (MD), purchase department (PD) and financial department (FD). There are eight users in workflow system: a, b, c, d, e, f, g and h, among which b is employee(ep) of manufacture department, h is the director(dr) of financial department, e is accountant(ac) of financial department, c and d are cashiers(csh) of financial department, a, f are employees(ep) of purchase department, and g is the director (dr)of purchase department.

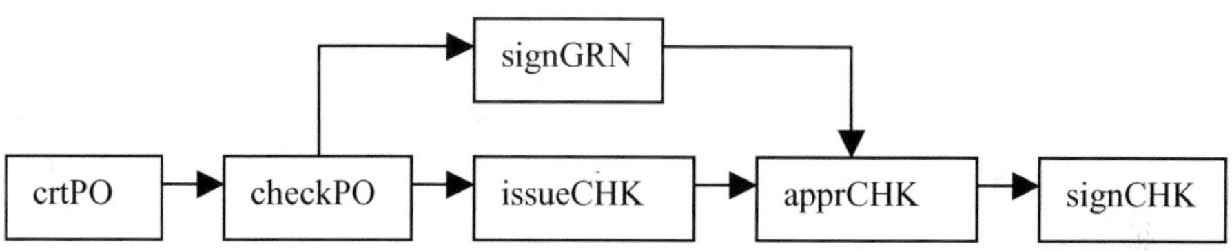

Fig. 2. A purchase workflow example

Assuming user a ask for executing task crtPO, because a assigned role of employee in organization unit PD, which inherit the common task from its topper organization implicitly, so approve the requirement, and add a to bod_blacklist (signGRN) because of the constraint $TT_{BOD}(crtPO, signGRN)$, and add a to sod_blacklist(checkPO) because of the constraint $TT_{SOD}(crtPO, checkPO)$; task checkPO can be assigned to a, f or g, due to a lies in the task's sod_blacklist, so only f and g can execute the task, and f prior to g to assign task checkPO due to the role of f (ep) is junior to the role of g (dr). Task issueCHK can be assigned to user c or d. About task signGRN, due to a existing in its bod_blacklist, so only a can execute the task. Task apprCHK is assigned to user h directly, so it should be executed by h, and task signCHK could be assigned to user e who assumes role ac in organization unit FD.

4 Conclusions

In this paper we propose an organization and task based access model for workflow. By introducing organization and task into traditional RBAC, we have shown that our model could reflect enterprise's organization structure, and could define role in which organization unit has permissions to execute a task. When there are same roles in different, this is necessary. A dynamic access control algorithm has also been presented. By adding blacklist data structure to each task, the algorithm can realize SOD or BOD constraints in workflow dynamically. The shortcoming of our model is that it lacks supporting for task delegation which is necessary when users are vacant, e.g., for disease or being on business trip, we will make efforts to solve this problem in our future work.

References

1. Sandhu, R., Coyne, E.J., Feinstein, H.L., et al.: Role-Based Access Control Models. IEEE Computer, 29, 38-47 (1996)
2. Huang, W.-K., Atluri, V.: Secureflow: A Secure Web-based Workflow Management System. In: Proc. of the 4th ACM Workshop on Role-based Access Control Fairfax, VA, USA, pp. 83-94 (1999)
3. Bertino, E., Ferrari, E., Atluri, V.: A Flexible Model Supporting the Specification and Enforcement of Role-based Authorizations in Workfiow Management Systems. In: Proc of the 2nd ACM Workshop on Role-based Access Control, pp. 1–12. ACM Press, New York (1997)
4. Kandala, S., Sandhu, R.: Secure Role-based Workflow Mode. In: Proc of the 15th Annual Working Conference on Database and Application Security, Dordrecht, pp. 1–14. Kluwer Academic Publishers, Norwell, MA, USA (2001)
5. Fan, H., Guang-Lin, X.: A Family of RBAC-based Workflow Authorization Models. Wuhan University Journal of Natural Sciences 10, 324–328 (2005)
6. Crampton, J.: XACML and role-based access control, DIMACS Workshop on Security of Web Services and E-Commerce (2005)
7. Crampton, J.: A Reference Monitor for Workflow Systems with Constrained Task Execution. In: Proc. of the 10th ACM Symposium on Access Control Models and Technologies, pp. 38-47 (2005)

A Method of Web Services Composition Based on Service Alliance

Chunming Gao[1,2], Liping Wan[1], and Huowang Chen[2]

[1] College of Mathematics and Computer Science, Hunan Normal University
410081 Changsha, China
[2] School of Computer Science, National University of Defense Technology
410073 Changsha, China
gcm_powerman@163.com, wanle2000@tom.com

Abstract. The method of QoS-based composition generally doesn't consider trade alliance of the service providers, so it can't be applied to the scenario of service alliance based on modern supply chain. In this paper, we study a service composition model composed of the business service process and the service relational network, and use genetic algorithm with penalty function to solve the 0-1 nonlinear programming problem. We can conclude from the simulation results that the optimal model considering the service alliance can obtain better results in the average total number of QoS attributes and objective-function values as well as better commercial benefits.

Keywords: service composition, service alliance, 0-1 nonlinear programming, genetic algorithm.

1 Introduction

Web Services system is a dynamically integrated, autonomous and loose-coupling distributed system. There has been wide concern and research about web service composition based on QoS-driven[1,2,3,4]. Along with development of the web service application, qualities of service composition become an important factor in the composition methods.

However, most of the composition methods don't consider business alliance relationship among the service providers that lie in the modern supply chain. Literature [2] and [4] only consider global optimization composition. Literature [3] presents a Web Services Outsourcing Manager framework, and gives the concept of service alliance relationship, but it does not present a model or solution farther.

In this paper, we construct a double-hierarchy relational network structure and consider the economic effect on service composition by the business alliance among the service providers. we consider the statistic property of control structure, adopt probabilistic method and provide a global optimization model based on service alliance. Since it is a complex 0-1 nonlinear programming problem that can't be linearized, the normal modified linear programming method [4] can't be used, so we choose genetic algorithm to accomplish service compositions. The rest of this paper is

K.C. Chang et al. (Eds.): APWeb/WAIM 2007 Ws, LNCS 4537, pp. 491–496, 2007.

organized as follows: In section 2, we construct a double-hierarchy relational network structure and provide the global optimization model in section 3. The solution of global optimization model is presented in Sections 4. The experiment results are in section 5. In section 6 we conclude the paper and describe the future work.

2 Double-Hierarchy Network Based on Service Alliance

We use a Directed Acyclic Graph (DAG) to represent the execution paths of the state chart, i.e., the upper hierarchy network structure in Fig.1, which just considers the properties of control structure in the state chart. It includes three basic control structures: Flow, Switch and Sequence. With cycle structure, DAG will be transformed to multi-sequence structure according to average executive times that are statistical value from Log file.

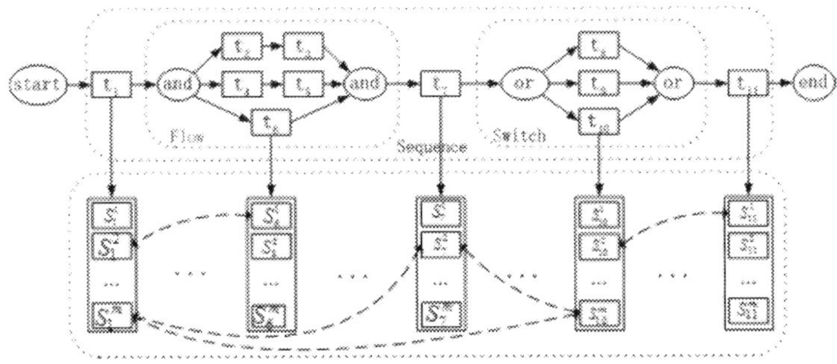

Fig. 1. Double-hierarchy state chart of composite service based on service alliance

According to the properties of control structure and Log file statistical value, in Structure Flow and Sequence, all the task executive probability is 1. Single task of Structure Switch is less than 1, and the total of all tasks' executive probability is 1.

In an execution path, task t_k has a candidate services set $S_k = \{S_k^1, S_k^2, \cdots, S_k^m\}$ that can execute task t_k. These services are provided by different Service Providers. We take the following general quality metrics into account: execution price, execution duration, successful execution rate, availability[2]. By merging the quality vectors of all these candidate services, a matrix $Q = (Q_{ijk;} 1 \le i \le m, 1 \le j \le 4, 1 \le k \le n)$ is built. We scale the quality matrix according to the method proposed in Literature [2].Then, we obtain the normalized matrix $V = (V_{ijk}; 1 \le i \le m, 1 \le j \le 4, 1 \le k \le n)$. We select a service from each set so that global objective function is optimized and all quality metrics constraints are satisfied[2][4].

In business application, Service Providers will gradually form lots of business alliances[1] which are service relational networks (see lower network in Fig.1). If the candidate services belonging to a service alliance are selected, the service composition can get better quality of service in total. In order to allow for service alliance in global QoS-optimized model, we define alliance preferential rates matrix and service alliance adjacency list as follows:

Definition: If preferential rates of service alliance i are $P_i = (p_{i1}, p_{i2}, p_{i3}, p_{i4})$, price, execution duration, successful execution rate and availability respectively has preferential rate $p_{i1}, p_{i2}, p_{i3}, p_{i4}$ of all services in the service alliance. Service alliance i exists if and only if $P_i > 0$, at least one $p_{ik} > 0, (k = 1,2,3,4)$.

We are using adjacency list to store service alliance for the sake of depositing and inquiring efficiency. Fig.2 is adjacency lists containing four service alliances.

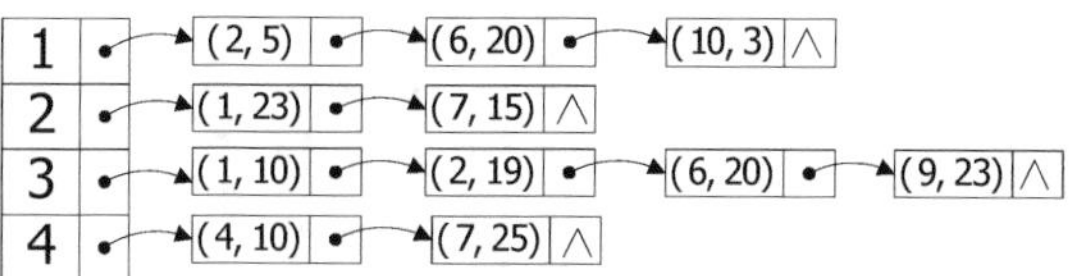

Fig. 2. Adjacency list for service alliance

From Fig.2 we can see, Service alliance 1 includes candidate Service $S_2^5, S_6^{20}, S_{10}^3$, corresponding alliance preferential rates is P_1. If these services are selected to execute corresponding task, then qualities of service have preferential rates $p_{ik}, (k = 1,...,4)$.

3 Global QoS-Optimized Model Based on Service Alliance

Global optimization computation of service composition is to select a service sequence that includes service alliance under the condition of global optimized objective function. In all, we define the objective function and constraints as follows.

(1) Objective function
After quality matrix is scaled, it is then weighted. Thus the multi-objective programming is transformed to a single-objective programming. According to executive probability of task, all tasks of the multi-path process are aggregated. So the objective function of the global optimization is:

$$Max\ F(x) = Max\left(\sum_{k=1}^{n}\sum_{i=1}^{m}\sum_{j=1}^{4} x_{ik}\rho_k w_j v_{ijk} + \sum_{l=1}^{u_{all}} (x_{r_{l1}} x_{r_{l2}} \ldots x_{r_{lm_l}} \sum_{j=1}^{4} w_j \sum_{k\in m_l} p_{lj}\rho_{r_{lj}} v_{r_{ljm_l}}) \right)$$

where $\rho_{r_{lj}}, \rho_k$ is the selected executive probability of service r_{lj}, k, and

$$\begin{cases} \rho_{r_{lj}}, \rho_k < 1, & service\ r_{lj}, k\ corrsponding\ task\ in\ Structure\ switch \\ \rho_{r_{lj}}, \rho_k = 1, & other \end{cases}$$

where $x_{ik}, x_{r_{lm_l}} \in \{0,1\}, \sum_{i=1}^{m} x_{ik} = 1, k \in \{1,2,\ldots,n\}$. w_k is the weight of k*th* quality metrics provided by the user, $\sum_{k=1}^{4} w_k = 1$ .n is the total number of tasks, m is the number of candidate service sets and u_{all} is the number of service alliance sets. m_l is the number of services in each service alliance set.

(2) QoS constraints
According to user's requirement for QoS, the properties of the control structure and service alliance, the QoS constraints table can be shown in Table 1.

Table 1. QoS constraints

Constraints	Illustration
$\sum_{k\in n}\sum_{i\in m} x_{ik}\rho_k c_{ik} - \sum_{l=1}^{u_{all}} x_{r_{l1}} x_{r_{l2}} \ldots x_{r_{lml}} p_{l1}(\sum_{k\in m_l} \rho_{r_{lk}} c_{r_{lk}}) \le B_c$	$c_{ik}, c_{r_{ik}}$ are price, B_c is the price budget set by the user.
$\sum_{k\in A}\sum_{i\in m} x_{ik}\rho_k z_{ik} - \sum_{l=1}^{u_{all}} x_{r_{l1}} x_{r_{l2}} \ldots x_{r_{lml}} p_{l2}(\sum_{k\in m_l} \rho_{r_{lk}} z_{r_{lk}}) \le B_d$ $e_{next} - (p_k + e_k) \ge 0, \forall t_k \to t_{next}, k, next \in A,$ $B_{du} - (p_k + e_k) \ge 0, \forall k \in A$	$z_{ik}, z_{r_{ik}}$ are corresponding to critical task. A is the critical task set. B_{du} is the execution duration budget set by the user. e_k, p_k are the expected start time and expected duration[2].
$\sum_{k\in n}\sum_{i\in m} x_{ik}\rho_k \ln(a_{ik}) + \sum_{l=1}^{u_{all}} x_{r_{l1}} x_{r_{l2}} \ldots x_{r_{lml}} p_{l3}(\sum_{k\in m_l} \rho_{r_{lk}} \ln(a_{r_{lk}})) \ge B_{rat}$	$a_{ik}, a_{r_{ik}}$ are successful execution rate, B_{rat} is successful execution rate budget set by the user.
$\sum_{k\in n}\sum_{i\in m} x_{ik}\rho_k \ln(b_{ik}) + \sum_{l=1}^{u_{all}} x_{r_{l1}} x_{r_{l2}} \ldots x_{r_{lml}} p_{l4}(\sum_{k\in m_l} \rho_{r_{lk}} \ln(b_{r_{lk}})) \ge B_{av}$	$b_{ik}, b_{r_{ik}}$ are the availability, B_{av} is the availability budget set by the user.

Objective function $\max F(x)$ and QoS constraints form the global optimization model of service composition considering service alliance. The model is a complex 0-1 nonlinear programming problem and can't be linearized any more. So the normal linear programming method[4] can't be applied here.

4 Global Optimization Solution Algorithm

Non-linear 0-1 programming, which is a typical NP problem, may be difficult to be solved in the traditional optimizing methods. Therefore, we adopt genetic algorithm [5] to solve the service composition problem based on alliance. Considering the solution framework of genetic algorithm and feature of this problem, our algorithm is shown as follows:

(1) Encoding Design and Creation of Initial Population
Take *N* chromosomes into account. The length of each chromosome is *n* and every gene is encoded by natural numbers in [1,*m*] randomly produced to compose initial

population. N is the number of population, n is the number of task in service composition flow and m is the number of candidate services of every task.

(2) Stop Rule

The maximum evolution generation configured previously is the stop condition.

(3) Fitness Function Design

Penalty function is to combine the global constraint sets and objective function. The fitness function Fit is:

$$Fit = F - \lambda(\sum_{j=1}^{con} \frac{\Delta D_j}{C_{j\max} - C_{j\min}}),$$

where $C_{j\max}, C_{j\min}$ are maximum and minimum in the j_{th} constraint condition respectively, con is the number of constraint, λ representing penalty factor is a empirical value, ΔD_j is the value exceeding the j_{th} constraint threshold, and $\Delta D_j = 0$ when chromosomes satisfy the constraints.

(4) Selection ,Crossover and Mutation

We use roulette wheel selection with elitist policy. Crossover operation chooses single-point crossover. Mutation operation adopts random perturbation. The selected genes are replaced by a natural number selected randomly in [1,m].

5 Simulation and Experiment Results

In our performance study, we implement a simulation system in java of JDK1.5.1. Experiments are performed on an Intel Pentium IV 2.6G, 512M RAM, with Windows XP. The population size is 50, crossover probability is 0.7, mutation probability is 0.01, and penalty factor is 0.1. Every time the experiment data is the same. Simulator randomly generates some candidate services and service quality matrix. Thus the alliance relationship adjacency matrix and its preferential rates matrix are created.

The flow includes 11 tasks in Fig.1. Every task has 40 candidate services. Suppose 15 service alliances. Here relaxation coefficient is set as 0.7[4]. The other parameters are set as W = (0.5,0.3,0.1,0.1), $\rho_8 = 0.4$, $\rho_9 = 0.3$, $\rho_{10} = 0.3$. The result of fitness compared with different genetic generation by running 50 times is in Fig.3. A is a non-alliance model, while B is a service alliance model.

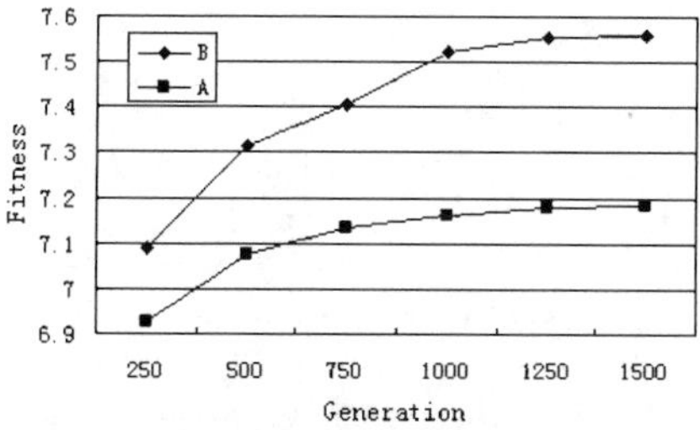

Fig. 3. Fitness Contrast of Different Genetic Generation

In Fig.3, the line above is the simulation results considering service alliance, while the below represents the results not considering service alliance. Obviously, fitness of the above line is larger.

Under the situation that we consider service alliance model (A) and do not consider model (B), the complexity of genetic generation is different, and this is reflected in Table 2 that shows the contrast results of QoS metrics. From the table, it is obvious that the cost and response time of optimization model considering service alliance B is smaller than A in each genetic generation, but its successful rate and reliability is larger. Model B obtains service composition of higher quality. Thus users can gain more profits if provided better service composition scheme with service alliance. From Table 2 we can see that the average computation time of Model B is longer than that of Model A, for Model B is a non-linear model. However, its computation time is within the tolerable range.

Table 2. Contrast Results of Average Total of Different Genetic Generation (A:B)

	250 Generation	500 Generation	750 Generation	1000 Generation
Cost	107.9 :103.5	104.5 : 97.5	103.4 : 95.1	102.7: 91.7
Response Time	113.6: 111.5	110.8 : 106.9	108.8:105.4	108.3:102.9
Successful Rate	40.5:42.6	40.5 : 44.2	40.6:46.62	40.6: 47.23
Reliability	40.3: 41.8	40.3 : 42.4	40.3:42.9	40.3: 43.4
Computation Time(ms)	176 : 1189	360 : 2366	525: 3556	688 : 4740

6 Conclusions

Through studying global optimization service composition deeply, we consider the service alliance in commercial application and construct a double-hierarchy relational network structure. Based on probabilistic method, we build a global optimization model with service alliance. We select genetic algorithm to accomplish the optimization of web services composition. The future work contains improving genetic algorithm to get better optimization solutions and higher efficiency on this kind of nonlinear programming problems.

References

1. Sayah, J.Y., Zhang, L.-J.: On-demand business collaboration enablement with web services. Decision Support Systems 40, 107–127 (2005)
2. Zeng, L., Benatallah, B., Dumas, M., Kalagnanam, J., Chang, H.: QoS-Aware Middleware for Web Services Composition[C]. In: Proc. IEEE transactions on software engineering 30(5), 311–327 (May 2004)
3. Zhang, L.-J., Li, B.: Requirements Driven Dynamic Services Composition for Web Services and Grid Solutions. Journal of Grid. Computing 2, 121–140 (2004)
4. Berbner, R., Spahn, M., Repp, N., Heckmann, O., Steinmetz, R.: Heuristics for QoS-aware Web Service Composition. In: ICWS'06, pp. 72–82 (2006)
5. Srinivas, M., Patnaik, L.M.: Genetic algorithm: A survey. IEEE Computer 27(6), 17–26 (1994)

Toward a Lightweight Process-Aware Middleware

Weihai Yu

University of Tromsø, Norway
weihai@cs.uit.no

Abstract. Process technology is widely adopted for composition of web services. On-demand composition of web services by a large number of small businesses or even end-users, however, calls for a peer-to-peer approach for process executions. We propose a lightweight process-aware component of middleware for peer-to-peer execution of processes. The approach is of continuation-passing style, where continuations, or the reminder of executions, are passed along with messages for process execution. Conducting the execution of a process is a series of local operations rather than global coordination. Two continuations are associated with an execution: a success continuation and a failure continuation. Recovery plans for processes are automatically generated at runtime and attached to failure continuations.

Keywords: Web services composition, peer-to-peer, workflow, continuation.

1 Introduction and Related Work

Process technology is widely adopted for web services composition. A composite service is a process that uses other services (sub-processes) in some prescribed order. WS-BPEL [5], or simply BPEL, is becoming a de facto standard for process-based web services composition. However, process technology today can hardly be adopted by a large number of small businesses or even end-users for on-demand compositional use of available services. One particular reason is that process executions are conducted by heavyweight engines running on dedicated servers. They are generally to costly for small end-users.

In the process research area, there are proposed approaches to decentralized process execution that do not involve a central process engine (see for example [1]-[4][6]). With these approaches, a process specification is typically analyzed before execution, and proper resources and control are pre-allocated in the distributed environment. These approaches inevitably allocate resources even for the part of the process that is not executed. This conflicts with the design goals of the individual service providers. To achieve high scalability, most service providers choose their implementations to be as stateless as possible. To these service providers, allocating resources only for some unpredictable chance of future use is an unbearable burden.

In [7], we proposed a peer-to-peer approach to process execution that does not involve static process instantiation. It is lightweight in the sense that there is less housekeeping of runtime states and hold of resources than the other approaches. The approach is of continuation-passing style, which is a common practice in the

K.C. Chang et al. (Eds.): APWeb/WAIM 2007 Ws, LNCS 4537, pp. 497–503, 2007.

programming language community. Basically, a continuation represents the rest of execution at a certain execution point. By knowing the continuation of the current execution, the control can be passed to the proper subsequent processing entities without the involvement of a central engine. In addition, the approach supports automatic recovery of workflow. To achieve this, two continuations are associated with any particular execution point. The success continuation represents the path of execution towards the successful completion of the process. The failure continuation represents the path of execution towards the proper compensation for committed effects after certain failure events.

In this paper, we propose a process-aware component of a general purpose middleware based on this work.

2 Process Container for Continuation-Passing Messaging

Basically, a message tells a processing entity, known as an *agent* here, what to do next. If a message also contains a continuation, the agent can figure out the execution plan that follows up. In our approach, conducting the execution of a process is the sequences of sending and interpreting messages that contain continuations. New continuations are dynamically derived from the current status and continuations.

We use BPEL as the process model and SOAP as the underlying messaging protocol, though our approach is not restricted to these. A SOAP message consists of an optional header element and a mandatory body element wrapped in an envelope element. The body element of a SOAP message for process execution consists of several sub-elements: a control activity, a success continuation and a failure continuation, as shown below:

```
<a:A> activity </a:A>
<a:Ks><a:A> activity </a:A> ... <a:A> activity </a:A></a:Ks>
<a:Kf><a:A> activity </a:A> ... <a:A> activity </a:A></a:Kf>
```

The first sub-element <a:A> represents the activity, called *control* activity, to be executed immediately. It is either an original BPEL activity or an *auxiliary activity*. Auxiliary activities are automatically generated during execution for management tasks (see some examples in the next section). The sub-elements <a:Ks> and <a:Kf> represent the *success* and *failure continuations*, one of which is to be executed after the control activity. A continuation is represented as a stack of activities.

An agent is realized using a process-aware component, called *process container*, of a general purpose middleware. A process container runs at each site where processes are to be executed. The structure of the process container is shown in Figure 1. A process container conducts the executions of processes at its site according to the SOAP messages in the message queue, as described below.

Requests for process executions are put in the *message queue* (1). A *process interpreter* is a pool of threads that interpret the messages in the message queue. A thread dequeues a message from the message queue (2) and decides the next action according to the control activity of the message. There are two possibilities here: either can the process move on with local processing, or it is dependent on some other messages that are not available yet, such as a `receive` activity dependent on an incoming `invoke` message. In the former case, the thread invokes (3, 4) some local programs. In the latter case, the current message is put in the *pending message pool* (5). Hence a message in the pending message pool represents a branch synchronized with a dependent activity of another branch. This message will be used later (6) when a dependent message is available (2 again). After the execution of the local procedures, new messages are either put in the pending message pool (5) or sent to a remote site (7).

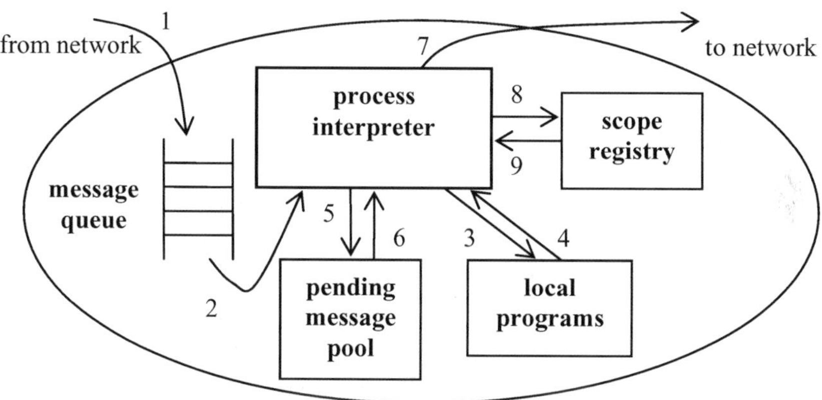

Fig. 1. Structure of process container

A BPEL scope provides a boundary for fault handling and recovery (a top-level process is a top-level scope). Every scope is managed by an agent. An agent maintains the states of the scopes in its charge in the *scope registry* (8, 9). Basically, the scope state contains the current locations of all active parallel branches within the scope. The location of a branch changes when a message is sent to a remote site. To keep this location state up to date, when an agent sends a message to a remote site (7), it also notifies the management agent of the immediate enclosing scope. To terminate a scope, the scope agent asks the agents of all active branches within the scope to stop the corresponding local activities by deleting the corresponding entries in the message queue, the pending message pool and the scope registry. To rollback a scope, all these agents run the respective parts of failure continuations.

3 Process Execution

To illustrate, we use an example process that helps some party organizers collect dancing music according to every organizer's favorites.

Figure 2 sketches the BPEL-like process. The event handler of the process interrupts the process and stops all activities. The primary activity of the process spawns several parallel branches: one for each party organizer to propose and collect his/her favorite music (for example, based on his/her own collections and free online downloading). If, after the parallel branches are done, there are still some pieces of proposed music unavailable, a borrowOrder request at LocalLibrary is issued. The compensation operation of borrowOrder is borrowCancel.

A process is started with an initial SOAP message sent to the process agent AlicePC. The message has the following body element.

```
<a:A><process> ... </process></a:A>
<a:Ks></a:Ks> <a:Kf></a:Kf>
```

```
<process name="PartyMusic" agent="AlicePC">
  <eventHandlers><onEvent operation="stop">
                  <throw faultName="Any" /> </onEvent>
  </eventHandlers>

  <sequence>
    <receive operation="collectMusic"
   createInstance="yes"
             outputVariable="musicLists" />

    <forEach counterName="numberOfOrganizers"
                 parallel="yes">
      Each organizer proposes and collects music
    </forEach>

    <invoke partnerlink="LocalLibrary"
            operation="borrowOrder" ... />
      <compensationHandler>
            <invoke partnerlink="LocalLibrary"
                    operation="borrowCancel" ... />
      </compensationHandler>
    </invoke>

    <reply operation="collectMusic" ... />
  </sequence>
</process>
```

Fig. 2. Example process: collecting party music

In the message, the control activity is the entire process; the success and failure continuations are empty. This message will be dequeued and be put in the pending message pool, waiting for an invocation on operation collectMusic.

Upon dequeuing the corresponding invoke message, the process container at AclicePC instantiates the process by inserting an entry in the scope registry and creating two new messages, one for the event handler and one for the primary activity.

```
<a:A><onEvent> <throw ... /> </onEvent></a:A>
<a:Ks></a:Ks> <a:Kf></a:Kf>

<a:A><sequence> ... </sequence></a:A>
<a:Ks><a:A><eos> ... </eos></a:A></a:Ks>
<a:Kf><a:A><eosf> ... </eosf></a:A></a:Kf>
```

The message for the event handler is put into the pending message pool. Later, when a matching stop event (i.e., an invoke message on operation stop) arrives, the throw activity will be executed.

The message for the primary activity now has two auxiliary activities eos and eosf (end of scope) pushed to the success and failure continuations. They encapsulate sufficient information for proper normal or abnormal scope termination. The new control activity is a sequence activity (without the instantiating receive activity). It is processed by setting the first activity of the sequence as the new control activity and pushing the rest of the activities into the success continuation. That is, they are to be executed after the successful execution of the first one.

```
<a:A><forEach ... parallel="yes"> ... </forEach></a:A>
<a:Ks>
   <a:A><invoke operation="borrowOrder"> ... </invoke></a:A>
   <a:A><reply operation="collectMusic" /></a:A>
   <a:A><eos> ... </eos></a:A>
</a:Ks>
<a:Kf><a:A><eosf> ... </eosf></a:A></a:Kf>
```

When multiple parallel branches are created using parallel forEach, multiple messages are created, each representing a branch, as below.

```
<a:A> propose and collect music </a:A>
<a:Ks>
   <a:A><join agent="AlicePC">
           <joincondition> ... </joincondition></join></a:A>
   <a:A><invoke operation="borrowOrder"> ... </invoke></a:A>
   <a:A><reply operation="collectMusic" /></a:A>
   <a:A><eos> ... </eos></a:A>
</a:Ks>
<a:Kf>
   <a:A><join agent="AlicePC">
           <joincondition> ... </joincondition></join></a:A>
   <a:A><eosf> ... </eosf></a:A>
</a:Kf>
```

Here two auxiliary join activities are pushed to both the success and failure continuations. The join condition in the success continuation states that all branches are successfully done. The join condition in the failure continuation states that the branches either have not succeeded or their committed effects are successfully compensated for.

The messages of the branches are then sent to the agents of the organizers. These agents are registered at the scope registry at AlicePC. When a branch terminates, the join activity in the success continuation will become the control activity in a new SOAP message. This message will be sent to AlicePC. The join is successful when the join messages of all branches are delivered to the join agent.

After a successful join, the new control activity becomes an invocation of the borrowOrder operation. This message, when delivered to LocalLibrary and matches a receive message in its pending message pool, will activate the service provided by LocalLibrary. So the PartyMusic process (service) now is composed of a sub-process (service) at LocalLibrary, which is now registered at the scope registry of `AlicePC`. When the `borrowOrder` operation finishes, two messages are generated: a `receive` message for the installed compensation operation at `LocalLibrary`'s pending message pool, and a return message in which the compensation operation `borrowCancel` is pushed into the failure continuation.

```
<a:A><receive operation="borrowCancel" /></a:A>
<a:Ks><a:A><reply operation="borrowCancel" /></a:A></a:Ks>
<a:Kf></a:Kf>
```

```
<a:A><reply operation="collectMusic" /></a:A>
<a:Ks><a:A><eos> ... </eos></a:A></a:Ks>
<a:Kf>
   <a:A><invoke operation="borrowCancel" /></a:A>
   <a:A><eosf> ... </eosf></a:A>
</a:Kf>
```

If a fault is thrown, for example, by the event handler of the process, all running activities are stopped and the corresponding fault handler is executed (here a default fault handler runs `compensate` and then `rethrow`). The `compensate` activity will cause the failure continuation to be applied, i.e., the installed `borrowCancel` operation will be invoked.

A process or scope is terminated upon `eos` or `eosf`. This will stop all current active activities such as the event handlers. The installed compensation handlers are only stopped if the current scope is a top-level process.

4 Conclusion

Our contribution is a lightweight process-aware component of general purpose middleware to support peer-to-peer execution of processes (or composite services). Without the reliance on a costly central engine, small businesses and end-users can now dynamically compose available services as processes. It does not unnecessarily pre-allocate resources or involve extensive housekeeping of runtime states. The approach is of continuation-passing style. This makes the conduction of process control flow as local operations rather than global coordination. Furthermore, our approach supports automatic process recovery by automatically generating recovery plans into failure continuations. The approach is verified by our current working prototype. Future work includes security and performance study.

References

[1] Barbara, D., Mehrotra, S., Rusinkiewicz, M.: INCAs: Managing Dynamic Workflows in Distributed Environments, Journal of Database Management, Special Issues on Multidadatabases, 7(1) (1996)

[2] Chafle, G., Chandra, S., Mann, V.: Decentralized Orchestration of Composite Web Services. In: 13th international World Wide Web conference (Alternate track papers and posters), pp. 134–143 (May 2004)

[3] Gokkoca, E., Altinel, M., Cingil, I., Tatbul, N., Koksal, P., Dogac, A.: Design and Implementation of a Distributed Workflow Enactment Service. In: 2nd IFCIS International Conference on Cooperative Information Systems (CoopIS 97), pp. 89–98 (June 1997)

[4] Muth, P., Wodtke, D., Weißenfels, J., Dittrich, A.K., Weikum, G.: From Centralized Workflow Specification to Distributed Workflow Execution. Journal of Intelligent Information Systems 10(2), 159–184 (1998)

[5] WS-BPEL, Web Services Business Process Execution Language Version 2.0, public review draft, QISIS Open http://docs.oasis-open.org/wsbpel/2.0/wsbpel-specification-draft.pdf (August 2006)

[6] Yan, J., Yang, Y., Raikundalia, G.: Enacting Business processes in a Decentralised Environment with p2p-Based Workflow Support. In: 4th International Conference on Web-Age Information Management (WAIM 03). LNCS, vol. 2762, pp. 290–297. Springer, Heidelberg (2003)

[7] Yu, W., Yang, J.: Continuation-passing Enactment of Distributed Recoverable Workflows, the 2007 ACM SIGAPP Symposium on Applied Computing (SAC 2007), pp. 475–481 (March 2007)

Automatic Generation of Web Service Workflow Using a Probability Based Process-Semantic Repository

Dong Yuan[1], Miao Du[2], Haiyang Wang[3], and Lizhen Cui[4]

School of Computer Science and Technology, Shandong University,
250061 Jinan, China
[1] yd0116@gmail.com, [2] dm0119@163.com, [3] why@sdu.edu.cn,
[4] clz@dareway.com.cn

Abstract. Workflow management system has been utilized in the web service environment to make mass of different web services work cooperatively. In this paper, the authors propose a novel automatic generation method of web service workflow and build an e-travel system, which could generate travel process automatically according to passenger's requirements. The e-travel system is a dynamic and self-adaptive virtual travel agency, which uses a probability based process-semantic repository as the foundation. This paper will carefully discuss the process-semantic repository and the automatic generation algorithm of web service workflow based on it. At last, some experiments are presented to evaluate the efficiency of the e-travel system with the process-semantic repository.

1 Introduction

Nowadays there are tremendous web services added on the internet everyday, this is rapidly changing the way of internet using. Many research projects about the workflow model in the web service environment were appeared, such as METEOR-S [1]. But in the service oriented Internet environments, the traditional workflow model is facing new challenges:

The users of workflow model, in the internet era, are the common internet users from various social circles. They are in very large number and usually have individual requirements. Both defining a universal process for all users and designing specific processes for every user are impossible and unrealistic. So we need a method that could automatically generate business processes to satisfy the users. Some early researches use AI planning algorithm to achieve the automatic web service composition [3]. Future more, some researches, using semantic technology [4] and ontology to define workflow, proposed a workflow ontology language [5] OWL-WS based on OWL-S [2]. And It is also reported [6] the using of the agent technology to reasoning. All the methods above are based on the pre-designed rules. But the reasonable and specific rules are always hard to design, such as in tourism industry that are often related to many domains.

Considering the limitations discussed above, the authors decide to use the uncertainty theory to achieve automatic generation of web service workflow. Some researches already accrued. Doshi et al [7] introduced a web service composition

K.C. Chang et al. (Eds.): APWeb/WAIM 2007 Ws, LNCS 4537, pp. 504–509, 2007.

method using MDP (Markov decision process). Canfora et al [8] used genetic algorithm to solve this problem. But these algorithms have a high time complexity, as the number of services is huge.

Here we propose an automatic generation method of web service workflow based process-semantic repository and build an e-travel system, which has the following novel features:

- In the e-travel system, the business process is automatically generated instead of being pre-defined by the designer or users.
- We propose a probability based process-semantic repository, which is the foundation of the generation method. The transfer probability dynamically changed during the run-time. This mechanism takes full advantage of experience sharing among users.
- Our e-travel system uses the meta-service ontology to classify the numerous web services on the Internet. It is greatly simplify the complexity of the process generation and service discovery.

2 E-Travel System with the Process-Semantic Repository

2.1 Meta-service and Meta-flow

In practical, travel services can be classified into categories according to their functions. Services in a same category have so similar functions that it is very sound to be seen as a virtual service.

Definition 1: Meta-service
A meta-service is a set of services with similar operating properties. It can be seen as a delegate of some similar services.

We use ontology to organize the meta-services in different concept levels. Using this way we could clearly represent the relationship between meta-services and could easily relate meta-service to the service ontology (OWL-S).

Definition 2: Meta-flow
A meta-flow is an abstract process that composed by states, where a state is composed by one or more meta-services in the meta-service ontology except two special states: *Start* and *End*.

A meta-flow describes which kind of web service the process needs and what kind of control flow the process is. After services matching, a meta-flow becomes a travel process that can be executed in the workflow engine.

2.2 Probability Based Process-Semantic Repository

We propose a probability based process-semantic repository, in order to achieve automatic generation of travel process in the e-travel system. In the repository we use a preference n-tuple and transfer probability to represent the process semantic.

The preference n-tuple, also can be named preference dimensions, is some predefined dimensions by the expert that represent the process semantic. As in e-travel system, the n-tuple is *cost, time, travel, fallow, excite* and so on.

The process-semantic repository also stores the transfer probability. The transfer probability is changing dynamically and automatically during the run-time based on the adjusting methods of the process-semantic repository. The higher probability denotes that this transfer is always selected in the meta-flow and the meta-service is always satisfied by the user.

The process-semantic repository is the foundation of the meta-flow generate module in e-travel system. As a dynamic self-adaptive system, the process-semantic repository must have some functions for modification and adjustment.

addTransfer(s, s'), is the function that could add a new transfer of state s to s' in the process-semantic repository.

addState(s), is the function that could add the new state s to the process-semantic repository.

delTransfer(s, s'), is the function that could delete the transfer state s to s' from the process-semantic repository.

delState(s), is the function that could delete the state s from the repository.

posAdjust(s, s') is to positively adjust the transfer probability of state s to s'.

negAdjust(s, s') is the negative adjust function.

The functions listed above will be used in the generation method of travel process, and the process-semantic repository is the foundation for automatic generation of meta-flow in e-travel system.

2.3 Structure of the E-Travel System

The e-travel system is a virtual travel agency in the web service environment. Web service providers register their services to our system, and the system could pick up the right services to generate a workflow for the tourists based on their requirements.

Fig. 1. shows the overall structure of the e-travel system. It uses three main steps to generate a travel process for the tourist, namely User interactive module, Meta-flow generate module, and Service discover module.

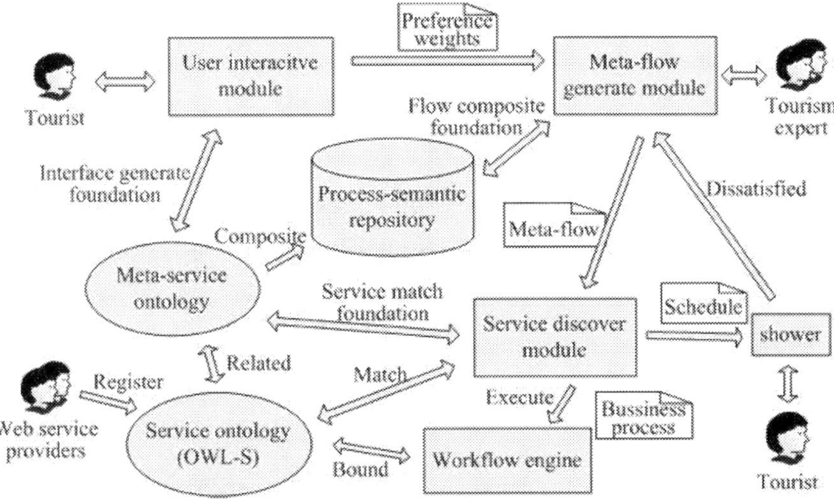

Fig. 1. Structure of the E-travel System

3 Automatic Generation Method

In this section, we propose an automatic generation method for travel process, based on the process-semantic repository.

First, we design the algorithm to pick up the most satisfied state of the meta-flow. It calculates the best state for the current state to transfer to of one step.

```
[Algorithm1] nextState(ω,s)
[Input]      ω                 /* the preference weight */
             s                 /* the current state */
[Output]     s'        /* the state transfer to */
Begin:       ϖ = Σ_{i=1}^{n} ω_i
      Loop (for every s_i in S):  /* S is the state set of s */
             If ( Σ_{i=1}^{n} f_i(s,s_i) * ω_i > Σ_{i=1}^{n} f_i(s,s_i) * ϖ ) then do
                        Add s_i to S'        /* S' is the satisfied states set */
      End Loop
      If (S'≠Φ) then do          s'={s_j | max p(s,s_j) ∩ s_j ∈ S'}
             return s'
      Else   return Φ
End
```

The main generation method in the meta-flow generation module is as follow.

```
[Algorithm2] generateMF(ω, MS)
[Input]      ω                 /* the preference weight */
             MS                /* the meta-service set of user */
[Output]     mf=<T, S>         /* a meta-flow */
Begin:
     Set:     mf=Φ             /* the meta-flow */;
             cur_s=start       /* the current state of mf */
             t_ms=Φ            /* the current meta-service set of mf */
     Loop (t_ms ⊂ MS and mf < MAXLength):
             s=nextState(ω,cur_s);
             If (s≠Φ) then do
                        add s to mf        /* as next state */
                        cur_s=s
                        update t_ms        /* with meta-service in s */
             Else do
                        mf=MF  /* MF is the meta-flow defined by expert */
                        modifyRepository(mf)
                        break loop
     End Loop
     If (mf≧MAXLength) then do
             mf=MF              /* MF is the meta-flow defined by expert */
             modifyRepository(mf)
     Return mf
End
```

With the meta-flow, the last step to generate a travel process is service matching in the service discovery module. If the service matching succeeds, the schedule of the travel will be showed to the tourist. We will evaluate the performance of the algorithm through some experiment in the next section.

4 Experiment

In order to evaluate our algorithm, we do some experiments on our e-travel system. The PC configuration is P4 2.8G CPU, 512M memory and Windows 2003 Server.

We use 20 groups of predefined tourists' preference weights, and the random preference vectors in the process-semantic repository to test the performance of the e-travel system with the automatic generation method. The result is listed in Table 1.

Table 1. Scuccess rate of the meta-flow generation

State number	Transfer number	Scuceess generation	Scuceess rate
6	20	4	20%
10	50	9	45%
20	150	15	75%
50	600	18	90%
100	2000	19	95%

In the table, as the state number in the process-semantic pository increased, the success rate of meta-flow generation increased accordingly. The result obviously shows that the e-travel system is a dynamic and self-adptive system with the automatic generation method.

In order to evaluate the efficiency of the e-travel system, we compare our generation method using process-semantic repository with the MDP method [7]. We still use the 20 groups of predefined tourists' preference weights as the input data, and the average time cost of generation is shown in Fig. 2.

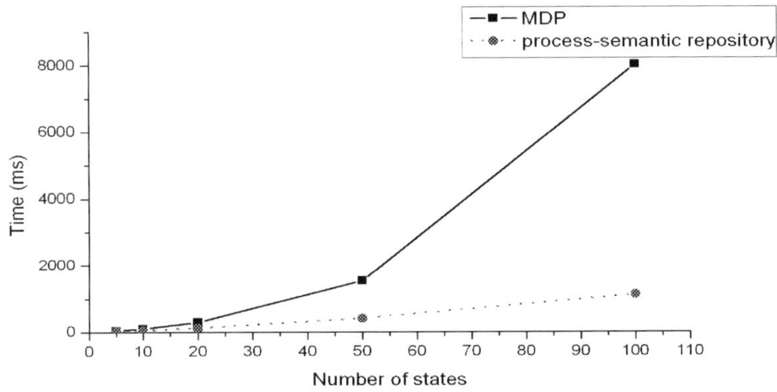

Fig. 2. Time cost of meta-flow generation

Fig. 2 shows that in the MDP generation, the time cost is strongly influenced by the number of states. Our generation method with the process-semantic repository has a linear relationship with the state number, so we could have more states in the repository and more meta-services in the meta-service ontology.

5 Conclusion and Future Work

This paper proposes an automatic generation method of web service workflow using a process-semantic repository. Based on this method we build an e-travel system that could generate travel process for the tourists.

In the future work, some veracious validation algorithm will be proposed to evaluate the satisfaction of the tourist about the travel process.

References

[1] http://lsdis.cs.uga.edu/projects/meteor-s

[2] http://www.daml.org/services/owl-s

[3] Medjahed, B., Bouguettaya, A., Elmagarmid, A.K.: Composing web services on the semantic web. The VLDB Journal, vol. 12(4) (November 2003)

[4] Yao, Z., Liu, S., Pang, S., Zheng, Z.: A Workflow System Based on Ontology. In: Proceedings of the 10th International Conference on Computer Supported Cooperative Work in Design 1-4244-0165 (2006)

[5] Beco, S., Cantalupo, B., Giammarino, L.: OWL-WS: A Workflow Ontology for Dynamic Grid Service Composition. In: Proceedings of the First International Conference on e-Science and Grid Computing (e-Science'05) (2005)

[6] Korhonen, J., Pajunen, L., Puustjarvi, J.: Automatic Composition of Web Service Workflows Using a Semantic Agent. In: Proceedings of the IEEE/WIC International Conference on Web Intelligence 0-7695-1932-6/03 (2003)

[7] Doshi, P., Goodwin, R., Akkiraju, R., Verma, K.: Dynamic Workflow Composition using Markov Decision Processeses. In: proceedings of the IEEE International Conference on Web Services (ICWS2004) (2004)

[8] Canfora, G., Di Penta, M., Esposito, R., Villani, M.L.: An approach for QoS-aware service composition based on genetic algorithms. GECCO, pp. 1069–1075 (2005)

A Three-Dimensional Customer Classification Model Based on Knowledge Discovery and Empirical Study

Guoling Lao and Zhaohui Zhang

School of Information Management and Engineering, Shanghai University of Finance and Economics, Shanghai, 200433, P.R. China
gllao@shufe.edu.cn

Abstract. It is important for stockjobbers to carry out customer segmentation and find out the high-valued customers. This article focuses on the main factors that act on customer lifecycle value (CLV) and customer potential contribution value (CPV). At the basis of analyzing some key factors which the stockjobbers largely depend on from the classical CLV model, a three-dimensional customer classification and CPV estimate model is put forward. This model is convinced of feasible and reasonable by an empirical study with factual data from one stockjobber. It solves out the problem of looking for quantitative approach to estimating customer's level of CPV.

1 Introduction

Finding out the right customers, knowing their needs and offering right services at the right time are the main goal for stockjobbers, and these are also the key to success for them. It is needed for them to predict, find out and track the customers who account for most of their future profits. Stockjobbers have accumulated large amounts of securities exchange data and customer information during the past years. The problem is how they find out valuable information from these data.

KDD (Knowledge Discovery based on Database) is the nontrivial process of identifying valid, novel, potentially useful, and ultimately understandable patterns in data. Data mining is a step in the KDD process and refers to algorithms that are applied to extract patterns from the data. The extracted information can then be used to form a prediction or classification model, identify trends and associations, refine an existing model, or provide a summary of the database being mined [1] [2]. Thomas H. Davenport gave the definition of knowledge: knowledge is a combination of structured experiences, value viewpoint, relevant information and expert opinion and so on [3]. Through the definition of knowledge, we can find that knowledge always exists as follows: opinions, Rules, Regularities, Patterns, Constraints, Visualization. The knowledge can help people make decisions [4]. It is obvious that knowledge discovery is not only KDD or data mining, the process of acquiring concepts, rules, and models, is also knowledge discovery. This article sheds lights on the process of discovering the knowledge like customer segmentation and customer potential value estimate through data mining technology.

K.C. Chang et al. (Eds.): APWeb/WAIM 2007 Ws, LNCS 4537, pp. 510–515, 2007.

2 Customer Classification and Customer Potential Value Estimate Model for Stockjobbers

The most popular customer lifecycle value model which is recognized by many specialists and scholars is the net current value evaluating system put forward by Frederick F. R in 1996 [5]. Kotler developed the theory and thought that customer lifecycle value is the current value of all profits which customers contribute to companies during the whole lifetime [6]. At the basis of the five lifecycle phase model which put forward by Dwyer in 1987 [7], Mingliang Chen divides customers into four groups according to two dimension: customer potential value (CPV) and customer current value (CCV) [8] [9] which is also discussed by Jingruan Han [10]. These studies on the two-dimensional model of customer segmentation stop only at theoretical research for the difficult of customer potential value estimate. Finding out the reasonable and feasible quantitative approach to estimating customer potential value is vital for the two-dimensional model being put into practice.

Customer value in retail industry means that customers' purchase brings profits to the companies [11], while customer value in securities industries means comes from customers' commission and interest margin. If the CLV model[6][12][13] is used for stockjobbers, it illustrates as follows: d represents discount rate, $C_t \times x_t$ represents contribution of commission, and C_t represents trading volume, x_t represents commission ratio; $I_t \times r_t$ represents interest margin, I_t represents average deposit in the t year, r_t represents the gap of interest rate; K(t) represents the customer service cost in t year. Then the CLV model for stockjobbers is:

$$CLV = \sum_{t=t_0}^{T} [C_t \times x_t + I_t \times r_t - K(t)] \times (\frac{1}{1+d})^t \tag{1}$$

From the CLV model above, we can find out that CLV depends on C_t, x_t, I_t, r_t, K(t), d, t. Variable x_t, d and r_t can be regarded as fixed because of the government's limits. The customer service cost is the same for customers in the same group, and it is controlled by the stockjobber. Then the main factors which act on CLV is trading volume C_t and average deposit I_t. But I_t is in direct proportion to the customer's fund which he wants to invest, C_t has relationship with customer's activity degree, ability of gaining profits, risk preference and stock market's weatherglass. three main factors: customer's fund, customer's activity degree and ability of gaining profits which have direct relevant to CLV are focused on, as stock market's weatherglass is the same to every customer and risk preference act on CLV through customer's exchanging behavior which is represented by customer's activity degree and ability of gaining profits here. The value of the three indexes: customer's fund, customer's activity degree and ability of gaining profits, is easy to get from customer database and securities exchange database. High or low degree of the three indexes combination divides all customers into eight groups. The customer three-dimensional model is illustrated as the following chart:

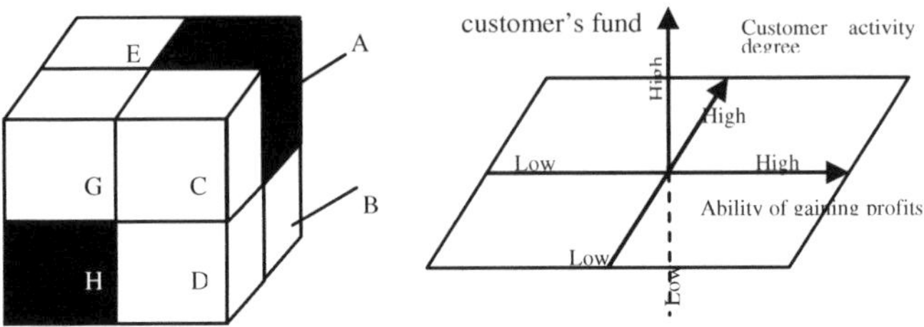

Fig. 1. Three-dimensional customer classification model

Customers in Group A (High, High, High) have high degree of customer activity, high degree of ability of gaining profits and high degree of customer's fund, they are the most valuable customers currently and in the future for the stockjobbers. No doubt they are the customers who account for most of the stockjobbers' profits. Customers in Group B (High, High, and Low) have less money investing into the stock market compared customers in Group A, but they exchange frequently, and they are high potential valued customers. Customers in Group C (High, Low, and High) also contribute large amounts of profits to the stockjobbers, but they lack of ability to earn money in stock markets. Customers in Group D (High, Low, and Low) exchange frequently, but the profits they contributed to stockjobbers are very limited. Customers in Group E (Low, High, High) are typical long line investor. Though they have large volume of capital for investing and earn good money from the stock market, their contribution to stockjobber is very little. Customers in Group F (Low, High, and Low) also account for very little of stockjobbers' profits. Customers in Group G (Low, Low, and High) have large amounts of money for investment, but they don't know where they should invest. If customization services are introduced, when they find the objective for investing, they will become a high-valued customer. Customers in Group H (Low, Low, and Low) bring very little profits to stockjobbers. Their value is low both in short term and in the long term.

At the basis of analysis above the three-dimensional customer classification model, it is reasonable that customers in Group A, B, C have high level of customer potential value in the future. Customers in Group F and H have low level of customer potential value. Customers in Group D, G and E probably have high level of customer potential value if right customization services offered to them.

3 Empirical Studies on Looking for High CPV Customers

As CLV in securities industries mostly depends on customer's fund, customer's activity degree and ability of gaining profits, we can find out the three indexes' value combination to represent features of customers who have the high level of customer contribution value to stockjobbers through their historical security exchange data.

3.1 Objective and Approach of Empirical Study

The purpose of this empirical study is to find out the relationship between profits customers contributed to stockjobbers and the value combination of the three indexes.

Then other customer's CPV can be estimated and predicted by their value combination of the three indexes.

The actual data used for this empirical study came from one of the famous stockjobbers in China. All customers, who opened an account in this stockjobber in the last month of 2000 and the account is still valid in October in 2006, is selected. We screen out customer's exchange data from January 1st, in 2001 to December 31st, in 2005. After data pretreatment, value of four indexes is obtained: the levels of customer's activity degree (Cust_Act), customer's fund (Cust_fund), customer's ability of gaining profits (Cust_Pro), and profits which customer contributed to stockjobber (Cust_Cv)during the five years. The three indexes and a target variable are all divided into two levels: high and low accord to the rule of 20% and 80%, or stockjobber and experts' critical value which actual be used.

3.2 Modeling

This empirical study sets up a decision tree model with the help of SAS, as SAS is one of the most famous and recognized data analysis tools. SAS offers a perfect and systemic data mining approach, which is known as SEMMA (Sample, Explore, Modify, Model, Assess). After the phase of data selecting, data pretreatment and preparing, import source data into SAS. Use data Partition tool random select 40% data as train set, 30% data as validation set, 30% data as test set. Set Cust_Cv as target variable, then SAS output results as follows (shown in fig. 2):

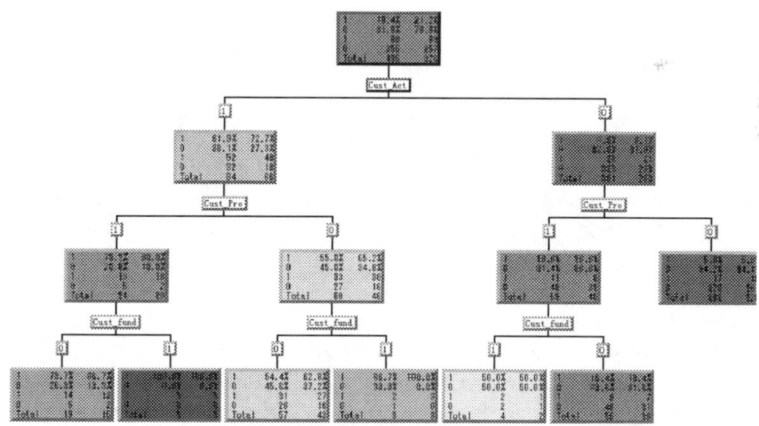

Fig. 2. Output result——Decision tree model

There are seven leaves in the decision tree; the rule of the decision tree is explained as follows from left to right:

(1) IF (Cust_Act=1, Cust_Pro=1, Cust_fund=0), THEN the probability of Cust_Cv=1 is 73.7%. It means that if one customer has high levels of customer's activity degree (Cust_Act) and customer's ability of gaining profits (Cust_Pro), but has low level of the amount of money used for investment (Cust_fund), then we estimate this customer's level of customer contribution value is high with the precision

73.7%. (2) IF (1, 1, 1), THEN the probability of Cust_Cv=1 is 100%. (3) IF (1, 0, 0), THEN the probability of Cust_Cv=1 is 54.4%. (4) IF (1, 0, 1), THEN the probability of Cust_Cv=1 is 66.7%. (5) IF (0, 1, 1), THEN the probability of Cust_Cv=1 is 50%. (6) IF (0, 1, 0), THEN the probability of Cust_Cv=1 is 16.4%, and the probability of Cust_Cv=0 is 83.6%. (7) IF (0, 0,1) and (0, 0, 0), THEN the probability of Cust_Cv=1 is 5.8%,the probability of Cust_Cv=0 is 94.2%.

3.3 Assessment and Modeling Results Analyzing

In SAS model, Validation data set is selected for assessment. From the assessment report of SAS, it is convinced that 10%of the customers has a high level of customer contribution value with the precision more than 80%, and 20%of the customers has a high level of customer contribution value with the precision more than 60%, as the proportion of high level customers goes up, the precision goes down.

It is concluded that if one customer belongs to Group A, stockjobber can predict that the level of his customer contribution value is high with the precision 100%, from the results of the decision tree. If one customer belongs to Group B and C, stockjobber can predict his level of customer contribution value is high with the precision 73.7% and 66.7% respectively. So when stockjobber estimate customer's CPV, if the customer belongs to Group A, B and C, stockjobber have enough precision to predict their level of CPV is high. It is also concluded that not all big customer (like customers in Group E, G) have high level of customer contribution value, while some ordinary customer such as customers in Group B have high level of customer contribution value.

3.4 Modeling Results Application

Most stockjobbers in China classify their customers into three main groups: big customer, secondary or ordinary customer, according to the customers' possessions they declared. But this classification approach is not reasonable and effective enough[14], as not all wealthy customers bring high contribution value to the stockjobber while some customers who do not have a high level of investment capital account for large amounts of stockjobber's profits. It is obvious that the decision tree model from above can solve out the problem of looking for quantitative approach to estimating customer's level of CPV in Mingliang C's two-dimensional customer segmentation model [8]. The result of our empirical study can also put the two-dimensional model of customer segmentation into practice. Such as, according to customer's activity degree, customer's fund, customer's ability of gaining profits, and profits which customer contributed to stockjobber which come from customers' information and their historical security trading data, stockjobber can segment their customer into four groups (High CCV and High CPV, Low CCV and High CPV, High CCV and Low CPV, Low CCV and Low CPV) for customization services strategy.

4 Conclusions

The approach of customer segmentation according to the customers' fund for investing they declared which is used by most stockjobbers in China is not reasonable and effectual. It is very important to find out the customers who account for most of stockjobber's profits and the customers who will bring high level of potential contribution value. This is feasible and useful if customers' information and their historical security trading data are best used.

References

1. Fayyad, U.M., Piatetsky- Shapiro, G., Smyth, P., Uthurusamy, R.: Knowledge Discovery and Data Mining: Towards a Unifying Frame-work. In: Proc. of KDD, Menlo park, CA, pp. 82–88. AAAI Press, California (1996)
2. Kantardzic, M.: Data Mining Concepts, Models, Methods and Algorithms[M] (Copyright@). IEEE Press, New York (2002)
3. Thomas, H.: Davenport haurence Prusak.: Working Knowledge. Harvard Business School Press, Boston (1998)
4. Weijin, L., Fansheng, K.: Data Warehouse and Knowledge Discovery. Computer engineering and applications, vol. 10 (2000)
5. Frederick, F.R.: The loyalty effect: the hidden force behind growth, profits, and lasting value [M]. Harvard Business School Press, Boston, Massachusetts (1996)
6. Kotler, P., Armstrong, G.: Principles of marketing [M], 7th edn. Prentice-Hill, Englewood Cliffs (1996)
7. Dwyer, F.R., Schurr Paul, H., Sejo, O.: Developing Buyer - Seller Relations [J]. Journal of Marketing 51, 11–28 (1987)
8. Mingliang, C., Huaizu, L.: Study on Value Segmentation and Retention Strategies of Customer. Group Technology and Production Modernization, vol. 4 (2001)
9. Mingliang, C.: A Study of the Customer Life Cycle Model. Journal of Zhejiang University (Humanities and Social Sciences), vol. 6 (2002)
10. Jingyuan, H., Yanli, L.: An Approach of Customer classification based on customer value. Economics forum, vol. 5 (2005)
11. Jiaying, Q., Shu, H.: Assessing, modeling and Decision making of Customer value. Beijing University of Posts and Telecommunications Press (2005)
12. Jingtao, W.: Segmentation of Customer life-time value and strategy of customer relationship. Journal of Xi'an finance and economics college, vol. 2 (2005)
13. Paul, D.: Customer lifetime value: Marketing models and applications[J]. Journal of Interactive Marketing 12(1), 17–30 (1998)
14. Xiaodong, J.: Index system based on Securities Companies' CRM. Statistics and Decision, vol. 9 (2003)

QoS-Driven Global Optimization of Services Selection Supporting Services Flow Re-planning

Chunming Gao[1,2], Meiling Cai[1], and Huowang Chen[2]

[1] College of Mathematics and Computer Science, Hunan Normal University
410081 Changsha, China
caimeiling418@sina.com
[2] School of Computer Science, National University of Defense Technology
410073 Changsha, China
gcm_powerman@163.com

Abstract. In most cases, QoS-driven global optimization is a non-linear 0-1 programming. Genetic Algorithms are superior in non-linear 0-1 programming and multi-objective optimization. However, encoding methods that many Genetic Algorithms (GAs) adopts are too complex or simple to apply to services selection. A novel Tree-coding Gas (TGA) is presented for QoS-driven service selection in services composition. Tree-coding schema carries the messages of static model of service workflow, which qualifies TGA to encode and decode chromosomes automatically, and keeps the medial results for fitness computing. Tree-coding can also support the services composition flow re-planning at runtime effectively. The experiment results show that TGA run faster than one-dimensional Genetic Algorithms when the optimal result is the same, furthermore the algorithm with Tree-coding is effective for re-planning.

Keywords: Service Composition; Quality of Service; Genetic Algorithms; Tree-coding; re-planning.

1 Introduction

At the stage of service composition, because of increasing homological and similar services in functionality, it is very important for us to select services based on QoS (Quality of Service) attached to every service and this problem is NP hard[5].

The GAs is a novel Web Services selection method based on QoS attributes[2,3,4]. Zhang et al.[2] and Canfora et al.[3] use one-dimensional chromosome-encoded method. Each gene of chromosome in [2] represents candidate service. Since each task can only select a service from its numerous candidate services, the readability of the chromosomes is fairly weak. Different from Zhang in [3], each gene represents the task of the services composition, and then the length of the chromosome is evidently shorter than that in [2]. Because the methods employed in [2,3] adopt one-dimensional chromosome-encoded, the chromosome can't take semantics of constructed logic of service composition, and encoding method, crossover and mutation operation also can't represent the composition relationships and so on. A

K.C. Chang et al. (Eds.): APWeb/WAIM 2007 Ws, LNCS 4537, pp. 516–521, 2007.

GAs based on relational matrix encoding method is designed in [4]. But this encoding method is too complex for operations, the crossover or mutation of which may generate illegal individuals frequently, thus it needs checking the validity of the crossover and mutation operations frequently, which will lower the efficiency. Since the GAs above does not take account of fault-tolerance of service composition at runtime, the optimization approaches with encoding above can't support re-plan of service composition.

This paper proposes a Tree-coding GAs (TGA) to solve global optimization of service composition. The tree-encoding mode carries static-model structure messages of the process, allowing automatic encoding and decoding of the chromosomes and supporting runtime re-planning of service composition. TGA run faster than the one-dimension GAs due to tree-coding storing the medial results for fitness calculating.

2 Services Selection Based on Tree-Coding Genetic Algorithm

(1) QoS Aggregation

When computing the QoS of services composition in service selection, we consider price, execution time, successful execution rate and availability as the criteria[1]. The aggregate QoS of Web Service Composition depends on the QoS of component services. Table 1 provides aggregation functions. Note that a loop structure with k iterations of task t is equivalent to a sequence structure of k copies of t.

Table 1. The model for computing the QoS of Services Composition.

QoS Attr.	Price	Time	Successful Rate	Availability
Sequence	$\sum_{i=1}^{n} q_{price}(S_i)$	$\sum_{i=1}^{n} q_{resp}(S_i)$	$\prod_{i=1}^{n} q_{rel}(S_i)$	$\prod_{i=1}^{n} q_{av}(S_i)$
Concurrence	$\sum_{i=1}^{p} q_{price}(S_i)$	$MAX_{I\in\{1..p\}}\{q_{resp}(S_i)\}$	$\prod_{i=1}^{p} q_{rel}(S_i)$	$\prod_{i=1}^{p} q_{av}(S_i)$
Choice	$\sum_{i=1}^{m} p_{ai} * q_{price}(S_i)$	$\sum_{i=1}^{m} p_{ai} * q_{resp}(S_i)$	$\sum_{i=1}^{m} p_{ai} * q_{rel}(S_i)$	$\sum_{i=1}^{m} p_{ai} * q_{av}(S_i)$
Loop	$k * q_{price}(inner)$	$k * q_{resp}(inner)$	$(q_{rel}(inner))^k$	$(q_{av}(inner))^k$

(2) Tree-Coding for Chromosome

Web Services Composition represented by the structural process can be decomposed to substructure. The substructures can combine together following four kinds of relationships that are sequence, choice, concurrence and loop. Recursively, the decomposition can act on the substructures until all substructures are decomposed to atomic activities. Therefore, the structural process model can be equivalently represented as a tree, where the leaf nodes and inner nodes represent atomic activities and combination relationships of activities respectively. As shown in Fig.1, "|", "$\cap$" and "$\cup$" separately expresses sequence, concurrence and choice structural activity. N1~N6 are inner nodes of the tree, indicating the combination relationships; R1~R9 are leaf nodes, indicating the task nodes. We called the tree representing constructed logics of Web Service Composition as Process Tree(PTree). According to the tree's recursive definition, the PTree is a recursive structure starting from root node defined as TNode, which satisfies the definitions below:

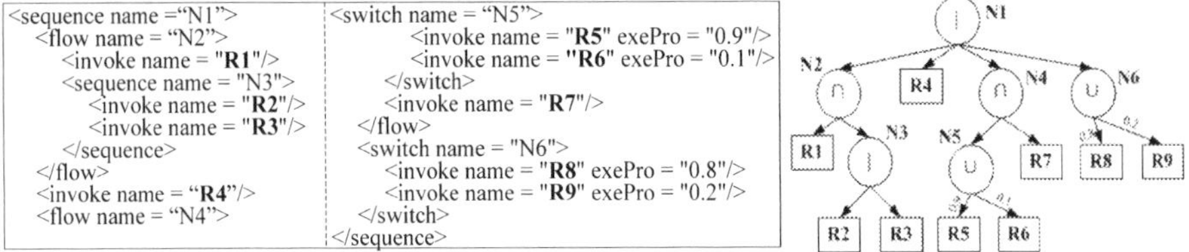

Fig. 1. Tree expression of process

(1) *type* ∈ {*seq,cos,cho,loop,inv*},types of TNode, including sequence, concurrence, choice,loop and invoke(atomic activity);
(2) *parent*:TNode,the parent of TNode;
(3) *childrenList*: {child_1, child_2,..., child_n:TNode }, the children list of TNode;
(4) *task*=t_k, TNode is responsible for task t_k,the structural node has no task to do;
(5) *service* ∈ $S_k=\{S_k^1,S_k^2,\cdots,S_k^m\}$,TNode will execute *service* to complet task t_k;
(6) *exePro* ∈ R^+,the weight of node,represents TNode's execution probability relative to its parent. p_i is the i_{th} child's execution probability of choice node , and $\sum_{i=1}^{n} p_i = 1$; execution probability of child in loop is *itNum*, which is iterations of loop body; and the probability is 1 in sequence and concurrence;
(7) *QoSVector*= $(q^{(p)}(p),q^t(p),q^s(p),q^a(p))$ is a QoS vector, each member of which separately expresses the price, time, success rate and availability. The task node's QoS vector is *service* QoS vector, while the inner node's QoS vector is gotten through calculating its child nodes' QoS vector according to Table 1.

The PTree containing static constructed model messages and dynamic statistical messages provides a good data structure for QoS computation of specific Composition Plan. Since each node in PTree is attached by a QoS vector representing sub-structures' aggregation QoS, PTree is traversed in post order from the leaf nodes and at the same time QoS of non-leaf nodes is calculated according to Table 1. Thus the QoS of root node is the QoS of service composition.

The chromosome is represented as a serial of nodes gotten through traversing the PTree in post order. For example, the PTree in fig1 can be represented as $R_1R_2R_3N_3N_2R_4R_5R_6N_5R_7N_4R_8R_9N_6N_1$. Since each node stores messages of its children and father, the serial nodes uniquely correspond to a PTree. Each gene of the serial nodes, namely a subtree, corresponds to a node of PTree, and the last gene denotes the root and the leaf nodes represent tasks. The chromosome is concrete representation of the PTree's leaf nodes. Therefore, Tree-coding scheme possesses dynamic properties (i.e. evolving of concrete Web Service and QoS messages of nodes) and static properties (i.e. process model structure).

(3) Fitness and the Strategies of Selction, Crossover and Mutation

Fitness: the overall quality score for Process Plan p $F(p)$ is designed as:
$F(p)=\sum w(x)*V^{(x)}(p)\ (x=p,t,s,a;0\le w(x)\le 1;\sum w(x)=1)$, where $w(x)$ represents the weight

on QoS criterion, $V^{(x)}(p)$ is the standardized QoS of p as expatiated in[1]. And we introduce the penalty model in [3].

Select operator: Using the roulette wheel selection with the optimal approach.

Crossover operator: First of all, two parent individuals are selected and later the crossover point is selected randomly, and then the parents exchange their subtree, whose root lies at the crossover point of intersection, to reproduce two descendants.

Mutation operator: adopting random disturbance, selecting another candidate service in candidate services set as *service* for every task node at a certain probability.

After successfully crossover and mutation, update QoS of nodes in the path from the crossover or mutation point's parent node to the root node.

(4) The Ability of Tree-Coding

The abilities of tree-coding are listed below:

1) Simultaneously expressing multiple kinds of composite relationships and supporting automatically encoding and decoding of chromosomes.
2) The ability to re-plan of process flow at runtime

```
CAlFN(ROOT, LIST, FN)
// ROOT the root of tree LIST the list of stop nodes, FN the current failure node
   if FN is ROOT, return FN;
   LIST.remove(FN);
   switch FNP'type
     case CONCURRENCE
        FN=New CONCURRENCE;
       for each child of FNP
         clist = sublist(child, LIST);
         FN.add(CALFN(child, clist, first node of clist));
     case SEQUENCE:
        FN = new SEQUENCE;
        for each child of FNP from FN
      FN.add(child);
     case CHOICE:
        FN = Root.replace(FNP, FN);
     case LOOP:
       newFN = new SEQUENCE;newFN.add(FN);newLoop = new LOOP,
       newLoop.setItNum(estNum- completedNum), newLoop.setChild(FNP.child),
       newFN .add(newLoop);FN = newFN;
    CALFN(ROOT, LIST, FN);
```

Fig. 2. Re-planning slice computing

Web Services run autonomously within a highly changeable Internet environment. As a result, during execution of composite service, the component services become unavailable or QoS of the component services changes significantly. Consequently, a re-planning procedure is triggered in order to ensure that the QoS of the composite service execution remains optimal. Re-planning includes two phases. The first step obtains a sub process containing all the nodes that haven't yet executed. The current failure node (FN) determines a tree whose root is FN. The algorithm in Fig.2 calculates a new FN from FN's parent node (FNP), and takes it as the current failure node in next recursion. The algorithm ends until the FN is the root of the PTree. Then FN is the slice needed to be re-planned. The second phase finds a sub optimization on the sub PTree gotten just now. Due to the ability of automatic encoding and decoding, TGA is quite appropriate for the re-planning at runtime.

3) TGA Running Faster than Traditional GAs.

In one-dimensional GAs, all the medial results used to get fitness have to be calculated repeatedly every generation. Due to the Tree-coding carrying the control structure messages of workflow, the QoS vector is exchanged when crossover, thus when it goes to compute the aggregate QoS, it only needs to update the QoS vector of nodes in path from crossover point to the root. Mutation is aimed at the task nodes, and once successfully mutated, it only needs to update the QoS vector of nodes in path from the leaf point to the root. As it shows below, TGA shows good performance with much less overhead compared with one-dimensional algorithm.

3 Simulation Analysis

The simulation analysis for coding scheme and re-planning experiments are implemented. All experiments were conducted on a test-bed of Windows PCs with Pentium IV 2.2 processor and 512M RAM. The population size of Genetic system is 50, evolution maximum is 1000, crossover probability is 0.7, and mutation probability is 0.01. Service composition is constrained on execution price and execution time.

(Experiment 1) Coding Scheme Comparison
The experiment is based on the randomly generated Web Services Composition, i.e., constructing topology of PTree and execution probability of nodes at random with certain size of tasks. The task size varies from 9 to 27 with 3 as step length.

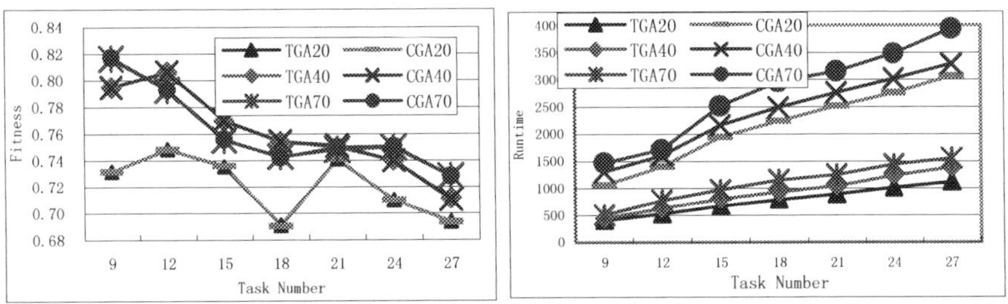

Fig. 3. Average fitness and runtime of CGA and TGA in different Web Service Compositions

We compare the average fitness and run time of TGA with one-dimensional coding algorithms [3](named CGA). For each test case, we run 100 times for both algorithms and compute the average results. Figure 3(a) plots the average optimal fitness of CGA and TGA where candidate service size is 20, 40 and 70. In all the test cases, the optimal fitness of TGA and CGA are almost the same as it shows. This proves that Tree-coding is effective for services composition optimization. Figure 3(b) plots the average run time of CGA and TGA where candidate service size is 20, 40 and 70. As it shows, TGA run faster than CGA about 40%. This is because TGA stores the medial results for fitness computation, thus it saves much time.

(Experiment 2) Re-planning at Runtime
The second experiment takes the process in fig1 as example. The size of candidate service is 70. Row 2 in Table 2 is initial optimum solution by TGA. We investigate the effect of our re-planning approach. In the experiment, every candidate services

keep available at its availability. Then the 54th candidate service of Task 4(R_4) is not available. So when composition process runs to R_4(marked as time T_4), the re-planning is triggered. The first step of re-planning is to compute the sub process needed to be re-planned according to Algorithm in Fig.2, as illustrated in Row 4.

The second step: select available services for the sub process which is figured out just now. The sub process plan of newly selected services must satisfy the constraints (Row 4, 5) all the same. In the end the workflow completes, the price is 70.31, increasing 35.04% to initial optimum(52.07); execution time is 551.18, increasing 2.67% to initial optimum(536.87). However, the execution effect still satisfies the constraints(Row1). Compared with the global optimization at time T_4, the execution price decreases 3.33%, while the execution time is consistent (the difference owes to estimated value of choice activities). Result shows the re-planning's effectiveness.

Table 2. Re-planning of composition process at runtime

1	All tasks	R1	R2	R3	R4	R5	R6	R7	R8	R9	$cons^{(p)}$:482.72	$cons^{(t)}$: 683.86
2	Initial process	66	24	64	54	58	54	7	64	3	52.07	536.87
3	Executed partial flow	66	24	64							26.00	220.88
4	Tasks to re-plan				R4	R5	R6	R7	R8	R9	$cons^{(p)}$: 456.72	$cons^{(t)}$:462.98
5	Re-planning result				45	58	41	62	23	37	44.94	330.27
6	Final result	66	24	64	45	58	□	62	23	□	70.31(35.04%)	551.18(2.67%)
7	Global optimization at T4	66	24	64	45	58	41	14	23	37	72.73(-3.33%)	551.15(0%)

4 Conclusion and Future Work

A new Tree-coding Genetic Algorithms named TAG was proposed. Through the tree encoding, we can express the composite relationships among the tasks and support automatic encoding and decoding of the chromosomes. A PTree is expression of Web Service Composition, and it provides better computation model for QoS of the services selection and solves the dynamic re-planning efficiently. The simulation experiments confirm that the tree-encoding has excellent capability and reasonability in services selection, and TGA run faster than CGA. The future work is improving TGA to get better optimization solutions at fitness computation.

References

1. Zeng, L., Benatallah, B., Ngu, A.H.H., Dumas, M., Kalagnanam, J., Chang, H.: QoS-Aware Middleware for Web Services Composition. IEEE TRANSACTIONS ON SOFTWARE ENGINEERING 30(5), 311–327 (2004)
2. Liang-Jie, Z., Bing, L.: Requirements Driven Dynamic Services Composition forWeb Services and Grid Solutions. Journal of Grid. Computing 2, 121–140 (2004)
3. Canfora, G., Penta, M.D., Esposito, R., Villani, M.L.: An approach for QoS-aware service composition based on Genetic Algorithms. In: Genetic and Evolutionary Computation Conference(GECCO 2005), Washington DC,USA (2005)
4. Cheng-Wen, Z., Sen, S., Jun-Liang, C.: Genetic Algorithm on Web Services Selection Supporting QoS. Chinese Journal of Computers 29(7), 1029–1037 (2006)
5. Garey, M., Johnson, D.: Computers and Intractability: a Guide to the Theory of NP-Completeness. W.H. Freeman (1979)

SOA-Based Collaborative Modeling Method for Cross-Organizational Business Process Integration

Hongjun Sun, Shuangxi Huang, and Yushun Fan

Department of Automation, Tsinghua University, 100084 Beijing, P.R. China
sunhj05@mails.tsinghua.edu.cn, {huangsx, fanyus}@tsinghua.edu.cn

Abstract. Business process modeling is a key technology for cross-organizational business process integration. However, current modeling methods always fall short in describing complex collaboration relationship existing in business process integration in SOA-based collaborative environment. On the basis of existing multi-views business process model, a collaborative business process modeling method is presented to meet the above requirement in this paper. According to the analysis of inter-enterprise collaborative behavior, multi-enterprises collaborative meta-model integrating process, role, service and data is put forward. Then, on the foundation of collaborative meta-model, adopting model mapping method, existing business process model is transformed into multi-views collaborative business process model. The proposed method lays a solid foundation for cross-organizational business process integration.

Keywords: Business process, collaboration, modeling, meta-model, mapping.

1 Introduction

With the rapid development of business environment and the global economic integration, different enterprises have to cooperate to face the high competition [1]. Cross-enterprise integration becomes more important. Four integration methods have been put forward [2], and business process integration is the most important one. It enables an enterprise to respond with flexibility and speed to altering business conditions by integrating its business processes end to end across the company [3].

Furthermore, currently, the business collaboration within and across enterprise is becoming increasingly frequent. However, information, resource and service owned by enterprise has the characteristics of heterogeneity, distribution, dynamic, loose coupling, even autonomy. It is important that each industry department cares about how to integrate these IT resources. SOA (Service Oriented Architecture) provides the solution of this problem. The research of SOA-based collaborative management system (CMS) is becoming a hot topic. Business process modeling is the basis for business process integration and collaboration [4]. In CSM, running through the full lifecycle of model, the collaboration within and across enterprise can be achieved based on business-driven management method and MDA. Here, model is classified into business model, platform independent collaborative business process model,

K.C. Chang et al. (Eds.): APWeb/WAIM 2007 Ws, LNCS 4537, pp. 522–527, 2007.

platform dependent collaborative business process model, service model. Furthermore, collaborative model bridges business and IT. It can direct business optimization and is the precondition to actualize cross-organizational business process integration.

However, facing such environment, there are many problems on current business process modeling methods. Firstly, they mainly meet intra-enterprise integration and ignore inter-enterprise integration. Second, they can't describe complex collaborative relationships between any pair of process, role, service and data thoroughly. Furthermore, service oriented environment puts forward new requirements. In order to address these problems, a new business process modeling method is needed.

2 Collaborative Meta-model

The motivation of meta-model are to help to establish an environment in which business knowledge can be captured and business rules can be traced from their origin [5]. It is the foundation of business process model. Business object is the mechanism denoting abstractly business entity, and business state machine is used to describe its behavior characteristic. In order to support collaborative business process modeling, traditional state machine need to be extended [6]. The method is using state to describe status of business system, using activity to describe variance of business status and using the communication among state machines to describe inter-activity collaboration. Furthermore, Business rule along with business activity are the driven of state transition. Therefore, in order to get collaborative model, it is necessary to study collaborative meta-model based on MOF and business state machine.

2.1 Requirement

Business collaboration mainly refers to four elements: process, role, service and data. Accordingly, collaboration can be classified into collaboration between process and process, service and services, role and role, role and process, data and data and etc, as illustrated in Fig. 1. Obviously, the relationships among cross-organizational business cooperation are anfractuous, and current business process model can't support them.

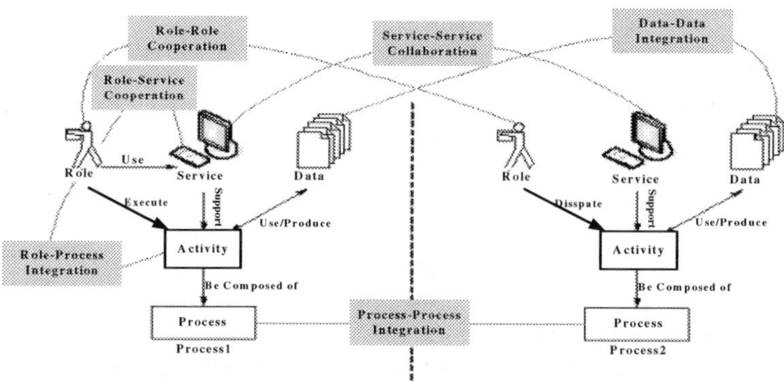

Fig. 1. Collaborative relationship between any pair of elements

2.2 The Framework of Collaborative Meta-model

Fig. 2 shows the framework of collaborative meta-model. It consists of an abstract basic class (Model Element) and six sub-models, including process, event, role, service, data, and state machine sub-model.

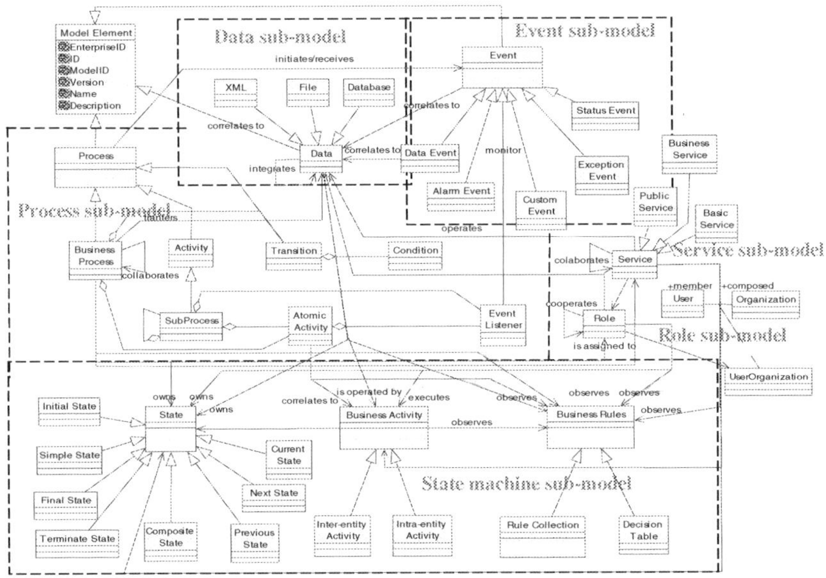

Fig. 2. The framework of collaborative meta-model

2.3 Sub-model

Reference [7] has described process sub model and event sub model in detail. Here, we mainly introduce class and associated relationship in other sub-models.

2.3.1 State Machine Sub Model

State machine sub-model comprises three basic classes that are State, Business Activity and Business Rule, and their deprived classes. State is used to describe entity status that participates in collaboration. The relationship between Business Activity and State is interlaced. State enables or prohibits the execution of Business activity. And the execution of Business Activity results in the transfer form a state to another. Business rule is an atomic piece of business logic, specified declaratively [8].

2.3.2 Data, Role and Service Sub Models

Data class is the abstract of business information and is used to sustain the collaboration between process and data, role and data, as well as service and data. Role class represents any entity that has the ability to initiate actions on other objects and is used where a person, organization, or program needs to be associated with others. Service class is used to abstract and organize enterprise information resources.

2.3.3 Relationship

Correlation relationship embodies the collaboration between any pair of process, role, service and data. Here, formal language is used to define these relationships. "M" represents the collection composed of all classes of meta-model.

1) The description of collaboration relationships between any pair of Business Process, Role, Data and Service can be defined as follows:

$$R_{collaborates}=\{<x, y>|x\in M, y\in M, \text{x and y cooperate and interact in order to finish the same task}\}. \quad (1)$$

2) Business Process, Role, Data and Service are described based on business state machine. It comprises three elements: State, Business Activity and Business Rule. The relationships between any pair of them can be described as follows:

$$R_{owns}=\{<x, y>|x\in M, y\in M, \text{y abstracts the state information x comprises}\}. \quad (2)$$

$$R_{observes}=\{<x, y>|x\in M, y\in M, \text{y abstracts the rules that x need to abide by}\}. \quad (3)$$

$$R_{behavirs}=\{<x, y>|x\in M, y\in M, \text{y describes the behaviors that exist in x}\}. \quad (4)$$

3) Other important relationships can be defined according to the following description.

$$R_{initiates/receives}=\{<x, y>|x\in M, y\in M, \text{y is initiates or receives by x}\}. \quad (5)$$

$$R_{correlates\ to}=\{<x, y>|x\in M, y\in M, \text{y is the correlative data of x}\}. \quad (6)$$

$$R_{is\ assigned\ to}=\{<x, y>|x\in M, y\in M, \text{y uses x when y executes}\}. \quad (7)$$

3 Collaborative Business Process Modeling

In collaborative environment, collaborative business process model need not only to support complex collaboration relationships, but also reflect enterprise business requirement completely. So, in order to keep the consistency between collaborative business process model and business requirement, model mapping method is used.

3.1 Mapping Rules

Current business process model [4] [9] is used to describe business requirement. It is depicted in the form of a group of enterprise models. The conjunction of these views is achieved through process view. Here, in order to get collaborative business process model based on collaborative meta-model, these views will be abstracted into the elements of collaborative model. Among the rest, business process and interaction between any pair of them can be modeled into business state machine, information view element can be modeled in business object, organization view elements can be modeled into role and others such as function view can be mapped into service. Based on these rules, business requirement can be transformed into collaborative model.

3.2 Mapping Process

Collaborative business process reflects all kinds of collaborative scenes. The mapping processes from business requirement to collaborative model are as follows.

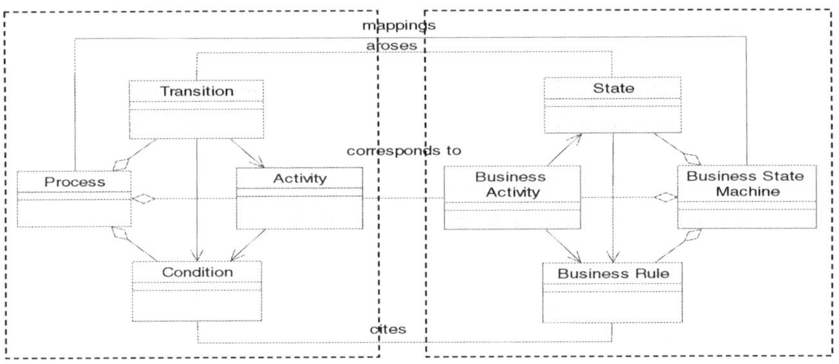

Fig. 3. Process model mapping description based on meta-model

1. Mapping between process and business state machine: Fig. 3 shows the mapping relationship. The left is traditional process meta-model, and the right is the process meta-model that is described in the form of business state machine.
2. Mapping between information view and business object: Business object is the abstract of heterogeneous data. XML can bridge data elements and business object.
3. Mapping between organization view and role: Role is the executer of task and it need to be taken on by users. The correlation between role and user is built.
4. Mapping between function view and service: The function can be encapsulated into service by service description, composition and etc.
5. Relationship: In order to realize collaboration between any pair of process, service, role and data, business state machine is used to describe them. Accordingly, collaboration can be achieved between business state machines.

3.3 Modeling Characteristics

Collaborative business process modeling has some new characteristics in contrast to others. Firstly, it includes process, role, service and data, and can describe complex collaboration relationship. Secondly, business function and achievement are encapsulated into services. Furthermore, it reflects enterprise business requirement. These characteristics indicate that collaborative model can well support cross-organizational business process integration in SOA-based collaborative environment.

4 Conclusions and Future Work

In this paper, a new business process modeling method is put forward, which supports the description of complex collaboration relationships that exist in cross-organizational business process integration in SOA-based collaborative environment. The proposed method extends the traditional enterprise business process model and lays a solid foundation for cross-organizational business process integration in SOA-based collaborative management environment. In the future, the research should centralize in mapping consistency between current business process model and collaborative business process model. Moreover, the research on run evaluation of collaborative business process model also should be carried out.

Acknowledgments. The work published in this paper is funded by the National Natural Science Foundation of China under Grant No. 60504030 and the National High Technology Research and Development Program of China (863 Program) under Grant No. 2006AA04Z166. Also, the work is supported by the project of IMPORTNET under Contract No. 033610.

References

1. Chakraborty, D., Lei, H.: Extending the Reach of Business Processes, 37(4) 78–80 (2004)
2. Khriss, I., Brassard, M., Pitman, N.: GAIL: The Gen-it@ Abstract Integration Layer for B2B Application Integration Solutions. In: the 39th International Conference and Exhibition on Technology of Object-Oriented Languages and Systems, pp. 73–82 (2001)
3. Kobayashia, T., Tamakia, M., Komodab, N.: Business process integration as a solution to the implementation of supply chain management systems, Information and Management pp. 769–780 (2003)
4. Lin, H.P., Fan, Y.S., Wu, C.: Integrated Enterprise Modeling Method Based on Workflow Model and Multi-Views. Tsinghua Science and Technology 6(1), 24–28 (2001)
5. Yu, Y., Tang, Y., L.L., Feng, Z. S.: Temporal Extension of Workflow Meta-model and Its Application. In: the 8th International Conference on Computer Supported Cooperative Work in Design Proceedings. pp. 293–297 (2003)
6. Thomas, D., Hunt, A.: State Machines, IEEE SOFTWARE, pp. 10–12 (2002)
7. Lin, H.P., Fan, Y.S.,Wei, T.: Interactive-Event-Based Workflow Simulation in Service Oriented Computing. In: 5th International Conference on Grid and Cooperative Computing, Hunan, China. pp. 177–184 (2006)
8. Ross, R.: The Business Rule Book: Classifying, Defining and Modeling Rules, 2nd edn, (Ross Method, version 4.0). Business Rule Solutions, Inc, Houston, Texas (1997)
9. Li, J.Q., Fan, Y.S., Zhou, M.C.: Performance Modeling and Analysis of Workflow. IEEE Transactions on Systems, Man, and Cybernetics-Part A:Systems and Humans, pp. 229–242 (2004)

Model Checking for BPEL4WS with Time

Chunming Gao[1,2], Jin Li[1], Zhoujun Li[2], and Huowang Chen[2]

[1] College of Mathematics and Computer Science, Hunan Normal University
410081 Changsha, China
lj_mvp@sina.com
[2] School of Computer Science, National University of Defense Technology
410073 Changsha, China
gcm_powerman@163.com

Abstract. The mobile ambient is a formal model for mobile computation, but the real-time property of the mobility has not been well described. We extend mobile ambient with time, and then present discrete time mobile ambient calculus (DTMA). We also propose a modal logic for DTMA, and then give a model checking algorithm for DTMA on a subset of its logic formulas. Based on DTMA, we investigate the modelling and model checking for web service composition orchestration that has time constraint. Our work is a foundation for the model checking of the real-time mobile computation.

Keywords: Timed Mobile Ambient, modal logic, model checking, BPEL.

1 Introduction

The process algebra[1][3] has a stronger ability to model mobile computation. For example the π-calculus[1][6] has the characteristic of describing mobile computation, but it can't describe the process behaviour in traversing spatial region. Mobile ambient calculus[5] can not only describe mobile process, but also describe the ambient of mobile process and the mobility of the ambient. However, by now there isn't any calculus that can describe behaviours of both traversing spatial region and real time constraint. We propose discrete time mobile ambient calculus(DTMA) to solve the problem in describing properties for spatial region and real time of mobile process.

At present the researches about process algebra in the service composition have met some difficulties, among which main obstacle is how to formalize the property of service composition with time constraint. Someone thought it is very difficult to give descriptions about time, fault and compensation handling[2].

In this paper, we investigate the modelling for web service composition orchestration that has some mobility and time constraint, based on DTMA, and model the BPEL4WS[4] by DTMA. The rest of this paper is organized as follows: In section 2 and section 3 we presented the syntax and semantics of DTMA. In the section 4 we introduced the DTMA logic for describing the properties of the mobile processes. In the section 5 we propose model checking algorithm. In section 6 we encoded basic actions of BPEL4WS by DTMA. Finally conclusions and future work are presented.

K.C. Chang et al. (Eds.): APWeb/WAIM 2007 Ws, LNCS 4537, pp. 528–533, 2007.

2 DTMA Syntax

For convenience, we adopt non-negative integral set $N^{\geq 0}$ as time range. Clock variable is defined on time set $N^{\geq 0}$, We use interval to state time slice, for example $[e_1, e_2]$ denotes duration from time e_1 to time e_2 .

The table 1 describes the syntax of DTMA calculus.

In the DTMA calculus, the operators such as $\text{in\_M}_e^{e'} @t$, $(\text{n})_e^{e'} @t.P$, $\text{<M>}_e^{e'} @t$ etc., restrict the time variable t , which is called bound time variable. In table 1, $(\text{n})_e^{e'} @t.P$ and $\text{<M>}_f^{f'} @t$ denote input and output processes, which restrict the durations of the input and output actions in $[e,e'],[f,f']$.

Table 1. The syntax of DTMA calculus

Def(*DTMA*)							
$P \in \pi ::=$	*Process*			M::=	Message		
nil	inactivity	$(vn)P$	restriction	$t \in \Phi(C) \cup \{\infty\}$			
P|P	parallel	M[P]	ambient	n	name	$\text{in\_M}_e^{e'} @t$	entry capability at t
!P	replication	M.P	exercise a capability	e	time expression	$\text{out\_M}_e^{e'} @t$	exit capability at t
$(\text{n})_e^{e'} @t.P$	input locally at t, and e $\leq$ t $\leq$ e'			ε	empty path	$\text{open\_M}_e^{e'} @t$	open capability at t
$\text{<M>}_e^{e'} @t$	output locally at t, and e $\leq$ t $\leq$ e'			M.M'	composite path		

3 DTMA Semantic

For the requirement of the real time system, we define the DTMA semantics by the reduction relation $P \to Q$ and the delay relation $P \mapsto Q$. The delay $\mapsto_u$ is used to describe the system delaying time u and doesn't do any actions; the reduction $\to_u$ is used to describe the system delaying time u and completes the reduction. The structural congruence[7], $P \equiv Q$, is an auxiliary relation used in the definition of the reduction. The DTMA semantics is defined as table2.

Table 2. DTMA semantics

$$n[in\_m_e^{e'} @t.P \mid Q] \mid m[R] \longrightarrow_u m[n[P \mid Q] \mid R]\{u/t\} \ , \ e \leq u \leq e'$$

$$n[in\_m_e^{e'} @t.P \mid Q] \mid m[R] \mapsto_u n[in\_m_{e\text{-}u}^{e'\text{-}u} @t.P \mid Q] \mid m[R]\{t+u/t\} \ , \ u \leq e'$$

$$m[n[out\_m_e^{e'} @t.P \mid Q] \mid R] \longrightarrow_u n[P \mid Q] \mid m[R]\{u/t\} \ , \ e \leq u \leq e'$$

$$m[n[out\_m_e^{e'} @t.P \mid Q] \mid R] \mapsto_u m[n[out\_m_{e\text{-}u}^{e'\text{-}u} @t.P \mid Q] \mid R]\{t+u/t\} \ , \ u \leq e'$$

$$m[open\_m_e^{e'} @t.P] \mid Q \longrightarrow_u P \mid Q\{u/t\} \ , \ e \leq u \leq e'$$

$$m[open\_m_e^{e'} @t.P] \mid Q \mapsto_u (m[open\_m_{e\text{-}u}^{e'\text{-}u} @t.P] \mid Q)\{t+u/t\} \ , \ u \leq e'$$

$$(n)_e^{e'} @t.P \mid\!< M >_f^{f'} @t' \to_u \{M/n\}\{(u,u)/(t,t')\} \ , \max(f,e) \leq u \leq \min(f',e')$$

$$(n)_e^{e'} @t.P \mid\!< M >_f^{f'} @t' \mapsto_u (n)_{e \odot u}^{e'-u} @t.P \mid\!< M >_{f \odot u}^{f'-u} @t'\{(t+u,t+u)/(t,t')\} \ , u \leq f', u \leq e'$$

$$P \to_u Q \ , \ P \equiv P', \ Q \equiv Q' \Rightarrow P' \to_u Q'$$

$$P \to_u P', Q \to_u Q' \Rightarrow P \mid Q \to_u P' \mid Q' \quad P \to_u Q \Rightarrow (vn)P \to_u (vn)Q \quad P \to_u Q \Rightarrow n[P] \to_u n[Q] \quad P \to_u P' \Rightarrow !P \to_u !P'$$

$$P \mapsto_u P', Q \mapsto_u Q' \Rightarrow P \mid Q \mapsto_u P' \mid Q' \quad P \mapsto_u P' \Rightarrow !P \mapsto_u !P' \quad P \mapsto_u Q \Rightarrow (vn)P \mapsto_u (vn)Q \quad nil \mapsto_u nil$$

4 The Logic of DTMA

The modal logic[7] for Mobile Ambient is used to describe the properties of mobile computation[5]. In order to characterize the real time properties, we need to redefine the modal logic for DTMA, including its syntax and the satisfactions. The syntax of the modal logic for DTMA is shown below:

Definition 1. Syntax of the logic for the DTMA

η	*a name n or a variable x*				
$A\ B ::=$	*formula*	$n[A]$	*location*	$A@n$	*location adjunct*
T	*ture*	$\neg A$	*negation*	$\eta^R A$	*revelation*
$A \vee B$	*disjunction*	$A^v\eta$	*revelation adjunct*	0	*void*
$\Diamond A$	*sometime modality*	$A \mid B$	*composition*	ΩA	*somewhere modality*

Formula A is closed, if it does not contain free variable. T, $\neg A$,are proposition logic 0 , $A|B$, $\eta[A]$ are used to describe the space.

The satisfactions of logic for DTMA are below:

Definition 2. Satisfaction

$P \models T$

$P \models A \vee B$	*iff*	$P \models A \vee P \models B$	$P \models \neg A$	*iff*	$\neg P \models A$
$P \models n[A]$	*iff*	$\exists P' : \Pi . P \equiv n[P'] \wedge P' \models A$	$P \models 0$	*iff*	$P \equiv 0$
$P \models \exists x.A$	*iff*	$\exists m : \Lambda . P \models A\{x \leftarrow m\}$	$P \models A \mid B$	*iff*	$\exists P', P'' : \Pi . P \equiv P' \mid P'' \wedge P' \models A \wedge P'' \models B$
$P \models \eta^R A$	*iff*	$\exists P' . P \equiv (vn)P' \wedge P' \models A$	$P \models A @ n$	*iff*	$n[P] \models A$
$P \models \Omega A$	*iff*	$\exists P' : \Pi . P \downarrow^* P' \wedge P' \models A$	$P \models A^v\eta$	*iff*	$(vn)P \models A$
$P \models A \rhd B$	*iff*	$\forall P' : \Pi . P' \models A \Rightarrow P \mid P' \models B$	$P \models \Diamond A$	*iff*	$\exists P' : \Pi . P \overrightarrow{\mapsto}^* P' \wedge P' \models A$
$P \models <\mapsto, u> A$	*iff*	$\exists P' : \Pi . P \mapsto_u P' \wedge P' \models A$	$P \models <\rightarrow, u> A$	*iff*	$\exists v : T . v \leq u \wedge P \rightarrow_v P' \wedge P' \models A$

By now, we have given the DTMA calculus and its logic. In the fifth chapter, the algorithms for model checking of DTMA will be introduced.

5 Model Checking Algorithm for DTMA

Model checking[8] is a method to decide whether a process can satisfy a given formula. In MA, some process and some formulas can not be checked[10]. Due to DTMA is an extended MA with time, only a subset of the processes and logic formulas for DTMA can be checked as MA. In the subset of DTMA and its logic, the DTMA contains constraint operation, and in order to deal with the constraint operation, we impose a function "separate" that is derived from paper[8]. We define two functions below:

$$Reachable(N,P) \triangleq \{< N',P' >|< N,P > \overrightarrow{\mapsto}^* < N',P' >$$

$$Sublocations(N,P) \triangleq \{< N,P > \downarrow^* < N,P' >\}$$

We write $\downarrow^*$ for the reflexive and transitive closure of $\downarrow$, and $\overrightarrow{\mapsto}^*$ for that of $\rightarrow$ and $\mapsto$. For convenience, we define the time prefix (e)E, which presents delays time e then the process active is the same as process E.

The algorithm for model checking of DTMA is below:

DTMA Model-checking Algorithm

$Check(N,P,T) \triangleq T$

$Check(N,P,\neg A) \triangleq \neg Check(N,P,A)$

$Check(N,P,A \vee B) \triangleq Check(N,P,A) \vee Check(N,P,B)$

$Check(N,P,0) \triangleq T \quad if \quad P \equiv 0 \quad , \quad F \quad otherwise$

$Check(N,P,A \mid B) \triangleq \vee_{N_1 \cup N_2 = N} \vee_{P_1 \mid P_2 = P} Check(N_1,P_1,A) \wedge Check(N_2,P_2,B) \wedge fn(P_1) \cap N_2 = \phi \wedge fn(P_2) \cap N_1 = \phi$

$Check(N,P,n[A]) \triangleq n[Q] \wedge n \notin N \wedge Check(N,Q,A)$

$Check(N,P,A @ n) \triangleq Check(N,n[P],A)$

$Check(N,P,n^R A) \triangleq \vee_{m \in N} Check(N - \{m\}, P\{m \leftarrow n\}, A) \vee (n \notin fn(P) \wedge Check(N,P,A))$

$Check(N,P,A^{\text{v}} n) \triangleq Check(N \cup \{n\}, P, A)$

$Check(N,P,\Diamond A) \triangleq \vee_{<N',P'> \in Reachable(N,P)} Check(N',P',A)$

$Check(N,P,\Omega A) \triangleq \vee_{<N',P'> \in Sublocations(N,P)} Check(N,P',A)$

$Check(N,P,\exists x.A) \triangleq \vee_{n \in fn(N,P) \cup fn(A)} Check(N,P,A\{x \leftarrow n\}) \vee Check(N,P,A\{x \leftarrow n_0\}) \quad let\ n_0 \notin N \cup fn(P) \cup bn(P) be\ a\ fresh\ name$

$Check(N,P,<\mapsto,u> A) \triangleq P \equiv (u)Q \wedge Check(N,Q,A)$

$Check(N,P,<\rightarrow,u> A) \triangleq \vee_{v \le u} P \rightarrow_v Q \wedge Check(N,Q,A)$

6 Modeling BPEL4WS Activities

In this section, we primarily discuss the modeling on the basic activities of BPEL4WS. In Sect. 6.1 we encode the basic program structures. Section 6.2 models the basic activities of BPEL4WS.

For some actions that haven't time constraint, we will write them for short without time variables and expressions. For example, $M_0^{\infty} @ t.P$ can write as $M.P$.

6.1 Encoding the Basic Program Structures

1) Encoding Control Structure of Sequence
With the help of implication of locks[5], we can get the sequence control between two processes of ambient. We denote the operator of sequence control with $\$^l$, where 'l' is the name of lock. We encode the control structure of sequence as $P_1 \$^{l_1} P_2 \$^{l_2} P_3 \triangleq (P_1 \$^{l_1} P_2) \$^{l_2} P_3$.

2) Encoding Channels
By requirement for the actual scenarios, the input process can't forever wait for an output in an ambient. So it is necessary to model channel with timed restriction using DTMA.

$$buf\ n \triangleq n[!open\ io] \quad , \quad n\langle M\rangle_e^{e'} @t \triangleq io[in\ n.\langle M\rangle_e^{e'} @t] \ , \quad n(x)_f^{f'} @t.P \triangleq (vp)(io[in\ n.(x)_f^{f'} @t.p[out\ n.P]] \mid open\ p)$$

The above channel with timed restriction has the reduction as follows:

$$buf\ n \mid n\langle M\rangle_e^{e'} @t_1 \mid n(x)_f^{f'} @t_2.P \rightarrow_u buf\ n \mid P\{x \leftarrow M\}\{u/t_2\} \quad \max(e,f) \le u \le \min\{e',f'\}$$

3) Encoding Match
The semantic of match is that when the name n is equal to the name y, a process P should be executed, otherwise the process 0. Encoding the match as follows:

$$[n=y]_{t_1}^{t_1+e} @t.P \triangleq <n> \mid (x)_f^{f'} @t_0.(x[] \mid open\_y_{t_1}^{t_1+e} @t_1.P\{t_1/t\})$$

4) Encoding Summation
Summation (or internal choice), it denotes that a behavior will be activated due to the system's choice. Encoding the summation based on DTMA as follows:

$$P + Q \triangleq (vx)(open\ x \mid x[P] \mid x[Q]) \quad (x \notin fn(P) \cup fn(Q))$$

6.2 Modeling the Basic Activities for BPEL4WS

We think that the specification of BPEL4WS has offered the mechanism such as pick structure for expressing time[8].

For convenience, we define the function T, which map the string of time to non-negative integer. The T can be implemented easily in practice. The mapping from the basic activities of BPEL4WS to their corresponding expressions of DTMA as showed in table 3. note: the name of channels consists of the standard attributes of activities in series.

Table 3. The modeling for basic activities of BPEL4WS based on DTMA

Activities	BPEL4WS Activities	DTMA terms
Receive	<receive ...variable="var" > </receive>	$P_{receive} \triangleq n(var) \mid buf\ n$
Reply	<reply...variable="var" > </reply>	$P_{reply} \triangleq n\langle variable \rangle \mid buf\ n$
Invoke	<invoke ...inputVariable="input" outputVariable="out"> </invoke>	$P_{invoke} \triangleq$ $buf\ n \mid n<inputVariable>\$^{l}\, n(outputVariable)$
Invoke (one way)	<invoke ...outputVariable="out"> </invoke>	$P'_{invoke} \triangleq n\langle outputVariable \rangle \mid buf\ n$
Flow	<flow > activity1... activityk </flow>	$P_{flow} \triangleq n[\prod_{i=1}^{k} P_i.release\ l_i.0 \mid acquire\ l_1$ $.acquire\ l_2.L.acquire\ l_k.out\ n.open\ n.0]$
Sequence	<sequence> activity1 ... activityk </sequence>	$P_{sequence} \triangleq$ $(\nu l_1,L,l_k)n[P_1\$^{l_1}P_2\$^{l_2}L\$^{l_{k-1}}P_k\$^{l_k}$ $(go(out\ n).l[\ ])] \mid open\ l.open\ n.0$
Pick	<pick> <onMessage... variable="variable"> activity1 </on Message> <onAlarm for="timeExpress"> activity2 < /onAlarm > < /pick >	$P_{pick} \triangleq buf\ n \mid n(variable)_t^{t+T(timeExpress)}@t_1.$ $release\ l.P_{activity1} \mid open\ l_{t+T(timeExpress)}^{t+T(timeExpress)}@t_2.$ $open\ secret.0 \mid secret[$ $out\ secret_{t+T(timeExpress)+\varepsilon}^{t+T(timeExpress)+\varepsilon}@t_3.P_{activity2}]$
Switch	<switch> <case condition='cond1'> activity1</case> <case condition="cond2"> activity2 </case> <otherwise> activity 3 </otherwise> </switch>	$P_{switch} \triangleq [n=("cond1","cond2")]_{t}^{t+3\varepsilon}$ $@t.((P_{activity1},P_{activity2}),P_{activity3}))$

There is an example about business process is described as follows:

Client agent sends a request to engine, which is accomplished in 2 hours. Then engine invokes the service according to the request, which is also accomplished in 2 hours. The corresponding BPEL resource file is as follows:

```
<process name="ModelProcess" ... >

  <sequence>
 <pick createInstance="yes">
  <onMessage portType=" " partnerLink=" " operation=" " variable=" ">
     </onMessage>
```

```
  <invoke name="M" partnerLink="" portType="" operation="" outputVariable="">
    </invoke>
   <onAlarm for="P0DT2H">
   </onAlarm>
  </pick>
 </sequence>
......
</process>
```

According to the above rules the process is modeled as

$$\begin{aligned} P &= l(r)_2^4 @ t.(\nu n)(m[n < r >_4^6 @ t.0]) \mid buf\ n) \mid < \alpha >_2^4 @ t \mid buf\ l \\ &= (\nu\ m)(\ io[\ in\ l.\ (r)_2^4 @ t.\ q[out\ l.\ q]]) \mid io[\ in\ l.\ < M >_2^4 @ t.0] \mid l[open\ io \mid open\ io.0] \end{aligned}$$

The reachability of the service is described by the DTMA logic as $A = l^R <\to, 2> l[n^R <\to, 2> n[T]]$. According to the model checking algorithm in section 5, the reachability can be checked as $Check(N, P, A) = T$. The conclusion is that the service can be invoked and give a corresponding response.

7 Summary and Future Work

In this paper, we extend mobile ambient calculus with time, introduce new reduction semantics, and then present discrete time mobile ambient calculus (DTMA). We analyze and model the basic actions of BPEL4WS and an instance of service composition with time restriction. The future work includes implementing model checking algorithm for DTMA based on replication-free processes and close logic without composite adjunct, applying the model checking of DTMA to analyze the behaviours of the decentralized orchestration with mobility and time restriction.

References

1. Milner, R.: Communicating and Mobile Systems:The π-Calculus. Cambridge University Press, Cambridge (1999)
2. Koshkina, M., van Breugel, F.: Verication of business processes for web services. Technical Report CS-2003-11, Department of Computer Science, York University (2003)
3. Chen, L.: Timed Processes: Models, Axioms and Decidability. Laboratory for Foundations of Computer Science (Theses and Dissertations), University of Edinburgh (1992)
4. Andrews, T., Curbera, F., et al.: Business process execution language for web services, version1.1 (2003), www-128.ibm.com/developerworks/library/specification/ws-bpel
5. Cardelli, L., Gordon, A.D.: Mobile ambients. In: Nivat, M. (ed.) Proc. FOSSACS'98, LNCS, vol. 1378, pp. 140–155. Springer, Heidelberg (1998)
6. Jing, C.: Study of Real-time Value-passing and Real-time Mobile System. PhD thesis. Institute of Software, Chinese Academy of Science. Beijing, China (2003)
7. Cardelli, L., Gordon, A.D.: Anytime,anywhere?Modal logics for mobile ambients. In: Proceedings POPL'00, pp. 365–377. ACM, New York (2000)
8. Charatonik, W., Talbot, J.M.: The Decidability of Model Checking Mobile Ambients. In: FOSSACS 2004. LNCS, Springer, Heidelberg (2004)

A Version Management of Business Process Models in BPMS

Hyerim Bae[1], Eunmi Cho[1], and Joonsoo Bae[2,*]

[1] Department of Industrial Engineering
Pusan National University, San 30, Jangjeon-dong, Geumjeong-gu,
Busan, Korea, 609-735
hrbae@pusan.ac.kr, oldlace@naver.com
[2] Department of Industrial and Information Systems Engineering
Chonbuk National University, 664-14, Dukjin-dong, Duckjin-gu,
Jeonju, Korea, 561-756
jsbae@chonbuk.ac.kr

Abstract. BPM system manages increasing number of business processes and the necessity of managing processes during whole process lifecycle from process modeling to process archiving has been emerged. Despite of wide use of the BPM system and the maturing of its technology, the main focus has been mainly on correctly executing process models, and convenient modeling of business processes has not been considered. In this paper, a new method of versioning business processes is developed in order to provide users with an easy modeling interface. Version management of a process enables a history of the process model to be recorded systematically. In our method, an initial version and changes are stored for each process model, and any version can be reconstructed using them. We expect that our method enhances the convenience of process modeling in an environment of huge number of business processes, and thereby assists the process designer. In order to verify the effectiveness of our method, a prototype system is presented in this paper.

Keywords: Business Process Management, Version Management, XML.

1 Introduction

A business process is represented as a flow of tasks, which are either internal or external to the enterprise. Business Process Management (BPM) is an integrated method of managing processes through their entire lifecycle. The BPM system is a software system that models, defines, controls and manages business processes [6, 7]. By contrast with the simple, linear versioning methods of existing systems, the version management method presented in this paper allows parallel versioning, automatic detecting of changes and the option of keeping track of the change history

[*] Corresponding author.

K.C. Chang et al. (Eds.): APWeb/WAIM 2007 Ws, LNCS 4537, pp. 534–539, 2007.

with a graphical tool. With its exactingly developed functions, our method also improves user convenience and minimizes space used to store process models. Version management, in its broad sense, is a method of systematically handling with temporal features and changes in objects over time. A version is defined as a semantically meaningful snapshot of an object at a point in time [5]. A version model is used to represent a history of changes to an object over time, or simply records a version history of that object. The history of change is usually described using a version graph, such as that shown in Figure 1. In a version graph, a node is a version, and a link between two neighboring nodes is a relationship between them. For example, version $v_2(o)$ was created by modifying a previous version, $v_1(o)$.

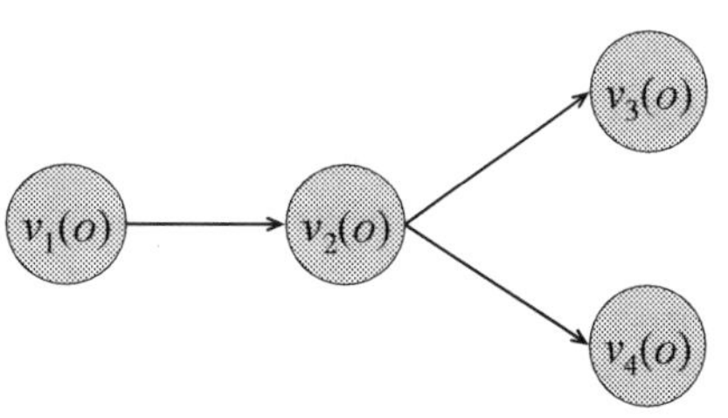

Fig. 1. Version Graph

2 Business Process Model

In this chapter, we define the process model, which is a target object of our version management. A business process model used in the BPM system is usually composed of basic objects such as tasks, links and attributes[8]. Attributes describe features of the objects. We define objects as the elements of a process model.

Definition 1. *Business Process Model*
A process model p, which is an element of a process model set P, consists of tasks, links, and attributes. That is, $p= (T, L, A)$.

- A set of tasks: $T = \{t_i \mid i= 1,\ldots, I\}$, where, t_i represents i-th task and I is the total number of tasks in p.
- A set of links: $L = \{l_k= (t_i, t_j) \mid t_i, t_j \in T, i \neq j \}$, where, l_k represents a link between two tasks, t_i and t_j. A link represents a precedence relation between the two tasks. That is, the link (t_i, t_j) indicates that t_i immediately precedes t_j.
- A set of attributes: A is a set of task attributes or link attributes.
  1) $A_i = \{t_i.a_s \mid s =1,\ldots, S_i\}$ is a set of task attributes, where $t_i.a_s$ represents s-th attribute of task t_i and S_i is the total number of t_i's attributes.
  2) $A_k = \{l_k.a_s \mid s =1,\ldots, S_k\}$ is a set of link attributes,where $l_k.a_s$ represents s-th attribute of link l_k and S_k is the total number of l_k's attributes.

3 Version Management of Business Process

This chapter explains our method of process version management. We first define a version graph by introducing concepts of object, a version, and change types. Then, we present a version management algorithm using check-in/check-out algorithms.

3.1 Version Graph

In the BPM system, process design procedure is a work of defining all the process elements introduced in Section 2 by using a design tool. In this procedure, it may be impossible to prepare a perfect process model at once. Therefore, business requirements for reusing the previous models have been always raised. A user can modify a previous model to make it more complete one. After the user modifies a process model, change of the process is defined as a set of changes to component objects. A component object o can be a task, a link or an attribute. That is, $o \in T$, $o \in L$ or, $o \in A$.

A process version results from user's modifications in designing a process model. Versions are recorded using a version graph. In the version graph, a node represents a version of a process p, which is denoted as $v(p)$. A modification of a process includes changes to the objects in it, and each of the changes is represented using δ. We define a process modification as a set of object changes $\Delta = \{\delta_q(o) \mid q=1,\ldots, Q \text{ and } o \in T, L, A\}$. Applying the changes to a previous version to create a new version is represented by '$\cdot$' (change operation). If the n-th version of p is derived from the m-th version of p by applying the changes Δ_{mn}, we represent that $v_n(p)$ is equal to $v_m(p) \cdot \Delta_{mn}$. A version graph is defined as follows.

Definition 2. ***Version Graph, VG***
A version graph is used to record the history of a single process. Let p denote a process, and $v_m(p)$ the m-th version of the process. A version graph (VG) of p is a directed acyclic graph $VG = (V, E)$, where V and E is a set of nodes and arc respectively.

- $V = \{v_m(p) \mid m= 1, \ldots,M\}$
- $E = \{(v_m(p), v_n(p)) \mid v_m(p) \in V, v_n(p) \in V, v_n(p) = v_m(p) \cdot \Delta_{mn}\}$

In a version graph, if a version $v_n(p)$ can be derived from $v_m(p)$ by modifying $v_m(p)$ repetitively($\{(v_m(p), v_k(p)), (v_k(p), v_{k+1}(p)), \ldots, ((v_{k+l}(p), v_n(p))\} \subset E$), we say that $v_n(p)$ is 'reachable' from $v_m(p)$.

3.2 Combination of Changes

In general, a change can be classified into three types: adding, modifying or deleting. In Figure 1, we consider a component object o, which is $t_{order}.a_{due}$. If a value is added to the object, and then the value is modified into another value, the change can be represented as follows.

$$\delta_1(o) = ADD\ t_{order}.a_{due}(\text{“2005-12-24”})$$
$$\delta_2(o) = MOD\ t_{order}.a_{due}(\text{“2005-12-24”, “2006-01-24”})$$

While designing a process, it is usual that the process is modified repetitively, and multiple changes are created. These multiple and repetitive changes for the same object can be represented by a single change. For example, the two changes δ_1 and δ_2 in the above can be combined using a combination operator, '$\circ$'.

$$\delta_{new} = \delta_1(o) \circ \delta_2(o) = ADD\ t_{order}.a_{due}\ (\text{“2006-01-24”})$$

The combination of changes is calculated by using our rules, which are summarized in Table 1. Reverse change (δ^{-1}) is used for δ and empty change ($\delta_\emptyset$) is used for the rules. Based on the combination of object changes, we can extend and apply to combinations between sets of changes. A combination of two change sets Δ_1, Δ_2 ($\Delta_1 \circ \Delta_2$) is defined as a set of changes. All of the object changes are included as elements of the set, and two changes from the two sets for the same object should be combined into one element.

Table 1. Combination of changes

reverse change		combined change	
ADD operation	$\delta_{ADD}^{-1} = \delta_{DEL}$	ADD and DEL operation	$\delta_{ADD} \circ \delta_{DEL} = \delta_\emptyset$
		ADD and MOD operation	$\delta_{ADD} \circ \delta_{MOD} = \delta_{ADD}'$
DEL operation	$\delta_{DEL}^{-1} = \delta_{ADD}$	DEL and MOD operation	$\delta_{DEL} \circ \delta_{ADD} = \delta_{MOD}$
MOD operation	$\delta_{MOD}^{-1} = \delta_{MOD}'$	MOD and MOD operation	$\delta_{MOD} \circ \delta_{MOD}' = \delta_{MOD}''$
		MOD and DEL operation	$\delta_{MOD} \circ \delta_{DEL} = \delta_{DEL}$

When we apply combination operators, the following axioms are used.

- Communicative law is not valid. ($\delta' \circ \delta'' \neq \delta'' \circ \delta'$)
- Associative law is valid. ($\delta' \circ (\delta'' \circ \delta''') = (\delta' \circ \delta'') \circ \delta'''$)
- De Morgan's law is valid. ($(\delta' \circ \delta'')^{-1} = \delta''^{-1} \circ \delta'^{-1}$)
- All changes are unaffected by empty change. ($\delta' \circ \delta_\emptyset = \delta_\emptyset \circ \delta' = \delta'$)
- Associative operation between the change and reverse change is equal to empty change. ($\delta' \circ \delta^{-1} = \delta^{-1} \circ \delta' = \delta_\emptyset$).

3.3 Version Management Procedure

Our version management method is based on two important procedures; check-in and check-out[2, 3]. When a user wants to make a new process model from an existing model, he can request a previous version of the process to be taken out into his private work area. This is called a ‘check-out’ procedure. After a user finishes modifying the process, he may want to store it in a process repository as a new process version. We call this a ‘check-in’ procedure. That is, check-out transfers a process model from public storage to an individual workplace, and check-in returns the model to the public storage.

If all of the versions of a process are stored whenever a new version is created, storage space might be wasted. To avoid such a waste of space, we use the modified delta method [4]. The modified delta method is implemented using our check-in/out algorithms. The modified delta method uses the combination operators. It stores only a root version of a process and changes, and reconstructs any version when a user retrieves that version.

Check-out is invoked when a user requests a certain version of a process, $v_m(p)$. It first searches the changes from database, which are required to reconstruct the version and combines them into a single change. Then, by applying the change to a root version $v_0(p)$, the requested version can be reconstructed and sent to the user.

Conversely, check-in is invoked when a user returns the modified version of a process. First, it identifies which objects were changed. Then it detects the change types (add, modify, delete). The changes are stored as a change set, and finally the process version graph is updated. In Figure 2, we provide flows of the two procedures.

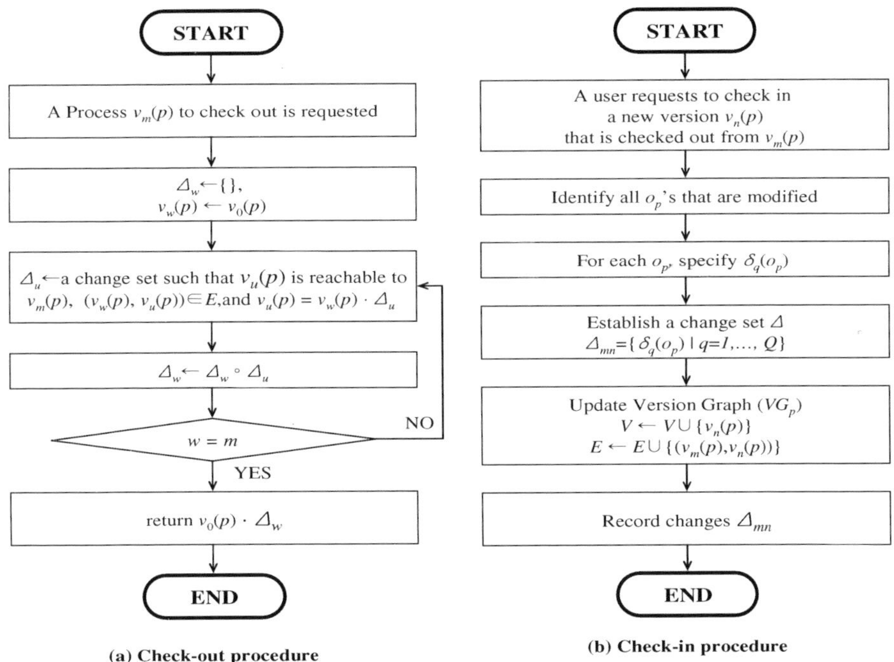

Fig. 2. Flowcharts of check-out/in procedures

3.4 Prototype System

The proposed method is implemented as a module of process designer, which implementation is a build-time function of our BPM system, called ILPMS (Integrated Logistics Process Management System) [1]. A user designs the process models with a process design tool and the designed process model is stored in a process DB. The user can easily modify and change the process using the modeling tool supported by our version management. When a user requests a process version, the system, with the check-out function, automatically generates the requested process version. After modifying the version delivered to the user, he checks in the newly created version.

4 Conclusions

In this paper, we propose a new method of process version management. Our method enables BPM users to design process models more conveniently. Though the BPM system is becoming increasingly essential to business information system, the difficulty of process modeling is a significant obstacle to employing the system. Consequently, beginners have not been able to easily design business processes using the BPM design tool. For this reason, we presented process models that use XML technology. If a user modifies a process model, our system detects the changes in the XML process definition. Then, the changes are recorded and the version graph is updated. With the version graph, we can manage history of process model change systematically. Any version of a process can be reconstructed, once its retrieval has been requested, by combining the changes and applying them to the initial version. We expect that our method can be easily added to the existing BPM system and, thereby, can improve the convenience of process modeling in an environment where a huge number of process models should be dealt with.

Acknowledgements. This work was supported by "Research Center for Logistics Information Technology (LIT)" hosted by the Ministry of Education & Human Resources Development in Korea.

References

1. Bae, H.: Develpment Integrated Logistics Process Management System (ILPMS) based on XML. PNU-Technical Paper-IS-2005-03, Pusan National University. (2005)
2. Conradi, R., Westfechtel, B.: Version Models for Software Configuration Management. ACM Computing Survey 30(2), 232–282 (1998)
3. Dittrich, K.R., Lori, R.: A. Version Support for Engineering Database Systems. IEEE Transaction on Softwate Engineering 14(4), 429–437 (1988)
4. Hunt, J. J., Vo, K. P., Ticky, W. F.: An Emperical Study of Delta Algorithms. In: Proceedings of ICSE'96 SCM-6 Workshop, LNCS, vol. 1167 (1996)
5. Katz, R.H.: Toward a Unified Framework for Version Modeling in Engineering Database. ACM Computing Surveys 22(4), 375–408 (1990)
6. Smith, H.: Business process management - the third wave: business process modeling language (bpml) and its pi-calculus foundations. Information and Software Technology 45(15), 1065–1069 (2003)
7. van der Aalst, W.M.P., Weijters, A.J.M.M.: Process mining: a research agenda. Computers in Industry 53(3), 231–244 (2004)
8. WfMC: Workflow Management Coalition the Workflow Reference Model. WfMC Standards, WfMC-TC00-1003 (1995), http://www.wfmc.org

Research on Architecture and Key Technology for Service-Oriented Workflow Performance Analysis

Bo Liu and Yushun Fan

Department of Automation, Tsinghua University, Beijing 100084, China
liubo03@mails.tsinghua.edu.cn, fanyus@tsinghua.edu.cn

Abstract. With the advent of SOA and Grid technology, the service has become the most important element of information systems. Because of the characteristic of service, the operation and performance management of workflow meet some new difficulties. Firstly a three-dimensional model of service is proposed. Then the characteristics of workflow in service-oriented environments are presented, based on which the workflow performance analysis architecture is described. As key technologies, workflow performance evaluation and analysis are discussed, including a multi-layer performance evaluation model and three kinds of performance analysis methods.

Keywords: workflow; service; performance evaluation; performance analysis.

1 Introduction

With the advent of new computing paradigms like Grid[1], SOA(Service-Oriented Architecture)[2] and P2P Computing[3], the service has become a vital element of information systems. Service and workflow are close related: workflow can be constituted with service, and workflow itself can be encapsulated into service as well. Thus it appears a new tendency that combines web service and workflow together. Because of the loosely coupled, autonomic and dynamic characteristics of service, Service-Oriented Workflow (SOWF) represents some new difficulties.

[4] regarded every activity in workflow as service. [5] discussed the interaction mechanism between web service and business process. [6] proposed a conceptual model of Web services workflow. [7] studied the modeling and implementation of organization centered workflows in the Web service environment. IBM developed a workflow management system “intelliFlow” based on SOA[8]. Grid workflow has also become a hotspot. Accordingly, the research of workflow performance management in service-oriented environments has evoked a high degree of interest.

2 Definition of Service

A new field Service Science has become the focus in recent years, but there still lacks a uniform definition of the service. In this paper, a ***service*** is defined as an IT-enabled

K.C. Chang et al. (Eds.): APWeb/WAIM 2007 Ws, LNCS 4537, pp. 540–545, 2007.

or IT-innovated functionality involving certain business process or activity, which is offered by a provider. A three-dimensional service model is shown in figure 1, which describes service from three views.

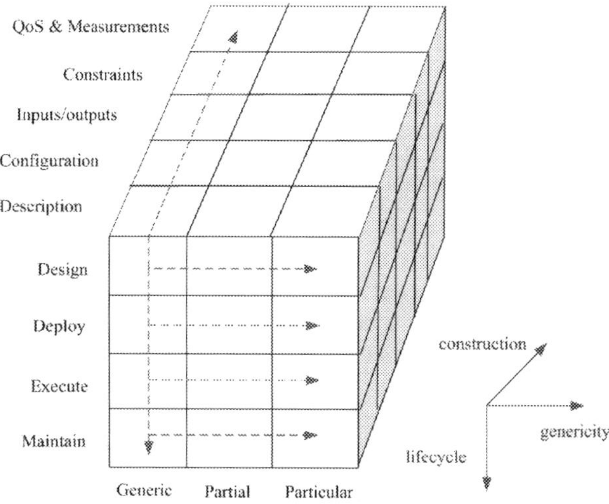

Fig. 1. Three-dimensional service model

1. *Construction dimension* consists of five parts. *Description* contains the name, identifier, domain of the service. *Configuration* includes related information used for configuring service, such as specific coherences. *Inputs/Outputs* describe the input and output parameters. *Constraints* mean pre- or post-conditions, rules, etc. *QoS & Measurements* provide the performance indicators of the service.

2. *Lifecycle dimension* involves four periods. During *Design* period, the service providers define the basic construction of the service. When the service is ready to release to the Network, its coherences need to be confirmed according to the user's demand. After *Deploy* period, the service is realized during *Execute* period. Finally in *Maintain* period, the service is monitored and managed, upgraded or withdrawn.

3. *Genericity dimension* encompasses three levels. *Generic* level comprises a collection of services that have the widest application in the representation of service domains. *Partial* level contains sets of partial services, each one being applicable to a specific domain. *Particular* level is concerned solely with one particular service domain. It should embody all necessary information in a way that can be used directly for its implementation. Three levels are ordered, in the sense that *Partial* level is a specialization of *Generic* level and *Particular* level is a specialization of *Partial* level.

In this model, service is constructed, realized and specialized gradually, and the characteristics of service are shown across-the-board.

3 Workflow in Service-Oriented Environments

Because of the loosely coupled, autonomic and dynamic characteristics of services, workflow in service-oriented paradigm also presents many new characteristics:

1. Services are implemented by workflow; Workflow is another kind of services.
2. Multiple processes interact with event/messages and share the resource or data.
3. The processes change dynamically along with the change of services. It requires ensuring the usability of services and selecting service components in real time which also result in the difficulty in evaluating the workflow performance.

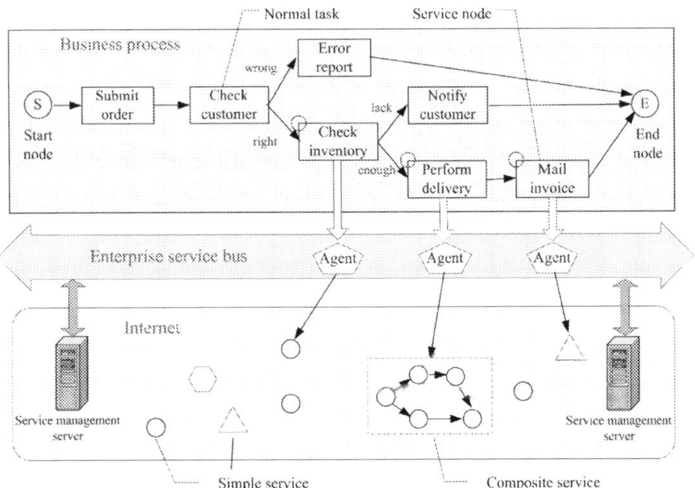

Fig. 2. Business process in service-oriented environments

Figure 2 illustrates the scenario of a business process in service-oriented environments. There are two kinds of activities in business processes, *normal task* and *service*. Each service node has a corresponding agent in Enterprise Service Bus, which is responsible for the execution of the service through querying the service management server. Consequently a simple service or composite service (constructed by composing several simple services according to certain regulation) in Network is selected to match the requirement of specific service.

4 Workflow Performance Analysis Architecture

In the highly autonomous, distributed environment, performance issue is of great importance. For example, composition of processes is according to performance requirements; selection and execution of processes is based on performance metrics; monitoring of processes assures compliance with initial performance requirements. So the evaluation and analysis of workflow performance have attracted great attention.

Figure 3 represents the architecture of business process performance analysis system. There are five layers shown as follows:

1. *Business operation layer* builds process model using modeling tools, and saves models in model DB (database) for process execution or simulation. Workflow execution data and log are stored in instance DB and log DB, while workflow simulation data and log are stored in simulation DB and log DB.

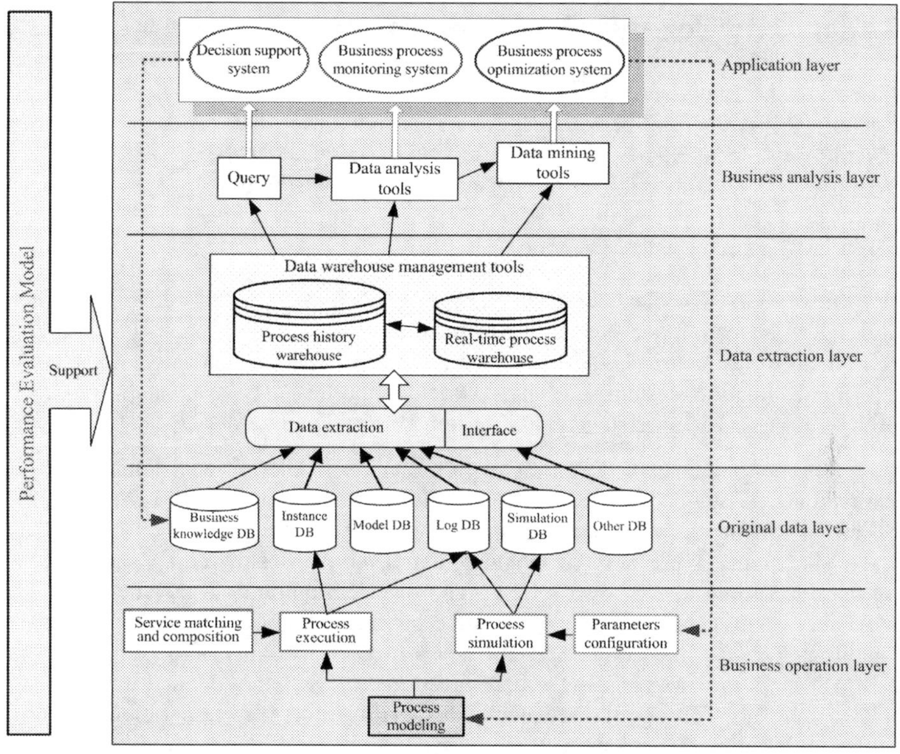

Fig. 3. Architecture of the business process performance analysis system

2. *Original data layer* includes several DBs which are sources of data analysis and data mining.

3. *Data extraction layer* extracts interesting information from original data by ETL (extract, transform and load) tools, and stores the information in business process information data warehouse. Meanwhile, considering the different data formats of data sources, an interface is added for transformation of data formats. Data warehouse management tools are responsible for maintaining data.

4. *Business analysis layer* has three kinds of tools for different purposes. Query obtains related data from data warehouse and generates reports for users. Data analysis tools utilize the function of OLAP, and provide services on data analysis and decision support in Application layer. Data mining tools operate deep analysis, predict future development tendency, and discover relationships and rules among data.

5. *Application layer* includes three systems facing end users. The result of business process optimization feeds back to process modeling and simulation. Business process monitoring system monitors the operation status and notifies exception to users. Decision support system assists decision-making activities and summarizes business rules and knowledge which are fed back to business knowledge DB.

In addition, the operation of business process performance analysis system is under the direction of business process performance evaluation model. The KPIs (Key Performance Indicator) in evaluation model are the basis of business analysis.

5 Key Technologies for Workflow Performance Analysis

Based on the above architecture, it can be concluded that workflow modeling, performance evaluation, performance analysis, monitoring and decision-making are several key technologies for workflow performance analysis, among which performance evaluation and performance analysis are most important.

5.1 Performance Evaluation Model

Workflow performance includes several aspects: time, cost, quality, reliability, agility etc. The evaluation of single performance indicator is obviously unilateral. Performance evaluation should be toward multi-indicators synthetically. Considering existing evaluation systems, a service-oriented performance evaluation model is proposed. The business system and IT system are divided into four layers.

1. The bottom layer is IT Infrastructure layer, and the corresponding KPIs are throughput, delay, bandwidth, etc. that reflect the performance of the network, operating system, facility and so on.
2. The higher layer Service Composition layer is used to composite required services according to this layer's KPIs. The service requirements include functional indicators which measure the function of the service and non-functional indicators (or Quality of Service, QoS) which reflect the non-functional quality of the service.
3. Business Process layer is the core layer, and its KPIs are divided into process-related indicators and activity-related indicators including cost, duration, resource utilization, waiting queue length etc.
4. The top layer Business Strategy layer faces end users and managers. Strategic goals vary from user to user, including maximum profit, minimum business costs, increased adaptability, flexibility, and efficiency, lower risk of system implementation, better governance and compliance, better customer satisfaction, etc.

The two bottom layers are from the IT view, while the other two from the business view. Each layer maps to neighbor layer with regard to KPIs' mapping.

Most existing performance evaluation models merely consider the mapping from business strategy layer to business process layer, or merely consider the performance of Network in IT layer, thus resulting in the disjoint of business and IT systems. The above performance evaluation model takes into account both systems synthetically and introduces Service Composition layer to present the particularity of services.

5.2 Performance Analysis Method

As far as performance analysis approach, there are mainly three methods at present.

Model analysis mainly utilizes different kinds of stochastic Petri-nets to build corresponding Continuous Time Markov Chain model or Queue Theory Model, based on which the performance parameters of the system can be obtained.

Workflow simulation could use special simulation tools for business process, simulation tools based on Petri-net, or discrete-event dynamic system simulation tools. Simulation in service-oriented environments must deal with multiple processes sharing common resources and organizations, with interaction of messages and events. Related research contains: simulating mechanism of business process with multi-entrance, resource scheduling algorithm, interaction between process model with organization model and resource model, etc.

Data analysis is based on history data or runtime data in model DB, instance DB and others. Because they are all relational databases, data warehouse and data mining technologies are used to analyze data. The mining and analysis of history data could master the rules of business operation, and the monitoring and analysis of run time data could master the operation status of business process in time. Existing data analysis methods mostly appear the deficiency of explicit delay of performance feedback, so we need to research real-time data analysis. Related studies involve: designing data model of the data warehouse, selecting or developing a most suitable data mining algorithm, the visualization of mining result etc. Considered data mining algorithms include classification, estimation, prediction, affinity grouping or association rules, clustering and statistics.

6 Conclusions and Future Work

Service-oriented workflow shows many new characteristics in aspects of execution mechanism and performance evaluation. In this paper, a three-dimensional service model is presented, the business process performance analysis architecture is proposed, and the key technologies for workflow performance analysis are discussed, especially workflow performance evaluation model and workflow performance analysis methods. It can be foreseen that service-oriented workflow will become the next generation of workflow and there are many fields deserve attention.

References

1. Foster, I., Kesselman, C. (eds.): The Grid: Blueprint for a New Computing Infrastructure, 2nd edn. Morgan-Kaufmann, Washington (2003)
2. Huhns, M.N., Singh, M.P. Service-Oriented Computing: Key Concepts and Principles. IEEE Internet Computing 9(1), 75–81 (2005)
3. Loo, A.W.: The Future of Peer-to-Peer Computing. Communications of the ACM 46(9), 56–61 (2003)
4. Zhou, X., Cao, J., Zhang, S.: Agent and Workflow Integration Based on Service. Computer Integrated Manufacturing Systems 10(3), 281–285 (2004)
5. Leymann, F., Roller, D., Schmidt, M.-T.: Web Services and Business Process Management. IBM Systems Journal 41(2), 198–211 (2002)
6. Xiao, Y., Chen, D., Chen, M.: Research of Web Services Workflow and its Key Technology Based on XPDL. In: Proc, IEEE International Conference on Systems, Man and Cybernetics (2004)
7. Zhao, X., Liu, C.: Supporting Relative Workflows with Web Services. In: Proc. 8th Asia-Pacific Web Conference 2006, LNCS 3841, pp. 680–691 (2006)
8. http://www-900.ibm.com/cn/software/websphere/solution/solu_intelliflow.shtml

The Study on Internet-Based Face Recognition System Using Principal Component Analysis

Myung-A Kang and Jong-Min Kim[1]

Dept. Computer Science & Engineering, KwangJu University, Korea
[1] Computer Science and Statistic Graduate School, Chosun University, Korea
makang@gwangju.ac.kr, mrjjoung@chosun.ac.kr

Abstract. The purpose of this study was to propose the real time face recognition system using multiple image sequences for network users. The algorithm used in this study aimed to optimize the overall time required for recognition process by reducing transmission delay and image processing by image compression and minification. At the same time, this study proposed a method that can improve recognition performance of the system by exploring the correlation between image compression and size and recognition capability of the face recognition system. The performance of the system and algorithm proposed in this study were evaluated through testing.

1 Introduction

The rapidly growing information technology has fueled the development in multimedia technique. However, demand for techniques involving searching multimedia data in a large scale database efficiently and promptly is still high. Among physical characteristics, face image is used as one of the reliable means of identifying individuals. Face recognition system has a wide range of applications such as face-based access control system, security system and system automation based on computer vision. Face recognition system can be applied to a large number of databases but requires a large amount of calculations. There are three different methods used for face recognition: template matching approach, statistical classification approach and neural network approach[1].Elastic template matching, LDA and PCA based on statistical classification approach are widely used for face recognition[2, 3]. Among these methods, statistical classification-based methods that require a small amount of calculations are most commonly used for face recognition. The PCA-based face recognition method identifies feature vectors using a Kahunen-Loeve transform. Given the proven feasibility of PCA as face recognition method, this study used PCA along with Kenel-based PCA[4, 5] and 2D-PCA[6].

The real-time face recognition system proposed in this study will be available in a network environment such as Internet. Each client is able to detect face images and forward detected images to remote server by compressing the images to reduce file size. However, the compression of facial images poses a critical risk because of the possibility of undermining image quality. This study investigated the effects of image compression and image size on recognition accuracy of the face recognition systems

K.C. Chang et al. (Eds.): APWeb/WAIM 2007 Ws, LNCS 4537, pp. 546–553, 2007.

based on PCA, KPCA, 2D_PCA algorithms and came up with the most effective real-time face recognition system that can be accessed across the Internet.

2 Internet-Based Face Recognition System

Based on the assumption that multiple variations of the face improves recognition accuracy of face recognition system, multiple image sequences were used. To reduce transmission delay, the images were compressed and minimized in the proposed system Fig 1.

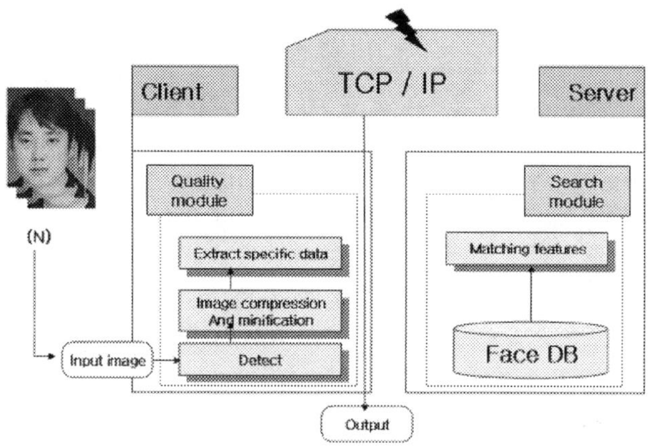

Fig. 1. Composition of the Proposed Face Recognition System

3 Face Recognition Algorithms

The real-time recognition accuracy was evaluated using PCA, KPCA and 2DPCA-based algorithms.

3.1 PCA (Principal Component Analysis)

The PCA-based face recognition algorithm calculates basis vectors of covariance matrix (C) of images in the following equation.

$$C = \frac{1}{M}\sum_{i=1}^{M}(X_i - m)(X_i - m)^T \tag{1}$$

Where x_i represents 1D vector converted from the i th image in a sequence of images in the size of $m \times n$. m indicates average of total M images of training face. Maximum number of eigenvectors ($m \times n$) of covariance matrix (C) of images are also calculated. Top K number of eigenvectors are selected according to descending eigenvalues and defined as basisvector (U)[7].

Feature vectors(w) of input image(x) are distributed as basis vectors in the vector space according to the following equation (2):

$$w = U^T (x - m) \tag{2}$$

3.2 2DPCA

While covariance matrix is computed from 1D images converted from input images for PCA, covariance matrix (G) is computed from 2D images and the average image for 2DPCA in the following equation (3) [6].

$$C = \frac{1}{M} \sum_{i=1}^{M} (A_i - E(A))(A_i - E(A))^T \tag{3}$$

Eigenvalues/eigenvectors of covariance matrix of images are calculated. Top k number of eigenvectors according to descending values are defined as basis vectors (U). Feature vector (w_i) of the i th image in the image sequence of face (A) are extracted in the equation (4). Characteristics of the face $B = [w_i, \ldots\ldots, w_k]$ can be extracted from w_i .

$$w_i = Au_i, \qquad i = 1,2,\ldots, K \tag{4}$$

Compared with covariance matrix used for PCA analysis, the covariance matrix derived from input images for 2DPCA analysis is smaller. This means that 2DPCA has the advantage of requiring less learning time [6].

3.3 KPCA (Kernel Principal Component Analysis)

KPCA face recognition algorithm involves converting input data on a face image into an image using nonlinear functions Φ . The converted images are reproduced as eigenvectors of the covariance matrix calculated for a set of nonlinear functions Φ and coefficients obtained during this process are used for face recognition in KPCA analysis.

For PCA, the covariance matrix can be efficiently computed by using kernel internal functions as the elements of the matrix [8,9]. In the equation (5), nonlinear function $\Phi(x)$ is substituted for input image x , and F was substituted for the feature space R^N .

$$\Phi : R^N \rightarrow F, x_k \rightarrow \Phi(x_k) \tag{5}$$

The training matrix and covariance matrix of images in the nonlinear space are presented in the equations (6) and (7). The nonlinear function $\tilde{\Phi}$ in the equation (6) must be mean zero by meeting the requirements for normalization.

$$C^{\Phi} = \frac{1}{l} \sum_{k=1}^{l} \tilde{\Phi}(x_k) \tilde{\Phi}(x_k)^T \tag{6}$$

$$\sum_{k=1}^{l} \tilde{\Phi}(x_k) = 0 \tag{7}$$

4 Face Recognition Rate

4.1 Changes in Recognition Rates with Image Compression

Image compression is essential to shorten transmission time through the Internet. However, compression approach also has a downside as it may hinder image quality. As presented in Fig 2, data file size was reduced but subjective image quality deteriorated as quantization parameters of compressed images increased. As a result, the recognition performance of the system is expected to decline.

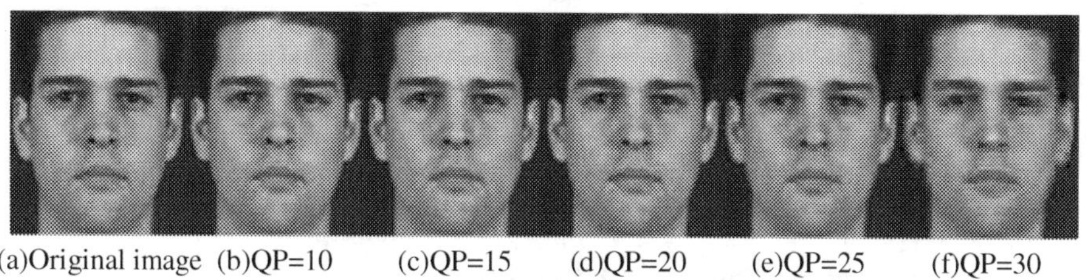

(a)Original image (b)QP=10 (c)QP=15 (d)QP=20 (e)QP=25 (f)QP=30

Fig. 2. Changes in visual image with QP value

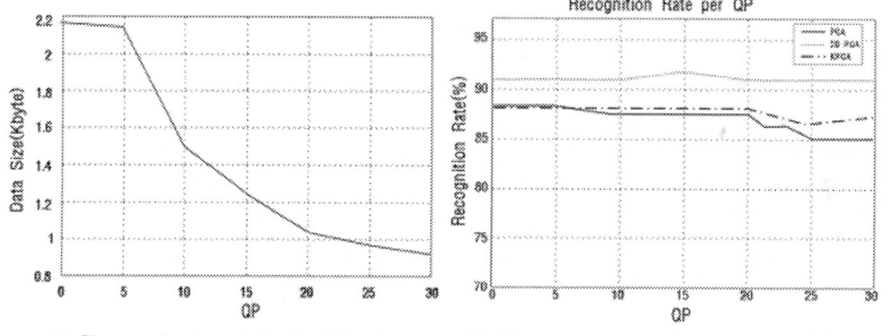

(a) Changes in data size with QP value (b) Changes in recognition rate with QP value

Fig. 3. Effects of image compression on data size and recognition performance

It is however found that recognition rate was not varied by the value of quantization parameters at the original image size of 92*122 pixels, as shown in Fig. 4 (b). Such a phenomenon was also confirmed by the distance between the original image and compressed image. Changes to the distance between the original image and compressed image with the value of quantization parameters are presented in Fig 4. There was a positive correlation between the distance and QP value.

In other words, the distance between the original image and compressed image (d_3) increased to 59, 305 and 689 as the value of QP reached 5, 15 and 30, respectively. However, changes to the distance is meager, so the effects of compression can be eliminated, meaning almost no changes in recognition performance but a significant reduction in data file size to be forwarded. In conclusion, transmission time can be reduced without affecting recognition performance of the system.

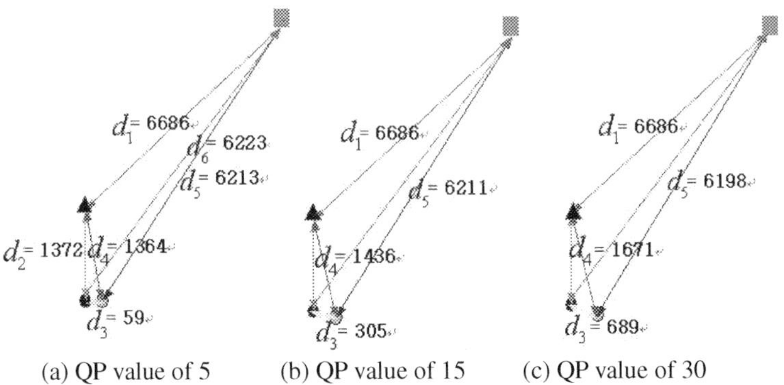

(a) QP value of 5 (b) QP value of 15 (c) QP value of 30

Fig. 4. Changes to the distance between the original image and compressed image with QP value. (Triangle in an image with the minimum distance, rectangular indicates the closest image to other class, green represents the original image and blue represents an compressed image.).

4.2 Changes in Recognition Rates with Image Size

The size of image has an impact on transmission time and computational complexity during the recognition process. Images were passed through a filtering stage to get the low-low band using wavelet transform. Image size (S_R) is defined in the following equation:

$$S_R = \frac{S_{origin}}{4^{(R)}} \tag{8}$$

For instance, the original image is reduced to 25% of its actual size when R equals to 1. Effects of image filtering are presented in Fig 5.

Effects of image size on time required for learning and recognizing images and recognition performance are presented in Fig 6. As shown in Fig 6 (a) and (b), the time required for learning and recognizing images drastically fell as the size of the image was reduced. The recognition rate also dropped when R was less than 4 but stopped its decline and remained almost unchanged when R was 4 or above. In fact, it is difficult to recognize features of the image with eyes when the image size became smaller. This is due to the fact that image size reduction involves reducing the number of faces in original images and the size of coefficient vectors.

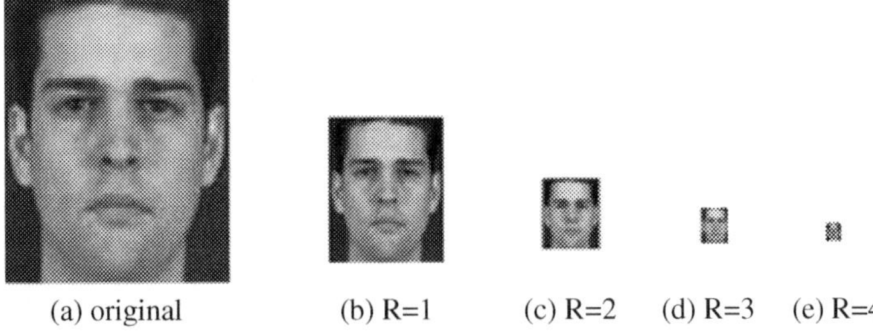

(a) original (b) R=1 (c) R=2 (d) R=3 (e) R=4

Fig. 5. Effects of image filtering

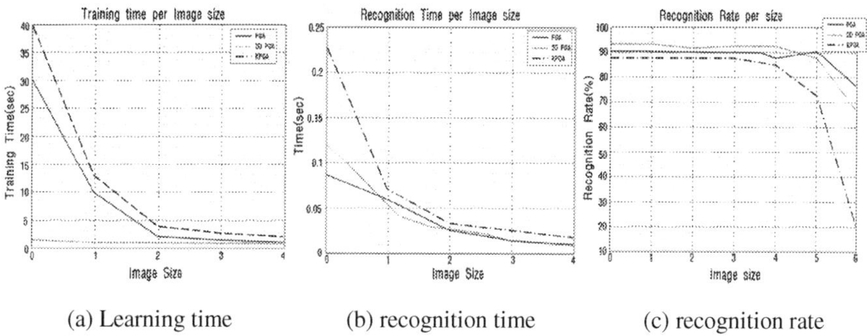

(a) Learning time (b) recognition time (c) recognition rate

Fig. 6. Effects of image size on time required for learning and recognizing images and recognition rate

5 Majority-Making-Decision Rule

The present study found that recognition rates remained almost unchanged in response to certain degrees of compression and minification of images. Based on these findings, the study proposes a face recognition algorithm capable of improving recognition performance of the system. The algorithm is designed to calculate a recognition rate based on the majority, composition and decision-make rules when multiple input images are used. A theoretical estimation of recognition rate (P_m) can be calculated in the following equation on the condition that more than half of transmitted images were matched with image models.

$$P_m = \sum_{k=\lfloor n/2 \rfloor}^{n} \binom{n}{k} p_s^k (1 - p_s)^{n-k} \tag{9}$$

Where n is the number of images forwarded, P_s is the average probability of face recognition $\binom{n}{k}$ is possible number when k number of features are recognized among n number of features. $\lfloor x \rfloor$ is a fixed number that is smaller than x but the largest value. For instance, when P_s is 0.94 and three images are forwarded, the value of P_m is 0.99 based on the majority, composition and decision-making rules.

The proposed algorithm was tested with 3, 5 and 7 images in PCA-, KPCA- and 2DPCA-based real-time face recognition systems. According to the equation (9), the saturation estimation was achieved when P_s was roughly larger than 0.7 and n equaled to 5. Five images were therefore used for the test of the proposed system. Test results are presented in Fig 7.

6 Experimental Results

The composition of the proposed system is presented in Fig 1. For the test, Chosun-DB (50 class, 12 images in the size of 60*1 20) and Yaile DB were used. The test was performed in a internet environment, and the optimized value of quantization parameters was applied. Experimental results are presented in Table 1.

The performance of real-time face recognition system is measured by the length of time required for learning and recognizing face images, total amount of data transmitted and recognition rate. The KPCA-based proposed system increased recognition rate by 14% and reduced the time required for recognizing images by 86%. The time required for learning images was reduced when smaller sizes of images were used. The 2DPCA-based proposed system showed the recognition rate of 95.4%, compared with 91.3% of the existing 2DPCA-based system. Besides, a 78% decrease was observed in learning time and a 24% decrease in recognition time in the same system. The amount of data transmitted was reduced to 3610 bytes from 19200 bytes, leading to a 81 reduction in transmission delay.

Table 1. Comparison of performance between proposed and existing systems

Algorithm	Recognition Rate(%)	Training Time	Recognition Time(sec)
PCA	88.7	28	1.5
Proposed system(PCA)	92.0	16	1.0
2D PCA	91.3	27	0.5
Proposed system(2D PCA)	95.4	6	0.38
KPCA	79.0	2.4(hour)	36
Proposed system(KPCA)	93.5	0.33(hour)	5

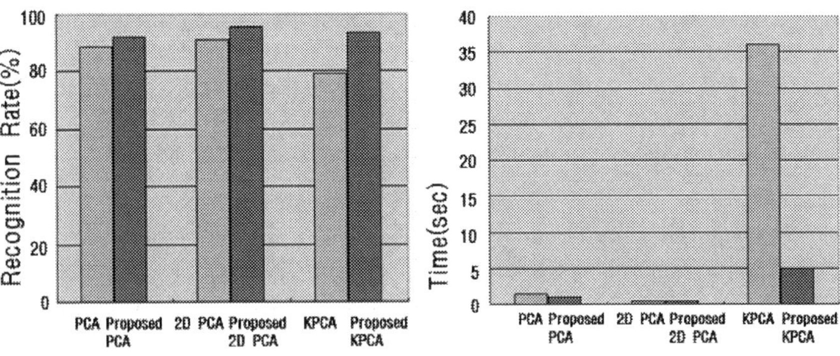

Fig 7. Effects of image compression and minification on recognition rate and time required for the recognition process of five images

7 Conclusion

This study proposed a real-time face recognition system that can be accessed across the internet. The test of the proposed system demonstrated that image filtering and image compression algorithms reduced transmission delay and the time required for learning and recognizing images without hindering recognition accuracy of the system. This study used multiple input images in order to improve the recognition performance of the system, and the proposed real-time face recognition system proved robust on the internet. Although the system was based on PCA algorithms, it can be integrated with other face recognition algorithms for real-time detection and recognition of face images.

References

1. Jain, A.K., Duin, R.W., Mao, J.: Statistical pattern recognition: a review. IEEE Trans.on Pattern Analysis and Machine Intelligence 22(1), 4–37 (January 2000)
2. Yambor, W.: Analysis of PCA based and Fisher Discriminant-based Image Recognition Algorithms. Technical Report CS-00-103, Computer Science Department Colorado State University, 7 (2000)
3. Murase, H., Nayar, S.K.: Visual Learning and Recogntion 3-Dobject from appearance, international journal of Computer Vision, vol. 14 (1995)
4. Zhang, Y., Liu, C.: Face recognition using kernel principal component analysis and genetic algorithms. In: Proceedings of the, 12th IEEE Workshop on Neural Networks for Signal Processing, pp.337–343 (2002)
5. Yang, J., Zhang, D., Frangi, A.F., Yang, J.Y.: Two-Dimensional PCA: A New Approach to Appearance-Based Face Representation and Recognition, IEEE Trans. Pattern Analysis and Machine Intelligence, vol.26(1) (January 2004)
6. Bourel, F., Chibelushi, C.C., Low, A.A.: Robust facial expression recognition using a state-based model of spatially localised facial dynamics. In: Proceedings of Fifth IEEE International Conference on Automatic Face and Gesture Recognition, pp.106–111 (2002)
7. Georghiades, A.S., Belhumeur, P.N., Kriegman, D.J.: From Few to Many: Illumination Cone Models for Face Recognition under Variable Lighting and Pose. IEEE Trans. Pattern Analysis and Machine Intelligence 23(6), 643–660 (June 2001)
8. Viola, P., Jones, M.: Rapid Object Detection using a Boosted Cascade of Simple Features, Computer Vision and Pattern Recognition, Vol. 1, pp. 511–518, 12 (2001)
9. Yang, H.-S., Kim, J.-M., Park, S.-K.: Three Dimensional Gesture Recognition Using Modified Matching Algorithm, Lecture Notes in Computer Science LNCS3611 pp. 224–233 (2005)
10. Belhumeur, P.N., Hepanha, J.P., Kriegman, D.J.: Eigenfaces vs. Fisherfaces: Recognition Using Class Specific Linear Projection. IEEE Trans. Pattern Analysis and Machine Intelligence 19(7), 711–720 (July 1997)
11. Georghiades, A.S., Belhumeur, P.N., Kriegman, D.J.: From Few to Many: Illumination Cone Models for Face Recognition under Variable Lighting and Pose. IEEE Trans. Pattern Analysis and Machine Intelligence 23(6), 643–660 (June 2001)

Semantic Representation of RTBAC: Relationship-Based Access Control Model*

Song-hwa Chae[1] and Wonil Kim[2,**]

[1] Korea IT Industry Promotion Agency, Seoul, Korea
shchae@software.or.kr
[2] College of Electronics and Information, Sejong University, Seoul, Korea
+82-2-3408-3795
wikim@sejong.ac.kr

Abstract. As Internet expands, many enterprise systems require managing security policies in a distributed environment in order to complement any authorization framework. The eXtensible Markup Language (XML) allows the system to represent security policy properly in a heterogeneous, distributed environment. In access control model, the security problem exists not only on subject but also on object side too. Moreover, when the system is expanded to ubiquitous computing environment, there are more privacy invasion problems than current Internet services. Proper representation of relationship in access control mechanism can be a solution for privacy invasion problem. In this paper, we develop XML Document Type Definition (DTD) and XML schema for representing the schema of the relationship-based access control model. This model supports object privacy since it introduces a new constraint called *relationship* between subject and object. It supports more constraints on object's policy than current Role-based Access Control Model (RBAC) does.

1 Introduction

Recently, many enterprises are growing toward a heterogeneous, distributed environment. This has motivated many enterprises to adopt their computing services for efficient resource utilization, scalability and flexibility. The eXtensible Markup Language (XML) is a good solution for supporting the policy enforcement in a heterogeneous, distributed environment. In addition, as wireless networking has become more common, ubiquitous computing begins to receive increasing attention as Internet's next paradigm [1]. Invisible and ubiquitous computing aims at defining environments where human beings can interact in an intuitive way with surrounding objects [2]. The ubiquitous service should consider more frequent movement than current Internet services since the user can use various services anytime, anywhere. For these services, ubiquitous system needs to control a lot of information, which it leads to lots of privacy invasion problems.

* This Paper is supported by Seoul R& BD Program.

** Corresponding author.

K.C. Chang et al. (Eds.): APWeb/WAIM 2007 Ws, LNCS 4537, pp. 554–563, 2007.

Even though several security mechanisms are suggested for user privacy, none is yet de facto standard model. In access control model, this privacy invasion problem exists not only on subject but also on object. The representing relationship in Access control mechanism can be a solution for privacy invasion problem. Most access control decisions depend on subject, *who you are*. The object's access control policy has been ignored in the most of access control models including RBAC. For example, in a hospital model, normally doctors can read patient's record even though the patient does not want that all doctors access his/her record. In order to preserve the patient privacy, only the doctor who has permission from the patient is able to access the patient record. Another example is in university model, a professor should be able to use read-student-record service only if the professor is advisor of the student. In this paper, we develop XML Document Type Definition (DTD) and XML schema for representing the schema of the Relationship-based Access Control Model (RTBAC) by introducing a relation between subject and object. This method presents the privacy by asking not only who you are but also what your relationship is. It supports more constraints on object's policy than current RBAC. Moreover, the proposed model has the strength value of relationship for dynamic privacy service.

This paper is organized as follows. Chapter 2 surveys related works. Chapter 3 discusses the XML DTD and schema for Relationship-based Access Control Model for ubiquitous computing environment. In Chapter 4, RTBAC is applied to RBAC model and shows various example scenarios. Chapter 5 concludes with future works.

2 Related Works

Access control refers to controlling access to resources on a computer or network system. There have been many researches on access control mechanism in information security. Access control mechanism is categorized into three areas, such as mandatory access control (MAC), discretionary access control (DAC) and role-based access control (RBAC). MAC is suitable for military system, in which data and users have their own classification and clearance levels respectively. This policy indicates which subject has access to which object. This access control model can increase the level of security, because it is based on a policy that does not allow any operation not explicitly authorized by an administrator. DAC is another access control method on objects with user and group identifications. A subject has complete control over the objects that it owns and the programs that it executes. RBAC has emerged as a widely acceptable alternative to classical MAC and DAC [4] [5]. It can be used in various computer systems. RBAC is one of famous access control model [7]. RBAC has shown to be policy neutral [8] and supports security policy objectives as the least privilege and static and dynamic separation of duty constraints [9]. In order to protect abusing rights, the user must have the least privilege. For that reason, some researchers extend RBAC model that has constraints - time and location are suggested for the least privilege service such as TRBAC (Temporal RBAC) [8] and SRBAC (Spatial RBAC) [10]. However, these extended models do not consider privacy service. Recently, Byun et al [6] suggested Purposed Based Access Control model for privacy protection. This model focused on access purpose for complex data. It defined purpose tree and the system defined access permission depend on this tree and access purpose. There are many RBAC implementations in commercial products. Therefore, the access control model for ubiquitous computing environment should consider these factors such as location, time, role and relationship. For the reasons, it is clear that current RBAC model is not suitable for ubiquitous computing environment and new access control model should be developed.

In most systems, simple access control is maintained via access control lists (ACLs) which are lists o f users and their access rights. This use of ACLs has problem for a large system. It has a large number of users, and each one has many permissions. The system implies a very large number of user/permission associations that have to be managed. Thus, when a user takes on different responsibilities within the enterprise, reflecting these changes entails a thorough review, resulting in the selective addition or deletion of user/permission associations on all servers. The larger the number of user/permission associations to be managed, the greater the risk of maintaining residual and inappropriate user access rights. [11].

3 The Relationship-Based Access Control Model (RTBAC)

3.1 Relationship and Relationship Hierarchy

There are many common examples where access decisions must include other factors such as user attributes, object attributes, user relationships to other entities. The relationships among entities associated with an access decision are often very important. Roles of RBAC can be used to represent relationships. However, using roles to express relationships may be inefficient and/or counter intuitive. When roles cannot be used to represent relationships, it is common to program access decision logic directly into an application [3]. Therefore, the relationship component should be included in the access control model.

Relationship (RT) of the relationship-based access control model is the relation between Subject Entity (SE) and Object Entity (OE). For example, the Relationship (RT) may be friend, classmate, club-member, tutor and attending doctor. This Relationship (RT) is determined by OE and administrator. Subject Entity (SE) is subjects such as users and devices that have a right to use some services in ubiquitous computing environment. Object Entity (OE) is objects such as users and devices that are targets of service. Figure 1 shows the relation among subject entity, object entity and relationship.

Fig. 1. The relation among Subject Entity (SE), Object Entity (OE) and Relationship (RT)

Definition 1. (*Relationship*) Relationship *rt* is represented through a symbolic formalism and can be expressed as $rt = \{ rt_1, rt_2, ..., rt_i\}$ where *i* is integer.

Example 1. Let *rt* = *{friend, classmate, club-member, tutor}*. It represents the set of relationships such as friend, classmate, club-member and tutor.

The Relationship is able to represent in hierarchical structure. It helps easy representing of policy in complex environment. Figure 2 shows an example relationship hierarchy. In this figure, relationship 'Friends' has tree different relationships such as 'School Friends', 'Social Friends' and 'Child Friends'. 'School Friends' is also has tree different relationships. For instance, let an object Alice want to permit accessing her current location to her friends in school. In set expression that is normally used in ubiquitous computing model, the expression is {Middle School Friends, High School Friends, University Friends}. The elements of relationship set increases when the number of permitted relationship is increased. However, hierarchical expression is able to represent this in simple way. In the proposed access control model, Alice's object policy expression is {School Friends}. It reduces the number of expression and operation.

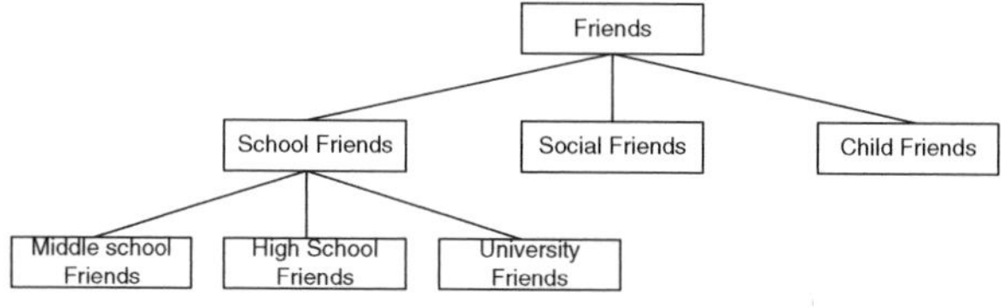

Fig. 2. An example of relationship hierarchy

Definition 2. (*Relationship Hierarchy*) A relationship hierarchy is defined as a 2-tuple $[RT, \leq]$, where RT is a set of relationships and $\leq$ is a partial order defined over RT. Let $rt_i, rt_j \in RT$. We can say that rt_i is a specialized relationship of rt_j if $rt_i \leq rt_j$.

Figure 3 and 4 show XML DTD and schema to represent relationship hierarchy in Relationship-base access control model. The example relationship hierarchy is converted to XML in Appendix A.

```
<!ELEMENT Relationship_Tree (Relationship)*>
<!ELEMENT Relationship (Name, Cardinality?, (Parent_relation?)*,
(Child_relation?)*)>
<!ELEMENT Name (#PCDATA)>
<!ELEMENT Cardinality (#PCDATA)>
<!ELEMENT Parent_relation (#PCDATA)>
<!ELEMENT Child_relation (#PCDATA)>
```

Fig. 3. Relationship_Tree.dtd

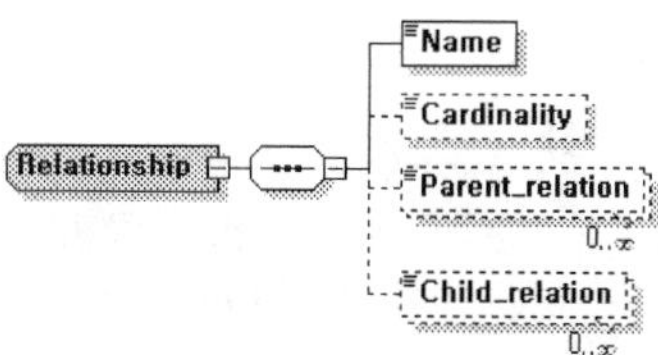

Fig. 4. XML Schema of Relationship Hierarchy

3.2 The Strength Value of the Relationship

In the access control model, a relationship rt_i is the same meaning to everyone. For example, all subjects have the same permission if subjects are members of object's relationship set 'Social Friends'. However, in a real environment, every person is not same relationship with object even if they are friends. For representing various relationships, the proposed model represents the strength value of the relationship. Thus, a relationship (RT) should be represented from strong relationship to weak relationship. We defined strength value of relationship that is number expression from 0 to 1. If the strength value is 0, it means week relationship. If the strength value is 1, it is strong relationship.

Definition 3. (*Relationship strength value*) A strength value of relationship is defined as a 2-tuple [*RT, S*], where *RT* is a set of relationships and *S* is a strength value of relationship in [0, 1]. 0 is week and 1 is strong.

4 Semantic Representation of RTBAC

4.1 Access Control Mechanism

RBAC model is well-known access control model therefore the RTBAC model is applied to RBAC model. In order to represent relationship among entities, new concepts such as Object Entity (OE) and Relationship (RT) are newly defined in RBAC model. We expand the meaning of subject (user or devices) to Subject Entity (SE) that has a right to use some services in ubiquitous computing environment. Object Entity (OE) is objects such as users and devices that are targets of service.

When a user tries to access some resource or use some services, the system makes an access-permit decision depending on relationship policy. OE has relationship policy table that consist of member, relationship and strength value of relationship. The system has a default role-permission table, which shows a default access control policy of role. For the more complex and specific access control, the OE can have another table that represents permitted role condition table.

Definition 4. (*Relationship Policy*) An OE has a relationship table that has members, relationship and strength value of relationship. One record is defined as a 3-tuple [*M, RT, S*], where *M* is a set of subject, *RT* is a set of relationships and *S* is a strength value of relationship in [0, 1].

Definition 5. (*Default Access Control Policy*) A default access control policy for a role is defined as a 3-tuple [*R, P, RT, SE*], where *R* is a set of roles, *P* is a set of permissions (or services), *RT* is a set of relationships and *SE* is a strength value of relationship expression. *SE* is consisted of operation and strength value. The operation are { ==, !=, >=, <=, < , > , and , or }.

Definition 6. (*Specific Relationship Policy*) An OE has a specific relationship table that has permission, relationship and strength value of relationship expression. One record is defined as a 3-tuple [*P, RT, SE*], where *P* is a set of permissions (or services), *RT* is a set of relationships and *SE* is a strength value of relationship expression. *SE* is consisted of operation and strength value. The operation are { ==, !=, >=, <=, < , > , and , or }. If a specific relationship policy is [find-friends, school friends, >= 0.8], we can say that OE only permits the find-friends service where relationship is 'School Friends' and the strength value of relationship is equal or lager than 0.8.

XML DTD and Schema for Relationship Policy, Default Access Control policy and Specific Access Control Policy are illustrated in Figure 5 and 6, Figure 7 and 8 and Figure 9 and 10 respectively. The system defines *Default Access Control Policy* and OE also has *Specific Relationship Policy*. Some cases, *Default Access Control Policy* conflicts with *Specific Relationship Policy*. The proposed model checks *Default Access Control Policy* first and then *Specific Relationship Policy*. Therefore, *Specific Relationship Policy* is normally specific subset of *Default Access Control Policy*. If the *Specific Relationship Policy* is in collision with *Default Access Control Policy*, this access is denied.

```
<!ELEMENT Relationship_Set (Relationship)+>
<!ELEMENT Relationship (Name, Member+)>
<!ELEMENT Name (#PCDATA)>
<!ELEMENT Member (Mem_name,Strength)>
<!ELEMENT Mem_name (#PCDATA)>
<!ELEMENT Strength (#PCDATA)>
```

Fig. 5. Relstionship_policy.dtd

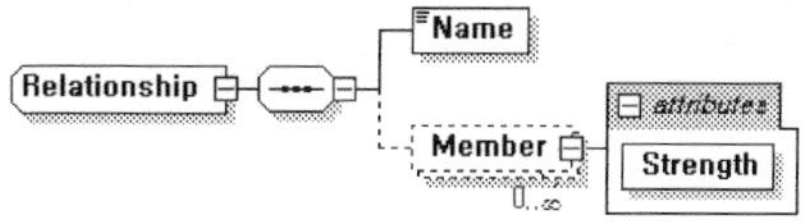

Fig. 6. XML Schema of Relationship policy

```
<!ELEMENT Default_Policy_Set (Default_Policy)+>
<!ELEMENT Default_Policy (Permission_Relationship+)>
<!ATTLIST Default_Policy name ID #REQUIRED>
<!ELEMENT Permission_Relationship ((Relationship_name+,Conditions)+)>
<!ATTLIST Permission_Relationship name ID #REQUIRED>
<!ELEMENT Relationship_name (#PCDATA)>
<!ELEMENT Conditions (Operation+)>
<!ATTLIST Conditions Lp (And|Or) #IMPLIED>
<!ELEMENT Operation (Value)>
<!ATTLIST Operation Op (equal|Not|GTequal|LSequal|GT|LS) #REQUIRED>
<!ELEMENT Value (#PCDATA)>
```

Fig. 7. Default_Access_Control_Policy.dtd

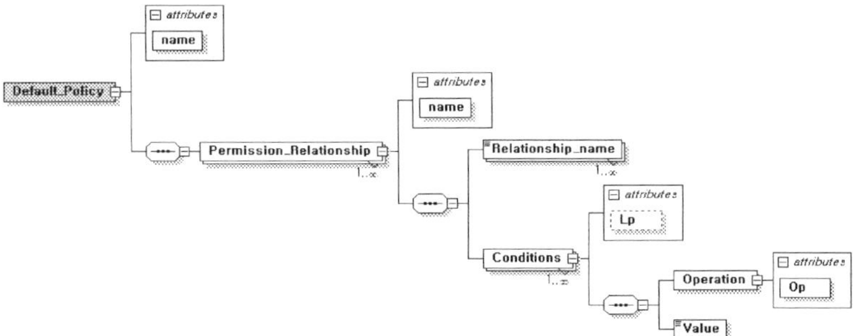

Fig. 8. XML Schema of Default Access Control Policy

```
<!ELEMENT Specific_Policy_Set (Specific_Policy)+>
<!ELEMENT Specific_Policy (Permission_Relationship+)>
<!ELEMENT Permission_Relationship ((Relationship_name,Conditions)+)>
<!ATTLIST Permission_Relationship name ID #REQUIRED>
<!ELEMENT Relationship_name (#PCDATA)>
<!ELEMENT Conditions (Operation+)>
<!ATTLIST Conditions Lp (And|Or) #IMPLIED>
<!ELEMENT Operation (Value)>
<!ATTLIST Operation Op (equal|Not|GTequal|LSequal|GT|LS) #REQUIRED>
<!ELEMENT Value (#PCDATA)>
```

Fig. 9. Specific_Access_Control_Policy.dtd

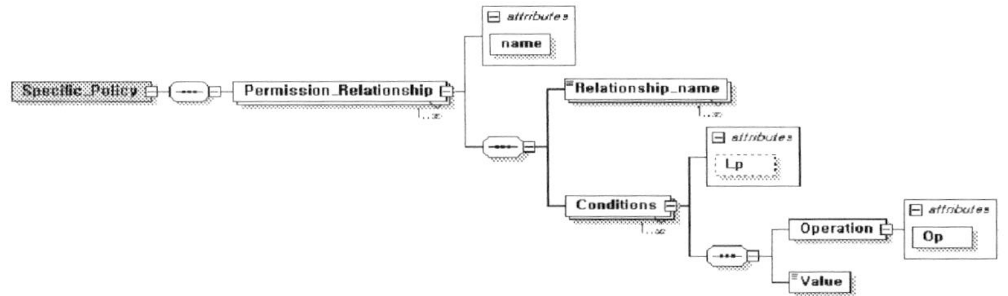

Fig. 10. XML Schema of Specific Access Control Policy

4.2 Example Cases

In the proposed access control model, the role is enabled or disabled according to relationship and strength value of relationship. Many SE can have the same role but the role activation may be different depending on their relationship and strength value of relationship. It also normally happens in the following cases in ubiquitous computing environment as university and hospital.

First case is in university model. There are many ubiquitous computing services in university such as read-student-record, mobile campus, mobile club and others. For instance, read-student-record service is a service that it returns students record, thus professor role has read-student-record service permission. It means that every professor can access all of student record. It makes privacy problem. In order to solve this problem, the RTBAC model has new concepts such as relationship and strength value of relationship. In this example, the object is able to produce *Relationship Policy* and *Specific Relationship Policy*. Figure 11 shows an example policy of this case. . Appendix B shows XML documents for above examples.

Alice's policy

- *Default Access Control Policy* : [*Professor, read-student-record, Professor-student, >=0*]
- *Relationship Policy* : [*Bob, Professor-student, 0.9*], [*Chris, Professor-student, 0.3*]
- *Specific Relationship Policy* : [*read-student-record, Professor-student, >= 0.5*]

Fig. 11. An example policy for find-friends service

In Figure 11, if Bob, who is advisor of Alice, tries to access Alice's record, the system permits it since his role and his relationship with Alice are fully satisfied with *Relationship Policy*, *Default Access Control Policy* and *Specific Relationship Policy*. However, if Chris, who is also professor, tries to use this service, the system denies it because his relationship policy is not satisfied with *Specific Relationship Policy*.

Second case is in a hospital model. There are various ubiquitous computing services in hospital such as mobile medical treatment, medical examination, prescription and others. For these services, the user should be able to access patient's medical record. However, the medical record is very private data, the system supports protecting privacy problem. For instance, read-Patient-Record service is only permitted to the attending physician. The attending physician can be defined either by the patient or hospital. Figure 12 shows an example policy of this case.

David's policy

- *Default Access Control Policy* : [*Doctor, read-Patient-Record, Doctor-Patient >=0*]
- *Relationship Policy* : [*Emily, Doctor-Patient, 0.8*], [*Frank, Doctor-Patient, 0.3*]
- *Specific Relationship Policy* : [*read-Patient-Record, Doctor-Patient, >= 0.7*]

Fig. 12. An example policy for read-Patient-Record service

In Figure 12, a patient David wants his patient record to be read only by attending doctor. Even though Emily and Frank are doctors in hospital, Frank is not allowed to read David's record because he is not an attending doctor. It is not satisfied with *Specific Relationship Policy*

5 Conclusion

In this paper, we developed XML Document Type Definition (DTD) and XML schema for the Relationship-based Access Control Model. This model solves a privacy invasion problem, which is more frequently involved than current Internet services, in ubiquitous computing environment. This model defines new concepts such as relationship and the strength value of relationship. The relationship is relation between subject and object. The strength value of relationship is degree of strength relation between subject and object. It supports more constraints on object's policy than current RBAC. It is able to represent complex and specific policy. Our specification language provides compact representation of access control policy to protect object privacy. We applied the RTBAC model to RBAC model that is well-known access control model and developed example XML documents.

References

1. Stajano, F., Anderson, R.: The Resurrecting Duckling: Security Issues for Ubiquitous Computing, IEEE security and Privacy (2002)
2. Bussard, L., Roudier, Y.: Authentication in Ubiquitous Computing, UbiCom2002 (2002)
3. Barkley, J., Beznosov, K., Uppal, J.: Supporting Relationships in Access Control Using Role Based Access Control. In: Proceedings of the Fourth ACM Workshop on Role-Based Access Control, pp. 55–65 (1999)
4. Choun, E.H.: A Model and administration of Role Based Privileges Enforcing Separation of Duty. Ph.D. Dissertation, Ajou University (1998)
5. Ahn, G., Sandhu, R.: Role-Based Authorization Constraints Specification. ACM Transactions on Information and System Security 3(4), 207–226 (2000)
6. Byun, J., Bertino, E., Li, L.: Purposed based access control of complex data for privacy protection.CERIAS Tech Report 2005, 12 (2005)
7. Ahn, G., Sandhu, R.: Role-Based Authorization Constraints Specification. ACM Transactions on Information and System Security 3(4), 207–226 (2000)
8. Bertino, E., Bonatti, P.A., Ferrari, E.: A Temporal Role-Based Access Control Model. ACM Transactions on Information and System Security 4(3), 191–223 (2001)
9. Ferraiolo, D.F., Sandhu, R., Gavrila, E., Kuhn, D.R., Chandramouli, R.: Proposed NIST Standard for Role-Based Access Control. ACM Transactions on Information and System Security 4(3), 224–274 (2001)
10. Hengartner, U., Steenkiste, P.: Implementing Access Control to People Location Information. In: proceedings of 9th ACM Symposium on Access Control Models and Technologies, pp. 11–20 (2004)
11. Ferraiolo, D.F., Barkley, J.F., Kuhn, D.R.: A Role-Based Access Control Model and Reference Implementation Within a Corporate Intranet. ACM Transactions on Information and System Security 2(1), 34–64 (1999)
12. eXtensible Markup Language, http://www.w3.org/XML/

Appendix: A


```
<?xml version="1.0"?>
<!DOCTYPE Relationship_Tree SYSTEM "Relationship_Tree.dtd">
<Relationship_Tree>
<Relationship>   <Name> Friends</Name>
   <Cardinality>3</Cardinality>
   <Child_relation>School Friends</Child_relation>
   <Child_relation>Social Friends</Child_relation>
   <Child_relation>Child Friends</Child_relation> </Relationship>
<Relationship> <Name>School Friends</Name>
   <Cardinality>3</Cardinality>
   <Parent_relation>Friends</Parent_relation>
   <Child_relation>Middle School Friends</Child_relation>
   <Child_relation>High School Friends</Child_relation>
   <Child_relation>University Friends</Child_relation> </Relationship>
```

```
<Relationship> <Name>Social Friends</Name>
   <Parent_relation>Friends</Parent_relation> </Relationship>
<Relationship> <Name>Child Friends</Name>
   <Parent_relation>Friends</Parent_relation> </Relationship>
<Relationship> <Name>Middle School Friends</Name>
   <Parent_relation>School Friends</Parent_relation> </Relationship>
<Relationship> <Name>High School Friends</Name>
   <Parent_relation>School Friends</Parent_relation> </Relationship>
<Relationship> <Name>University Friends</Name>
   <Parent_relation>School Friends</Parent_relation> </Relationship>
</Relationship_Tree>
```

Appendix: B

```
<!--Relationship policy for find-friends example --!>
<?xml version="1.0"?>
<!DOCTYPE Relationship_Tree SYSTEM "Relationship_Policy.dtd">
<Relationship_Set>
<Relationship> <Name>Professor-student</Name>
        <Member>
                <Mem_name>Bob</Mem_name>
                <Strength>0.9</Strength>
        </Member>
        <Member>
                <Mem_name>Chris</Mem_name>
                <Strength>0.3</Strength>
        </Member> </Relationship>
</Relationship_Set>

<!--Default access control policy for find-friends example --!>
<?xml version="1.0"?>
<!DOCTYPE Relationship_Tree SYSTEM "Default_Access_Control_Policy.dtd">
<Default_Policy_Set>
   <Default_Policy name="Professor">
        <Permission_Relationship name="read-student-record">
                <Relationship_name>Professor-student</Relationship_name>
                <Conditions> <Operation Op="GTequal">
                                <Value>0.0</Value>
                        </Operation> </Conditions>
        </Permission_Relationship>
   </Default_Policy>
</Default_Policy_Set>

<!--Specific access control policy for find-friends example --!>
<?xml version="1.0"?>
<!DOCTYPE Relationship_Tree SYSTEM "Specific_Access_Control_Policy.dtd">
<Specific_Policy_Set>
   <Specific_Policy>
        <Permission_Relationship name="read-student-record">
                <Relationship_name>Professor-student</Relationship_name>
                <Conditions> <Operation Op="GT">
                                <Value>0.5</Value>
                        </Operation>
                </Conditions>
        </Permission_Relationship>
   </Specific_Policy>
</Specific_Policy_Set>
```

A Code-Based Multi-match Packet Classification with TCAM

Zhiwen Zhang and Mingtian Zhou

College of Computer Science and Engineering,
University of Electronic Science and Technology of China,
ChengDu 610054, P.R. China
zwenzhang@tom.com, mtzhou@uestc.edu.cn

Abstract. Packet classification, especially multi-match packet classification has become a challenging problem in Network Intrusion Detection Systems(NIDSs). Because of the speed limitation of software based packet classification algorithm, it is necessary to apply a hardware or software with hardware assist solution. Ternary Content Addressable Memory(TCAM) is being used more often to solve packet classification problem. However, commercially available TCAM is more expensive, consumes more power and dissipates more heat, compared to tranditional memory. In this paper, to mitigate the disadvantages of TCAM, we describe a code-based multi-match packet classification scheme, which transmits the product of rule's fields into summation and requires less number and narrow width of TCAM entries. The classify speed of our scheme is deterministic, which is determined by the dimension of the classifier. The simulation results show that our scheme is superior in saving memory to the schemes presented previously, such as, MUD[5], SSA[6] and Geometric Intersection-based[4] solutions.

Keywords: Code-based, Multi-match Packet Classification, TCAM, NIDS.

1 Introduction

Multi-match packet classification is indispensable to Network Intrusion Detection Systems(NIDSs), in which all matching rules need to be provided. Traditional software based packet classification schemes, such as, Grid of Tries[1], FIS-tree[2] and Tuple Space Search[3], etc., can not catch up with the increase in link bandwidth, and lead to a bottleneck in NIDSs. Because of their massively parallel, highly deterministic search characteristics, Ternary Content Addressable Memory(TCAM) can be used effectively to solve the problem of multi-match packet classification in NIDSs.

However, TCAM is much more costly than conventional memory. In addition, TCAM consumes more power and dissipates more heat. Thus, it is crucial to compact packet classification table size so that a smaller number of TCAM can be applied to reduce its power consumption, heat dissipation and cost. Several solutions have been presented on TCAM-based multi-match classification.

K.C. Chang et al. (Eds.): APWeb/WAIM 2007 Ws, LNCS 4537, pp. 564–572, 2007.

F.Yu et al. proposes a Geometric Intersection-based solution[4], which produces multi-match classification results with only one TCAM lookup. This approach is not efficient on memory when rules have many intersections. Since it needs all filter intersections inserted as new filters in the TCAM, and consumes most TCAM memory. The MUD scheme[5] proposed by Lakshminarayanan et al. applies extra unused bits in each TCAM entry to support multi-match lookup. The MUD does not increase the number of TCAM entries needed, but needs extra search cycles. The total number of cycles taken for finding k matches is k. This results long processing time and high power consumption. To solve the problem of Geometric Intersection-based solution, the SSA scheme[6] splits the filters into several sets and performs separate TCAM lookups for each set. If SSA splits rules into K sets, then K TCAM lookups are needed to find all matching filters. The amount of TCAM memory needed by SSA is also greater than or at least equal to the original number of rules.

The TCAM memory consumption of previously presented solutions is considerably large, i.e., at least equal to the original number of rules. With the increasing number of rules and the widening of TCAM width, such as IPV6 rules, this has become a challenging problem. In this paper, we present a code-based scheme that provides a solution for the multi-match problem in NIDSs. Through less number and narrower width of TCAM entries, our scheme reduces the consumption of TCAM space significantly. This allows a smaller TCAM to be used to reduce cost, power consumption and thermal dissipation.

2 A Code-Based Scheme

The idea behind our scheme is based on the simple observation that for two or more positive integers named n_1, n_2, $\cdots$, n_d, the product of them $n_1 \times n_2 \times \cdots \times n_d$ increases more rapidly than the summation of them $n_1 + n_2 + \cdots + n_d$ when the integers increase. Unfortunately, multi-field classifiers based on TCAM technology typically concatenate all fields into a single search key, which just resembles the product of integers $n_1 \times n_2 \times \cdots \times n_d$, results in a large TCAM width and requires lot amount of TCAM entries. This leads to more TCAM memory occupancy. To reduce memory consumption, a straightforward method is to change the product of rule's fields into summation.

In the next sections, we will use the example classifier in Table 1, which is a multi-matching classifier with four rules in four fields labeled as F_1, F_2, F_3 and F_4, to demonstrate our method.

2.1 Terminology

In this section, we introduce some terms and definitions used in next sections.

R_c: All rules in the classifier.
N_d: The dimension of the classifier.
RF_i: All sub sets of classifier located at $i-th$ field.

Table 1. An example classifier

Rule	F_1	F_2	F_3	F_4
R_1	101*	111*	7	8
R_2	00*	01*	11	*
R_3	1*	0*	16	11
R_4	*	*	*	11

under: A rule set is called under another rule set if and only if the rule set is one of the divided sub set of that rule set.

overlap: The relationship between two field if and only if one is a prefix of the other, for example $10*$ overlaps with $1010*$, but not with $010*$.

$O(f_1, f_2)$: A boolean function, if the field f_1 overlaps field f_2 it return true, otherwise return false.

$NF_i(R)$: A function which return the number of different $i-th$ field in rule set R.

$S_f(R)$: The sub sets under R divided by field f.

$F_i(r)$: A function return the $i-th$ field of rule r.

2.2 Assigning of Code

The method of assigning code to field is considerably simple. All different entries of each dimension are sorted with TCAM compatible order, and store in TCAM. At first, the entries of 1-th field are ordered on TCAM compatible order. Then, from the begin to the end we assign code to them sequentially, i.e., the first one assigned code 0, the second one assigned code 1, and so on. This procedure is repeated to all dimension of the classifier. For the example classifier, we have four tables(i.e.,F_1,F_2,F_3 and F_4) shown as Table 2 to store in TCAM, their corresponding sub-codes is stored in attached SRAM memory on TCAM.

Table 2. Field tables stored in TCAM and their sub-codes

field	F_1				F_2				F_3				F_4		
entry	101⋆	00⋆	01⋆	⋆	111⋆	01⋆	0⋆	⋆	16	11	7	⋆	11	8	⋆
sub-code	00	01	10	11	00	01	10	11	00	01	10	11	00	01	10

2.3 Dividing of Rules

We describe our scheme briefly as follows. Through a preprocessing step, which is offline, we split the rules of a multi-field classifier into independent sets. These subsets are stored in SRAM memory, and indexed by pointers built from returned codes during TCAM searches. Each subset is the multi-match result of a packet. The same fields of rules, for example, the first field source IP, the second field destination IP, and so on, are ordered in TCAM compatible order, each of them is given a code simultaneously. The field with its correspond code constructs a independent table stored in TCAM. When a packet is extracted from network link, the corresponding fields are used as search keys one by one, and obtained the assigned code. After every code of the fields are obtained, we mix them to form a pointer, which is used to index the multi-match rules table. So we can get all matched rules of a packet.

There are d times needed to split the rules, where d is the dimension of the classifier. In the first times, all rules in the classifier are divided by the first filed, and $NF_1(R_c)$ rule sets are produced *under* R_c. For every field f1 from 1 to $NF_1(R_c)$, a comparison is made between the field f_1 and that of every rule in R_c. If the field f_1 *overlaps* the corresponding field of the rule, the rule is put into the set corresponding field f_1, otherwise skip it. Simultaneously, the sub-code returned through searching the table described in section 2.2 is assigned to the rule. After having done this, we have gotten $NF_1(R_c)$ subsets of rules under R_c.

In the second times, we process the rule sets obtained from the first times processing. Every rule set R in RF_1 is divided by the second field, and $NF_2(R)$ rule sets are produced under R. This processing is similar as the first step. The procedure is repeated until d dimensions are consumed. Finally, we can get the splitting rule sets and its corresponding codes. The pseudo code is shown as fallows:

```
div_rule(R_c)
{
    RF_0 = R_c;
    for every field d = 1 ··· N_d {
            for every rule set R under RF_{d-1} {
                for every field f = 1 ··· NF_d(R) {
                    F_f(R) = ø;
                for every rule r ∈ R {
                    if(f == F_d(r)) {
                        S_f(R) = S_f(R) + {r};
                    }else if(O_f(f, F_d(r)) {
                        S_f(R) = S_r(R) + r;
                    }else
                        continue;
                    }
                }
            }
        }
    }
}
```

The splitting result of the example classifier is shown as Fig. 1.

2.4 Mapping Between Code and List of Rules

During packet classification, every packet will produce d sub-codes. The number of bits needed to represent every sub-code is determined by the number of entries

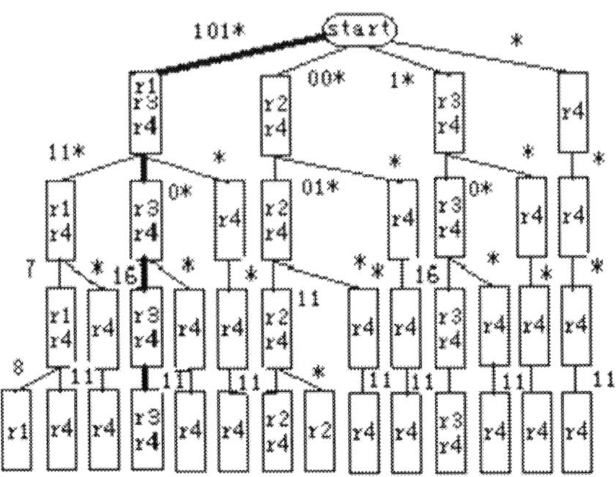

Fig. 1. The tree of splitting rules into subsets

in the corresponding filed. The number of bits of field f can be determined by $\lceil \log_2 N_f \rceil$, where N_f is the number of different entries in $f - th$ field. For the example classifier, the bits needed by F_1, F_2, F_3 and F_4 are calculated as follows:

$$b_1 = \lceil \log_2 4 \rceil = 2$$

$$b_2 = \lceil \log_2 4 \rceil = 2$$

$$b_3 = \lceil \log_2 4 \rceil = 2$$

and

$$b_4 = \lceil \log_2 3 \rceil = 2$$

Combining them together, so it needs 8 bits to represent a code, which is mapped to a list of multi-match rules. The map is shown as Table 3 .

Table 3. Map between codes and rule sets

Code	Rule set
00001001	r_1
00001000, 00001100, 00101100 00111100, 01011100, 01111100 10101100, 10111100, 11111100	r_4
01010100	r_2, r_4
00100000, 10100000	r_3, r_4

2.5 Classifying of Packets

When a packet is extracted from network link, we fetch the first field of the packet and used it as a key to search the TCAM table of this field. If there is any entry in the TCAM table matching the field, a code is returned from the search, otherwise there is no any rule matched this packet, and can drop it safely. Given that the dimension of a classifier is d, this procedure is repeated until all d fields is consumed, and the corresponding code is obtained. Finally, all returned d codes are mixed to form a pointer, which is used as index to the table of matched rules. As shown as Fig. 2. To better understand the

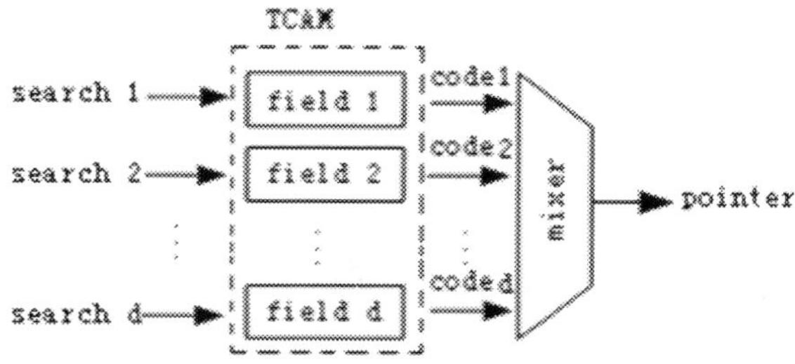

Fig. 2. The procedure of packet Classifying

multi-match packet classification algorithm, we use an example packet to demonstrate our algorithm. For example, there is a packet with source IP address 160.2.135.8, destination IP address 10.16.2.2, source port 16 and destination port 11. In order to find all matched rules, firstly we use source IP address 160.2.135.8 as the key to lookup TCAM table of field 1, and code 00 is returned. Then, destination IP address 10.16.2.2 is used as the key to search TCAM table filed 2, and code 10 is returned. So do to field 3 and field 4, and code 3 and code 4 are returned. The returned sub-codes are used to form a pointer 00100000b, and map to the beginning of multi-match rules list of this packet. So we know rule r_3 and r_4 matched this packet, this is the result what we want. The bold line in Fig. 1 matching process and the result of this packet.

3 Evaluation

In this section, we firstly give some simulation results of our scheme, such as throughput, power consumption and memory occupancy. Secondly, we compare our scheme with three previously proposed TCAM-based approaches: MUD, Geometric Intersection-based approaches and SSA. In the same way, we also use the SNORT[7] rule database as the test sets for multi-match classification. The environment of simulation is composed of the Intel developer workbench(v3.5), which is used to provide simulation environment of IXP2400 network processor, and the IDT75KTA062134200 kit[8], which is used to provide simulation environment of TCAM.

3.1 Throughput

The current version(v2.4.0) of Snort[7] rule database is applied , which contains 3859 rules and has 491 unique rule headers, to test query speed of our scheme.

The number of TCAM lookups of our schemes is deterministic, i.e., the dimension d of the classifier. In our simulation, the classifier contains four fields: destination IP address, source IP address, destination port number, and source port number, with 32, 32, 16, and 16 bits in each field, respectively. So it needs four TCAM lookups to classify an incoming packet. We examine the effect of microengine number on the throughput of our scheme in simulation environment. Packet streams with shortest bytes, for example, the shortest 64 bytes Ethernet packet, are constructed using Intel developer workbench. The packet classification task is allocated running on various number of microengines, for example, 1, 2, 3, and 4. For the different microengine number, the values of throughputs are measured from the Intel developer workbench. The simulation results are shown as Fig. 3 (a).

In the figure, we can find that at first, when the number of microengines on which the packet classification task running increases one times, the throughout will increase one times accordingly. But when we increase microengine number further, the query speed increases slowly, or achieves a saturated value. The reason is that too many query command issues blocks the query channel, and leads to return query result with long latency.

3.2 Power Consumption

We examine the effect of number of microengines and packet length on the power consuming of TCAM in simulation environment. The version of SNORT rule database is same as section 3.1. The consumed power of TCAM under various combinations of different parameters is examined. The simulation results are shown as Fig. 3(b).

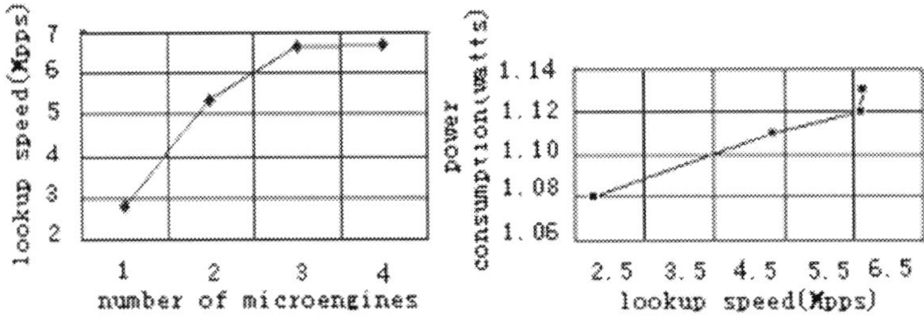

Fig. 3. (a)Lookup speed vs. number of microengines (b)Power consumption vs. lookup speed

In the figure, we can find that the power consumption is linear with the query speed. The reason is that when the query speed increases, during a given time interval the number of commands issued increases correspondingly and accesses more TCAM space. So more heat is dissipated.

3.3 Memory Occupancy

Suppose we have a classifier, whose dimension is d, each field with N_i different entries and E_i bits length. If we configure the TCAM width to w bits, which is bigger than or at last equal to the max bit length of field, then the total TCAM space required to accommodate all dimension entries is $w \times \sum_{i=1}^{d} N_i$. The efficient space is calculated as $w \times \sum_{i=1}^{d} E_i$. The percentage of the efficient utilization of TCAM space is calculated as $(w \times \sum_{i=1}^{d} E_i)/(w \times \sum_{i=1}^{d} N_i) \times 100\%$.

Two versions of SNORT rule database are used to test our scheme, i.e. V2.3.3 and V2.4.3. The number of the rule set is 3370 and 3859. Many rules share a same rules header, and the unique rule headers is 447 and 491. Considering that the number of ranges in SNORT rule database is small, and our scheme have change the product of fields into summation, we use the simple method to expand range fields, for example, range [1,4] is expanded as 1, 1*, 100. The result is shown as Table 4 .

Table 4. Memory consumption of two rule databases

version	entries	width	space	efficient	percentage
2.3.3	267	36	9612	4672	48.61
2.4.0	283	36	10188	4928	48.37

3.4 Comparison to Other Schemes

In this section, we give some comparison to other schemes proposed previously memory occupancy and query speed. Since memory occupancy and query speed have significant relationship with power consumption. Similarly, four versions of SNORT rule databases are used to the comparison, i.e., v2.0.0 v2.0.1 v 2.1.0 and v2.1.1. The original rule size is used as the comparison base. The comparison results are shown as Table 5. The Geometric Intersection-based solution consumes the most number of TCAM entries, which is about 16 times the original rule size. MUD and SSA use about the same number of original rule size of TCAM entries. The width of TCAM required by MUD, SSA and Geometric Intersection-based solutions is same, i.e. 144 bits. The Geometric Intersection-based solution has the fastest query speed, which just needs one TCAM lookup. But the number of TCAM lookups needed by MUD varies from packet to packet. The lookup number of SSA is deterministic, which is equal the number of divided rule sets. Our scheme significantly decreases consumption of TCAM space. We use the least number of TCAM entries, and require the narrowest width of TCAM, i.e. 36 bits. Our scheme uses about 0.22 times the TCAM space used by the original rule set, and the classify speed is deterministic, which is determined by d, where d is the dimension of classifier.

Table 5. Comparison results to other various schemes

version	original	Geometric Intersection-based			MUD			SSA			code-based		
2.0.0	240	3693	144	15.39	240	144	1	241	144	1.00	207	36	0.22
2.0.1	255	4009	144	15.72	255	144	1	256	144	1.00	217	36	0.21
2.1.0	257	4015	144	15.62	257	144	1	257	144	1	218	36	0.21
2.1.1	263	4330	144	16.46	263	144	1	263	144	1	221	36	0.21

4 Conclusion

Multi-match packet classification is a crucial task required by NIDS. Commercially available TCAM is suitable to offload this task from operation of network processor. In this paper, we present a code-based scheme to solve the multi-match classification problem with TCAM. The solution reports all matching results with d TCAM lookups, where d is the dimension of a classifier. Comparing to several TCAM-based solutions, our scheme possesses less TCAM memory and consumes lower power. Apart from that, our scheme has several advantages as follows:

1. Our scheme requires less number and narrow width of TCAM entries. This makes our scheme can support large rule set classifier.
2. The classify speed is deterministic, which is determined by the dimension of classifier.
3. The search of fields taken from a packet is performed on independent tables, and can be taken place in parallel.

References

1. Srinivasan, V., Suri, S., Varghese, G., Waldvogel, M.: Fast and Scalable Layer four Switching. In: Proceedings of ACM Sigcomm pp. 203–214 (September 1998)
2. Feldman, A., Muthukrishnan, S.: Tradeoffs for packet classification. Proceedings of Infocom 3, 1193–1202 (March 2000)
3. Srinivasan, V., Suri, S., Varghese, G.: Packet Classification using Tuple Space Search. In: Proceedings of ACM Sigcomm pp.135–146 (September 1999)
4. Yu, F., Katz, R.H.: Effixient multi-match packet classification with TCAM. Hot Interconnects (August 2004)
5. Lakshminarayanan, K., Rangajan, A., Venkatachary, S.: Algorithms for advanced packet classification with ternary CAM. In: Proc. ACM Sigcomm (2005)
6. Yu, F., Lakshman, T.V., Motoyama, M.A., Katz, R.H.: SSA: A power and memory efficient scheme to multi-match packet classification. Technical Report, No. UCB/CSD-5-1388 (May 2005)
7. http://www.snort.org. Snort system
8. Integrate Device Technology, Inc, Integrate IP Co-Processor(IIPC) with QDR Interface IDT75k52134 /IDT75k62134 User Manual (2002)

Home Network Device Authentication: Device Authentication Framework and Device Certificate Profile

Yun-kyung Lee, Deok Gyu Lee, Jong-wook Han, and Kyo-il Chung

Electronics and Telecommunications Research Institute,
161 Gajeong-dong, Yuseoung-gu, Daejeon, Korea
neohappy@etri.re.kr

Abstract. As home network service is popularized, the interest in home network security is going up. Many people interested in home network security usually consider user authentication and authorization. But the consideration about home device authentication almost doesn't exist. In this paper, we describes home device authentication which is the basic and essential element in the home network security. We propose home device authentication framework, home device registration and certificate issuing method, and home device certificate profile. Our home device certificate profile is based on the internet X.509 certificate. And our device authentication concept can offer home network service users convenience and security.

Keywords: home network security, home device authentication, home device certificate.

1 Introduction

As the home devices have various functions and have improved computing power and networking ability, the importance of home device authentication is increasing for improving of home network users' security. In using home network service, user authentication and authorization technology are applied to home network services only for authorized persons to use the home network services. But It has some problems: the leakage of user authentication information by user's mistake, usage of guessable authentication information, and finding of new vulnerability about existing authentication method. So it is necessary that home network service user can be served the secure home network service by only using credible device. This means that home device authentication besides user authentication and authorization is essential to the secure home network service. Also, the unauthorized accessing possibility for our home network is very high by the device included in neighbor home network because of the home network characteristic; various wired/wireless network devices is used in the home network. This is an additional reason about the necessity of device authentication.

Finally, we think that the secure relationship among home network devices is very important factor because home network service evolves into more convenient one; user's role in receiving home network service is minimized and the service served by cooperation among home devices is maximized.

K.C. Chang et al. (Eds.): APWeb/WAIM 2007 Ws, LNCS 4537, pp. 573–582, 2007.

Device authentication ensures that only specific authorized devices by specific authorized credential is compromised, the security between two parties is still protected as long as the authorized device is not used. Besides this, the device authentication is a mandatory technology that enables emerging context-aware services providing service automatically through device cooperation without user intervention, and DRM systems also need the device authentication [1, 2].

So far, several mechanisms have been proposed for this purpose. Some industries suggest hardware fingerprint based approach [3,4] that extract the secret information from the unique hardware fingerprint and trust the device by verifying the secret. Bluetooth [5] and Zigbee [6] provide device authentication mechanism based on shared symmetric key, and CableLab [7] also provides PKI based one. Personal CA [8, 9] provides localized PKI model. However, to the best of our knowledge none of them are applicable for multi-domain environment for several reasons [10].

This paper describes device authentication. In section 2, we describe the reason for using PKI in device authentication and our device authentication framework. In section 3, we propose device certificate profile. Finally, our paper concludes with section 4.

2 Home Device Authentication Using PKI

2.1 Why PKI?

Generally, authentication methods are grouped into three categories: 1) what you know (e.g. password); 2) what you have (e.g., token, certificate); and 3) who you are (e.g., biometric) [11]. But, on the other point of view on authentication methods, we can categorize them into two types: public key authentication and private key authentication. In our paper, we describe home device authentication methods using device certificate. It belongs to "what you have" in the former and "public key authentication" in the latter. Also, the classification of the former is more applicable to the user authentication method. But it prefers device authentication methods to divide into the used key; public key and private key.

In the authentication method using private key, the main issue is key sharing and key management problem. That is, it is most important that the both authentication parts securely share and securely store a private key. But the operation of private key cipher algorithms (for example: AES, DES, etc.) is more easy than that of the public key cipher algorithm. In other word, it needs less computing power and takes a few operation times. On the other hand, the public key authentication method has a merit that is free from key sharing and key management.

The authentication method using private key is appropriate to the model which authenticates small number of devices and authenticates only between server and end-device. But it is not adequate for our model because our model considers the authentication between server and end-device and between end-device and end-device. Also, fixed home appliances at home and portable devices of the home members are our considering devices. And we consider device authentication between our home device and our neighbors' home device. This means that each device must store and manage so many private keys to participate in home network. So, we think home device authentication using public key is adequate for our model.

2.2 Global PKI vs. Personal PKI

Global PKI can be defined as follows: it has some public root CAs, each root CA administrates subordinate CAs and they issue certificates to end entities and manage them. While personal PKI can be defined as follows: it has many private CAs, each private CA issues certificate to limited numbers of devices and manages them. These two PKIs are identical in the point that an end-entity kept the valid certificate is authenticated by it's CA and uses safe services in the administrative area of the CA. However, a difference is in the administrative area of CA.

In the Global PKI, it has one or a few numbers of root CA. every end entity can be authenticated and authenticates each other using an valid certificate by a convention between these root CAs. However, in the personal PKI, it has great many root CAs. These root CAs have limited administrative area. There are many restrictions in the convention between root CAs. Also, if each family has one personal CA and it issues certificates to all devices in the house, then each family have to bear the cost of personal CA maintenance and a limit follows the use of home network service through an authentication between my house device and the other home device.

Home network service includes the indoor service through an authentication between my house devices in my home, outdoor service through an authentication between my house device and outdoor service server, and the indoor and outdoor service using portable device. Therefore, the device authentication system for the home network service seems to do according to the global PKI.

2.3 Home Device Authentication Framework

As described in section 2.1., this paper proposes home device authentication mechanism using PKI. It covers intra-home device authentication and inter-home device authentication. We consider not personal CA [8, 9] but public CA. The use of personal CA [8, 9] may be proper solution if only device authentication in the intra-PAN (Personal Area Network) is considered. But if we consider inter-home network, public CA is more proper. Figure 1 shows our home device authentication framework.

In figure 1, our home device authentication framework has hierarchical PKI (Public Key Infrastructure) structure. That is, root CA (Certificate Authority) manages it's subordinate CAs and CAs manage home devices and HRA(Home Registration Authority). HRA is a home device which has enough computing power for public key operation, communication ability with other home devices and user interface equipment (for example, monitor, keypad, etc.). And it functions as RA (Registration Authority) and has more authority and requirement.

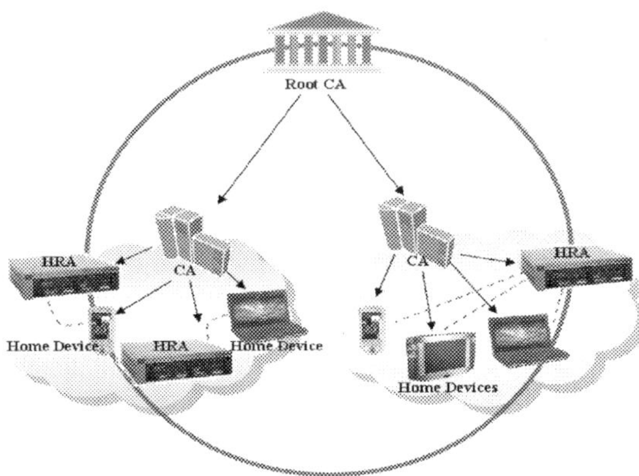

Fig. 1. Home device authentication framework

The devices in the figure 1 means home devices included in the home network. They can communicate with each other and have basic computing ability. That is, internet-microwave, internet-refrigerator, digital TV such as IPTV, internet-washing machine, PDA, notebook computer, wall-pad, PC, cellular phone, etc. are included in our home device. Many home devices are used in everyday life. And more and more home devices will be developed.

Device certification path will be root CA -> CA1 -> CA2-> … -> HRA/device. And it will be different if the devices are included different CAs. In this case, home devices are authenticated by using CA's trust list which is made by agreement between the CAs.

2.4 Home Device Registration and Certificate Issuing

This section describes home device registration and device certificate issuing process. Figure 2 shows home device registration and certificate issuing process.

Home device registration and certificate issuing process need user intervention. In figure 2, (1) and (7) processes expressed by broken line specially are off-line processes by user. Home device registration and certificate issuing processes are as follows;

(0) Buy home device with home networking ability and bring it home.
(1) Register the home device through HRA at home. In this time, user must inp ut device identity information and other information which is necessary for certificate issuing.
(2) TLS channel is established between HRA and device manufacturer portal. HRA requests device manufacturer portal to verify the validity of that devic e by forwarding the device identity through the TLS channel.
(3) Device manufacturer portal checks whether the device is his product or not through the received device identity.

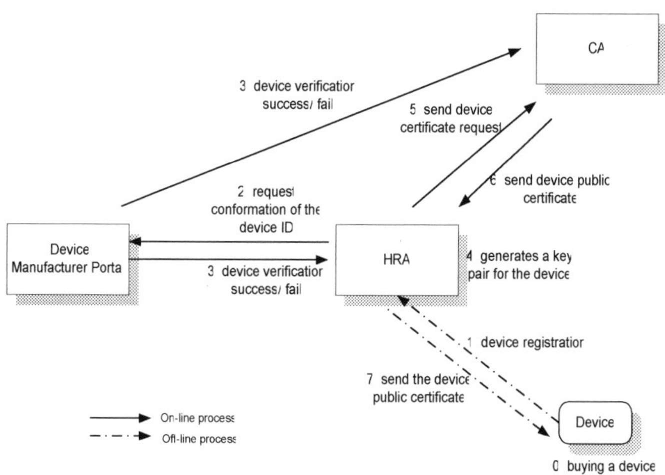

Fig. 2. Issuing process of home device public certificate

(4) If HRA receives 'verification success' message from device manufacturer portal, then HRA generates a key pair: public key and private key for the de vice.

(5) HRA sends the request of the device certificate issuing to CA.
If CA receives 'verification success' message from device manufacturer portal and 'certificate request' message from HRA, then CA issues a certificate of the home device. If CA doesn't receive 'verification success' message from device manufacturer portal, then CA rejects the certificate request. And CA can reject the certificate request if the device is already registered and is included in a report of the lost devices.

(6) HRA sends the received certificate of the home device and generated key p air to the device. This process needs user intervention. Maybe it is processe d by off-line method for security.

Home device identity referred before is a factor which can identify a device. It can be a new device identity system or existing information such as device serial number, barcode, or MAC address, etc.

Our HRA verifies the certificate contents and the identity of the device like RA (Registration Authority) in general PKI. Two RA models exist in general PKI. In the first model, the RA collects and verifies the necessary information for the requesting entity before a request for a certificate is submitted to the CA. the CA trusts the information in the request because the RA already verified it. In the second model, the CA provides the RA with information regarding a certificate request it has already received. The RA reviews the contents and determines if the information accurately describes the user. The RA provides the CA with a "yes" or "no" answer [12]. Our HRA is similar to the first model of general RA, but it is not CA had public trust but a home device of the kind. It is a device that has the same or more computing power, memory, and data protection module. So, HRA generates key pair and requests and receives certificates for other home devices.

3 Home Device Certificate Profile

Home device certificate follows the basic form of internet X.509 certificate [13]. That is, it is the same with X.509 version 3 certificate, but it adds some other extensions about home device authentication. Whatever they has different target: our home device certificate authenticates home devices, but internet X.509 certificate

Table 1. Basic device certificate profile

version
serialNumber
signature
issuer
validity
* subject
subjectPublicKeyInfo
* extensions
signatureAlgorithm
signature

Table 2. Extensions of home device certificate

Extensions	Explain
*Device information	Home device manufacturer and device identity
*HRA information	The location of HRA(Home Registration Authority)
*Device ownership	The information of home device owner and whether the device is HRA or not
*Device description	Description about the basic function of home device
Authority key identifier	Provides a means for identifying certificates signed by a partic ular CA private key
Subject key identifier	Provides a means for identifying certificates containing a parti cular public key
Subject alternative name	Additional information about home device
Issuer alternative name	Additional information about CA
Basic constraints	Maximum number of subsequent CA certificates in a certificat ion path Where it is end device or not
CRL distribution points	Acquisition method of CRL information
Authority information access	The method of accessing CA information and services (LDAP location)

authenticates human, enterprise, server, router, and so on. It is more efficient that home device certificate is implemented based on X.509 certificate because of popularity of the X.509 certificate. It means that implementation of our home device authentication frame work can be easier and spread of our mechanism can be faster.

Table 1 and 2 show our home device certificate profile.

In Table 1, subject and extensions fields signed with '*' are different with those of X.509 certificate. In table 2, four extensions signed with '*' are newly added in our home device certificate.

Now, we describe home device certificate fields which are different with X.509 certificate fields.

3.1 Signature

This field contains an algorithm identifier, and it identifies the digital signature algorithm used by the certificate issuer to sign the certificate. It is a copy of the signature algorithm contained in the signature algorithm field. In X.509 certificate, RSA algorithm is recommended. In home device certificate, but, ECDSA(Elliptic Curve Digital Signature Algorithm) signature algorithm is recommended.

3.2 Subject

Fundamentally, subject field of our certificate follows that of X.509 certificate. Subject field of CA certificate is the same with that of X.509 certificate. But subject field of end-device certificate has some difference. In other words, 'detail-locality', 'city', and 'state' attribute is added to the naming attributes of the subject field and we recommend 'locality' attribute is filled with detailed postal address and 'common name' attribute is filled with the kind of home device(for example, refrigerator, PDA, TV, microwave, notebook, and so on). If there are two TVs at home, they can be distinguished with appended number: TV1 and TV2. 'detail-locality' attribute is filled with concrete location of the device; bed room, living room, porch, kitchen, and so forth. For example:

Country = KR, city = Daejeon, locality = 101-302, Hankook apartment, Yuseong-gu, common name = notebook1, serial number = 1, pseudonym = father's favorite device, detail-locality = study room.

Country = KR, state = Kyunggi-do, city = ilsan, locality = 1102-507, Donghwa apartment, common name = refrigerator1, detail-locality = kitchen.

3.3 Device Information Extension

Device information extension describes the information of the home device. This extension consists of 'manufacturer' attribute and 'device recognition number' attribute. 'manufacturer' attribute fills with the name of manufacturer. 'device recognition number' attribute means unique number of the device; it is determined by the manufacturer. It can be serial number or MAC address.

This extension is useful in identifying of home device and deciding whether device manufacturer serves after-sales service or not.

3.4 HRA (Home Registration Authority) Information Extension

HRA information extension describes the location of HRA related with the device. The location of HRA is filled with IP address of the HRA and postal address of the home. If we lost a home device and notice it to the CA, this extension can help taking back it.

3.5 Device Ownership Extension

Device ownership extension describes the device owner's information. This extension and HRA information extension give the information of the device owner and give the owner legal and moral responsibility about using home service through the device. And this extension describes whether the device is HRA or not.

Device ownership extension consists of 'hRA', 'sharing', and 'owner' attributes. 'hRA' attribute means whether the device is HRA or not, 'sharing' attribute means whether the device owner is one person or not, and 'owner' attribute is the real name or role in the home(i.e. father, mother, son, daughter, grand-parents, and so on.) of the device owner.

If 'hRA' attribute is TRUE (this means the device is HRA), then 'sharing' attribute must be FALSE (this means the owner of the device is one person) and 'owner' attribute must be the real name of representative of the home. Also, it must be verified by credible agency. If 'hRA' attribute is FALSE (this means the device is general end-device), then there is no restriction. But, if 'sharing' attribute is TRUE (it means this device is shared by two or more persons), then "OWNER_GROUP" of the 'owner' attribute must be "public". ASN.1 syntax of this extension is as follows;

```
ownerShipInfo ::= SEQUENCE{
                      hRA     BOOLEAN DEFAULT FALSE,
                      sharing BOOLEAN DEFAULT FALSE,
                      owner   Owner }
Owner         ::= SEQUENCE{
                      OwnerGroup    OWNER_GROUP OPTIONAL,
                      Real_name     IA5String        OPTIONAL }
OWNER_GROUP ::= CHOICE{
                      Public        [0],
                      Father        [1],
                      Mother        [2],
                      Son           [3],
                      Daughter      [4],
                      Guest         [5] }
```

3.6 Device Description Extension

Device description extension identifies the basic function of home device. This extension offers 4 device characteristics: 'device control', 'home appliance', 'simple UI media' 'PCs'. The 4 device description choices are as follows:

device control. 'device control' indicates that this device may be used to control home devices in remote. i.e. remote controller.

home appliance. 'home appliance' indicates that this device may be home appliance such as refrigerator, washing machine, microwave, audio, and so forth.

simple UI media. 'simple UI media' indicates that this device may have simple user interface equipment, such as wall pad, PMP(portable media player), DMB player, and so on. These devices have not powerful computing ability like personal computer but have basic computing ability and simple user interface.

PCs. 'PCs' indicates that this device may be PC, notebook computer, PDA, etc.

We can expect the basic function, computing power, and communication ability of the device through this extension. And this extension can be used in home device access control.

4 Conclusions

This paper described the necessity of home device authentication. It needs to provide home network security and user convenience. And this paper proposed home device authentication method using PKI. We described the process of home device registration and the issuing process of home device certificate. Finally, we proposed home device certificate profile based on internet X.509 certificate.

Home device certificate differs from internet X.509 certificate in some fields of certificate. They are subject, device information extension, HRA information extension, device ownership extension, device description extension. That is, device sort and the main detail-location(i.e. bedroom, living room, study room, etc.) are included in the subject field value of home device certificate. Device information extension includes device manufacturer information and device identity information. HRA extension includes the postal address of the home which is subordinated by the HRA and IP address of the HRA. It is possible to find out the lost home device, and to relate HRA and the home device. Device ownership extension indicates whether the home device is personal possession or not. If the device is possessed by one person, then the person can use simple home network service only by device authentication. Finally, device description extension can provide the information about computing power of the device and the accessible home service. It is useful in device access control.

References

1. Lee, J., et al.: A DRM Framework for Distributing Digital Contents through the Internet. ETRI Journal 25(6), 423–436 (December 2003)
2. Jeong, Y., Yoon, K., Ryou, J.: A Trusted Key Management Scheme for Digital Right Management. ETRI Journal 27(1), 114–117 (February 2005)
3. Device Authentication. http://www.safenet-inc.com
4. TrustConnector 2. http://phoenix.com

5. Bluetooth Core Specification v2.0 (2004), http://www.bluetooth.org/spec/
6. ZigBee Specification v1.0, (December 2004) http://www.zigbee.org/en/spec_download/
7. OpenCable Security Specification (2004), http://www.opencable.com/specifications/
8. Gehrmann, C., Nyberg, K., Mitchell, C.J.: The personal CA-PKI for a personal area network, IST Mobile and Wireless Telecommunications Summit, pp. 31–5 (2002)
9. Intermediate specification of PKI for heterogeneous roaming and distributed terminals, IST-2000-25350-SHAMAN (March 2003)
10. Hwang, J.-b., Lee, H.-k., Han, J.-w.: Efficient and User Friendly Inter-domain Device Authentication/Access control in Home Networks. In: Sha, E., Han, S.-K., Xu, C.-Z., Kim, M.H., Yang, L.T., Xiao, B. (eds.) EUC 2006. LNCS, vol. 4096, Springer, Heidelberg (August 2006)
11. O'Gorman, L.: Comparing Passwords, Tokens, and Biometrics for User Authentication,In: Proceedings of the IEEE, Vol. 91, No. 12 (December 2003)
12. Planning for PKI: Best Practices Guide for Developing Public Key Infrastructure, John Wiley & Sons, Inc. (2001)
13. Housley, R., Polk, W., Ford, W., Solo, D.: Internet X.509 Public Key Infrastructure Certificate and Certificate Revocation List(CRL) Profile,' RFC 3280, April, 2002.Baldonado, M., Chang, C.-C.K., Gravano, L., Paepcke, A.: The Stanford Digital Library Metadata Architecture. Int. J. Digit. Libr. vol. 1, pp. 108–121 (1997)

P-IDC: Information Security and Consideration in Building Internet Data Centers for Pervasive Environment

Jae-min Yang[1], Jong-geun Kim[2], and Jong-in Im[1]

[1] Center for Information Security of Technologies (CIST), Korea University, Seoul, Korea
{jaeminy,jilim}@korea.ac.kr
[2] IDC Service Team, Hanwha S&C
jkkim@hanwha.co.kr

Abstract. Internet Data Center (IDC) offers servers and networks, and manages all contents in an enterprise. All servers should run 24 hours at no failure because the servers are very sensitive to keep their business continues. IDC has launched and operated already in many countries. IDC offers the enterprises to achieve their business goals and provides faster and safer service by converging all data services to one place for physical handling. In this paper, we propose Pervasive Internet Data Center (P-IDC) for preservation of high security on confidential information for large enterprises. The proposed model is appropriate for an enterprise which plans to build IDC and has many family companies in different business field. Definitely P-IDC should provide the solution to the problems ordinary IDC have.

Keywords: Pervasive Internet Data Center (P-IDC), Enterprise Internet Data Center (E-IDC), Multi-point operating system, Single-point operating system, and Risk analysis.

1 Introduction

IDC business was developed in 2000, a year known for its high-speed Internet boom [1]. As for a high-speed Internet service provider, IDC was required for developing secured web system, and efficient maintenance and operation.

To meet all the requirements of providing secured high-speed internet service, huge amount of investment needed to be involved. Duplication of investigation is a good example causing a very costly system [2]. When IDC market became very harsh, many IDC operators had to improve the service quality rather than expanding number of meaningless IDC.

Generally, there are two types of Internet Data Center (IDC), providing total network-connected environment, and managing any size of customer-business related system [3]. In this paper, we will only discuss enterprise IDC (E-IDC) that provides network managements, server managements, and system managements for combining all family companies. Although IDC has been evolved through many years, it still

K.C. Chang et al. (Eds.): APWeb/WAIM 2007 Ws, LNCS 4537, pp. 583–592, 2007.

suffers from couple weaknesses and vulnerabilities. Our researches were focused on how to make current IDC more secured and safer.

By developing E-IDC, an enterprise can expect well-defined business continuity through four achievements: increased purchasing power, prevention of duplicated investigation, efficient management of human power, and reduced business complexity by avoiding over purchases. Meanwhile, following five matters should be considered: disclosure information from database, illegal action under several related regulation, threatening for terror, physical weakness, and losing all data. Notice that several physical weaknesses were inherited from previous multi-point operating system when changed to a new single-point operating system.

Enterprise in large scale usually consists of ten family companies, and they are all in different business fields such as finance, manufacturing, national defense, medical, or commercial service. Because each family company has different types of confidential information and for more effective problem evaluations, development of E-IDC should be considered in an information security's point of view.

When the database operators are merged into one team, customer data should be handled more carefully since any unauthorized access by different database administrators are prohibited under the law of Financial Supervisory Service. This paper compares three IDC models consisting of IDC for a regular multi-point operation, IDC without a backup center, and IDC with a backup center. We propose IDC with a backup center that fulfills these three requirements; requisitions of a financial banking agency, reporting and getting permission from the finance supervisory service, reviewing of security deliberation by finance supervisory service.

Although, the enterprise puts its priority on efficient data management, the operators often do not know which method to be used to handle data efficiently. And this is the reason why IDC with proper data management is preferred. We propose Pervasive Internet Data Center (P-IDC) under the considerations of separation of duty, geographically backup center, and secure transferring human resources under the appropriate law. In order to set a strategy for the enterprise's vision and improve the quality of customer service, the present data center needs to be expanded since current ones are designed only for old equipments, utilities, and facilities. In a strategy of an enterprise, there are advanced Information Technology (IT) and superiority of competition, expanding its' business for future.

When a large enterprise operates the E-IDC, it is necessary to consider service methods for IDC. As an example of operating human resource, people should consider E-IDC operating system model in three methods, full outsourcing, selective outsourcing, and in-house operation [4]. According to the analysis on policy performance, in using lease based resource sharing, long lease times hurt high priority workloads [5]. Besides, E-IDC users pay for the services customers obtain, so they want them to be delivered according to established service-level agreements [6].

Here we present a summary of the policy performance analysis [7]. Policy performance is the different policies impact the ability to meet service level objectives in different ways. The Dater Center administrators and system designers need to understand the effects of a given set of policies on different workloads [7].

In Section 2, we present requirements for building P-IDC. In Section 3, we introduce P-IDC model for the secure and efficient IDC. Finally, the conclusion and future works are discussed in Section 4.

2 Requirements for Building P-IDC

In order to protect from all threats and possible problems, we have to consider following requisitions. Firstly, a regular database server and sensitive database server with confidential contents should not be sharing data together. For example, the national defense database server or hospital database server definitely could have much sensitive information as confidential. For the same reason, financial database server and regular database server should not be sharing data, either. Secondly, in case of getting serious physical damage from attacks by terrorists, everything would be stopped running the business. It could be worse than operating separately for each family company. These kinds of threats are not only the terrorist, but also earthquake, flooding, fire, and so forth. Furthermore, when we think about having an integrated IDC, we should consider many different regulations such as financial laws and information & communication acts.

In consequence to overcome these problems, the following requirements must be met with the P-IDC model.

***Separation of Duty for Independent Database*:** Make a role to separate of duty so that only authorized operator can access. If an enterprise has database servers for a financial business company and regular business company, the database administrators (DBA) must be separated with different duty. Therefore, the DBA from the regular business company cannot access to database at the financial business company. In the same token, it is equally applied between regular business company and national defense company.

***Backup Center for Overcome of Single-Point Operation*:** To solve the problem when single E-IDC physically failed, make sure there is a backup center in case of something happened. Before making E-IDC, they used to operate separately. Therefore, even if one of family companies was failed physically, the other family company would not be affected at all. However, E-IDC stands alone as a single point physically. Therefore, in order to get over the physically weakness, have the backup center in different place. The backup center should have a system which is able to do business normally at the minimum.

***Transferring Human Resources under the Financial Law*:** Make sure there is no violation under the financial law. If there is transferring human resources inevitably from a regular business company to a financial company and vice versa, it must be check if any violation occurs under the financial law. If there is any violation occurred, do not transfer human resource and employ not even one through a screening test.

***Transferring Human Resources under Information & Communication Acts*:** In the same way, this is equally considered as the financial law. Make sure there is no illegal transfer under information & communication acts. If there is transferring human resources inevitably between two companies, it must be check any violation occurs under Information & communication acts. When any violation occurred, do not transfer human resources through a screening test.

3 Proposed Scheme of P-IDC

Even though many enterprises have built E-IDC, these following the restriction are missed. Many of enterprises have constructed E-IDC compatibly without consideration of information security. In this paper, we would like to propose scheme of Pervasive Internet Data Center (P-IDC) with these next considerations.

3.1 Proposed P-IDC System Architecture

Proposed Pervasive Internet Data Center protects contents and information of each family company's database. It is very strictly applied under each field law. This P-IDC system architecture is aimed to take advantage of the ideal features provided by advances in IDC [8].

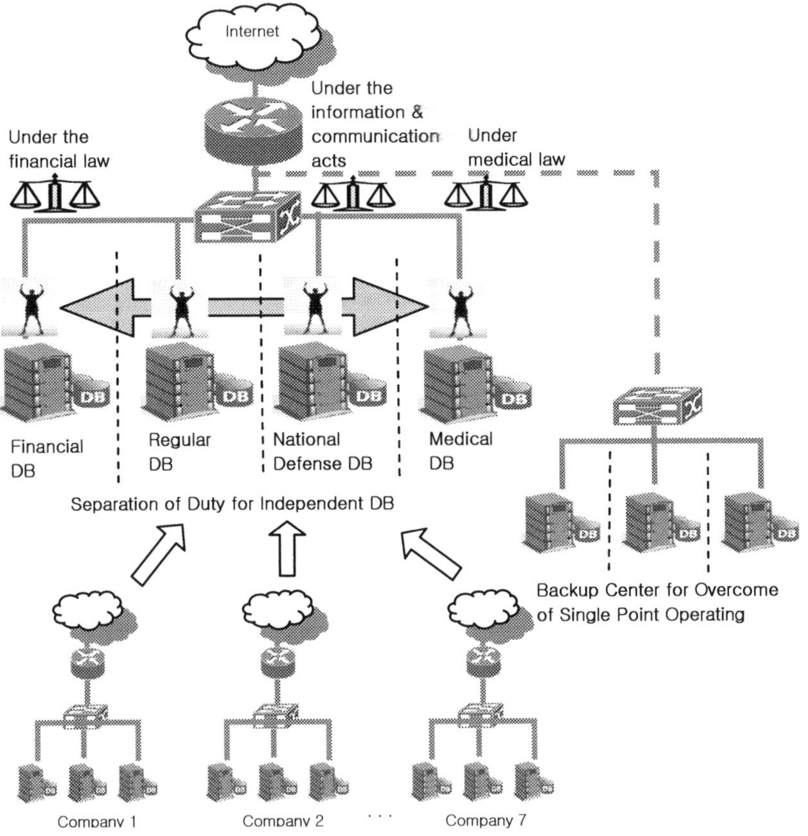

Fig. 1. Proposed Pervasive Internet Data Center (P-IDC)

Figure 1 shows company 1 through company 7 integrated P-IDC as a single-point operating. Herein, the database for financial companies such as banks, stock companies, and insurance companies is referred as financial DB. Similarly, the database used at companies providing national defense for each different countries is referred as national defense DB. In addition, we call regular DB as a common business database excepting financial and national defense DB. Medical DB indicates all of information and records about patients. P-IDC has separation of duty for independent database such as financial DB, regular DB, national defense DB, and medical DB. In addition, it shows the backup center for overcome of single-point operating. In case of force majeure such as fires, earthquakes or strikes, the operation switches over to the backup center in order to recover the business process as much as possible. There is transferring human resource under the financial and information & communication acts. When they transfer human resource between them, make sure there is no leaking out about customer personal data and national defense information.

3.2 Characteristics of Proposed Scheme

This proposed scheme is satisfied with all of requirements we listed in previous Section. There are five considerations are disclosure information from database, violence under the related regulation, threatening for terror, physical and geographically weakness, and losing all data. Those considerations are already mentioned in requirement for building P-IDC above. Disclosure information from each company's database is covered by Separation of Duty for Independent Database. Violence under the related regulation is made up by Transferring Human Resources under the Financial Law and Information & Communication acts. Threatening for terror, physical and geographical weakness, and losing all data should be taken cared of a backup center for overcome of single-point operating. Furthermore, there are efficient directions following the restriction for integration as the P-IDC.

At first, requisitions of a financial banking agency are that physical security facilities are required. When there are transferring human resources and properties occur from financial business area, the requisition must be followed as well.

Second of all, it is necessary to report and get permission from the Finance Supervisory Service about IT service outsourcing before the enterprises make a decision to have outsourcing by Third Party Company.

Third of all, a finance company related with security deliberation needs to be reviewed by Finance Supervisory Service after submit a request of security deliberation form. Whether or not transferring properties when the P-IDC is built, the enterprise needs to consider a price for taking over and tax. After integration from each family company as the P-IDC, access control of the each other's properties have to meet financial law and information & communication acts. Furthermore, transferring human resource is recommended by guideline of Finance Supervisory Service to save the costs for manpower by using outsourcing.

Last of all, considering for constructing of the backup center is recommended to build a long distance from the P-IDC to prevent for any disaster such as earthquake, flooding, fire, lighting, mud sliding, tornado, riot and so on.

3.3 Analysis of P-IDC

In this section, we compare current IDC system and proposed IDC system. From this proposed scheme, we show advantage of the scheme. Suppose that all three types of IDC have same assets category in this scheme. There are six categories that are information, documentation, software, hardware, human resource, and facilities & utilities. However, only five categories are showed for easy and better understanding. Let's make an assumption that each IDC has same assets categories. In the categories, there are information, documentation, software, hardware, facility and utility. In addition, all of main information processing devices and network lines are designed as a high availability (HA).

Table 1. Notation of the Risk Analysis for P-IDC Model

The meaning of following abbreviations is listed bellow:

- Components (C: Confidentiality, I: Integrity, A: Availability)
- CI: Customer information is leaked outbound
- DD: Possible to disclosure valued-documentation from third party employees
- FF: Damaged by flood or fire
- CD: Compromised or deleted by unauthorized access
- LU: Limitation of UPS by blackout
- Asset value = (C + I + A) * Support value
- Asset value = 1 ~ 9 : L, 10 ~ 18 : M, 19 ~ 27 : H
- Risk = (Occurrence frequency + Influence) * Asset value
- Risk = 1 ~ 54 : L, 55 ~ 108 : M, 109 ~ 162 : H

Asset value is multiplication by support value and sum of confidentiality, integrity, and availability. Support value represents simply how much an asset supports their business in good faith. Risk is multiplication by Asset value and sum of Occurrence frequency and Influence. Occurrence frequency shows how often this risk happens. Influence indicates how much business continuance gets impact by this risk.

Table 2. Operating with None IDC

Asset	Components			Support value	Asset value	Possible threat	Occurrence frequency	Influence	Risk
	C	I	A						
Information	3	3	2	3	24(H)	CI	2	2	96(M)
Document	3	2	1	3	18(M)	DD	3	2	90(M)
Software	2	2	3	3	21(H)	CD	3	3	126(H)
Hardware	1	2	3	3	18(M)	FF	2	2	72(M)
Facility & Utility	1	1	3	3	15(M)	LU	3	2	45(L)
									Total Risk : 429

N-IDC indicates an environment of an enterprise doesn't have IDC. Let an enterprise has couple of family companies which have different types of business.

Therefore, if they want to construct network infrastructure of their company, they have to purchase same network devices on same purpose. For example, seven family companies pay money for seven routers and firewalls separately. They might not have any discount for that purchase because they have no purchasing power than other two IDC kinds. Moreover, N-IDC might have duplication of investigation.

Table 2 represents what possible threats N-IDC gets. And how much total risk of N-IDC shows from fixed asset value, possible threat, occurrence frequency, and influence. As we mentioned about asset above, three types of IDC have same assets. Therefore, components from CIA values are same. Support and asset value are same of course. CD of software occurs possibly a lot more than the other two because many single-point operators exit at that rate. Therefore, occurrence frequency is high. Altogether, N-IDC total risk is 429 which are high.

Table 3. Enterprise-IDC

Asset	Components			Support value	Asset value	Possible threat	Occurrence frequency	Influence	Risk
	C	I	A						
Information	3	3	2	3	24(H)	CI	3	2	120(H)
Document	3	2	1	3	18(M)	DD	2	2	72(M)
Software	2	2	3	3	21(H)	CD	2	2	84(M)
Hardware	1	2	3	3	18(M)	FF	2	3	90(M)
Facility & Utility	1	1	3	3	15(M)	LU	2	3	45(L)
									Total Risk : 411

Risk of Information represents worse than N-IDC does. Customer information is leaked outbound worse. It seems awkward that risk value increased although there is IDC existed. The reason why is there is no secure transferring human resources under the appropriate law. However, risk of software decreased because there is not many single-point operators exist.

Table 4. Pervasive- IDC

Asset	Components			Support value	Asset value	Possible threat	Occurrence frequency	Influence	Risk
	C	I	A						
Information	3	3	2	3	24(H)	CI	2	2	96(M)
Document	3	2	1	3	18(M)	DD	2	2	72(M)
Software	2	2	3	3	21(H)	CD	2	2	84(M)
Hardware	1	2	3	3	18(M)	FF	2	1	54(L)
Facility & Utility	1	1	3	3	15(M)	LU	1	1	30(L)
									Total Risk : 336

Table 4 represents P-IDC with a backup center that fulfills all requirements. Every category risk value downs even a bit. Especially, risk of hardware decreased a lot because of consideration for the backup center.

The analysis of P-IDC is based on requirements that we mentioned in Section 3. Comparison criterion for analysis of P-IDC simply is broken down into items, namely, separation of duty for independent database, a backup center for overcome of single point, transferring human resources under the financial law, the information & communication acts, and medical law [8]. In addition, it shows how each IDC works well on each item. We omit transferring human resource under the medical law in this table since it is same concept of rest of two transferring human resources.

Table 5. Comparison with the Operating with None IDC (N-IDC), Enterprise IDC (E-IDC), Pervasive IDC (P-IDC) Scheme (O: Good, △: Medium, ✕: Bad), Criteria for Good, Medium, and Bad is based on the risk analysis above. Information and software of risk value is related with SDID. Hardware risk value is related with BCOSP.

Item	Operating with None IDC	Enterprise-IDC	Pervasive-IDC
Separation of Duty for Independent Database (SDID)	O	✕	O
Backup Center for Overcome of Single-Point Operating (BCOSP)	✕	✕	O
Transferring Human resources under the Financial Law (THFL)	△	✕	O
Transferring Human resources under the Information & communication acts (THIC)	△	✕	O

Now, based on the risk analysis, we compare three IDC with requirements for building IDC. First of all, ***separation of duty for independent database (SDID)*** is satisfied with N-IDC and P-IDC in Table 5 above. N-IDC is totally separated each family company that are operated own themselves as well. It might be duplication of human resources for the N-IDC. E-IDC just merged database administrators (DBA) without consideration for SDID. It could be a problem of access control the database. SDID is contented with P-IDC which made of thoughtfulness with SDID.

Second of all, ***backup center for overcome of single-point operating (BCOSP)*** is only happy with P-IDC, not rest of them. N-IDC has almost no backup center for each family company. N-IDC is only focused on operating own themselves which mentioned in Figure 1. Moreover, E-IDC is also designed without a backup center generally.

Third of all, ***transferring human resources under the financial law (THFL)*** is only contented with P-IDC and okay with N-IDC, but not happy with E-IDC. N-IDC would take care of THFL. N-IDC is physically separated for their each company and their organization members are totally different. Therefore, N-IDC should take care of THFL sensitively. However, E-IDC does not have firm organization about information security because they work together at same place.

Last of all, ***transferring human resources under the information & communication acts (THIC)*** is fairly same as THFL. THFL is satisfied with P-IDC and acceptable with N-IDC on same reason.

To sum up, enterprise IDC might be very good at purchasing power, prevention for duplication of investigation, and using human power efficiently. However, construction IDC without consideration of all requirements like the E-IDC could have worse case than operating with none IDC has. Even though E-IDC is based on IDC environment, there is no definition for separation of duty and a backup center. E-IDC is not quite satisfied with transferring human resources under the financial law and information & communication acts since there is partially defined for that. Pervasive IDC is excellent for every items listed. P-IDC gets all the necessary requirements which are advanced of E-IDC as well and made up for all weak point of E-IDC. As it shows above, P-IDC satisfied with separation of duty for independent database, backup center, and transferring human resources.

4 Conclusion

There are certainly many efficient parts if the enterprise makes up for weak points and brings the merits from the IDC integration from each independent family company. Besides, if we consider security issue to meet a specialized the enterprise's goal for the operating human power, it will save costs for human resources. At last, it is vividly possible to build safe and secured IDC if physical vulnerability is covered by complement such as changing multi-point operation into single-point operation.

We proposed Pervasive Internet Data Center (P-IDC) which E-IDC does not cover all information security issues. In the P-IDC, there are separations of duty for independent database, a backup center for overcome of single-point operating, and transferring human resources under the financial Law and information & communication acts. Not only P-IDC is in accordance with basic requirements of protecting problems from E-IDC, but also P-IDC gets many benefits from the IDC integration. Therefore, P-IDC is more superior than current E-IDC systems for the reasons that are separation of duty, backup system, and transferring human resources.

Further work will be worked on P-IDC that integrates Grid computing which uses concurrently all of computer devices through all networks to get maximized availability.

References

1. Won, N.H.: A study on The Roles & Development direction of IDC (Internet Data Center) business in IT society. Sejong University (2005)
2. Kim, Y.: Understanding IDC and Business Trends, Samsung SDS IT Review (2000)

3. Managed Co-Location: Korea Internet Data Center, KIDC- Co-location, Server Hosting & Managed Services, Sunday December 10 2006 (2006). http:// www.kidc.net/ eng/ service/ managed_co/co.jsp
4. Rational News, Korea IBM-Rational Software, March 2006 (2006). http://www-903.ibm.com/kr/software/rational/news/news_content.jsp?Idx=177
5. Srivatsa, M., Rajamani, N., Devarakonda, M.: Multi-Site Resource Management, Policy Performance Analysis, IBM Research Report RC23497, January 2005 (2005). http://domino.watson.ibm.com/library/cyberdig.nsf/Home
6. Menasce, D.A.: Internet Computing, IEEE Vol. 8, Issue 3, Digital Object Identifier 10.1109/MIC.2004.1297280, pp. 94–96 (2004)
7. Devarakonda, M., Naik, V.K., Rajamanim, N.: Policy-based multi-datacenter resource management. Policies for Distributed Systems and Networks, Sixth IEEE International Workshop, pp. 247–250 (2005)
8. Balaji, P., Vaidyanathan, K., Narrula, S., Jin, H.-W., Panda, D.K.: Designing Next Generation Data-Centers with Advanced Communication Protocols and System Services. In: the workshop on NSF Next Generation Software (NGS) Program; held in conjunction with IPDPS, Rhodes Island, Greece (2006)
9. Kim, H., Nang, J.: A Security Scheme for Protecting Network Resources in IDC Environment (2001)
10. Kim, J., Lee, J.: Study in IDC/ISP Information Security Enhancement. Dong-kuk University (2003)

Improvement of an Authenticated Key Agreement Protocol

Yongping Zhang, Wei Wei, and Tianjie Cao

School of Computer Science and Technology,
China University of Mining and Technology, Xuzhou 221008, China
{ypzhang,weiwei,tjcao}@cumt.edu.cn

Abstract. Popescu proposed an authenticated key agreement protocol based on Diffie-Hellman key agreement, which works in an elliptic curve group. The author also presented a simpler authenticated key agreement protocol than the the proposed one and a multiple key agreement protocol which enables the participants to share two or more keys in one execution of the protocol. However, in this paper, we show that their protocols do not authenticate each other. We also present an improved version.

Keywords: Authentication, Key Agreement Protocol, Elliptic Curves.

1 Introduction

A key agreement protocol is the process by which two or more distributed entities establish a shared secret key. Subsequently, this secret key (or called session key) can be used to create a confidential communication channel among those entities. Since the Diifie-Hellman key exchange protocol proposed by Diffie and Hellman in 1976 [2], key establishment protocols who are based on the Diffie-Hellman problem in finite groups have been used extensively.

Miller [6] and Koblitz [3] firstly suggest to use elliptic curve groups for realizing public-key cryptography. The elliptic curve cryptosystems are based on the elliptic curve discrete logarithm problem(ECDLP) over a finite field. We give the definition of ECDLP as this: Let E be an elliptic curve defined over a finite field F_q and give a pair of points $[P, mP]$ where P is a point of order n. The elliptic curve discrete logarithm problem is to find the integer m where $0 \leq m \leq n-1$. It is widely believed that the ECDLP is difficult to solve (computationally infeasible) when the point P has a large prime order.

A authenticated key establishment protocol should achieve a security service with some basic security properties. At the end of this protocols run between entities A and B, only A and B (or perhaps an entity who is trusted by them) should know the shared secret key K, A and B should know that the other entity knows K, and know that K is fresh. Some other desirable security attributes of authenticated key agreement protocols have also been identified[1][4][5]:

1. **known-key security.** Each run of a key agreement protocol between two entities should produce a unique secret key(session key). A protocol still

K.C. Chang et al. (Eds.): APWeb/WAIM 2007 Ws, LNCS 4537, pp. 593–601, 2007.

achieves its goal in the face of an adversary who has learned some previous session keys.

2. **(perfect) forward secrecy.** If long-term secrets of one or more entities are compromised, the secrecy of previous session keys established by honest entities is not affected.
3. **unknown key-share.** Entity A cannot be coerced into sharing a key with entity B without A's knowledge, i.e., when A believes the key is shared with some entity $C \neq B$, and B (correctly) believes the key is shared with A.
4. **key-compromise impersonation.** Suppose A' s long-term private key is disclosed. Clearly an adversary that knows this value can now impersonate A, since it is precisely this value that identifies A. However, it may be desirable in some circumstances that this loss does not enable an adversary to impersonate other entities to A.
5. **key control.** Neither A nor B can predetermine any portion of the shared secret key being established.

In [8], Popescu proposed an authenticated key agreement protocol(AKAP) based on Diffie-Hellman key agreement, which works in an elliptic curve group. The author also presented a simpler authenticated key agreement protocol (SAKAP) than the first one and a multiple key agreement protocol which enables the participants to share two or more keys in one execution of the protocol. We can see clearly that their protocols meet the above attributes. However, Popescu's protocols does not meet the authentication attribute, namely their protocols could not authenticate each other.

All protocols described in this paper have been described in the setting of the group of points on an elliptic curve defined over a finite field. Suitable choices include the multiplicative group of a finite field, subgroups of Z_n^* where n is a composite integer, and subgroups of Z_p^* of prime order q. Elliptic curve groups offer equivalent security as the other groups with smaller key sizes and faster computation times. In this paper we use an elliptic curve E defined over a finite field F_q of characteristic p. And also we choose elliptic curve domain parameters for the elliptic curve cryptography(ECC): field size q where q is a prime power, two field elements $a, b \in F_q$ which define the equation of the elliptic curve E over F_q, two field elements x_p and y_p in F_q which define a finite point $P = (x_p, y_p)$ of prime order in $E(F_q)$ ($P \neq O$, where O denotes the point at infinity), and the order n of the point P. The elliptic curve domain parameters should be verified to meet some requirements to avoid kinds of attacks and ensure the security [4][5].

2 Brief View of the Authenticated Key Agreement Protocol

The AKAP is divided into two phases, the key generation and the protocol description. The operation of the key generation is as follows. First, choose a one-way hash function, H, such as SHA-1 [7]. Then, select two random integers s_A, s_B from the interval $[1, n-1]$. The value s_A is a secret key of the user A and

s_B is the secret key of the user B. Next, compute the points $Y_A = -s_AP$ and $Y_B = -s_BP$, which are the public key of a user A and B respectively. Finally, select ID_A and ID_B, which are the identity information of a user A and B respectively. Let x_i be the x-coordinate of i.

We describe AKAP between two distributed entities A and B in brief as follows:

1. $A \rightarrow B : -k_AP$
   A generates random integers r_A, k_A from the interval $[1, n-1]$ and computes QA, V_A, points on E, such that $Q_A = r_AP, V_A = -k_AP$. A sends the point V_A to B.
2. $B \rightarrow A : \langle -k_BP, H(xQ_B, x_{V_B}, x_{V_A}, ID_B, ID_A), r_B + e_Bk_B + e_Bs_B \rangle$
   B randomly selects integers r_B, k_B from the interval $[1, n-1]$ and computes Q_B, V_B, points on E where $Q_B = r_BP, V_B = -k_BP$. B computes $e_B = H(xQ_B, x_{V_B}, x_{V_A}, ID_B, ID_A)$ and $d_B = r_B + e_Bk_B + e_Bs_B$, where x_{Q_B} is the x-coordinate of Q_B, xVA is the x-coordinate of V_A and x_{V_B} is the x-coordinate of V_B. B sends V_B, e_B, d_B to A.
3. $A \rightarrow B : \langle H(x_{Q_A}, x_{V_A}, x_{V_B}, ID_A, ID_B), r_A + e_Ak_A + e_As_A \rangle$
   A computes the point U_B, such that $U_B = d_BP + e_B(V_B + Y_B)$ and checks if $e_B = H(x_{U_B}, x_{V_B}, x_{V_A}, ID_B, ID_A)$. If it does not hold, then A terminates the execution. Otherwise, A computes $e_A = H(x_{Q_A}, x_{V_A}, x_{V_B}, ID_A, ID_B)$ and $d_A = r_A + e_Ak_A + e_As_A$, where x_{U_B} is the x-coordinate of U_B, x_{V_B} is the x-coordinate of V_B, x_{Q_A} is the x-coordinate of Q_A and x_{V_A} is the x-coordinate of V_A. A computes the point K_A where $K_A = -k_AV_B$ and sends e_A, d_A to B.
4. B computes the point U_A, such that $U_A = d_AP + e_A(V_A + Y_A)$ and checks if $e_A = H(x_{U_A}, x_{V_A}, x_{V_B}, ID_A, ID_B)$. If it does not hold, then B terminates the execution. Otherwise, B computes $K_B = -k_BV_A$.

If the entities A and B run this regular protocol, the shared secret key between A and B is the point $K = K_A = K_B$.

3 Security Analysis of AKAP

In [8], the author proved that their protocols including AKAP have the security attributes described in Section 1. Unfortunately, we see obviously that through AKAP run, entities A and B could not authenticate each other as the following theorem.

Theorem 1. *Through the authenticated key agreement protocol(AKAP) run, entities who participates in the protocols could not authenticate each other.*

Proof. Assume that M is a vicious entity, he wants to pose B and then run the AKAP with A successfully.

M chooses two random integers r_1, $r_2 \in [1, n-1]$, and computes $Q_B = r_1P, V_B = r_2P - Y_B$. He computes the verifier $e'_B = H(xQ_B, x_{V_B}, x_{V_A}, ID_B, ID_A)$ and $d'_B = r_1 - r_2e_B$.

We can see that $U_B = d_B P + e_B(V_B + Y_B) = (r_1 - r_2 e_B)P + e_B(r_2 P - Y_B + Y_B) = r_1 P = Q_B$. The another entity A passes the verification with e'_B and d'_B, and then considers that she authenticates B already. □

Let us describe this attack scenario with protocol flows for easier understanding. M is an adversary in this attack who wants to poses the entity B. $M(B)$ denotes that M poses B and processes the protocol.

Setup 1a. $A \rightarrow M(B) : -k_A P$

A generates random integers r_A, k_A from the interval $[1, n-1]$ and computes Q_A, V_A, points on E, such that $Q_A = r_A P, V_A = -k_A P$. A sends the point V_A to B.

Setup 2a. $M(B) \rightarrow A : \langle r_2 P - Y_B, H(xQ'_B, x_{V'_B}, x_{V_A}, ID_B, ID_A), r_1 - r_2 e_B \rangle$

M intercept A's message, $M(B)$ randomly selects integers r_1, r_2 from the interval $[1, n-1]$ and computes Q'_B, V'_B, points on E, such that $Q'_B = r_1 P, V'_B = r_2 P - Y_B$. $M(B)$ computes $e'_B = H(xQ'_B, x_{V'_B}, x_{V_A}, ID_B, ID_A)$ and $d'_B = r_1 - r_2 e_B$, where $x_{Q'_B}$ is the x-coordinate of Q'_B, x_{V_A} is the x-coordinate of V_A and $x_{V'_B}$ is the x-coordinate of V'_B. $M(B)$ sends V'_B, e'_B, d'_B to A.

Setup 3a. $A \rightarrow M(B) : \langle H(x_{Q_A}, x_{V_A}, x_{V'_B}, ID_A, ID_B), r_A + e_A k_A + e_A s_A \rangle$

A computes the point U_B, such that $U_B = d'_B P + e'_B(V'_B + Y_B)$ and checks if $e'_B = H(x_{U_B}, x_{V'_B}, x_{V_A}, ID_B, ID_A)$. If it does not hold, then A terminates the execution. Otherwise, A computes $e_A = H(x_{Q_A}, x_{V_A}, x_{V'_B}, ID_A, ID_B)$ and $d_A = r_A + e_A k_A + e_A s_A$ where x_{U_B} is the x-coordinate of U_B, $x_{V'_B}$ is the x-coordinate of V'_B, x_{Q_A} is the x-coordinate of Q_A and x_{V_A} is the x-coordinate of V_A. A computes the point K_A, such that $K_A = -k_A V'_B$, and sends e_A, d_A to $M(B)$.

As we have observed, A do not detect that M poses B. The adversary M poses the entity B communicate with A successfully. Similarly, M can also poses A to communicate with B.

4 Security Improvement on AKAP

In this section, we can take a secure ahuthentication scheme instead of the previous strategy to resist the above attack.

The Elliptic Curve Digital Signature Algorithm (ECDSA) is the elliptic curve analogue of the Digital Signature Algorithm (DSA). ECDSA was first proposed in 1992 by Vanstone [9] in response to NIST's (National Institute of Standards and Technology) request for public comments on their first proposal for the Digital Signature Standard(DSS). It was accepted as standard by many organizations, such as ISO(International Standards Organization) standard (ISO 14888-3) and so on. The ECDSA protocol consists of the three standard phases: Key Generation, Signature Generation and Signature Verification.

In ECDSA Key Generation phase, the user A selects a random integer s from $[1, n-1]$, compute $Q = sP$. The public and private keys of A are Q and s, respectively. A wants to sign a message m, he does a series of steps:

1. Selects a random integer r from $[1, n-1]$;
2. Computes $rP = (x_1, y_1)$ and $t = x_1 \bmod n$. If $t = 0$ then go to step 1 as any points selected will not generate lines that intercept the elliptic curve.
3. Computes $r^{-1} \bmod n$.
4. Computes $v = r^{-1}(H(m) + st) \bmod n$. Where H is the secure hash algorithm, such as SHA-1 [7]. If $v = 0$ go to Step 1. As s represents the gradient of generated lines, there is a strong chance of no intercepts.
5. The signature for the message m is the pair of integers (t, v).

Another user B wants to verify the A's signature (t, v) on the message m, he does the following steps:

1. Compute $c = v^{-1} \bmod n$ and $H(m)$.
2. Compute $u_1 = H(m)c \bmod n$ and $u_2 = tc \bmod n$.
3. Compute $u_1P + u_2Q = (x_0, y_0)$ and $h = x_0 \bmod n$.
4. Accept the signature if $h = t$.

We give an improved protocol by using ECDSA. Similarly, This improved version is divided into two phases, i.e. key generation and protocol description. In key generation phases, A and B choose s_A and s_B randomly from $[1, n-1]$ as their secret key respectively, then they compute $Y_A = s_AP$ and $Y_B = s_BP$ as their public key. Finally, select ID_A and ID_B, which are the identity information of a user A and B respectively.

We describe the improved authenticated key agreement protocol with protocol flows.

1. $A \to B : V_A$
   A generates random integers k_A from the interval $[1, n-1]$ and computes V_A, points on E, such that $V_A = -k_AP$. A sends the point V_A to B.
2. $B \to A : \langle V_B, x_{U_B}, W_B \rangle$
   B randomly selects integers k_B from the interval $[1, n-1]$ and computes V_B, points on E where $V_B = -k_BP$. B also chooses a random integer $r_B \in [1, n-1]$, computes $U_B = r_BP$, and ensures $x_{U_B} \neq 0 \bmod n$. B computes $W_B = r_B^{-1}(H(-k_BV_A \| ID_A) + s_Bx_{U_B}) \bmod n$. If $x_{W_B} = 0$, B selects another random integer $r_B \in [1, n-1]$. Finally, B sends V_B, x_{U_B}, W_B to A.
3. $A \to B : \langle x_{U_A}, W_A \rangle$
   A computes $c_A = W_B^{-1} \bmod n$ and $H(-k_AV_B \| ID_A)$, then computes $h_{A1} = H(-k_AV_A \| ID_A)c_A \bmod n$ and $h_{A2} = x_{U_B}c_A \bmod n$. If $x_{h_{A1}P + h_{A2}Y_B} \neq x_{U_B} \bmod n$, then A terminates the execution. Otherwise, A chooses a random integer $r_A \in [1, n-1]$, computes $U_A = r_AP$, and ensures $x_{U_A} \neq 0 \bmod n$. A computes $W_A = r_A^{-1}(H(-k_AV_B \| ID_B) + s_Ax_{U_A}) \bmod n$. If $x_{W_A} = 0$, A selects another random integer $r_A \in [1, n-1]$. Finally, A sends x_{U_A}, W_A to B.
4. B computes $c_B = W_A^{-1} \bmod n$ and $H(-k_BV_A \| ID_B)$, then computes $h_{B1} = H(-k_BV_A \| ID_B)c_B \bmod n$ and $h_{B2} = x_{U_A}c_B \bmod n$. If $x_{h_{B1}P + h_{B2}Y_A} \neq x_{U_A} \bmod n$, then B terminates the execution.

If the entities A and B run this protocol regularly, the shared secret key between A and B is the point $K = -k_AV_B = -k_BV_A = k_Ak_BP$.

We proof the soundness of this protocol. If V_B, X_{U_B} and W_B were indeed generated by B, $W_B r_B^{-1}(H(-k_B V_A \| ID_A) + s_B X_{U_B}) \bmod n$. Then, we get the equation:

$$\begin{aligned} r_B &= W_B^{-1}(H(-k_B V_A \| ID_A) + s_B x_{U_B}) \\ &= W_B^{-1} H(-k_B V_A \| ID_A) + W_B^{-1} s_B x_{U_B} \\ &= h_{A1} + h_{A2} s_B \bmod n. \end{aligned}$$

Thus, $h_{A1}P + h_{A2}Y_B = h_{A1}P + h_{A2}s_B P = (h_{A1} + h_{A2}s_B)P = r_B P \bmod n$. We can see the correctness of $x_{h_{A1}P+h_{A2}Y_B} = x_{U_B} \bmod n$. Similarly, we can derive the correctness of $x_{h_{B1}P+h_{B2}Y_A} = x_{U_A} \bmod n$.

5 Security of Our Improved Protocol

In this section, we discuss which of the security attributes described in Section 1 the proposed improved protocol possess under the assumption that the elliptic curve discrete logarithm problem is secure.

5.1 Basic Security Properties

Obviously, our improved protocol achieves the following three security properties: only A and B should know the shared secret key K, A and B should know that the other principal knows K, and A and B should know that K is fresh.

For the secret key K involves the random integers newly generated by A and B, only A and B know the shared secret key K, and know K is fresh. Through the protocol run, in Step 2 and Step 3, A and B authenticate the other entity respectively and know the other principal knows K.

5.2 Known-Key Security

An adversary M may learn previous session keys. The known-key security demands that no adversary be able to learn any information about the session key generated by a fresh protocol run even when the adversary learns other session keys. If the two entities A and B execute the improved protocol run, then they clearly share their unique session key K with their one-off random numbers as above.

5.3 (Perfect) Forward Secrecy

Intuitively, the improved protocol proposed achieve forward secrecy. Assume that an adversary M may capture the entities' private keys s_A or s_B. But, during the computation of the session key K for each entity, the random integers r_A, k_A, r_B, k_B still act on it. An adversary would have to extract the random integers k_A, k_B from V_A, V_B and extract the random integers r_A, r_B from ECDSA to know the previous or next session key between them. Obviously, this is the elliptic curve discrete logarithm problem.

5.4 Key-Compromise Impersonation

It is obvious that our improved protocol could provide key-compromise impersonation attribute. Now, suppose that the long-term private key s_A of the entity A is disclosed. An adversary M who knows this private key s_A can clearly impersonate A (Also, M can impersonates B to A if M knows B's long-term private key s_B). For the success of the impersonation, the adversary M must know A's random numbers r_A and k_A. In this case, the adversary M would have to extract k_A from A's ephemeral public value V_A and extract r_A from A's signature on message V_A to generate the same session key K with A. we see clearly that this also is the elliptic curve discrete logarithm problem.

5.5 Unknown Key-Share

Suppose an adversary M tries to make A believe that the session key is shared with B, while B believes that the session key is shared with M. To launch the unknown key-share attack, the adversary M would have to set his public key to be certified even though he does not know his correct private key. In this sense, the adversary M uses the public values (points) Y_A, Y_B and P. Let $f_t(R_1, ..., R_l) = \sum_{i=1}^{l} t_i R_i$, where R_i are points on E and $t = (t_1, ..., t_l)$ are integers from the interval $[1, n-1]$. Then M should set his public key Y_C as $Y_C = f_t(Y_A, Y_B, P)$. Suppose C got the value Y_C certified as his public key and suppose the following generalized model for unknown key-share attack. Suppose that $V_C = f_p(Y_A, Y_B, P, V_B)$ and $V_C' = f_m(Y_A, Y_B, P, V_A)$, where $p = (p_1, ..., p_l)$ and $m = (m_1, ..., m_l)$ are integers from the interval $[1, n-1]$. For C to launch the unknown key-share attack successfully, he should force A and B to share the same secret session key $K = K_A = K_B$ through the protocol run.

In practice, through the protocol run, A and B get their session key K_A and K_B respectively as those in the following: $K_A = -k_A V_C$, $K_B = -k_B V_C'$ The adversary C does not know s_A, s_B, k_A, k_B even though C can control the integer values t_i, p_i, m_i. The adversary C can force the equation $K_A = K_B$ to hold for many values of k_A and k_B. Now we can consider the following equation as an identical one for the variables k_A and k_B: $k_A V_C = k_B V_C'$ We can change this equation as the form $aP = O$, by unfolding the values V_A, Y_C, V_C, V_C' with respect to P. Then we can not solve this equation for t_i, p_i, m_i, since we do not have sufficient information on s_A, s_B, k_A, k_B.

5.6 Key Control

We could also see that our improved protocol meet the key control attribute. The key-control attack is impossible for the third party. The only possibility of key-control attack may be brought out by the participant of the protocol. But, for the entity B to make the entity A generate the session key K_B which is pre-selected value by B, for example B should solve the equation $K_B = -k_B V_A$. This is the elliptic curve discrete logarithm problem.

6 Remarks on SAKAP and Multiple Key Agreement Protocol

The SAKAP is a simpler version of AKAP. In SAKAP, the entities A and B share the long term secret key $K_S = -s_B Y_A = -s_A Y_B = s_A s_B P$. A generates a random integer $k_A \in [1, n-1]$, computes $V_A = -k_A P, e_A = H(x_{V_A}, x_{K_S})$ and sends V_A and e_A to B. B randomly selects an integer $k_B \in [1, n-1]$, computes $V_B = -k_B P, e_A = H(x_{V_B}, x_{K_S})$ and sends V_B, e_B to A. As a result, A and B achieve the same shared secret key $K = -k_A V_B = -k_B V_A = k_A k_B P$. We can show that this protocol also do not provide the authentication. this simpler version could suffer from message replay attack and reflection attack.

In a message replay attack, an adversary M has previously recorded an old message from a previous run of a protocol and now replays the recorded message in a new run of the protocol. We show the message replay attack against SAKAP in details. A sends V_A, e_A to B, where $V_A = -k_A P$ and $e_A = H(x_{V_A}, x_{K_S})$. The adversary M intercepts this message, poses B and send the message $V'_A = -k'_A P, e'_A = H(x_{V'_A}, x_{K_S})$ recorded from a previous run of this protocol to A.

In a reflection attack, when one entity A in the protocol sends to an intended communication partner B a message for the latter to perform a cryptographic process, M intercepts the message and simply sends it back to the message originator. M has manipulated the identity and address information which is processed by a lowerlayer communication protocol so that A will not notice that the reflected message is actually one created by her. The reflection attack against SAKAP can be described as this: A sends V_A, e_A to B, where $V_A = -k_A P$ and $e_A = H(x_{V_A}, x_{K_S})$. The adversary M intercepts this message, and sends this message V_A, e_A to A.

In those two attacks, A believes that the message received is from the intended parter B. We can see that M poses B successfully by message replay attack and reflection attack.

The multiple key agreement protocol proposed by the author in [8] is similar to AKAP. An adversary M can also impersonates one of the entities and deceives the another one as we have showed is Section 3. And also we can give an improved version which is similar to the protocol described in Section 4.

7 Conclusion

We have pointed out the drawback of a secure protocol(AKAP) for authenticated key agreement based on Diffie-Hellman key agreement proposed by Popescu. An adversary can poses one of the participants and run protocol with the other participant successfully. Considering the security, we have proposed an improved version in this paper. We also see that this drawback exist in the multiple key agreement protocol and point out this has a similar improved version. Popescu proved a simpler authenticated key agreement protocol than AKAP. Unfortunately, this protocol can not resist the message replay attack and reflection attack. Because of the faultiness in their authentication method, all their protocols could not provide the authentication to the participant entities.

Acknowledgment

This work was supported by the Science and Technology Foundation of CUMT.

References

1. Blake-Wilson, S., Johnson, D., Menezes, A.: Key Agreement Protocols and Their Security Analysis. In: Proceedings of Sixth IMA International Conference on Cryptography and Coding, Cirencester, UK, pp. 30–45 (1997)
2. Diffie, W., Hellman, M.: New directions in cryptography. IEEE Transactions on Information Theory, IT 22(6), 644–654 (1976)
3. Koblitz, N.: Elliptic curve cryptosystems. Mathematics of Computation 48, 203–209 (1987)
4. Law, L., Menezes, A., Qu, M., Solinas, J., Vanstone, S.: An efficient Protocol for Authenticated Key Agreement, Technical Report CORR98-05, Department of CO, University of Waterloo (1998)
5. Law, L., Menezes, A., Qu, M., Solinas, J., Vanstone, S.: An efficient Protocol for Authenticated Key Agreement. Designs, Codes and Cryptography 28(2), 119–134 (2003)
6. Miller, V.: Uses of elliptic curves in cryptography. In: Proceedings of Crypto'85, Santa Barbara, USA, pp. 417–426 (1986)
7. National Institute of Standards and Technology, Secure Hash Standard (SHS), FIPS Publication pp. 180–181 (1995)
8. Popescu, C.: A Secure Key Agreement Protocol Using Elliptic Curves. International Journal of Computers and Applications 27(3), 147–152 (2005)
9. Vanstone, S.: Responses to NIST's Proposal. Communications of the ACM 35, 50–52 (1992)

A Mechanism for Securing Digital Evidences in Pervasive Environments

Jae-Hyeok Jang[1], Myung-Chan Park[1], Young-Shin Park[1], Byoung-Soo Koh[2], and Young-Rak Choi[1]

[1] Department of Computer Engineering, Daejeon University, Yongun-Dong, Dong-Gu, Daejeon, Korea
{changjh1002,hparkmc,good4u23}@nate.com, yrchoi@dju.ac.kr
[2] DigiCAPS Co. Ltd., Jinjoo Bldg. 938-26 Bangbae-Dong, Seocho-Gu, Seoul, Korea
bskoh@digicaps.com

Abstract. In pervasive environment, for the mal functions in appliances and system errors, the unaccepted intrusion should be occurred. The evidence collecting technology uses the system which was damaged by intruders and that system is used as evidence materials in the court of justice. However the collected evidences are easily modified and damaged in the gathering evidence process, the evidence analysis process and in the court. That's why we have to prove the evidence's integrity to be valuably used in the court. In this paper, we propose a mechanism for securing the reliability and the integrity of digital evidence that can properly support the Computer Forensics. The proposed mechanism shares and manages the digital evidence through mutual authenticating the damaged system, evidence collecting system, evidence managing system and the court(TTP: Trusted Third Party) and provides a secure access control model to establish the secure evidence management policy which assures that the collected evidence has the corresponded legal effect.

1 Introduction

The intrusion to the system is dangerous and makes bad influence to pervasive environments where all digital appliances and PC's are connected all personal data each other. Such an unaccepted intrusion to personal data invades one's private life even more produce a powerful bad effect on companies lives.

Privacy is an important consideration for pervasive environment, because an individual's data contain a large amount of personal information. Therefore, technical advanced country like United State of America concentrate on studying for developing forensics mechanism that the major functions are colleting the digital evidence, analyzing the digital evidence if the data have modified[1,2,3]. Also, they are developing the forensics tools[4,5] as the national strategic industry to provide safety business communication. So the computer forensic tool developing area is one of the most preferred jobs.

The commonly used forensic tools use the damaged system to collect the digital evidence after an attack has been occurred[6,7]. But, the digital evidences used in the

K.C. Chang et al. (Eds.): APWeb/WAIM 2007 Ws, LNCS 4537, pp. 602–611, 2007.

court are not proved if the evidence data was collected from the damaged system. That's why we have to prove the integrity and the confidentiality of the collected digital evidences[8].

In this paper we proposed a mechanism to guarantee digital evidence's integrity which can prevents and checks the system errors and unaccepted intrusion to the system and collect the digital evidence from the damaged system to submit in the court. Also the mechanism uses media independent property of digital evidence to have more reliability and effective in the court of justice.

2 A Mechanism for Guarantee Digital Evidence's Integrity

2.1 System Environment and Organization

To apply in the existing network environment, the proposed mechanism for guarantee digital evidence's integrity has comprised in digital evidence management system (DEM), digital evidence collecting system (DEC) and the trusted third party (TTP).

The proposed mechanism is comprised, according to the inspection process of digital evidence, identify evidence (1 step), collect evidence (2nd step), preserve evidence (3rd step), send and store (4th step). The first, second and third are executed in DEC, the fourth step is establish in DEM. And the creation of identifier and storing the profile for preserving the evidence is executed in TTP. Fig 1 demonstrates the procedures for how to secure the evidence's integrity.

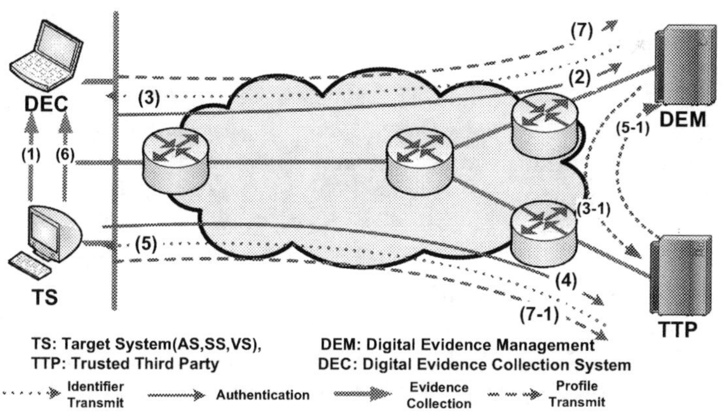

Fig. 1. System organization for guarantee the integrity

(1) DEC accesses to TS and extracts system information (SI). The extracted information use H/W and S/W property to identify TS. Then, DEC creates CN_{DEC} (Case Number) by using SI, (2) request the identifier to DEM. Also, DEC creates CN's identifier CN_{DEC}, (3) and transmit to DEC and (3-1) TTP.

DEC preserves the transmitted CN_{DEC}, then, use the identifier required to generate the profile. (4) DEC request identifier and TT (Time-sTamp) to TTP. TT is used as the current execution program standard time. TTP compares the received CN_{DEC} with

CN_{DEC} which is received from DEC and if the received CN_{DEC} are the same, creates TTP's system value CN_{TTP} (5) and sends to DEC (5-1) then, send identifier to DEM.

DEC creates DEO (Digital Evidence Object) by collecting the digital data from TS. Then, creates PI (Profile Info) by using the collected DEO and stored CN, and generates DEI (Digital Evidence Item) by combining PI and DEO. In this time, PI writes the information about whom, when and where the digital evidence was created. The created DEI is sent to DEM and is stored as digital evidence, and PI is sent and stored to TTP (7-1).

DEM generate the report by using DEI and is submitted to the court of justice, and assures the digital evidence's integrity by using the profile. And whenever the court requests for the profile's integrity, DEM request PI value to guarantees the profile's integrity.

2.2 Digital Evidence Collecting System

DEC is installed in TS to analyze the submitted security intrusion criminal in DEM and verifies the digital data stored in TS to extract digital evidence. The function of DEC is request the identifier from DEM and TTP to guarantee digital evidence's integrity and create the profile using extracted identifier and finally send that profile to DEM and TTP. The detailed operation policy of DEC, TS, DEM and TTP are as follow.

(1) DEC and TS

DEC and TS obtain system information and create CN according to the TS's environment, and the created CN is used to identify the TS. DEC accesses to TS and analyze the digital data. In this time, TS can be a on-line and off-line status which can access to the network and system and collect the information if TS's status in on-line.

The collected information can be a network information and system information. Network information are comprised in TS's original MAC address and IP address, and system information are comprised in CPU's identifier number, HDD S/N and BIOS information. Then the collected information is used as a guaranteed data that the collected data was created from TS.

DEC extract the identifier required in the profile from DEM and TTP. Then, accesses to TS and collects and analyses the digital data and create DEO. The created DEO establish profiling by using the identifier, as a result, creates DEI. Finally, PI contains the information about who, when, where and what make the modification of digital evidence.

(2) DEC and DEM

Fig 2 demonstrates the identifier creation procedure between DEC and DEM. According to the TS's environment obtain system information and create $CN_{DEC,}$ (the created CN_{DEC} establish TS's identifying method) then send to DEM. DEM create CN_{DEM} based on CN_{DEC} value and transmit securely. To secure the transmission path, it is used PKI based certificate.

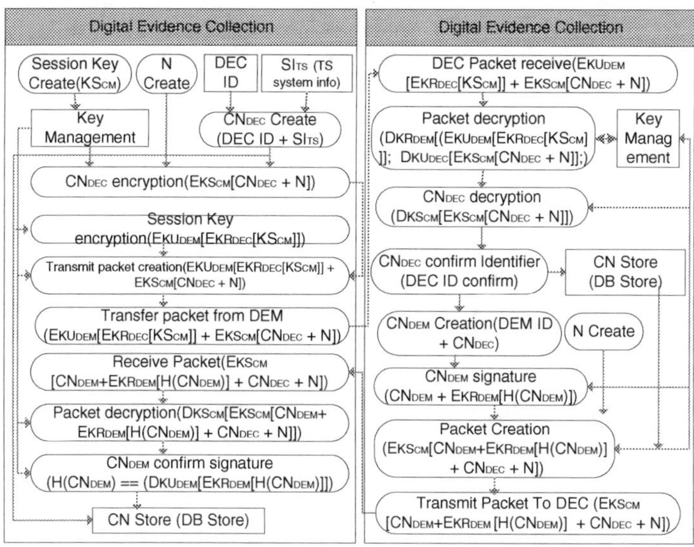

Fig. 2. DEC and DEM's identifier creation procedure

(3) DEC and TTP

Fig 3 shows the identifier creation procedures between DEC and TTP.

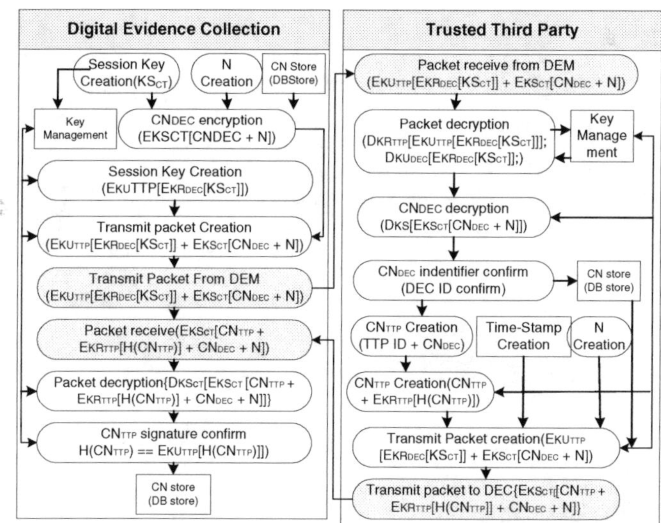

Fig. 3. Identifier creation procedures between DEC and TTP

First of all, according to the TS's condition, creates CN_{DEC} by obtaining the system information and make sure to be delivered to TTP. Secondly, TTP generates CN_{TTP}

using CN_{DEC} value and send that value to other entity. Also, uses to secure the transmission route the certificate which is based on public key cryptography. Lastly, DEC sends PI to TTP.

2.3 Digital Evidence Management System

DEM's functions are classified in DEC's user authentication, digital evidence storing and evidence providing functions. DEC, user authentication module, is installed in TS by DEM module and authenticates the user if that user has correspondent rights to DEM using TS's system information. The storing function receives digital evidence through trusted interface from DEC, and stores and manages the received digital evidence in DB. Then, collects and keeps the digital evidence profile from the system information. Finally, storing function provides reporting service to supply to the court.

(1) DEM identifier authentication

DEM reads CN_{DEC} and CN_{DEM} from CN DB and sends to TTP (Fig 4). First of all, we can prove that KS is sent securely from DEC to TTP using $EKU_{TTP}[EKR_{DEC}[KS_{MT}]]$. Also, the encrypted $EKS_{MT}[CN_{DEC} + CN_{DEM} + N]$ by KS guarantees the integrity itself.

TTP obtains KS using own public key $(DKR_{TTP}[EKU_{TTP}[EKR_{DEC}[KS_{MT}]]])$ and DEC's public key $(DKU_{DEC}[EKR_{DEC}[KS_{MT}]])$, and gets CN through decryption $(DKS_{MT}\ [EKS_{MT}[CN_{DEC} + CN_{DEM} + N\ ]])$.

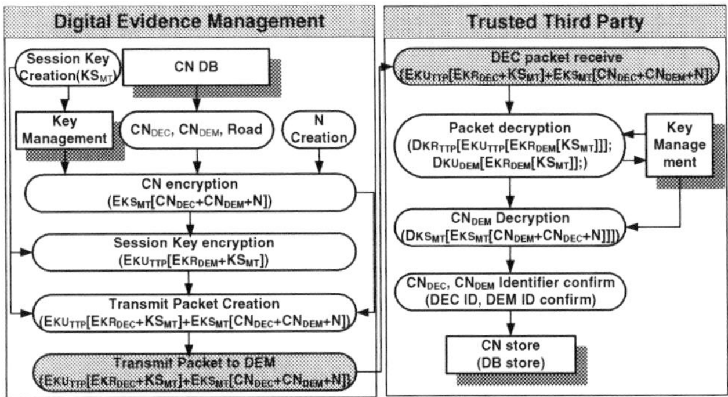

Fig. 4. DEM identifier authentication

(2) TTP identifier authentication

Fig 5 demonstrates TTP identifier authentication process. TTP reads CN_{DEM}, CN_{TTP} stored in CN DB and sends to DEM encrypted with KS. The transmitted packet is decrypted$(DKS_{MT}[EKS_{MT}[CN_{DEM} + CN_{TTP} + N]])$ with KS, then verifies CN_{DEM} which is stored in CN DB and CN_{DEM} which is transmitted from TTP.

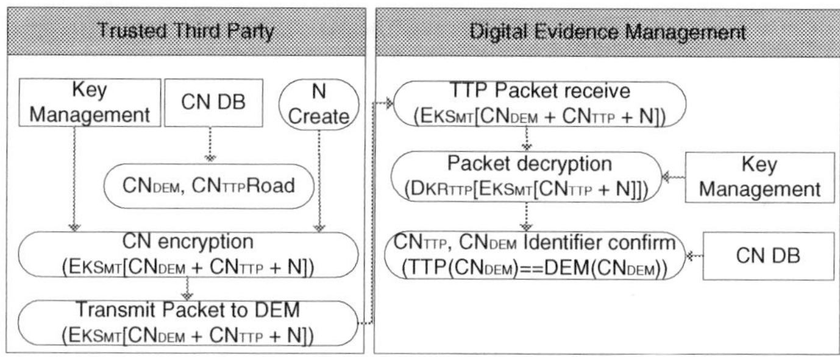

Fig. 5. TTP identifier authentication

2.4 Trusted Third Party

The major functions of TTP are creation and transmission of identifier, profile management and TimeStamp creation. First of all, in the identifier creation function generates the identifier using CN_{DEC} and TT to guarantee the confidentiality of digital evidence. The profile management function stores the profile in DB which is received from DEC, and transmits the profile whenever the court request for the verification of integrity of digital evidence. Finally, TimeStamp creation function provides the standard time.

TTP can be installed in the system that DEM can manipulates directly but it is recommend to be installed in the system out of reach of DEM and it is required to establish the access control policy to prevent un-granted user or organization.

3 Proposed Mechanism Evaluation

3.1 Evaluation Requirements

It is required the requirements of the confidentiality in the damaged system, commit system, pass through system, and collect/store the digital evidence. The proposed system that provides the confidentiality guaranteeing services has the evaluation requirements of digital evidence integrity guaranteeing, digital evidence objectivity and access control for guarantee the intrusion evidence's integrity.

In this section, we provide the validation of the proposed model analyzing the evaluation requirements separately in each items and we compared with the existing forensics tools.

3.2 Analysis of Digital Evidence's Integrity Guarantee Mechanism

(1) Guarantee of Digital Evidence's Integrity

Guaranteeing the integrity of digital evidence means that the profile of digital evidence must be unchanged and keep in a secured place.

The user who has to prove the integrity can use the digital evidence profile validation mechanism. That is, the user can prove the integrity by comparing the stored profile in DB since the profile which is created by DEC and the digital evidence are stored securely in TTP and DEC. The contents stored in DEM are as follow.

- DEM = DEI
- DEI = PI || DEO
- $PI_{DEM} = CN_{DEC} || CN_{DEM} || CN_{TTP} || TT || H(DEO)$

The contents stored in TTP are as follow.

- $PI_{TTP} = H(CN_{DEC} || CN_{DEM} || CN_{TTP} || TT || H(DEO))$

In DEM, we store the collected profile of digital evidence and the digital evidence from DEC, and we can prove the integrity of digital evidence through PI. But the stored PI in DEM may have the confidentiality problem. So, we have to ask and compare the PI values (PI_{DEM} and PI_{TTP}) which are stored in TTP to prove the integrity of digital evidence.

- Guarantee integrity $\rightarrow$ if($H(PI_{DEM}) == PI_{TTP}$) CDI

(2) Objectivity of Digital Evidence

Before, to confiscate the investigation related to digital evidence we have to stand upon to the media. But, confiscate investigation makes big privacy problems since confiscate means take over all the individual's properties. Therefore, we need the media independent access methodology which proves that the digital evidence was not modified.

In media independent access method, the collected digital evidence must be objectified to produce the profiles. We must separate the profile from the identifier to prevent the profile from the identifier to be modified. And we have to prove that mechanism by technologically and legally. To have the objectivity function we must use the subject management objectivity, time creation objectivity, contents objectivity, writing method objectivity and storing method objectivity.

a) Subject Objectivity

The subject that manages the profile must be separated from identifier and let the subject manage the profile. In this paper, we use the identifier DEC, manager DEM and TTP.

- DEC: analyze digital evidence and create DEI
- DEM: manage DEI
- TTP: manage PI

The created DEI must be divided into identifier and manager to manage and to objectify the subject.

b) Time creation objectivity

It is effort to distinguish the produced evidence by comparing the digital evidence creation time and profile creation time. And it is method to prevent the modification of the digital evidence by objectifying the digital evidence creation time.

- PI: $CN_{DEC} || CN_{DEM} || CN_{TTP} || TT || H(DEO)$

In this paper, we objectify the required TimeStamp which it is used to create the evidence and case investigation. And TTP provides TT with identifier whenever there is a request from DEC.

c) Content objectivity
To prevent the modification of the content of digital evidence, we used one-way hash function.

- $PI = CN_{DEC} \parallel CN_{DEM} \parallel CN_{TTP} \parallel TT \parallel H(DEO)$

PI creates identifier which is included in DEO.

d) Write and store method objectivity
It is used to do not modify the digital evidence once it has been written and stored at any reason in the case of write and store of the digital evidence. This is the method for preventing the modification of digital evidence using the append-only method and read-only method.

In this paper, we provide the write and store method by using the IEB security model characteristic. That is, the subject that has the rights to create a new object but can not modify any object.

(3) Access control for integrity of guarantee the intrusion evidence
If the collected digital evidence be valuable evidence in the forensics system, we must provide the method to prove the integrity of the original digital evidence. And, if the intrusion signs used in the court be usefulness must be unchanged and deleted. It is obvious that the low security level system can not access the high security level system.

However, even though the object which is granted to be accessed by others object, we must prevent the write and modification of the digital evidence to be used in computer forensics. A set of subject(S), a set of object (O), subject's security level(C(s)), object's security level(C(o)) have the following security policy to protect intrusion evidence.

- Read security property: subject S grants object O when C(s) >= C(o)
- Write security property: object O do not have write permission to subject S
- Create security property: if subject S has the write permission to object O, when C(s) = C(o) subject S can create a new Object

(S: a set of subject, O: a set of object, C(s); C(s): subject's security level, (Co): object's security level)

3.3 Evaluation of Digital Evidence Integrity

The proposed digital evidence integrity guarantee model's evaluation is based on the requirements of the Fig 4.1 and it is compared with the existing forensics tools. The method creates the digital evidence's profile based on who, when, where and what and with this method, we could analyzed the digital evidence integrity with media independent method. That means, we could provide the integrity of the collected digital evidence by using the profile stored in TTP.

We analyzed the validation requirement in each item separately and we could reach that our model has better result comparing to existing forensics tools. The validation items are as follow:

- trust of computer forensics tool
- guarantee of digital evidence integrity
- objectivity of digital evidence

Table 1 demonstrates the compared result with the proposed model and existing forensics tool.

Table 1. The compared result with the proposed model and existing forensics tool

Requirement	Existing method	Proposed method	Comparison
Trust validation in system	X	Δ	TTP
Trust validation of compute r forensics tool	Δ	O	TTP
Providing usability service	X	O	Media independent
Providing integrity service o f digital evidence	Δ	O	Objectivity of digital evid ence
Subject objectivity	X	O	Duty separation
Objectivity of time creation	X	O	Use TTP's TimeStamp
Contents objectivity	O	O	Hash function
Write method objectivity	Δ	O	Apply IEB
Store method objectivity	Δ	O	Apply IEB

(O: support, Δ: partially support, X: not support)

In the existing method, there is no mechanism to validate the system and the program. Only, some of the program manufactures provide the basic value itself. The proposed method generates the profile and stores in TTP and is managed from the trusted administrator.

In the case of guarantee of digital evidence's integrity, the existing method provides the integrity of the gathered evidence through comparing the collected evidence in the time of attack with the stored original digital evidence. But, in the proposed mechanism, we can collect the digital evidence by using the media independent method. The proposed model creates profile to guarantee digital evidence, and store the created profile to TTP and is used to guarantee the digital evidence's integrity.

The proposed model which is comprised in DEC, DEM and TTP administrates the rights in create and manage rights. Also, provides contents objectivity through profiling and time creation objectivity using the standard time which TTP generates. That means, the proposed model guarantees the reliability of the collected and created digital evidence from the attacked system which it is accomplished applying the media independent in the existing system environments.

4 Conclusion

In this paper we proposed a mechanism to guarantee digital evidence's integrity in the pervasive environments which can prevents and checks the system errors and unaccepted intrusion to the system and collect the digital evidence from the damaged system to be used in the court. Also the mechanism uses media independent property of digital evidence to have more reliability in the court.

The proposed mechanism contributes to strength the information security in the pervasive environments while we can improve the reliability of collected digital evidence in the damaged systems which are effectively used in the court. Also, we can collect the digital evidence periodically even though there is no intrusion activities and when the unaccepted intruder modify or attack the system, we can use the stored original digital evidence to get which information was modified and prove the integrity of the digital evidence. And also, we can contribute in the field of electronic commerce since our model guarantees the integrity of the digital data.

Hereafter, if the proposed mechanism for guarantee the digital evidence's integrity to be used widely in the future, it is necessary that the government must build the trusted third party. Also, it is required more research on the reliability guarantee mechanism to protect the reliability of main program and the system that generates the digital evidence.

References

1. Kruse ii, W.G., Heiser, J. G.: COMPUTER FORENSICS: Incident Response Essentials, Addison Wesley
2. Marcella Jr, A.J.: Greenfield, S., (ed.) Cyber Forensics: A Field Manual for Collecting, Examining, and Preserving Evidence of Computer Crimes, (2002)
3. Casey, E.: Handbook of Computer Crime Investigation: Forensic Tools & Technology (2001)
4. Farmer, D., Venema, W.: The Coroner Toolkit (TCT) v1.11, (September 2002) Available at: http://www. Porcupine.org/ forensics/tct.thml
5. TCT: The Coroner's Toolkit, http://www.fish.com/tct
6. Carrier, B.: Defining Digital Forensic Examination and Analysis Tools Using Abstraction Layers, International Journal of Digital Evidence (2003)
7. Palner, G.: A Road Map for Digital Forensics Research. Technical Report DTR-T001-01. DFRWS. Report From the Fiest Digital Forensic Reserch Workshop(DFWS) (November 2001)
8. Garber, L.: Encase: A Case Study in Computer-Forensics Technology, IEEE Computer Magazine (January 2001)

A Secure Chaotic Hash-Based Biometric Remote User Authentication Scheme Using Mobile Devices

Eun-Jun Yoon[1] and Kee-Young Yoo[2,*]

[1] Faculty of Computer Information, Daegu Polytechnic College,
San 395 3-Manchon-Dong, Susung-Gu, Daegu 706-711, South Korea
ejyoon@tpic.ac.kr

[2] Department of Computer Engineering, Kyungpook National University,
1370 Sankyuk-Dong, Buk-Gu, Daegu 702-701, South Korea
Tel.: +82-53-950-5553; Fax: +82-53-957-4846
yook@knu.ac.kr

Abstract. Traditional remote user authentication methods mainly employ the possession of a token (magnetic cards, cell phones, personal digital assistant (PDA), and notebook computers, etc.) and/or the knowledge of a secret (password, etc.) in order to establish the identity of an individual. In 2006, Khan et al. proposed an efficient and practical chaotic hash-based fingerprint biometric remote user authentication scheme on mobile devices. The current paper, however, demonstrates that Khan et al.'s scheme is vulnerable to a privileged insider's attacks and impersonation attacks by using lost or stolen mobile devices. Also, we present an improvement to their scheme in order to isolate such problems.

Keywords: Chaotic hash, Biometric authentication, Cryptanalysis, Mobile devices.

1 Introduction

With the large scale proliferation of mobile technology, remote user authentication in e-commerce and m-commerce has become indispensable for accessing precious resources. Remote authentication is a mechanism to authenticate remote users over insecure communication networks. It is evident that, with the passage of time, the volume of mobile user authentication is increasing because of ease in accessing resources from any remote location. Traditional remote user authentication methods mainly employ the possession of a token (magnetic cards, cell phones, personal digital assistant (PDA), and notebook computers, etc.) and/or the knowledge of a secret (password, etc.) in order to establish the identity of an individual. To spread this technology, commercial companies are providing remote authentication of mobile users to access their resources remotely (e.g. online banking and mobile commerce). A typical representation of a mobile user remote authentication system is depicted in Figure 1 [2].

* Corresponding author.

K.C. Chang et al. (Eds.): APWeb/WAIM 2007 Ws, LNCS 4537, pp. 612–623, 2007.

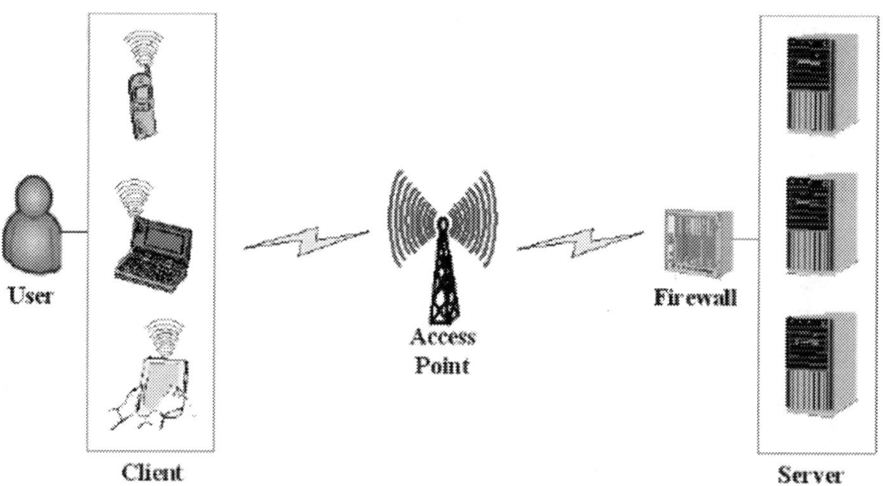

Fig. 1. Remote user authentication using mobile devices

A token, however, can be lost, stolen, misplaced, or willingly given to an unauthorized user; and a secret can be forgotten, guessed, or unwillingly-or willingly-disclosed to an unauthorized user. The science of biometrics has emerged as a powerful tool for remote user authentication systems. Since it is based on the physiological and behavioral characteristics of an individual, biometrics does not suffer from disadvantages found in traditional methods [1]. Also, biometrics and tokens have the potential to be a very useful combination. First, the security and convenience of biometrics allow for the implementation of high-security applications regarding tokens. Second, tokens represent a secure and portable way of storing biometric templates, which would otherwise need to be stored in a central database. Among the various biometric technological tools in use today, fingerprint recognition seems to be particularly suitable for token systems.

Recently, Khan et al. [2] proposed an efficient and practical chaotic hash-based fingerprint biometric remote user authentication scheme using mobile devices (e.g. cell phone and PDA). Their scheme is based on a new family of one-way collision free chaotic hash functions [3], which are more efficient than modular exponentiation-based authentication schemes (e.g. RSA). However, their scheme is vulnerable to a privileged insider's attacks and impersonation attacks by using lost or stolen mobile devices [4]. Therefore, the current paper demonstrates that Khan et al.'s scheme is still vulnerable to such attacks. Also, we present an improvement to their scheme in order to remove such problems.

This paper is organized as follows: Section 2 briefly reviews Khan et al.'s chaotic hash-based fingerprint biometric remote user authentication scheme using mobile devices; then Section 3 discusses its weaknesses. Our proposed scheme is presented in Section 4, while Section 5 discusses the security and efficiency of the proposed scheme. Our conclusions are presented in Section 6.

2 Review of Khan et al.'s Biometric Authentication Scheme

This section briefly reviews Khan et al.'s chaotic hash-based biometric authentication scheme [2]. Khan et al.'s scheme is composed of four phases: registration, login, authentication, and password change.

2.1 Registration Phase

Figure 2 shows the registration phase of Khan et al.'s scheme. In the registration phase, user U_i chooses his/her ID_i and password pw_i, and interactively submits these to the registration center. U_i also imprints his/her fingerprint impression at the sensor, and then registration system performs the following operations.

1. Computes $A_i = h_c(ID_i \oplus x)$, where x is the private key of the remote system, $\oplus$ is a bit-wise exclusive-or operation and $h_c(\cdot)$ is a collision free one-way chaotic hash function [3].
2. Computes $V_i = A_i \oplus h_c(pw_i \oplus S_i)$, where S_i is the extracted fingerprint template of the user.
3. The remote system personalizes the secure information $\{ID_i, A_i, V_i, S_i, h_c(\cdot)\}$ and saves it into the system of the U_i.

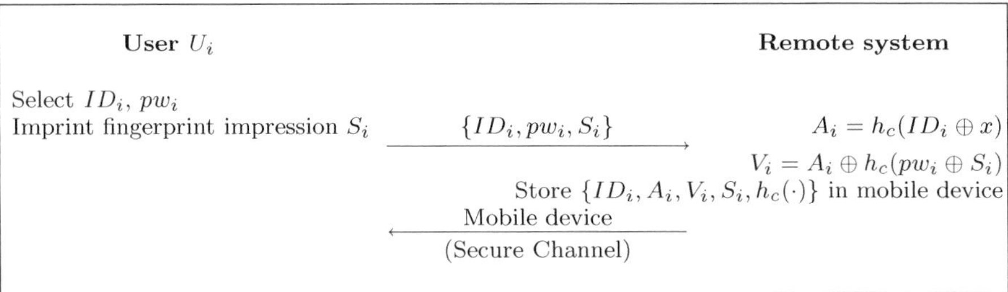

Fig. 2. Registration phase of Khan et al.'s scheme

2.2 Login Phase

Figure 3 shows the login and authentication phase of Khan et al.'s scheme. If U_i wants to login the system, he or she opens the login application software, enters ID_i and pw_i^*, and imprints a fingerprint biometric at the sensor. If U_i is successfully verified by his/her fingerprint biometric, a mobile device will perform the following operations.

1. Computes $B_i = V_i \oplus h_c(pw_i^* \oplus S_i)$, and verifies whether B_i is equal to the stored A_i or not. If the two are equal, the user's device performs further operations; otherwise it terminates the operation.
2. Computes $C_1 = h_c(B_i \oplus T_u)$, where T_u is the current timestamp of the device.
3. At the end of the login phase, U_i sends the login message $m = \{ID_i, C_1, T_u\}$ to the remote system over an insecure network.

2.3 Authentication Phase

In the authentication phase, the remote system receives the message from the user and performs the following operations.

1. Checks if the format of ID_i is invalid or if $T_s = T_u$, where T_s is the current time stamp of the remote system, then rejects the login request.
2. If $(T_s - T_u) > \Delta T$, where ΔT denotes the expected valid time interval for transmission delay, then the remote system rejects the login request.
3. Computes $C_1^* = h_c(h_c(ID_i \oplus x) \oplus T_u)$. If C_1^* is equal to the received C_1, it means the user is authentic, and the remote system accepts the login request and performs step 4; otherwise the login request is rejected.
4. For mutual authentication, the remote system computes $C_2 = h_c(h_c(ID_i \oplus x) \oplus T_s)$ and then sends a mutual authentication message $\{C_2, T_s\}$ to the U_i.
5. Upon receiving the message $\{C_2, T_s\}$, the user verifies that either T_s is invalid or $T_u = T_s$, then the user U_i terminates this session; otherwise performs step 6.
6. U_i computes $C_2^* = h_c(B_i \oplus T_s)$ and compares $C_2^*? = C_2$. If they are equal, the user believes that the remote party is an authentic system, and the mutual authentication between U_i and the remote system is completed; otherwise U_i terminates the operation.

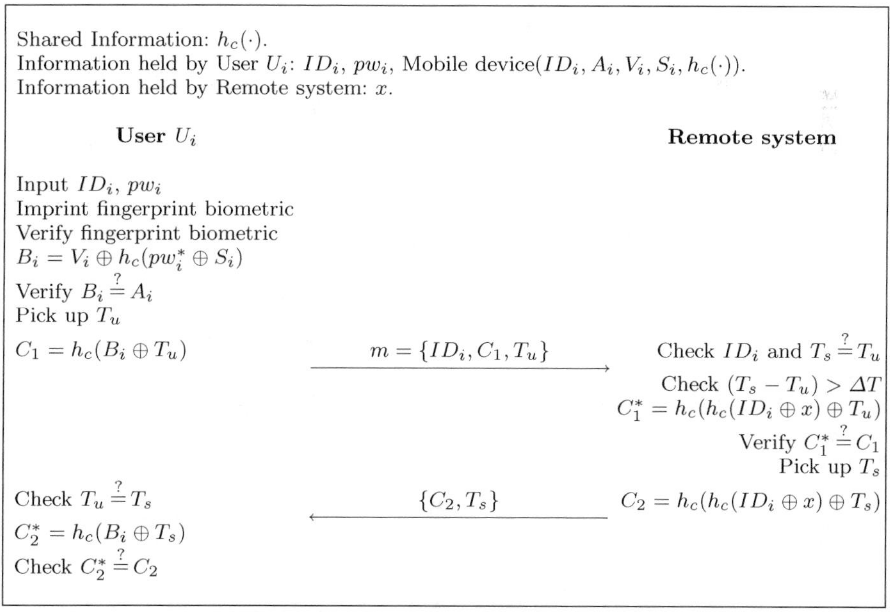

Shared Information: $h_c(\cdot)$.
Information held by User U_i: ID_i, pw_i, Mobile device$(ID_i, A_i, V_i, S_i, h_c(\cdot))$.
Information held by Remote system: x.

User U_i		**Remote system**
Input ID_i, pw_i		
Imprint fingerprint biometric		
Verify fingerprint biometric		
$B_i = V_i \oplus h_c(pw_i^* \oplus S_i)$		
Verify $B_i \stackrel{?}{=} A_i$		
Pick up T_u		
$C_1 = h_c(B_i \oplus T_u)$	$m = \{ID_i, C_1, T_u\}$ →	Check ID_i and $T_s \stackrel{?}{=} T_u$
		Check $(T_s - T_u) > \Delta T$
		$C_1^* = h_c(h_c(ID_i \oplus x) \oplus T_u)$
		Verify $C_1^* \stackrel{?}{=} C_1$
		Pick up T_s
Check $T_u \stackrel{?}{=} T_s$	← $\{C_2, T_s\}$	$C_2 = h_c(h_c(ID_i \oplus x) \oplus T_s)$
$C_2^* = h_c(B_i \oplus T_s)$		
Check $C_2^* \stackrel{?}{=} C_2$		

Fig. 3. Login and authentication phase of Khan et al.'s scheme

2.4 Password Change Phase

Whenever U_i wants to change or update his/her old password pw_i to the new password pw_i', he or she opens the login application on his/her mobile device and enters his/her ID_i and pw_i^*, and also imprints a fingerprint biometric at the sensor. If Ui is successfully verified, the mobile device performs the following operations without any help from the remote system:

1. Computes $B_i = V_i \oplus h_c(pw_i^* \oplus S_i) = h_c(ID_i \oplus x)$.
2. Verifies whether B_i is equal to the stored A_i or not. If the two are equal, the mobile device performs further operations; otherwise it terminates the operation.
3. Computes $V_i' = B_i \oplus h_c(pw_i' \oplus S_i)$.
4. Stores V_i' on the user's mobile device and replaces the old value of V_i. Next, the new password is successfully updated, and this phase is terminated.

3 Cryptanalysis of Khan et al.'s Scheme

This section shows that Khan et al.'s scheme is vulnerable to privileged insider attacks and impersonation attacks using lost or stolen mobile devices.

3.1 Privileged Insider Attack

In the registration phase of Khan et al.'s scheme, U_i's password pw_i will be revealed to the remote system because it is directly transmitted to the system. In practice, users offer the same password to access several remote servers for their convenience. Thus, a privileged insider of the remote system may try to use pw_i to impersonate U_i to login to the other remote systems that U_i has registered with outside this system. If the targeted outside remote system adopts the normal password authentication scheme, it is possible that the privileged insider of the remote system could successfully impersonate U_i to login to it by using pw_i. Although it is also possible that all the privileged insiders of the remote system can be trusted and that U_i does not use the same password to access several systems, the implementers and the users of the scheme should be aware of such a potential weakness.

3.2 Impersonation Attack by Using Lost or Stolen Mobile Device

Khan et al.'s scheme is vulnerable to impersonation attacks using lost or stolen mobile devices. Namely, a user can be authenticated to a remote system even if he or she does not have the valid password pw_i. Precisely, if an attacker gets a user's mobile device and extracts secure value A_i from the mobile device, then he or she can simply be authenticated by using A_i without the user's password. Needless to say, this is a serious problem. Figure 4 depicts the message transmission of an impersonation attack using a lost or stolen mobile device.

Public Information: $h_c(\cdot)$.
Information held by Attacker: ID_i, A_i.
Information held by Remote system: x.

Attacker)		Remote system
Pick up T_u		
$C_1 = h_c(A_i \oplus T_u)$	$m = \{ID_i, C_1, T_u\}$ $\longrightarrow$	Check ID_i and $T_s \stackrel{?}{=} T_u$
		Check $(T_s - T_u) > \Delta T$
		$C_1^* = h_c(h_c(ID_i \oplus x) \oplus T_u)$
		Verify $C_1^* \stackrel{?}{=} C_1$
		Pick up T_s
Check $T_u \stackrel{?}{=} T_s$	$\longleftarrow$ $\{C_2, T_s\}$	$C_2 = h_c(h_c(ID_i \oplus x) \oplus T_s)$
$C_2^* = h_c(A_i \oplus T_s)$		
Check $C_2^* \stackrel{?}{=} C_2$		

Fig. 4. Impersonation attack by using a lost or stolen mobile device in Khan et al.'s scheme

4 Proposed Biometric Authentication Scheme

This section proposes an improvement of Khan et al.'s scheme. The proposed scheme is also composed of four phases: registration, login, authentication, and password change. These phases are presented in the following subsections.

4.1 Registration Phase

Figure 5 shows the registration phase of the proposed scheme. User U_i chooses his/her ID_i, password pw_i and a random nonce n, and interactively submits $\{ID_i, pw_i \oplus n\}$ to the registration center. U_i also imprints his/her fingerprint impression at the sensor, and then the registration system performs the following operations.

1. Computes $A_i = h_c(ID_i \oplus x)$ and $X_i = h_c(A_i)$, where x is the private key of the remote system and $h_c(\cdot)$ is a collision free one-way chaotic hash function [?].
2. Computes $V_i = A_i \oplus h_c(pw_i \oplus n \oplus S_i)$, where S_i is the extracted fingerprint template of the user.
3. The remote system personalizes the secure information $\{ID_i, X_i, V_i, S_i, h_c(\cdot)\}$ and saves it into the system of the U_i.
4. U_i enters n into his/her mobile device.

4.2 Login Phase

Figure 6 shows the login and authentication phase of the proposed scheme. If U_i wants to login the system, he or she opens the login application software,

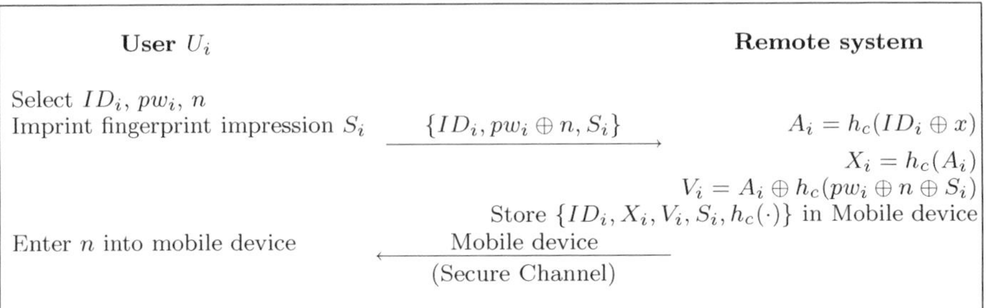

Fig. 5. Registration phase of proposed scheme

enters ID_i and pw_i^*, and imprints a fingerprint biometric at the sensor. If U_i is successfully verified by his/her fingerprint biometric, the mobile device performs the following operations.

1. Computes $B_i = V_i \oplus h_c(pw_i^* \oplus n \oplus S_i)$, and verifies whether $h_c(B_i)$ is equal to the stored X_i or not. If the two are equal, the user's device performs further operations; otherwise it terminates the operation.
2. Computes $C_1 = h_c(B_i, T_u)$, where T_u is the current timestamp of the device.
3. At the end of the login phase, U_i sends the login message $m = \{ID_i, C_1, T_u\}$ to the remote system over an insecure network.

4.3 Authentication Phase

In the authentication phase, the remote system receives the message from the user and performs the following operations.

1. Checks if the format of ID_i is invalid. If so, it then rejects the login request.
2. If $(T_s - T_u) > \Delta T$, where T_s is the current time stamp of the remote system and ΔT denotes the expected valid time interval for transmission delay, then the remote system rejects the login request.
3. Computes $C_1^* = h_c(h_c(ID_i \oplus x) \oplus T_u)$. If C_1^* is equal to the received C_1, it means the user is authentic, and the remote system accepts the login request and performs step 4; otherwise the login request is rejected.
4. For mutual authentication, the remote system computes $C_2 = h_c(h_c(ID_i \oplus x) \oplus C_1)$ and then sends the mutual authentication message C_2 to the U_i.
5. Upon receiving the message C_2, user U_i computes $C_2^* = h_c(B_i \oplus C_1)$ and compares $C_2^*? = C_2$. If they are equal, the user believes that the remote party is an authentic system, and the mutual authentication between U_i and the remote system is completed; otherwise U_i terminates the operation.

4.4 Password Change Phase

Whenever U_i wants to change or update his/her old password pw_i to a new password pw_i', he or she opens the login application on his/her mobile device

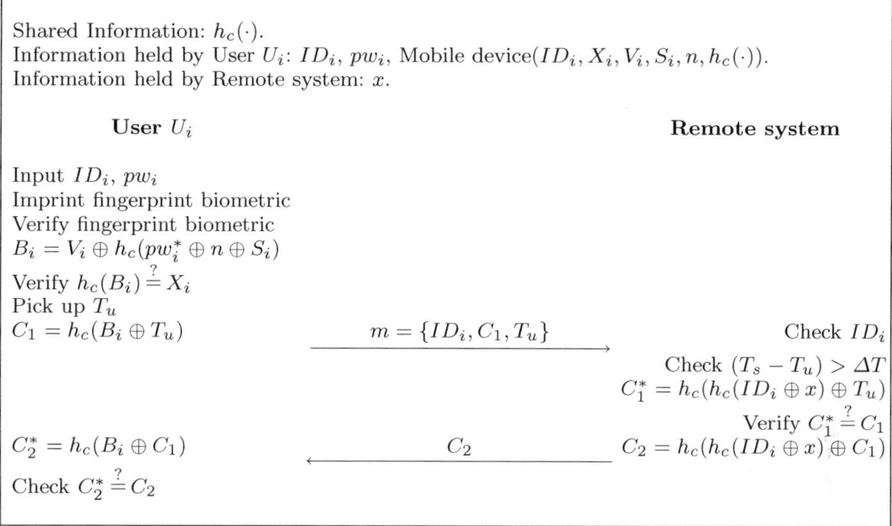

Fig. 6. Login and authentication phase of proposed scheme

and enters his/her ID_i and pw_i^*, and also imprints a fingerprint biometric at the sensor. If Ui is successfully verified, the mobile device performs the following operations without any help from the remote system.

1. Computes $B_i = V_i \oplus h_c(pw_i^* \oplus n \oplus S_i) = h_c(ID_i \oplus x)$.
2. Verifies whether $h_c(B_i)$ is equal to the stored X_i or not. If the two are equal, the mobile device performs further operations; otherwise terminates the operation.
3. Computes $V_i' = B_i \oplus h_c(pw_i' \oplus n \oplus S_i)$.
4. Stores V_i' on the user's mobile device and replaces the old value of V_i. Next, a new password is successfully updated, and this phase is terminated.

5 Security and Efficiency Analysis

This section discusses the security and efficiency features of the proposed scheme.

5.1 Security Analysis

This subsection provides the proof of correctness of the proposed scheme. First, the security terms [5] needed for the analysis of the proposed scheme are defined as follows:

Definition 1. *A weak password (pw_i) has a value of low entropy, which can be guessed in polynomial time.*

Definition 2. *A strong secret key (x) has a value of high entropy, which cannot be guessed in polynomial time.*

Definition 3. *A secure chaotic one-way hash function $y = h_c(x)$ is where given x to compute y is easy and given y to compute x is hard.*

Given the above definitions, the following analyzes the security of the proposed authentication scheme:

1. *The proposed scheme can resist replay attacks:* For replay attacks, the replay of an old login message $m = \{ID_i, C_1, T_u\}$ in the authentication phase will not work, as it will fail in step 2 due to the time interval $(T_s - T_u) > \Delta T$.
2. *The proposed scheme can resist guessing attacks:* Due to the fact that a chaotic one-way hash function is computationally difficult to invert, it is extremely hard for any an attacker to derive x from $h_c(ID_i \oplus x)$. Assume that attacker intercepts U_i's login request message $m = \{ID_i, C_1, T_u\}$ and the remote system's response message C_2 over a public network, due to the one-way property of a secure chaotic one-way hash function, the attacker cannot derive secure value B_i from C_1 and C_2.
3. *The proposed scheme can prevent a parallel session attack and reflection attack:* Because of the different message structures between $C_1 = h_c(B_i \oplus T_s)$ and $C_2 = h_c(B_i \oplus C_1)$, the proposed scheme prevents a parallel session attack and reflection attack.
4. *The proposed scheme can resist an insider attack:* Since U_i registers to the remote system by presenting $pw_i \oplus n$ instead of pw_i, an insider of the remote system using an off-line password guessing attack cannot obtain pw_i without knowing the random nonce n.
5. *The proposed scheme can resist an impersonation attack using lost or stolen mobile devices:* If legal users lose their mobile devices or an attacker steals a mobile device for a short duration and makes a duplicate of it. The attack cannot pass the mobile device verification process, step 1 of the authentication phase, because the attacker does not know the legal user's password pw_i and the mobile device checks $h_c(B_i) = X_i$. Furthermore, even if an attacker extracts all values $\{ID_i, X_i, V_i, S_i, n, h_c(\cdot)\}$ from the mobile device, the attacker cannot get A_i from X_i and V_i without the user's password pw_i. Thus, the proposed scheme prevents an impersonation attack using lost or stolen mobile devices to which Khan et al.'s scheme is vulnerable.
6. *The proposed scheme can achieve mutual authentication:* In step 3 of the authentication phase, the remote system can authenticate U_i. Also, U_i can authenticate the remote system in step 5 of the authentication phase because only a valid remote system can compute $h_c(B_i \oplus C_1)$. Therefore, the proposed scheme can achieve mutual authentication.
7. *The proposed password change scheme is secure:* In the password change phase, the user has to verify himself or herself by a fingerprint biometric, and it is not possible to impersonate a legal user because the biometric is unique. Furthermore, the value of $h_c(B_i)$ is also compared with the value of X_i in the mobile device. If these two values are not the same, the user is not

allowed to change the password. Moreover, if the mobile device (e.g. PDA or cell phone) is stolen, unauthorized users cannot change the password. Hence, our scheme is protected from the denial-of-service attack through a stolen device [6].

The security properties of Khan et al.'s scheme and of the proposed scheme are summarized in Table 1. In contrast with Khan et al.'s scheme, the proposed scheme is more secure.

Table 1. Comparison of security properties

	Khan et al.'s Scheme [2]	Proposed Scheme
Replay attack	Secure	Secure
Guessing attack	Secure	Secure
Parallel session attack	Secure	Secure
Reflection attack	Secure	Secure
Insider attack	Insecure	Secure
Impersonation attack	Insecure	Secure
Mutual authentication	Provide	Provide
Password change	Secure	Secure

5.2 Efficiency Analysis

Comparisons between Khan et al.'s scheme [2] and our proposed scheme are shown in Table 2. To analyze the computational complexity of the proposed scheme, we define the notation T_h, which is the time for computing the chaotic one-way hash function.

In the registration phases, Khan et al.'s scheme requires a total of two hashing operations, and the proposed scheme requires a total of three hashing operations. In the login and authentication phases, Khan et al.'s scheme requires a total of six hashing operations, and the proposed scheme requires a total of seven hashing operations. In the password change scheme, Khan et al.'s scheme requires a total of two hashing operations. For the same scheme, the proposed scheme requires a total of three hashing operations. The number of hash operations is increased by only three in the proposed scheme compared with Khan et al.'s scheme. The proposed scheme also does not add many additional computational costs.

Also, the proposed scheme uses a minimum communication bandwidth, unlike Khan et al.'s scheme. Among the four transmitted messages $\{ID_i, C_1, T_u, C_2\}$, one is the user's identifier (80 bit); one is a timestamp (80 bit); and two are hash output bits (128 bit such as a chaotic one-way hash function [3]). These are very low communication messages. Finally, the proposed scheme does not require server timestamp T_s to prevent reflection attacks, unlike Khan et al.'s scheme. Therefore, the proposed scheme is not only efficient but also enhances security.

Table 2. A comparison of computation costs

	Khan et al.'s scheme [2]	Proposed scheme
Registration phase	$2 \times T_h$	$3 \times T_h$
Login phase	$2 \times T_h$	$3 \times T_h$
Authentication phase	$4 \times T_h$	$4 \times T_h$
Password change scheme	$2 \times T_h$	$3 \times T_h$
Server timestamp	Required	Not Required
Communication costs	$\approx$ 496 bits	$\approx$ 416 bits

6 Conclusions

The current paper demonstrated that Khan et al.'s chaotic hash-based biometric authentication scheme is vulnerable to privileged insider's attacks and impersonation attacks using lost or stolen mobile devices. This paper also presented an improved scheme in order to solve such problems. As a result, in contrast to Khan et al.'s scheme, the proposed scheme is able to provide greater security.

Acknowledgements

This research was supported by the MIC of Korea, under the ITRC support program supervised by the IITA (IITA-2006-C1090-0603-0026).

References

1. Rila, L., Mitchell, C.J.: Security protocols for biometrics-based cardholder authentication in smartcards, 2846th edn. LNCS, vol. 2846, pp. 254–264. Springer-Verlag Heidelberg, Heidelberg (2003)
2. Khan, M.K., Jiashu, Z., Wang, X.M.: Chaotic hash-based fingerprint biometric remote user authentication scheme on mobile devices. Chaos, Solitons & Fractals, Elsevier Science, doi:10.1016/j.chaos.2006.05.061 (in press)
3. Wang, X.M., Jiashu, Z., Wenfang, Z.: Keyed hash function based on composite nonlinear autoregressive filter. Acta Phys Sinica;54:5566-73 (in Chinese) (2005)

4. Ku, W.C., Chuang, H.M., Tsaur, M.J.: Vulnerabilities of Wu-Chieus improved password authentication scheme using smart cards. IEICE Trans. Fundamentals E88–A(11), 3241–3243 (2005)
5. Menezes, A.J, Oorschot, P.C, Vanstone, S.A.: Handbook of applied cryptograph. CRC Press, Boca Raton, FL (1997)
6. Yoon, E.J., Ryu, E.K., Yoo, K.Y.: An improvement of Hwang.Lee.Tang's simple remote user authentication scheme. Comput. Secur. 24, 50–56 (2005)

Adapting Web Services Security Standards for Mobile and Wireless Environments

Nelly A. Delessy and Eduardo B. Fernandez

Dept. of Computer Science and Eng.,
Florida Atlantic University, Boca Raton, FL 33431, USA
ndelessy@fau.edu, ed@cse.fau.edu

Abstract. Web services are an important way for enterprises to interoperate. They are also becoming important for user access to services that depend on location and they are appearing in mobile devices. We consider the security standards needed for the use of web services in wireless networks. Web services security standards are used for the secure design of the communications between a web service and a mobile client and for the storage of the web service and its data. However, because those standards are designed to be flexible, they are also complex and verbose, and most often difficult to understand and implement. In addition, wireless devices have specific technological constraints and their own standards. We show here the use of patterns as a way to adapt web services security standards to the wireless environment. We also present a new pattern for the Liberty Alliance PAOS service.

Keywords: Mobile systems security, software patterns, web services security, web services standards, wireless devices.

1 Introduction

Web services are becoming an important way for enterprises to interoperate. They are platform-independent, language-independent services that expose an XML-based interface (WSDL) and can be registered in a UDDI registry. They communicate using XML messages in compliance with the SOAP standard. They usually use HTTP as a transport layer. They can be automatically discovered by their potential clients, and support loosely-coupled interactions between systems. Currently, they are mainly used to integrate existing heterogeneous systems. But they also enable the creation of new applications through web services composition, thus implementing a Service-Oriented Architecture (SOA). This type of application may involve a number of web services providers, possibly from different organizations that may not even know each other in advance, and could discover each other on the fly. The benefit of using web services for an organization is obviously to target a broader spectrum of customers without need of prior interactions with them. These goals are achieved in particular through standardization.

Web services are also becoming important for user access to services that depend on location and they are appearing in mobile devices. The concept of dynamic access

K.C. Chang et al. (Eds.): APWeb/WAIM 2007 Ws, LNCS 4537, pp. 624–633, 2007.

to web services allied with the flexibility of wireless accesses makes it possible to envisage a new type of applications, where the mobility of the user supplies the application with context elements. We can imagine applications in the field of disaster management, location services, advertising, etc.

We consider the security standards needed for the use of web services in wireless networks. Web services security standards are used for the secure design of the communications between a web service and a mobile client and for the storage of the web service and its data. However, because those standards are designed to be flexible, they are also complex and verbose, and most often difficult to understand and implement. On their part, wireless devices have specific technological constraints as well as their own standards. To use web services standards in mobile devices, it is necessary to adapt these standards to consider the limitations of portable devices. We show here the use of patterns as a way to adapt web services security standards to the wireless environment. We do not consider the security issues that come with the use of wireless networks alone, or web services alone. In the rest of the paper, we restrict our study to cellular networks and WLANs, since their usage is widespread. Ad hoc wireless networks or satellite networks are not considered here because they bring their own set of issues. Our approach uses security patterns. A security pattern solves a security problem in a given context, e.g. the XML Firewall pattern describes a mechanism to filter commands so as to prevent some security problems. This type of pattern is becoming accepted by industry.

There is a large variety of standards for web services security [10]. We will consider here XACML and the Liberty Alliance Identity Federation. XACML, (eXtensible Access Control Markup Language) is a security standard defined by OASIS [xac]. It includes policy and access decision languages. The Liberty Alliance is a consortium whose numerous industry members develop suites of standards concerning identity management and web services, such as the Liberty Alliance Identity Federation specification. We are proposing two approaches for adapting standards to wireless systems. XACML will be used to show one possible approach: *standard subsetting*, where parts of the standard which are considered too wasteful of resources or unnecessary for this environment are left out. Another approach is to start from the functions themselves and see which ones are the ones needed for some targeted applications and adapt the standard to these functions. As an illustration for this latter approach we show the PAOS pattern, a wireless-oriented version of the traditional HTTP binding for SOAP for web services. Since we have already shown a pattern for XACML [4], we only summarize it. The pattern for PAOS is new and is another contribution of this paper. There is little work on this subject, the most relevant appearing to be [2], which uses a lightweight service container, a quality of use model, and an adaptive algorithm to configure applications. Our approach can be complementary in that it can be used to guide its configurations.

The paper is organized as follows: Section 2 considers possible architectures to implement wireless web services. Section 3 is a summary of our XACML pattern while Section 4 discusses how to adapt three web services standards to wireless communications. We end with some conclusions in Section 5.

2 Architectures

Web services rely heavily on XML. This language is quite verbose and requires much computational power and memory space to parse it into usable information. We consider three possible architectures for the use of web services over wireless networks.

2.1 Use of a Gateway

Figure 1 illustrates this architecture. This is used where portable devices are particularly limited in memory and computational power. Furthermore, the connection bandwidth and reliability of the wireless connection are limited compared to wired connections. This is typical of most basic cellular phones. Web services are not delivered directly to the portable device but transformed in a gateway [6]. The portable device is a user of the web service. The gateway is in charge of transforming the SOAP messages into a compressed form that can be used by the mobile device. The gateway can also implement cache functions. Other examples of reductions include lower quality of images (resolution, size, color levels), minimization of user inputs, etc.

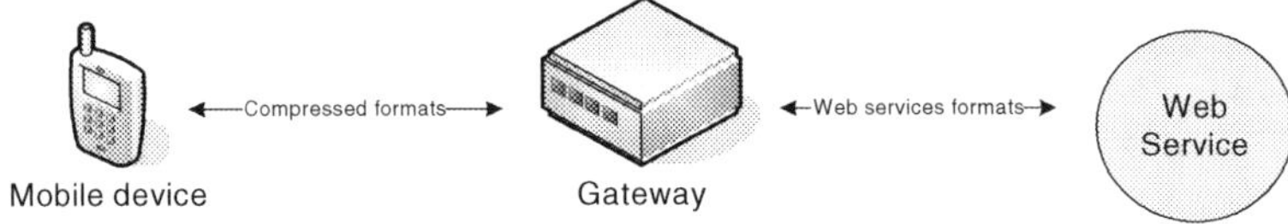

Fig. 1. Gateway architecture

2.2 Direct Consumer or Provider of Web Services

Figure 2 illustrates this architecture. In this case, portable devices must have built-in implementations of the web services technologies. This concerns the high end market segment, and includes smart phones, PDAs, and laptops. Hardware and operating systems security are important issues in this configuration. In particular, since the device, that is now a consumer of web services, can run client applications from different providers, a strong level of security is needed, including some type of authorization system. When the device is a provider of web services we also need it to control access to its services.

Fig. 2. Direct consumer architecture

2.3 Use of Mobile Agents

Figure 3 illustrates this architecture. This approach is suggested in [1]. The agents act as proxies on behalf of clients. This approach is used because using a web service can imply multiple passes between client, server and third parties (for security purposes for example) while the wireless link is not reliable and the bandwidth can be limited. Besides, agents can work even when the coverage is weak, or when the connection is lost.

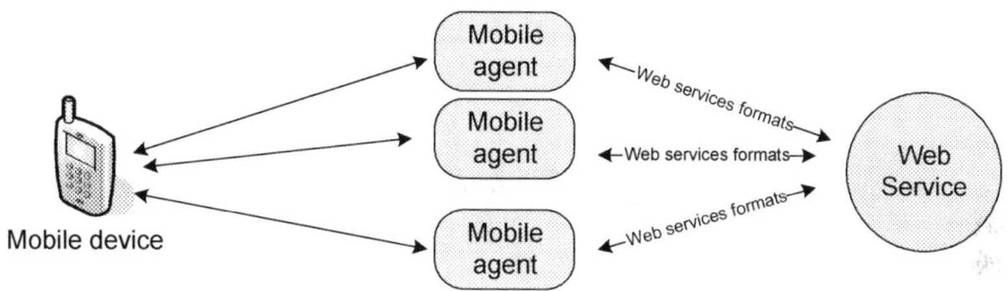

Fig. 3. Mobile agent architecture

3 Security Patterns

A pattern is a solution to a problem in a given context [11]. A security pattern solves a security problem, defining a generic solution to control some type of attacks. Our work and the work of others have shown that the use of patterns appears as a good approach to develop secure systems [7]. We have developed patterns to describe security mechanisms for web services, e.g. XML Firewalls [3], to describe web services security standards such as XACML [4] and SAML [8], to describe identity management for web services [Del 07], and to compare web services standards [8]. We are developing a catalog of security patterns that can be used to guide the design and evaluation of systems using web services. These are high level patterns that can abstract architectural solutions useful for securing web services deployed in a standalone manner or composed with other services. We describe below a complementary pair of our patterns [4] as a reference to show the typical complexity of some web services standards.

3.1 XACML Authorization

The intent of this pattern is to write all policies in a common language using a standard format, generic enough to implement some common high level policies or models (open/closed systems, extended access matrix, RBAC, multilevel). In addition, it defines a way to compose policies so that when several policies apply to one access, it is possible to render one unique decision: The policies are defined with an embedded combining algorithm. Figure 4 shows its class model.

In Figure 4 the **Subject** is authorized to access the **Resource**, and the **Environment** of the access defines conditions for the access. The Environment represents the characteristics of an access that are independent of the Subject or Resource and they could include the current date, time, or other environmental properties. A **Rule** is a basic unit of **Policy** and it has the usual meaning. In the access matrix model, it defines a set of **Subjects**, **Resources** (i.e. protection objects), and **Actions** (i.e. access types). However, in this pattern, a Rule associates not only one, but a set of Subjects, with a set of Resources, and a set of Actions. It also includes a set of Environments to which the rule is intended to apply, a condition and an effect ("Permit" or "Deny"). The condition refines the rule by imposing constraints on the subjects, the resources, or the environment' attributes. Policies are structured according to a Composite Pattern [11], where a **PolicySet** is the composite element. This indicates that policies have a tree structure. Each **PolicyComponent** may include an obligation that defines an operation that should be performed after enforcing the access decision. For example, an obligation could be an audit operation or a notification to an external client.

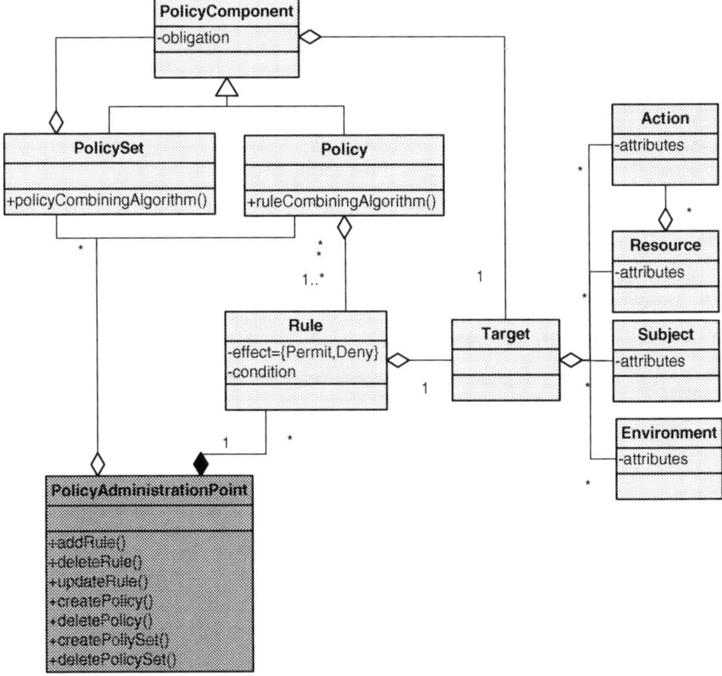

Fig. 4. Class diagram for XACML Authorization

3.2 XACML Access Control Evaluation

The idea of this pattern is to enforce the authorization rules defined by XACML Authorization using a **Policy Enforcement Point** (Figure 5). All access requests are submitted to a unique Policy Decision Point in a common format. This Policy

Decision Point returns an **Access Response** (decision), based on the **Applicable Policy Set** corresponding to the access's context. A Subject can access a Resource in the current Environment only if an XACML Access Response authorizes it to do so. The Subject, Resource and Environment are described through their attributes.

The **Policy Enforcement Point** requests an access decision to the Policy Decision Point through a **Context Handler**, which is an adapter between any specific enforcement mechanism and the XACML Policy Decision Point. The Policy Decision Point is responsible for deciding whether or not an access should be permitted, by locating the Applicable Policy Set, applying it to the **XACML Access Request**, and issuing a corresponding **XACML Access Response**. The Context Handler can also get attributes from a **Policy Information Point**, which is responsible for obtaining attributes from the subject.

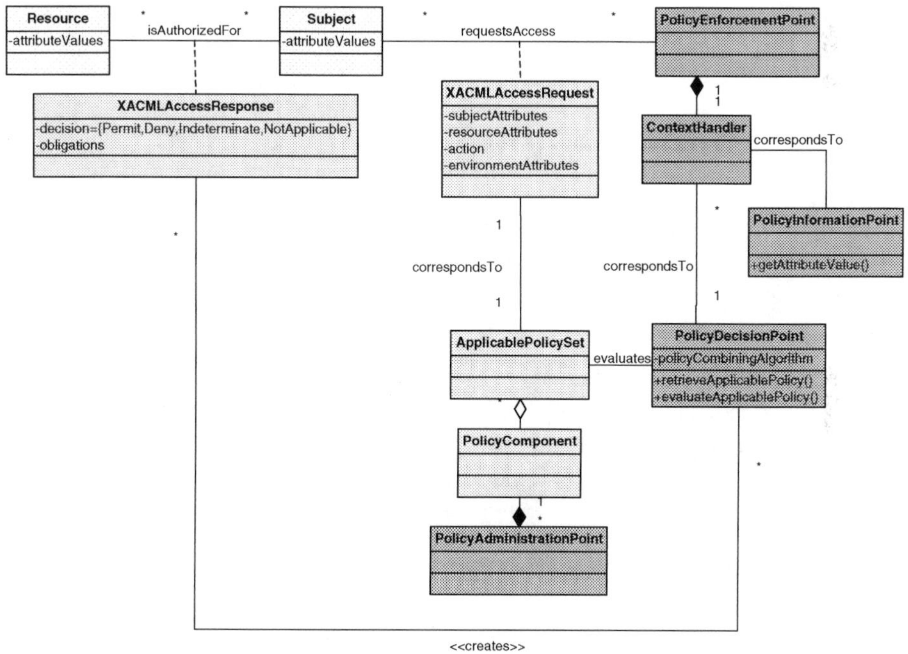

Fig. 5. Class diagram for the XACML Access Control Evaluation pattern

It is desirable that access control is realized by the web service. In this configuration, the resource is simply the web service endpoint, and the subject is the user. If this is realized according to the XACML access control evaluation pattern, the wireless operator, which already manages the users' identities, payment information, etc, can act as the PolicyInformationPoint, that is, as an attribute provider. XACML does not detail how to protect the communication between the PolicyInformationPoint and the PolicyEnforcementPoint. We need to emphasize that this communication channel must be protected, and in order to do that, a trust relationship must be developed.

4 Wireless Standards

All these standards are complex and require large amounts of memory and computation. We need to simplify them to use them in portable devices. Looking at a standard such as XACML, shown in the last section, we can see that its model for authorization rules and policies is too complex for the simple authorizations needed in wireless web services. In particular, a portable device only needs to control access to its own services and to the XML data stored in it. From this observation, we can eliminate the Composite pattern of policies and allow only simple policies. Environments and actions may not be needed either, as well as negative rules. Access to the web service is controlled by a set of rules. If we describe the rules according to the XACML Authorization pattern, we have a large flexibility to describe the user, which corresponds to a possible subject in the rule. User attributes may include phone number, SIM, name, location, IP address, etc, depending on the type of network she is using, and her privacy restrictions. In any case, a set of attributes can be defined further to define the context in which the user operates. This indicates that these aspects should be kept in the wireless reduced standard. A systematic analysis of each class and its attributes can lead us to remove or simplify other features.

Another approach to adaptation is to include minimal functionality; for example, Nokia includes identity services (based on Liberty alliance standards), and support for XML and SOAP. In addition to sending/receiving web services this allows authentication of the client or server. More advanced standards can be added later.

We present here a pattern that describes the Liberty Reverse HTTP Binding for SOAP specification which is an adaptation of the traditional HTTP binding for SOAP to wireless communications. This is done by binding a SOAP request to an HTTP response and a SOAP response to an HTTP request. The idea behind the reverse binding is to implement some kinds of call back service on a mobile device, so that a regular service can pull information from the user's device directly. Such information typically includes its location or some elements from one of its profiles. We use a common template to describe the sections of this pattern.

4.1 Liberty Alliance PAOS Service Pattern

An adapted version of the traditional HTTP binding for SOAP, suitable for mobile devices.

Context. Mobile devices with limited memory and computational power, equipped with HTTP clients (browsers) accessing web services that can be personalized. Rich information about the user, such as its location and its profiles may be available from the mobile device.

Example. Consider a user who wants to access a personalizable web service such as a restaurant directory from its mobile device. The personalizable web service could use the location information about the user to propose a list of nearby restaurants, but it does not know how to locate the user, and the user does not know precisely where he is and how to specify it to the web service.

Problem. Mobile devices usually do not have a practical interface which is convenient for manually entering information such as text, figures, etc. Accessing a web service from such a client should be not more as cumbersome than accessing the service from a traditional web service client. Besides, the mobile device can hold information about the user such as its location.

How to communicate personal information from the mobile device in a user-friendly fashion? The solution to this problem is affected by the following force:

- Mobile devices can be from different types (PDAs, laptops, smart phones, etc), and may run different platforms.
- They are usually limited in processing power.

Solution. The mobile device exposes PAOS-based services which can be accessed by the personalizable web services. The mobile device advertises for its PAOS-based services when accessing a personalizable service.

Structure. This is described in Figure 6.

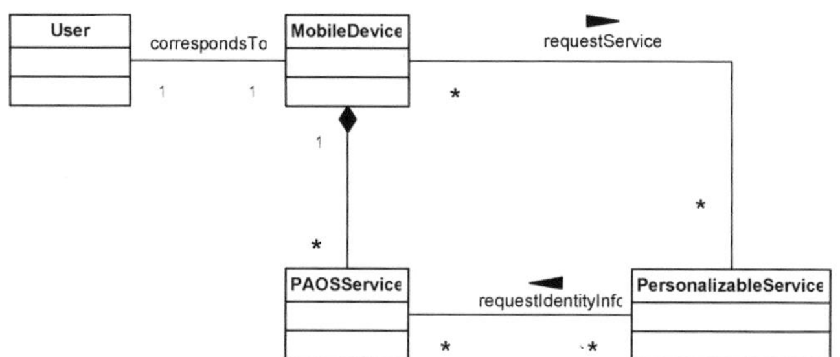

Fig. 6. Class diagram for Liberty Alliance PAOS

Dynamics. We describe the dynamic aspects of the Liberty Alliance PAOS pattern showing a sequence diagram for the use case User accesses a Personalizable Service.

Use case: User accesses Personalizable Service

Summary: A User accesses a Personalizable Service and its PAOS Service is advertised in the request. Upon detection of such an advertisement, the Personalizable Service sends back a request for some elements to the PAOS Service. Then the Personalizable Service can personalize the response to the service.

Actors: User, PAOS Service, Personalizable Service

Preconditions: The PAOS Service is running on the Mobile Device.

Description:

1. A User requests a service to a Personalizable Service.
2. The request is transparently augmented with an advertisement for the PAOS Service.
3. Upon detection of such an advertisement, the Personalizable Service sends back a request for elements to the PAOS Service.
4. The Personalizable Service personalizes the response using the elements and sends it back to the user.

Postcondition: The response to the user's request has been personalized using elements from its PAOS Service.

Example resolved. The user's mobile device can host a PAOS service that exposes the user's location. When accessing the personalizable restaurant directory service, the mobile device advertises for the existence of such a PAOS service. The restaurant directory service can therefore request the user's location and use this information for preparing the list of nearby restaurants that it will send back.

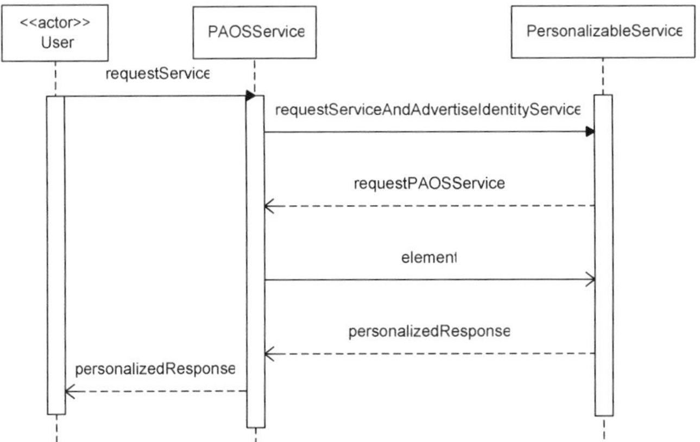

Fig. 7. Sequence diagram for Liberty Alliance PAOS Service

Consequences. This pattern presents the following advantages:

- It is possible to personalize services for each type of mobile device.
- Services have the ability to locate the user.
- The operation is transparent for the user.

Its possible liabilities include:

- The solution requires multiple messages between the mobile device and the service. This can increase the response time for the service.
- Possible privacy violations, since identity information is exposed to personalizable services.

5 Conclusions

Web services have the potential to offer consumers a variety of services. As the technology on mobile devices such as mobile phones or PDAs matures, web service providers will offer more advanced services to consumers. With this focus increasing, new security issues arise. Such issues involve identity standards, encryption/ decryption standards, as well as authorization. As we have seen some of the vendors of mobile devices are starting to adopt the security standards used for wired web services, although in reduced variations. We have made a case to use security patterns to adapt web services standards to wireless devices. We have also presented a pattern for a reduced web services standard as an example of adaptation. This pattern has value on its own for use in secure software development [7]. Patterns are valuable in these two approaches because of the graphic nature of their solutions; we can see by inspection what can be deleted or simplified or how to tailor the pattern for some application. If the technology changes, removed classes may be reintroduced or unnecessary classes can be removed. Our approach is simple and based only in understanding of the applications but it can be combined with more sophisticated approaches such as [2], which uses adaptive algorithms to configure applications.

It is important that wireless devices use security models based on standards. This approach allows the incorporation of wireless devices into a unified enterprise policy. In this way, policies for wireless devices can be designed as part of the overall design as we do in our methodology [7], and not in an ad hoc way. Approaches such as [12] that use separate policy models for wireless devices result in systems that will not be interoperable with other systems.

References

1. Bellavista, P., Corradi, A., Montanari, R., Stefanelli, C.: Context-aware middleware for resource management in the wireless Internet. IEEE Transactions on Software Engineering 29(12), 1086–1099 (December 2003)
2. Chu, H., You, C., Teng, C.: Challenges: Wireless Web Services. In: Procs of the 10th International Conference on Parallel and Distributed Systems (ICPADS '04) (2004)
3. Delessy, N., Fernandez, E.B., Rajput, S., Larrondo-Petrie, M.M.: Patterns for application firewalls. In: Procs. of Pattern Languages of Programs (PLoP 2004) http:// hillside.net/ plop/2004/final_submissions.html
4. Delessy, N., Fernandez, E.B.: Patterns for XACML, In: Procs. of Pattern Languages of Programs (PLoP 2005), (2005), http://hillside.net/plop/2005/proceedings.html
5. Delessy, N., Fernandez, E.B., Larrondo-Petrie, M.M.: A pattern language for identity management. In: Procs.of the 2nd IEEE Int. Multiconference on Computing in the Global Information Technology (ICCGI 2007) March 4-9, Guadeloupe, French Caribbean (2007)
6. Elkarra, N.: A Web Services Strategy for Mobile Phones, XML.com (2003), http://webservices.xml.com/pub/a/ws/2003/08/19/mobile.html
7. Fernandez, E.B., Larrondo-Petrie, M.M., Sorgente, T., VanHilst, M.: A methodology to develop secure systems using patterns. In: Mouratidis, H., Giorgini, P. (eds.) Ch. 5 in Integrating security and software engineering: Advances and future vision, IDEA Press, pp. 107–126 (2006)
8. Fernandez, E.B., Delessy, N.: Using patterns to understand and compare web services security products and standards. In: Procs of the IEEE Int. Conf. on Web Applications and Services (ICIW'06), Guadeloupe (February 2006)
9. Fernandez, E.B., Delessy, N.A., Larrondo-Petrie, M.M.: Patterns for web services security. In: Skar, L., Bjerkestrand, A.A., (eds.) Best Practices and Methodologies in Service-Oriented Architectures, 29-39, part of OOPSLA, the 21st Int. Conf. on Object-Oriented Programming, Systems, Languages, and Applications, Portland,OR, ACM, October 22-26 (2006)
10. Fernandez, E.B., Sorgente, T., Larrondo-Petrie, M.M., Delessy, N.: Web services security: Standards, industrial practice, and research issues, submitted for publication
11. Gamma, E., Helm, R., Johnson, R., Vlissides, J.: Design Patterns: Elements of Object-Oriented Software. Addison-Wesley, Boston, MA (1994)
12. Jansen, W., Karygiannis, T., Iorga, M., Gavrila, S., Korolev, V.: Security policy management for handheld devices. In: Procs. of the Int. Conf. on Security and Management (SAM'03) (June 2003)
13. http://www.13ia.com/13ia/0,8764,56843,00.html
14. Schumacher, M., Fernandez, E.B., Hybertson, D., Buschmann, F., Sommerlad, P.: Security Patterns: Integrating security and systems engineering. Wiley, Chichester, UK (2006)

Coexistence Proof Using Chain of Timestamps for Multiple RFID Tags*

Chih-Chung Lin[1], Yuan-Cheng Lai[1], J.D. Tygar[2], Chuan-Kai Yang[1], and Chi-Lung Chiang[1]

[1] Department of Information Management, National Taiwan University of Science and Technology
D9409106@mail.ntust.edu.tw, laiyc@cs.ntust.edu.tw,
[2] University of California, Berkeley
doug.tygar@gmail.com

Abstract. How can a RFID (Radio Frequency Identification Devices) system prove that two or more RFID tags are in the same location? Previous researchers have proposed *yoking-proof* and *grouping-proof* techniques to address this problem – and when these turned out to be vulnerable to replay attacks, a new *existence-proof* technique was proposed. We critique this class of existence-proofs and show it has three problems: (a) a race condition when multiple readers are present; (b) a race condition when multiple tags are present; and (c) a problem determining the number of tags. We present two new proof techniques, *a secure timestamp proof* (secTS-proof) and a *timestamp-chaining proof* (chaining-proof) that avoid replay attacks and solve problems in previously proposed techniques.

Keywords: RFID, coexistence proof, timestamp, computer security, cryptographic protocol, race condition.

1 Introduction

Radio Frequency Identification Devices (RFID), are supported by systems comprised of wireless readers and tags, and allows objects to be identified and, in some cases, tracked. [1] The most commonly used tags, passive tags, are inexpensive devices powered by radio signals from readers. They have sharply limited memory and processing capabilities. Some argue that RFID tags allow less expensive inventory management capabilities. [2] One important issue for RFID systems is generation of coexistence proofs demonstrating two or more RFID tags are simultaneously located. The key contribution of this paper is a critique of previous RFID coexistence proofs and a set of new proofs avoiding previous shortcomings.

* This research was supported in part with funding from the iCAST project NSC95-3114-P-001-002-Y02 in Taiwan, from the US National Science Foundation, from the US Air Force Office of Scientific Research, and from the Taiwanese National Science Council. The opinions expressed in this opinion are solely those of the authors and do not necessarily reflect the opinions of any funding sponsor.

K.C. Chang et al. (Eds.): APWeb/WAIM 2007 Ws, LNCS 4537, pp. 634–643, 2007.

Here are some motivating examples:

1) *Medical care*: a medical professional can prove that a set of correct drugs, blood products, or other medical materials are brought together for patient needs. The proof can be retained in case of dispute or for insurance purposes. [3]

2) *Transportation*: a transport or logistic firm can prove that items are always stored together in a safe box, if RFID tags are on the box and contents. [3]

3) *Forensics*: if RFID tags are on phones or other personal devices, law enforcement can use it to identify witnesses to a crime.

Here is a brief summary of prior work: Juels defined *yoking-proofs* which use a random number independently generated by tags to produce a coexistence proof. [3] Saito and Sakurai observed that yoking-proofs are vulnerable to replay attacks and proposed *grouping-proofs* that request timestamps from a trusted server. [4] Piramuthu observed that grouping-proofs were still vulnerable to replay attacks and proposed *existence-proofs* that keep a random number in the tag memory and sets the inputs of one tag to information generated by a second tag. [5] We can group these techniques into those using an off-line verifier (yoking-proofs) and those using an online verifier (grouping-proofs and existence-proofs).

We show existence-proofs have three problems:

- A race condition when multiple readers are present;
- A race condition when multiple tags are present; and
- Difficulties in determining the number of tags.

We give two new proof techniques:

- a *secure timestamp proof* (secTS-proof) – a proof technique based on a secure online verifier that issues secure timestamps; and
- a *timestamp-chaining proof* (chaining-proof) – a proof technique based on a secure off-line verifier. [6]

Both schemes avoid the replay attack, and can scale up to support proofs of arbitrary number of tags in a simultaneous environment.

Notation:

- *OV*: a trusted online verifier
- *FV*: a trusted off-line verifier
- *TSD*: a trusted timestamp database, which stores timestamps and message authentication codes from readers
- *TS*: a timestamp
- x: a symmetric key
- r: a random number
- *MAC*: message authentication code.
- $MAC_x[m]$: the MAC of message m under key x
- $SK_x[m]$: the encryption of m under key x
- P_{AB}: a proof that tags A and B were scanned simultaneously
- $P_{1\sim n}$: a proof that tags 1, 2, 3, …, n were scanned simultaneously

2 Related Work

2.1 Yoking-Proof

In yoking-proofs, the reader interacts with two RFID tags, T_A and T_B, and an off-line verifier (*FV*). T_A and T_B share secret keys x_A and x_B with *FV* and generate random numbers r_A and r_B, respectively, in every session. Figure 1 gives the protocol. After receiving P_{AB}, the reader forwards P_{AB}, r_A and r_B to *FV*.

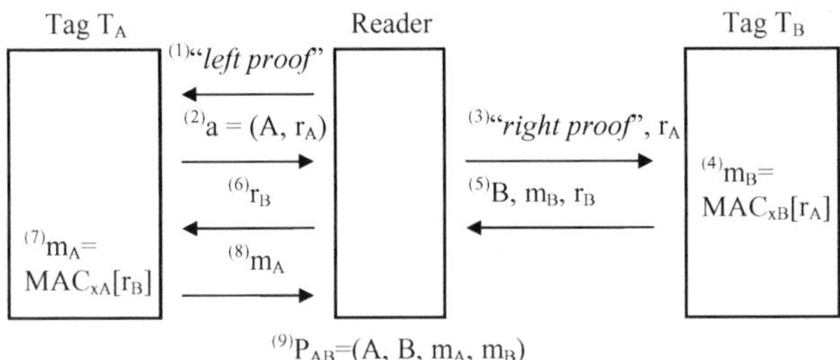

Fig. 1. Yoking-Proof

Because each tag uses a random number to compute a MAC, an adversarial reader can perform a replay attack by reusing previously generated random values. Saito and Sakurai showed a replay attack on T_A (see Figure 2 – the dashed line indicates the reader interacts with T_A and T_B at different times) [4] and Piramuthu showed a replay attack on T_B [5]. The yoking-proof technique cannot be repaired since the adversarial reader can send P_{AB}, r_A and r to *FV* for verification and ignores r_B.

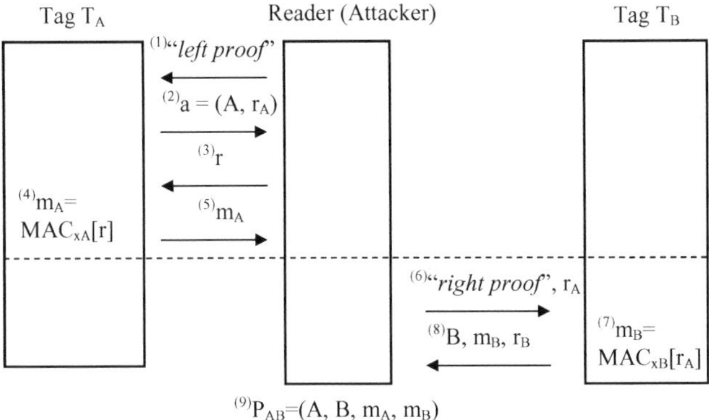

Fig. 2. Replay attack against Yoking-Proof

2.2 Grouping-Proof

Saito and Sakurai give a *grouping-proof* technique with the intention of avoiding replay attacks. [4] The reader acquires a timestamp (*TS*) from an on-line verifier (*OV*) and sends it to T_A and T_B. T_A and T_B individually compute m_A and m_B using the secret keys x_A and x_B – see Figure 3. After receiving proof P_{AB}, the reader sends P_{AB}, A, and B to OV. Grouping-proofs rely on a timeout mechanism – if the OV receives P_{AB} at time more than $TS+\Delta$, it rejects the proof.

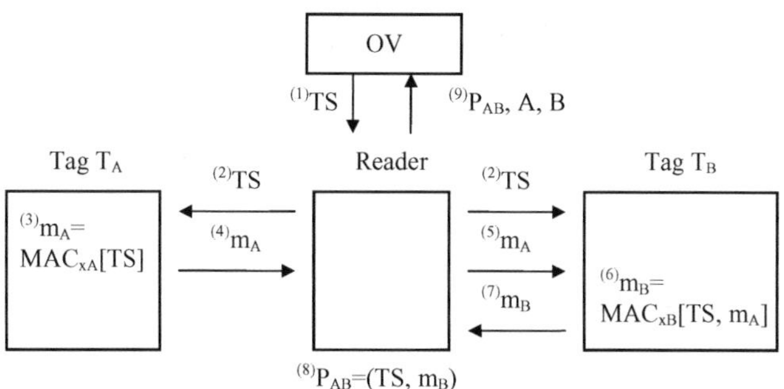

Fig. 3. Grouping-Proof

Piramuthu shows grouping-proofs are vulnerable to replay attacks (see Figure 4.). [5] An adversarial reader repeatedly transmits different future timestamps to tag T_A, generating different (TS, m_A) pairs. At the future time, the adversarial reader transmits these combinations to the tag T_B and then sends P_{AB} to OV. Note that the adversarial reader formally acquires TS from OV.

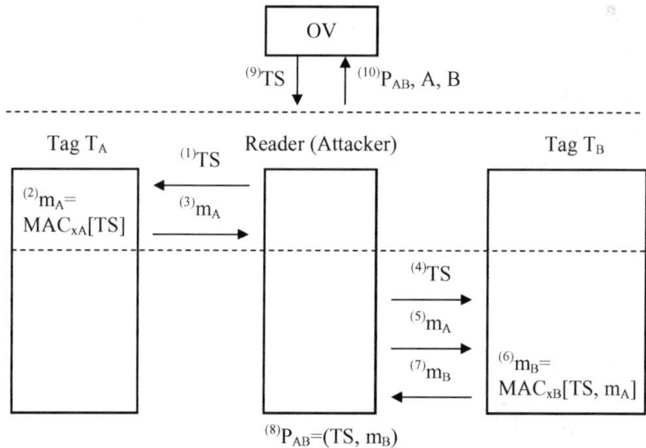

Fig. 4. Replay attack against Grouping-Proof

Saito and Sakurai extend their scheme to prove coexistence of multiple tags (Figure 5). Their scheme needs two types of tags, *product tags* and *pallet tags*. Product tags function similarly to the T_A or T_B discussed above. Pallet tags can compute symmetric key encryption and have larger memory stores than product tags.

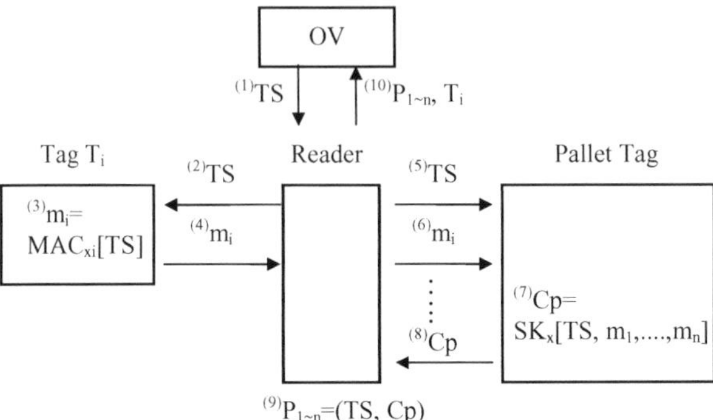

Fig. 5. Grouping-Proof for multiple tags

Product tags and a pallet tag share their secret keys with *OV*. The reader gathers n MACs from the product tag T_i ($1 \leq i \leq n$) and sends them to the pallet tag. The pallet tag encrypts n MACs m_i and *TS* to generate the ciphertext *Cp*. After the reader receives *Cp* from the pallet tag, it sends $P_{1\sim n}$ and all T_i ($1 \leq i \leq n$) to *OV*. *OV* first checks whether $P_{1\sim n}$ are within the timeout range. *OV* decrypts *Cp* using x to get m_i. *OV* verifies m_i using x_i. Note that this approach is also vulnerable to replay attacks.

2.3 Existence-Proof

Piramuthu proposes *existence-proofs* with the intention of avoiding replay attacks. [5] His idea is to ensure that inputs to a tag depend on information generated by other tags. Figure 6 shows his approach:

- The reader requests random number r from *OV*, which in term is a seed for generating r_A and r_B by tags T_A and T_B.
- T_B generates m_B which depends on both r and r_A.
- T_A uses m_B and r_A to generate m_A.

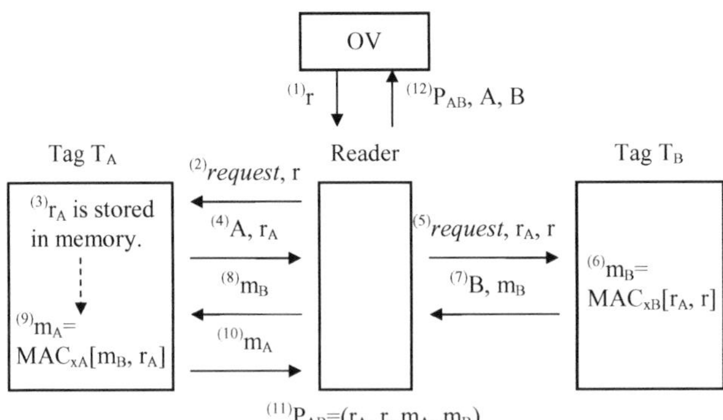

Fig. 6. Existence-Proof

Since both T_A and T_B rely on values generated by the other, it is robust against replay attacks. This scheme also uses a timeout mechanism to ensure the freshness of proofs.

3 Problems of Existence-Proofs

While existence-proofs avoid replay attacks, they have other problems. First, when tag T_A interacts with multiple readers, a race condition can occur (Figure 7). Reader1 sends r_1 to T_A and Reader2 sends r_2 to T_A almost simultaneously. T_A generates r_{1A} and r_{2A}, stores them in the memory, and transmits them to Reader1 and Reader2. After Reader1 and Reader2 individually interact with their other tags, i.e. T_{1B} or T_{2B}, they send m_{1B} and m_{2B} to T_A almost simultaneously. T_A does not know which r_A (r_{1A} or r_{2A}) should be used with m_{1B} or m_{2B} to generate m_{1A} or m_{2A}, causing a race condition.

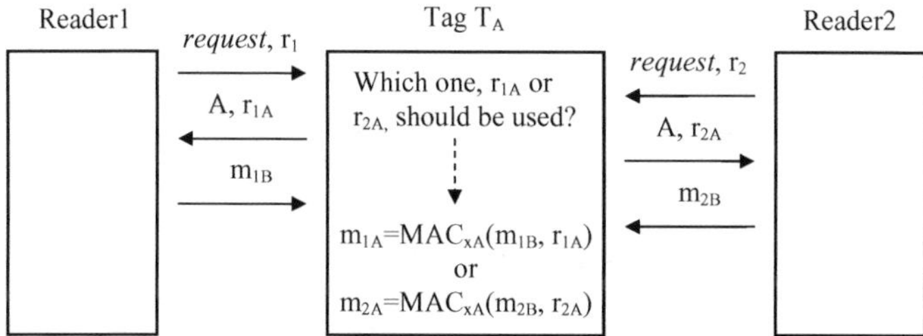

Fig. 7. Race condition for multiple readers

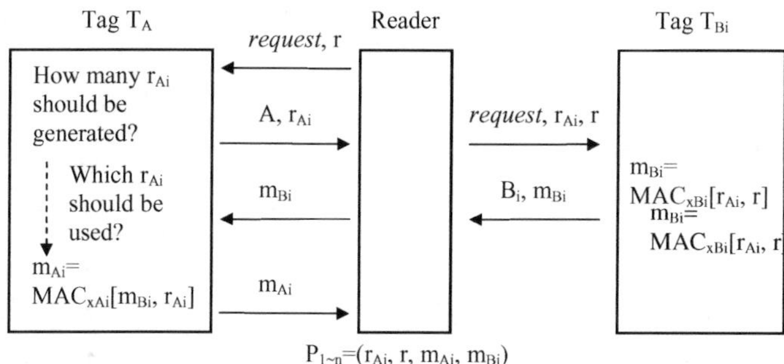

Fig. 8. Race condition for multiple tags and determining the number of tags

Piramuthu designed existence-proofs to scale to handle multiple tags. Tag T_A generates r_{Ai} (i=1, … , n-1, where n is the number of tags) by partitioning r into n-1 parts and the reader generates m_{Bi} according to each r_{Ai}. In the end, $P_{1\sim n}$ is composed of r_{A1}, r_{A2}, …, $r_{A(n\text{-}1)}$, r, m_{A1}, m_{A2}, …, $m_{A(n\text{-}1)}$ and m_{B1}, m_{B2}, …, $m_{B(n\text{-}1)}$. However, because T_A does not know how many r_{Ai} should be generated, it cannot determine the number of tags (see Figure 8.) Even if the reader can give the number of T_{Bi} to T_A, T_A

does not know which r_{Ai} is used with m_{Bi} to generate m_{Ai}, causing a different type of race condition, which we call the race condition for multiple tags.

4 Two Proposed Coexistence-Proofs

Above, we distinguished between systems with online verifiers and off-line verifiers. Below we give two proposed proof types, *secTS-proof* (with an online verifier) and *chaining-proof* (with an off-line verifier.)

Figure 9 shows the secTS-proof. To prevent adversarial readers from generating bogus timestamps, when the reader requests *OV*, *OV* generates a random number *r* and uses its secret key *x* to encrypt *TS* and *r* to create a unique *S*. *OV* also checks the freshness of proofs – if a proof is submitted after $TS+\Delta$, it is rejected.

Figure 10 shows the chaining-proof. In this scheme, the reader can issue the timestamp by itself. Because there is not an on-line verifier to monitor the reader's behavior, an attacker may issue a bogus timestamp. We use Haber-Stornetta timestamps [6] to avoid attack; each new timestamp is formed by taking a hash and using the hash value and MAC from previous timestamps. Because the reader does not have the tag's secret key, the timestamp can be generated until the last timestamp has been obtained. To complete verification, the reader must report the last timestamp, tag id, and timestamp MAC computed by the tag to an offline trusted third party (timestamp database *TSD*). When *TSD* receives the timestamp, it marks the timestamp information with a trusted time value.

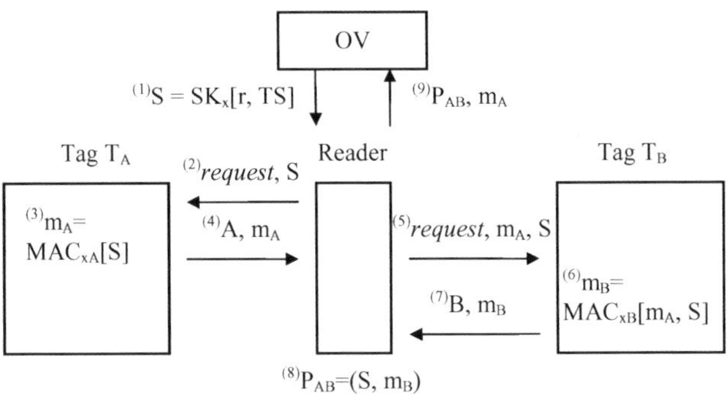

Fig. 9. SecTS-Proof

In the chaining-proof, each tag shares its secret key x_i with *FV*. Furthermore, the reader reports timestamps $ms_i=(T_i, TS_i(h(ms_{i-1})), m_i)$ to *TSD*, where *i* indicates the sequence that the reader scans the tags. ms_0 is a random number *r* acquired from *TSD*. Once *TSD* receives one ms_i, *TSD* stores and combines ms_i with a time (RT_i) which means when *TSD* receives this ms_i. The procedure of chaining-proof for multiple tags is described as follows:

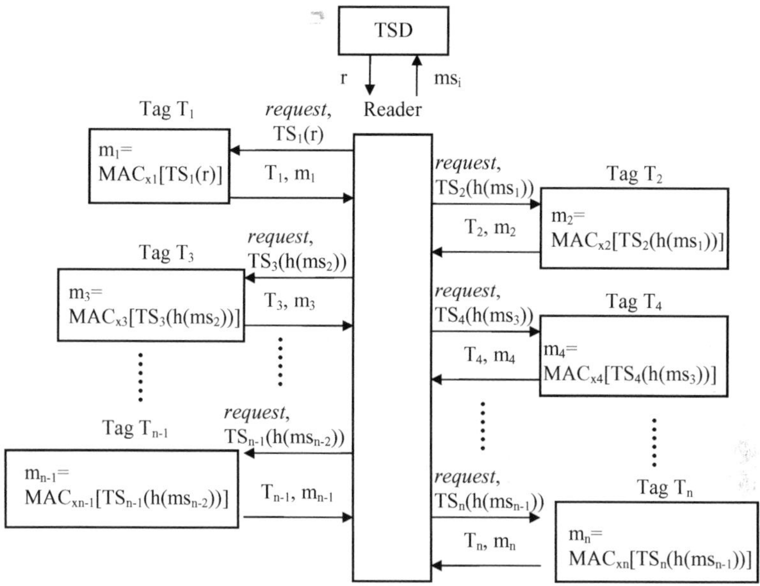

Fig. 10. Chaining-Proof

1. *TSD* gives a random number r to the reader and combines r with a time, RT_0.
2. The reader issues TS_1 including a random number r to the tag T_1.
3. T_1 generates m_1 by applying x_1 to TS_1, and sends its id T_1 and m_1 to the reader.
4. The reader reports $ms_1=(T_1 \| TS_1(r) \| m_1)$ to *TSD*, and *TSD* stores and combines it with a time, i.e. RT_1.
5. The reader sends TS_2 including $h(ms_1)$ to the tag T_2. $h(\ )$ is a one-way hash function.
6. T_2 use x_2 on TS_2 to generate m_2, and submits its id T_2 and m_2 to the reader.
7. The reader reports $ms_2=(T_2 \| TS_2(h(ms_1)) \| m_2)$ to *TSD*, and *TSD* stores and combines it with a time, i.e. RT_2.
8. The remainder tags and the reader repeat until the reader gets all MACs of nearby tags.
9. When verifying, *FV* receives the proof $P_{1\sim n}$ from the reader.
10. *FV* extracts ms_i from $P_{1\sim n}$ and sequentially acquires RT_i from *TSD* according to ms_i.
11. *FV* checks that the duration between RT_i and RT_{i-1} is less than Δ, a predefined time threshold.
12. After *FV* checks all time durations, if no any problem exists, *FV* verifies each MAC m_i by using the corresponding secret key x_i and $TS_i(h(ms_{i-1}))$.
13. Finally, *FV* verifies whether each TS_i is located between RT_0 and RT_{n-1}. If so, TS_i are accepted. If no problem exists, the proof is.

5 Sketch of Security Proof

Space does not permit a full proof of security; so here we just provide a sketch of security for the chaining-proof. (SecTS-proofs are substantially simpler to show security for.) We assume an adversary can control one or more reader, but not a verifier or *TSD*. We assume that our cryptographic functions observe standard requirements (see [6] for a fuller discussion of security of the timestamp mechanism.). Now the sketch of the proof is straightforward. Refer to Figure 10. If the tag T_2 is not in range of an adversarial reader the attacker cannot get m_2 from T_2. Without m_2, the adversary cannot compute $ms_2=(T_2 \| TS_2(h(ms_1)) \| m_2)$ and cannot submit ms_2 to the on-line timestamp database (*TSD*). Alternatively, if the adversary waits for the tag T_2, *TSD* will mark ms_2 with the later time timestamp (RT_2) – and if this exceeds timeout range Δ, the key will be discarded. Replay attacks against T_1, T_3, T_4, …, T_n are not possible for parallel reasons.

Note that because the reader uses a random number r from *TSD* and this value is combined with a trusted time in *TSD*, *FV* can discover if an adversarial tries to collect information by interacting individually with each tag using bogus timestamps at some later time.

Note further, if a proof $P_{1234}=(r, T_1, T_2, T_3, T_4, TS_1(r), TS_2(h(ms_1)), TS_3(h(ms_2)), TS_4(h(ms_3)))$ is already verified and is valid. If an adversarial reader attempts to duplicate the same proof by using a valid r and insert existence evidence of T_5 , i.e. $(T_5, TS_5(h(ms_4))$, in this proof, the attack will be detected, since each timestamp inside the proof chains is reported to *TSD*.

6 Conclusion

We showed three RFID co-existence proof types (yoking-proofs, grouping-proofs, and existence-proofs) suffer from a number of problems: replay attacks, race conditions, and ambiguity in the number of tags. We proposed two novel proof types: secTS-proofs and chaining-proofs. SecTS-proof is applied on the environment having an online verifier while chaining-proof is used on the environment having an off-line verifier. Our two schemes successfully avoid all known attacks, including replay attacks.

References

1. Shepard, S.: RFID: Radio Frequency Identification. McGraw-Hill, New York (2005)
2. Juels, A.: Strengthening EPC tags against cloning. In: Proceedings of the 4th ACM Workshop on Wireless Wecurity, pp. 67–76 (2005)
3. Juels, A.: Yoking-proofs for RFID tags. In: Proceedings of the Second IEEE Annual Conference on Pervasive Computing and Communications Workshops. 2004, pp. 138–143 (2004)

4. Saito, J., Sakurai, K.: Grouping proof for RFID tags. In: Proceedings of the 19th IEEE International Conference on Advanced Information Networking and Applications, pp. 621–624 (2005)
5. Piramuthu, S.: On existence proofs for multiple RFID tags. In: Proceedings of the ACS/IEEE International Conference on Pervasive Services, pp.317–320 (2006)
6. Haber, S., Stornetta, W.: How to time-stamp a digital document. Journal of Cryptology 3(2), 99–111 (1991)

A Design of Authentication Protocol for Multi-key RFID Tag

Jiyeon Kim[1], Jongjin Jung[2], Hoon Ko[3], Boyeon Kim[4], Susan Joe[2], Yongjun Lee[2], Yunseok Chang[2], and Kyoonha Lee[1]

[1] Inha University
[2] Daejin University
[3] Chungnam National University
[4] Hanyang University
jini_69@naver.com, khlee@inha.ac.kr,
{jjjung,sjoe3,eyongjun,cosmos}@daejin.ac.kr,
skoh21@cnu.ac.kr, bykim@hanyang.ac.kr

Abstract. RFID systems are spreading in the area of the industry and marketplace faster and now this time it is coming up to usual area covering individual life and environment as a core element of the ubiquitous technologies. But the RFID system has RF device communicates each other by radio frequency and it could make some serious problems on the important area such as privacy violation and information security. In this paper, we propose multi-keys RFID tag scheme including tag structure and authentication protocol which can support efficient security on the tag information by using SEED algorithm. The multi-key RFID tag could maintain multiple object IDs for different applications in a single tag and allow simultaneous access for their pair applications. The authentication protocol for multi-key RFID tag is designed to overcome the security and privacy problems and has enough robustness against the various attacks on the RFID system with low cost. We compared the security capability of the schemes and the evaluation results showed the proposed scheme has better efficiency than existing schemes.

Keywords: RFID, multi-key, tag structure, authentication protocol.

1 Introduction

Recently, the information technology has evolved toward the ubiquitous environment accessible to the network everywhere and every time. The ubiquitous environment can provide easy access to the devices and make one's economical benefit also. The RFID system is an important core technology in ubiquitous environment. RFID system consists of contactless devices to communicate each other by radio frequency. It provides technologies of automatic object identification in invisible range, read/write function and adaptability against various circumstances. These advantages make RFID systems to be applied in various fields and expect one of the big markets in the area of human life such as traffic card system, toll gate system, logistics, access control, etc. As RFID tags identify many different types of objects, it is going to

K.C. Chang et al. (Eds.): APWeb/WAIM 2007 Ws, LNCS 4537, pp. 644–653, 2007.

increase the tags that have to be carried in individual life. But it is uneasy for a person to control many tags in a hand because traditional RFID systems have restriction that is one tag per each object and it is difficult to distinguish tags without some kind of effort. That is why a tag is used to store identifying information just for a single object in common RFID applications. In this paper, we proposed multi-key RFID tag structure which is shared by many different applications. A key identifies a corresponding object in an application and we call it object ID. Therefore, multi-key RFID tag means that a tag has more than one key for multiple objects stored in a tag memory. The proposed tag structure can integrate different keys for different applications into data memory in a same tag.

We also proposed an efficient authentication protocol for multi-key RFID tag structure. RFID system often raises seriously violation of privacy caused by various attacks because contactless devices communicate each other by radio frequency. Various proposals have been made to prevent these kinds of problems. They are divided into two types of solutions. One is the physical scheme as kill-command scheme [2], Faraday cage scheme [6] and blocker-tag scheme [5]. The other is encryption scheme as hash lock scheme [1], randomized hash lock [2], hash chain [3] and variable ID scheme [4]. However, most schemes still have weakness against various attacks such as spoofing, traffic analysis or location tracking, etc. Moreover, some of these schemes are not applicable because they are not suitable for low cost RFID systems which are being developed. Others are not practicable on account of their demand for circuit size and operation power. In this paper, we design an efficient authentication protocol to overcome the problems of security and privacy. The proposed protocol is designed by considering of robustness against various attacks in low cost RFID systems.

We described RFID system technologies and requirements in section 2 and reviewed several existing schemes in section 3. In section 4, we proposed multi-key RFID tag scheme and evaluated the efficiency of our scheme comparing with other schemes in section 5 consequently. Finally, we concluded in section 6.

2 Basic Requirements

The standardization organizations for RFID tags are JTC1 (Joint Technical Committee 1) by ISO/IEC and EPC Global by GS1 (Global Standard 1). Tag identifier is a number to distinguish a tag from others in communication with readers. International standard observes the structure of the number [8]. Permanent unique ID named as chip ID or tag ID is written (masked) within a tag by tag producer according to the ISO/IEC 15963 standard [7]. Another identifier is item ID which is used to distinguish the tagged objects. Tag memory could have the item ID under user's decision and standardization on item ID is still under discussing. Therefore, we would like to use the item ID on multiple objects access in this paper. Fig. 1 shows the typical example of the multiple item IDs in a same tag.

RFID systems often make serious violation of privacy and security caused by various attacks through the weakness of their wireless interface. Vulnerabilities to eavesdropping, location tracking, spoofing, message losses or replay attack can threaten RFID components. These attacks may affect individual privacy and

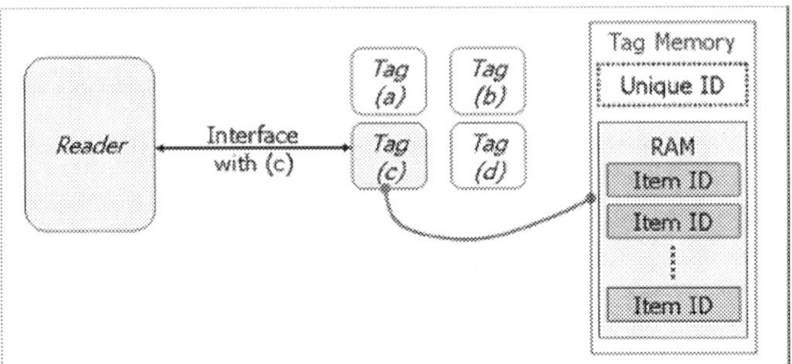

Fig. 1. Concept of tag identifier

information security. Therefore, efficient security mechanism against attacks must be considered when RFID application systems are designed.

Unprotected tags could be read by unauthorized readers. Attackers are used to eavesdrop on the general information stored in a tag such as serial number or code number from readers or tags though they can't understand the meaning of the number. The attackers retransmit the eavesdropping data to the server and find out the critical information. With the eavesdropping data, the attackers also detect the tag location and analyze traffics. Therefore, RFID system has to be designed with data encryption and authentication protocol for preventing the attackers from replay of tapped data and location tracking. This reason makes the authentication protocol of RFID system may change the response of corresponding tag to the query of reader after each session. Spoofing is another type of attacks and it means that attacker joins in authentication protocol in the disguise of authorized tag or reader. To prevent the spoofing attack, a RFID system has to control the authority of tag access in authentication protocol and must block the attacker' illegal information gathering. In addition to the above attacks, the attackers often intercept communications between reader and tag. This behavior causes loss of data which is important to authentication process. Therefore, the safe and reliable RFID systems should detect the interference coming from various kind of attack as many as it can.

3 Related Works

Several papers proposed the solution schemes against the attacks of privacy threat. There are two types of scheme: One is the physical scheme such as kill-command scheme, Faraday cage scheme and blocker-tag scheme. The other is the encryption scheme such as hash lock scheme, randomized hash lock, hash chain and variable ID scheme. In this study, we have focused on the encryption methods which are hot among researchers lately. The hash lock scheme is simple access control mechanism based on one-way hash function. It uses metaID to process authentication between reader and tag. The metaID is a temporary ID generated from one-way hash function with single key. Both the reader and tag store the metaID separately and match the key with metaID during the authentication process. A tag responds to all queries with only its metaID and decides whether the tag offers it or not. This scheme only requires a hash function on the tag and key management on the back-end database. So, it would be the best one of the cost-efficient solutions in the near future. Based on the difficulty of inverting a one-way function, this scheme also prevents unauthorized readers from reading tag contents. Maybe, spoofing attempts may be detected under this scheme but not prevented. The hash lock scheme can't prevent replay attack and location tracking either because the metaID has constant value. Randomized hash lock scheme improves the hash lock scheme and uses variable metaID. It can generate a different metaID with random number generator in every session. Therefore, a tag would not respond to queries from unauthorized readers. This scheme can prevent RFID tag from tracking but not replay attack and spoofing. Hash chain scheme uses two different hash functions to change the response message for reader. This scheme can prevent tag tracking and replay attack but still has weakness against spoofing. And actually, it is not practicable on account of their demand for circuit size and operation power because a tag has to keep two hash functions. Variable ID scheme changes tag ID by a random value in every session. For the replay attack, the scheme keeps transaction ID (TID) and last successful transaction (LST) in each session. However, this scheme may allow adversaries to track when LST was not updated on occurring message loss during the session.

4 Multi-key RFID Tag Scheme

4.1 Tag Structure

Multi-key RFID tag structure is a new type of RFID tag structure that has multiple keys for multiple RFID readers. As increasing the RFID systems applies on individual life, people are expected to have several tags to identify the different types of objects. It causes serious problem on managing different tags in one's pocket and needs a new method to handle multiple objects in a simple way. But it is not easy to control many tags because traditional RFID systems can handle only one tag per each object. In ubiquitous computing environment, information integrating and device sharing have been increased on the various application areas and now we need an efficient solution to simplify the control and management method. Therefore, if there is a kind of method to share one tag for many RFID applications, it can be suitable for the ubiquitous computing environment in many ways. In this paper, we propose a multi-key RFID tag structure which can be shared by different RFID applications.

In usual RFID systems, a key saved in a RFID tag is used to identify an object in a specific application. We call this key as *object ID*. Multi-key RFID tag for multiple objects means that a single tag has more than a key for each distinct object with tag ID. This tag structure integrates different keys for different applications into data memory in a tag. In actual cases, multiple object application causes security problem in multiple key accesses and needs some appropriate method to tie up between peer key and object with security level. To satisfying this specific need, the tag ID and each object ID are encrypted with SEED encryption algorithm [9]. We call the results of encryption as *encrypted IDs*. The SEED is one of the famous block cipher algorithm developed by the KISA (Korean Information Security Agency) and broadly used throughout South Korean public and industry field. These encrypted ID and object ID peers are stored in a tag memory and back-end database respectively. Fig. 2 simply shows the concept of multi-key tag structure.

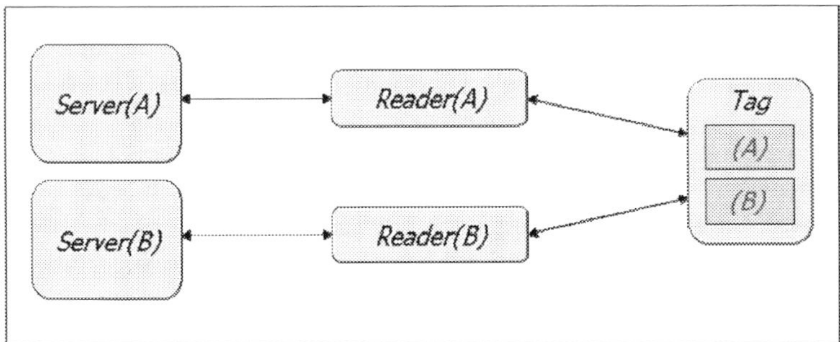

Fig. 2. Proposed multi-key tag structure

When a certain type of reader (*A*) transmits a query for tag connection, the tag searches corresponding encrypted ID matches the reader's object type. Then, the tag makes a variable ID with a random value and the inquired encrypted ID responding to the reader. At this point, corresponding reader transmits the random value when it queries to the tag. In RFID application on multi-key tag structure, some type of data is necessary to identify authorized components in authentication protocol. We define several types of data which should be stored in tag memory and back-end database in server respectively as shown in Table 1.

Table 1. Data types in the server and tag

RFID components	Data types
Server	tag ID, object ID, encrypted ID, partial ID
Tag	object ID, encrypted ID, partial ID

The meanings of each stored data type are as follows:

- *Tag ID* is a unique identifier of tag. It is used as the key for creating encrypted ID by SEED encryption algorithm. In the multi-key RFID tag structure, only one tag is used to identify multiple objects. This means that a tag should maintain multiple tag IDs for multiple objects. But instead of maintaining multiple tag IDs, encrypted IDs take place the roles of tag IDs and stored in tag memory for distinguishing each distinct object. Different encrypted IDs can be generated by SEED algorithm using object IDs even if they have the same tag ID.
- *Object ID* is an identifier to distinguish different objects in multi-key RFID system. It is stored in tag memory and server database respectively. When a reader

transmits a query to a specific tag, the query includes an object ID that is used to search peer encrypted ID in tag memory.

- *Encrypted ID* is a key value to identify a certain object. It is correspondent to object ID and generated by SEED encryption algorithm using both object ID and tag ID. The encrypted ID is stored in tag memory and server database respectively when system is initialized. According to the object type, two or more encrypted IDs can be stored in a tag.
- *Partial ID* is a temporary identifier of encrypted ID which is generated by hash function embedded in a tag. It is a transformation or variation of encrypted ID made by a tag using random value transmitted from a reader. Partial ID is changed and becomes a different value at every session. The last successful partial ID is maintained in database and a tag. If the authentication process ends successfully, the tag and the server update the existing partial ID to prevent attack's threat. In the multi-key RFID tag system, the tag or the server can perceive illegal adversaries' spoofing or replay attack by comparing the stored partial ID with the attacker's.

4.2 Authentication Protocol for Multi-key RFID Tag

To ensure the security and safety between tag and reader, we also designed an authentication protocol for multi-key RFID tag structure. As mentioned above, server and tags store the encrypted IDs generated by SEED encryption algorithm respectively at the point of RFID system initialization. The SEED algorithm takes both object ID and tag ID as input keys. Since a decryption of the result of SEED encryption algorithm without input keys is very hard and expensive work, physical replication of tag isn't worth for attacks [9]. These characteristics of SEED provide a high level security and reliability in RFID systems. Therefore, we employ SEED algorithm on key encryption to identify a corresponding object. We also designed an authentication protocol by considering of robustness against privacy threats such as location tracking, spoofing, re-play and message loss. The Fig. 3 shows the basic scheme of the authentication protocol for multi-key RFID tag structure.

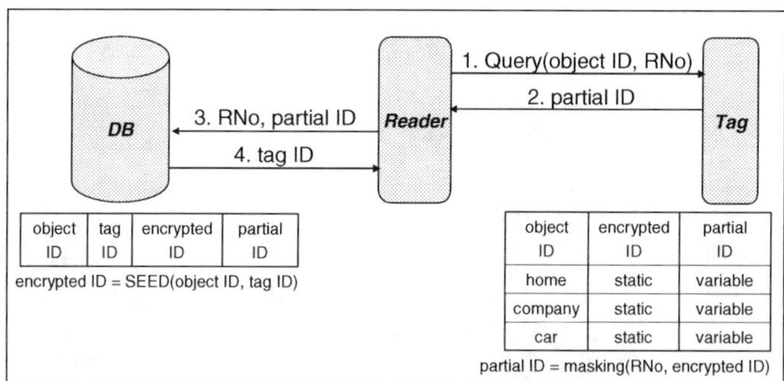

Fig. 3. Basic scheme of the authentication protocol for multi-key RFID tag structure

The partial ID is the most important key value in our scheme. A temporal one-time key of partial ID can keep tag from spoofing attack and location tracking. The last successful partial ID can also prevent RFID components from illegal adversaries' spoofing or replay attack. Since the partial ID can be created by bit masking operation in a simple way, it can be easily processed in most of the low-cost RFID systems. The authentication proceeds through the following steps.

① The reader sends a query to the tag. The query includes a pair of values *(object ID, RNo)*. *RNo* is taken as a random number to make different response from the tag and has two separate values in expression (1):

$$RNo = R_{Forward\_no} \| R_{Backward\_no} \quad (1)$$

② The tag searches the encrypted ID corresponding to the object ID. It creates a partial ID by bit masking operation for encrypted ID using *RNo*. If the tag memory has 128bits block size and holds the pair of (object ID, encrypted ID) for an object in the same block, the size of encrypted ID can be *(128 – size of object ID)* bits. We decide bi-directional masking points ($R_{Forward}$, $R_{Backward}$) for encrypted ID in the following expression (2) and (3).

$$R_{Forward} = R_{Forward\_no} \text{ mod (size of encrypted ID / 2)} \quad (2)$$
$$R_{Backward} = R_{Backward\_no} \text{ mod (size of encrypted ID / 2)} \quad (3)$$

$R_{Forward}$ is the starting point of forward masking in encryption ID and $R_{Backward}$ is the starting point of backward masking in encryption ID. The tag creates a partial ID in two steps. First step, it makes two 32 bits strings by bit masking operation using both $R_{Forward}$ and $R_{Backward}$. Next, it creates a 64 bits partial ID by concatenating two masking operation results. These operations are shown in Fig. 4.

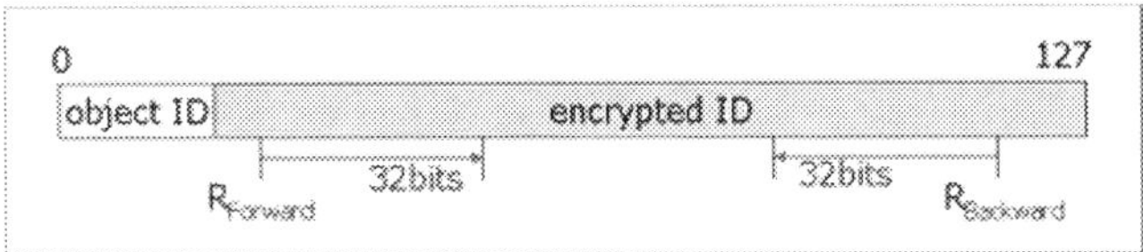

Fig. 4. Creation of partial ID

Since the tag maintains the partial ID generated through the last successful authentication process in its memory, the reader may be an illegal adversary in current session if partial ID in current session is equal to the stored value in tag memory. When the partial ID does not match to the one stored in tag memory, the tag recognizes the reader as legal and send the partial ID. If the illegal reader had obtained the message at the step ① in the former session, it may disguise as an authorized reader and retransmit the obtained message as replay attack. However, the tag can perceive replay attack by comparing stored partial

ID with present one. In that case, the tag wouldn't transmit the value to the illegal reader because the partial ID doesn't be changed.

③ After the reader receives the response message from the tag, it sends a message to the corresponding server. This message contains both the random value *RNo* and the partial ID received from the tag.

④ Server retrieves proper encrypted ID with the *RNo* and partial ID pair in database. If there is a match, the server decrypts the encrypted ID by SEED decryption algorithm and resolves tag ID. At the end of this session, the server starts service to the reader.

5 Evaluation

5.1 Evaluation of Security

The authentication protocol for multi-key RFID tag structure can keep its security for various kinds of attack. To prove this guarantee, we had evaluated the security of our scheme against the attacks as follows:

- Eavesdropping: The information tapped by attackers can be used for spoofing or location tracking. To solve this problem, a query message of reader should contain random value and then the tag should respond to the reader with variable value using this random value in the proposed scheme. In addition, this scheme will not let an attacker have any security information because the response message of variable partial ID is not complete value but the partial masking value of encrypted ID.
- Location tracking and replay attack: Since the tags respond differently for each query in our scheme, attackers can't catch the source of ID as well as the original value of ID. Therefore our scheme guarantees anonymity and supports blocking replay attacks.
- Spoofing: In our scheme, unauthorized readers have no way of joining in authorization process since readers should make a specified random value and transmit it to the tags. If an unauthorized reader uses the eavesdropped value again in current session, tags or server can be aware of abnormal status by comparing current partial ID with the stored one. On the other hand, if an unauthorized tag catches a random value from the reader, it can't create the correct partial ID without encrypted ID either. In addition, though an unauthorized tag catches the partial ID by eavesdropping, the tag is unable to reuse it because the random value for partial ID is changed in every session.
- Message loss: The encrypted ID is non-volatile and fixed in the tag memory. Therefore, even if there is some data loss during the authentication process by transmission interference, the tag doesn't have to recover the data in the proposed scheme.

Table 2 shows the comparison of the multi-key RFID tag scheme with the several existing schemes from the viewpoint of security.

Table 2. Comparison in security (o: strong, x: weak)

	Location Tracking	Spoofing	Replay Attack	Message Loss
Hash Lock	x	x	x	x
Randomized Hash Lock	o	x	x	o
Hash Chain	o	o	x	o
Variable ID	o	x	o	x
Proposed Scheme	o	o	o	o

5.2 Evaluation of Efficiency

Almost existing schemes are not practicable on account of their demand for circuit size and computing power. Those schemes often require heavy operation in tag but the proposed scheme requires small operation. In the multi-key RFID tag scheme, a tag computes bits masking once only for encrypted ID. This operation is simply compared with the other complicated hashing operations. On the other side, the server processes the tag ID encryption with SEED algorithm for each tag in the proposed scheme when the system is initialized. The server also executes bit masking operations of n/2(n: numbers of ID in database) times on an average using message of (*RNo*, partial ID) to retrieve exact encrypted ID. If there is matched one, the server processes decryption of the retrieved encrypted ID once only by SEED algorithm. As the result, the proposed scheme guarantees more privacy and security than the existing schemes. Table 3 shows the comparison of the schemes from the viewpoint of efficiency.

Table 3. Comparison in efficiency

	Tag	Reader	Server
Hash Lock	hashing:1	-	-
Randomized Hash Lock	randomizing:1 hashing:1	hashing:n/2	-
Hash Chain	hashing:2	-	hashing:(n/2)*i (i:update count)
Variable ID	hashing:3	randomizing:1	randomizing:1 hashing:3
Proposed Scheme	bit masking:1	Randomizing:1	bit masking:n/2

6 Conclusion

We proposed a new RFID scheme including the multi-key tag structure and the authentication protocol to give more privacy than existing schemes. The multi-key tag structure can maintain more than one object ID for different applications in a tag and allows applications to access them simultaneously. So, each application can share a tag on the multi-key and it can result many tags in one tag. The multi-key RFID tag scheme can also keep the RFID components from various attacks without heavy system load. Especially, the multi-key RFID tag could pay small cost against security threaten during authentication process. We evaluated the security and efficiency of the multi-key RFID tag for several types of attacks. The evaluation results show that the proposed scheme has better performance in security and efficiency than existing schemes.

The multi-key tag scheme is just at the beginning in our study and we proposed basic concept on multi-key tag structure in this paper. For the deep research and implementation, RFID reader has to be physically redesigned to support the proposed multi-key tag structure. We are going to design the RFID components fit with the proposed authentication protocol based on multi-key tag with embedded SEED algorithm in further study and make update such as performance evaluation results for the proposed scheme in real experimental environment.

Acknowledgement. This work was supported by the Korea Science and Engineering Foundation(KOSEF) grant funded by the Korea government(MOST) (No. R01-2006-000-10280-0).

References

[1] Weis, S.A.: Security and Privacy in Radio-Frequency Identification Devices, Masters thesis, MIT (2003), http://theory.lcs.mit.edu/ sweis/masters.pdf

[2] Weis, S.A., Sarma, S.E., Rivest, R.L., Engels, D.W.: Security and Privacy Aspects of Low-Cost Radio Frequency Identification Systems, First International Conference on Security in Pervasive Computing, In: Hutter, D., Müller, G., Stephan, W., Ullmann, M. (eds.) Security in Pervasive Computing. LNCS, vol. 2802, pp. 201–202. Springer, Heidelberg (2003)

[3] Ohkubo, M., Suzuki, K., Kinoshita, S.: Cryptographic Approach to Privacy-Friendly Tags, RFID Privacy Workshop (2003), http://www.frid.edu.com

[4] Saito, J., Sakurai, K.: Variable ID Scheme of Anonymity in RFID tags. The 2004 Symposium on Cryptography and Information Security 1, 713–718 (2004)

[5] Juels, A., Rivest, R.L., Szydlo, M.: The Blocker Tag: Selective Blocking of RFID Tags for Consumer Privacy. In: Proceedings of 10th ACM Conference on Computer and Communications Security, CCS 2003, pp. 103–111 (2003)

[6] mCloak: Personal/corporate management of wireless devices and technology (2003), http://www.mobilecloak.com

[7] ISO/IEC/JTC 1/SC 31, Information technology–Radio frequency identification for item management–Unique identification for RF tags, ISO/IEC 15963 (2004)

[8] Telecommunication Technology Association (TTA), Technical Report on Numbering an RFID tag, TTA Technical Report TTAR-06.0013 (2006)

[9] IETF, The SEED Encryption Algorithm, IETF RFC 4269 (2005)

[10] Garfinkel, S., Rosenberg, B. (eds.): RFID-Applications, Security and Privacy. Addison Wesley, New York (2005)

An Efficient Fragile Watermarking for Web Pages Tamper-Proof

Chia-Chi Wu[1], Chin-Chen Chang[1,2], and Shang-Ru Yang[2]

[1] Department of Computer Science and Information Engineering, National Chung Cheng University, Chiayi 621, Taiwan, R.O.C.
[2] Department of Information Engineering and Computer Science, Feng Chia University, Taichung 40724, Taiwan, R.O.C.
{wcc,ccc}@cs.ccu.edu.tw

Abstract. The Internet is the most popular and convenient mode of communication for people. An incredible amount of information and many services are provided through the Internet. However, many famous websites often suffer from tampering. Therefore, website security has become an important issue. Recently, Zhao and Lu proposed a PCA-based watermark scheme for web pages, claiming that their process would make web pages tamper-proof and more efficient. However, the PCA algorithm is time-consuming in processing the features of the web page. Therefore, we design a fast fragile watermark for web pages based on hash functions, because hash functions can generate rapidly a digest for the web page. According to our experimental results, our scheme is up to 100 times faster than the PCA algorithm, and, at the same time, our scheme improves case insensitivity and the malfunctions of HTML tags that are drawbacks of Zhao and Lu's scheme.

1 Introduction

With the rapid popularization of the Internet, many government organizations and enterprises provide announcements and services through Internet web servers. Also, the general population also obtains information from the Internet. Therefore, the security of the web server and web pages has become a very important issue. Because the Internet is an open environment, it offers a convenient channel for service providers and users alike. If the web pages are altered by a malicious attacker, panic and transaction disputes will be the result.

To deal with this problem, the web manager usually deploys a Firewall or an Intrusion Detection System (IDS) in place to detect and prevent malicious attackers. However, because updates of operating systems cannot occur continuously, those with malicious attackers can often find new ways to tamper with the content of web pages.

Watermarking schemes [1, 2, 3, 4, 5] have recently become a topic of widespread discussion by enterprises and the academic community as a feasible means for protecting digital intellectual property. The watermarking scheme can embed logotypes into a digital object (e.g., digital image, digital text, and digital multimedia) to protect ownership of the digital property. Generally, the watermarks can be divided

K.C. Chang et al. (Eds.): APWeb/WAIM 2007 Ws, LNCS 4537, pp. 654–663, 2007.

into three kinds according to functionality. First, the Robust watermark [3, 8, 13] can resist image processing, e.g., filtering, lossy compression, and sharpening. Second, the Fragile watermark [2, 7] can detect the smallest alterations of the content. Third, the Semi-fragile watermark is moderately robust for lossy compression such as JPEG.

Fragile watermarks are used to protect the integrity of a digital object. If the object has been altered, the fragile watermark will be changed from the original. In 1996, Craver et al. [6] defined a "non-invertible watermark" concept in which it is impossible or unfeasible to counterfeit a forgery watermark for passing the test. Afterwards, Zhao and Lu [14] proposed a modified version from this concept for application to web page watermarks. We briefly review their concept below.

D denotes the original document, and to protect the integrity of D, we compute a fragile watermark W for D with a secret key K:

$$W=G(D, K), \tag{1.1}$$

where $G(*)$ means the watermark-generating algorithm. Then W can be embedded in the original document to generate a watermarked document, designated D_W, which is defined as shown in Equation 1.2:

$$D_W=E(D, W), \tag{1.2}$$

where $E(*)$ represents the watermark-embedding algorithm.

D_W can satisfy the following equation, when the document has not been altered:

$$W= G(D_W, K). \tag{1.3}$$

Since D_W is a public document, attackers can easily get it. However, an attacker can obtain the published document, D_W, only a counterfeit key K^F is used to generate a forgery watermark, W^F, since the original key K is secret. Then, the attacker would have to further alter D_W to make counterfeit documents, D^F and D_W^F, to satisfy Eq. (1.4) and Eq. (1.5).

$$W^F = G(D^F, K^F). \tag{1.4}$$

$$D_W^F = E(D^F, W^F). \tag{1.5}$$

$$W^F = G(D_W^F, K). \tag{1.6}$$

If Eq. (1.6) is also valid, the attacker can show the counterfeit documents D_W^F and W^F, to cheat people who think that D_W^F is a legal document. This kind of watermarking scheme, called invertible watermarking, is deficient. Otherwise, it is called non-invertible.

In this paper, we design a fast fragile watermarking scheme for web pages that can satisfy non-invertible properties and ensure low computation cost as well.

The rest of this paper is organized as follows. In Section 2, we briefly review Zhao and Lu's scheme [14], so as to offer more background information and discuss its drawbacks in detail. In Section 3, we present our proposed watermarking scheme, and this is followed in Section 4 by a discussion of how the proposed scheme ensures web page security. In Section 5, we demonstrate our experimental results. Finally, in Section 6, our conclusions are presented.

2 Related Works

In this section, we introduce the processes of Zhao and Lu's web page watermarking scheme [14] and discuss the scheme in detail.

2.1 Review of Zhao and Lu's Web Page Watermarking Scheme

Zhao and Lu's web page watermarking uses the principle component analysis (PCA) [1, 14] to retrieve features of the original web page. PCA is a multivariate technique, which can transform a number of related variables to other uncorrelated variables. These uncorrelated variables can represent the original characters before the transformation took place. Usually, PCA is used to retrieve features and recognize patterns in images. PCA has two main properties: (1) the principle vectors (PVs) are projection axes of the original data and (2) the projected vectors can express the most features of the original data.

We further explain Zhao and Lu's watermarking scheme as follows.

(1) They retrieved an integer matrix H from source codes of the original web page, in which English characters are mapped from "0" to "25", with both "a" and "A" mapped to "0", "b" and "B" mapped to "1" in the corresponding row vector of H, and so forth. For example, the text line "</form> </tR> </Table> </td>" is mapped to (5 14 17 12 19 17 19 0 1 11 4 19 3). Thus, the coding scheme is case insensitive. However, since the number of characters of each row may be not the same, they take "cyclic filling" to expand the length of every row vector to the longest one. For instance, if the above row vector is the longest one and another shorter one, such as (8 11 14 21 4 24 14 20) exists, then the shorter one will be expanded to (8 11 14 21 4 24 14 20 8 11 14 21 4). Thus, they can get the alphabet-document matrix H. Suppose that $H \in F^{R \times C}$, where F represents either the real number domain or the complex number domain.

(2) They used a matrix H^t to amplify the changes in H to get another matrix $D \in F^{R \times R}$ by $D=HH^t$, where t represents the transpose operation.

(3) They used a random sequence as a key K; assume it is an $N \times N$ matrix. Then they used Shannon's Diffusion [12] to convolute the D matrix, as shown in the following equation:

$I=D \otimes K$,

where '⊗' denotes the operation of convoluting and $I \in F^{(R+N-1) \times (R+N-1)}$.

(4) PCA is applied to the matrix I, and they choose the first R principle components to generate watermarks. They performed steps as follows:

(a) They computed the covariance matrix V according to the equation:

$$V = \sum_{i=1}^{N} (I_i - \overline{I}_R)^t (I_i - \overline{I}_R), \tag{2.1}$$

where I_i is the ith row vector in I, t represents the transpose operation, and $\overline{I}_R \in F^{1 \times N}$ is the average vector of the row vectors in I, it is computed by

$$\overline{I}_R = \frac{1}{N}\sum_{i=1}^{N} I_i \,. \tag{2.2}$$

(b) The eigen decomposition (ED) was applied to V: $V=ULU^{-1}$, (2.3)
where U^{-1} is the inverse matrix of U. L denotes a diagonal matrix with eigenvalue of V, whose diagonal elements are $\lambda_1, \lambda_2,\ldots, \lambda_N$, and they assumed that these elements have been sorted in descending order ($\lambda_1 \geq \lambda_2 \geq \ldots \geq \lambda_N$), and the columns of U, $u_1, u_2,\ldots, u_N$, are the eigenvectors of V.

(c) They used a subset of $u_1, u_2,\ldots, u_N$ as basis vectors of a feature space S:

$$S=\text{span}\ (u_1, u_2,\ldots, u_m),\ \ m \leq N. \tag{2.4}$$

(d) They computed some feature vectors by projecting them into S using the following equation:

$$Z_i = (I_i - \overline{I}_R)\cdot\left[u_1 u_2 \cdots u_m\right], i = 1, 2, ..., N, \tag{2.5}$$

where $Z_i \in F^{1\times m}$ can be taken as the coordinates of the original data in the feature space S. Z_i's can be called "the principle components."

(e) They chose the first R principle components as $Z_1, Z_2,\ldots, Z_R \in F^{1\times R}$, and converted them into binary form, i.e. a sequence of '0' and '1'. Let z_{ij} denote the jth element of Z_i, i=1,2,…, R, j=1,2,…, R. Thus, each z_{ij} is converted to binary form β_{ij}, and all of them are connected to a binary sequence $W_i = \beta_{i1}\beta_{i2}\ldots\beta_{iR}$, which was taken as the watermark for the ith text line. Thus, the watermark, W, of the whole web page is composed of all W_i's, i=1,2,…,R

(5) In 2000, Katzenbeisser et al. [9] designed a web page embedding scheme through changing Tab space. However, that will increase the web page size. Therefore, Zhao and Lu embedded watermarks through modifying the case of letters in HTML tags (called Upper-Lower Coding, ULC). For example, the watermark W_i is a binary bits string that contains 0's and 1's, and they modified the case of the letters in the HTML tags in the ith text line T_i: the jth letter was changed to lower case if the jth element of W_i is '0', otherwise upper case. If the number of letters is larger than the length of W_i, the letters index is modular by the length of W_i. For example, W_i= (1001) and T_i = "<title> Web Page Watermarking</title>," after being watermarked, T_i will be transformed into T_i^W = "<TitLE> Web Page Watermarking</tiTLe>."

(6) To validate whether a watermarked web page has been maliciously altered, they first generated a watermark W_G by the aforementioned embedding scheme. Then, they extracted another watermark W_E from the letters in the HTML tags of the web page. The upper case letter is retrieved as '1,' and the lower case letter as '0.' If $W_G = W_E$, then the integrity of the web page is intact; otherwise, the page has been altered.

2.2 Drawbacks of Zhao and Lu's Scheme

According to the above-mentioned description, the main advantage of Zhao and Lu's scheme is that the watermarked web page will not increase file size. They claimed that their scheme is feasible and efficient; however, we found some drawbacks in their scheme.

(1) The PCA algorithm has heavy computational burdens.
According to the above-mentioned PCA algorithm, we find that it generates the principle components, z_{ij}, via complex matrix computation while requiring the use of floating point computation, since they are generated from Eq. (2.1) to Eq. (2.5). Therefore, the PCA algorithm will consume a lot of computation time and computer memory.

(2) Their scheme makes parts of HTML functions lose efficacy.
In their essay, they did not take into consideration that directory systems of many web servers are case sensitive. They showed their watermarked web page tag fragment,
"<A Ref="HTtp://Www.CS.Sjtu.EDu.CN/EngLISH/engLISh_INdEx.HTM#">". We tested and found that this sample can work well, but many popular web sites in which the directory system is case sensitive (except domain name), will incur an original hyperlink malfunction if this modification is made. For example, the tag, "<a href="http://www.opensource.org/licenses/W3C.php">" is transformed into "<a hReF="http://www.opensource.org/liCeNses/W3c.pHp">".

(3) Case-insensitive property cannot provide integrity protection of the original web page.
According to their scheme, both "a" and "A" are mapped to "0," and "b" and "B" are mapped to "1," and so on. Therefore, their scheme cannot guarantee the integrity of the watermarked web page without case-sensitive property.

(4) The convolution operation is time-consuming.
In the conclusions of their essay, they mentioned this comment, especially for large web pages. It will make their scheme inefficient.

3 The Proposed Scheme

In this section, we shall propose a fast, fragile watermarking scheme to improve Zhao and Lu's scheme and provide better efficiency. We think that the PCA algorithm can generate some multi-dimensional vectors to represent the features of images or multimedia, but the content of the web page substantially is text, which is composed of letters and symbols. The PCA algorithm might not be applicable to extract features from web pages. The well-known, one-way hash functions, such as SHA-1[10] and MD5 [11], are still the most efficient methods to generate a digest from the text. In addition, the traditional web page validation method usually stores a mirror file or a digest of the web page on another server or storage device, and this redundancy consumes resources unnecessarily. Therefore, we use one-way hash functions to design a watermarking scheme without redundant backup or the need for a digest of the web page. The administrator of web server can detect the web page tamper-proof through this embedded watermark.

3.1 The Watermark Generation Process

The input to the watermark generator is a web page D. To avoid malfunctioning of HTML tags, we especially keep the double-quotes parts in the HTML tags, even if the

HTML tags are case insensitive in parsing by the browser. Assume that, starting with the original web page, D, we first transform all letters of HTML tags into lower case to create a new web page, D_L, but the double-quotes parts will not be changed. Then, we choose a long length (128 bits), random-number key K (can be text or a binary bitmap but must be secret) to generate a watermark W=H(D_L||K), where H(*) is one of the above-mentioned, famous one-way hash functions. Generally, W is a fixed length bit string irrespective of the original web page and key size. In MD5, W is a 128-bit string, whereas in SHA, the length of W is 160 bits. In addition, if D_L and K are altered in any way, the result of the hash computation will be different from W. Therefore, the digest can be treated as a fragile watermark for the original web page to protect integrity.

3.2 The Watermark Embedding Process

We embed the bit string of W in HTML tags of the lowercased web page D_L, except for the double-quotes parts. The bit "0" will keep the lower case letter of the HTML tags, which would otherwise be upper case, until all bits are embedded in D_L. If there are tag letters remaining, we can embed the watermark redundancy. We call that the watermarked web page D_W.

3.3 The Watermark Validation Process

To validate the integrity of the watermarked web page D_W, we input this web page and the secret key K into the mentioned watermark generation process, and then it generates a watermark W. Afterwards, we extract a watermark W' from HTML tags of this web page according to the watermark embedding process, i.e., the upper case letter is extracted as '1' and the lower case letter is extracted as '0.' If W and W' are equivalent, the integrity of the web page can be confirmed; otherwise, it has been altered.

Our scheme overcomes the drawbacks of Zhao and Lu's watermarking scheme for the following reasons:

(1) Hash functions are much faster than the PCA algorithm.
We deeply investigated the computational processes of hash functions, which use non-linear logic computation to scramble data, i.e. XOR, shift, and modular computation without floating point. However, the PCA algorithm possesses multiplication, division, and subtraction computation with floating point. According to our experimental results presented in Section 5, it is evident that hash functions are much faster than the PCA algorithm.
(2) Our scheme will not influence HTML tags function, since we preserve the integrity of the double-quotes.
(3) Our scheme is case sensitive.
Since we only change the HTML tags into lower case to generate a watermark, the content of web page will not be changed. Therefore, our fragile watermark is case sensitive.
(4) Our scheme does not need convolution operation.
Since our scheme uses a key and a hash function to generate the watermark, there is no need for the convolution operation. Thus, our scheme is efficient.

4 Analysis and Discussion

If the watermarked web page has undergone a tampering attempt by a hacker, there are three possible cases as follows:

(1) The hacker damaged watermarks, but did not embed any new watermark.
(2) The hacker destroyed watermarks and embedded a counterfeit watermark in the web page.
(3) Watermarks have not been broken.

In Cases 1 and 3, we can easily to find tampering. However, we must further validate whether our scheme can withstand the counterfeit watermark attack in Case 2. That can be divided into two questions:

(a) Can the hacker recover the secret key K?
(b) Does the counterfeit key K^F satisfy Eq. (1.4) and Eq. (1.6)?

We shall analyze these two questions to explain that the proposed scheme has non-invertible property as follows.

1. Given D_W, the attacker cannot get the correct secret key K. Because we use a one-way hash function to generate W from W=H(D_L||K), even if the attacker extracts the W from D_W, he still gets no information about K. In addition, it is infeasible computationally for the attacker to derive K by the guessing attack because the length of K is at least 128 bits.
2. Without K, the attacker cannot counterfeit a watermarked web page. We apply some popular hash functions, which have collision free character. Even if the attacker attempts to generate a counterfeit watermark W^F=H(D_L^F ||K^F) to satisfy Eq. (1.4), the attempt will fail, since (D_L^F ||K^F) is different from (D_L||K). According to collision free character, it is not possible that H(D_L||K) and H(D_L^F ||K^F) could be equivalent.
3. Without K, the attacker can generate a counterfeit web page D_W^F =$E(D^F, W^F)$ to satisfy Eq. (1.5), but it can be detected easily, because the attacker must use a forgery watermark, W^F= H(D_L^F ||K^F), that will be not be equal to H(D_L^F ||K).
4. If we input the correct K to validate a embedded a forgery watermark web page that generate the watermark W_K^F = H(D_W^F ||K), this watermark cannot possibly be the same as the forgery watermark, W^F= H(D_W^F ||K^F), due to the collision-free character of hash functions. Therefore, our scheme cannot satisfy Eq. (1.6).

5 Experimental Results and Interpretation

We used both MD5 and SHA to implement our scheme in C++ language. Our original web page source code fragment is shown in Fig. 1, and the watermarked web page source code fragment is shown in Fig. 2. In addition, we list execution time in Table 1. Obviously, our method is much faster than Zhao and Lu's experimental data [14], no matter which one hash function we used.

Afterwards, we randomly test 200 web pages in Internet to draw Fig. 3. In that figure, we drew the executing time needed by using our scheme. Since we redundantly embed the watermark, the executing time of embedding is more than the validating time. Furthermore, we mark the coordinates of Zhao and Lu's experimental data in Fig. 4. We can find that executing times of our scheme are lower increasing than Zhao and Lu's scheme, when the web page size extension. The main reason is that our method does not perform complex floating computation.

```
><A
HREF="jdbc-callproc.html" tppabs="http://twpug.net/docs/pgsqldoc-7.4-big5/jdbc-callproc.html"
>Prev</A
></TD
><TD
WIDTH="10%"
ALIGN="left"
VALIGN="top"
><A
HREF="jdbc.html" tppabs="http://twpug.net/docs/pgsqldoc-7.4-big5/jdbc.html"
>Fast Backward</A
></TD
><TD
WIDTH="60%"
ALIGN="center"
```

Fig. 1. The original web page source code fragment

```
><A
Href="jdbc-callproc.html" TPpABs="http://twpug.net/docs/pgsqldoc-7.4-big5/jdbc-callproc.html"
>Prev</a
></td
><TD
WiDth="10%"
AliGn="left"
VAlign="top"
><A
hrEf="jdbc.html" TppaBs="http://twpug.net/docs/pgsqldoc-7.4-big5/jdbc.html"
>Fast Backward</A
></td
><tD
WIDTh="60%"
aLign="center"
```

Fig. 2. The watermarked web page source code fragment

Table 1. The comparisons of our scheme and Zhao and Lu's scheme

	Time(s)					
	Embed			Validate		
File size(KB)	SHA	MD5	PCA	SHA	MD5	PCA
0.828	0.009158	0.009282	0.1625	0.000564	0.000574	0.1609
2.65	0.010347	0.010274	0.6844	0.001092	0.000661	0.6891
4.58	0.01161	0.011506	0.9578	0.00092	0.000911	0.9578
6.71	0.012584	0.012506	1.7562	0.001168	0.001167	1.6125
9.86	0.014583	0.014292	1.9406	0.001665	0.001992	2.0031
16.4	0.017695	0.016516	7.1547	0.002623	0.002535	7.0656
19.3	0.018308	0.018389	16.1625	0.002898	0.002949	16.1781
21.5	0.018651	0.018927	39.7235	0.003375	0.003439	39.7703

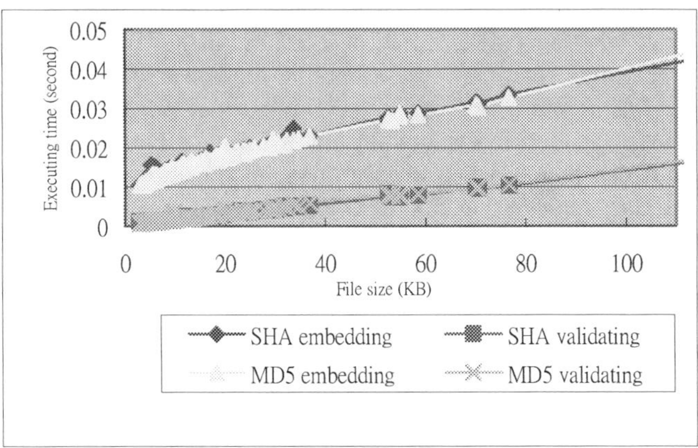

Fig. 3. The executing time by the proposed scheme

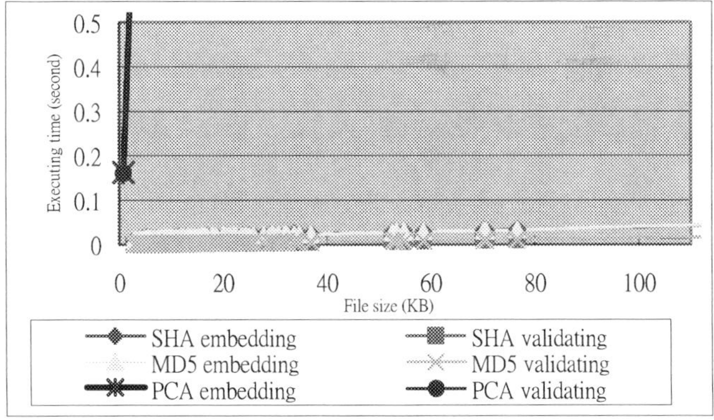

Fig. 4. The executing time by PCA and our scheme

6 Conclusions

In this paper, we propose a fast fragile watermark scheme for web pages. The proposed scheme can fulfill non-invertible security requirements, while keeping computation loading very light. Furthermore, to validate the efficiency benefits of our proposed scheme, we implemented two famous hash functions to compare with the PCA algorithm method. The comparison showed that the proposed scheme is approximately 100 times faster than the PCA algorithm method. Especially, our scheme increasingly improved performance over the PCA method as web page size increases. Therefore, our conclusion is that our proposed scheme provides a fast fragile watermark scheme that guarantees security for web pages while reducing computational costs. Moreover, the proposed approach is more practical and is easier to implement.

References

[1] Chang, C.C., Chan, C.S.: A Watermarking Scheme Based on Principal Component Analysis Technique. Informatica 14(4), 431–444 (2003)
[2] Chang, C.C., Wu, W.C., Hu, Y.C.: Public-Key Inter-Block Dependence Fragile Watermarking for Image Authentication Using Continued Fraction. Informatica 28, 147–152 (2004)
[3] Chang, C.C., Tsai, P.Y., Lin, M.H.: SVD-based Digital Image Watermarking Scheme. Pattern Recognition Letters 26(10), 1577–1586 (2005)
[4] Chang, C.C., Hu, Y.S., Lu, T.C.: A Watermarking-Based Image Ownership and Tampering Authentication Scheme. Pattern Recognition Letters 27, 439–446 (2006)
[5] Cox, I.J., Miller, M.L., Bloom, J.A.: Digital Watermarking. Morgan Kaufmann publishers, Seattle, Washington, USA (2002)
[6] Craver, S., Memon, N., Yeo, B., Yeung, M.: Can Invisible Watermarks Resolve Rightful Ownership? IBM Research Division, Technical Report RC 20509 (1996)
[7] Hwang, M.S., Chang, C.C., Hwang, K.F.: A Watermarking Technique Based on One-way Hash Functions. IEEE Transactions on Consumer Electronics 45(2), 286–294 (1999)
[8] Hwang, M.S., Chang, C.C., Hwang, K.F.: Digital Watermarking of Images Using Neural Networks. Journal of Electronic Imaging 9(4), 548–555 (October 2000)
[9] Katzenbeisser, S., Petitcolas, A.P.: Information Hiding Techniques for Steganography and Digital Watermarking. Artech House, Boston (2000)
[10] NIST FIPS PUB 180-1, Secure Hash Standard, National Institute of Standards and Technology, Available (April 1995), at http://www.itl.nist.gov/fipspubs/fip180-1.htm
[11] Rivest, R.: The MD5 Message-Digest Algorithm. RFC 1321, Internet Activities Board, Internet Privacy Task Force (1992)
[12] Stallings, W.: Cryptography and Network Security Principles and Practice. Prentice-Hall Inc, Englewood Cliffs, New Jersey (1999)
[13] Wu, H.C., Chang, C.C.: A Novel Digital Image Watermarking Scheme Based on the Vector Quantization Technique. Computers & Security 24(6), 460–471 (September 2005)
[14] Zhao, Q., Lu, H.: PCA-based Web Page Watermarking. Pattern Recognition, doi: 10.1016/j.patcog.2006.04.047 (2006)

Classification of Key Management Schemes for Wireless Sensor Networks

Hwaseong Lee*, Yong Ho Kim, Dong Hoon Lee, and Jongin Lim

Center for Information Security Technologies (CIST),
Korea University
{hwaseong,optim,donghlee,jilim}@korea.ac.kr

Abstract. For secure wireless sensor networks, a pair-wise key between nodes has to be established securely and efficiently. It is common to evaluate the security and efficiency of a (symmetric) key management scheme with those of other schemes without considering adversary's ability. In this paper, we defined attack models according to adversary's ability and classify existing key management schemes according to the attack models. We also suggested a guideline on selecting a proper key scheme with a given attack model. We expect our classification provides a certain criteria to fairly evaluate key management schemes in the same attack model.

Keywords: security, key management, attack model, guideline, wireless sensor networks.

1 Introduction

Wireless sensor networks are well recognized as a new paradigm for future communication. Sensor networks consist of a huge number of battery-powered and low-cost devices, called (sensor) nodes. As soon as nodes are deployed in a field, they monitor an event and transmit sensing data to the base station [2,5].

Wireless sensor networks are used in diverse fields - military, environment, and industry - because of cheap network configuration, data collection in wide range, and self-organization without infrastructure. There are many requirements. To satisfy such requirements, the first consideration is a guaranteed security and an inexpensive cost. In general, both we may 'trade-off' security and cost in designing a key management scheme. Because the price of a node is expected to be cheap, a node cannot support temper-resistance. (Symmetric) key management schemes have been designed to organize secure networks. However, it is not wise to emphasize only security and apply extremely tight security requirements to commodity sensor networks. It is reasonable to apply a little loose security to the application which requires efficiency and performance rather than security. Consequently, different security levels, not a uniform one, should be applied to

* "This research was supported by the MIC (Ministry of Information and Communication), Korea, under the ITRC (Information Technology Research Center) support program supervised by the IITA (Institute of Information Technology Advancement)".

K.C. Chang et al. (Eds.): APWeb/WAIM 2007 Ws, LNCS 4537, pp. 664–673, 2007.

different applications. Hence, a reasonable criterion, not yet studied, to evaluate and select a key scheme considering application is required.

In our paper, we first define attack models according to an adversary's ability in the practical environment where nodes are deployed. Moreover, we classify key schemes in the literature with attack models and suggest a guideline on selecting a key scheme considering an application environment.

The rest of the paper is organized as follows. Section 2 shows applications of wireless sensor networks. We define attack models in Section 3. In Section 4, key schemes are classified based on the attack models defined in Section 3. We suggest a guideline on the key schemes in Section 5. Finally, we conclude our paper in Section 6.

2 Applications in Wireless Sensor Networks

As we mentioned, wireless sensor networks are used in many applications with varied purposes. The applications are usually divided by sensing types : detection application, tracking application, and periodic measurement application. Detection application is used in military and EFOS, Environment Forecast and Observation System, to launch a warning when sensing data is over a pre-defined threshold or special value. Tracking application chases the movement of an object. It is helpful in animal or enemy tracking and traffic tracking. The application monitoring objects and collecting data at regular intervals belongs to periodic measurement application. Examples are invention system, health-care monitoring, and agriculture.

Even though the sensing type of application is identical, a security level may be not equal since each application is to face a distinct adversary. Accordingly, to choose an effective key scheme, applications should be classified according to an adversary's ability.

Generally, some applications may need strong confidentiality and high integrity. These applications have to adopt a key management scheme which is secure under the assumption an adversary is very active and powerful. However, in other applications, it can be enough to regard the power of an adversary as less strong than in other applications. Even though some key management schemes cannot thwart all possible attacks, they may satisfy a loose security suited to some applications and organize more efficient networks. Consequently, determining an attack model suitable for an application environment may result in maximizing efficiency as well as fully satisfying necessary security.

3 Attack Model

3.1 Criterion in Attack Model

Attack models can be defined by attack factors which can be a threat on key establishment. The attack factors generally mentioned are how to attack and when to begin attacks.

Attack Type: The goal of an adversary in a key scheme is to illegally obtain a pair-wise key between nodes. It is a *passive* adversary who can get no physical access to node and just obtain some information through overhearing and analyzing communication and detect the existence of unknown nodes or any suspicious activities such as RF signals. In practice, the case that physical access is hard or second shield exists is included in this attack type (e.g. a mine field and Smart Home). Besides overhearing, the adversary capable of launching active attack such as node capture is distinguished into an *active* adversary. Node capture means secure information leakage through the physical access to node [3]. In general, attacks affecting secure key establishment are usually overhearing and node capture [9,3,10]. Hence, we focus on such attacks.

Attack Time: Once deployed, for security, nodes establish a pair-wise key at short time so that it is crucial whether the phase of key setup is exposed to an adversary or not. An adversary can get ready to attack in advance before key setup. This adversary can analyze communication between nodes or get the physical access to node during key setup. This adversary is regarded as strong and intensive. It means the application requiring a high security level must design a key scheme as assuming a prepared adversary. On the contrary, to make an application more flexible and usable, a real attack model can be defined as [1]. In this attack model, after a key is established, an attack is possible whether it is active or passive. Practically, a key is established in a short time after node deployment. It is hard that an adversary not knowing deployment time and unable to access to deployment place tries an attack at key setup. This attack is a very real case despite loose attack. On the application where loose or no attacks during key setup are launched, it is reasonable to design a key scheme to improve efficiency and scalability as providing only loose security.

3.2 Attack Model Definition

Attack Model 1: We define as attack model 1 the environment where an adversary can overhear after key setup. An attack is not launched like node capture all along the network life. This attack model expects the weakest adversary.

Attack Model 2: We define as attack model 2 the environment where attacks are almost possible after key setup. In other words, during key setup, active attacks such as node capture are not in place and overhearing hardly exists. After key setup, an adversary is capable of overhearing and obtaining secret information through node capture.

Attack Model 3: We define as attack model 3 the environment where an adversary can only overhear communication just when nodes are deployed and, after key establishment, she is prepared for all the attacks including node capture.

Attack Model 4: Attack model where an active adversary always waits for node deployment is defined as attack model 4. It means that overhearing and node capture happen already in the phase that nodes are deployed. This attack model, consisting of the strongest adversaries, is a general assumption but requires an expensive cost.

Generally, if an adversary is able to launch active attack including node capture, she is considered to have the enough ability to overhear transmitted data. Accordingly, other attack models need not be considered : the case node capture is always possible but overhearing is practical after key setup and the case all the attacks except overhearing are possible only after key setup. Moreover, the higher an attack model level is, the stronger an adversary is. If a key scheme is secure under the attack model of high level, it is also secure in the attack model of low level. The attack models are as Figure 1. The examples of each attack model are explained in Section 5.

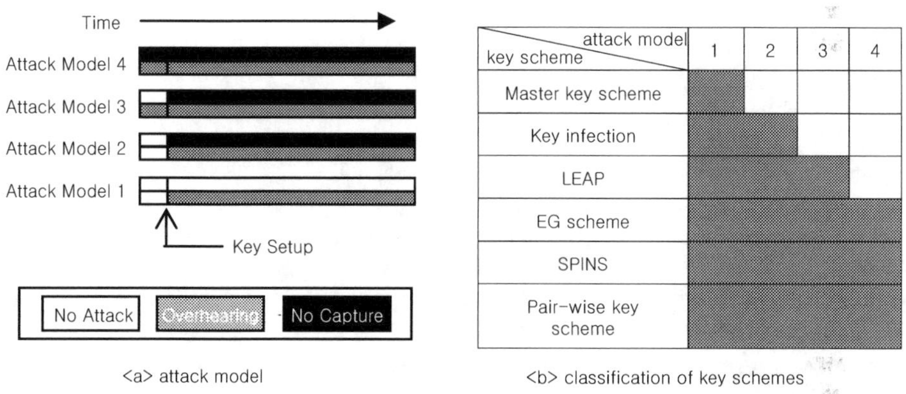

Fig. 1. Attack model and classification of key schemes by attack model

4 Classification of Key Schemes

4.1 General Classification of Key Schemes

Before categorizing key schemes into four attack models, we classify them by mechanism used in key establishment. In each mechanism, many schemes have been proposed to improve a basic scheme. Our explanation is basic scheme-oriented in each mechanism.

Pair-Wise Key Pre-distribution: A pair-wise key between a pair of nodes is directly stored, pre-distributed, in each node before node deployment (hereafter Pair-wise key scheme). Since each node in this scheme stores its pair-wise keys, it has perfect resilience against node capture which means even if a node is captured, the keys of non-captured nodes are never compromised [3]. However, scalability is limited because network scale depends on the memory of node where potential keys are stored.

Master Key Based Pre-distribution: A pair-wise key is derived from both a random number exchanged between each node and a single master key pre-distributed into each node (hereafter Master key scheme [7]). It results in great key connectivity and a little memory required. However, resilience is very low since all the pair-wise keys can be compromised when the master key is exposed to an adversary. Unlike Master key scheme which does not erase a master key after key setup, in 'LEAP' a master key is erased completely after a pair-wise key is established [10]. Although resilience is improved by erasing a master key of deployed nodes, there is still the risk of compromising a master key during node addition since added nodes store a master key.

Base Station Participation: 'SPINS' is included in this mechanism [9]. In SPINS, each node is given its shared key with the base station. The base station directly transmits a pair-wise key respectively encrypted with each node's shared key. In other words, the base station intermediates in key setup. This scheme supports not only full connection but also perfect resilience. However, it is not scalable because of the terrible traffic volume resulting from intermediation.

Probabilistic Key Pre-distribution: For large networks, a probabilistic method is more efficient than a deterministic method. This mechanism results from the concept all the nodes in the entire networks are connected with the 0.9997 probability- almost fully connected- if the probability each node can establish a pair-wise key with its neighbor nodes is 0.33. A key ring is stored in each node before deployment (a key ring k is randomly selected from key pool P which is randomly selected from huge key space). A common key in both key rings of a pair of nodes is used as their pair-wise key. It guarantees enough resilience even though not perfect resilience, because the probability of breaking communication link is k/P. Moreover, it supports the large scale networks. The representative scheme is 'EG scheme' [6]. Its variants are proposed like a combination of the EG scheme and the the Blundo scheme and a combination of the EG scheme and the Blom scheme which significantly enhance the security [4,8].

No Key Pre-distribution: This mechanism is considering the reality of wireless sensor networks. If an adversary does not know where and when nodes are deployed, it is difficult to launch active attack at an early phase. It can be a good trade-off to improve efficiency instead of a little node loss due to attacks during key setup. 'Key infection' is a representive scheme. In Key infection, key setup is completed in a relatively short time through a few transmissions [1]. The advantage in this mechanism is the base station does not take part in a key setup so that it consumes relatively less energy. Unlike the pre-distribution schemes above, it need not load potential keys into a node, which results in the low cost of network organization. However, it is only strong when an adversary does not observe communication during key setup and it cannot add nodes since a pair-wise key is established through exchanged data during key setup.

4.2 Criterion in Key Schemes

The key schemes above have trade-off consequences. According to attack model of application, the advantage of a key scheme looks better than actually is while its disadvantage can not be harmful to security as expected. In order to choose a reasonable key scheme, key schemes are reclassified by a novel criterion, not by mechanism above.

Attack Model: Attack model is to expect the real attack in the field as like we classify in Section 3. This attack model makes it possible to design a key scheme where enough security is provided and efficiency is also maximized.

Secondary Factors: It consists of efficiency, scalability, and connectivity. Efficiency includes memory requirement, computational cost, and communication cost. Memory requirement affects the hardware price of nodes. Computational cost and communication cost influence the energy consumption of nodes during the network life. Scalability affects the entire network scale. Connectivity relates to the number of isolated nodes. Table 1. is the result analyzing the secondary factors of key schemes : (A)-Pair-wise key scheme, (B)-Master key scheme, (C)-LEAP, (D)-SPINS, (E)-EG scheme, and (F)-Key infection. Scalability in Table 1. is determined by whether (i) initial network scale is large or not and (ii) node addition is easy or not. The higher the rank is, the better the scalability is supported. More details are in full paper [11].

The first consideration is to determine attack model in an application environment. It is a basic and essential element in organizing secure networks. After it, secondary factors are deliberated to efficiently operate the network for a long time. The definition such as attack model and secondary factors help regulate security and efficiency in designing a key scheme.

Table 1. Comparison on Secondary Factors

Key Scheme	Efficiency			Scalability	Connectivity
	Memory	Computation Complexity	Communication Cost		
(A)	$2(N\text{-}1)$	Search	$O(1)$	1	1
(B)	1	$1\times$PRF	$O(1)$	2	≈ 1
(C)	2	$4\times$MAC	$O(1)$	2	≈ 1
(D)	1	$2\times$MAC + $1\times$ENC	$O(n)$	1	1
(E)	$2k$	Search	$O(k)$	3	≈ 1
(F)	0	$1\times$ENC	$O(1)$	2	≈ 1

4.3 Reclassification of Key Schemes

Now, we reclassify key schemes with attack model as Figure 1 and the key schemes in the same category is compared with secondary factors in Section 5.

Scheme under Attack Model 1: Master key scheme assumes the weakest adversary. During key setup, an adversary in this attack model is considered to hardly overhear communication and obtain no master key through node capture. Hence, Master key scheme is secure under attack model 1.

Scheme under Attack Model 2: Since Key infection establishes a pair-wise key through exchanging data instead of using pre-distributed key, it is secure against node capture while it is very vulnerable to overhearing communication during key setup. After all, this mechanism is classified into attack model 2.

Scheme under Attack Model 3: In LEAP, a pair-wise key is established from a pre-distributed master key. It means the security of the master key is a significant matter. In LEAP, it is assumed that a pair-wise key is established before node capture. Therefore, LEAP is classified into attack model 3 where no physical access happens during key setup. Accordingly, overhearing is hardly threat in LEAP while the attack to disrupt the security of the master key such as node capture is a very menacing attack to LEAP.

Scheme under Attack Model 4: Both SPINS and Pair-wise key scheme are secure under the most powerful attack model. The reason is that SPINS satisfies perfect resilience since SPINS establishes a pair-wise key through the base station. In addition, communication between the base station and each node is encrypted with a shared key between only both entities so that an adversary can obtain little information by overhearing. Pair-wise key scheme is also perfect resilience and secure against overhearing because a pair-wise key is pre-distributed. Hence, these schemes are secure under attack model 4. In addition, even though EG scheme does not provide perfect resilience, it guarantees high resilience. Its variants have the similar result [3,4,8]. Consequently, this mechanism is also categorized into attack model 4.

5 Guideline on Key Management

5.1 Guideline Suggestion

To select a key scheme suitable for a specific application, attack model must be firstly decided as Section 3. And then the key schemes under the same attack model are analyzed comparing secondary factors. In other words, scalability has to be considered for large scale networks. Efficiency is essential for low energy consumption. If non-isolated node is important, connectivity is significant. Below are a few examples analyzing secondary factors of key scheme under the same attack model.

5.2 Examples on Analysis of Secondary Factors

SPINS and Pair-wise key scheme can be utilized by attack model 4. In the scalablity's point of view, both schemes are suitable for small scale network and

SPINS has to consider communication cost while Pair-wise key scheme depends on memory requirement. In key connectivity, SPINS and Pair-wise key scheme support full connectivity. Finally, in computation complexity, SPINS requires much computation than Pair-wise key scheme requiring only search.

SPINS and EG scheme both are classified into attack model 4 but they have the difference on the secondary factors. At first, SPINS supports perfect resilience but has the limitation on network scale due to communication cost. On the contrary, even though EG scheme does not guarantee perfect resilience, it excels SPINS in scalability. Also, SPINS stores a single key before node deployment while EG scheme has to pre-distribute random keys as many as a key ring on each node. Finally, in computational complexity, EG scheme costs less than SPINS.

5.3 Application Examples

From now, key schemes defined above are applied into the various applications under four attack models.

Habitat Monitoring Application: Wireless sensor networks can be utilized to observe the endangered species. If information on animal object or location information is exposed to an adversary, there are threats like poaching. It is necessary to organize secure networks in order to prevent such threats. In this case, nodes are deployed in a wide and rough field so that it is difficult for an adversary to be fully prepared for any attack during key setup. It means this application is included in attack model 2. Therefore, Key infection can be recommended which is vulnerable for attacks during key setup but maximizes efficiency, scalability, and connectivity. However, a key scheme like SPINS or Pair-wise key scheme assuming a stronger attack than a real attack does not operate wireless sensor networks for the long period due to communication cost and computation complexity.

Smart Home/Office and Structure Health Monitoring Application: Recently, Smart Home/Office is being developed in accordance with life quality improvement. In this application, nodes are deployed in home or office but physical attacks from the outside can be tried. Since nodes under these two environments are deployed in places possible to access or in public places such as a bridge and a dam, nodes can be secure from node capture during key setup but might be insecure against node capture after key setup and overhearing all along the network life. These applications can be also considered to be placed under attack model 3. Consequently, selecting LEAP explained above makes both security and efficiency satisfied.

Military Application: In case that nodes are distributed into the enemy force, it is general to assume attack model in the field to be the strongest attack model 4. In attack model 4, a few schemes exist : SPINS, Pair-wise key scheme, and EG scheme. SPINS or Pair-wise key scheme is restrictive in network scale so that EG scheme is more effective when a large field has to be monitored.

However, it is practical that, during key setup, adversaries in the field consist of passive adversaries who can detect themselves to be monitored and then they can try more active attacks including node capture. In this environment, LEAP can be also efficient because even if both mechanisms - EG Scheme and LEAP - are suitable for a large scale network, LEAP goes through more light pre-distribution process. In other words, in LEAP a single common master key is pre-distributed into each node while in EG scheme k different keys have to be pre-distributed. It can cause the network organization delay in military requiring a prompt action.

Examples explained above is possible to be changed. LEAP, a key scheme of attack model 3, can be used in military application, generally expecting attack model 4, by considering an actual condition. In addition, Smart Home/Office application can select a key scheme of attack model 4, such as SPINS, in order to support tight security.

6 Conclusion

Many key schemes have been proposed since a key scheme has been studied in wireless sensor networks. Unfortunately, comparing the key schemes under a different attack model with a uniform criterion is not an accurate analysis. Our paper proposed a novel approach defining attack model and classifying key management schemes with attack model. Moreover, the guideline in our paper is expected to provide a clear criterion choosing a key scheme that application uses and become a fair criterion analyzing a key scheme. In conclusion, our paper helps application choose a key scheme to support enough security and maximize efficiency as well as considering attack model. Our future work is to evaluate public key management schemes, which are more recently being researched, according to security level.

References

1. Anderson, R., Chan, H., Perrig, A.: Key Infection : Smart Trust for Smart Dust. In: Proceedings of the 12th IEEE International Conference on Network Protocols, pp. 206–215 (October 2004)
2. Akyildiz, I.F., Su, W., Sankarasubramaniam, Y., Cayirci, E.: A survey on sensor networks. In: Proceedings of the IEEE Communications Magazine, vol. 40(8), pp. 102–114 (August 2002)
3. Chan, H., Perrig, A., Song, D.: Random key predistribution schemes for sensor networks. In: Proceedings of the 2003 IEEE Symposium on Security and Privacy, pp. 197–213 (May 2003)
4. Du, W., Deng, J., Han, Y.S., Varshney, P.K., Katz, J., Khalili, A.: A Pairwise Key Pre-distribution Scheme for Wireless Sensor Networks. In: Proceedings of the ACM Transactions on Information and System Security, pp. 228–258 (August 2005)
5. Djenouri, D., Khelladi, L., Badache, A.N.: A Survey of Security Issues in Mobile Ad Hoc and Sensor Netowkrs. In: Proceedings of Communications Surveys and Tutorials, Vol. 7(4), pp. 2–28 (2005)

6. Eschenauer, L., Gligor, V.D.: A key-management scheme for distributed sensor networks. In: Proceedings of the 9th ACM conference on Computer and communications security, pp. 41–47 (November 2002)
7. Lai, B., Kim, S., Verbauwhede, I.: Scalable session key construction protocol for wireless sensor networks. In: Proceedings of the IEEE Workshop on Large Scale RealTime and Embedded Systems LARTES (December 2002)
8. Liu, D., Ning, P., Li, R.: Establishing Pairwise Keys in Distributed Sensor Networks. In: Proceedings of the ACM Transactions on Information and System Security, pp. 41–77 (February 2005)
9. Perrig, A., Szewczyk, R., Wen, V., Cullar, D., Tygar, J.D.: "SPINS: Security protocols for sensor networks". In: Proceedings of the 7th Annual ACM/IEEE Internation Conference on Mobile Computing and Networking, pp. 189–199 (July 2001)
10. Zhu, S., Setia, S., Jajodia, S.: "LEAP: Efficient Security Mechanisms for Large-Scale Distributed Sensor Networks". In: Proceedings of the Tenth ACM conference on Computer and Communications Security, pp. 62–72 (October 2003)
11. http://protocol.korea.ac.kr/~puzzle/paper/classification.pdf

An Efficient Algorithm for Proportionally Fault-Tolerant Data Mining

Tianding Chen

Institute of Communications and Information Technology,
Zhejiang Gongshang University, Hangzhou 310035
chentianding@163.com

Abstract. The frequent pattern mining problem has been studying for several years, while few works discuss on fault-tolerant pattern mining. Fault-tolerant data mining extracts more interesting information from real world data which may be polluted by noise. However, those few previous works either not define the problem maturely or restrict the problem to finding those patterns tolerate fixed number of fault items. In this paper, the problem of mining proportionally fault-tolerant frequent patterns is discussed. Two algorithms are proposed to solve it. The first algorithm, applies FT-Apriori heuristic and performs the idea of finding all FT-patterns with all possible number of faults. The second algorithm, divides all FT-patterns into several groups by their number of tolerable faults, and mines the content patterns of each group respectively. The experiment result shows more potential fault-tolerant patterns are extracted by our approach. Our contribution is offering a different type of fault-tolerant frequent pattern, in those patterns, the number of tolerated faults is proportional to the length of patterns. This gives the user another choice when traditional fault-tolerant frequent pattern mining result can't satisfy them.

1 Introduction

Since information is the most powerful weapon in the knowledge economic age, knowledge discovering has become a popular research. Association rule mining was first investigated in [1], and it explored the relationship among data items.

The emphasis of association rule mining is put on mining frequent patterns. The task of frequent pattern mining problem is, given a minimum support threshold, to enumerate all the frequent itemsets in the given database. Meanwhile, frequent pattern is extended to all kind of patterns, such as sequential patterns[2], cyclic patterns[3], closed patterns[4], and maximum patterns[5][6][7], depending on application requirements. Generally speaking, approaches for finding association rules could be roughly parted into two categories, one is Apriori-based[8] algorithm. This influential algorithm generates candidate patterns according to the anti-monotone heuristic. The main problems of Apriori algorithm is it needs to scan database for k+1 times, where k is the length of maximum frequent pattern and generates too many candidate patterns. Therefore, in [9], hash table is used to store candidate patterns to promote the efficiency; [9][10] try to reduce the number of transactions scanned in future

K.C. Chang et al. (Eds.): APWeb/WAIM 2007 Ws, LNCS 4537, pp. 674–683, 2007.

iterations; [11] and [12] adopt the techniques of partitioning and sampling respectively; And [13] proposes the dynamic pattern counting method by adding candidate patterns at different points during a scan.

The second category is tree-based algorithm which was proposed as FP-tree (frequent pattern tree) in [14]. This algorithm scans database to find all frequent items, and compresses the database which representing frequent items into a FP-tree. Finally, all frequent patterns can be obtained by searching on the tree. When the database is large, it is sometimes unrealistic to construct a main memory-based FP-tree. Therefore, by extending the pattern growth concept, H-mine is proposed in [15]. Without maintaining FT-tree or creating physical database, H-mine designs a dynamic structure to adjust links dynamically. The motivation of this method is to preserve space, and loading transactions into memory initially. However, H-mine has to maintain a head table in each level and modify links to build a queue of the collection of transactions containing the same prefix before the pattern support was counted. Since each approach has its own advantages and limitations, [16] suggests opportunistically chooses between two different structures, array-based or tree-based, to represent projected transaction subsets, and heuristically decides to build unfiltered pseudo projection or to make a filtered copy according to features of the subsets.

2 Mining Proportionally Fault-Tolerant Frequent Patterns

2.1 Problem Definition

Let pattern $X = \{i_1,\ldots, i_n\}$ be a set of items, while the *length* of X is the cardinality of X, denoted as |X|. Moreover, X is called *|X|-pattern* since it contains |X| items. A transaction T=(tid, X) is a 2-tuple record, where tid is the transaction-id and X is a pattern. Transaction T=(tid, X) is said to contain pattern Y iff $Y \subseteq X$.

A transaction database TDB is a set of transactions. The number of transactions in TDB containing pattern X is called the *support* of X, denoted as sup(X). Given a transaction database TDB and a user defined support threshold *min_sup*>0, pattern X is a *frequent pattern* iff sup(X)≥min_sup. Moreover, a frequent pattern with length k is denoted as frequent-k pattern. Besides, in the process of frequent pattern mining, possible patterns are generated as *candidate* patterns and which will be tested if they are frequent. The problem of frequent pattern mining is to find the complete set of frequent patterns in a given transaction database with respect to a given support threshold.

Extending frequent pattern mining problem, fault-tolerant frequent pattern mining problem relaxes the definition of containing as *FT-containing*. In addition to mining those patterns exactly occurs with high frequency, we find those frequent patterns that tolerate some errors. In [17][18], FT-containing is defined as mismatching a fixed number of items in a pattern. However, as mentioned above, it is unfair for the patterns with different lengths tolerate the same number of faults. Therefore, the problem of proportionally fault-tolerant frequent pattern mining is proposed.

Definition 2.1 (Proportionally Fault-tolerant frequent pattern)
Let P be a pattern, a transaction T=(tid, X) is said to *FT-contain* pattern P with respect to a given *fault-tolerant parameter* δ ($0<\delta\leq 1$) iff there exists $P' \subseteq P$ such that $P' \subseteq X$ and $\frac{P'}{P} \geq \delta$. The number of transactions in a database FT-containing pattern P is called the *FT-support* of P, denoted as $sup^{FT}(P)$.

Let B(P) be the set of transaction FT-containing pattern P. given a frequent *item-support* threshold min_sup^{item} and a FT-support threshold min_sup^{FT}. A pattern P is called a fault-tolerant frequent pattern, or FT-pattern in short, iff

1. $sup^{FT}(P) \geq min\_sup FT$; and
2. for each item $p \in P$, $sup^{item}{}_{B(P)}(p) \geq min\_sup^{item}$, where $sup^{item}{}_{B(P)}(p)$ is the number of transactions in B(P) containing item p.

The above definition mostly extends [17], except the fault-tolerant parameter and the definition of FT-containing definition. The item-support threshold avoids the problem of sparse pattern happens in [19] by constraining the plenty occurrence of each item in a pattern. Besides, with the new definition of FT-containing, the number of fault items tolerable in a pattern is no more fixed; on the contrary, it will increase as the length of patterns growth. Fig. 1 shows the relation between the length of pattern X and #fault(|X|), where #fault(|X|) denotes the number of fault items tolerable in pattern X. According to definition 2.1, we have the following equation: $\#fault(|X|)=\lfloor (1-\delta)\times|X| \rfloor$. In the horizontal part of the stair showed in Fig. 1, our problem can be simplified to previous works, i.e., FT-Apriori can be extended as the following theorem to solve part of our problem.

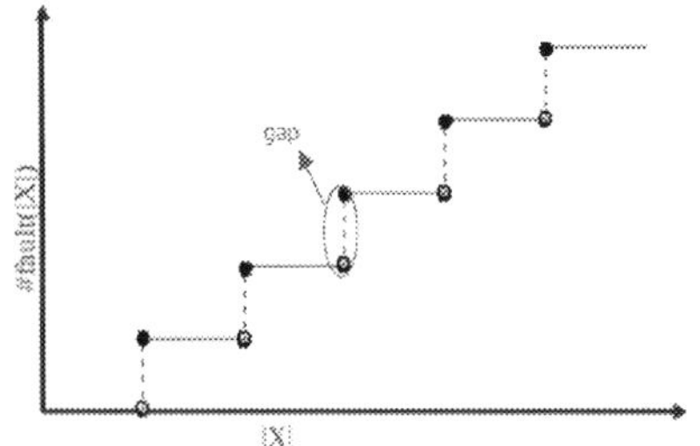

Fig. 1. Relation between |X| and #fault(|X|)

abc: (2: 2-2-2)	abcd: (2; 2-2-2-0)
abd: (4; 3-3-2)	abce: (2; 2-2-2-0)
abe: (4: 3-3-2)	abde: (2; 1-1-2-2)
acd: (4: 3-3-2)	acde: (2; 1-1-2-2)
ace: (4; 3-3-2)	bcde: (2: 1-1-2-2)
ade: (3: 1-1-3)	
bcd: (4: 3-3-2)	
bce: (4; 1-3-4)	
bde: (3: 1-1-3)	ab:(sup^{FT}(ab); $sup^{item}{}_{B(ab)}$(a)-$sup^{item}{}_{B(ab)}$(b))
cde: (3; 1-1-3)	

Fig. 2. supports of some patterns of DB1

Lemma 2.1 (Extended Fault-tolerant Apriori). If X is not a FT-pattern, then none of its superset with the same number of faults will be a FT-pattern.

However, we have a challenge in the gaps area, which presented as the vertical part of the stair on Fig. 1. In these areas, the anti-monotonic property does not exist. That is, if a pattern is not a frequent FT-pattern, its superset can still be a frequent FT-pattern. Therefore, solutions in previous works can not solve our problem. Furthermore, Fig. 2 listed all FT-support and item-support of Y's subset with length-3 and length-4. We got a bad news from this example, that is, the sets of frequent

FT-patterns parted by the gap are independent. Even though all sub-patterns of Y are not frequent FT-patterns, Y still has chance to be a frequent FT-pattern.

Fortunately, by observing the properties of patterns parted by the gap, we got the following lemma:

Lemma 2.2. Given a pattern X and the set of its sub-patterns set($X_{subpattern}$), where for all P$\in$ set($X_{subpattern}$), |P| = |X|-1. Moreover, let #fault(|X|)-1= #fault(|X|-1). (i.e., X and the considered subsets are parted by the gap), If X is not a frequent FT-pattern, then we have following two conditions:

case 1. if supFT(X)<min_sup^{FT}, then for all P$\in$ set($X_{subpattern}$), P can not be a FT-pattern.
case 2. else if $sup^{item}{}_{B(X)}(x_j)$<min_sup^{item} where x_j denotes an item contained by X, then none of patterns in set($X_{subpattern}$) which contains item xj can be a FT-pattern.

To improve the efficiency, the second algorithm, FT-LevelWise, is proposed. Several pruning properties are adopted in this algorithm to improve the performance. The algorithm FT-LevelWise is discussed in section 2.3.

2.2 FT-BottomUp Algorithm

The idea of the basic algorithm, FT-BottomUp, is to find all FT-patterns with number of faults from $(1-\delta)\times|X|$ to *MaxFault* in level-|X|, where MaxFault is the maximum possible number of faults. Let *DB'* be the pre-processed database which removed the infrequent items from the original database. We estimate MaxFault by maximum possible length of FT-patterns, denoted as *MaxPattern,* as the following equations:

MaxPattern=min(number of frequent-1 patterns, *length of longest transaction in DB'/δ*); MaxFault = #fault(|MaxPattern|).

Algorithm 1 (FT-BottomUp Algorithm)
Input: Transaction database DB
Frequent item-support threshold: min_sup^{item}
Frequent Fault-tolerant support threshold: min_sup^{FT}
Fault-tolerant parameter: δ
Output: FT-Patterns /* $F_{i,j}$ means the set of DT-patterns with length i and j faults */

```
Method:
1. Scan DB, find the set of frequent 1-patterns, denoted as F1,0;
Let DB'=DB∩frequent 1-patterns;
2. MaxPattern=min(length of longest transaction in DB'/δ, |F1,0|);
   MaxFault = #fault(MaxPattern);
3. for(int i = 2; i<MaxPattern; i++)
{   for(int j = #fault(i); j<MaxFault; j++)
        { generate Ci,j by Fi-1,j;   /* Ci,j is the candidate patterns for Fi,j */
          Fi,j=FT_frequent(Ci,j, min_sup^FT, min_sup^item);
            /* FT_frequent(Ci,j, min_sup^FT, min_sup^item) returns the set of patterns
               in Ci,j passed the two support thresholds*/
        }
}
output Fi,j;
```

Table 1. Frequent FT-patterns of DB2

i	$F_{i,0}$	$F_{i,1}$	$F_{i,2}$
2	{1 2},{1 3},{1 4},{1 5},{1 7},{2 3},{2 4},{2 5},{2 6},{2 7},{2 8},{3 4},{3 5},{3 7},{4 5},{4 7},{5 7}	All combination of 2 F1 items excepts{6 8}	All combination of 2 F1 items
3	{1 2 3},{1 2 4},{1 2 5},{1 2 7},{1 3 4},{1 3 5},{1 3 7},{1 4 5},{1 4 7},{1 5 7},{2 3 4},{2 3 5},{2 3 7},{2 4 5},{2 4 7},{2 5 7},{3 4 5},{3 4 7},{3 5 7},{4 5 7}	{1 2 3},{1 2 4},{1 2 5},{1 2 6},{1 2 7},{1 2 8},{1 3 4},{1 3 5},{1 3 7},{1 4 5},{1 4 6},{1 4 7},{1 5 7},{2 3 4},{2 3 5},{2 3 6},{2 3 7},{2 3 8},{2 4 5},{2 4 6},{2 4 7},{2 4 8},{2 5 6},{2 5 7},{2 5 8},{2 6 7},{2 7 8},{3 4 5},{3 4 7},{3 5 7},{4 5 7},{4 5 8},{4 6 7}	All combination of 3 F1 items
4	----------------	{1 2 3 4},{1 2 3 5},{1 2 3 7},{1 2 4 5},{1 2 4 6},{1 2 4 7},{1 2 5 7},{1 3 4 5},{1 3 4 7},{1 3 5 7},{1 4 5 7},{2 3 4 5},{2 3 4 7},{2 3 5 7},{2 4 5 7},{2 4 5 8},{2 4 6 7},{3 4 5 7}	{1 2 3 4},{1 2 3 5},{1 2 3 6},{1 2 3 7},{1 2 3 8},{1 2 4 5},{1 2 4 6},{1 2 4 7},{1 2 4 8},{1 2 5 6},{1 2 5 7},{1 2 5 8},{1 2 6 7},{1 2 6 8},{1 2 7 8},{1 3 4 5},{1 3 4 6},{1 3 4 7},{1 3 5 7},{1 4 5 6},{1 4 5 7},{1 4 5 8},{1 4 6 7},{1 5 6 8},{2 3 4 5},{2 3 4 6},{2 3 4 7},{2 3 4 8},{2 3 5 6},{2 3 5 7},{2 3 5 8},{2 3 6 7},{2 3 6 8},{2 3 7 8},{2 4 5 6},{2 4 5 7},{2 4 5 8},{2 4 6 7},{2 4 6 8},{2 4 7 8},{2 5 6 7},{2 5 6 8},{2 5 7 8},{2 6 7 8},{3 4 5 7},{3 4 5 8},{3 4 6 7},{4 5 6 7},{4 5 7 8},{5 6 7 8}
5	----------------	{1 2 3 4 5},{1 2 3 4 7},{1 2 3 5 7},{1 2 4 5 7},{1 3 4 5 7},{2 3 4 5 7}	{1 2 3 4 5},{1 2 3 4 6},{1 2 3 4 7},{1 2 3 5 7},{1 2 4 5 6},{1 2 4 5 7},{1 2 4 5 8},{1 2 4 6 7},{1 2 5 6 8},{1 3 4 5 7},{2 3 4 5 7},{2 3 4 5 8},{2 3 4 6 7},{2 4 5 6 7},{2 4 5 7 8},{2 5 6 7 8}
6	----------------	{1 2 3 4 5 7}	{1 2 3 4 5 7}
7	----------------	None	None
8	----------------	--------------------	None

The algorithm shows at the first $\left\lceil \frac{\delta}{1-\delta} \right\rceil$ levels, we adopt the basic Apriori method since patterns at those levels tolerate 0 items as faults, and FT-patterns with 0 faults are only required to pass the item-support threshold to be frequent. When the faults tolerable is greater than 0, the FT-Apriori is repeatedly used to generate candidate FT-patterns with the tolerable fault item number from that of this level to that of Maximum possible length.

Considering DB2, fault-tolerant parameter and support thresholds are set in Fig. 3. At the starting, frequent-1 items are extracted to reduce the database size. MaxFault is evaluated in Fig. 3. Table 1 shows the FT-patterns with every possible number of faults generated by FT-BottomUp.

$min\_sup^{FT} = 4$

$min\_sup^{item} = 2$

$\delta = 0.75$

➔ F1 :{ 1 2 3 4 5 6 7 8}

➔ MaxPattern = $\min(\frac{6}{0.75}, 8) = 8$

➔ MaxFault = #fault(8) = $\lfloor (1-0.75)\times 8 \rfloor = 2$

Fig. 3. Preprocess data before FT-pattern mining

From table 1, we know that most frequent FT-patterns we generate and check are not for the current level but for the future levels. This is doubtless time consuming and resource exhausting process. Therefore, we propose another algorithm to improve the efficiency in the following section.

2.3 FT-LevelWise Algorithm

The FT-LevelWise algorithm adopts the idea of partition the FT-patterns into MaxFault groups as each steps of the stair showed in Fig. 1.

For each group G_i, i=0 to MaxFault, i is the label of the group and also presents the number of faults tolerated by the patterns in the group. The shortest FT-patterns of group G_i are called the *head patterns* of G_i, denoted as $head_i$, while the longest ones are called *tail patterns* and denoted $tail_i$. The depth of group G_i , denoted $depth_i$, is the difference between the length of $tail_i$ and $head_i$, and can be evaluated by fault-tolerant parameter δ as $depth_i = \left\lceil \frac{1}{1-\delta} \right\rceil$ or $\left\lfloor \frac{1}{1-\delta} \right\rfloor$. After finding frequent-1 patterns, for each group G_i ($0 \le i \le$MaxFault), we first generate candidate patterns for $head_i$ by frequent-1 items, and scan database to check whether the candidates are frequent patterns. If $head_k$ contains no frequent patterns, group G_k can be deleted. Then, we can generate the candidates of $tail_i$ by $head_i$ since they tolerate the same number of fault items. Furthermore, according to lemma 2.2, some candidates of $tail_i$ can be pruned by the information collected in $head_{i+1}$. That is, a candidate pattern of $tail_i$ will never be a FT-pattern if its super-pattern do not existed in $head_{i+1}$ because of the small FT-support or because any item-support is not large enough and the candidate pattern contains the item as well.

Moreover, in the middle layers mid_{ij} (j is the difference between the length of mid_{ij} and $head_i$), we adopt lemma 2.1 from both sides. That is, candidate patterns of mid_{ij} are generated by existed longest sub-patterns of mid_{ij}, and if there existed any FT-patterns as super-pattern of candidate, this candidate is already set to be frequent and needless to be checked by scanning database.

Algorithm 2 (FT-LevelWise Algorithm)

```
Input: Transaction database DB
   Frequent item-support threshold: min_sup^item
   Frequent Fault-tolerant support threshold: min_sup^FT
   Fault-tolerant parameter: δ
Output: FT-Patterns
   Method:
   1. Scan DB, find the set of frequent 1-patterns, denoted as F1;
   Let DB'=DB∩frequent 1-patterns;
   2. MaxPattern = min(length of longest transaction in DB/δ, | F1|);
      MaxFault = #fault(MaxPattern);
   3. Construct MaxFault+1 groups: Gi, i=0 to MaxFault;
   For each group Gi
   {  Headi: generate candidate by F1;
             Check whether candidates are frequent FT-patterns;
      Taili:  generate candidate by Headi;
             Prune by the head of next group (Head(i+1));
             Check whether candidates are frequent FT-patterns;
      Midij: For j=1 to depthi/2
                 { Generate candidate Cij and Ci(depthi-j)) by Midi(j-1);
                   Prune{
                       If ( Candidate is the subset of Midi(depthi-j+1))
                          Set the candidate as frequent;
                          }
                   Check other candidate as frequent FT-patterns
                 }
   Output Midik, k = 0 to depthi
   }
```

Refer to table 1, the FT-LevelWise algorithm mining process is obviously. After scanning database once and evaluating the maximum possible FT-pattern length, we divide all FT-patterns into 3 groups by the number of faults they tolerated. For each group, the head FT-patterns are generated by frequent-1 patterns, while the tail FT-patterns are generated by the head and pruned by the head of next group. Then, FT-patterns are generated by the longest sub-pattern and pruned by the shortest super-pattern from both sides of the group.

3 Experiment Results

In this section, the performance of our two algorithms, FT-BottomUp and FT-LevelWise, is evaluated. These mining algorithms are implemented in JAVA, and all experiments are done on P-IV 2.4GHz with 512 MB RAM running Windows XP. The experimental datasets are generated by IBM synthetic data generator. Each dataset contains 1k different items, 10k transactions, 10 items in a transaction in average, and several potential frequent patterns with average length 8.

The parameters used in our simulation are described in table 2. The performance of FT-BottomUp and FT-LevelWise is compared by their execution time.

Table 2. Parameters considered in experiments

Parameter	Symbol	Range of values	Default value
Fault-tolerant parameter	δ	0.7~0.99	0.8
Minimum FT-support threshold	min_sup^{FT}	0.06~0.1	0.1
Minimum item-support threshold	min_sup^{item}	0.04~0.09	0.075

In the first simulation, the relation between fault-tolerant parameter and the number of FT-patterns is considered. It's trivial that the larger δ is, the less FT-patterns would be extracted. This situation is especially obvious in the range from δ=0.7 to 0.9 as the curve shown in Fig. 4. When δ is over 0.9, the patterns mined out are closed to traditional frequent patterns, i.e., exactly matched patterns without fault-tolerant property. Fig. 5 presents the total execution time of FT-BottomUp and FT-LevelWise. The later algorithm is plainly performs better than the former one.

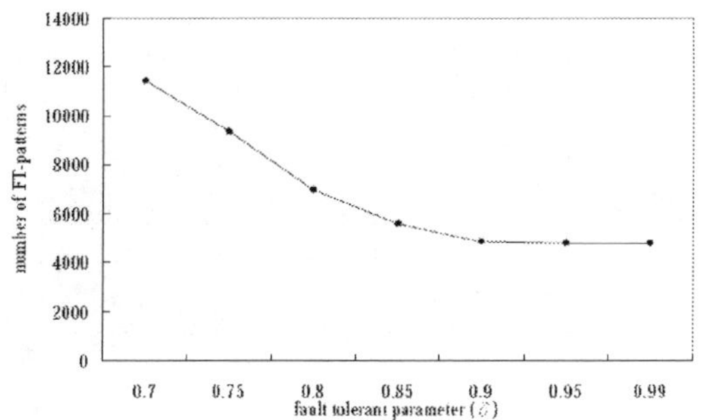

Fig. 4. Number of FT-patterns w.r.t. fault-tolerant parameter

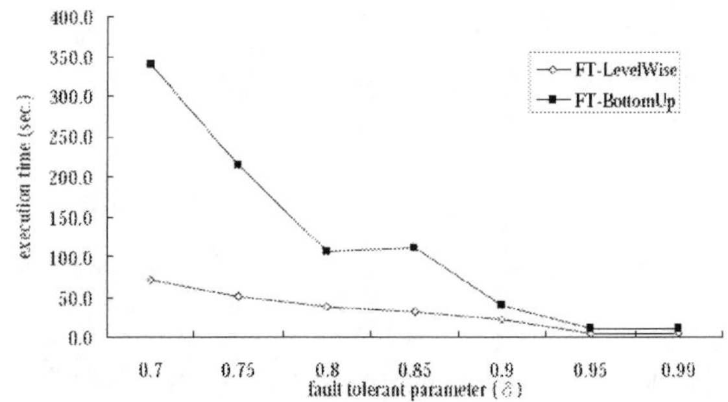

Fig. 5. Execution time of two algorithms w.r.t. fault-tolerant parameter

The second simulation sets δ as default value and studies the reciprocal effect of the two support thresholds. Fig. 6 shows the variation of number of FT-patterns when min_sup^{FT} and min_sup^{item} changes.

The scalability of FT-BottomUp and FT-LevelWise with respect to the two support thresholds are presented in Fig. 7. The whole performance of FT-LevelWise is still better than FT-BottomUp. Besides, with the same min_sup^{item}, the larger min_sup^{FT} is, the more execution time it spends.

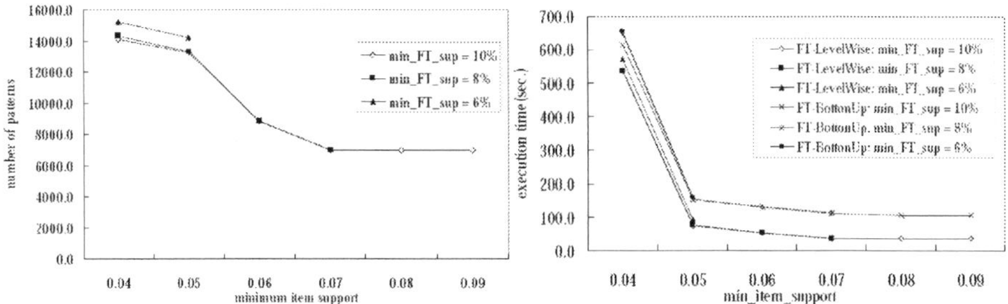

Fig. 6. Number of FT-patterns w.r.t. min_sup^{FT} and min_sup^{item}

Fig. 7. Execution time of our algorithms w.r.t. min_sup^{FT} and min_sup^{item}

4 Conclusion and Future Works

In this paper, the problem of mining proportionally fault-tolerant frequent patterns and two effective algorithms are proposed. Differ to previous works on fault-tolerant data mining, the number of faults tolerable in the FT-patterns found out by our approach is proportional to the length of pattern. By evaluating the maximum possible length of FT-pattern, the maximum number of tolerable faults can be obtained. The experiment also shows the scalability of FT-BottomUp and FT-LevelWise. Undoubtedly, FT-LevelWise performs much better than FT-BottomUp since it need not to generate so many candidate patterns and the DB scanning time is half of FT-BottomUp.

In the future, there are some problems which are worth studying. First, investigating more efficient algorithms to improve the scalability of mining proportionally fault-tolerant frequent patterns is important. Furthermore, it is interesting to apply proportional fault-tolerant data mining into the real world database such as medical data, marketing trade data and etc.

References

1. Agrawal, R., Imielinski, T., Swami, A.: Mining Association Rules between Sets of Items in Large Databases. In: Proc. 1993 ACM-SIGMOD Int. Conf. Management of Data(SIGMOD'93), pp. 207–216, Washington, DC (May 1993)
2. Agrawal, R., Srikant, R.: Mining sequential patterns. In: Proc. 1995 Int. Conf. Data Engineering, pp. 3–14, Taipei, Taiwan (March 1995)

3. Ozden, B., Ramaswamy, S., Silberschatz, A.: Cyclic association rules. In: Proc. 1998 Int. Conf. Data Engineering (ICDE'98), pp. 412–421, Orlando, FL (February 1998)
4. Pei, J., Han, J., Mao, R.: Closet: An efficient algorithm for mining frequent closed itemsets. In: Proc. of ACM SIGMOD DMKD Workshop, pp. 21–30 (2000)
5. Agrawal, R., Aggarwal, C., Prasad, V.: Depth first generation of long patterns. In: Proc. of ACM SIGKDD Conf. pp.108–118 (2000)
6. Burdick, D., Calimlim, M., Gehrke, J.: Mafia: A maximal frequent itemset algorithm for transactional databases. In: Proc. of ICDE Conf. pp. 443–452 (2001)
7. Gouda, K., Zaki, M.J.: Efficiently mining maximal frequent itemsets. In: Proc. of ICDM conf. pp. 163–170 (2001)
8. Agrawal, R., Srikant, R.: Fast Algorithm for Mining Association Rules. In: Proc. 1994 Int. Conf. Very Large Data Bases(VLDB'94), pp. 487–499, Santiago, Chile (September 1994)
9. Park, J.S., Chen, M.S., Yu, P.S.: An Efficient Hash-based Algorithm for Mining Association Rules. In: Proc. 1995 ACM-SIGMOD Int. Conf. Management of Data (SIGMOD'95), pp. 175–186, San Jose, Ca (May 1995)
10. Han, J., Fu, Y.: Discovery of Multiple-level Association Rules from Large Databases. In: Proc. 1995 Int. Conf. Very Large Data Bases (VLDB'95), pp. 420-431, Zurich, Switzerland (September 1995)
11. Savasere, A., Omiecinski, E., Navathe, S.: An Efficient Algorithm for Mining Association Rules in Large Databases. In: Proc, Int. Conf. Very Large Data Bases(VLDB'95), pp. 432–443, Zurich, Switzerland, (Septemper 1995)
12. Toivonen, H.: Sampling Large Databases for Association Rules. In: Proc. 1996 Int. Conf. Very Large Data Bases (VLDB'96), pp. 134–145, Bombay, India (September 1996)
13. Brin, S., Motwani, R., Ullman, J.D., Tsur, S.: Dynamic Itemset Counting and Implication Rules for Market Basket Analysis. In: Proc. 1997 ACM-SIGMOD Int. Conf. Management of Data(SIGMOD'97), pp. 255–264, Tucson, AZ (May 1997)
14. Han, J., Pei, J., Yin, Y.: Mining Frequent Patterns Without Candidate Generation, ACM SIGMOD Record. In: Proceedings of 2000 ACM SIGMOD International Conference on Management of data (May 2000)
15. Pei, J., Han, J., Lu, H., Nishio, S., Tang, S., Yang, D.: H-Mine: Hyper-Structure Mining of Frequent Patterns in Large Databases. In: Proc. of ICDM conf. pp. 441–448 (2001)
16. Liu, J., Pan, Y., Wang, K., Han, J.: Mining Frequent Item Sets by Opportunistic Projection, ACM SIGKDD '02 (July 23-26, 2002)
17. Pei, J., Tung, A.K.H., Han, J.: Fault-Tolerant Frequent Pattern Mining: Problems and Challenges, DMKD'01, Santa Barbara, CA (May 2001)
18. Wang, S.-S., Lee, S.-Y.: Mining Fault-Tolerant Frequent Patterns in Large Database, International Computer Symposium (December 2002)
19. Yang, C., Fayyad, U., Bradley, P.S.: Efficient discovery of error-tolerant frequent itemsets in high dimensions. In: Proceedings of the seventh ACM SIGKDD international conference on Knowledge discovery and data mining (August 2001)

SOGA: Fine-Grained Authorization for Self-Organizing Grid*

Ming Guo[1], Yong Zhu[1], Yuheng Hu[1], and Weishuai Yang[2]

[1] Department of Computer Science, Zhejiang University, City College
{guom, zhuyong, 3035211012}@zucc.edu.cn
[2] Department of Computer Science, Binghamton University
wyang@cs.binghamton.edu

Abstract. The potential of truly large scale Grids can only be realized with Grid architectures and deployment strategies that lower the need for human administrative intervention, and therefore open the Grid to wider participation from resources and users. Self-Organizing Grids (SOGs) was proposed to address this issue. Current general solutions for Grid authorization are not scalable enough, inflexible, inefficient, coarse-grained, or require too much administrative work, thus do not fit for the needs of SOGs. The arising Semantic Web can contribute to a solution. In this paper, we propose SOGA, a fine-grained authorization architecture for SOG environment. Its characteristics includes, policy based authorization, fine-grained access control, dynamic environment monitor and ontology driven access rights reasoning.

1 Introduction

A computational Grid has been defined as "a hardware and software infrastructure that provides dependable, consistent, pervasive and inexpensive access to high-end computational capabilities" [5]. Computational capabilities come from resources. Grid resources are provided by various organizations and are used by people from diverse sets of organizations.

Usually, the process of resource access control in grid is simplified using the concept of virtual organization (VO) [6] consists of users and resources (possibly geographically dispersed) working towards a common set of problems. The VO defines who its members are, assigns roles or credential to members, and also regulates the access to the resources. Authorization mechanisms are deviced to implement the policies to govern and manage the use of computing resources.

Specialty of Self-Organizing Grids (SOGs) [2] are that resource may come and go freely. Individual resources in SOG may have different access policies. We believe that a greater flexibility can be achieved, if in addition to the global access policy set by the VO, individual resource owners were also allowed to refine the access control policies for their resources, by means of their own local policies.

* This work is supported by National Science Foundation of China under Grant No. 60473025 and Zhejiang Provincial Natural Science Foundation of China under Grant No. Y106427.

K.C. Chang et al. (Eds.): APWeb/WAIM 2007 Ws, LNCS 4537, pp. 684–693, 2007.

Current general solutions for Grid authorization are not scalable enough, inefficient, inflexible coarse-grained, or require too much administrative work, thus do not fit for the needs of SOGs. In this paper, we propose SOGA, distributed fine-grained authorization architecture for SOG environment. Its characteristics includes, policy based authorization, fine-grained access control, dynamic environment monitor and ontology driven access rights reasoning.

The rest of the paper is organized as follows. Section 2 reviews the concept of Self-Organizing Grids (SOGs). Section 3 requirements for authorization system in SOG environment. Section 4 illustrates the architecture of SOGA. A prototype implementation is covered in section 5. Section 6 provides a classification of existing authorization mechanisms, modules and systems developed for or used in Grid environment. Section 7 concludes this paper by highlighting the major contribution of this research, and providing pointers for future research.

2 Preliminary: Self-Organizing Grids

The potential of truly large scale grids can only be realized with grid architectures and deployment strategies that lower the need for human administrative intervention, and therefore open the grid to wider participation from resources and users. Grids should be as simple to assemble and organize as peer to peer (P2P) and public resource computing (PRC) systems, without narrowing the kinds of applications they can host.

We believe that this "best of both worlds" goal can be achieved with self-organizing grids (SOGs) [2][11]. In the SOG model, participating resources self-organize to provide effective support for applications that are submitted against them. Essentially, selforganization replaces administrative intervention to facilitate the aggregation and use of grid resources, and the deployment of grid applications. Having truly large-scale grids will lead to more grid applications, and innovative uses for the grid, making grid computing genuinely and significantly different from and more enabling than-traditional distributed, enterprise, and cluster computing. A self organizing grid environment will have several defining characteristics that fundamentally differentiate it from current production grids, and that make necessary new solutions for a variety of different grid problems. First, the system will have a more dynamic and "scattered" structure. Some grid resources, because they are owned and controlled by individual users rather than organizations, will join and leave the grid more frequently and less predictably, and will not all be part of natural static clusters. Second, SOGs will be characterized by a wide range in the heterogeneity of resource capabilities, fault characteristics, degrees of trust, and more.Finally, the SOG model eliminates the feasibility of centralized administrative control of all grid resources, because there will be so many more "point of contact".

Thus, SOGs represent the intersection of P2P computing, grid computing, and autonomic computing, and can potentially offer the desirable characteristics of each of these models.

3 Requirements for SOG Authorization

To point out some access control challenges and requirements in SOG environment, we start by considering the spontaneous scenario of distance education in SOG among members from diverse locations. In the remainder of this paper,we use this scenario as a running example to illustrate the main access control issues and our solution guidelines. In this scenario, each participant may grant access to his resources to other participants, in order to enable teaching,cooperation and knowledge sharing. Access to these resources must be regulated in order to protect them from malicious access or misuse. Let us note some basic requirements for authorization in this scenario:

Distributed Mechanisms: Due to huge number of participating nodes, e.g. different students from diverse areas. Centralized solutions have single point failure problem and are not scalable. The security mechanisms must be fully distributed. This implies not only authorization decision should be made distributedly, but also user information should be stored distributedly.

Resource Control Autonomy: Different from regular Grid systems, nodes in SOGs are flexible to come and go. Resource owner needs higher level of control to the shared resources. In this scenario, access token only provides the role of the user, exactly what this role can do relies on resource owner's policy and requestor's context. External incentives may be needed to encourage resource owner to lower privilege requirements.

Fine-grained Access Rights and Policy Enforcement: Resources in SOGs span across multiple geographical locations, usually heterogeneous in nature and administered individually by different resource owners. Such an environment requires the possibility of different access control policies for each resource, instead of one global policy uniformly applied to all Grid resources. In our scenario,an access right is "fine-grained" if it denotes a specific right of a given user to a given object. To support fine grain access rights the enforcement mechanisms must be able to constrain the execution of services at an equally fine level of granularity.

Exact Role Assignment: In case a user has more than one role, least privileged role should be selected in delegated execution. This selectability increases overall system security as services can be executed with only the least access rights required.

Basic security requirements, such as the support for single sign-on and mutual authentication, have been omitted from this list since these requirements are well met by existing grid systems.

4 Architecture of SOGA

In this section, we present an extensible authorization model(SOGA) for fine-grained access control in SOG environments. Through the specific scenario, the process of SOGA is easy to follow, we would like to share our motivation and details of SOGA in the following part.

4.1 Policy Based Authorization Model

Models like role based access control (RBAC) [13] were widely used to protect the processes from unauthorized access in an adequate manner. Even these approaches have deficiencies in large distribute environment, since role definitions and hierarchies might vary across parties, thus making their interpretation difficult outside the specific boundaries of each organization. Specifically, changing access rights seems inflexible because of close coupling between roles and access rights.

In order to properly control access to resources, we claim the need for a more general and comprehensive approach that exploits not only identity and role information but also other contextual information, such as login time, ongoing activities, IP address, etc. In particular, we consider that it may be advantageous for each participant to define the access control policies for the managed resources simply according to the current condition of the requestor, i.e. request context and the resource, i.e resource context. For instance, the resource owner defines his own policy that requestors accessing to his resource depend on the pre-assigned maximum load, i.e. system load. If overloaded, the coming request will be denied. The request owner could also define other rules related to requestor's context, for example, the rule could be defined as " requestor only permitted to access to the studying resource during the teaching time". It is useful when resource owner have both education and entertainment materials.

The integration of policy-based authorization with contextual information make the access control more flexible and easy to regulate. Resource owner could defined granularly policies to better control or protect resources.

4.2 Proposed Architecture

We proposed a fine-grained policy based authorization architecture (SOGA) with semantic web technologies for SOG environment. We state that, instead of assigning permissions directly to the subjects and defining the context in which these permissions should be considered valid and applicable (like RBAC), the resource owner defines specific policies for his resources the contextual conditions that enable one to operate on it.

Figure 1 illustrates our architecture, the contextual information (both requestor's and resource's) is captured by environment monitor. Moreover, a policy evaluation point (PEP) is proposed to merge user's request and these contexts. Policy Administrator contributes to edit and deliver the external policies to context handler. Finally, the response to a requestor must be made according to the policies, request and context by a inference.

We would like to illuminate specific procedures in the following steps, according to which, the advantages of SOGA will be showed.

1. SOGA parses the request and collects basic information i.e., user ID, request time, resource name,etc. According to resource name, static parameters i.e. maxLoad, TimeSpan could be extract from resource repository. All these

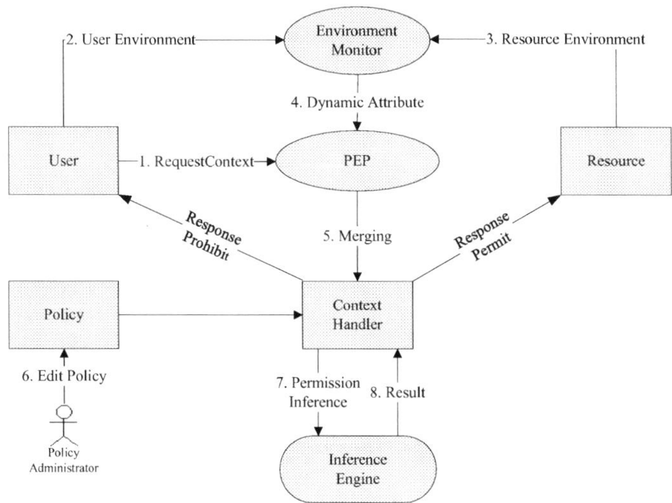

Fig. 1. Architecture of policy authorization model

basic information (from request's and resource's) could be encapsulated into RequestContext object and sent to PEP (Policy evaluation point). We assume the user have already passed authentication in our scenario.

2. Environment monitor contributes to capture dynamic information include the requestor context and resource context. As a result, dynamic variables e.g.,CPU usage, requestor's time(from requestor), resource load, connects, system time(from resource owner) will be recorded.
3. Environment monitor combines collected dynamic information and then sends it to PEP.
4. The PEP merges all static and dynamic information and forwards them to the context handler.
5. Policy administrator is in charges of editing the policies composed by different rules and sends these policies to the context handler.
6. Content handler transfers policies and situations provided by PEP to inference engine. According to the policies and situations, the result could be "Permit" or "Prohibit".
7. The result would be pushed back to the requestor via the context handler, if the result is "Permit", then the requestor is allowed to get the resource. If not, the request is denied.

5 SOGA Policies

In this section, we firstly explain the basic concepts of the Semantic Web and the Semantic Web Rule Language (SWRL), then we propose our implementation of SOGA that exploiting ontologies for modeling the requests and context. Finally, we present our policies.

5.1 Semantic Web and Ontology

An ontology [14] is capable of describing concepts in a certain domain and relationships among them. Implicit information in the data and relations between data can be made explicit by using inference engines. Complex inference rules can be realized with Semantic Web Rule Language (SWRL) [16], which based on the Rule Markup Language (RuleML).

5.2 SOGA Ontology

We use protege [15] to create the ontology for describing context, request, resource and exploring the implicit relationships between them. The followings are the basic classes in our ontology and the fundamental domain knowledge in SOGA.

- **Permission(P):** The variable P in class Permission(P) is the sign of whether the request is allowed or prohibited.
- **Resource(R):** In SOGs, files, service or computers are treated as a kind of resources. Specifically, in our scenario, we define two different resources: the computing resource and the files (include entertainment and studying). System maxLoad, resource spanTime and resource name are the static subclasses of Resource(R), which could be pre-assigned by resource owner.
- **Resource Environment(RE):** The environment monitor captures dynamic context from resource activities. For instance, resource load is the dynamic subclass of Resource Environment(RE).
- **User environment(UE):** Similarly, user activities are recorded as a instance of class "User environment".
- **User(U):** Variable U represent of the requestors.

In our application, relationships among classes are usually referred to as properties. A property links individuals from its domain to individuals of its range. We present the properties below. Besides, figure 2 shows the specific properties with classes in our ontology. For example, the hasResourceEnvironment property links the resource class to resource environment class.

- **hasResourceEnvironment(R, RE):** This predicate determines relationship between resource and resource environment: each resource has only a resource environment. This predicate could be refreshed while environment monitor captures the changes of resource.
- **hasUserEnvironment(U, UE):** This predicate determines relationship between resource and resource environment: each user has only a user environment. This predicate could be refreshed when environment monitor captures the changes of user.
- **hasPermit(U, P):** This predicate is true determines relationship between resource and user: It is true if the user is permitted to get the resource.
- **RequestContext(U, R):** This predicate combines static attributes from user's and resource's . Every user will have this predicate if he/she sends the request.

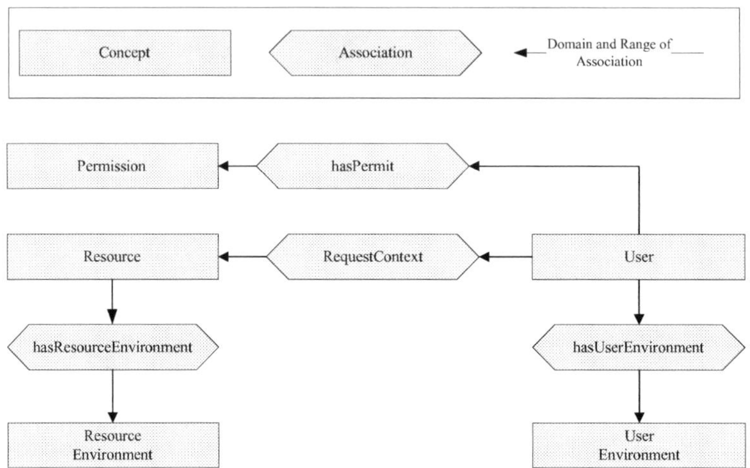

Fig. 2. Process ontology

5.3 SOGA Policies

In the SOGA framework, a policy is defined by set of rules that specify the request if prohibited or permitted. In our model, a policy have three basic outcomes: RequestContext, UserEnvionment, and ResourceEnvironment. Figure 3 and figure 4 are two different rules which compose the basic AC policy in our scenario.

The **Load Rule** present that if the resource load is less than the max load, the request is "Permitted" and the requestor is allowed to get the resource. On the contrary, the resource might have some risk and the request is automatically denied.

$$\begin{aligned}
&\text{User}(user) \wedge \text{request_context}(user,r) \wedge \text{hasResourceEnvironment}(r,re) \\
&\wedge \text{resource_load}(re,rl) \wedge \text{resource_max_load}(r,ml) \\
&\wedge swrlb: lessThanOrEqual(rl,ml) \wedge \text{Permission}(p) \\
&\Rightarrow \text{hasPermit}(user, p)
\end{aligned}$$

Fig. 3. Load Rule

$$\begin{aligned}
&\text{User}(user) \wedge \text{request_context}(user,r) \wedge \text{hasUserEnvironment}(r,ue) \\
&\wedge \text{application_time}(ue,at) \wedge \text{resource_system_start_Time}(r,st) \\
&\wedge \text{resource_system_end_Time}(r,et) \\
&\wedge swrlb: lessThanOrEqual\ (rl,ed) \wedge swrlb: greaterThanOrEqual(rl,st) \\
&\wedge \text{Permission}(p) \\
&\Rightarrow \text{hasPermit}(user, p)
\end{aligned}$$

Fig. 4. Time Rule

The **Time Rule** shows that if the request time is between resource start time and resource end time, e.g. teaching time is between 8:00AM to 16:00PM, the request for the specific entertainment resource is "Prohibit". This rule regulate the user must to access to the learning material only during the studying time.

All above rules are written in SWRL syntax and processed by JESS [17] policy inference engine.

6 Related Work

Virtual Organization Management Service (VOMS) [3], to some extent similar to the CAS [9], was developed by the European Data Grid Project. The user has to contact several VOMS servers, retrieve VOMS pseudo certificates containing his attributes (groups and roles) and include these pseudo certificates in a non-critical extension of a usual proxy certificate. The user pushes the proxy certificate to the resource Gatekeeper authorizing the submitted job on the attributes included in the certificate received.

Akenti [10] uses three types of certificates: attribute certificates binding an attribute-value pair to the principal of the certificate, use-condition certificates indicating lists of authoritative principals for user attributes and containing relational expressions of required user attributes for access rights and policy certificates consisting of trusted CAs and stakeholders issuing use-condition certificates and lists of URLs from where use-condition and attribute certificates can be retrieved. Akenti engine authenticates clients based on their X.509 certificate, gathers in a pull model certificates available with the authenticated identity (attribute certificates) and those of the accessed resource (use-condition certificates) and makes a decision regarding user rights at the resource.

Torsten Priebe et al. [12] presented an approach for simplifying the specification and maintenance of attribute-based access control policies by extending the attribute management with an ontology-based inference facility. This model enables policy administrators to concentrate on the properties they deem necessary from their point of view. A semantic mapping between different attributes can be performed via ontology. The model is based on the established XACML standard and features thorough use of open standards like RDF and OWL in the semantic extension of the architecture.

Besides listed above, research on grid authorization also include GridLab [4], PRIMA [7], etc.

7 Conclusion

In this paper, we introduced an efficient and capability based fine-grained authorization infrastructure, for SOG environment. It is based on user-level, peer-to-peer trust relationships, and used for building secure Web services and Grid services with the support of fine-grained authorization policy enforcement. Abstract entities and processes are defined and fundamental communication sequences for authentication and authorization requests and decisions are shown.

Since authentication and authorization are basic requirements for grid security, for future research, other aspects like accounting, integrity, confidentiality will be studied later.

Acknowledgments

We would like to thank Dr. Zhaohui Yang for his selfless help. We also want to thank the anonymous reviewers for their comments.

References

1. Oasis extensible access control markup language (XACML) committee specification (2002), http://www.oasisopen.org/committees/xacml/docs/s-xacmlspeci.cation-1.0-1.doc
2. Abu-Ghazaleh, N., Lewis, M.: Towards self-organizing grids. In: Proc. IEEE International Conference on High Performance Distributed Computing (HPDC-15) (2006)
3. eri, R.Al., Cecchini, R., Ciaschini, V., dell'Agnello, L., Frohner, A., Gianoli, A., Lorentey, K., Voms, S.F.: An authorization system for virtual organizations. In: European Across Grids Conference, pp. 33–40, (2003)
4. Allen, Davis, K., Dramlitsch, T., Goodale, T., Kelley, I., Lanfermann, G., Novotny, J., Radke, T., Rasul, K., Russell, M., Seidel, E., Wehrens, O.: The gridlab grid application toolkit. In: HPDC'02: Proceedings of the 11th IEEE International Symposium on High Performance Distributed Computing, p. 411. IEEE Computer Society, Washington, DC, USA (2002)
5. Foster, I., Kesselman, C.: The Grid2: Blue print for a New Computing Infrastructure. Morgan Kaufmann Publishers Inc, San Francisco, CA, USA
6. Foster, I., Kesselman, C., Tuecke, S.: The anatomy of the grid: Enabling scalable virtual organizations. Int. J. High Perform. Comput. Appl. 15(3), 200–222 (2001)
7. Lorch, M., Adams, D.B., Kafura, D., Koneni, M.S.R., Rathi, A., Shah, S.: The prima system for privilege management, authorization and enforcement in grid environments. In: GRID'03. Proceedings of the Fourth International Workshop on Grid Computing, p. 109. IEEE Computer Society Press, Washington, DC, USA (2003)
8. Novotny, J., Tuecke, S., Welch, V.: An online credential repository for the grid: Myproxy. In: HPDC'01. Proceedings of the 10th IEEE International Symposium on High Performance Distributed Computing, p. 104. IEEE Computer Society Press, Washington, DC, USA (2001)
9. Pearlman, L., Welch, V., Foster, I., Kesselman, C., Tuecke, S.: A community authorization service for group collaboration. In: POLICY'02. Proceedings of the 3rd International Workshop on Policies for Distributed Systems and Networks, p. 50. IEEE Computer Society Press, Washington, DC, USA (2002)
10. Thompson, M.R., Essiari, A., Mudumbai, S.: Certicate-based authorization policy in a PKI environment. ACM Trans.Inf.Syst.Secur. 6(4), 566–588 (2003)
11. Yang, W., Abu-Ghazaleh, N., Lewis, M.J.: Automatic clustering for self-organizing grids. In: Proc. of CLUSTER 2006 (2006)
12. Priebe, T., Dobmeier, W., Kamprath, N.: Supporting Attribute-based Access control with Ontologies. In: proceedings of the First International Conference on Availability, Reliability and Security (ARES'06) IEEE Computer Society

13. Ferraiolo, F., Sandhu, R., Gavrila, S., Kuhn, D., Chandramouli, R.: Proposed NIST standard for role-based access control. ACM transaction on Information and Systems Security. vol. 4(3) (August 2001)
14. Gruber, T.R.: A translation approach to portable ontologies. Knowledge Acquisition 5(2), 199–220 (1993)
15. Protege he official web site: http://protege.stanford.edu/
16. Horrocks, I., et al.: SWRL: A Semantic Web Rule Language Combining OWL and RuleML,plus 0.5em minus 0.4emW3C Member Submission (2004)
17. Jess, http://herzberg.ca.sandia.gov/jess

Permission-Centric Hybrid Access Control*

Sejong Oh

Dept. of Computer Science, Dankook University, San 29 Anseo-dong
Chonan-si, 330-714, South Korea
sejongoh@dankook.ac.kr
http://home.dankook.ac.kr/sejong

Abstract. Secure access control is a hot issue of large-scale organizations or information systems, because they have numerous users and information objects. They may adopt one of typical access control models such as ACL(access control list), MAC(mandatory access control), and RBAC(role-based access control). But an organization, in many cases, needs to apply multiple access control methods at the same time. In this paper, we proposed hybrid access control method for large organizations or information systems. In the proposed method, an organization or system can selectively choose one of access control model for each access event according to the nature of permissions.

1 Introduction

Access control is a central concern for information security in large organizations or information systems. The basic goal of access control is to offer the methodology that only authorized users (Subjects) can access information resources (Objects). Access is the ability to perform work such as reading, writing, and execution of system resources. Access control is the means to control the ability to perform above work. Access control of the computer system describes whether specific users or processors can access specific system resources or not, and their allowed access type.

Many researchers have developed access control models, such as access control list (ACL), discretionary access control (DAC), mandatory access control (MAC), and role-based access control (RBAC). Each access control model considers requirements of real world, and each organization has different requirements of access control. Therefore, there is no access control model satisfies all the organization. For example, MAC may be good for military environment, and RBAC may be good for enterprise environment.

The fact is that an organization needs several types of access control model at the same time. Let's suppose a company adopts RBAC model for it. In the RBAC model, many users belong to same role, and they shares authorities that assigned to the role. If a user U1, belongs to role R1, has an authority to read FILE1, all the users belong to R1 have the authority to read FILE1. Security officer may want to assign special authority only for U1, but it is impossible in the RBAC model. In this case, ACL

* This work was supported by Korea Research Foundation Grant funded by Korea Government (MOEHRD, Basic Research Promotion Fund) (KRF-2005-003-D00366).

K.C. Chang et al. (Eds.): APWeb/WAIM 2007 Ws, LNCS 4537, pp. 694–703, 2007.

model is required. In fact, most of relational DBMS products allow that DBA directly assign access rights for 'TABLE-A' to the user 'TOM'. They also allow indirectly assign access rights for 'TABLE-B' to 'TOM' through role 'CLERK'. As another case, let's suppose the company imports BBS (bulletin board system). Many BBS need to setup each user's authority level and sensitivity level of permission for each bulletin board. If a user U1's authority level is 70 and sensitivity level of '*can-file-attach*' permission for a bulletin board is 90, U1 cannot attach file when he write the bulletin board. RBAC cannot satisfy this situation; MAC model is required.

In this paper we propose hybrid access control method. 'Hybrid' means that several access control models can be applied at the same time. We does not propose new access control model. Instead, we propose how to merge several access control models to work at the same time. We assume many organizations need ACL, MAC, and RBAC models at the same time. Fig. 1 shows the concept of hybrid access control.

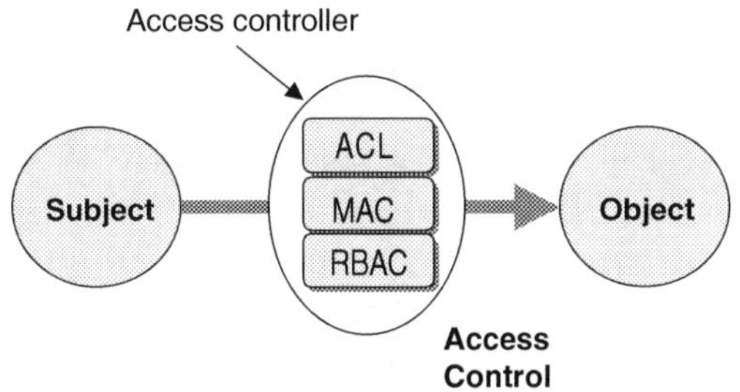

Fig. 1. Concept of Hybrid Access Control

This paper is organized as follows. Section 2 introduces our motivations. It contains concept of hybrid access control and nature of permission. Section 3 introduces typical access control models including ACL, MAC, and RBAC. Section 4 describes about hybrid access control method. First, we describe schema of authorization data for hybrid access control. Second, we describe decision-making algorithm for hybrid access controller. Third, we show an example of hybrid access control. And the paper is then conclusion in Section 5.

2 Motivations

Our motivation is that large organizations or information systems needs several types of access control model at the same time. They have several types of users and information resources. Single access control model may not satisfy them. They require plural access control mechanisms run concurrently.

In spite of rich research results about access control, we seldom find the research like this paper. Most of papers including hybrid access control describe about physical level access control [10][11]. Georgiadis et al. [12] proposed hybrid context and role-based access control model. It is a new access control model; context-based model and role-based model are fused into one. There is only one access control principle.

If an information system needs ACL, MAC, and RBAC model, how to integrate them? We try to answer for the question. The integration issue is related with following three questions:

- Which access control model should be assigned for specific access situation?
- Which access control model should be chosen when a user submits an access request?
- Which authorization data is maintained for the hybrid access control mechanism?

First question is about decision making for using three access control models. If we can use one of three access control models, we need to decide which access control model is assigned for specific access situation. There may be various principles for the decision. We propose a principle based on characteristics of permission. Permission is a pair of '(information object, operation)'. For example, permission '(file1, read)' means the access right of reading file1. The permissions may be assigned to users who need them according to their duty or job position. Some permissions are needed for a little user, and other permissions are needed for many users. Table 1 shows classification of the permissions and proper access control model for each group of permissions.

Table 1. Classification of Permissions

Group	Characteristics	Proper Access Control Model
A	very restricted users need them	ACL
B	well defined group users need them	RBAC
C	Large number of users need them, and the users or permissions are changed frequently	MAC

In general, ACL may be good if a permission should be owned by very restricted people. If a permission is shared among well defined group of users who have same job position or mission, RBAC may be good for it. If a permission shared among large number of people and it is difficult to group them, MAC may be efficient.

The solution of second question is related with first question. We can record proper access control model into a permission data. If user U1 submit access request '(U1, PERMSSION1)', access controller take the access control model that is recorded in ERMSSION1 data.

Third question is about schema design for authorization data. Each access control model needs authorization data including user, object, permission, role, etc. Access controller uses the data when it is working. Our hybrid access control method integrates ACL, MAC, and RBAC model. Therefore, integrated schema of the three models is required.

In the next section, we summary three access control model, and then we describe our hybrid access control method in section 4.

3 Summary of Typical Access Control Models

Proposed method merges three access control models, ACL, MAC, and RBAC. In this section, we briefly describe about the three models.

3.1 ACL (Access Control List)

ACL(access control list) [1-2] is a list of users and groups, with their specific permissions. There is one such list for each object, and the list shows all subjects who should have access to the object and the level of their access. Unix system uses ACL. Table 2 shows another way to store ACL authorization data. According to Table 2, 'David' has authorities to 'read file1' and 'execute file2'.

Table 2. An Example of ACL

User	Permissions
Jane	file1(r,w), file2(x), file3(r)
John	file1(r), file2(x), file3(r)
Sam	file1(r), file2(r,x), file5(r),
David	file1(r), file2(x)

3.2 MAC (Mandatory Access Control)

MAC(Mandatory access control)[3-4] means that access control policy decisions are made beyond the control of the individual owner of an object. A central authority determines what information is to be accessible by whom, and the users cannot change access rights. In the MAC model, users (Subjects) and information objects (Objects) have labels of their sensitivity level, and users are restricted access according to their sensitivity level. MAC protects against information disclosure. The MAC model can be implemented using a multilevel security mechanism that uses no read-up and no write-down rules, also known as Bell-LaPadula restrictions. These rules are designed to ensure that information does not flow from a higher sensitivity level to a lower sensitivity level. (*see Fig. 2*). To achieve information integrity, the access rules are formulated as no read-down and no write-up.

Remark. In the real world, there are various types of operation over 'read' and 'write'. For simplicity, we assume a user who has a higher sensitivity level owns all authority about objects that have lower sensitivity level.

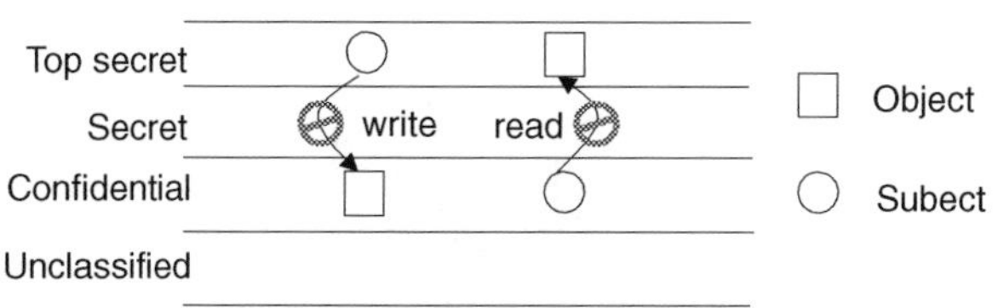

Fig. 2. MAC model

3.3 RBAC (Role-Based Access Control)

Role-based access control (RBAC)[5-9] has the central idea of preventing users from accessing company information at their discretion. Instead, access rights are associated with roles, and users are assigned to appropriate roles. The notion of role is an enterprise or organizational concept. Therefore, RBAC allows us to model security from an enterprise perspective, since we can align security modeling to the roles and responsibilities in the company. This greatly simplifies the management of access rights. Fig. 3 shows RBAC model.

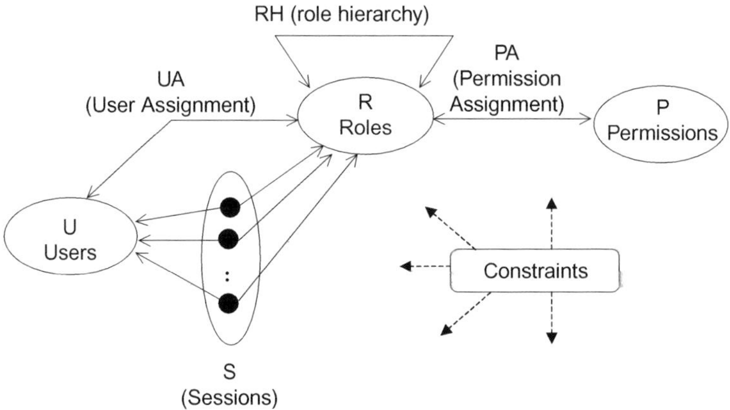

Fig. 3. RBAC model

4 Hybrid Access Control Method

Now, we describe hybrid access control method. As we mentioned before, it is not new type of access control model. We describe how to apply three typical access control model at the same time. Fig. 4 shows integrated model of ACL, MAC, and

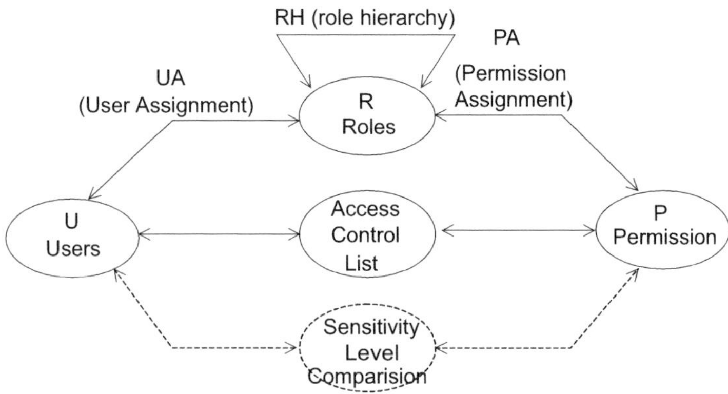

Fig. 4. Concept of Hybrid Access Control

RBAC model. RBAC part in Fig.4 indicates that users have a relation with permission through roles. ACL part indicates that users have a relation with permission through access-control-list. MAC part indicates that users and permissions have no direct relationship. Instead, users and permissions have indirect relationship through their sensitivity level.

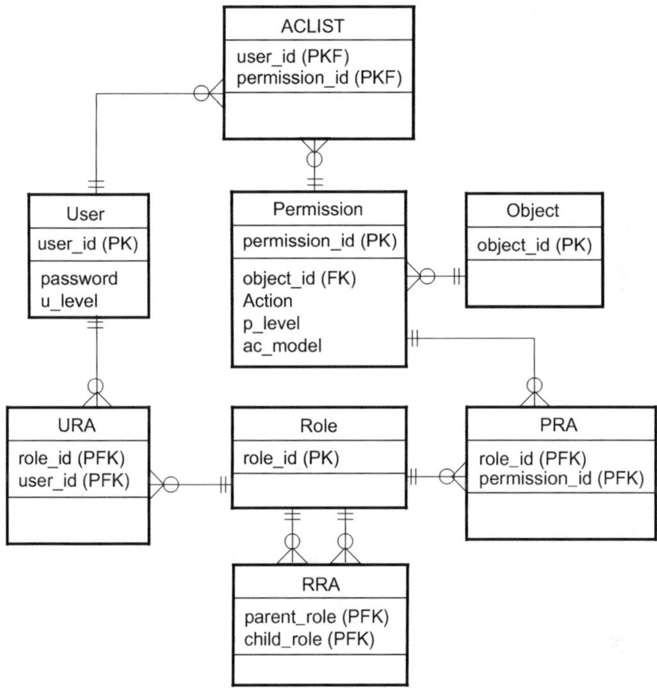

Fig. 5. ER Diagram for Schema of Hybrid Access Control

4.1 Schema for Authorization Data of Hybrid Access Control

If we want to apply three types of access control model at the same time, we should maintain authorization data that covers the three models. Therefore, we merge all the schema of three access control models. The result is as follows:

USER(user_id, password, u_level)
OBJECT(object_id)
PERMISSION(permission_id, object_id, action, p_level, ac_model)
ROLE(role_id)
URA(role_id, user_id)
PRA(role_id, permission_id)
RRA(parent_role, child_role)
ACLIST(user_id, permission_id)

In the *USER* information, *u_level* attribute means sensitivity level of a user for MAC model. In the *PERMISSION* information, *p_level* attribute means sensitivity level of a permission for MAC model, and *ac_model* attribute means access control model for the permission. Security administrator assigns one of 'ACL', 'MAC', or 'RBAC' to *ac_model* attribute based on Table 1. Fig.5 shows ER diagram for above schema.

4.2 Design of Hybrid Access Controller

Access controller is a software module that makes decision about users' access request. It allows or denies the access request based on authorization data described in section 4.1. Therefore, the core of access controller is decision-making algorism. The input of the algorism is an access request '(user_id, information_object, action)' where action means 'read', 'write' or 'execute'. The output of the algorism is 'True (allow access)' or 'False (deny access)'. Now we describe decision-making algorism for Hybrid access controller.

Let's suppose:

U = {*u* | *u* is a user with structure *USER*}
P = {*p* | *p* is a permission with structure *PERMISSION*}
R = {*r* | *r* is a role with structure *ROLE*}
UA = {*ua* | *ua* is a user-role assignment with structure *URA*}
PA = {*pa* | *pa* is a permission-role assignment with structure *PRA*}
RRA = {*rra* | *rra* is a role-role assignment with structure *RRA*}
ACL = {*a* | *a* is an authorization data with structure *ACLIST*}

$request(user_i, object\_id_j, action_k)$: access request of user $user_i$.

Then decision-making algorism is as follows:

```
If (user_i ∉ U)
   return False ;                                        // deny access
If ((object_id_j, action_k) is valid permission)
   take the permission id of (object_id_j, action_k) from P into perm_l ;
else
   return False ;                                        // deny access

If (perm_l.ac_model == 'ACL') {                          // case of ACL
   If ((user_i, perm_l) ∈ ACL)
      return True ;
   else
      return False ;
}

If (perm_l.ac_model == 'MAC') {                          // case of MAC
   If (user_i.u_level ≥ perm_l.p_level)
      return True ;
```

```
    else
        return False ;
}

If (perm_l.ac_model == 'RBAC') {                          // case of RBAC
    take the activated role for user_i from UA into set role_m ;
    take the child roles of role_m from RRA into set assigned_roles ;
    add role_m into assigned_roles ;

    while (assigned_roles != ϕ) {
        take a next role role_n from assigned_roles ;
        If ((role_n, perm_l) ∈ PA)
            return True;
    }
    return False;
}
```

4.3 An Example of Hybrid Access Control

In this section, we show how to work hybrid access control. In the example environment, authorization data is built following schema in section 4.1. Sample authorization data is shown in Fig. 6.

Now we discuss several cases of user requests. The requests may allowed or rejected by access controller depend on access control policy of information object.

Case 1. request (Jane, file1, execute). User 'Jane' requests execution of 'file1'. Permission p1 that contains (file1, execute) is managed by ACL and ACLIST contains (Jane, p1). Access controller allows the request because Jane owns required permission.

Case 2. requset (Jane, file2, execute). User 'Jane' requests execution of 'file2'. Permission p2 that contains (file2, execute) is managed by RBAC and 'Jane' belongs to role 'manager'. Access controller allows the request because 'manager' has permission p2.

Case 3. requset (Jane, file3, read). User 'Jane' requests reading of 'file3'. Permission p3 that contains (file3, read) is managed by MAC and her sensitivity level is 90. Access controller allows the request because 'Jane's sensitivity level is higher than permission p3.

Case 4. requset (John, file1, execute). User 'John' requests execution of 'file1'. Permission p1 that contains (file1, execute) is managed by ACL and ACLIST does not contain permission (John, p1). So access controller rejects the request.

Case 5. requset (John, file2, execute). User 'John' requests execution of 'file2'. Permission p2 that contains (file2, execute) is managed by RBAC and 'John' belongs to role 'manager'. Access controller allows the request because 'manager' has permission p2.

Case 6. requset (John, file3, read). User 'John' requests reading of 'file3'. Permission p3 that contains (file3, read) is managed by MAC and his sensitivity level is 70. Access controller rejects the request because sensitivity level of p3 is higher than 'John'.

USER

user_id	Password	u_level
Jane	(omit)	90
John	(omit)	70

OBJECT

object_id
file1
file2
file3
file4

PERMISSION

p_id	object_id	Action	p_level	ac_model
p1	file1	execute	null	ACL
p2	file2	execute	null	RBAC
p3	file3	Read	80	MAC
p4	file3	Write	90	MAC
p5	file4	execute	60	MAC

ROLE

role_id
manager
clerk

URA

role_id	user_id
Manager	Jane
Manager	John

PRA

role_id	p_id
Manager	p2

RRA

parent	child
manager	clerk

ACLIST

user_id	p_id
Jane	p1

Fig. 6. Authorization data for sample environment

5 Conclusion

Many organizations or information systems need hybrid access control. In this paper, we propose the way using three well-known access control models. The characteristics of permissions are basis for selecting access control model. In our method, access controller has ability to make decision based on three types of access control principles. In the proposed method, we use nature of permission for choosing access control model on specific access situation. In the web and pervasive environment (WPE), we may consider context information instead of nature of permission. Context information may also be used by access controller when it makes decision for user's access request. This is further research topic.

References

1. Pfleger, C.P.: Security in Computing, Prentice-Hall International Inc, Englewood Cliffs 2nd edn, pp. 244–250 (1997)
2. Russel, D., Gangemi, G.T.Sr.: Computer Security Basics, O'Reilly & Associates Inc. 67–77 (1991)

3. Amoroso, E.G.: Fundamental of Computer Security Technology, PTR, pp. 101–112. Prentice Hall, Englewood Cliffs (1994)
4. Joshi, J.B.D., Aref, W.G., Ghafoor, A., Spafford, E.H.: Security Model for Web-Based Applications. Communications of the ACM 44(2), 38–44 (2001)
5. Sandhu, R., Coyne, E.J., Feinstein, H.L., Youman, C.E.: Role-Based Access Control Method. IEEE Computer 29, 38–47 (1996)
6. Ferraio, D., Cugini, J., Kuhn, R.: Role-based Access Control (RBAC): Features and motivations. In: Proc. of 11th Annual Computer Security Application Conference (1995)
7. Sandhu, R.: Rationale for the RBAC96 Family of Access Control Models. In: Proc. of ACM Workshop on Role-Based Access Control, II, 1-8 (1995)
8. Gavrila, S.I., Barkley, J.F.: Formal Specification for Role Based Access Control User/Role and Role/Role Relationship Management. In: Proc. of the 3rd ACM workshop on Role-Based Access Control, pp. 81–90 (1998)
9. Sandhu, R., Bhamidipati, M.Q.: The ARBAC97 Model for Role-Based Administration of Roles. ACM Trans. on Information and Systems Security (TISSEC) 2, 105–135 (1999)
10. Human Recognition Systems, http://hrsltd.com/solutions/ access_control/access_control_hybrid.htm
11. Hybrid ALOHA: a novel medium access control, protocol (2005) http://www.egr.msu.edu/ wanghuah/HybridAloha.pdf
12. Georgiadis, C., Mavridis, I., Pangalos, G.: Context and Role Based Hybrid Access Control for Collaborative Environments. In: Proceedings of the Fifth Nordic Workshop on Secure IT Systems - Encouraging Co-operation (NORDSEC 2000) (2000)

Author Index

Lecture Notes in Computer Science

For information about Vols. 1–4431

please contact your bookseller or Springer

Vol. 4542: P. Sawyer, B. Paech, P. Heymans (Eds.), Requirements Engineering: Foundation for Software Quality. IX, 384 pages. 2007.

Vol. 4539: N.H. Bshouty, C. Gentile (Eds.), Learning Theory. XII, 634 pages. 2007. (Sublibrary LNAI).

Vol. 4538: F. Escolano, M. Vento (Eds.), Graph-Based Representations in Pattern Recognition. XII, 416 pages. 2007.

Vol. 4537: K.C.-C. Chang, W. Wang, L. Chen, C.A. Ellis, C.-H. Hsu, A.C. Tsoi, H. Wang (Eds.), Advances in Web and Network Technologies, and Information Management. XXIII, 707 pages. 2007.

Vol. 4534: I. Tomkos, F. Neri, J. Solé Pareta, X. Masip Bruin, S. Sánchez Lopez (Eds.), Optical Network Design and Modeling. XI, 460 pages. 2007.

Vol. 4531: J. Indulska, K. Raymond (Eds.), Distributed Applications and Interoperable Systems. XI, 337 pages. 2007.

Vol. 4530: D.H. Akehurst, R. Vogel, R.F. Paige (Eds.), Model Driven Architecture- Foundations and Applications. X, 219 pages. 2007.

Vol. 4529: P. Melin, O. Castillo, L.T. Aguilar, J. Kacprzyk, W. Pedrycz (Eds.), Foundations of Fuzzy Logic and Soft Computing. XIX, 830 pages. 2007. (Sublibrary LNAI).

Vol. 4526: M. Malek, M. Reitenspieß, A. van Moorsel (Eds.), Service Availability. X, 155 pages. 2007.

Vol. 4525: C. Demetrescu (Ed.), Experimental Algorithms. XIII, 448 pages. 2007.

Vol. 4524: M. Marchiori, J.Z. Pan, C.d.S. Marie (Eds.), Web Reasoning and Rule Systems. XI, 382 pages. 2007.

Vol. 4523: Y.-H. Lee, H.-N. Kim, J. Kim, Y. Park, L.T. Yang, S.W. Kim (Eds.), Embedded Software and Systems. XIX, 829 pages. 2007.

Vol. 4522: B.K. Ersbøll, K.S. Pedersen (Eds.), Image Analysis. XVIII, 989 pages. 2007.

Vol. 4521: J. Katz, M. Yung (Eds.), Applied Cryptography and Network Security. XIII, 498 pages. 2007.

Vol. 4519: E. Franconi, M. Kifer, W. May (Eds.), The Semantic Web: Research and Applications. XVIII, 830 pages. 2007.

Vol. 4517: F. Boavida, E. Monteiro, S. Mascolo, Y. Koucheryavy (Eds.), Wired/Wireless Internet Communications. XIV, 382 pages. 2007.

Vol. 4516: L. Mason, T. Drwiega, J. Yan (Eds.), Traffic Performance in Converged Networks. XXIII, 1191 pages. 2007.

Vol. 4515: M. Naor (Ed.), Advances in Cryptology - EUROCRYPT 2007. XIII, 591 pages. 2007.

Vol. 4514: S.N. Artemov, A. Nerode (Eds.), Logical Foundations of Computer Science. XI, 513 pages. 2007.

Vol. 4510: P. Van Hentenryck, L. Wolsey (Eds.), Integration of AI and OR Techniques in Constraint Programming for Combinatorial Optimization Problems. X, 391 pages. 2007.

Vol. 4509: Z. Kobti, D. Wu (Eds.), Advances in Artificial Intelligence. XII, 552 pages. 2007. (Sublibrary LNAI).

Vol. 4508: M.-Y. Kao, X.-Y. Li (Eds.), Algorithmic Aspects in Information and Management. VIII, 428 pages. 2007.

Vol. 4506: D. Zeng, I. Gotham, K. Komatsu, C. Lynch, M. Thurmond, D. Madigan, B. Lober, J. Kvach, H. Chen (Eds.), Intelligence and Security Informatics: Biosurveillance. XI, 234 pages. 2007.

Vol. 4505: G. Dong, X. Lin, W. Wang, Y. Yang, J.X. Yu (Eds.), Advances in Data and Web Management. XXII, 896 pages. 2007.

Vol. 4504: J. Huang, R. Kowalczyk, Z. Maamar, D. Martin, I. Müller, S. Stoutenburg, K.P. Sycara (Eds.), Service-Oriented Computing: Agents, Semantics, and Engineering. X, 175 pages. 2007.

Vol. 4501: J. Marques-Silva, K.A. Sakallah (Eds.), Theory and Applications of Satisfiability Testing – SAT 2007. XI, 384 pages. 2007.

Vol. 4500: N. Streitz, A. Kameas, I. Mavrommati (Eds.), The Disappearing Computer. XVIII, 304 pages. 2007.

Vol. 4496: N.T. Nguyen, A. Grzech, R.J. Howlett, L.C. Jain (Eds.), Agent and Multi-Agent Systems: Technologies and Applications. XXI, 1046 pages. 2007. (Sublibrary LNAI).

Vol. 4495: J. Krogstie, A. Opdahl, G. Sindre (Eds.), Advanced Information Systems Engineering. XVI, 606 pages. 2007.

Vol. 4494: H. Jin, O.F. Rana, Y. Pan, V.K. Prasanna (Eds.), Algorithms and Architectures for Parallel Processing. XIV, 508 pages. 2007.

Vol. 4493: D. Liu, S. Fei, Z. Hou, H. Zhang, C. Sun (Eds.), Advances in Neural Networks – ISNN 2007, Part III. XXVI, 1215 pages. 2007.

Vol. 4492: D. Liu, S. Fei, Z. Hou, H. Zhang, C. Sun (Eds.), Advances in Neural Networks – ISNN 2007, Part II. XXVII, 1321 pages. 2007.

Vol. 4491: D. Liu, S. Fei, Z.-G. Hou, H. Zhang, C. Sun (Eds.), Advances in Neural Networks – ISNN 2007, Part I. LIV, 1365 pages. 2007.

Vol. 4490: Y. Shi, G.D. van Albada, J. Dongarra, P.M.A. Sloot (Eds.), Computational Science – ICCS 2007, Part IV. XXXVII, 1211 pages. 2007.

Vol. 4489: Y. Shi, G.D. van Albada, J. Dongarra, P.M.A. Sloot (Eds.), Computational Science – ICCS 2007, Part III. XXXVII, 1257 pages. 2007.

Vol. 4488: Y. Shi, G.D. van Albada, J. Dongarra, P.M.A. Sloot (Eds.), Computational Science – ICCS 2007, Part II. XXXV, 1251 pages. 2007.

Vol. 4487: Y. Shi, G.D. van Albada, J. Dongarra, P.M.A. Sloot (Eds.), Computational Science – ICCS 2007, Part I. LXXXI, 1275 pages. 2007.

Vol. 4486: M. Bernardo, J. Hillston (Eds.), Formal Methods for Performance Evaluation. VII, 469 pages. 2007.

Vol. 4485: F. Sgallari, A. Murli, N. Paragios (Eds.), Scale Space and Variational Methods in Computer Vision. XV, 931 pages. 2007.

Vol. 4484: J.-Y. Cai, S.B. Cooper, H. Zhu (Eds.), Theory and Applications of Models of Computation. XIII, 772 pages. 2007.

Vol. 4483: C. Baral, G. Brewka, J. Schlipf (Eds.), Logic Programming and Nonmonotonic Reasoning. IX, 327 pages. 2007. (Sublibrary LNAI).

Vol. 4482: A. An, J. Stefanowski, S. Ramanna, C.J. Butz, W. Pedrycz, G. Wang (Eds.), Rough Sets, Fuzzy Sets, Data Mining and Granular Computing. XIV, 585 pages. 2007. (Sublibrary LNAI).

Vol. 4481: J. Yao, P. Lingras, W.-Z. Wu, M. Szczuka, N.J. Cercone, D. Ślęzak (Eds.), Rough Sets and Knowledge Technology. XIV, 576 pages. 2007. (Sublibrary LNAI).

Vol. 4480: A. LaMarca, M. Langheinrich, K.N. Truong (Eds.), Pervasive Computing. XIII, 369 pages. 2007.

Vol. 4479: I.F. Akyildiz, R. Sivakumar, E. Ekici, J.C.d. Oliveira, J. McNair (Eds.), NETWORKING 2007. Ad Hoc and Sensor Networks, Wireless Networks, Next Generation Internet. XXVII, 1252 pages. 2007.

Vol. 4478: J. Martí, J.M. Benedí, A.M. Mendonça, J. Serrat (Eds.), Pattern Recognition and Image Analysis, Part II. XXVII, 657 pages. 2007.

Vol. 4477: J. Martí, J.M. Benedí, A.M. Mendonça, J. Serrat (Eds.), Pattern Recognition and Image Analysis, Part I. XXVII, 625 pages. 2007.

Vol. 4476: V. Gorodetsky, C. Zhang, V.A. Skormin, L. Cao (Eds.), Autonomous Intelligent Systems: Multi-Agents and Data Mining. XIII, 323 pages. 2007. (Sublibrary LNAI).

Vol. 4475: P. Crescenzi, G. Prencipe, G. Pucci (Eds.), Fun with Algorithms. X, 273 pages. 2007.

Vol. 4474: G. Prencipe, S. Zaks (Eds.), Structural Information and Communication Complexity. XI, 342 pages. 2007.

Vol. 4472: M. Haindl, J. Kittler, F. Roli (Eds.), Multiple Classifier Systems. XI, 524 pages. 2007.

Vol. 4471: P. Cesar, K. Chorianopoulos, J.F. Jensen (Eds.), Interactive TV: a Shared Experience. XIII, 236 pages. 2007.

Vol. 4470: Q. Wang, D. Pfahl, D.M. Raffo (Eds.), Software Process Dynamics and Agility. XI, 346 pages. 2007.

Vol. 4468: M.M. Bonsangue, E.B. Johnsen (Eds.), Formal Methods for Open Object-Based Distributed Systems. X, 317 pages. 2007.

Vol. 4467: A.L. Murphy, J. Vitek (Eds.), Coordination Models and Languages. X, 325 pages. 2007.

Vol. 4466: F.B. Sachse, G. Seemann (Eds.), Functional Imaging and Modeling of the Heart. XV, 486 pages. 2007.

Vol. 4465: T. Chahed, B. Tuffin (Eds.), Network Control and Optimization. XIII, 305 pages. 2007.

Vol. 4464: E. Dawson, D.S. Wong (Eds.), Information Security Practice and Experience. XIII, 361 pages. 2007.

Vol. 4463: I. Măndoiu, A. Zelikovsky (Eds.), Bioinformatics Research and Applications. XV, 653 pages. 2007. (Sublibrary LNBI).

Vol. 4462: D. Sauveron, K. Markantonakis, A. Bilas, J.-J. Quisquater (Eds.), Information Security Theory and Practices. XII, 255 pages. 2007.

Vol. 4459: C. Cérin, K.-C. Li (Eds.), Advances in Grid and Pervasive Computing. XVI, 759 pages. 2007.

Vol. 4453: T. Speed, H. Huang (Eds.), Research in Computational Molecular Biology. XVI, 550 pages. 2007. (Sublibrary LNBI).

Vol. 4452: M. Fasli, O. Shehory (Eds.), Agent-Mediated Electronic Commerce. VIII, 249 pages. 2007. (Sublibrary LNAI).

Vol. 4451: T.S. Huang, A. Nijholt, M. Pantic, A. Pentland (Eds.), Artifical Intelligence for Human Computing. XVI, 359 pages. 2007. (Sublibrary LNAI).

Vol. 4450: T. Okamoto, X. Wang (Eds.), Public Key Cryptography – PKC 2007. XIII, 491 pages. 2007.

Vol. 4448: M. Giacobini et al. (Ed.), Applications of Evolutionary Computing. XXIII, 755 pages. 2007.

Vol. 4447: E. Marchiori, J.H. Moore, J.C. Rajapakse (Eds.), Evolutionary Computation,Machine Learning and Data Mining in Bioinformatics. XI, 302 pages. 2007.

Vol. 4446: C. Cotta, J. van Hemert (Eds.), Evolutionary Computation in Combinatorial Optimization. XII, 241 pages. 2007.

Vol. 4445: M. Ebner, M. O'Neill, A. Ekárt, L. Vanneschi, A.I. Esparcia-Alcázar (Eds.), Genetic Programming. XI, 382 pages. 2007.

Vol. 4444: T. Reps, M. Sagiv, J. Bauer (Eds.), Program Analysis and Compilation, Theory and Practice. X, 361 pages. 2007.

Vol. 4443: R. Kotagiri, P.R. Krishna, M. Mohania, E. Nantajeewarawat (Eds.), Advances in Databases: Concepts, Systems and Applications. XXI, 1126 pages. 2007.

Vol. 4440: B. Liblit, Cooperative Bug Isolation. XV, 101 pages. 2007.

Vol. 4439: W. Abramowicz (Ed.), Business Information Systems. XV, 654 pages. 2007.

Vol. 4438: L. Maicher, A. Sigel, L.M. Garshol (Eds.), Leveraging the Semantics of Topic Maps. X, 257 pages. 2007. (Sublibrary LNAI).

Vol. 4433: E. Şahin, W.M. Spears, A.F.T. Winfield (Eds.), Swarm Robotics. XII, 221 pages. 2007.

Vol. 4432: B. Beliczynski, A. Dzielinski, M. Iwanowski, B. Ribeiro (Eds.), Adaptive and Natural Computing Algorithms, Part II. XXVI, 761 pages. 2007.